Shell Mechanics

This book is devoted to shells, a natural or human construction, whose modelling as a structure was particularly developed during the 20th century, leading to current numerical models.

Many objects, either in industry or civil engineering, come under shell mechanics, so a good knowledge of their behaviour and modelling is essential to master their design.

This book highlights the very strong link between the deformation of geometric surfaces and the mechanics of shells. The theory is approached in a general formulation that can apply to any surface, and the applications bring the concepts and the methods of resolution to practical situations. It aims to understand the behaviour of shells and to identify the most important parameters, thus allowing a good interpretation of the numerical results. The reader will be able, with finite element software, to reproduce the proposed solutions.

The book is based on the knowledge acquired by the reader in structural mechanics and provides essential information on the geometry of surfaces. It is ideal for students in the fields of engineering using mechanics, as well as professionals wishing to deepen their knowledge of shells.

Philippe Bisch is a professor at the Ecole des Ponts ParisTech in France and has worked as a design engineer and consultant for EGIS Industries. He is former President of the European Association of Earthquake Engineering and of the French Association for Earthquake Engineering, and is now chairman of the Eurocode 8 project.

Shell Mechanics

Theory and Applications

Philippe Bisch

CRC Press
Taylor & Francis Group
Boca Raton London New York

CRC Press is an imprint of the
Taylor & Francis Group, an **informa** business

Cover image: Philippe Bisch

First published in English 2024
by CRC Press
2385 NW Executive Center Drive, Suite 320, Boca Raton FL 33431

and by CRC Press
4 Park Square, Milton Park, Abingdon, Oxon, OX14 4RN

CRC Press is an imprint of Taylor & Francis Group, LLC

Published in French by Presses des Ponts 2013

Library of Congress Cataloging-in-Publication Data

Names: Bisch, Philippe, author.
Title: Shell mechanics : theory and applications / Philippe Bisch.
Description: Boca Raton : CRC Press, 2023. | Includes bibliographical references and index.
Identifiers: LCCN 2022048147 | ISBN 9781138310599 (hbk) | ISBN 9781032451718 (pbk) | ISBN 9780429440403 (ebk)
Subjects: LCSH: Shells (Engineering)
Classification: LCC TA683.5.S4 B57 2023 | DDC 624.1/7762--dc23/eng/20221206
LC record available at https://lccn.loc.gov/2022048147

ISBN: 978-1-138-31059-9 (hbk)
ISBN: 978-1-032-45171-8 (pbk)
ISBN: 978-0-429-44040-3 (ebk)

DOI: 10.1201/9780429440403

Typeset in Sabon
by Deanta Global Publishing Services, Chennai, India

Contents

Preamble

This book is devoted to shells and to special cases of plates and membranes. It includes the establishment of general equations of motion and the presentation of particular methods and applications.

This is a book intended for students who have acquired a solid basic knowledge of Solid Mechanics, in particular of one-dimensional structures, and who wish to deepen their understanding of two-dimensional structures analysis. But this course is also, for engineers who already have experience in structural design, a way to understand the behaviour of shells, which differs according to geometry and situations.

The properties of shells, when assimilated to surfaces, cannot be considered as the simple reproduction in a two-dimensional space of the properties of one-dimensional structures: beams and arches. Indeed, surfaces have unique geometric properties that directly influence their mechanical behaviour. One of the goals of this book is to show this very strong link between geometry and mechanics, because it helps to better understand the behaviour of shells.

To achieve this, the development of the theory is done within the rigorous framework of the differential geometry of surfaces, based on tensorial analysis, which allows the setting of equations in a synthetic form independent of the shape of the shell considered, and avoids thus the long developments of calculation associated with each particular form. This method, if it requires a greater effort of understanding and formalisation at the beginning, then makes it possible to control the approximations granted and to approach any form of shell without having to reconstruct in each case the equilibrium equations on particular geometries. Reminders of mathematics are the subject of two chapters, so that the student with the usual basic notions, in particular of algebra and linear analysis, can acquire the tools necessary for the development of shell mechanics.

The finite element method, when the software used has the necessary finite elements and algorithms, makes it possible to deal with all the practical cases encountered by the design engineer. It is therefore not a question, in the applications that are developed here, of giving "exact" solutions, especially if they require a lot of calculations. These applications are given with the main purpose of describing the behaviours and highlighting the most important terms in the equations, and their physical meaning. That is why a number of them have been developed here, at least to the point of allowing useful interpretations; they therefore often include approximations; more exact analytical solutions can be found in the literature. It is therefore a question here of illustrating the theory and behaviour of shells, not of offering ready-made solutions to designers. A chapter dedicated to the finite elements of shells allows them to make the link between practical problem-solving and the underlying theory.

As regards the establishment of general equations, the two approaches by the Principle of Virtual Powers and by the General Theorems are explained and compared, which makes it possible to clearly highlight the differences between the resultants of stresses and the generalised stresses. The method for obtaining general equations by integrating local 3D equations is also presented. The foundations of shell theory are exposed in their generality, then the equations are applied to particular geometries.

A central question in shell mechanics is the behaviour of points located in the thickness. The two “classic” models are explained, taking into account or not the distortion of the material segment located on the normal vector to the mean surface. Openings are also made towards more complete models.

Only elasticity is considered for setting up the constitutive laws of shells, as well as for the applications. Nevertheless, the exposed method is general and can be used to introduce other types of behaviour.

Use was made, especially for the illustrations, of a few digital tools. Some are freely accessible on the Internet, others can be acquired at a reasonable cost. The reader, especially if he is a student, can therefore reproduce at low cost the numerical solutions that are presented.

▼▼▼

This book is the materialisation of a course taught at the Ecole des Ponts, initiated by Professor Yves Bamberger, and which is a continuation of his course in Structural Mechanics. Without his support, this course could not have been developed over the years. Given his fundamental contribution to the content of this course, the author wishes to pay him a special tribute.

The content of the course that gave birth to this book has been enriched over the years thanks to the many questions asked by the students. This contribution is in fact fundamental, because the questions very often flush out the dark areas that hide behind well-established truths.

Such a course, when it lasts several years, benefits from the contribution of colleagues, teachers, or professionals. Several of them have helped to clarify the delicate points, they are thanked. I would like to make a special mention to Alain Millard, in particular, because of his in-depth knowledge of finite elements.

Finally, I owe a very big thank you to Sébastien Brisard, who had the courage to read the text meticulously and to point out many of the errors that cannot fail to hide in editing so many equations. His insightful remarks also led to a very significant improvement in the presentation.

▼▼▼

Philippe BISCH

The figures, photos, and drawings intended to illustrate the text, but also to brighten it up, are by the author.

General Principles of Notation

Notations are based on the following general principles, which may not be respected in the event of conflict. In particular, it is unavoidable that the same letter may sometimes designate very different quantities. In these cases, the context and the text make it possible to remove any ambiguity.

Latin indices are relative to the ambient space (except in Chapter 9). They vary from *1* to *3*. Greek indices are relative to the middle surface or $\mathbb{R}^2$. They vary from *1* to *2*.

Numbers are in *italics* when it comes to indices. Otherwise, in particular for powers, they are in normal characters.

Quantities related to the deformed configuration (that is, at time t) are represented by lowercase letters; the corresponding quantities on a reference configuration are represented by the same letters, but in capital letters.

Quantities in $\mathbb{R}^3$ are highlighted with a bar, when there are corresponding quantities defined on the surface (Example: ρ is the surfacic mass, $\overline{\rho}$ is the volumic mass).

The total differentiation with respect to time (real velocity) is indicated by a dot on the letter: $\dot{f} = \dfrac{df}{dt}$.

The virtual rate of a quantity f is designated by a star * on the letter: $\overset{*}{f} = \underbrace{\lim}_{u \to 0} \dfrac{f(u)}{u}$.

Points m, P, p and other geometrical quantities (C, F, P) are in brush script font.

Energies $\mathcal{W}$, $\mathcal{F}$, $\mathcal{V}$ are in mural script font.

Endomorphisms are most often represented by letters in mural script font.

A circumflex on a letter representing an endomorphism designates the determinant of this endomorphism: $\hat{m} = \det\left(m_{\bullet}^{\ \bullet}\right)$.

If T denotes a tensor of variance 2, [T] denotes the matrix associated with this tensor, in a given basis.

To avoid confusion, the vectors in Latin letters are sometimes in bold right letters, but their components are always in light italic letters (Example: **t** is a tangent vector with components t^{α}, but time is also designated by t).

The following pages give a list of the most used or significant symbols. Other notations are introduced throughout the text.

The chapters are subdivided into paragraphs according to the usual decimal numbering. The figures and formulas are numbered sequentially by chapter. The reference to a paragraph, figure, or formula "nn" is of the form "(X.nn)", where "X" refers to the number of the chapter in Arabic numerals.

The equations are numbered extensively. An unnumbered equation is usually of less importance and corresponds to an intermediate calculation that helps to understand the reasoning.

MAIN NOTATIONS

Euclidean vector space

E_n, E	euclidean vector space of dimension n on $\mathbb{R}$
E^*	dual space of E
(e_i)	basis of E
(e^i)	basis of E^* dual of e_i
V, W	vectors of E
V^i	components of V
ℓ, h	linear one-forms on E
ℓ_i	components of ℓ
δ^i_j	Kronecker delta, components of endomorphism identity
$\mathcal{L}(E\times E,\mathbb{R})$	set of bilinear forms on E
$\mathfrak{b}$	bilinear form on E
$\mathfrak{b}_{ij}$	components of $\mathfrak{b}$
$\mathcal{L}(E,E)$	set of endomorphisms on E
ω, $\mathfrak{f}$, $\mathfrak{g}$	endomorphisms on E
$\mathcal{P}$	matrix of basis change in E
$\mathfrak{p}_I{}^i$	components of $\mathcal{P}$ and of the associated endomorphism
$\mathcal{Q}=\mathcal{P}^{-1}$	
t, u, v	any tensor in E
g	covariant tensor of type (0, 2) associated to the scalar product (metric tensor)
$\flat$	flat, mixed endomorphism associated to the scalar product, $\flat \in \mathcal{L}(E,E)$
$\# = \flat^{-1}$	sharp, $\# \in \mathcal{L}(E^*,E)$

GENERAL OPERATORS

$\bullet$	dot (scalar) product of two vectors or tensor contraction
$\times$	cross product of two vectors
$\otimes$	tensorial product of two tensors

AMBIANT EUCLIDIEN SPACE

$\mathbb{R}^3$	Euclidean tridimensional space on $\mathbb{R}$, provided with a scalar product
$E_3^\bullet$	ambiant space, affine space of dimension 3 on $\mathbb{R}$
$\mathcal{p}$, $\mathcal{O}$	points in the ambiant space

(e_i) basis vectors of a cartesian or a natural coordinate system of $E_3^\bullet$
$(\mathbf{i}, \mathbf{j}, \mathbf{k})$ basis vectors of an orthonormal cartesian coordinate system of $E_3^\bullet$
(x, y, z) or (z_i) coordinates associated to an orthonormal cartesian coordinate system of $E_3^\bullet$
$\tilde{e}_i$ unit basis vector associated to a vector e_i of a natural basis
$\tilde{\mathrm{e}}_{ijk}$ signature of a permutation (i, j, k) with respect to $(1, 2, 3)$
$\overline{\mathrm{e}}$ Levi-Civita tensor in $\mathbb{R}^3$
D covariant differentiation in $E_3^\bullet$
D_X differentiation along a vector X
$\overline{\Gamma}_{ij}^k$ Christoffel coefficients in $E_3^\bullet$
$\frac{\partial}{\partial x^i} = \partial_i$ partial differentiation with respect to x^i
$\overline{R}$ Riemann-Christoffel tensor of differentiation D in $E_3^\bullet$, of variance 4

GEOMETRY OF THE MIDDLE SURFACE

σ middle surface in the present configuration at time t
$\partial\sigma$ boundary of σ
d part of σ
∂d boundary of d
(x^α) curvilinear coordinates, $\alpha \in [1,2]$
m current point on the middle surface
T_m plane tangent to the middle surface at point m
a_α basis vector in plane T_m, associated to x^α
$\tilde{a}_\alpha$ unit basis vector in T_m associated to a_α
a_3 unit vector normal to middle surface at point m
$a_{\alpha\beta}$ components of the metric tensor field of the surface, first quadratic form a
$\hat{a}$ $= \det(a)$
$b_{\alpha\beta}$ components of the curvature tensor field of the surface, second quadratic form b
$c_{\alpha\beta} = b_\alpha{}^\gamma b_{\gamma\beta}$ components of the third quadratic form c
R_α principal radii of middle surface
$H = \frac{1}{2}\left(\frac{1}{R_1} + \frac{1}{R_2}\right)$ mean curvature
$K = \frac{1}{R_1 R_2}$ gaussian curvature
ℓ length, span
e antisymetric quadratic form on the surface
$\mathcal{A}$ area measured on the surface
$\mathcal{C}$ curve drawn on surface σ
s curvilinear abscissa along $\mathcal{C}$
R radius of curvature of $\mathcal{C}$

$\mathbf{t}$	unit vector tangent to curve $\mathcal{C}$
$\mathbf{n}$	unit vector normal to curve $\mathcal{C}$
ν	vector of T_{m} normal to $\mathbf{t}$ (geodesic normal vector)
ρ_n	normal curvature of curve $\mathcal{C}$
ρ_g	geodesic curvature of curve $\mathcal{C}$
τ_g	geodesic torsion of curve $\mathcal{C}$
∇	covariant differentiation (Levi-Civita) on σ
$\Gamma^{\gamma}_{\alpha\beta}$	Riemannian connection coefficients
$R_{\alpha\beta\gamma\delta}$	components of the Riemann-Christoffel tensor of differentiation ∇

PARTICULAR GEOMETRIC PARAMETERS OF THE MIDDLE SURFACE

R_0	(positive) radius of a circle or a curve at a particular point
θ, z	cylindric coordinates
$\theta, \varphi, \phi, \psi$	spheric coordinates
p, q, r	Monge notations for surfaces defined by cartesian coordinates

GEOMETRY IN THE SHELL THICKNESS

e	shell thickness
$\mathscr{p}$	point in shell thickness
x^3	transverse coordinate, perpendicular to the middle surface
z	transverse coordinate, perpendicular to the middle plane of a plate
$\mathscr{a}, \mathscr{b}$	transverse coordinates of the two shell faces, perpendicular to the middle surface
$\mathcal{P}$	plane
$\mathcal{S}$	surface generated by a segment normal to the surface, based on $\mathcal{C}$
$\mathcal{D}$	volumic portion of a shell, limited by $\mathcal{S}$
ω	volumic portion of a shell, defined by domains d or σ drawn on the middle surface
$\partial\omega$	boundary of ω
$\mathcal{F}$	facet drawn in the thickness of the shell
g_α	vector of the local basis associated to x^α, in the vicinity of the middle surface
$g_3 = a_3$	third vector of the local basis associated to (x^i), in the vicinity of the middle surface
g_{ij}	components of the metric tensor of $\mathbb{R}^3$ in basis (g_i)
$\hat{g}$	$= \det(g)$
m	endomorphism used to deduct the local basis (g_i) in the vicinity of the middle surface σ, from the local basis (a_i) of the middle surface
$\ell = m^{-1}$	endomorphism, inverse of m

KINEMATICS AND KINETICS OF THE MIDDLE SURFACE

$\mathcal{M}$	current point on the reference surface Σ
(A_i)	basis on the reference surface
$\mathfrak{f}$	transformation of the middle surface
J	Jacobian of the transformation
ξ	vector displacement of a point on Σ
$\xi = \xi^\alpha A_\alpha$	orthogonal projection of ξ on $T_{\mathcal{M}}$
ζ	component of ξ on a_3 normal to $T_{\mathcal{M}}$
(u, v, w)	components of the displacement vector in an orthonormal basis
$\boldsymbol{\omega}$	vector rotation of a_3
$\mathcal{A}_\alpha{}^\beta$	component of $a_\alpha - A_\alpha$ on A_β
$\mathcal{B}_\alpha$	component of $a_\alpha - A_\alpha$ on A_3
$\mathcal{N}^\alpha$	component of a_3 on A_α
$\cos\omega$	component of a_3 on A_3
Ξ	surface dilatation
ε	surface strain tensor in Lagrangian variables
κ	surface curvature variation tensor in Lagrangian variables
$\tilde{\varepsilon}$ and $\tilde{\kappa}$	tensors ε and κ expressed in an orthonormal basis (Lagrange)
τ	variation of the third quadratic form tensor (Lagrange)
e	surface strain tensor in Eulerian variables
k	surface curvature variation tensor in Eulerian variables
t	time
$v = \dot{\xi}$	velocity of a point on σ
$\dot{a}_i$	velocity of a basis vector a_i
Ω	instantaneous rotation velocity vector of a_3 at a point on σ
$\mathfrak{s}$	spin tensor
$\dot{e}$	strain rate vector
$\dot{k}$	rate of curvature variation tensor
γ	acceleration of a point on σ
$\overset{*}{\xi}$	virtual velocity of a point on σ
$\overset{*}{\Omega}$	virtual rotation velocity of the normal vector at a point on σ
$\overset{*}{e}$	virtual strain rate tensor
$\overset{*}{k}$	virtual curvature variation rate tensor
Φ	physical quantity representable by a tensor
ϕ	surfacic density of Φ
ρ	surfacic mass
I	mass inertia of the material segment normal to the middle surface
$\mathscr{d}$	director vector, characteristic of the kinematics in the shell thickness
δ	field of director variation
ϑ_i	components of the director distorsion
$\mathfrak{p}_{\alpha i}$	components of the material curvature tensor
$\rho_{\alpha i}$	components of the variation of material curvature tensor

KINEMATICS AND KINETICS IN THE THICKNESS OF THE SHELL

$\overline{\xi}$	displacement of a point in the shell thickness
$\overline{V}$	velocity of a point in the shell thickness
$\overline{\gamma}$	acceleration of a point in the shell thickness
$\overline{\rho}$	volumic mass
$\overline{\varepsilon}$	tridimensional strain tensor in the shell (Lagrange)
$\overset{*}{\overline{e}}$	virtual strain rate tensor (tridimensional)

ENERGIES

$\mathcal{C}$	kinetic energy
$\mathcal{W}_j$	inertial actions work
$\mathcal{W}_e$	external actions work
$\mathcal{W}_i$	internal actions work
$\mathcal{W}_f$	strain energy
$\overset{*}{\mathcal{W}}$	virtual power of an action
Φ	internal thermodynamic potential
ϕ	density of internal thermodynamic potential
$\mathcal{F}$	elastic potential
$\mathcal{F}_m$	membrane elastic potential
$\mathcal{F}_f$	flexural elastic potential
T	temperature
τ	temperature variation
Θ	mean temperature variation in the thickness of the shell
$\Delta\Theta$	difference of temperature variation between the two faces of the shell

EXTERNAL ACTIONS

p	pressure
p	surfacic density vector of external forces on σ
c	surfacic density vector of external moments on σ
q	linear density vector of external forces on $\partial\sigma$
m	linear density vector of external moments on $\partial\sigma$

INTERNAL ACTIONS

$\overline{\overline{T}}$	tridimensional stress tensor, in Euler variables
T^{ij}	components of $\overline{\overline{T}}$
$\overline{\overline{\sigma}}$	tridimensional stress tensor, in Lagrange variables
N	generalised stress tensor associated to ε
M	generalised stress tensor associated to κ
Q	generalised stress tensor associated to distorsion ϑ

N, $\bar{N}$	membrane tensor
M, $\bar{M}$	bending moment tensor
L	modified membrane tensor
Q , $\bar{Q}$	shear force vector
$\mathcal{R}$	resultant of general stresses on a curve
$\mathcal{M}$	resultant moment of general stresses on a curve
N	axial force in the tangent plane, applied to a facet $\mathcal{F}$ (per unit length)
T	shear force in the tangent plane, applied to a facet $\mathcal{F}$ (per unit length)
V	shear force normal to the tangent plane, applied to a facet $\mathcal{F}$ (per unit length)
M	bending moment applied to a facet $\mathcal{F}$ (per unit length)

Thermo-elastic constitutive law

α_t	thermal coefficient of expansion of an isotropic solid
$\bar{\alpha}$	tridimensional tensor of thermoelasticity coefficients
Λ	surfacic tensor of elasticity coefficients
$\bar{\Lambda}$	tridimensional tensor of elasticity coefficients
E	Young modulus
ν	Poisson coefficient
$K = \dfrac{Ee}{1-\nu^2}$	membrame shell stiffness
$D = \dfrac{Ee^3}{12\left(1-\nu^2\right)}$	flexural shell stiffness

Chapter 1

Introduction to the behaviour of shells

This chapter is a first approach to the behaviour of shells, where some essential but simple ideas are discussed, which are developed in the following chapters.

It is based on elementary notions encountered in Solid Mechanics and more precisely in the Mechanics of Structures.

After having defined the objectives of the theory of shells, the preponderant role of shell curvature in their behaviour is highlighted by simple examples.

The basic concepts commonly used in shell theory, including the nature of internal forces, are introduced by adopting "natural" assumptions.

1.1 SOME GENERAL IDEAS

1.1.1 Definitions

1.1.1.1 Shell theory

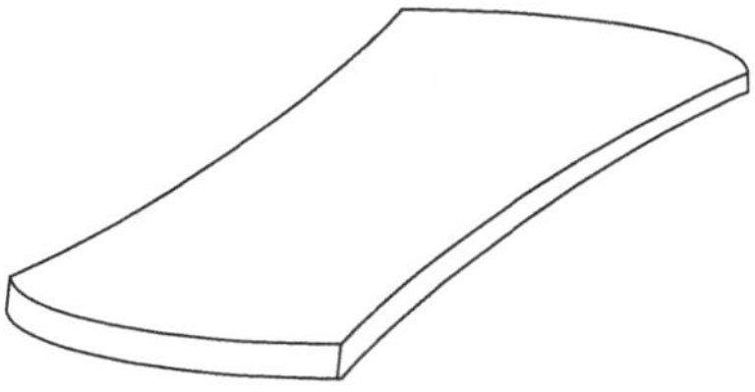

Figure 1.1

The term shell commonly refers to a solid with a dimension that is much smaller than the other two dimensions (Figure 1.1). This dimension is called **thickness**. This simple definition is however not sufficient to establish a shell theory.

Physically, according to the approach of the theory of structures, this fundamental property gives rise to hypotheses making it possible to simplify the study of the behaviour of the shell in its thickness, in order to reduce the study of the three-dimensional solid body to that of a body to two dimensions.

Mathematically, this reduction can only be of practical interest if the two-dimensional geometrical entity representing the shell has sufficient properties of regularity (continuity, piecewise differentiability): indeed, the equations of motion are partial differential equations. However, particular singularities, such as discontinuity of slope, can be the subject of specific treatment in the context of shell theory.

Shell theory is therefore intended for a family of three-dimensional media with particular properties:

- geometric dimensions,
- regularity,
- mechanical behaviour,

DOI: 10.1201/9780429440403-1

such as the model of analysis which results from the basic hypotheses gives satisfactory results in comparison with the three-dimensional theory and, of course, with the experimental results, which obviously gives it all its practical interest.

► In fact, a shell is not a natural object in itself, but is defined by the model(s) developed in the next chapters, in particular by its geometrical representation in its successive configurations. Just as in the case of beams, several types of models are possible, depending on the geometry (shape of the shell), kinematics (assumptions on the displacement), and mechanical behaviour assumptions retained. There is therefore no single shell model. Subsequent developments aim to provide a reasoning framework applicable to any shell model and to show the usual models.

Note: As a solid, the shell can obviously be studied by methods applicable to three-dimensional media. The shell models are intended, by a direct approach and following the methods of the theory of structures, to calculate the behaviour of such solids from simpler equations than the three-dimensional equations. As always in physics, a model is considered effective if it gives satisfactory results compared to the experimental observation. In the present case, it would also be interesting to ensure that a solution obtained by a shell model is a limited solution with respect to the three-dimensional solution, the difference between the two making it possible to evaluate the error introduced by the "shell" solution. This aspect, which is mathematically difficult, is not addressed in this document. Nevertheless, some known practical situations developed from Chapter 6 make it possible to discuss the validity of the different models studied (see also §1.2.2.2). In the case of the simplest shell models, the fields making it possible to represent the behaviour of the middle surface are in principle sufficient to represent the behaviour of the shell. Nevertheless, in many situations, this approach is insufficient and other parameters are needed. The appropriate methods are discussed in Chapter 5.

1.1.1.2 Middle surface and thickness

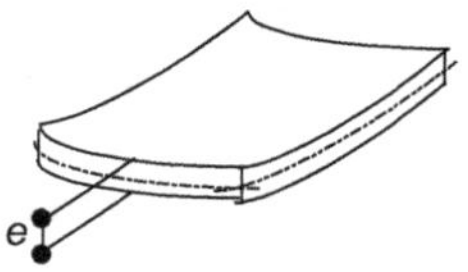

Figure 1.2

Originally, shell theory was built around the particular case, very important in practice, of thin shells, that is shells whose thickness is very small as compared to the minimum radii of surface curvature (or its span). The developments in the following chapters show that, as the thickness increases, the model to be used to correctly represent the behaviour becomes complicated and progressively loses interest in comparison to solving the three-dimensional equations (numerically, for example, using the finite element method).

► It is necessary to make the assumption, at the very base of shell theory, that there is a **middle** surface, from which the shell is built by thickening (Figure 1.2): the thickness is measured in all point perpendicular to the middle surface (along the segment normal to the middle surface, inscribed in thickness, called **normal segment** in the following); it can vary according to the point considered.

In order to completely characterise it, it should be noted that each point of the middle surface is the centre of mass of the **normal segment**. By designating the thickness of the shell and the dimension along the normal segment, measured from the middle surface, in common cases[1] :

$$\int_{-e/2}^{e/2} z\,dz = 0 \tag{1.1}$$

Note: This definition is satisfactory in the case of solid shells, the thickness of which varies slightly and regularly. It only covers approximately the cases where the thickness is very uneven, even discontinuous, as is the case for example in a ribbed floor: the set of centres of mass no longer constitutes a regular surface. The approximation provided by the shell theory is all the better as the ribs are closer together: it is a problem of scale and homogenisation. This question is considered in particular in Chapter 8.

► The definition of the thickness proposed above assumes the choice of a reference configuration: during a configuration change (in particular when the displacements can no longer be considered small), the material segment normal to the middle surface is deformed and the thickness varies: the thickness is not an invariable data *a priori.*

By generalising somewhat, the shell, considered a three-dimensional medium, can be parameterised using the parameterisation of the middle surface. In this case, if $\mathcal{P}$ is the current point in the thickness of the shell, its curvilinear coordinates (x^1, x^2, x^3) are such that:

$$x^3 \in \left[\mathcal{a}(x^1,x^2), \mathcal{b}(x^1,x^2)\right], \quad \text{with} \quad \mathcal{a} < 0 \quad \text{and} \quad \mathcal{b} > 0$$

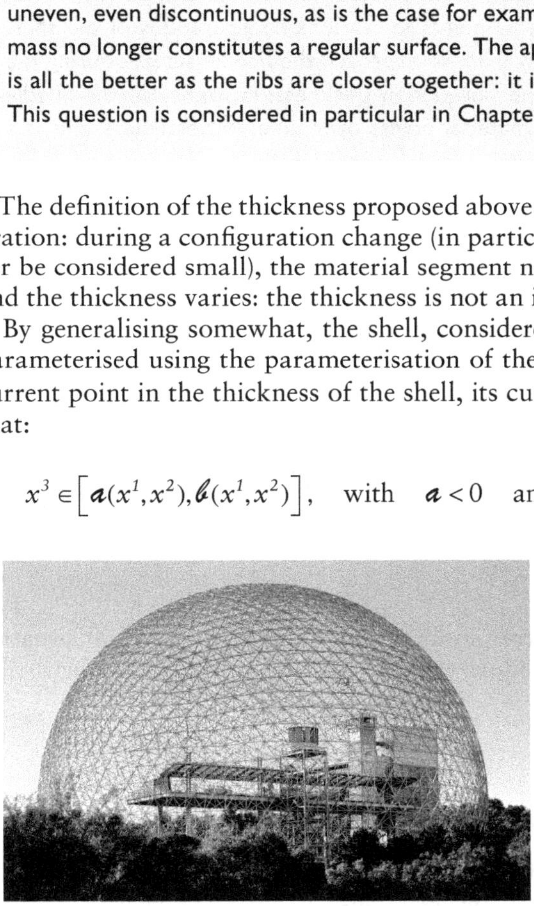

Photo 1.1 A three-dimensional structure made up of steel bars can be modelled as a shell vis-à-vis certain types of loads (Montreal, Canada)

is a given reference position. x^3 can be measured in a direction that is not tangent to the middle surface, but not necessarily in the normal direction. By convention, the surface $x^3 = 0$ refers to the middle surface.

Thus, in this reference position, the shell is the set of material surfaces $\mathcal{P}(x^1, x^2, x^3 = \text{cst})$, limited by both extreme surfaces $x^3 = \mathcal{a}(x^1, x^2)$ and $x^3 = \mathcal{b}(x^1, x^2)$. This parameterisation, although quite general, does not make it easy to represent situations where there are strong

[1] In the case of materials with tensile and compressive non-symmetrical behaviour, such as reinforced concrete, or composite materials asymmetrical along the thickness, the middle surface can be characterised differently. Guidance on this issue is given in Chapter 5. In such cases, "mean surface" is a more convenient wording than "middle surface", which will be used nevertheless as the most common designation.

discontinuities; it allows, on the other hand, to give a mathematical framework to the cases where:

- the thickness is variable and can not be considered very small,
- the "middle" surface is not really in the geometric middle of the thickness (anisotropic shells),
- it is desirable to use a mathematical representation free from the normal to the shell.

Note: From a physical point of view, the geometric surface defined by relation (1.1) is not necessarily constituted of the same material points for various configurations of the shell. Indeed, the material points can move away from or approach the middle surface during the evolution of the shell, which relates to the variation in thickness. In particular, the middle surface is not *a priori* made of the same material points over time. Nevertheless, this is a fundamental assumption in shell theory.

► Over time, a material point occupies successive positions that can be represented by a function $\mathcal{P}(x^1, x^2, x^3, t)$. This general parametric representation is often restricted in the context of particular behavioural assumptions (see Chapter 5 for example).

Such a coordinate system can be used in Lagrangian representation to describe the successive positions of the shell during its movement. Any material point is then set in the form: $\mathcal{P}\left[x^1(t), x^2(t), x^3(t)\right]$. It is not so obvious that the shell can be parameterised simply, in all circumstances, in Lagrange[2] variables, by its movements from a "natural" position: for example, what about the initial position of a balloon, inflated in final configuration? In such a case, an Eulerian representation is more appropriate.

1.1.1.3 Particular case of plates

When, in a given reference position (corresponding, for example, to the ideal situation where the structure is not loaded), the middle surface is a portion of a plane, the shell is called a **plate**.

1.1.2 Importance of shape

1.1.2.1 Spherical tank under pressure

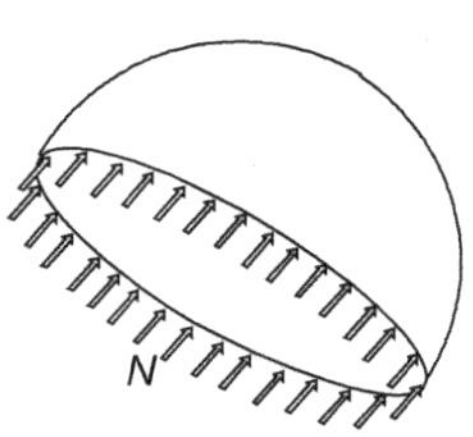

Figure 1.3

Consider a spherical tank of radius R_0 subjected to a constant internal pressure p. The shell consists of a homogeneous material forming a very thin envelope of which, locally, the bending stiffness is negligible (this notion is analysed in the next chapters).

The sphere is cut off along any diametrical plane (Figure 1.3); the stresses exerted on the edge of the cut, thus released, result in a force N per unit length.

Given the spherical symmetry, N is constant and normal to the plane of cut and balances the pressure applied on the half sphere

[2] Joseph Louis, comte de Lagrange, French (born Italian) mathematician and astronomer (1736–1813).

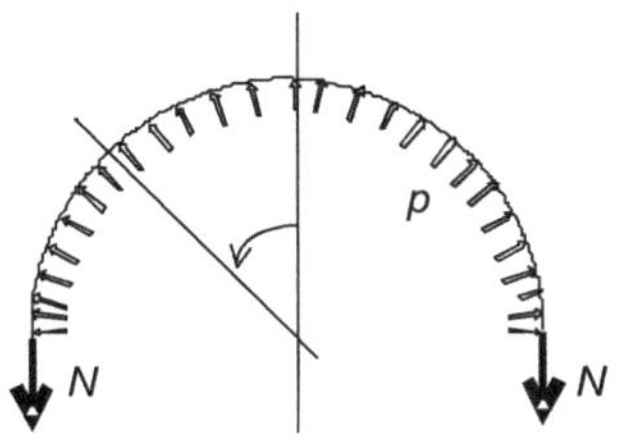

Figure 1.4

considered (Figure 1.4). By projection on the normal to the plane of cut becomes:

$$2\pi R_0 N = \int_0^{2\pi} R_0^2 d\theta \int_0^{\pi/2} p \cos\varphi \sin\varphi \;\, d\varphi = \pi R_0^2 p$$

Indeed, according to Stokes' theorem, the resulting force applied by the pressure on the hemisphere is the same as the force applied on the diametral plane. So:

$$\boxed{N = \frac{pR_0}{2}} \tag{1.2}$$

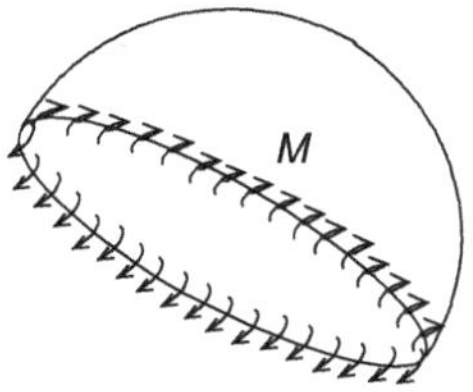

Figure 1.5

If the bending stiffness is not negligible, the stresses on the cut can have a resulting moment M per unit length (bending moment), but the resultant of this moment along the cut is zero (Figure 1.5); it cannot therefore be determined by simple equilibrium considerations. On the other hand, the out-of-plane shear force applied normal to the surface at the cut is zero, by symmetry.

1.1.2.2 Cylindrical tank under pressure

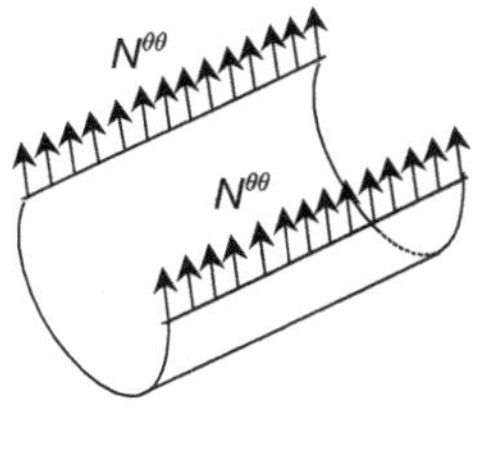

Figure 1.6

Now consider a cylindrical tank of circular section and infinite length; the distribution of the internal forces does not benefit from a symmetry similar to the spherical symmetry: in fact, a cut of the cylinder by a plane perpendicular to a generatrix does not release any force, in the event that the displacements are free in the direction of the generatrix.

On the other hand, a cut by a diametral plane parallel to a generatrix (Figure 1.6) releases stresses whose resultant $N^{\theta\theta}$ per unit length is constant and normal to the cutoff plane.

$N^{\theta\theta}$ is obtained by writing the equilibrium of a half-cylinder portion of unit length (Figure 1.7):

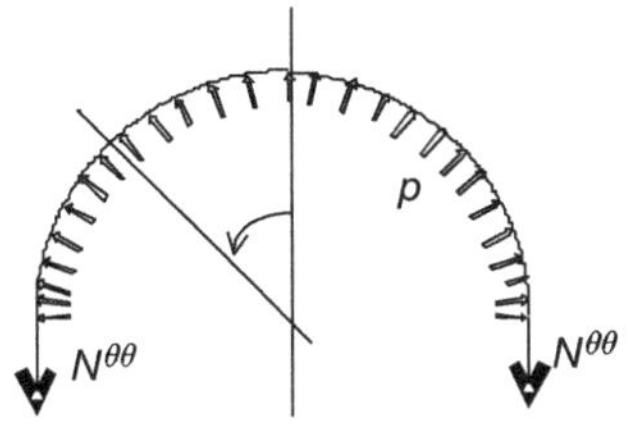

Figure 1.7

$$2N^{\theta\theta} = \int_{-\pi/2}^{\pi/2} pR_0 \cos\theta \; d\theta = 2pR_0$$

from which:

$$\boxed{N^{\theta\theta} = pR_0} \tag{1.3}$$

If the cylinder is of finite length and has bottoms (can), the pressure is also exerted on the bottoms (Figure 1.8). A cut by a perpendicular plane (for example in the middle of the can) releases a normal force per unit length, directed along the generatrices (bottom effect).

The balance of a half can be written as:

$$2\pi R_0 N^{zz} = \pi R_0^2 p$$

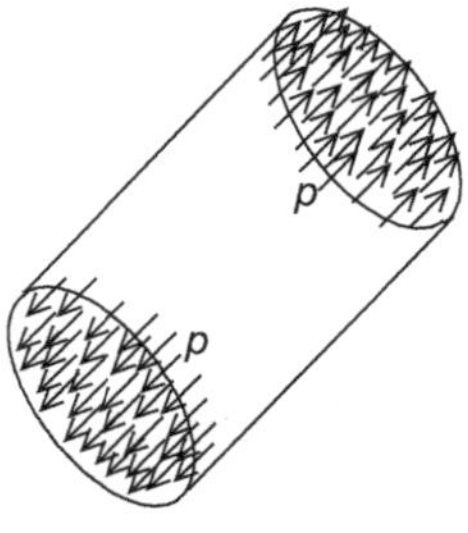

Figure 1.8

that is:

$$\boxed{N^{zz} = \frac{pR_0}{2}} \tag{1.4}$$

while this effort is zero when the cylinder is infinite and without a bottom or if the bottom effect cannot be transmitted (lyres of pipes).

> In the theory of structures in general, and shell theory in particular, internal forces are represented by generalised stresses (normal and shear forces, bending or twisting moments), which depend on the orientation of the facet on which they are evaluated

Photo 1.2 Spherical tank

Photo 1.3 Cylindrical tank

1.1.2.3 Conclusion

The comparison between the sphere and the closed cylindrical can show that:

- in the case of the sphere, the tensile force N is equal to $\frac{pR_0}{2}$ whatever the diametral plane of cut,
- in the case of the cylindrical barrel, N is equal to pR_0 or $\frac{pR_0}{2}$ according to whether the cut is parallel or perpendicular to the generatrices.

The internal forces are different, even if the internal volume and pressure are identical. What differentiates the two containers is their shapes. More precisely, it is highlighted in later chapters that the double curvature of the sphere makes it possible to better distribute the forces.

In the same way that an arc resists transverse forces by developing axial forces, thanks to its curvature, a shell resists by developing tangential forces in two orthogonal directions, thanks to its double curvature. Bending moments can also develop, for various reasons. They are necessary for equilibrium when the curvature effect does not exist (case of the plates).

1.1.3 Behaviour of the shell

By analogy with the case of a funicular arc and generalising the results obtained for spherical and cylindrical tanks under internal pressure, it is clear that a shell can withstand a load thanks to its curvature in two directions and, in many cases, without no significant flexure develops. Unlike bows and because of the double curvature, a shell of a given shape can withstand, under certain conditions, various load cases, without flexures necessary to ensure balance: a balloon, which does not resist bending, is not necessarily spherical under internal pressure. Such a state of stretching stresses where the flexion is neglected is called a **membrane. Membrane theory** is developed in Chapter 6.

Conversely, a plate subjected to forces perpendicular to the middle plane cannot, by definition, resist thanks to its curvature. It is therefore essential that the forces are transmitted to the supports in bending. The **theory of bending plates** is discussed in Chapter 7.

In the more general real conditions, where the hypotheses relating to the two particular cases above are no longer valid, the shell is simultaneously subjected to in-plane forces and bending moments.

Theories are then developed from the assumptions on the deformation of the shell according to the geometry. They are examined in particular in Chapters 4 and 5.

Another important feature of a shell is its open or closed character. For example, a cylinder is much less rigid to overall torsion along its axis if it is cut along a generatrix. A cylindrical barrel with bottoms is closed: there is no edge, so no boundary conditions. The same bottomless cylinder is semi-closed: the edges are the ends of the generatrices, where boundary conditions are to be expressed. If in addition the cylinder is cut along a generatrix, the shell is open and boundary conditions are to be expressed in both main directions. In a direction along which a shell is closed, there are periodicity conditions to be expressed, which restricts movement and make the shell more rigid in the direction considered. The rigidity is often a favourable character; however, it may not be the case if the shell is subjected to imposed displacements or deformations.

Thanks to their curvature, shells resist the loads imposed at least in part by in-plane (or membrane or stretching) forces, which in practice leads to low thickness to assure the resistance. But when the membrane state is compression in a given direction, there is a risk of instability that can lead to rupture. This is a risk, for example, for a submarine that is at a much greater depth than expected, as the external pressure creates a state of generalised compression of the shell that can cause instability, even if the strength of the metal is not primarily involved. **Stability of shells** is discussed in Chapter 8.

1.1.4 Methods used

Equilibrium equations and boundary conditions deal with generalised stresses. They can be obtained by two methods.

The first method consists in defining the generalised stresses resulting from the three-dimensional stresses in the thickness of the shell. Shell equilibrium equations are then obtained, with certain assumptions, by integrating three-dimensional equilibrium equations (see Chapter 5).

The second method is a direct approach where the shell is reduced to its middle surface. It consists in supposing that the strain energy of the shell can be expressed with a sufficient approximation starting from certain variables representing the deformation of the middle surface. Equilibrium equations and boundary conditions are then obtained by applying the Principle of Virtual Power (PVP).[3] This method makes it possible to specify the modes of deformation of the shell and it adapts, without any other difficulty than that of the calculations, to any hypothesis of deformation. From its clarity and power, it allows the concise development of the general theory (Chapters 4 and 5). An equivalent approach consists in writing the equilibrium of a shell portion assimilated to its middle surface, by the balance theorems, by introducing the screw resulting from the stresses in the thickness (Chapter 4).

Photo 1.4 Aircraft hangar (Marignane, France). This bold structure combining arches and a reinforced concrete shell 6 cm thick is due to Auguste Perret (architect) and Nicolas Esquillan (engineer)

1.2 NATURE OF INTERNAL FORCES

A general method of shell theory is to define on the middle surface the fields of scalars, vectors, and tensors[4] allowing to represent globally displacements, deformations, and actions to which the shell is subjected. These fields defined on the middle surface (so-called global fields) are the unknowns of the problem. They are then connected to three-dimensional fields of displacement, deformation, and stress (defined at any point in the volume occupied by the solid).

1.2.1 Generalised actions applied to the shell

Let σ a part of the middle surface be limited by a closed curve $\mathcal{C}$ (Figure 1.9). The three-dimensional medium $\mathcal{D}$ consisting of the shell portion bounded by the surface $\mathcal{S}$ generated by the normal segments relying on $\mathcal{C}$ is assimilated to σ; σ is subject to:

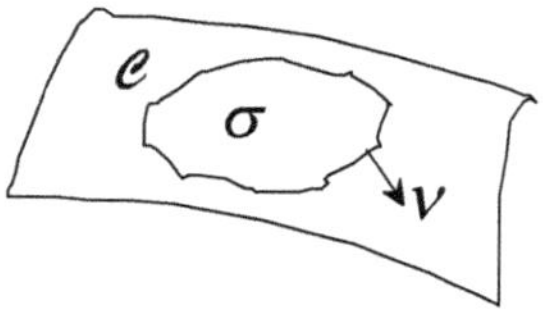

Figure 1.9

- external loads, in the form of moments and forces, concentrated or distributed, applied to $\mathcal{D}$, expressed on σ,
- contact actions applied on $\mathcal{S}$, expressed along $\mathcal{C}$, actions exerted by the outer portion to σ on σ; these actions can be represented by a screw per unit length on $\mathcal{C}$.

[3] or Principle of Virtual Works, according to the preferred notations.
[4] See Chapter 2.

At any point of $\mathcal{C}$, the screw resulting from the contact actions applied to σ is composed of two vectors (Figure 1.10):

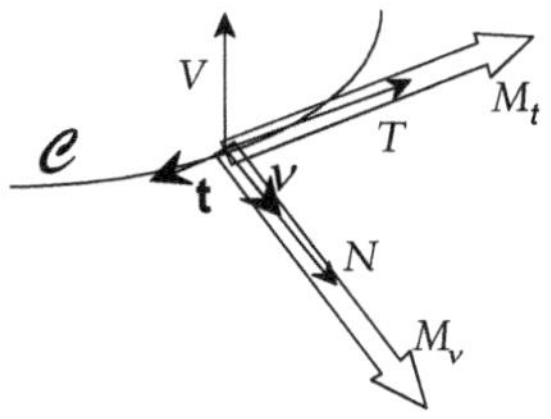

Figure 1.10

- a resultant (stretching) force per unit length with components:
  - component N along v is called **in-plane normal** (or **direct**) **force**,
  - component T in the plane tangent to σ and perpendicular to v is called **in-plane** (or **central**) **shear force**,
  - component V normal to σ is called **out-of-plane** (or **transverse**) **shear force.**
- A resulting moment per unit length with components:
  - component M_v along v is called the **twisting moment.** It should be noted that, while this definition appears to be consistent with that used in beam theory, it does not cover the same concept, since the effects caused by such a moment in terms of stresses are different from those caused by a torsional moment in a beam;
  - component M_t in the plane tangent to σ and perpendicular to v is called the **bending moment,** its effects are similar to those in a beam;
  - component normal to σ is due to shear in the plane tangent to σ; this component is generally not considered in usual shell theory as it is of little practical interest. However, there is no difficulty in introducing this component into the analysis of the shell behaviour (cf. for example §4.5).

Leaving aside this last component, contact actions have five components at every point of $\mathcal{C}$.

In membrane theory, the moment resulting from contact actions is neglected, whatever the point m and the direction v considered. In pure flexional theory, on the contrary, the resultant force with components (N,T), tangent to σ is neglected.

By analogy with the three-dimensional theory[5], by assimilating the shell to its middle surface, as a two-dimensional medium, a fundamental result[6] is that, at a given point of $\mathcal{C}$, the screw of the contact actions depends linearly on v. This makes it possible, in particular, to introduce a tensor of in-plane (or membrane or stretching) forces $\boldsymbol{N}$, defined on σ, such that:

$$\begin{pmatrix} N \\ T \end{pmatrix} = \boldsymbol{N} \cdot \mathrm{v}$$

and a vector $\boldsymbol{Q}$ such that:

$$V = \boldsymbol{Q} \cdot \mathrm{v}$$

Note: The above considerations relate to a particular configuration of the shell. It is highlighted that the notions of middle surface, thickness, and vector v depend on the configuration and evolution of the geometry during a transformation. In reality, the expression of the

[5] Cauchy postulate and theorem.
[6] Cf. §4.2.1.3.

resulting screw is related to the shell model used. The most common models are studied here, admitting the existence of the basic geometrical and mechanical entities defined above.

1.2.2 Link with three-dimensional stresses

1.2.2.1 Approximate values of stresses

Pressure vessels serve as an example again.

In the case of the spherical tank, if the thickness is small (low flexural stiffness), the stresses can reasonably be assumed to be uniformly distributed in the thickness of the shell. In this hypothesis, the bending moments are zero and the tensor of the local stresses $\overline{\overline{T}}$ (Eulerian[7]) can be written in the orthonormal spherical coordinate system $(\tilde{e}_\theta, \tilde{e}_\phi, e_r)$, given the results of §1.1.2:

$$\overline{\overline{T}} = \begin{pmatrix} \frac{pR_0}{2e} & 0 & 0 \\ 0 & \frac{pR_0}{2e} & 0 \\ 0 & 0 & T^{rr} \end{pmatrix} \tag{1.5}$$

T^{rr} may be neglected vis-à-vis the other two stresses, for reasons given in §1.2.2.2 below.

The stress tensor in the cylindrical tank with a bottom can be written approximately in the orthonormal cylindrical coordinate system $(\tilde{e}_\theta, e_z, e_r)$, T^{rr} being in fact negligible:

$$\overline{\overline{T}} = \begin{pmatrix} \frac{pR_0}{e} & 0 & 0 \\ 0 & \frac{pR_0}{2e} & 0 \\ 0 & 0 & T^{rr} \end{pmatrix} \tag{1.6}$$

At identical thickness and radius, the maximum stress developed in the cylinder is twice that developed in the sphere, because of the double curvature of the latter: the sphere is more resistant to pressure.

1.2.2.2 Comparison with the three-dimensional solution

The approximation given above is made clearer using the three-dimensional elastic solution. In the case of the cylinder, the expression of the stresses can easily be obtained, in small displacements, by solving the equation of elasticity in the orthonormal coordinates associated with the cylindrical coordinates; it depends on the boundary conditions along the axis of the cylinder:

[7] Leonhard Euler, Swiss mathematician and physicist (1707–1783).

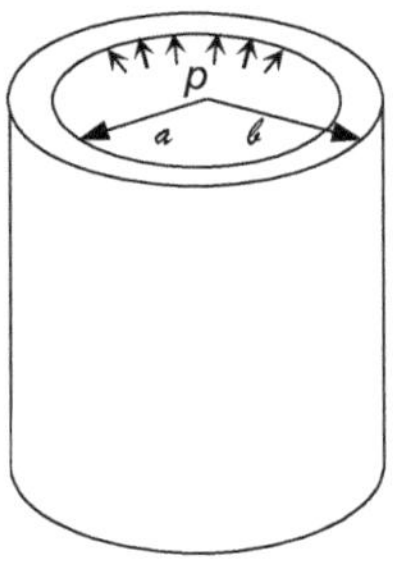

Figure 1.11

1. in plane deformations and small displacements, the displacement along z is assumed to be null, the Navier[8] equation is reduced to:

$$\text{grad div}\left[u(r)e_r\right]=0$$

with the boundary conditions (the stresses being expressed in Lagrange variables) – Figure 1.11:

$$\sigma^{rr}(a)=-p$$

$$\sigma^{rr}(b)=0$$

it becomes:

$$\sigma^{rr}=p\frac{a^2}{b^2-a^2}\left[1-\frac{b^2}{r^2}\right]$$

$$\sigma^{\theta\theta}=p\frac{a^2}{b^2-a^2}\left[1+\frac{b^2}{r^2}\right]$$

$$\sigma^{zz}=2\nu p\frac{a^2}{b^2-a^2}$$

and the radial displacement can be written as:

$$u(r)=p\frac{a^2}{b^2-a^2}\frac{1+\nu}{E}\left[(1-2\nu)r+\frac{b^2}{r}\right]$$

In this situation, the deformation ε_{zz} is zero (the cylinder is flanged longitudinally) and:

$$\sigma^{zz}=\nu\left(\sigma^{rr}+\sigma^{\theta\theta}\right)$$

This solution is slightly different from that obtained in §1.1.2.2 in the case of the cylinder not longitudinally restrained.

σ^{rr} is always negative (compression), while $\sigma^{\theta\theta}$ is always positive (traction) and maximum on the inner surface.

2. in plane stresses, the axial displacement w is not zero. This situation corresponds to the case of the non-restrained cylinder longitudinally examined in §1.1.2.2.

Starting from the three-dimensional constitutive law:

$$\sigma^{ij}=\frac{E}{1+\nu}\left(\frac{\nu}{1-2\nu}\text{TR}(\varepsilon)g+\varepsilon\right)$$

[8] Claude Louis Marie Henri Navier, French engineer, mathematician, and economist (1785–1836).

the condition of free deformation along the z-axis: $\sigma^{zz}=0$ (without the bottom effect, which can be added) gives the relation:

$$\varepsilon_{zz} = -\frac{\nu}{1-\nu}\left(\varepsilon_{rr}+\varepsilon_{\theta\theta}\right)$$

then the constitutive law in-plane stresses:

$$\sigma^{rr} = \frac{E}{1-\nu^2}\left(\varepsilon_{rr}+\nu\varepsilon_{\theta\theta}\right)$$

$$\sigma^{\theta\theta} = \frac{E}{1-\nu^2}\left(\varepsilon_{\theta\theta}+\nu\varepsilon_{rr}\right)$$

with $\varepsilon_{rr}=u'$ and $\varepsilon_{\theta\theta}=\frac{u}{r}$, where $u(r)$ is the radial displacement supposed not to depend on z. This constitutive law makes appear the rigidity of plane stresses $\frac{E}{1-\nu^2}$ whose fundamental role is subsequently developed.

The equation of equilibrium:

$$\frac{d\sigma^{rr}}{dr}+\frac{\sigma^{rr}-\sigma^{\theta\theta}}{r}=0$$

is rewritten:

$$\frac{d}{dr}\left(u'+\nu\frac{u}{r}\right)+\frac{1}{r}(1-\nu)\left(u'-\frac{u}{r}\right)=0$$

which integrates into:

$$u(r)=\frac{Ar}{2}+\frac{B}{r}$$

This relation is similar to that obtained in plane deformations, but the expressions of the displacements are slightly modified taking into account the plane stresses law. By expressing the boundary conditions on σ^{rr}, it finally becomes:

$$u(r)=\frac{p}{E}\frac{a^2}{b^2-a^2}\left[(1-\nu)\,r+(1+\nu)\frac{b^2}{r}\right]$$

$$w(z)=-2\nu\frac{p}{E}\frac{a^2}{b^2-a^2}z+w_0$$

and the expressions of σ^{rr} and $\sigma^{\theta\theta}$ are identical to those obtained in the case of plane deformations.

Noting: $b=R_0+\frac{e}{2}$ and $a=R_0-\frac{e}{2}$, where :

- R_0 is the mean radius of the cylinder,
- e its thickness,

it becomes:

$$\sigma^{\theta\theta}(a) = p\frac{R_0^{\ 2} + \frac{e^2}{4}}{eR_0}$$

$$\sigma^{\theta\theta}(b) = p\frac{R_0^{\ 2} + \frac{e^2}{4} - eR_0}{eR_0}$$

A median value of the stress is:

$$\overline{\sigma}^{\theta\theta} = \frac{1}{2}\left[\sigma^{\theta\theta}(a) + \sigma^{\theta\theta}(b)\right] = \frac{pR_0}{e} - \frac{p}{2} + \frac{pe}{4R_0}$$

and the variation between the two faces

$$\sigma^{\theta\theta}(a) - \sigma^{\theta\theta}(b) = p$$

is related to the variation of σ^{rr} by equilibrium equations.

If $e << R_0$ (thin tube), approximations:

$$\overline{\sigma}^{\theta\theta} = \frac{pR_0}{e} \quad ; \quad \frac{\sigma^{\theta\theta}}{\sigma^{rr}} = \frac{R_0}{e} \tag{1.7}$$

conclude that orthoradial stresses are much larger than radial stresses: the tube withstands the tensile pressure along the ring perpendicular to the axis of the cylinder. The stress $\sigma^{\theta\theta}$ is even larger than the radius is larger. The thin tube hypothesis is consistent with the approximation obtained in §1.2.2.1. These results are independent of the hypothesis taken (restrained or not) along the z-axis.

Note: Considering the expression of the displacement field (in either case), it is easy to verify that the radial displacements vary in thickness and that the average of the radial displacements of the two faces is not equal to the radial displacement of the middle of the thickness. This makes it possible to conclude on the one hand that the thickness varies during the transformation, on the other hand, that the material points of the middle surface do not remain on the middle surface and thus to confirm the comments of §1.1.1.2.

1.2.2.3 Resultant of three-dimensional stresses

The considerations above make it possible to clarify the notion of in-plane force in a "thin" shell. A facet contained in the plane of the cut and of which one dimension is the thickness of the shell is subjected to the screw resulting from the three-dimensional stresses applied to it. If the stress tensor is assumed of the form:

$$\overline{\overline{T}} = \begin{pmatrix} T_{11} & T_{12} & 0 \\ T_{12} & T_{22} & 0 \\ 0 & 0 & 0 \end{pmatrix} \tag{1.8}$$

where the pinch stress in the thickness (T^{rr} for the sphere or the cylinder) and the shear stress resulting from transverse shear forces are neglected, the resulting internal force is in the plane tangential to the middle surface of the shell.

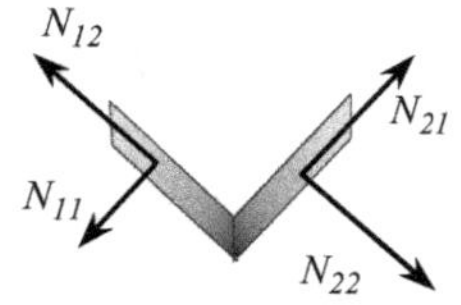

Figure 1.12

If the stresses are constant in the thickness, the state of stress is entirely determined by the plane-symmetric tensor (the symmetry is due to that of the stress tensor), obtained by multiplication of the components $T_{\alpha\beta}$ by thickness e:

$$\overline{\overline{N}} = \begin{pmatrix} N_{11} & N_{12} \\ N_{12} & N_{22} \end{pmatrix} \tag{1.9}$$

More generally, the tensor $\overline{\overline{N}}$ is introduced as a resultant of the stresses defined by (1.8) (even when these vary along the thickness):

- (N_{11}, N_{12}) are the components of the force resulting from stresses applied to the facet, which normally is parallel to the first basic vector (Figure 1.12).
  - is normal to the facet,
  - is tangent (in-plane shear).
- (N_{12}, N_{22}) are the components of the force resulting from stresses applied to the facet, which normally is parallel to the second basic vector.

In the case of the sphere under internal pressure:

$$N_{\theta\theta} = N_{\phi\phi} = \frac{pR_0}{2}; \qquad N_{\theta\phi} = 0$$

and in the case of the bottomless cylinder:

$$N_{\theta\theta} = pR_0 \quad ; \qquad N_{zz} = N_{z\theta} = 0$$

It is obvious, in the case of a thin shell in membrane theory, that the tensor $\boldsymbol{N}$, introduced in §1.2.1 from the generalised contact actions, is identical to the tensor $\overline{\overline{N}}$ introduced above as a resultant of stresses. Such an identity is less obvious in the case of less restrictive hypotheses (cf. §5.1.3).

1.2.3 Validity of the membrane theory

Now consider a cylindrical tank, closed at both ends by a hemisphere (Figure 1.13).

The result of the pressure on the bottoms is transmitted along the generatrices of the cylinder. Indeed, a cut by a plane perpendicular to the generators shows that $N^{zz} = \dfrac{pR_0}{2}$.

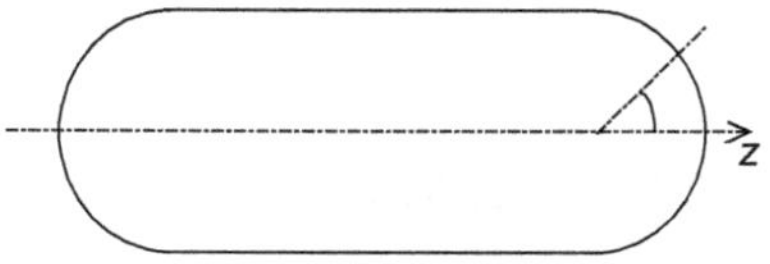

Figure 1.13

Examination of the connection zone between the cylinder and a hemisphere shows that N^{zz} is in the continuity of $N^{\phi\phi}$, that in-plane shear stresses $N^{\theta\phi}$ et $N^{\theta z}$ are null, but that $N^{\theta\theta}$ is discontinuous: this is the consequence of the discontinuity of the curvature along θ at the connection.

There are necessarily in this zone bending moments, the deformations due to the normal forces being not compatible between the spherical part and the cylindrical part (cf §6.1.2.3), which is obvious in the – simple – hypothesis where the orthoradial deformation is proportional to $N^{\theta\theta}$. A membrane state of equilibrium is therefore not always possible, even in a shell with double curvature, because of a lack of kinematic continuity. Other situations limit the membrane theory (see Chapter 6). In general, a shell simultaneously develops in-plane forces and bending moments. In these cases, the stresses cannot be uniformly distributed in the thickness.

1.2.4 Contribution of global balances

Simple equilibrium considerations associated with possible symmetries sometimes make it possible to determine certain internal actions. This has already been highlighted in §1.1.2.

1.2.4.1 Infinite plane in cylindrical bending

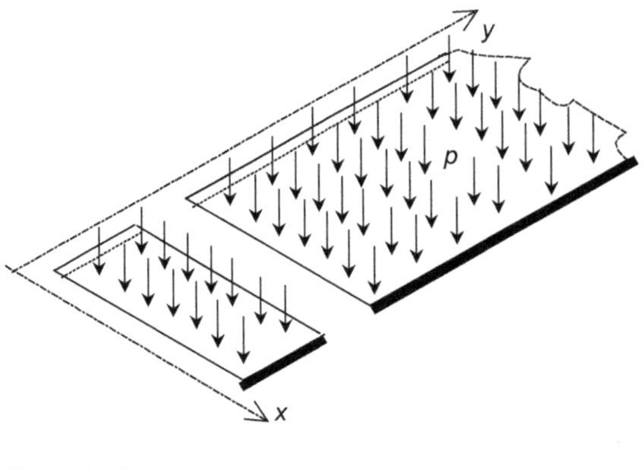

Figure 1.14

Consider an infinite plate in one direction, simply resting on the two finite-distance edges (Figure 1.14). This plate is uniformly loaded by a density of load p. Due to the infinite dimension along y, the deformation is independent of y. Thus, the plate can be cut into parallel "slices" of unit width deforming identically.

A "slice" of the plate, bounded by two straight lines, can be likened to a beam, which makes it possible to easily determine the transverse shear force V^x and the bending moment M^{xx} in the main direction of bending of the slice. In particular, the bending moment is parabolic and is at most:

$$M = \frac{p\ell^2}{8}$$

where ℓ is the span of the plate in the direction x.

Moreover, in small displacements, the curvature of each slice along x is equal to $\frac{d^2w}{dx^2}$, where w refers to the transverse displacement (deflection), independent of y. In the thickness of the plate, at z with respect to the middle surface, the deformation $\overline{\varepsilon}_{xx}$ is calculated in accordance with beam theory (Bernoulli[9]–Euler hypothesis):

[9] Daniel Bernoulli, Swiss mathematician and physicist (1700–1782).

$$\bar{\varepsilon}_{xx} = -z\frac{d^2w}{dx^2}$$

On the other hand, there is no curvature along y and the condition of cylindrical bending imposes that $\varepsilon_{yy} = 0$.

In the case of a homogeneous and isotropic elastic plate, the local deformations are calculated from the stresses, in the event that $\sigma^{zz} = 0$ (plane stress elastic constitutive law):

$$\begin{cases} \bar{\varepsilon}_{xx} = \dfrac{\sigma_{xx}}{E} - \nu\dfrac{\sigma_{yy}}{E} \\ \bar{\varepsilon}_{yy} = \dfrac{\sigma_{yy}}{E} - \nu\dfrac{\sigma_{xx}}{E} = 0 \end{cases}$$

hence the stresses, in Lagrange variables:

$$\begin{cases} \sigma_{xx} = -\dfrac{Ez}{1-\nu^2}\dfrac{d^2w}{dx^2} \\ \sigma_{yy} = \nu\sigma_{xx} \end{cases}$$

The relationship between the curvature and the bending moment is obtained by integration, for a unit width slice:

$$M_{xx}(x) = -\int_{-e/2}^{e/2} \sigma_{xx}\, z\, dz = \frac{Ee^3}{12\left(1-\nu^2\right)}\frac{d^2w}{dx^2} = D\frac{d^2w}{dx^2}$$

where D is the bending stiffness of the plate, which differs a little from the bending stiffness of the beam of unit width $EI = \dfrac{Ee^3}{12}$ (cf. §5.4.3.3). The moment $M_{yy} = \nu\, M_{xx}$ is not zero.

1.2.4.2 Spherical dome

Consider a spherical dome subjected to its own weight. In the hypothesis of a state of membrane equilibrium, a cut by a horizontal plane releases a force tangent to the shell (Figure 1.15). Angle φ being the angle of the cut with the vertical, writing the equilibrium of the upper cap, in vertical projection, makes it possible to obtain the relation:

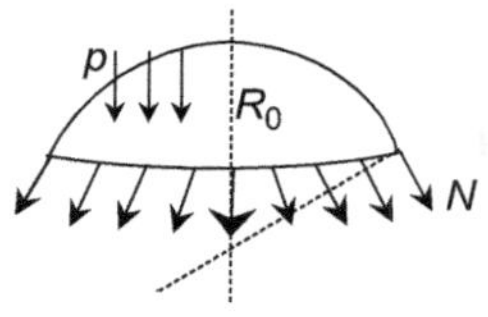

Figure 1.15

$$2\pi R_0 N \sin\varphi + 2\pi\int_0^{\varphi} pR_0^2 \sin\psi\, d\psi = 0$$

from where:

$$N = -\frac{pR_0}{1+\cos\varphi}$$

This does not make it possible to completely determine the state of stress, the orthoradial forces not being determined at this stage (see §6.1.3.1).

Photo 1.5 Church with spherical cupolas (Beirut, Lebanon)

Photo 1.6 Mosque with spherical cupolas (Beirut, Lebanon)

EXERCISES

Exercise 1.1

A balloon having the shape of an ellipsoid of revolution of axis z (Figure 1.16) is under constant internal pressure p. Determine the membrane forces applied along the cut defined by the intersection of the plane $z = 0$ with the ellipsoid, then with a plane z = cst, as a function of z. What is the value at the top?

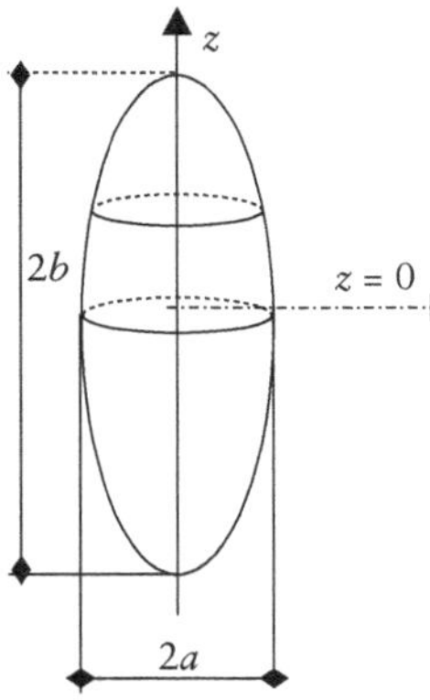

Figure 1.16

Answer:

- Given the symmetry of revolution, along the median cut, the membrane shear force is zero and the membrane normal force is constant along the cut. It is worth:

$$N^{zz} = \frac{pa}{2}$$

- Along a cut by a plane perpendicular to the z-axis, the forces are of the same nature and the normal constant force is:

$$N^{zz} = \frac{pa}{2}\sqrt{1+\left(\frac{a^2}{b^2}-1\right)\frac{z^2}{b^2}}$$

- and $N^{zz} = \dfrac{pa^2}{2b}$ at the top.

Exercise 1.2

A conical tank of revolution with axis z (Figure 1.17) and angle α, suspended from an upper ring, is filled with water of density ρ. What is the action exerted by the cone on the ring which supports it and the axial force which results from it in the ring? Determine

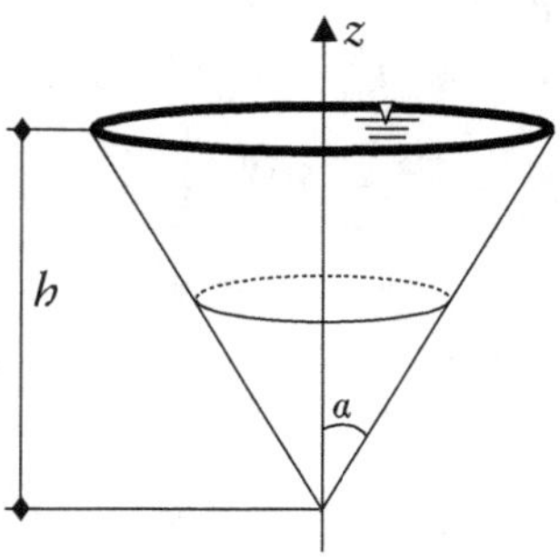

Figure 1.17

in-plane (membrane) forces along any cut by a horizontal plane z = cst, due to the action of water alone.

Answer: The ring supports the weight of the water in the tank, equal to $\frac{\pi}{3}\rho g h^3 \tan^2\alpha$. The inclined action exerted by the cone on the ring is equal to: $N^{zz}(z=h)=\frac{\rho g h^2 \sin\alpha}{6\cos^2\alpha}$, which results in compression in the ring equal to $\frac{\rho g h^3 \sin^2\alpha}{6\cos^3\alpha}$.

Given the symmetry of revolution, along a cut perpendicular to the axis of revolution, the in-plane shear force is zero and the in-plane normal force is constant along the cut. It is worth, at a height z from the bottom, writing the equilibrium of the lower part of the tank:

$$N^{zz}=\frac{\rho g h^2 \sin\alpha}{6\cos^2\alpha}\left(\frac{z}{h}\right)^2\left(3-2\frac{z}{h}\right)$$

This in-plane force vanishes near the bottom.

Exercise 1.3

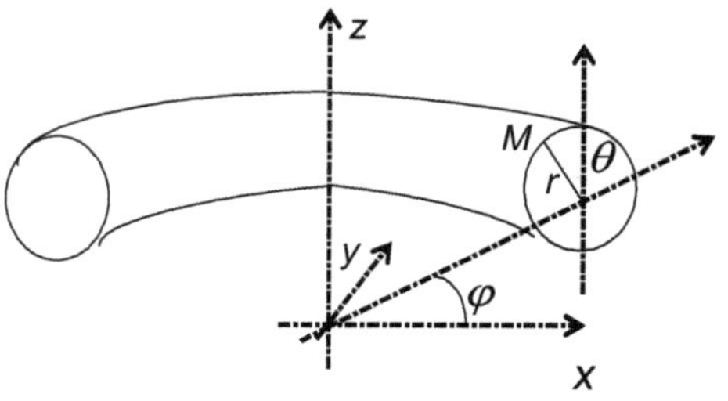

Figure 1.18

A buoy under constant internal pressure p has the shape of a torus of circular section (Figure 1.18). The radius of the small circle constituting the section of the torus is r and the radius of the great circle described by the centre of the small circle in rotation around the axis of revolution is R. What information on the internal forces is obtained by practising cuts by a diametral plane (separating the buoy into two identical half-tori) or by radial planes passing through the axis of revolution?

Answer:
In this problem, the forces and displacements do not depend on φ. The cut by a diametral plane perpendicular to the z-axis consists of two concentric circles, along which the normal forces are constant, but different between the two circles. Calling them N_1 (outer circle) and N_2 (inner circle), the equilibrium of the cutoff plane leads to:

$$N_1(R+r)+N_2(R-r)=2pRr$$

By cutting the half-torus into two identical pieces by the plane (x, z), the equilibrium expressed in a moment with respect to the x-axis leads to the following relation:

$$N_1(R+r)^2+N_2(R-r)^2=2pr\left(R^2+\frac{r^2}{3}\right)$$

These two relations make it possible to obtain the two values, which are valid only on the diametrical cut ($\theta = 0$ and π):

$$N_1 = \frac{R + r/3}{R + r} pr \; ; \; N_2 = \frac{R - r/3}{R - r} pr \quad (N_1 < N_2)$$

The same reasoning can be done for any value z of the cut-off plane, but the calculations are more difficult. Then, $N^{\theta\,\theta}$ can be fully determined.

The cut by a radial plane makes it possible to obtain the average N of the normal ortho-radial forces $N^{\varphi\varphi}$:

$$N = \frac{pr}{2}$$

this force varies as a function of θ around this mean value.

A complete solution to this problem is given in Chapter 6.

Chapter 2

Elements of tensorial algebra and analysis

This chapter is devoted to reminders of linear algebra and analysis, limited to the knowledge strictly necessary for a good understanding of the rest of the document and the practice of calculations. It can be omitted by the reader possessing a solid culture in these fields. Reading it (possibly fast) is nevertheless recommended for a good understanding of the adopted notations. Most demonstrations that are omitted are simple and can therefore be used as an exercise. The objective here is "utilitarian". Readers wishing a more general approach can refer to mathematical works.

First, the tensors on a vector space are introduced as multilinear applications.

Then, Euclidean tensors are defined by considering the metric of vector space.

To finish, the differentiation of the tensor fields is approached, which makes it possible to establish the differentiation formulas on a natural basis, then the definition of the differential operators used in mechanics are recalled to define the curvature of space.

2.1 TENSORS ON A VECTOR SPACE

2.1.1 Vector space and dual space

Let E_n a vector space of **finite dimension** n on $\mathbb{R}$ and (e_i), $i \in [1,n]$ any basis of E_n. Let V a vector of E_n expressed on the basis (e_i) by the relation:

$$V = \sum_{i=1}^{n} V^i e_i$$

where V^i denotes the components of V in the basis (e_i). The **Einstein**[1] **summation convention** allows V to be written in the form:

$$\boxed{V = V^i e_i} \tag{2.1}$$

where the index i (said dummy) is summed from 1 to n when it appears twice (one superscript, one subscript), unless otherwise indicated. The position of the index is not indifferent, as is established below. From now on, E_n is designated by E, no ambiguity being possible within the framework of this chapter.

A linear one-form ℓ on E is such that:

$$\forall V \in E: \quad \ell(V) = \ell\left(V^i e_i\right) = V^i \ell(e_i) = V^i \ell_i \in \mathbb{R} \tag{2.2}$$

[1] Albert Einstein, German physicist (1879–1955).

DOI: 10.1201/9780429440403-2

The linear one-form[2] ℓ is entirely determined by n scalars ℓ_i such as:

$$\ell_i = \ell(e_i)$$

Let e^i the linear form be as:

$$e^i(e_j) = \delta^i_j \tag{2.3}$$

where δ^i_j denotes the Kronecker[3] symbol. The components ℓ_i of the form are distinguished from those V^i of the vector by the position of the index.

So:

$$\forall V \in E : e^i(V) = e^i(V^j e_j) = V^j e^i(e_j) = V^j \delta^i_j = V^i \tag{2.4}$$

The form associates a vector V with its i-th component V^i on e^i. The set of linear forms on E is generated by the forms (e^i), $i \in [1,n]$. Indeed, the previous relations allow it to be written as:

$$\forall \ell, \quad \forall V \in E: \quad \ell(V) = V^i \ell(e_i) = \ell_i \, e^i(V) \tag{2.5}$$

The n forms e^i are independent, which is easy to verify. They thus form a basis of the space of the linear forms on E, denoted E^* and called **dual of** E. E^* is of dimension n. The family of forms (e^i) is the **dual basis** of (e_i).

The n scalars ℓ_i are the components of ℓ in basis (e^i). Thanks to (2.5), it is easy to establish a canonical bijection (that is independent of the bases) between E and E^{**}, the dual of E^* and bidual of E, so that the two spaces can be assimilated: $E \equiv E^{**}$. This identification makes it possible to write as:

$$\forall V \in E, \; \forall \ell \in E^* : \quad \ell(V) = V(\ell) = V^i \ell_i \in \mathbf{R}$$

and $\langle \ell, V \rangle$ (duality pairing) then designates the common value obtained by this equality.

2.1.2 Bilinear forms on E

Let $\mathcal{L}(E \times E, \mathbf{R})$ be the set of bilinear forms on E:

$$\begin{aligned} &\mathfrak{b} \in \mathcal{L}(E \times E, \mathbf{R}): \quad \forall (V, W) \in E \times E: \\ &\qquad \mathfrak{b}(V, W) = \mathfrak{b}(V^i e_i, W^j e_j) = V^i W^j \; \mathfrak{b}(e_i, e_j) \in \mathbf{R} \end{aligned} \tag{2.6}$$

[2] To simplify, a one-form is simply called "form" in the following.
[3] Leopold Kronecker, German mathematician (1823–1891).

ℓ and h being two linear forms, let $\ell \otimes h$ be the bilinear form defined by:

$$\forall (V,W) \in E \times E: \quad \ell \otimes h(V,W) = \langle \ell, V \rangle \langle h, W \rangle = \ell_i h_j V^i W^j \tag{2.7}$$

A bilinear form such that $\ell \otimes h$ is called **"decomposed"**. The sign $\otimes$ (tensorial) is introduced here to represent an operation between the two forms ℓ and h underlying the definition (2.7). The order of the indices is not indifferent: it determines the order of the operations carried out, therefore the nature of the bilinear form. A particular case of such decomposed forms is obtained by taking for ℓ and h the forms of the dual basis. In this case, the relation (2.7) takes the particular form:

$$\forall (V,W) \in E \times E: \quad e^i \otimes e^j (V,W) = \langle e^i, V \rangle \langle e^j, W \rangle = V^i W^j \tag{2.8}$$

The set of forms $e^i \otimes e^j$ begets $\mathcal{L}(E \times E, \mathbf{R})$; indeed:

$$\begin{aligned} &\forall b \in \mathcal{L}(E \times E, \mathbf{R}): \ \forall (V,W) \in E \times E: \\ &\quad b(V,W) = V^i W^j b(e_i, e_j) = V^i W^j \, b_{ij} = b_{ij} \, e^i \otimes e^j (V,W) \end{aligned} \tag{2.9}$$

where it is noted:

$$b_{ij} = b(e_i, e_j)$$

It is easy to demonstrate that the forms $e^i \otimes e^j$ are independent and therefore form a basis for $\mathcal{L}(E \times E, \mathbf{R})$. The n^2 scalars b_{ij} are the components of b in this basis:

$$b = b_{ij} \, e^i \otimes e^j \tag{2.10}$$

(a generalisation of the Einstein convention to two indices), from which:

$$\dim\left(\mathcal{L}(E \times E, \mathbf{R})\right) = n^2$$

The n^2 scalar b_{ij} make it possible to fully define the bilinear form b and is usually stored in a square matrix $[b]$. The sign $\otimes$ represents the **tensor product**. Any bilinear form may not be decomposed, but it is a linear combination of decomposed forms. A decomposed form is of rank one.

Now, let the space $\mathcal{L}(E, E^*)$ of functions that associate a linear form to a vector:

$$f \in \mathcal{L}(E, E^*), \qquad \forall V \in E: \ f(V) \in E^*$$

Let ℓ and h two forms on E and their product (noted provisionally T) defined by:

$$\forall V \in E: \quad \ell T h(V) = \langle \ell, V \rangle h \in E^*$$

$\ell T h$ is a decomposed element of $\mathcal{L}(E, E^*)$ and T designates the composition operation of the two forms associated with this definition. In the same way as above, it is easy to show that the decomposed elements $e^i T e^j$ de $\mathcal{L}(E, E^*)$ such that:

$$\forall V \in E : \quad e^i T e^j (V) = < e^i, V > e^j$$

are in number n^2 and form a basis of $\mathcal{L}(E, E^*)$.

The application from $\mathcal{L}(E \times E, \mathbb{R})$ to $\mathcal{L}(E, E^*)$ where the decomposed element $\ell \otimes h$ matches the decomposed element $\ell T h$ is a canonical isomorphism. $\ell \otimes h$ and $\ell T h$ are two representatives of the same equivalence class called a **tensor.** This leads to identifying both spaces and noting $E^* \otimes E^*$ (tensor product of E^* by itself) the set of such tensors. Then confusing the applications $\otimes$ and T under the common name $\otimes$, $\ell \otimes h$ designates the **tensor product** of ℓ and h, an element of $E^* \otimes E^*$. The family $(e^i \otimes e^j)$ is a basis of $E^* \otimes E^*$.

Note: Similarly, the set $E \otimes E$ can be introduced from $\mathcal{L}(E^* \times E^*, \mathbb{R})$, a set of bilinear forms on E^*, generated by the forms $e_i \otimes e_j$ such that:

$$\forall (\ell, h) \in E^* \times E^* : \quad e_i \otimes e_j (\ell, h) = \langle e_i, \ell \rangle \langle e_j, h \rangle = \ell_i h_j \tag{2.11}$$

Then:

$$\forall f \in \mathcal{L}(E^* \times E^*, \mathbb{R}) : \qquad f = f^{ij} e_i \otimes e_j \tag{2.12}$$

2.1.3 Endomorphisms on E

Let $\ell \in E^*$ and $V \in E$, from which endomorphism is formed $\ell \otimes V \in \mathcal{L}(E, E)$ such that:

$$\forall W \in E : \quad \ell \otimes V(W) = \langle \ell, W \rangle V = \ell_i W^i V^j e_j \tag{2.13}$$

$\ell \otimes V$ is a decomposed element of $\mathcal{L}(E, E)$. The particular decomposed elements $e^i \otimes e_j$ such that:

$$\forall W \in E : \quad e^i \otimes e_j (W) = \langle e^i, W \rangle e_j = W^i e_j \tag{2.14}$$

are independent and generate $\mathcal{L}(E, E)$. Indeed:

$$\forall f \in \mathcal{L}(E, E) :$$

$$f(W) = W^i f(e_i) = W^i f_i^{\,j} e_j = f_i^{\,j} \langle e^i, W \rangle e_j = f_i^{\,j} e^i \otimes e_j (W)$$

They, therefore, form a basis of space $\mathcal{L}(E, E)$, which is of dimension n^2. An element such that $\ell \otimes V$ is called a tensor product of ℓ and V and is a decomposed element (decomposed tensor) of $\mathcal{L}(E, E)$. The decomposed tensors generate the whole space.

Similarly, let the decomposed elements $\ell \otimes V$ of $\mathcal{L}(E \times E^*, \mathbb{R})$, such that:

$$\forall (W, h) \in E \times E^* : \qquad \ell \otimes V(W, h) = \langle \ell, W \rangle \langle V, h \rangle$$

It is easily established, by an approach similar to that carried out in §2.1.2, that there is a canonical isomorphism between $\mathcal{L}(E,E)$ and $\mathcal{L}(E\times E^*,\mathbb{R})$ matching the decomposed elements two by two. The two spaces are then merged and the common space thus formed is noted $E^* \otimes E$.

Note: Similarly, let $E\otimes E^*$ the space of endomorphisms $\mathcal{L}(E^*,E^*)$ on E^*:

$$\forall \ell \in E^*: \quad e_i \otimes e^j(\ell) = \langle e_i,\ell\rangle e^j = \ell_i e^j$$

Then:

$$\forall f \in \mathcal{L}(E^*,E^*): \quad f = f^i{}_j \, e_i \otimes e^j \tag{2.15}$$

This space is also identifiable to $\mathcal{L}(E^* \times E,\mathbb{R})$.

2.1.4 Tensor product space on E

By generalising the approach of §2.1.2 and 2.1.3, that is by relying on the highlighting of applications of rank one (decomposed elements), it is possible to identify between them multilinear application spaces on E. The classes of sets thus identified are called **tensor product spaces** on E. So, for example:

$$E^* \otimes E^* = \mathcal{L}(E\times E,\mathbb{R}) \equiv \mathcal{L}(E,E^*)$$

$$E\otimes E = \mathcal{L}(E^*\times E^*,\mathbb{R}) \equiv \mathcal{L}(E^*,E)$$

$$E^* \otimes E = \mathcal{L}(E\times E^*,\mathbb{R}) \equiv \mathcal{L}(E,E)$$

$$E\otimes E^* = \mathcal{L}(E^*\times E,\mathbb{R}) \equiv \mathcal{L}(E^*,E^*)$$

and also:

$$E^* \otimes E^* \otimes E^* = \mathcal{L}(E\times E\times E) \equiv \mathcal{L}(E\times E,E^*) \equiv \mathcal{L}(E,\mathcal{L}(E,E^*)) \equiv \text{etc}$$

$$= E^* \otimes (E^* \otimes E^*) = (E^* \otimes E^*) \otimes E^*$$

The tensor product is associative.

Any multilinear application on E and E* represents a tensor element of a tensor product space of E and E*

The number of multilinear spaces identified in a class increases rapidly according to the number of elements of the tensor product, the number of factors of which is called **order**.

The elements of a tensor product space are called tensors. **Decomposed tensors** are rank one applications. These are the tensor products of elements of E and E^*. For example, decomposed endomorphism $\ell \otimes V$ is the tensor product of the form $\ell \in E^*$ and of the vector $V \in E$.

Following the same process, it is easy to introduce multilinear applications on a tensor product space. So, if $\mathfrak{f}$ is an endomorphism of E and $\mathfrak{g}$ an application of $\mathcal{L}\left(E^*,E\right)$, $\mathfrak{f} \otimes \mathfrak{g}$ is a tensor of $E^* \otimes E \otimes E \otimes E$ such that:

$$\forall V \otimes \ell \in E \otimes E^* : \quad \mathfrak{f} \otimes \mathfrak{g}\,(V \otimes \ell) = \mathfrak{f}(V) \otimes \mathfrak{g}(\ell) \in E \otimes E \tag{2.16}$$

2.1.5 Change of basis

Let (E_I) be a second basis of E. The passage of the basis (e_i) to the basis (E_I) is done by means of a change-of-basis matrix $\mathcal{P}$ with components $\mathfrak{p}_I^{\ i}$ representing an endomorphism. The basis change formula is written in tensorial notations:

$$\boxed{E_I = \mathfrak{p}_I^{\ i} e_i} \tag{2.17}$$

Let $V = v^i e_i = V^I E_I$ be a vector in E:

$$v^i e_i = V^I \mathfrak{p}_I^{\ i} e_i$$

from which:

$$v^i = \mathfrak{p}_I^{\ i} V^I$$

This relationship is reversed by involving the matrix $\mathcal{Q} = \mathcal{P}^{-1}$ such that:

$$\mathfrak{q}_i^{\ I} \mathfrak{p}_J^{\ i} = \delta_J^{\ I}$$

$$\mathfrak{p}_I^{\ i} \mathfrak{q}_j^{\ I} = \delta^i_{\ j}$$

It becomes:

$$\boxed{V^I = \mathfrak{q}_i^{\ I}\, v^i} \tag{2.18}$$

In a basis change, the components of a vector vary in contrast to the vectors of the basis. They are said to be **contravariant.**

Let $\left(e^i\right)$ the basis of E^* dual of $\left(e_i\right)$ and $\left(E^J\right)$ the dual basis in $\left(E_I\right)$, be linked by the change of basis relationship: $E^J = \mathfrak{r}_j^{\ J} e^j$. But:

$$\left\langle E^J, E_I \right\rangle = \mathfrak{r}_j^{\ J} \mathfrak{p}_I^{\ i} \left\langle e^j, e_i \right\rangle = \mathfrak{r}_j^{\ J} \mathfrak{p}_I^{\ i} \delta^j_{\ i} = \mathfrak{r}_j^{\ J} \mathfrak{p}_I^{\ j} = \delta^J_{\ I}$$

This last relationship shows that $\mathcal{P}$ and the matrix $\mathcal{R}$ with components $\left(\mathfrak{r}_j^{\ J}\right)$ are inverse, then $\mathcal{R} \equiv \mathcal{Q}$:

$$\boxed{E^J = \mathfrak{q}_j{}^J\, e^j} \tag{2.19}$$

The dual basis (E^J) varies in opposition to the basis (E_I).

Let $\ell = \ell_i e^i = \mathcal{L}_I E^I$ be a form of E^*:

$$\ell_i\, e^i = \mathcal{L}_I\, \mathfrak{q}_i{}^I\, e^i$$

it becomes:

$$\boxed{\mathcal{L}_I = \mathfrak{p}_I{}^i\, \ell_i} \tag{2.20}$$

Entities varying as the basis have been noted with a subscript (covariant index), while the inverse-variant entities, such as the components of the vectors, have a superscript (contravariant index)

In a change of basis, the components of a form vary as the vectors of the basis. They are called a **covariant**. This convention is used generally in this book. The Einstein summation is always made on the same index, once in a covariant position, once in a contravariant position.

Let $t = t_{ij}\, e^i \otimes e^j \in E^* \otimes E^*$ be a bilinear form on E. Expressing the change of basis:

$$t = T_{IJ}\, E^I \otimes E^J = T_{IJ}\, \mathfrak{q}_i{}^I\, \mathfrak{q}_j{}^J\, e^i \otimes e^j$$

then by inverting the resulting relationship, it becomes:

$$\boxed{T_{IJ} = \mathfrak{p}_I{}^i\, \mathfrak{p}_J{}^j\, t_{ij}} \tag{2.21}$$

where it appears that the change of basis is made on each of the covariant indices. In matrix notation:

$$\left[T_{IJ}\right] = \left[\mathcal{P}\right]^T \left[t_{ij}\right] \left[\mathcal{P}\right] \tag{2.22}$$

For an endomorphism $t = t_i{}^j\, e^i \otimes e_j = T_I{}^J\, E^I \otimes E_J$, the change of basis is expressed by:

$$t = t_i{}^j \left(\mathfrak{p}_I{}^i E^I\right) \otimes \left(\mathfrak{q}_j{}^J E_J\right)$$

from which :

$$\boxed{T_I{}^J = \mathfrak{p}_I{}^i\, \mathfrak{q}_j{}^J\, t_i{}^j} \tag{2.23}$$

Or, in matrix form:

$$\left[T_I{}^J\right] = \left[\mathcal{Q}\right] \left[t_i{}^j\right] \left[\mathcal{P}\right] \tag{2.24}$$

The change of basis is expressed using the matrix $\mathcal{P}$ for a covariant index, or matrix $\mathcal{Q}=\mathcal{P}^{-1}$ for a contravariant index.

In summary, a bilinear form on E is a twice-covariant tensor, an endomorphism is a mixed tensor (once contravariant, once covariant).

The process described above is generalised to multilinear applications on E and E^*. For a tensor $(p,q)-p$ times covariant, q times contravariant – the rule for a change of basis is then obvious: $\mathcal{P}$ is used for every covariant index, $\mathcal{Q}$ for each contravariant index. A vector is a (0,1) tensor, a form is a (1,0) tensor.

The formulas for the change of basis above characterise tensors (which is not demonstrated here): if two families of numbers are linked by such a formula of change of basis, they are the components, in two different bases, of a tensor with an order defined by the formula (according to the number of endomorphisms $\mathcal{P}$ and $\mathcal{Q}$ used). It is a criterion of tensoriality.

Example: Let the tensor be $t=t_i{}^j{}_k\, e^i \otimes e_j \otimes e^k$. In a change of basis $e_i \to E_I$, its components in the new basis become:

$$T_I{}^J{}_K = \mathfrak{p}_I{}^i \mathfrak{q}_j{}^J \mathfrak{p}_K{}^k t_i{}^j{}_k$$

2.1.6 Contraction

Contraction is a formal linear operation that applies to mixed tensors (that is, including at least one contravariant index and one covariant index). It consists in operating one of the basic forms on one of the basic vectors.

Consider for example the tensor (2,1): $t=t_i{}^j{}_k\, e^i \otimes e_j \otimes e^k$. The contraction of the first and the second index is the linear application:

$$t \;\rightarrow\; t_i{}^j{}_k \left\langle e^i, e_j \right\rangle e^k = t_i{}^i{}_k\, e^k \in E^*$$

(summation on the index i in first and second positions). But the contraction of the second and third indices is also possible and gives a different result.

Similarly, for the tensor (2,2): $t=t_i{}^j{}_k{}^l\, e^i \otimes e_j \otimes e^k \otimes e_l$, the contraction of the first and fourth indices is the application:

$$t \rightarrow\; t_i{}^j{}_k{}^l \left\langle e^i, e_l \right\rangle e_j \otimes e^k = t_i{}^j{}_k{}^i\, e_j \otimes e^k \;\in\; E \otimes E^*$$

Contraction is an application of the tensor space (p,q) in a space $(p-1,q-1)$. Except in the trivial case of tensors (1,1), there are several possible contraction operations on a mixed tensor, which do not all have the same interpretation. Therefore, it is necessary to specify the chosen contraction. It is an **intrinsic** operation, that is to say, independent of the basis in which the tensors are expressed. In practice, the contraction consists of summing the components of t which two indices to contract are made equal: the indices in question are said to be **saturated**.

$\underline{\otimes}$ denotes a contracted tensor product. Where there is a risk of confusion, the indices contracted are indicated, for example $\overset{3}{\underset{1}{\otimes}}$.

The contraction gathers in fact in a single form all the multilinear operations applied to the tensors, in particular:

- Applying a form to a vector $\ell(V)$:

$$\ell \otimes V \rightarrow \ell \underline{\otimes} V = \ell_i V^i = \ell(V) \tag{2.25}$$

- The trace of an endomorphism ω:

$$\omega_i^{\ i} = \mathrm{TR}(\omega) \tag{2.26}$$

- Applying an endomorphism $\mathfrak{f}$ to a vector V:

$$\mathfrak{f} \otimes V \rightarrow \mathfrak{f} \overset{3}{\underset{1}{\otimes}} V \ (\text{or}\ \mathfrak{f} \underline{\otimes} V) = \mathfrak{f}_i^{\ j} V^i e_j = \mathfrak{f}(V) \tag{2.27}$$

(the contraction is made between the covariant index of $\mathfrak{f}$ and the contravariant index of V).

- The composition of two endomorphisms $\mathfrak{f}$ and $\mathfrak{g}$:

$$\mathfrak{f} \otimes \mathfrak{g} \rightarrow \ \mathfrak{f}_i^{\ j}\, \mathfrak{g}_j^{\ k}\, e^i \otimes e_k = \mathfrak{h} \in \mathcal{L}(E, E) \tag{2.28}$$

Applying the result $\mathfrak{h}$ to a vector V:

$$\mathfrak{h}(V) = \mathfrak{h}_i^{\ k}\, V^i\, e_k = \mathfrak{f}_i^{\ j}\, \mathfrak{g}_j^{\ k}\, V^i\, e_k$$

from which:

$$\mathfrak{h} = \mathfrak{g} \circ \mathfrak{f} = \mathfrak{f} \overset{2}{\underset{3}{\otimes}} \mathfrak{g} \tag{2.29}$$

2.1.7 Symmetry and antisymmetry

For example, a twice-covariant tensor: $t = t_{ij}\, e^i \otimes e^j$. t is symmetrical if, regardless of the couples (i, j):

$$\boxed{t_{ij} = t_{ji}} \tag{2.30}$$

In the same way, t is **antisymmetric** if, whatever the couples (i, j):

$$\boxed{t_{ij} + t_{ji} = 0} \tag{2.31}$$

Similar definitions apply to a twice-contravariant tensor. For a tensor of order greater than two, symmetry and antisymmetry are defined with respect to two covariant indices or two contravariant indices. This definition can be generalised for a set of indices of the same position (lower or upper), but this notion is of little use in the following (see nevertheless the symmetry of the tensor of elasticity, cf. 5.4.2.1).

The properties of symmetry and antisymmetry with respect to two indices of the same position are intrinsic properties, that is, independent of the chosen basis. On the other hand, this property is not intrinsic for two indices in different positions and therefore does not make it possible to define the symmetry or the antisymmetry of a mixed tensor.

The fully contracted product of an antisymmetric tensor and a symmetric tensor is zero.

Let us build the product $u \underline{\otimes} v = u^{ij}\, v_{ij}$ where u is symmetrical and v is antisymmetrical. The indices are dummy, so this product can be rewritten: $u^{ji}v_{ji}$. By using the properties of symmetry and antisymmetry, this last expression is rewritten: $-u^{ij}v_{ij}$, which demonstrates the property.

2.1.8 Euclidean tensors

E here is a Euclidean[4] space, that is to say, provided with a dot product noted $\bullet$. The dot product is a bilinear form on E, there is an associated tensor (2, 0) noted g:

$$\boxed{\forall (V,W) \in E \times E : \quad V \bullet W = g(V,W) = g_{ij} V^i W^j \quad \in \mathbf{R}} \tag{2.32}$$

with:

$$g_{ij} = g\left(e_i, e_j\right) = e_i \bullet e_j = g_{ji}$$

The tensor g is symmetric because of the symmetry of the dot product. The dot product makes it possible to associate with any vector V a bilinear form $\flat(V)$– **flat** – such that:

$$\boxed{\forall W \in E : \quad \left\langle \flat(V), W \right\rangle = V \bullet W} \tag{2.33}$$

The application $\flat$ thus defined is a canonical isomorphism, the dot product being definited.

$\flat \in \mathcal{L}\left(E, E^*\right)$ is identifiable to $g \in \mathcal{L}\left(E \times E, \mathbf{R}\right)$, because they constitute the same tensor (cf. §2.1.2). Indeed, let $\flat\left(e_i\right) = \flat_{ij} e^j$. Then:

$$\left\langle \flat\left(e_i\right), e_k \right\rangle = \flat_{ij} \left\langle e^j, e_k \right\rangle = \flat_{ik} = e_i \bullet e_k = g_{ik}$$

So, in a matrix form:

$$\begin{aligned} &\forall (V,W) \in E \times E : \quad V \bullet W = \left[V\right]^T \left[g\right]\left[W\right] \\ &\text{and} \quad \flat(V) = \left[V\right]^T \left[g\right] \end{aligned} \tag{2.34}$$

[4] Euclid, Greek mathematician (Mid-4th century BC, Mid-3rd century BC)

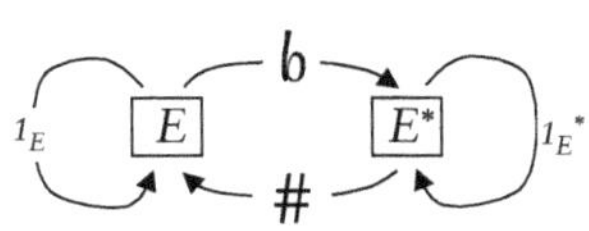

Figure 2.1

The inverse isomorphism is called # (**sharp**). Its representative matrix is, therefore $[g]^{-1}$, which is also symmetrical. Thanks to the metric, the spaces E and E^* are thus connected by a canonical isomorphism and it is therefore possible to identify vectors and forms by the isomorphisms represented in Figure 2.1.

Tensor products of applications such as $\flat$ and # can be applied to tensors on E. For example, a tensor $t = t^{ij} e_i \otimes e_j$ and the application noted $\mathfrak{f}_1 \otimes \mathfrak{f}_2$ applied to t such that:

$$\mathfrak{f}_1 \otimes \mathfrak{f}_2(t) = t^{ij}\, \mathfrak{f}_1(e_i) \otimes \mathfrak{f}_2(e_j)$$

where $\mathfrak{f}_1$ and $\mathfrak{f}_2$ are either the isomorphism $\flat$ or the identity 1_E. For example:

$$1_E \otimes \flat(t) = t^{ij} 1_E(e_i) \otimes \flat(e_j) = \left(t^{ij}\, e_i\right) \otimes \left(g_{jk}\, e^k\right) = t^{ij}\, g_{jk}\, e_i \otimes e^k \;\in\; E \otimes E^*$$

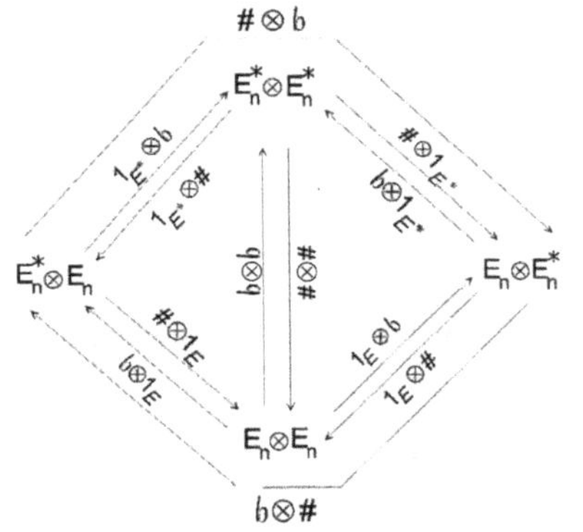

Figure 2.2

These applications make it possible to correspond in pairs two tensors whose order (= covariance + contravariance) is the same. By construction, these applications are canonical isomorphisms, which are summarised in the diagram of Figure 2.2.

A **Euclidean tensor** is the class of equivalence of the tensors which are images of each other by such isomorphisms; these tensors are noted with the same letter. In the example of Figure 2.2, there are four such tensors, called **representatives** of the Euclidean tensor.

Property:

$\flat$ (or g), # (or g^{-1}), 1_E and 1_{E^*} are the four representatives of the same Euclidean tensor called **metric tensor**, noted g.

Indeed:

$$\# \otimes 1_{E^*}(g) = g_{ij}\, \#\left(e^i\right) \otimes e^j = g_{ij} \left(g^{-1}\right)^{ki} e_k \otimes e^j = \delta_j^{\ k} e_k \otimes e^j = 1_{E^*}$$

$$1_{E^*} \otimes \#(g) = g_{ij} e^i \otimes \#\left(e^j\right) = g_{ij} \left(g^{-1}\right)^{kj} e^i \otimes e_k = \delta_i^{\ k} e^i \otimes e_k = 1_E$$

$$\# \otimes \#(g) = g_{ij} \# \left(e^i\right) \otimes \# \left(e^j\right) = g_{ij} \left(g^{-1}\right)^{ki} \left(g^{-1}\right)^{lj} e_k \otimes e_l$$

$$= \delta_j^k \left(g^{-1}\right)^{lj} e_k \otimes e_l = \left(g^{-1}\right)^{lk} e_k \otimes e_l = g^{-1}$$

The notation:

$$\left(g^{-1}\right)^{lk} = g^{lk} \tag{2.35}$$

may be used, since g is a Euclidean tensor.

Example: Let $t^{\bullet\bullet}$ a tensor twice contravariant, so belonging to $E \otimes E$. It is a matter of calculating its representative in $E \otimes E^*$, that is to say $t^{\bullet}{}_{\bullet}$, obtained by:

$$t^{\bullet}{}_{\bullet} = 1_E \otimes \flat\left(t^{\bullet\bullet}\right) = 1_E \otimes \flat\left(t^{ij} e_i \otimes e_j\right) = t^{ij} 1_E\left(e_i\right) \otimes \flat\left(e_j\right) = t^{ij} e_i \otimes \left(g_{jk} e^k\right)$$

hence the components of $t^{\bullet}{}_{\bullet}$, then, proceeding in the same way for the first index, those of $t_{\bullet\bullet}$:

$$t^i{}_k = \left[1_E \otimes \flat(t)\right]^i{}_k = t^{ij} g_{jk}$$

$$t_{kl} = \left[\flat \otimes \flat(t)\right]_{kl} = t^{ij} g_{ik} g_{jl}$$

Note the points in indices of t. They identify the type of representative.

Similarly, using #:

$$t^{kl} = g^{ik} g^{jl} t_{ij}$$

$$t^k{}_j = g^{ik} t_{ij} \quad \text{etc.} \ \blacksquare$$

The metric tensor is the index elevator: it allows movement up or down from one representative to another of the same Euclidean tensor.

It can be demonstrated without difficulty that if a twice-covariant representative is symmetric (or antisymmetric), the two contravariant representative of the same Euclidean tensor is **symmetric** (or **antisymmetric**). The Euclidean tensor is then called **symmetric** (or **antisymmetric**). This gives no particular property to the components of the mixed representative, except in an orthonormal frame.

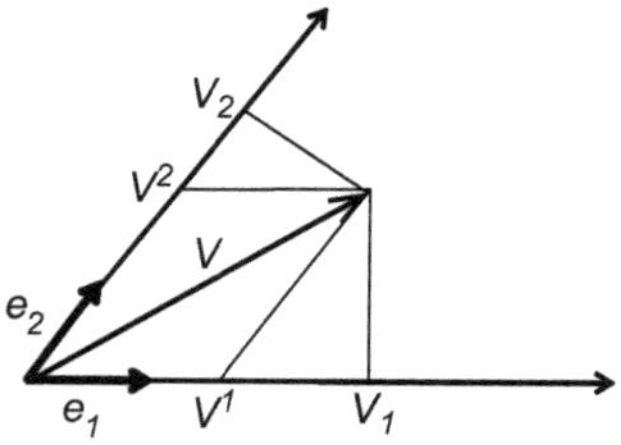

Figure 2.3

Note: If the basic vectors are unitary, the covariant components of a vector V are the orthogonal projections of the vector on the axes (Figure 2.3). The contravariant and covariant components are equal only if the basis is orthonormal (in such a basis, the components of g are the Kronecker symbols and the representative matrix of g is reduced to the unit matrix).

2.1.9 Levi-Civita tensor and cross product

In this paragraph, E is of dimension 3. Let $\tilde{e}_{ijk}$ the signature of the permutation (i, j, k) be compared to $(1, 2, 3)$, sometimes called the **generalised Kronecker symbol.** $\hat{g}$ denoting the determinant of the metric tensor, the family of numbers $\bar{e}_{ijk}$ defined by:

$$\bar{e}_{ijk} = \hat{g}^{1/2}\tilde{e}_{ijk} \tag{2.36}$$

is a tensor of $\mathbb{R}^3$ (see the demonstration in 3.1.3.2). It is called the **Levi-Civita[5] tensor.**
The cross product is defined from its expression on the basic vectors:

$$e_i \times e_j = \bar{e}_{ij}{}^k e_k$$

That is to say:

$$\boxed{V \times W = V^i W^j \,\bar{e}_{ij}{}^k e_k} \tag{2.37}$$

The covariant components of $\bar{e}$ are equal to the mixed product:

$$\bar{e}_{ijk} = \left(e_i \times e_j\right) \bullet e_k$$

The purely contravariant components of $\bar{e}$ are given by:

$$\bar{e}^{ijk} = g^{il}g^{jm}g^{kn}\bar{e}_{lmn} = g^{il}g^{jm}g^{kn}\sqrt{\hat{g}}\,\tilde{e}_{lmn} = \tilde{e}_{ijk}\sqrt{\hat{g}}\,\det\left(g^{-1}\right) = \frac{\tilde{e}_{ijk}}{\sqrt{\hat{g}}} \tag{2.38}$$

Other relationships using the tensor $\bar{e}$ are given in 3.1.3.2 in the two-dimensional case.

2.2 DIFFERENTIATION

2.2.1 Local basis in an affine space

Let E an Euclidian vector space, $E^\bullet$ an affine space on E. To each point m of $E^\bullet$ is associated with a vector space E_m isomorphic to E. A basis of E allows expression of the vectors attached to the point m. In practice, two kinds of local bases can be distinguished, for convenience: Cartesian[6] bases and natural bases:

[5] Tullio Levi-Civita, Italian mathematician (1873–1941)
[6] René Descartes, French mathematician, physicist and philosopher (1596–1650).

- a **Cartesian basis** is obtained by translation $\mathcal{O}\mathcal{m}$ of a basis defined at the origin $\mathcal{O}$.
- a **natural basis** is associated with any coordinate system $\left(x^i\right)$ of $\mathcal{m}$ and is defined by:

$$e_i = \frac{\partial \mathcal{m}}{\partial x^i} \tag{2.39}$$

These two types of local bases are illustrated in Figure 2.4. The vectors of a basis constitute a constant vector field in Cartesian coordinates, a variable in any coordinates. A Cartesian basis is a special case of a natural basis.

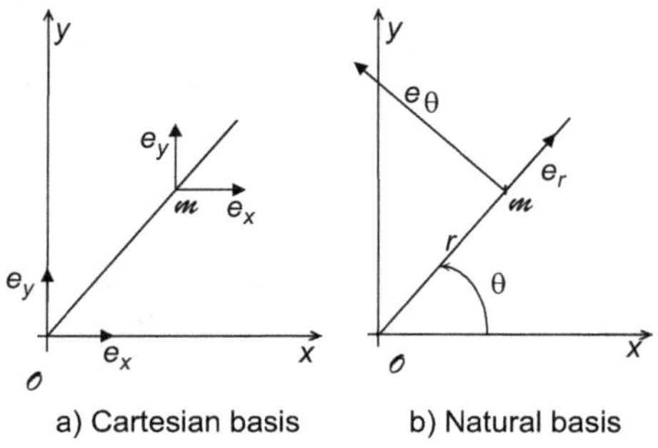

Figure 2.4

When the natural basis is orthogonal, it is convenient to introduce the associated orthonormal physical basis, obtained by normalising the vectors of the natural basis. In a physical basis, the components of a physical quantity represented by a tensor have the same dimension as the quantity itself.

2.2.2 Tensor fields

A tensor product space built on the attached space $E_{\mathcal{m}}$ can be defined at each point $\mathcal{m}$ of $E^\bullet$, by identification with the corresponding product space on E. A tensor field is an application:

$$\mathcal{m} \rightarrow T(\mathcal{m})$$

where $T(\mathcal{m})$ is a tensor with a given order on the attached vector space $E_{\mathcal{m}}$ at point $\mathcal{m}$. Components of T are expressed on the vectors and forms of the local basis. In any coordinates, the components of T in the local basis are not necessarily constant if the tensor is constant (a tensor is constant if its derivative is null).

2.2.3 Differentiation of the vectors of the basis

Whereas in Cartesian coordinates the vectors of the basis are constant because they are deduced from each other by translation (Figure 2.4(a)), they are variable in curvilinear coordinates (for example in polar coordinates in the plane, Figure 2.4(b)). Their derivative is not zero. At a point $\mathcal{m}$, this derivative is expressed in the natural basis in the form:

$$\boxed{\frac{\partial e_i}{\partial x^j} = \overline{\Gamma}_{ij}^k e_k} \tag{2.40}$$

Example: in polar coordinates in the plane, the vectors of the basis obtained by (2.39) are the vectors (e_r, e_θ), cf. Figure 2.4 (b), which are decomposed on the orthonormal Cartesian basis (e_x, e_y):

$$e_r = \frac{\partial \boldsymbol{m}}{\partial r} = \begin{vmatrix} \cos\theta \\ \sin\theta \end{vmatrix} \quad ; \quad e_\theta = \frac{\partial \boldsymbol{m}}{\partial \theta} = \begin{vmatrix} -r\sin\theta \\ r\cos\theta \end{vmatrix}$$

e_r and e_θ verify the relations obtained by differentiation:

$$\frac{\partial e_r}{\partial r} = 0 \quad ; \quad \frac{\partial e_r}{\partial \theta} = \frac{e_\theta}{r}$$

$$\frac{\partial e_\theta}{\partial r} = \frac{e_\theta}{r} \quad ; \quad \frac{\partial e_\theta}{\partial \theta} = -re_r$$

Therefore:

$$\bar{\Gamma}^r_{rr} = \bar{\Gamma}^\theta_{rr} = \bar{\Gamma}^r_{r\theta} = \bar{\Gamma}^r_{\theta r} = \bar{\Gamma}^\theta_{\theta\theta} = 0$$

$$\bar{\Gamma}^\theta_{r\theta} = \bar{\Gamma}^\theta_{\theta r} = \frac{1}{r} \quad ; \quad \bar{\Gamma}^r_{\theta\theta} = -r$$

The physical basis associated with the polar coordinates is obtained by normalising the vector e_θ, that is $\tilde{e}_\theta = \frac{e_\theta}{r}$. It must be emphasized that the differentiation rules (and in particular the calculation of the coefficients $\bar{\Gamma}$) are established using a natural basis, because of the definition (2.39) and cannot be directly applied in a physical basis for which it would not be possible to find an associated parameter setting.

Coefficients $\bar{\Gamma}$, called **Christoffel[7] coefficients**, are symmetrical with respect to the subscripts. Indeed:

$$\frac{\partial e_i}{\partial x^j} = \bar{\Gamma}^k_{ij} e_k = \frac{\partial^2 \boldsymbol{m}}{\partial x^i \partial x^j} = \frac{\partial^2 \boldsymbol{m}}{\partial x^j \partial x^i} = \frac{\partial e_j}{\partial x^i}$$

Coefficients $\bar{\Gamma}$ are not the components of a tensor.

The differentiation of a form e^j of the dual basis is expressed then, by differentiation of the duality pairing:

$$\frac{\partial}{\partial x^k}\left\langle e^j, e_i \right\rangle = \left\langle \frac{\partial}{\partial x^k} e^j, e_i \right\rangle + \left\langle e^j, \frac{\partial}{\partial x^k} e_i \right\rangle = 0$$

Therefore:

$$\left\langle \frac{\partial e^j}{\partial x^k}, e_i \right\rangle = -\left\langle e^j, \bar{\Gamma}^l_{ik} e_l \right\rangle = -\bar{\Gamma}^j_{ik}$$

[7] Elwin Bruno Christoffel, German mathematician and physicist (1829-1900).

It becomes:

$$\boxed{\frac{\partial e^j}{\partial x^k} = -\overline{\Gamma}^j_{kl} e^l} \tag{2.41}$$

Finally:

$$\boxed{\overline{\Gamma}^k_{ij} = \left\langle e^k, \partial_i e_j \right\rangle = -\left\langle e_j, \partial_i e^k \right\rangle} \tag{2.42}$$

2.2.4 Differentiation of tensor fields

D denotes the differentiation in the affine space $E^\bullet$, D_X the differentiation along a vector X defined by:

$$D_X f = \lim_{h \to 0} \frac{1}{h} \left[f\left(m + hX \right) - f(m) \right] \; , \quad h \in 0$$

The differentiation along a vector e_i, noted also $D_i f$, is equal to the partial derivative with respect to parameter x^i:

$$D_i f = D_{e_i} f = \frac{\partial f}{\partial x^i} \tag{2.43}$$

Notation D avoids confusion between the component of the derivative and the (partial) derivative of a component, as it appears hereinafter. By linearity of the differentiation:

$$D_X f = X^i \, D_{e_i} f = X^i \frac{\partial f}{\partial x^i} \tag{2.44}$$

In a Cartesian basis, the vectors of the basis are constant and their derivative is zero. In this case, let T a tensor field, for example:

$$T = T_i{}^j{}_k \; e^i \otimes e_j \otimes e^k$$

Then:

$$\frac{\partial T}{\partial x^l} = D_l T = \partial_l T_i{}^j{}_k \; e^i \otimes e_j \otimes e^k$$

with the notation:

$$\partial_l T_i{}^j{}_k = \frac{\partial T_i{}^j{}_k}{\partial x^l} \tag{2.45}$$

In the case where a natural basis associated with curvilinear coordinates is used, it is also necessary to differentiate the vectors of the basis. By applying the rule of differentiation of a product:

$$\frac{\partial T}{\partial x^l} = D_l T$$

$$= \partial_l T_i{}^j{}_k \, e^i \otimes e_j \otimes e^k + T_i{}^j{}_k \, \partial_l \left(e^i\right) \otimes e_j \otimes e^k$$

$$+ T_i{}^j{}_k \, e^i \otimes \partial_l \left(e_j\right) \otimes e^k + T_i{}^j{}_k \, e^i \otimes e_j \otimes \partial_l \left(e^k\right)$$

from which:

$$D_l T_i{}^j{}_k \;\; e^i \otimes e_j \otimes e^k = \left(\partial_l T_i{}^j{}_k - \bar{\Gamma}^m_{li} T_m{}^j{}_k + \bar{\Gamma}^j_{lm} T_i{}^m{}_k - \bar{\Gamma}^m_{lk} T_i{}^j{}_m\right) e^i \otimes e_j \otimes e^k$$

Here $D_l T_i{}^j{}_k = \left(D_l T\right)_i{}^j{}_k$ denotes the components of the tensor $D_l T$, which differ in general from $\partial_l T_i{}^j{}_k$. It becomes the following relation, which allows us to calculate them:

$$\boxed{D_l T_i{}^j{}_k = \partial_l T_i{}^j{}_k - \bar{\Gamma}^m_{li} T_m{}^j{}_k + \bar{\Gamma}^j_{lm} T_i{}^m{}_k - \bar{\Gamma}^m_{lk} T_i{}^j{}_m} \tag{2.46}$$

The tensor field DT, with components $D_l T_i{}^j{}_k$, a derivative of the tensor field T, has one more covariance unit than T. It verifies the relationships:

$$\begin{aligned} DT &= e^i \otimes D_i T \\ D_X T &= DT \bullet X \end{aligned} \tag{2.47}$$

where $\bullet$ denotes the product contracted between the contravariant index of the vector X and the covariant index of the differentiation.

The calculation rule (2.46) is easily generalised to any tensor, noting that, to express the components of the differentiated tensor:

- the tensor components are differentiated;
- each of the vectors of the basis (or of dual forms) is differentiated successively; the basic vectors are differentiated with the coefficients $\bar{\Gamma}$, the forms of the dual basis with coefficients $-\bar{\Gamma}$; care should be given to the position of the dummy indices which differs between the two cases;
- the other indices (non dummy) are present in the same position (upper or lower) on the left and on the right sides of the equality.

For example:

$$\begin{aligned} D_i T^k &= \partial_i T^k + \bar{\Gamma}^k_{ij} T^j \\ D_i T_j &= \partial_i T_j - \bar{\Gamma}^k_{ij} T_k \\ D_i T^{jk} &= \partial_i T^{jk} + \bar{\Gamma}^j_{im} T^{mk} + \bar{\Gamma}^k_{im} T^{jm} \\ D_i T^j{}_k &= \partial_i T^j{}_k + \bar{\Gamma}^j_{im} T^m{}_k - \bar{\Gamma}^m_{ik} T^j{}_m \end{aligned} \tag{2.48}$$

2.2.5 Expression of Christoffel coefficients versus metric

Coefficients $\bar{\Gamma}$ can be expressed according to the derivatives of the metric tensor g. Indeed, g is a constant tensor (it is easy to demonstrate it in Cartesian coordinates), so its derivative in any direction is zero:

$$D_i g_{jk} = \partial_i g_{jk} - \bar{\Gamma}^m_{ij} g_{mk} - \bar{\Gamma}^m_{ik} g_{jm} = 0 \tag{2.49}$$

By circular permutation:

$$\partial_j g_{ki} - \bar{\Gamma}^m_{jk} g_{mi} - \bar{\Gamma}^m_{ji} g_{km} = 0$$

$$\partial_k g_{ij} - \bar{\Gamma}^m_{ki} g_{mj} - \bar{\Gamma}^m_{kj} g_{im} = 0$$

Given the symmetries of coefficients $\bar{\Gamma}$, by adding the first two relationships and then subtracting the third, it becomes:

$$\boxed{\bar{\Gamma}^k_{ij} = \frac{1}{2} g^{mk} \left(\partial_j g_{im} + \partial_i g_{jm} - \partial_m g_{ij} \right)} \tag{2.50}$$

So the derivation of the basis vectors depends only on the metric. If the basis is orthogonal, the coefficients $\bar{\Gamma}$ equal to zero when the three indices are different.

2.2.6 Differential operators

f being a scalar field defined on $E^\bullet$, Df is a linear form of components $\partial_i f$. The associated vector $\#(Df)$ is called a **gradient** of f and noted grad f. Its components are:

$$\boxed{(\text{grad } f)^i = g^{ij} \partial_j f = D^i f} \tag{2.51}$$

X is a vector field defined on $E^\bullet$, DX is an endomorphism which trace is called **divergence** of the vector X and noted div X:

$$\boxed{\text{div } X = \text{TR}(DX) = D_i X^i} \tag{2.52}$$

f being a scalar function, the **scalar Laplacian**[8] of f noted Δf is defined by the relation:

$$\Delta f = \text{div}(\text{grad } f) = D_i \left(D^i f \right) = g^{ij} D_i \left(D_j f \right) \tag{2.53}$$

Now let X a vector field of $E_3^\bullet$. Let A the bilinear antisymmetric form associated with DX, defined by:

$$AX = DX_{\bullet\bullet} - DX_{\bullet\bullet}{}^T$$

[8] Pierre-Simon de Laplace, French mathematician and physicist (1749–1827).

with components:

$$AX_{ij} = D_i X_j - D_j X_i$$

Applying a vector Y to AX gives a linear form, to which a vector $\#(AX \bullet Y)$ can be associated. There is a single vector field called **curl** of X or **rotational** of X and noted curl X or rot X[9] such as, for any vector field Y:

$$\#(AX \bullet Y) = g^{jk}\left(D_i X_j - D_j X_i\right) Y^i e_k = \text{rot } X \times Y \tag{2.54}$$

Indeed, taking as Y the vector e_i:

$$\left(\text{rot } X\right)^j \,\overline{e}_{ji}{}^{k} = g^{jk}\left(D_i X_j - D_j X_i\right) = g^{jk}\left(\partial_i X_j - \partial_j X_i\right)$$

where $\overline{e}$ denotes the Levi-Civita tensor (§1.9); it becomes:

$$\left(\text{rot } X\right)^i \,\overline{e}_{ijk} = \partial_j X_k - \partial_k X_j$$

Or, multiplying by $\tilde{e}_{ijk}$ (generalised Kronecker symbol):

$$\boxed{\left(\text{rot } X\right)^i = \frac{1}{2}\frac{\tilde{e}^{ijk}}{\sqrt{g}}\left(\partial_j X_k - \partial_k X_j\right) = \frac{1}{2}\overline{e}^{ijk}\left(\partial_j X_k - \partial_k X_j\right)} \tag{2.55}$$

which defines the rot X vector in a unique way. By transforming the last relation, rot X can also be written:

$$\text{rot } X = g^{ij}\frac{\partial X^k}{\partial x^i} e_j \times e_k \tag{2.56}$$

Indeed: $\overline{e}^{ijk} = \overline{e}^{jki}$, then:

$$\text{rot } X = \frac{1}{2} g^{mj} g^{nk}\left(\partial_j X_k - \partial_k X_j\right) \overline{e}_{mn}{}^{i} e_i$$

$$= \frac{1}{2}\left(g^{mj}\partial_j X^n - g^{nk}\partial_k X^m\right) e_m \times e_n$$

Still in $E_3^{\bullet}$, the **Laplacian vector** of an X vector field is defined by:

$$\Delta X = D_i\left(D^i X\right) = D_i\left(g^{ij} D_j X\right) = g^{ij} D_i\left(D_j X\right) \tag{2.57}$$

ΔX verifies the relation (which is easily demonstrated in an orthonormal Cartesian coordinate system):

$$\boxed{\Delta X = \text{grad}\left(\text{div } X\right) - \text{rot}\left(\text{rot } X\right)} \tag{2.58}$$

[9] Both notations rot X and curl X may be used. In this book, the former is used as given in ISO international standard.

If f and X are respectively a scalar field and a vector field in E, the following relationships are easily found:

$$\boxed{\operatorname{rot}(\operatorname{grad} f) = 0} \tag{2.59}$$

$$\boxed{\operatorname{div}(\operatorname{rot} X) = 0} \tag{2.60}$$

2.2.7 Riemann–Christoffel tensor associated with a differentiation

D designating any differentiation, the antisymmetric part, with respect to the differentiation indices, of the second derivative of a vector U is noted:

$$D_i D_j U^k - D_j D_i U^k = \bar{R}_m{}^k{}_{ij} U^m \tag{2.61}$$

It is admitted here (which the reader can easily verify) that $\bar{R}$ defines a tensor, called **Riemann**[10]**–Christoffel tensor** (or **tensor of curvature of the differentiation** D)

The Riemann–Christoffel tensor serves in particular to establish the link between metric and curvature tensors on surfaces (cf. 3.2.2).

Writing that:

$$D_i D_j U^k = \partial_i\left(\partial_j U^k + \bar{\Gamma}^k_{jm} U^m\right) - \bar{\Gamma}^l_{ij}\left(\partial_l U^k + \bar{\Gamma}^k_{lm} U^m\right) + \bar{\Gamma}^k_{il}\left(\partial_j U^l + \bar{\Gamma}^l_{jm} U^m\right)$$

it becomes:

$$D_i D_j U^k - D_j D_i U^k = \left[\partial_i\left(\bar{\Gamma}^k_{jm}\right) - \partial_j\left(\bar{\Gamma}^k_{im}\right)\right] U^m + \left(\bar{\Gamma}^k_{li}\bar{\Gamma}^l_{jm} - \bar{\Gamma}^k_{lj}\bar{\Gamma}^l_{im}\right) U^m$$

hence the expression of the components of $\bar{R}$:

$$\bar{R}_m{}^k{}_{ij} = \partial_i\left(\bar{\Gamma}^k_{jm}\right) - \partial_j\left(\bar{\Gamma}^k_{im}\right) + \bar{\Gamma}^k_{il}\bar{\Gamma}^l_{jm} - \bar{\Gamma}^k_{jl}\bar{\Gamma}^l_{im} \tag{2.62}$$

Which shows that $\bar{R}$ depends only on the metric g. So there is a direct link between the second derivative (the curvature) and the metric.

Applying the definition (2.61) to a vector of the basis:

$$\begin{aligned} &D_i D_j e_k - D_j D_i e_k = \bar{R}_k{}^m{}_{ij}\, e_m \\ &\bar{R}_{kmij} = \left(D_i D_j e_k - D_j D_i e_k\right) \cdot e_m. \end{aligned} \tag{2.63}$$

[10] Georg Friedrich Bernhard Riemann, German mathematician (1826–1866).

Tensor $\bar{R}$ allows to express the antisymmetric part of the second derivative of a tensor of any kind, in particular:

$$\begin{aligned} D^2_{jk}T_i - D^2_{kj}T_i &= \bar{R}^m_{\ ijk}\, T_m \\ D^2_{kl}T_{ij} - D^2_{lk}T_{ij} &= \bar{R}^m_{\ ikl}\, T_{mj} + \bar{R}^m_{\ jkl}\, T_{im} \end{aligned} \tag{2.64}$$

In a Euclidean space, the relation (2.62) can be applied in a Cartesian coordinate system, where all $\bar{\Gamma}$ are zero, which leads immediately to:

$$\bar{R} = 0 \tag{2.65}$$

(this condition is also necessary and sufficient for the space to be Euclidean).

The following relationships are shown from (2.63):

$$\begin{aligned} &\bar{R}_{mijk} = -\bar{R}_{imjk} = -\bar{R}_{mikj} = \bar{R}_{jkmi} \\ &\bar{R}_{mijk} + \bar{R}_{mjki} + \bar{R}_{mkij} = 0 \end{aligned} \tag{2.66}$$

The number of independent components of $\bar{R}$ is equal to $\frac{1}{12}n^2\left(n^2-1\right)$, n being the dimension of the space.

When $n = 2$, there is only one independent component, $\bar{R}_{1212}$ for example.

When $n = 3$, there are six independent components, for example:

$$\bar{R}_{1212}, \bar{R}_{3112}, \bar{R}_{3221}, \bar{R}_{1313}, \bar{R}_{2323}, \bar{R}_{1323}$$

Photo 2.1 : shells are used for constructions of very diverse natures

MAIN RESULTS

TENSORS ON A VECTOR SPACE

Einstein summation convention $V = \sum_{i=1}^{n} V^i e_i = V^i e_i$

Change of basis on tensors $E_I = p_I{}^i e_i \quad T_I{}^J = p_I{}^i q_j{}^J t_i{}^j \; ; \; \left[T_I{}^J\right] = \left[\mathcal{Q}\right]\left[t_i{}^j\right]\left[\mathcal{P}\right]$

Symetry $T_{ij} = T_{ji} \quad T^{ij} = T^{ji}$
Antisymetry $T_{ij} = -T_{ji} \quad T^{ij} = -T^{ji}$

EUCLIDEAN TENSORS

$$\left\langle b(V), W \right\rangle = V \cdot W \qquad \left(g^{-1}\right)^{lk} = g^{lk}$$

$$t^{kl} = g^{ik} g^{jl} t_{ij} \qquad t^k{}_j = g^{ik} t_{ij}$$

$$\overline{e}_{ijk} = \hat{g}^{1/2} \tilde{e}_{ijk} \qquad V \times W = V^i W^j \, \overline{e}_{ij}{}^k e_k$$

DIFFERENTIATION OF TENSOR FIELDS

$$\frac{\partial e_i}{\partial x^j} = \overline{\Gamma}_{ij}^k e_k \; ; \qquad \overline{\Gamma}_{ij}^k = \frac{1}{2} g^{mk} \left(\partial_j g_{im} + \partial_i g_{jm} - \partial_m g_{ij} \right) ;$$

$$\frac{\partial e^j}{\partial x^k} = -\overline{\Gamma}_{kl}^j e^l$$

$$D_l T_i{}^j{}_k = \partial_l T_i{}^j{}_k - \overline{\Gamma}_{li}^m T_m{}^j{}_k + \overline{\Gamma}_{lm}^j T_i{}^m{}_k - \overline{\Gamma}_{lk}^m T_i{}^j{}_m$$

DIFFERENTIAL OPERATORS

$$\text{grad } f = \#(Df) \qquad \text{div } X = \text{TR}(DF) = D_i X^i$$

$$\Delta f = \text{div}\left(\text{grad } f\right) = D_i\left(D^i f\right) = g^{ij} D_i\left(D_j f\right)$$

$$\Delta X = \text{grad}\left(\text{div } X\right) - \text{rot}\left(\text{rot } X\right)$$

$$\text{rot } X = g^{ij} \frac{\partial X^l}{\partial x^i} e_j \times e_l$$

$$\text{rot}\left(\text{grad } f\right) = 0$$

$$\text{div}\left(\text{rot } X\right) = 0$$

EXERCISES

Exercise 2.1

Show the decomposition in a basis e_i of a tensor the representative of which is in space $\mathcal{L}\left(E^*, \mathcal{L}\left(E^*,E\right)\right)$ and give at least one other representative of the same tensorial space.

Answer:

$$\boldsymbol{t} \in \mathcal{L}\left(E^*,\mathcal{L}\left(E^*,E\right)\right): \ \boldsymbol{t} = t^{ij}{}_k e_i \otimes e_j \otimes e^k$$

$$\mathcal{L}\left(E^*,\mathcal{L}\left(E^*,E\right)\right) \equiv E \otimes E \otimes E^* \equiv \mathcal{L}\left(E^* \times E^* \times E\right) \equiv \mathcal{L}\left(E^* \times E^*, E\right) \equiv \text{etc...}$$

Exercise 2.2

Let the tensor (1, 2): $\boldsymbol{t} = t^{ij}{}_k \, e_i \otimes e_j \otimes e^k$

Using the basis change formulae, what are the components of $\boldsymbol{t}$ in another basis E_I?

Answer:
With the notations of §2.1.5: $\boldsymbol{t} = \mathrm{q}_i{}^I \mathrm{q}_j{}^J \mathrm{p}_K{}^k t^{ij}{}_k \, E_I \otimes E_J \otimes E^K$

Exercise 2.3

In an orthonormal basis (e_i), a mixed tensor $\boldsymbol{t}$ (1, 1) has components:

$$t_1{}^1 = 1; \ t_2{}^2 = -1; \ t_1{}^2 = 1; \ t_2{}^1 = 2$$

A new basis (E_I) is defined from (e_i) in Figure 2.5.

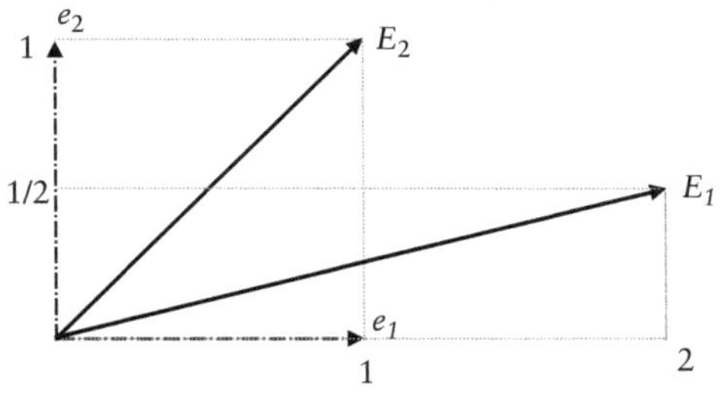

Figure 2.5

Numerically calculate the following entities:

- representatives of the metric tensor in the basis (E_I),
- $\mathcal{P}$ and $\mathcal{P}^{-1}$,
- the components T of $\boldsymbol{t}$ in the basis (E_I),
- the components of the other representatives of the Euclidean tensor associated with $\boldsymbol{t}$ in the base (E_I).

Answer:

$$G_{..} = \begin{pmatrix} \frac{17}{4} & \frac{5}{2} \\ \frac{5}{2} & 2 \end{pmatrix}; \quad G^{..} = \begin{pmatrix} \frac{8}{9} & -\frac{10}{9} \\ -\frac{10}{9} & \frac{17}{9} \end{pmatrix}$$

$$\mathcal{P} = \begin{pmatrix} 2 & p_1^{\ 2} = \frac{1}{2} \\ p_2^{\ 1} = 1 & 1 \end{pmatrix} \qquad \mathcal{Q} = \mathcal{P}^{-1} = \frac{1}{3}\begin{pmatrix} 2 & q_1^{\ 2} = -1 \\ q_2^{\ 1} = -2 & 4 \end{pmatrix}$$

$$T_1^{\ 1} = 1; \; T_1^{\ 2} = 1; \; T_2^{\ 1} = 2; \; T_2^{\ 2} = -1$$

$$T_{11} = \frac{27}{4}; \; T_{12} = \frac{9}{2}; \; T_{21} = 6; \; T_{22} = 3$$

$$T^1{}_1 = -\frac{2}{3}; \; T^2{}_1 = \frac{23}{6}; \; T^1{}_2 = \frac{2}{3}; \; T^2{}_2 = \frac{2}{3}$$

$$T^{11} = -\frac{4}{3}; \; T^{12} = 2; \; T^{21} = \frac{8}{3}; \; T^{22} = -3$$

Exercise 2.4

A vector field is defined in the plane, in polar coordinates, by:

$$X = \frac{e_\theta}{r}$$

Calculate the tensor derived from X in the natural coordinate system associated with the polar coordinates.

Answer:

$$DX = -\, e^\theta \otimes e_r$$

Chapter 3

Deformation of surfaces

This chapter is devoted to the study of some properties of surfaces immersed in the ambient space $E_3^{\bullet}$, which is useful in the development of shell theory, including properties that serve to specify the deformation of surfaces. The necessary notions of geometry not discussed here are introduced in later chapters, as and when needed.

In the first section, the fundamental forms of the surface (metric and curvature) are introduced. The first form makes it possible to measure distances and angles and remains internal to the surface. The second form makes it possible to express the shape of the surface in the ambient space and to establish a geometric frame that is relevant to work in the vicinity of the middle surface of a shell.

The differentiation of the fields defined on the surface are examined in the second section, which leads to considering an internal differentiation of the surface fields, called Levi-Civita differentiation.

The transformation of surfaces is studied in the third section and it is shown that their deformation is characterised by variations of metric and curvature.

In addition, the geometry of the curves plotted on a surface is recalled in the fourth section, with a view to express the boundary conditions on the edge of a shell.

3.1 FUNDAMENTAL FORMS ON A SURFACE

3.1.1 Introduction to surface geometry

The geometry of the surfaces was analysed in depth by Gauss[1] (1828), who showed that the surface can be considered from two points of view:

a) as a non-Euclidean mathematical object with two dimensions. Indeed, two coordinates are enough to position a point on the surface. It is possible to draw a line between two points on the surface and measure the distance between the two points along the line; this distance can withstand a change of shape of the surface. For example, a line segment can be drawn on a flat sheet of paper to join two points. If the sheet is wound without stretching to form a cylinder, the surface and the line change shape, but the distance between the two points, measured along the line, is not changed. A similar remark can be made about an angle drawn on the sheet. This reflects an **intrinsic** property of the surface, the measurement of the length or angle on the surface being independent of the position of the points of the sheet in the ambient space. Mechanically, if a shell can be likened to a surface, the changes in

[1] Johann Carl Friedrich Gauß, German mathematician (1777–1855)

DOI: 10.1201/9780429440403-3

length and angle of this two-dimensional mechanical object clearly bring into play a strain energy (**membrane**).

b) as a mathematical object immersed in an affine three-dimensional space. A change in shape such as the winding passage of the sheet of plane paper to the cylinder can be evidenced only by measurements in the ambient three-dimensional space. The shape (the curvature) is not therefore a property of the surface alone as a two-dimensional object because it implies its environment. It is an **extrinsic** property. On a mechanical point of view, a change of shape is associated with a strain energy (**flexional**).

One of the strengths of Gauss' analysis, which has primary consequences in shell mechanics, is to have demonstrated the link between intrinsic and extrinsic properties of surfaces via the **Gaussian curvature.** This property is discussed in detail in §3.2.3.

3.1.2 Definitions, local basis

3.1.2.1 Parameterisation of the surface

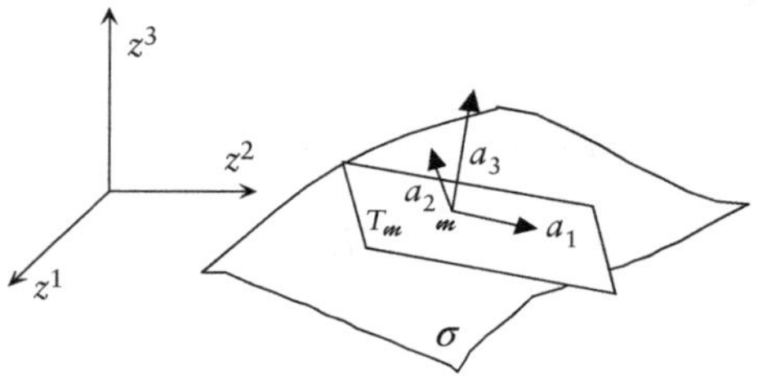

Figure 3.1

A surface σ is a manifold of dimension 2 plunged into an "ambient" affine space of dimension 3 (Figure 3.1). In the following, the Latin indices are relative to the ambient space and vary from *1* to *3*, while the Greek indices relate to the surface and vary from *1* to *2*.

The ambient space is the Euclidian affine space $E_3^{\bullet}$ on $\mathbb{R}$, equipped with a rectilinear coordinate system $\left(z^i, i \in [1,3]\right)$ in a Cartesian coordinate system associated with an orthonormal basis (e_i).

The surface is defined in parametric form by a one-to-one application[2] from $a_3 \cdot a_\beta = 0$ in $E_3^{\bullet}$, that is the parametric equation:

$$\left(x^1, x^2\right) \in \mathbb{R}^2 \rightarrow m\left(x^1, x^2\right) \in \sigma \subset E_3^{\bullet} \tag{3.1}$$

Parameters x^1 and x^2 are the **curvilinear coordinates** of the surface.

3.1.2.2 Tangent plane

In the same way that, in an affine space, an isomorphic vector space of the reference Euclidian vector space is attached to any point of the space, it is useful to attach a vector space to each point of the surface, especially to express vectors such as the displacement of the point.

The surface is supposed to be **differentiable** in all points; two independent vectors can then be defined at any point m on the surface:

[2] In some cases, the surface can usefully be defined by a set of such applications, called "maps".

$$\boxed{a_\alpha = \frac{\partial \boldsymbol{m}}{\partial x^\alpha} = \partial_\alpha \boldsymbol{m}} \tag{3.2}$$

such that $a_1 \times a_2 \neq 0$.

Vectors (a_α) generate a plane $T_{\boldsymbol{m}}$, called the **tangent plane**, of which they form a natural basis. $T_{\boldsymbol{m}}$ is independent of the curvilinear coordinates (x^α). Indeed, let a vector $V = \mathrm{v}^\alpha a_\alpha$ of $T_{\boldsymbol{m}}$ and (X^β) another parameterisation of the surface be linked to the first by the parameter change formula $X^\beta = X^\beta(x^\alpha)$, such that $\det\left(\frac{\partial X^\beta}{\partial x^\alpha}\right) \neq 0$ to maintain regularity. It comes, by differentiation:

$$V = \mathrm{v}^\alpha a_\alpha = \mathrm{v}^\alpha \frac{\partial \boldsymbol{m}}{\partial x^\alpha} = \mathrm{v}^\alpha \frac{\partial \boldsymbol{m}}{\partial X^\beta}\frac{\partial X^\beta}{\partial x^\alpha} = \mathrm{v}^\alpha \frac{\partial X^\beta}{\partial x^\alpha} A_\beta \tag{3.3}$$

where (A_β) refers to the natural basis associated with the curvilinear coordinates X^β. The proposition is demonstrated by considering the particular vectors $V = a_\alpha$. Equation (3.3) also provides the local basis change formula:

$$\boxed{V^\beta = \mathrm{v}^\alpha \frac{\partial X^\beta}{\partial x^\alpha}} \tag{3.4}$$

3.1.2.3 Normal vector to the surface

As it appears later, it is necessary to involve the ambient space $E_3^\bullet$ in the expression of the behaviour of the surface. For example, moving a point of the surface during a transformation belongs to $\mathbb{R}^3$.

The basis of the tangent plane is completed in $\mathbb{R}^3$ by the unit normal vector to the tangent plane, defined by:

$$\boxed{a_3 = \frac{a_1 \times a_2}{\left\| a_1 \times a_2 \right\|}} \tag{3.5}$$

a_3 is unitary and independent of the parameterisation, with the exception of its orientation which depends on the orientation of the angle (a_1, a_2), so of the order (arbitrary!) of the parameters.

A vector field on σ is an application of $\mathbb{R}^3$ in which at any point $\boldsymbol{m}$ of σ matches a vector $X \in T_{\boldsymbol{m}}$ attached to $\boldsymbol{m}$.

3.1.3 Metric

3.1.3.1 Metric tensor

The measurement of the deformation of the surface during the transformation involves measuring distances and angles on the surface. The quantity used for this measurement in a Euclidean affine space is the metric tensor. It is therefore a question of extending this notion to the case of the surface.

The ambient space $E_3^\bullet$ being Euclidean, it is equipped with a metric, represented by the metric tensor field g, of components: $g_{ij} = e_i \cdot e_j$ (where the dot $\cdot$ denotes the dot product), expressed for example in a Cartesian reference system. In this way, a Euclidean structure of the tangent plane $T_{\mathcal{m}}$ is defined at every point $\mathcal{m}$ of the surface, by restriction to $T_{\mathcal{m}}$ of the ambient Euclidean structure. This structure leads to defining on the surface a metric tensor field $a(\mathcal{m})$ which components are, at every point $\mathcal{m}$:

$$a_{\alpha\beta} = a_\alpha \cdot a_\beta \tag{3.6}$$

a is a twice-covariant symmetric tensor field. This is the **first fundamental form** of the surface. The structure of the surface is then called **Riemannian.**

(dx^α) designating the basis of $T_{\mathcal{m}}^*$ dual of the natural basis (a_α), the metric tensor is noted in the form:

$$a = a_{\alpha\beta} dx^\alpha dx^\beta \tag{3.7}$$

omitting the symbol ⊗ of the tensor product. This notation makes it possible to use the usual differential calculus:

$$d\mathcal{m} = \frac{\partial \mathcal{m}}{\partial x^\alpha} dx^\alpha = a_\alpha dx^\alpha$$

is a linear form and:

$$d\mathcal{m}^2 = (a_\alpha dx^\alpha) \cdot (a_\beta dx^\beta) = a_{\alpha\beta}\, dx^\alpha dx^\beta = a$$

is the metric tensor, also noted ds^2.

If vectors a_1 et a_2 are orthogonal ($a_{12} = 0$):

$$ds^2 = A^2 dx_1{}^2 + B^2 dx_2{}^2$$

where the coefficients A and B, called **Lamé[3] coefficients,** are the lengths of the basic vectors. It is then easy to associate with the natural coordinate system an orthonormal system (called **physical,** because the quantities expressed in this coordinate system have their usual physical dimension), such as:

$$\tilde{a}_1 = \frac{a_1}{A}\,;\; \tilde{a}_2 = \frac{a_2}{B}$$

Let $\hat{a}$ be the determinant of the symmetric matrix of components $a_{\alpha\beta}$; it verifies the relationship:

$$\|a_1 \times a_2\| = \sqrt{\hat{a}} \tag{3.8}$$

a quantity that measures the area of the parallelogram built on the vectors of the basis.

The cross product of two vectors u and v of the tangent plane is then written:

$$u \times v = (u^\alpha a_\alpha) \times (v^\beta a_\beta) = (u^1 v^2 - u^2 v^1)\sqrt{\hat{a}}\; a_3$$

[3] Gabriel Lamé de la Droitière, French mathematician (1795–1870).

3.1.3.2 Fundamental antisymmetric tensor

Let $\tilde{e}$ be the family of antisymmetric numbers (signature of the permutation (α,β) with respect to $(1,2)$), defined in the basis (a_α) by:

$$\tilde{e}_{11} = \tilde{e}_{22} = 0 \qquad \tilde{e}_{12} = 1 \qquad \tilde{e}_{21} = -1 \tag{3.9}$$

The sign of $\tilde{e}$ is determined by the orientation of the basis. The family $\tilde{e}$ does not verify the change of basis formulas of a tensor.

Let e be the antisymmetric family of numbers whose components are, in any basis (a_α):

$$e_{\alpha\beta} = \sqrt{\hat{a}}\, \tilde{e}_{\alpha\beta} \tag{3.10}$$

e is a tensor. Indeed, its components verify the change of basis formulas of a tensor; in a change of basis $X^\Gamma(x^\alpha)$, it comes successively:

$$A_\Gamma = \frac{\partial x^\alpha}{\partial X^\Gamma} a_\alpha$$

$$\hat{A} = \det(A_{\Gamma\Delta}) = \det\left(\frac{\partial x^\alpha}{\partial X^\Gamma}\frac{\partial x^\beta}{\partial X^\Delta} a_{\alpha\beta}\right) = \left[\det\left(\frac{\partial x^\alpha}{\partial X^\Gamma}\right)\right]^2 \hat{a}$$

and, if $\mathcal{E}_{..}$ designates the components of e in basis (A_Γ), using the change of basis formulas of a twice-covariant tensor:

$$\begin{aligned} \mathcal{E}_{\Gamma\Delta} &= \frac{\partial x^\alpha}{\partial X^\Gamma}\frac{\partial x^\beta}{\partial X^\Delta} e_{\alpha\beta} = \left(\frac{\partial x^1}{\partial X^\Gamma}\frac{\partial x^2}{\partial X^\Delta} - \frac{\partial x^2}{\partial X^\Gamma}\frac{\partial x^1}{\partial X^\Delta}\right)\sqrt{\hat{a}} \\ &= \det\left(\frac{\partial x^\alpha}{\partial X^\Gamma}\right)\tilde{e}_{\Gamma\Delta}\sqrt{\hat{a}} = \tilde{e}_{\Gamma\Delta}\sqrt{\hat{A}} \end{aligned} \tag{3.11}$$

e is therefore a tensor whose components are expressed by (3.10) in all the bases. It is called the **2-fundamental form** of the surface or **fundamental antisymmetric tensor.**

This tensor is very convenient because, as in a Euclidean space, it makes it possible to easily express certain antisymmetric quantities, for example:

$$\begin{aligned} a_\alpha \times a_\beta &= e_{\alpha\beta}\, a_3 \\ a_3 \times a_\beta &= e_{\beta\lambda} a^{\lambda\mu} a_\mu = e_\beta{}^\mu a_\mu \end{aligned} \tag{3.12}$$

Noting $\underline{\varpi}$ the endomorphism associated with e:

- if X and Y are two vectors of T_m:

$$X \times Y = X^\alpha Y^\beta e_{\alpha\beta} a_3 = \underline{\varpi}(X)\cdot Y\, a_3 \tag{3.13}$$

- if X and Y are two vectors of $\mathbb{R}^3$:

$$X \times Y = X^\alpha Y^\beta e_{\alpha\beta} a_3 + \left(X^3 Y^\alpha - Y^3 X^\alpha\right) e_\alpha{}^\beta a_\beta \tag{3.14}$$

The following relation relates to the contravariant representative of e :

$$e^{\alpha\beta} = a^{\alpha\lambda} a^{\beta\mu} e_{\lambda\mu} = \tilde{e}^{\alpha\beta} / \sqrt{\hat{a}} \tag{3.15}$$

where $\tilde{e}^{\alpha\beta}$ is the antisymmetric family defined in a similar way to $\tilde{e}_{\alpha\beta}$.

The tensor field e defined on the surface σ is a constant field (cf. §3.2.1.5).

The following relations are obtained by expressing that the components of the contravariant representative of the metric tensor are the terms of the inverse matrix of the covariant representative (and conversely):

$$a^{\alpha\beta} = e^{\alpha\lambda}\ e^{\beta\mu}\ a_{\lambda\mu} \tag{3.16}$$

$$a_{\lambda\mu} = e_{\lambda\alpha}\ e_{\mu\beta}\ a^{\alpha\beta} \tag{3.17}$$

e and $\tilde{e}$ also make it possible to express the determinants, for example:

$$\det\left(m_\alpha{}^\beta\right) = \frac{1}{2} e^{\alpha\gamma}\ e_{\beta\delta}\ m_\alpha{}^\beta\ m_\gamma{}^\delta \tag{3.18}$$

$$\det\left(m_{\alpha\beta}\right) = \frac{\hat{a}}{2} e^{\alpha\gamma}\ e^{\beta\delta}\ m_{\alpha\beta}\ m_{\gamma\delta} \tag{3.19}$$

in particular:

$$\hat{a} = \frac{1}{2} \tilde{e}^{\alpha\gamma}\ \tilde{e}^{\beta\delta}\ a_{\alpha\beta}\ a_{\gamma\delta} \tag{3.20}$$

then the inverse of a square matrix M representing a bilinear form:

$$\left(M^{-1}\right)^{\lambda\mu} = \frac{2\,\tilde{e}^{\lambda\gamma}\ \tilde{e}^{\mu\delta}\ M_{\gamma\delta}}{\tilde{e}^{\alpha\gamma}\ \tilde{e}^{\beta\delta}\ M_{\gamma\delta}\ M_{\alpha\beta}}$$

3.1.4 Curvature

3.1.4.1 Definition

It is a matter of establishing a quantity which translates the shape of the surface, that is to say, its curvature, in common sense. The way the normal vector "turns" when moving on the surface is a good way to measure the curvature in a given direction.

D denotes the differentiation in the ambient Euclidean space $E_3^{\bullet}$. Let X a vector field on σ at a point m. X is a vector of the tangent plane T_m. The differentiation with respect to X of the relation $\left(a_3\right)^2 = 1$:

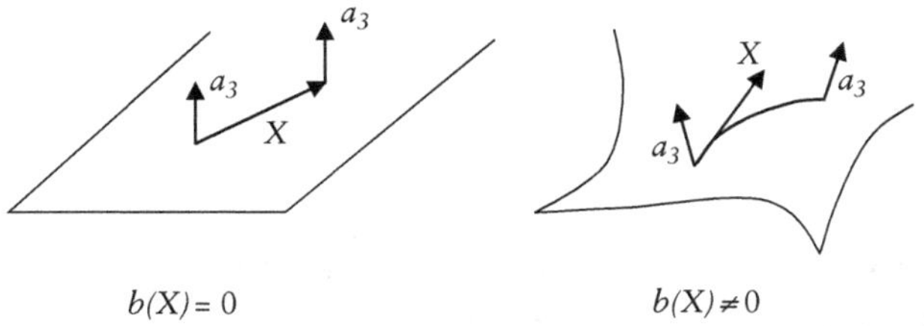

Figure 3.2

$$D_X a_3 \cdot a_3 = 0 \tag{3.21}$$

shows that $D_X a_3$ belongs to $T_{\mathcal{m}}$. The application:

$$X \rightarrow D_X a_3$$

is an endomorphism of $T_{\mathcal{m}}$ (the linearity is due to the linearity of the differentiation). Let $-b$ the associated tensor:

$$D_X a_3 = -b(X) = -X^\alpha b_\alpha{}^\beta a_\beta \tag{3.22}$$

Tensor b is called **curvature tensor** of the surface σ at point $\mathcal{m}$. As an endomorphism of $T_{\mathcal{m}}$, it belongs to the tensor product $T_{\mathcal{m}}^* \otimes T_{\mathcal{m}}$.

Geometrically, the tensor b expresses the rotation of the normal vector; indeed, if $D_X a_3$ is zero, a_3 remains parallel to itself in a displacement in the direction of X and the associated curvature is zero (Figure 3.2). If $b(X)$ is zero whatever the vector X at $\mathcal{m}$, the surface is locally a plane (cf. §3.1.3.2).

Denoting ∂_α the partial differentiation with respect to x^α $\left(\partial_\alpha = D_{a_\alpha}\right)$, the following relationships are inferred from the definition of b:

$$\boxed{\partial_\alpha a_3 = -b_\alpha{}^\beta a_\beta} \tag{3.23}$$

$$\boxed{b_{\alpha\beta} = -a_\beta \cdot \partial_\alpha a_3} \tag{3.24}$$

(Weingarten[4] formulas). As a bilinear form, b is the **second fundamental form** of the surface.

The calculation of b is generally done through its twice covariant components obtained by (3.24), a new expression of which is obtained by deriving the identity $a_3 \cdot a_\beta = 0$ with respect to a_α:

$$\partial_\alpha a_3 \cdot a_\beta + a_3 \cdot \partial_\alpha a_\beta = 0$$

Either, with relations (3.2) and (3.24):

$$\boxed{b_{\alpha\beta} = a_3 \cdot \partial_\alpha a_\beta = a_3 \cdot \partial^2_{\alpha\beta} \mathcal{m}} \tag{3.25}$$

---

[4] Julius Weingarten, German mathematician (1836–1910).

then, with (3.5):

$$\boxed{b_{\alpha\beta} = \frac{1}{\sqrt{\hat{a}}}\left(a_1, a_2, \partial^2_{\alpha\beta}\boldsymbol{m}\right)} \tag{3.26}$$

The covariant components of the curvature tensor are the projections of the second derivatives of the current point of the surface on the normal vector a_3

(3.26) shows that the tensor b is symmetrical. For any pair of vectors *(X, Y)* in $T_{\boldsymbol{m}} \times T_{\boldsymbol{m}}$:

$$D_X a_3 \cdot Y = D_Y a_3 \cdot X \tag{3.27}$$

from which:

$$\boxed{b(X,Y) = b(Y,X)} \tag{3.28}$$

In differential form:

$$-d\boldsymbol{m}\cdot da_3 = -\left(a_\alpha dx^\alpha\right)\left(\partial_\beta a_3\, dx^\beta\right) = \left(a_\alpha dx^\alpha\right)\left(b_\beta{}^\gamma a_\gamma\, dx^\beta\right) = b_{\alpha\beta}\, dx^\alpha dx^\beta = b \tag{3.29}$$

3.1.4.2 Geometric interpretation

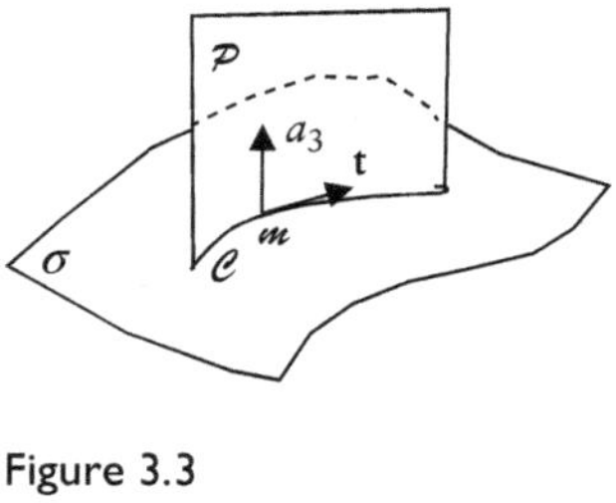

Figure 3.3

Let $\mathcal{P}$ a plane containing the normal vector a_3 at $\boldsymbol{m}$ and $\mathcal{C}$ the plane curve passing through $\boldsymbol{m}$, trace of $\mathcal{P}$ on the surface σ (Figure 3.3). Let **t** the vector tangent to the curve (which belongs to $T_{\boldsymbol{m}}$) and **n** the vector normal to the curve, which is identical[5] to a_3 provided that the curvilinear abscissa is properly oriented on $\mathcal{C}$.

Under these conditions, s denoting the curvilinear abscissa and R the radius of curvature of the curve, taking into account the relationship:

$$\mathbf{t} = \frac{d\boldsymbol{m}}{ds} = \frac{\partial \boldsymbol{m}}{\partial x^\alpha}\frac{dx^\alpha}{ds} = t^\alpha a_\alpha$$

it becomes:

$$\begin{aligned} \frac{\mathbf{n}}{R} &= \frac{d\mathbf{t}}{ds} = \frac{d}{ds}\left(\frac{dx^\alpha}{ds}a_\alpha\right) \\ &= \frac{d^2x^\alpha}{ds^2}a_\alpha + \frac{dx^\alpha}{ds}\frac{dx^\beta}{ds}\partial_\beta a_\alpha \end{aligned} \tag{3.30}$$

[5] In general, the normal to a curve drawn on a surface is not identical to the normal to the surface (see §3.1.3.4). The situation considered here is therefore specific, due to the particular plane considered.

$t^\alpha = \dfrac{dx^\alpha}{ds}$ being the components of the tangent vector **t** in basis (a_α), (3.25) and (3.30) get the relationship:

$$\frac{1}{R} = \mathbf{n} \cdot \frac{d\mathbf{t}}{ds} = b_{\alpha\beta} t^\alpha t^\beta = b(\mathbf{t}, \mathbf{t}) \tag{3.31}$$

3.1.4.3 Examples

- Let a cylinder parameterised in polar coordinates by:

$$(\theta, z) \quad \rightarrow \quad \mathcal{m} \begin{cases} R_0 = \text{cst} \\ \theta \\ z \end{cases}$$

The vectors of the basis are a_θ (with length R_0) and a_z ; $a_3 = e_r$ is the normal vector (Figure 3.4).

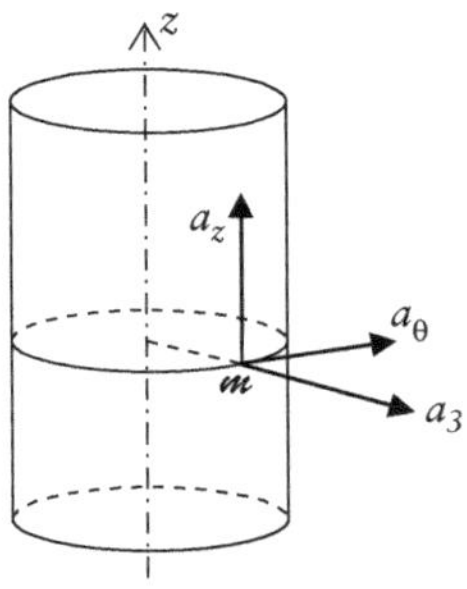

Figure 3.4

The expression of the curvatures is obtained from (3.25):

$$b_{\theta\theta} = a_3 \cdot \frac{\partial^2 \mathcal{m}}{\partial \theta^2} = -R_0 \; ; \; b_{\theta z} = a_3 \cdot \frac{\partial^2 \mathcal{m}}{\partial \theta \, \partial z} = 0 \; ; \; b_{zz} = a_3 \cdot \frac{\partial^2 \mathcal{m}}{\partial z^2} = 0$$

The transformation to the physical components, in the associated orthonormal coordinate system $\left(\tilde{a}_\theta = \dfrac{a_\theta}{R_0}, \tilde{a}_z = a_z \right)$, allows to find the usual expression of curvatures:

$$\tilde{b}_{\bullet\bullet} = \tilde{b}_{\bullet}^{\bullet} = \begin{pmatrix} -\dfrac{1}{R_0} & 0 \\ 0 & 0 \end{pmatrix}$$

$-\dfrac{1}{R_0}$ is the curvature of the circle, section of the cylinder by the plane z = cst, which contains the normal to the cylinder.

- Similarly, for a sphere parameterised in spherical coordinates (Figure 3.5):

$$(\theta, \varphi) \quad \rightarrow \quad \mathcal{m} \begin{cases} R_0 = \text{cst} \\ \theta \\ \varphi \end{cases}$$

The natural basis is $(a_\theta, a_\varphi, a_3 = e_r)$. a_θ and a_φ are tangent to the coordinate lines φ =cst and θ = cst respectively.

The metric tensor is written:

$$a_{\bullet\bullet} = \begin{pmatrix} {R_0}^2 \cos^2 \varphi & 0 \\ 0 & {R_0}^2 \end{pmatrix}$$

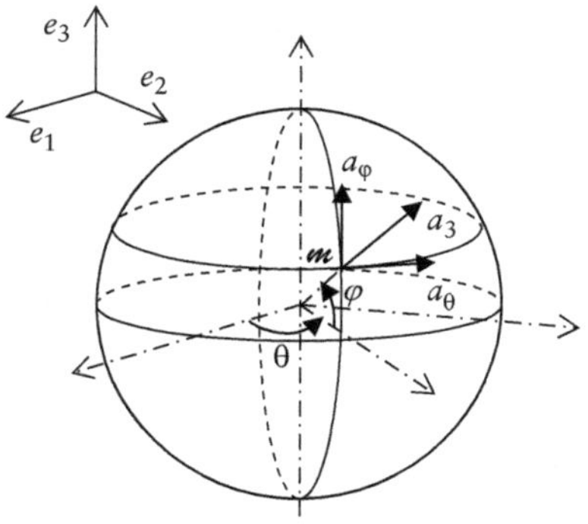

Figure 3.5

Using (3.25) :

$$b_{\theta\theta} = a_3 \cdot \frac{\partial^2 \boldsymbol{m}}{\partial \theta^2} = -R_0 \cos^2 \varphi$$

$$b_{\theta\varphi} = a_3 \cdot \frac{\partial^2 \boldsymbol{m}}{\partial \theta \, \partial \varphi} = 0 \,; \quad b_{\varphi\varphi} = a_3 \cdot \frac{\partial^2 \boldsymbol{m}}{\partial \varphi^2} = -R_0$$

also, in the associated orthonormal coordinate system:

$$\tilde{b}_{..} = \tilde{b}_{.}^{\,.} = \begin{pmatrix} -\dfrac{1}{R_0} & 0 \\ 0 & -\dfrac{1}{R_0} \end{pmatrix}$$

which interpretation is obvious ($-\dfrac{1}{R_0}$ curvature of the great circles).

3.1.4.4 Meusnier[6] theorem

Let (Figure 3.6):

- $\mathcal{C}$ a straight section at $\boldsymbol{m}$ as defined in §3.1.3.2, that is the intersection of σ with a plane $\mathcal{P}$ containing a_3 at $\boldsymbol{m}$;
- $\mathcal{C}'$ a tilted section, trace of a plane $\mathcal{P}'$ on σ, with the same tangent vector **t** than $\mathcal{C}$ at $\boldsymbol{m}$. $\mathcal{P}'$ makes an angle with respect to $\mathcal{P}$ at $\boldsymbol{m}$;
- ν the vector normal to $\mathcal{C}'$. $\mathcal{P}'$ contains **t** and ν.

(Meusnier) **Theorem**
The orthogonal projection of the centre of curvature of the straight-section $\mathcal{C}$ on the inclined plane $\mathcal{P}'$ is the centre of curvature of the tilted section $\mathcal{C}'$.

Proof: Let C and R (resp. C' and R') the centre and the radius of curvature of $\mathcal{C}$ (resp. $\mathcal{C}'$). If s' is the curvilinear abscissa along $\mathcal{C}'$, according to (3.30):

[6] Jean-Baptiste Marie Charles Meusnier de la Place, French engineer (1754–1793)

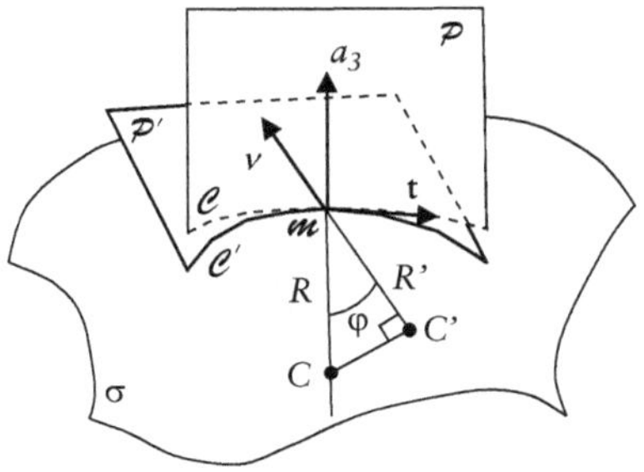

Figure 3.6

$$\frac{\nu}{R'} = \frac{d^2x^\alpha}{ds'^2} a_\alpha + \frac{dx^\alpha}{ds'} \frac{dx^\beta}{ds'} \partial_\beta a_\alpha$$

from where:

$$\frac{\nu \cdot a_3}{R'} = b_{\alpha\beta} t^\alpha t^\beta = \frac{1}{R} \tag{3.32}$$

or:

$$(R\, a_3) \cdot \nu = R'$$

that is: $mC \cdot \nu = mC'$

3.1.4.5 Principal curvatures

The curvature tensor being symmetrical, it has two real eigenvalues $\frac{1}{R_1}$ and $\frac{1}{R_2}$, R_1, and R_2 are called the **principal radii of curvature** of the surface. When R_1 and R_2 are different, two orthogonal eigenvectors (**principal directions of curvature**) are associated with them. If they are equal, any direction is principal.

In the orthonormal basis associated with the principal directions (principal basis of curvature), the vector **t** tangent to a straight section $\mathcal{C}$, as defined in §3.1.3.2, has components $(\cos\alpha, \sin\alpha)$. The relation (3.31) is then written:

$$\frac{1}{R} = \frac{\cos^2\alpha}{R_1} + \frac{\sin^2\alpha}{R_2} \tag{3.33}$$

R_1 and R_2 are the extreme values of the radii of curvature of the straight sections. In particular, if $\frac{1}{R_1} \le \frac{1}{R_2}$:

$$\frac{1}{R_1} = \inf_{X \in T_m} \frac{b(X,X)}{a(X,X)} \qquad \frac{1}{R_2} = \sup_{X \in T_m} \frac{b(X,X)}{a(X,X)}$$

Note: In all points, the principal directions and the principal radii of curvature can be determined by drawing the Mohr circle representing the curvature tensor.

When $R_1 = R_2$, the curvature is the same in all directions and any direction is principal; the point is called **ombilic.** In this case, two arbitrary directions (or in the continuity of the principal directions of the points of the neighbourhood) are chosen as support of the principal basis of curvature. a_1 and a_2 denoting the two orthogonal unit vectors of the principal basis of curvature at the point considered, the traces $\mathcal{C}_1$ and $\mathcal{C}_2$ of the planes (a_1, a_3) and (a_2, a_3) respectively on surface σ are plane curves which radii of curvature are respectively R_1 and R_2. This makes it possible to define two centres of curvature C_1 and

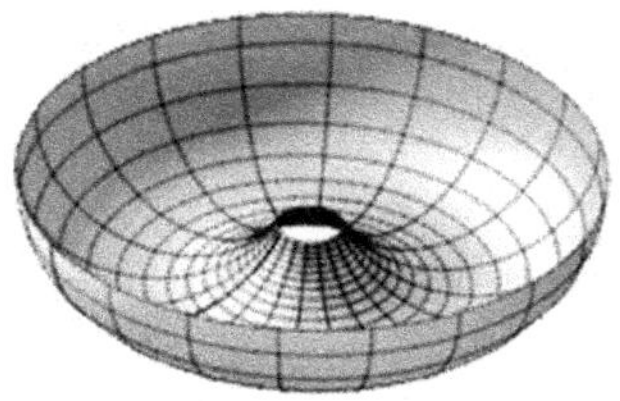

C_2 positioned on the normal axis containing a_3, given the sign of R_α. The two centres of curvature are on the same side of the surface if the two radii have the same sign. The curvilinear abscissa s_1 (resp. s_2) along $\mathcal{C}_1$ (resp. $\mathcal{C}_2$) is such that $R_1 = \dfrac{ds_1}{d\alpha_1}$, α_1 (resp. α_2) denoting the angle of rotation around C_1 (resp. C_2), the curve $\mathcal{C}_1$ (resp. $\mathcal{C}_2$) being plane.

The envelope curves of the principal directions of curvature are called **principal lines of curvature**. They usually differ from the previous curves $\mathcal{C}_1$ and $\mathcal{C}_2$. Their curvature is not necessarily equal to a principal curvature, except when the vectors normal to the curve and to the surface are identical.

The Meusnier theorem, relationship (3.32), allows in particular to calculate one of the principal curvatures of a surface of revolution (Figure 3.7): the small circle with radius r is the tilted section $\mathcal{C}'$, the plane $\mathcal{P}'$ being taken perpendicular to the axis of revolution. The centre of curvature C is the intersection of the axis bearing a_3 and of the axis of revolution. If φ is the angle (e_r, a_3):

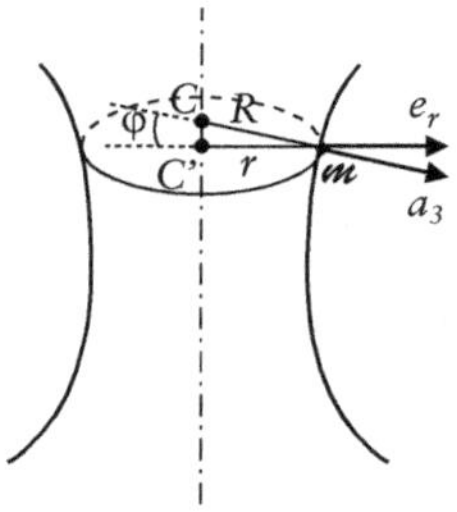

Figure 3.7

$$R = \frac{r}{\cos\varphi}$$

Note: A classical mistake is to take r as the second radius of curvature. One way to remember the good result is to follow the evolution of the normal vector a_3 along the small circle of radius r: it is clear that a_3 revolves around point C.

3.1.4.6 *Invariants of the curvature tensor*

The first and second invariants of the curvature tensor are calculated using R_1 and R_2:

$$H = \frac{1}{2} b_\alpha{}^\alpha = \frac{1}{2}\left(\frac{1}{R_1} + \frac{1}{R_2}\right) \tag{3.34}$$

$$K = \det\left(b_\alpha{}^\beta\right) = \frac{1}{R_1 R_2} = \frac{1}{2}\delta^{\alpha\beta}_{\lambda\mu}\, b_\alpha{}^\lambda\, b_\beta{}^\mu \tag{3.35}$$

where δ denotes the generalised Kronecker tensor. H is called **mean curvature** and K **Gaussian curvature**. The following non-reciprocal property is given without proof:

(Meusnier – 1776) **Property**
Of all the regular surfaces based on a given closed edge, the one with the minimum area has zero mean curvature.

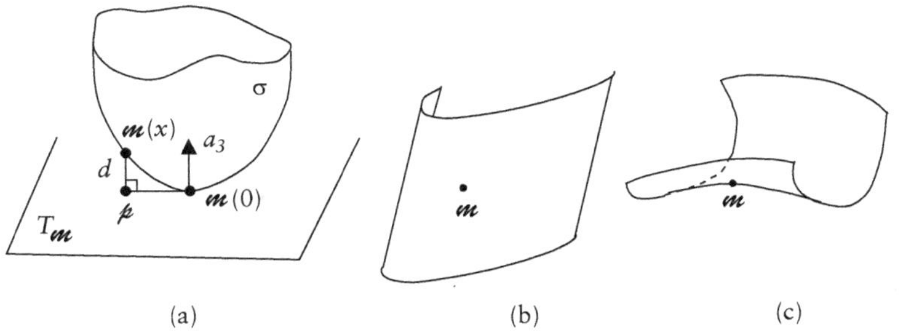

Figure 3.8

The main curvatures are roots of the characteristic polynomial:

$$-x^2+2Hx-K \tag{3.36}$$

The area element $d\mathcal{A}$ on the surface is equal to $ds_1\, ds_2$, that is to say $R_1R_2\, d\alpha_1\, d\alpha_2$. $d\psi = d\alpha_1\, d\alpha_2$ is the **solid angle** element intercepting the area element $d\mathcal{A}$. This leads to another property of Gaussian curvature, as introduced by Gauss in 1828:

$$\frac{1}{K}=\frac{d\mathcal{A}}{d\psi}$$

which generalises to two dimensions the relation on a plane arc: $R=\dfrac{ds}{d\phi}$.

The Gaussian curvature allows specifying of the local shape of the surface. Let x^1 and x^2 the surface parameters, be equal to zero in $\mathcal{m}(0)$. A Taylor[7] expansion of the surface is carried out in the vicinity of $\mathcal{m}(0)$:

$$\mathcal{m}(x)=\mathcal{m}(0)+\partial_\alpha \mathcal{m}(0)x^\alpha+\frac{1}{2}\frac{\partial^2\mathcal{m}}{\partial x^\alpha \partial x^\beta}(0)x^\alpha x^\beta+O\left[(x)^3\right]$$

Point $\mathcal{p}=\mathcal{m}(0)+a_\alpha(0)x^\alpha$ belongs to the tangent plane $T_{\mathcal{m}}$ at $\mathcal{m}(0)$. Let a_3 the normal vector at $\mathcal{m}(0)$. The distance d from $\mathcal{m}(x)$ to $T_{\mathcal{m}}$ is worth, neglecting third-order terms:

$$\begin{aligned} d &= a_3\cdot\left[\mathcal{m}(x)-\mathcal{m}(0)-a_\alpha(0)x^\alpha\right] \\ &= \frac{1}{2}a_3\cdot\frac{\partial^2\mathcal{m}}{\partial x^\alpha\partial x^\beta}(0)x^\alpha x^\beta \\ &= \frac{1}{2}b_{\alpha\beta}(0)x^\alpha x^\beta \end{aligned}$$

[7] Brook Taylor, British mathematician (1685–1731).

Photo 3.1. For reasons that are developed in Chapter 6, the Gaussian curvature of this stretched membrane is negative at all points (Montreal, Canada).

The curvature tensor thus defines the osculating quadric to σ at $m(0)$, that is, the local shape of the surface; if the Gaussian curvature $K = \dfrac{1}{R_1 R_2}$ is positive, the point is elliptic (Figure 3.8a). If it is negative, the point is hyperbolic (Figure 3.8c). If it is zero, the point is parabolic (Figure 3.8b).

If a surface is hyperbolic at one point, the principal curvatures being of opposite signs, there is a direction according to which the curvature is null (asymptotic direction).

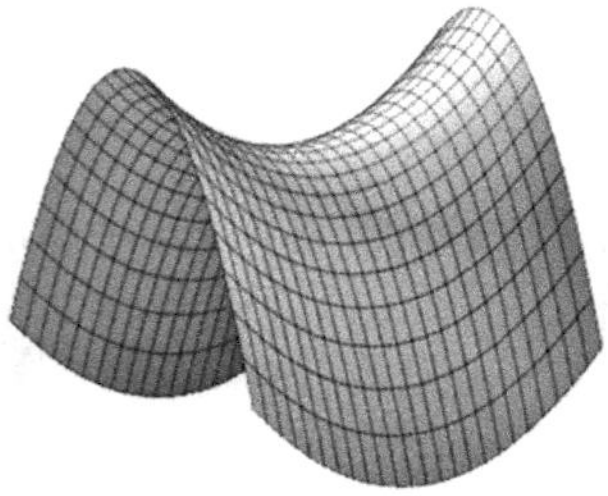

Example: If the surface is a hyperbolic quadric, its equation can be written (expressing it in the principal directions):

$$z = \frac{x^2}{\alpha^2} - \frac{y^2}{\beta^2}$$

where the two principal radii of curvature are of opposite signs: $R_1 = -\beta^2$, $R_2 = \alpha^2$, α and β being two constants.

Following the direction:

$$\frac{x}{\alpha} - \frac{y}{\beta} = \text{cst} = A$$

the equation of the curve inscribed on the surface is:

$$z = A\left(2\frac{x}{\alpha} - A\right)$$

So it is a straight line. Similarly, following the direction:

$$\frac{x}{\alpha}+\frac{y}{\beta}=\text{cst}=B$$

the curve drawn on the surface is also a straight line. In this way, two networks of parallel straight lines are drawn on the surface.

3.1.4.7 Third fundamental form

Let c be the third fundamental form of the surface defined by:

$$\forall(X,Y)\in T_{\mathcal{m}}\times T_{\mathcal{m}}: \qquad c(X,Y)=D_X a_3 \bullet D_Y a_3 \tag{3.37}$$

c is a symmetrical bilinear form on $T_{\mathcal{m}}$, so a tensor field twice covariant as:

$$c_{\alpha\beta}X^\alpha X^\beta=\left(-X^\alpha b_\alpha{}^\lambda a_\lambda\right)\left(-Y^\beta b_\beta{}^\mu a_\mu\right)$$

which components are:

$$\boxed{c_{\alpha\beta}=b_\alpha{}^\lambda b_{\beta\lambda}=c_{\beta\alpha}} \tag{3.38}$$

The three fundamental forms a, b, c are not independent. Indeed, according to Cayley[8]–Hamilton[9]'s theorem, the value of the characteristic polynomial of any endomorphism is zero. Applying this property to the curvature tensor, it becomes:

$$-b^2+2bH-aK=0 \tag{3.39}$$

that is to say:

$$\forall(X,Y)\in T_{\mathcal{m}}\times T_{\mathcal{m}}: \qquad \left(-c_{\alpha\beta}+2Hb_{\alpha\beta}-Ka_{\alpha\beta}\right)X^\alpha Y^\beta=0 \tag{3.40}$$

3.1.5 Normal coordinates in the neighbourhood of the surface

The objective is to build a parameterisation of $E_3^\bullet$ in the neighbourhood of σ and the associated natural coordinate system. Indeed, a shell is represented by its middle surface σ, but has a thickness on both sides of the middle surface and it is necessary to have a local 3D parameterisation based on that of σ.

3.1.5.1 Naturel coordinate system in the neighbourhood of the surface

The natural coordinate system at any point on the surface makes it possible to define a system of coordinates $\left(x^i\right)$ of $E_3^\bullet$ in the neighbourhood D of σ (Figure 3.9). Let $\mathcal{p}$ be a

[8] Arthur Cayley, British mathematician (1821–1895).
[9] Sir William Rowan Hamilton, Irish mathematician and physicist (1805–1865).

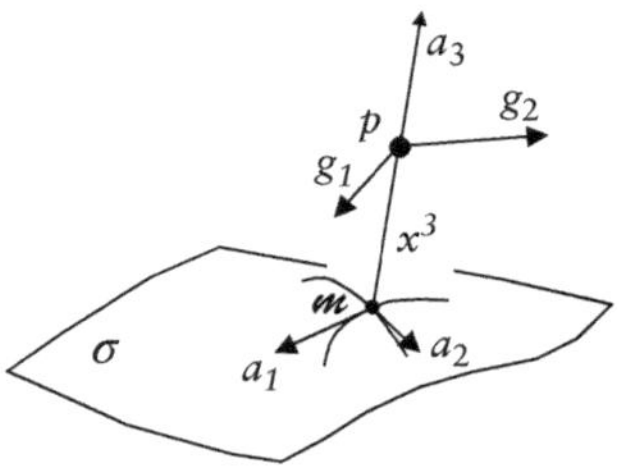

Figure 3.9

point of the neighbourhood of σ and m its orthogonal projection on σ.

p can be located using the normal coordinate system:

$$p(x^i) = m(x^i) + x^3 a_3(m) \tag{3.41}$$

(x^α, x^3) is a parametrisation of $E_3^\bullet$ in the neighbourhood of σ.

The natural basis (g_i) is associated with this normal coordinate system:

$$\begin{cases} g_\alpha = \dfrac{\partial p}{\partial x^\alpha} = \dfrac{\partial m}{\partial x^\alpha} + x^3 \dfrac{\partial a_3}{\partial x_\alpha} = a_\alpha - x^3 b_\alpha{}^\beta a_\beta \\ g_3 = a_3 \end{cases}$$

g_1 and g_2 are in the plane parallel to T_m. The curvature modifies g_1 and g_2 with respect to a_1 and a_2. If the surface is locally a plane, g_1 and g_2 are respectively equal to a_1 and a_2. g_3 remains equal to a_3, unitary and orthogonal to the other two vectors, which are in a plane parallel to T_m. Then:

$$\boxed{\begin{array}{l} m_\alpha{}^\beta = \delta_\alpha^\beta - x^3 b_\alpha{}^\beta \\ m_\alpha{}^3 = 0 \\ m_3{}^\alpha = 0 \\ m_3{}^3 = 1 \end{array}} \tag{3.42}$$

which allows it to be written as:

$$\boxed{g_i = m_i{}^j a_j} \tag{3.43}$$

The family m defines an endomorphism of $\mathbf{R}^3$ and is therefore of a tensorial nature. More precisely, m is defined on $T_m \times \mathbf{R}$, with values in $\mathbf{R}^3$. The endomorphism m is reduced to identity on the surface.

The metric tensor in p is written as:

$$g_{ij} = m_i{}^k m_j{}^l a_{kl} \tag{3.44}$$

with $g_{33} = 1$, $g_{3\alpha} = 0$ and

$$g_{\alpha\beta} = a_{\alpha\beta} - 2x^3 b_{\alpha\beta} + (x^3)^2 b_\alpha{}^\gamma b_{\gamma\beta}$$

Let the Levi-Civita tensor $\overline{\mathrm{e}}$ of $\mathbf{R}^3$ (cf. §2.1.9) expressed in the local coordinate system (g_i). The following relation is easily obtained according to (3.18):

$$\overline{e}_{\alpha\beta 3} = m_\alpha{}^\gamma m_\beta{}^\delta \; e_{\gamma\delta} = \hat{m} \; e_{\alpha\beta} \tag{3.45}$$

On the surface, the twice covariant tensor field with component $\overline{e}_{\alpha\beta 3}$ and the tensor field e are equal.

3.1.5.2 Principal curvature coordinate system

In the case where the basis (a_α) is that of principal curvature:

$$\begin{aligned} g_1 &= \left(1 - \frac{x^3}{R_1}\right) a_1 \\ g_2 &= \left(1 - \frac{x^3}{R_2}\right) a_2 \end{aligned} \tag{3.46}$$

g_1 and g_2 are also parallel to a_1 and a_2 respectively, but are not normalised. In this coordinate system:

$$g_{\alpha\alpha} = \left(1 - \frac{x^3}{R_\alpha}\right)^2 \tag{3.47}$$

There must be no singularity by (3.43), so that the normal coordinates are in bijection with $\mathcal{D}$, which implies that:

$$g_1 \times g_2 = \hat{m} \; a_1 \times a_2$$

is defined continuously, from which:

$$\hat{m} = \det\left(m_i{}^j\right) > 0$$

(must not change sign when x^3 increases).

From (3.44):

$$\hat{m} = \left(\hat{g} / \hat{a}\right)^{1/2} \tag{3.48}$$

In addition:

$$\begin{aligned} \det\left(m_i{}^j\right) &= \det\left(m_\alpha{}^\beta\right) = \det\left(\delta_\alpha^\beta - x^3 b_\alpha{}^\beta\right) = \left(1 - \frac{x^3}{R_1}\right)\left(1 - \frac{x^3}{R_2}\right) \\ &= 1 - 2x^3 H + \left(x^3\right)^2 K \end{aligned} \tag{3.49}$$

is zero when $x^3 = R_1$ or R_2, but remains positive as long as:

$$\left|x^3\right| < \min\left(\left|R_1\right|,\left|R_2\right|\right) \tag{3.50}$$

Inequality (3.50) is supposed to be verified in what follows, which in practice requires only one projection of $\not{p}$ on σ is of interest for the representation. If (3.50) was not verified for a sphere, for example, it would mean that its half thickness is greater than its radius, that is that the sphere is full.

3.1.5.3 Inverse endomorphism of $\mathfrak{m}$

When condition (3.50) is satisfied, the matrix $\left(\mathfrak{m}_\alpha{}^\beta\right)$ is reversible; ℓ refers to the inverse endomorphism.

Starting from (3.43) projected on a_α:

$$g_\gamma \cdot a_\alpha = g_\gamma \ell_\alpha{}^\beta g_\beta = g_{\gamma\beta}\ell_\alpha{}^\beta = \mathfrak{m}_{\gamma\alpha}$$

ℓ, therefore, has for components:

$$\begin{aligned} &\ell_\alpha{}^\beta = \left(\mathfrak{m}^{-1}\right)_\alpha{}^\beta = g^{\beta\gamma}\mathfrak{m}_{\gamma\alpha} \\ &\ell_3{}^\beta = 0;\ \ell_\alpha{}^3 = 0;\ \ell_3{}^3 = 1 \end{aligned} \tag{3.51}$$

The relation (3.44) leads, by raising the indices, to the expression:

$$g^{ij} = a^{im}a^{jn}\mathfrak{m}_m{}^k\mathfrak{m}_n{}^l a_{kl} = \mathfrak{m}^{ik}\mathfrak{m}^{jl}a_{kl} = \mathfrak{m}^i{}_k\mathfrak{m}^j{}_l a^{kl}$$

which is reversed in:

$$a^{ij} = \ell_k{}^i \ell_l{}^j g^{kl}$$

ℓ, as $\mathfrak{m}$, depends on the curvature tensor b and can be developed in the Taylor series of b. It can be shown that:

$$\ell_\alpha{}^\beta = \sum_{\mathrm{n}=0}^{\infty}\left(x^3\right)^{\mathrm{n}}\left(b^{\mathrm{n}}\right)_\alpha{}^\beta \tag{3.52}$$

with: $b^0 = a$; $b^1 = b$; $b^2 = c$

b^{n}, n-th power of b, is defined by the recurrence relation:

$$\left(b^{\mathrm{n}}\right)_\alpha{}^\beta = b_\alpha{}^\gamma\left(b^{\mathrm{n-1}}\right)_\gamma{}^\beta = b_\gamma{}^\beta\left(b^{\mathrm{n-1}}\right)_\alpha{}^\gamma$$

3.1.5.4 Surface parallel to the middle surface

Let σ' a surface parallel to the reference surface $\sigma\left(x^1,x^2\right)$. σ' is the set of points $\not{p}$ defined by their normal coordinates according to (3.41), such that $x^3 = \mathrm{cst}$. σ' is then parameterised by x^1 and x^2. The vectors of the plane tangent to σ' in this parametrisation are given by (3.43). It becomes:

$$\partial_\alpha g_\beta = -x^3\left(\partial_\alpha b_\beta{}^\gamma \cdot a_\gamma + b_\beta{}^\gamma\, \partial_\alpha a_\gamma\right) + \partial_\alpha a_\beta$$

From where the curvature b' of σ':

$$b'_{\alpha\beta} = g_3 \cdot \partial_\alpha g_\beta = \left(\delta_\beta{}^\gamma - x^3 b_\beta{}^\gamma\right) b_{\alpha\gamma}$$

b' is expressed here in basis(g_α), b in basis (a_α).

σ' has the same principal directions of curvature as σ. If (a_α) is the principal curvature coordinate system (orthonormal), the basis (g_α) is orthogonal but not normalised. Given (3.47):

$$\frac{1}{R'_\alpha} = \frac{1}{R_\alpha}\left(1 - \frac{x^3}{R_\alpha}\right) \Big/ \left(1 - \frac{x^3}{R_\alpha}\right)^2 = \frac{1}{R_\alpha - x^3}$$

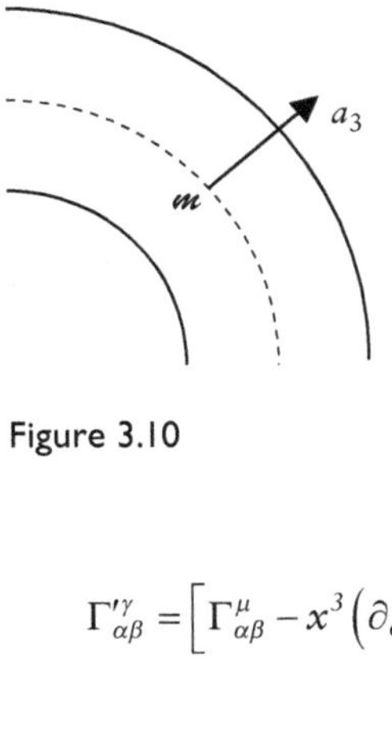

Figure 3.10

If R_α is negative (Figure 3.10), the radius of curvature increases in absolute value when x^3 increases. The metric of σ' is defined by (3.47), x^3 remaining constant.

The differentiation on the surface is determined by the connection coefficients. These are obtained from their definition:

$$\begin{aligned}\partial_\alpha g_\beta &= \Gamma^\mu_{\alpha\beta} a_\mu + b_{\alpha\beta} a_3 - x^3 \partial_\alpha b_\beta{}^\mu a_\mu - x^3 b_\beta{}^\lambda \Gamma^\mu_{\alpha\lambda} a_\mu - x^3 b_\beta{}^\mu b_{\alpha\mu} a_3 \\ &= \left(\Gamma^\mu_{\alpha\beta} - x^3 \partial_\alpha b_\beta{}^\mu - x^3 b_\beta{}^\lambda \Gamma^\mu_{\alpha\lambda}\right) a_\mu + b'_{\alpha\beta} a_3\end{aligned}$$

from which:

$$\Gamma'^\gamma_{\alpha\beta} = \left[\Gamma^\mu_{\alpha\beta} - x^3\left(\partial_\alpha b_\beta{}^\mu + b_\beta{}^\lambda \Gamma^\mu_{\beta\lambda}\right)\right] \ell_\mu{}^\gamma \tag{3.53}$$

3.2 DIFFERENTIATION ON THE SURFACE

3.2.1 Levi-Civita differentiation

3.2.1.1 Definitions

A tensor field (p, q) on σ is an application that at any point m of σ associates a tensor (p, q) on T_m. The purpose of this paragraph is to define a differentiation of such tensor fields; this differentiation is an internal operation on the surface.

Let X and Y be two vector fields of σ (Figure 3.11). The derivative of Y with respect to X in $E_3^\bullet$ is:

$$D_X Y = X^\alpha \left\{\partial_\alpha Y^\beta \cdot a_\beta + Y^\beta \partial_\alpha a_\beta\right\} \tag{3.54}$$

It appears by (3.54) that $D_X Y$ is not tangent to σ, because of the term $\partial_\alpha a_\beta$ (of which it was seen previously that, when the surface is curved, it has a component on a_3). $D_X Y$ is not an internal operation on σ.

In order to define a differentiation which is an internal operation on all tensor fields of the surface, it is natural to pose:

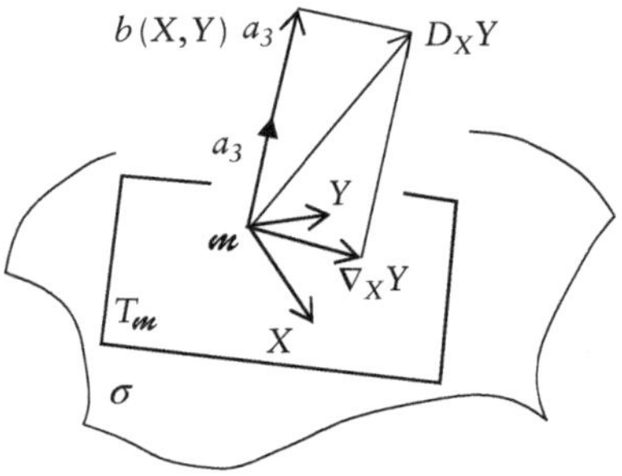

Figure 3.11

$$\nabla_X Y = \underbrace{\text{proj}}_{T_m // a_3}\left(D_X Y\right) \tag{3.55}$$

which leads to the linearity of ∇ and allows to write $D_X Y$ in the form:

$$D_X Y = \nabla_X Y + b(X,Y)a_3 \tag{3.56}$$

In the same way, for a form ℓ of $T_m{}^*$:

$$\nabla_X \ell = \underbrace{\text{proj}}_{T_m^* // a_3}\left(D_X \ell\right) \tag{3.57}$$

(where a_3 equates to its dual a^3).

To be a differentiation, ∇ must also verify the rule of differentiation of a tensor product (contracted or not), for example:

$$\nabla(u \otimes v) = \nabla u \otimes v + u \otimes \nabla v \tag{3.58}$$

where u and v belong to T_m or $T_m{}^*$. This property completes the definition of ∇. It is equivalent to a rule of projection of the derivative $D(u \otimes v)$ on a tensor product space of T_m.

Under these conditions, ∇ possesses the properties of a differentiation of the tensor fields on the surface; it is called a **covariant differentiation** on the surface or **Levi-Civita differentiation.**

The surface operators **gradient, divergence** and **scalar Laplacian** can be defined on the surface using differentiation ∇ as in a Euclidean space (see §2.2.6).

3.2.1.2 Differentiation formulas

By definition, $\nabla_\alpha a_\beta$ belongs to T_m. The coefficients $\Gamma^\gamma_{\alpha\beta}$ introduced by the relation:

$$\nabla_\alpha a_\beta = \Gamma^\gamma_{\alpha\beta} a_\gamma \tag{3.59}$$

analogous for ∇ to the Christoffel coefficients for the differentiation D (see §2.2.3), are called **Riemannian connection** (or Levi-Civita connection) **coefficients.**

Given this definition and the results of §3.1, especially the relationship (3.25):

$$D_\alpha a_\beta = \partial_\alpha a_\beta = \nabla_\alpha a_\beta + b_{\alpha\beta} a_3 = \Gamma^\gamma_{\alpha\beta} a_\gamma + b_{\alpha\beta} a_3 \tag{3.60}$$

(Gauss formula).

For any two vectors of $T_{\mathcal{m}}$:

$$X \in T_{\mathcal{m}}, Y \in T_{\mathcal{m}}: \qquad \nabla_X Y = X^\alpha \left\{ \partial_\alpha Y^\beta a_\beta + \Gamma^\gamma_{\alpha\beta} Y^\beta a_\gamma \right\} \tag{3.61}$$

and:

$$\boxed{\nabla_\alpha Y^\beta = \partial_\alpha Y^\beta + \Gamma^\beta_{\alpha\gamma} Y^\gamma} \tag{3.62}$$

For a form of the dual basis, the following relation is easily verified:

$$D_\alpha a^\beta = \nabla_\alpha a^\beta + b_\alpha{}^\beta a^3 = -\Gamma^\beta_{\alpha\gamma} a^\gamma + b_\alpha{}^\beta a^3 \tag{3.63}$$

3.2.1.3 Expression of the connection coefficients in $E_3^\bullet$

In $\mathbb{R}^3$, in the neighbourhood of σ, it is natural to use the normal coordinates, for which the following differentiation relationships are verified:

$$\begin{cases} \partial_3 g_3 = 0 \\ \partial_\alpha g_3 = -b_\alpha{}^\beta a_\beta = -b_\alpha{}^\beta \,\ell_\beta{}^\gamma \, g_\gamma \end{cases} \tag{3.64}$$

$$\begin{cases} \partial_3 g_\beta = \partial_3 \left(\delta^\alpha_\beta - x^3 b_\beta{}^\alpha \right) a_\alpha = -b_\beta{}^\alpha a_\alpha = -b_\beta{}^\alpha \,\ell_\alpha{}^\gamma \, g_\gamma \\ \partial_\alpha g_\beta = \partial_\alpha \left(m_\beta{}^\delta a_\delta \right) = a_\delta \, \partial_\alpha m_\beta{}^\delta + m_\beta{}^\delta \Gamma^\gamma_{\alpha\delta} a_\gamma + m_\beta{}^\delta \, b_{\alpha\delta} a_3 \\ \qquad = \left(\partial_\alpha m_\beta{}^\gamma + m_\beta{}^\delta \, \Gamma^\gamma_{\alpha\delta} \right) \ell_\gamma{}^\lambda \, g_\lambda + m_\beta{}^\delta \, b_{\alpha\delta} g_3 \end{cases} \tag{3.65}$$

Considering that:

$$\partial_\alpha m_\beta{}^\gamma + m_\beta{}^\delta \Gamma^\gamma_{\alpha\delta} = \nabla_\alpha m_\beta{}^\gamma + m_\delta{}^\gamma \Gamma^\delta_{\alpha\beta}$$

the Christoffel coefficients of $E_3^\bullet$ are thus expressed in the coordinate system (g_i) by:

$$\boxed{\begin{aligned} &\bar{\Gamma}^3_{33} = \bar{\Gamma}^\alpha_{33} = \bar{\Gamma}^3_{3\beta} = 0 \\ &\bar{\Gamma}^\gamma_{\alpha 3} = -b_\alpha{}^\beta \ell_\beta{}^\gamma \\ &\bar{\Gamma}^3_{\alpha\beta} = m_\beta{}^\gamma b_{\alpha\gamma} \\ &\bar{\Gamma}^\gamma_{\alpha\beta} = \Gamma^\gamma_{\alpha\beta} + \ell_\lambda{}^\gamma \, \nabla_\alpha m_\beta{}^\lambda \end{aligned}} \tag{3.66}$$

The non-zero coefficients are reduced on σ to:

$$\boxed{\bar{\Gamma}^\beta_{\alpha 3} = -b_\alpha{}^\beta; \qquad \bar{\Gamma}^3_{\alpha\beta} = b_{\alpha\beta}; \qquad \bar{\Gamma}^\gamma_{\alpha\beta} = \Gamma^\gamma_{\alpha\beta}} \tag{3.67}$$

Coefficients $\Gamma^{\gamma}_{\alpha\beta}$ also verify the following relation, similar to the relation (2.50) established for the Christoffel coefficients:

$$\boxed{\Gamma^{\gamma}_{\alpha\beta} = \frac{1}{2} a^{\gamma\delta} \left(\partial_\alpha a_{\delta\beta} + \partial_\beta a_{\alpha\delta} - \partial_\delta a_{\alpha\beta} \right)} \tag{3.68}$$

The covariant differentiation depends only on the metric tensor a (and therefore not the curvature tensor b). It is **intrinsic.** The relation (3.68) shows that if the natural coordinate system (a_α) is orthonormal in all points, its connection coefficients are zero.

By (3.60) and by differentiation of the relation $a^\gamma a_\beta = \delta^\gamma_\beta$, it also becomes:

$$\boxed{\Gamma^{\gamma}_{\alpha\beta} = a^\gamma \partial_\alpha a_\beta = -a_\beta \, \partial_\alpha a^\gamma} \tag{3.69}$$

Note: formulas established here have meaning only in a natural coordinate system, that is a coordinate system in which the vectors of the basis are defined by (3.2).

If the coordinate system is simply orthogonal, the only non-zero coefficients are:

$$\Gamma^{\alpha}_{\alpha\alpha} = \frac{1}{2} a^{\alpha\alpha} \partial_\alpha a_{\alpha\alpha} \quad \text{(ns)*}$$

$$\text{if } \alpha \neq \gamma: \qquad \Gamma^{\gamma}_{\alpha\alpha} = -\frac{1}{2} a^{\gamma\gamma} \partial_\gamma a_{\alpha\alpha} \quad \text{(ns)}$$

$$\Gamma^{\gamma}_{\gamma\beta} = \frac{1}{2} a^{\gamma\gamma} \partial_\beta a_{\gamma\gamma} \quad \text{(ns)}$$

* (ns) indicates that there is no summation here on the index α or γ.

For example: $\Gamma^1_{12} = \dfrac{1}{2} a^{11} \partial_2 a_{11}$.

The above expressions then make it possible to write the connection coefficients as functions of the Lamé coefficients A and B:

$$\Gamma^1_{11} = \frac{1}{A^2} A \, \partial_1 A \, ; \; \Gamma^2_{11} = -\frac{1}{B^2} A \, \partial_2 A$$

$$\Gamma^1_{22} = -\frac{1}{A^2} B \, \partial_1 B \, ; \; \Gamma^2_{22} = \frac{1}{B^2} B \, \partial_2 B$$

$$\Gamma^1_{12} = \frac{1}{A^2} A \, \partial_2 A \, ; \; \Gamma^2_{12} = \frac{1}{B^2} B \, \partial_1 B$$

3.2.1.4 Rules of differentiation of any tensor field

The rule of differentiation by of any tensor on the surface are the same as the rules of differentiation by D in $E_3^\bullet$. For example:

$$\nabla_\alpha\left(T_\beta{}^\gamma a^\beta \otimes a_\gamma\right) = \partial_\alpha T_\beta{}^\gamma a^\beta \otimes a_\gamma + T_\beta{}^\gamma \nabla_\alpha a^\beta \otimes a_\gamma + T_\beta{}^\gamma a^\beta \otimes \nabla_\alpha a_\gamma$$

from where:

$$\nabla_\alpha T_\beta{}^\gamma = \partial_\alpha T_\beta{}^\gamma - \Gamma^\delta_{\alpha\beta} T_\delta{}^\gamma + \Gamma^\gamma_{\alpha\delta} T_\beta{}^\delta \tag{3.70}$$

The connection coefficients also make it possible to express the derivative of any vector field $U \in \mathbb{R}^3$ defined on the surface with respect to a vector of $T_{\mathcal{M}}$:

$$X \in T_{\mathcal{M}},\, U \in \mathbb{R}^3\text{: } D_X U = \nabla_X\left(U^\alpha a_\alpha\right) + b_{\alpha\beta} U^\alpha X^\beta a_3 + X^\alpha \partial_\alpha U^3 a_3 - U^3 X^\alpha b_\alpha{}^\beta a_\beta$$

$$\boxed{D_X U = X^\alpha\left[\left(\partial_\alpha U^\beta + \Gamma^\beta_{\alpha\gamma} U^\gamma - b_\alpha{}^\beta U^3\right) a_\beta + \left(b_{\alpha\beta} U^\beta + \partial_\alpha U^3\right) a_3\right]} \tag{3.71}$$

where it is recalled that:

$$\nabla_\alpha U^\beta = \partial_\alpha U^\beta + \Gamma^\beta_{\alpha\gamma} U^\gamma$$

This relation is generalised to any tensor, thanks to relations (3.66), for example:

$$D_\alpha T_\beta{}^\gamma = \nabla_\alpha T_\beta{}^\gamma + b_{\alpha\beta} T_3{}^\gamma - b_\alpha{}^\gamma T_\beta{}^3$$

$$D_\alpha T_3{}^\gamma = \nabla_\alpha T_3{}^\gamma - b_\alpha{}^\beta T_\beta{}^\gamma - b_\alpha{}^\gamma T_3{}^3$$

$$D_\alpha T_\beta{}^3 = \nabla_\alpha T_\beta{}^3 + b_{\alpha\gamma} T_\beta{}^\gamma + b_{\alpha\beta} T_3{}^3$$

In these expressions, it is noted:

- $\nabla_\alpha T_3{}^\gamma = \partial_\alpha T_3{}^\gamma + \Gamma^\gamma_{\alpha\lambda} T_3{}^\lambda$, where T is considered as a vector of components $T_3{}^\lambda$;
- $\nabla_\alpha T_\beta{}^3 = \partial_\alpha T_\beta{}^3 - \Gamma^\lambda_{\alpha\beta} T_\lambda{}^3$, where T is considered as a form of components $T_\lambda{}^3$.

3.2.1.5 Differentiation of fundamental forms

The three-dimensional metric tensor g is constant (this result is obvious when the derivative of g is expressed in a cartesian coordinate system). Expressed near the surface in normal coordinates, this property allows it to be written as:

$$D_\gamma g_{\alpha\beta} = 0$$

That is:

$$D_\gamma g_{\alpha\beta} = \partial_\gamma g_{\alpha\beta} - \overline{\Gamma}^\lambda_{\gamma\alpha} g_{\lambda\beta} - \overline{\Gamma}^\lambda_{\alpha\beta} g_{\alpha\lambda} - \overline{\Gamma}^3_{\gamma\alpha} g_{3\beta} - \overline{\Gamma}^3_{\gamma\beta} g_{\alpha 3} = 0$$

Brought back to $x^3 = 0$, this relationship becomes:

$$D_\gamma a_{\alpha\beta} = \partial_\gamma a_{\alpha\beta} - \Gamma^\lambda_{\gamma\alpha} a_{\lambda\beta} - \Gamma^\lambda_{\alpha\beta} a_{\alpha\lambda} = \nabla_\gamma a_{\alpha\beta} = 0$$

giving:

$$\boxed{\nabla a = 0} \tag{3.72}$$

The covariant differentiation ∇ respects the metric.

The derivative ∇ of the fundamental antisymmetric tensor $\mathfrak{e}$ is calculated via the total derivative of its extension $\overline{\mathfrak{e}}$ in $\mathbf{R}^3$. $\overline{\mathfrak{e}}$ is a constant tensor (its components are constant in a Cartesian coordinate system), so:

$$D_i \overline{\mathfrak{e}}_{jkl} = 0$$

from which:

$$D_\gamma \overline{\mathfrak{e}}_{\alpha\beta 3} = \partial_\gamma \overline{\mathfrak{e}}_{\alpha\beta 3} - \overline{\Gamma}^i_{\gamma\alpha} \overline{\mathfrak{e}}_{i\beta 3} - \overline{\Gamma}^i_{\gamma\beta} \overline{\mathfrak{e}}_{\alpha i 3} - \overline{\Gamma}^i_{\gamma 3} \overline{\mathfrak{e}}_{\alpha\beta i} = 0$$

By expressing this relation on the surface, considering (3.45) and (3.67), it becomes:

$$D_\gamma \overline{\mathfrak{e}}_{\alpha\beta 3}\left(x^3 = 0\right) = \nabla_\gamma \mathfrak{e}_{\alpha\beta} - b_{\gamma\alpha} \overline{\mathfrak{e}}_{3\beta 3} - b_{\gamma\beta} \overline{\mathfrak{e}}_{\alpha 33} + b_\gamma{}^\delta \overline{\mathfrak{e}}_{\alpha\beta\delta} = 0$$

but:

$$\overline{\mathfrak{e}}_{3\beta 3} = \overline{\mathfrak{e}}_{\alpha 33} = \overline{\mathfrak{e}}_{\alpha\beta\delta} = 0$$

Indeed, the triplet (α, β, δ) cannot be a permutation of (*1*, *2*, *3*), since it contains only Greek indices. So:

$$\nabla_\gamma \mathfrak{e}_{\alpha\beta} = 0 \tag{3.73}$$

Tensor $\mathfrak{e}$ is constant:

$$\boxed{\nabla \mathfrak{e} = 0} \tag{3.74}$$

Expressing this property using the connection coefficients:

$$\nabla_\gamma \mathfrak{e}_{\alpha\beta} = \partial_\gamma \mathfrak{e}_{\alpha\beta} - \Gamma^\delta_{\gamma\alpha} \mathfrak{e}_{\delta\beta} - \Gamma^\delta_{\gamma\beta} \mathfrak{e}_{\alpha\delta} = \partial_\gamma \left(\tilde{\mathfrak{e}}_{\alpha\beta} \sqrt{\hat{a}}\right) - \sqrt{\hat{a}} \left(\Gamma^\delta_{\gamma\alpha} \tilde{\mathfrak{e}}_{\delta\beta} + \Gamma^\delta_{\gamma\beta} \tilde{\mathfrak{e}}_{\alpha\delta}\right) = 0$$

For $(\alpha, \beta) = (1, 2)$:

$$\partial_\gamma \left(\sqrt{\hat{a}}\right) - \sqrt{\hat{a}} \left(\Gamma^1_{\gamma 1} + \Gamma^2_{\gamma 2}\right) = 0$$

from which:

$$\boxed{\Gamma^\alpha_{\gamma\alpha} = \frac{\partial_\gamma \left(\sqrt{\hat{a}}\right)}{\sqrt{\hat{a}}}} \tag{3.75}$$

which is the restriction on the surface of an equivalent relation in $E_3^\bullet$. This expression notably makes it possible to simplify the calculation of the surface divergence:

$$\mathrm{div}X = \nabla_\alpha X^\alpha = \partial_\alpha X^\alpha + \Gamma^\alpha_{\alpha\beta} X^\beta = \frac{1}{\sqrt{\hat{a}}}\partial_\beta\left(\sqrt{\hat{a}}\ X^\beta\right) \tag{3.76}$$

3.2.1.6 Example: cylindrical coordinates on a cylinder

On a cylinder parameterised by the cylindrical coordinates (θ, z), the second derivatives of the current point are written, starting from the relations established in §3.1.4.3:

$$\frac{\partial^2 \boldsymbol{m}}{\partial\theta^2} = \partial_\theta a_\theta = -R_0 a_3\ ; \quad \frac{\partial^2 \boldsymbol{m}}{\partial\theta\partial z} = \partial_\theta a_z = \partial_z a_\theta = 0\ ; \quad \frac{\partial^2 \boldsymbol{m}}{\partial z^2} = \partial_z a_z = 0$$

all the coefficients Γ are zero. The natural basis is here assimilable to a Cartesian one.

3.2.1.7 Example: spherical coordinates on a sphere

The parameters are (θ, φ) (see §3.1.4.3). It comes, e_r designating the unit vector normal to the circle $\varphi = \mathrm{cst}$:

$$\frac{\partial^2 \boldsymbol{m}}{\partial\theta^2} = \partial_\theta a_\theta = -R_0 a_3 + R_0 \sin\varphi\ e_r = -R_0 a_3 + R_0 \sin\varphi\left(a_3 \sin\varphi + \frac{a_\varphi}{R_0}\cos\varphi\right)$$

$$\frac{\partial^2 \boldsymbol{m}}{\partial\theta\partial\varphi} = \partial_\theta a_\varphi = \partial_\varphi a_\theta = -a_\theta \tan\varphi$$

$$\frac{\partial^2 \boldsymbol{m}}{\partial\varphi^2} = \partial_\varphi a_\varphi = -R_0 a_3$$

hence the non-zero coefficients Γ:

$$\Gamma^\varphi_{\theta\theta} = \sin\varphi \cos\varphi$$

$$\Gamma^\theta_{\theta\varphi} = \Gamma^\theta_{\varphi\theta} = -\tan\varphi$$

3.2.1.8 Example: surfaces defined in Cartesian coordinates

The usual notations are used: x, y, z are Cartesian coordinates of $E_3^\bullet$ in an orthonormal coordinate system (e_x, e_y, e_z), also noted $(\boldsymbol{i}, \boldsymbol{j}, \boldsymbol{k})$; the surface is defined by the function $z = f(x, y)$.

The position of a point $\boldsymbol{m}$ of the surface being $\boldsymbol{m}(x, y, z = f(x, y))$, using the **Monge[10] notations:**

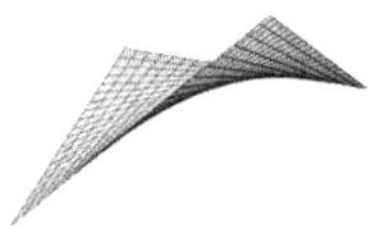

$$p = \partial_x f\ ;\ q = \partial_y f\ ;\ r = \sqrt{1 + p^2 + q^2}$$

[10] Gaspard Monge, French mathematician (1746–1818)

It becomes, by differentiation:

$$a_x = e_x + p\, e_z$$

$$a_y = e_y + q\, e_z$$

$$r\, a_3 = a_x \times a_y = e_z - p\, e_x - q\, e_y$$

from which:

$$e_z = \frac{1}{r^2}\left(p\, a_x + q\, a_y + r\, a_3\right)$$

On the other hand:

$$\partial_x a_x = \partial^2_{xx} f \cdot e_z$$

$$\partial_x a_y = \partial^2_{xy} f \cdot e_z = \partial_y a_x$$

$$\partial_y a_y = \partial^2_{yy} f \cdot e_z$$

and:

$$a_{xx} = 1 + p^2 \,; \qquad a_{xy} = pq; \qquad a_{yy} = 1 + q^2 \,; \qquad \hat{a} = r^2$$

$$b_{xx} = \frac{1}{r}\partial^2_{xx} f \,; \qquad b_{xy} = \frac{1}{r}\partial^2_{xy} f \,; \qquad b_{yy} = \frac{1}{r}\partial^2_{yy} f$$

$$\Gamma^x_{xx} = \frac{p}{r^2}\partial^2_{xx} f \;; \qquad \Gamma^x_{xy} = \frac{p}{r^2}\partial^2_{xy} f \;; \qquad \Gamma^x_{yy} = \frac{p}{r^2}\partial^2_{yy} f$$

$$\Gamma^y_{xx} = \frac{q}{r^2}\partial^2_{xx} f \,; \qquad \Gamma^y_{xy} = \frac{q}{r^2}\partial^2_{xy} f \,; \qquad \Gamma^y_{yy} = \frac{q}{r^2}\partial^2_{yy} f$$

3.2.2 Mainardi–Codazzi and Gauss relations

The purpose of this paragraph is to show that, except in special cases, it is not possible to modify the shape (curvature) of a surface without modifying its metric. This property has very important consequences for the mechanical behaviour of shells.

3.2.2.1 Riemann–Christoffel surface tensor field associated with the Levi-Civita differentiation

Let the Riemann–Christoffel tensor field R (see §2.2.7), defined on the surface, associated with the differentiation , by:

$$\forall Y \in T_{\mathcal{M}}\,, \qquad \nabla_\alpha \nabla_\beta Y^\gamma - \nabla_\beta \nabla_\alpha Y^\gamma = R_\delta{}^\gamma{}_{\alpha\beta} Y^\delta \tag{3.77}$$

R satisfies relations (2.62) to (2.66), with the exception of (2.65), the space being non-Euclidean.

Since the surface is of dimension 2, R has only an independent component R_{1212}. It is now necessary to express the relations existing between tensors R and b. Starting from the relation:

$$D_\beta a_\alpha = \nabla_\beta a_\alpha + b_{\alpha\beta} a_3$$

differentiating:

$$\begin{aligned} D_\lambda D_\beta a_\alpha &= D_\lambda \nabla_\beta a_\alpha + a_3\, \partial_\lambda b_{\alpha\beta} - b_{\alpha\beta} b_\lambda{}^\mu a_\mu \\ &= \nabla_\lambda \nabla_\beta a_\alpha + \left(b_{\lambda\gamma}\Gamma^\gamma_{\alpha\beta} + \partial_\lambda b_{\alpha\beta}\right) a_3 - b_{\alpha\beta} b_\lambda{}^\mu a_\mu \end{aligned} \tag{3.78}$$

But, in $E_3^\bullet$, $D_\lambda D_\beta a_\alpha$ is symmetric with respect to λ and β, provided that the function $\boldsymbol{m}(x_1, x_2)$ is of class C^3 because:

$$\partial^2_{\lambda\beta}\partial_\alpha \boldsymbol{m} = \partial^2_{\beta\lambda}\partial_\alpha \boldsymbol{m}$$

This symmetry makes it possible to calculate R by (3.77), taking into account (3.78):

$$\left(R_\alpha{}^\mu{}_{\lambda\beta} + b_{\alpha\lambda} b_\beta{}^\mu - b_{\alpha\beta} b_\lambda{}^\mu\right) a_\mu + \left(\nabla_\lambda b_{\alpha\beta} - \nabla_\beta b_{\alpha\lambda}\right) a_3 = 0 \tag{3.79}$$

3.2.2.2 Relations between metric and curvature

Mainardi–Codazzi[11] equations are obtained by multiplying the equality (3.79) by a_3:

$$\boxed{\nabla_\lambda b_{\alpha\beta} - \nabla_\beta b_{\alpha\lambda} = 0} \tag{3.80}$$

which are identically satisfied for $\lambda = \beta$. It remains the two relations:

$$\nabla_1 b_{\alpha 2} = \nabla_2 b_{\alpha 1}$$

(3.80) shows that the three-fold covariant tensor b is totally symmetric.

Note: These relationships can also be obtained by writing:

$$\partial_\alpha \partial_\beta a_3 = \partial_\beta \partial_\alpha a_3$$

Indeed:

$$\partial_\beta \partial_\alpha a_3 = \partial_\beta\left(-b_\alpha{}^\gamma a_\gamma\right) = -\partial_\beta\left(b_\alpha{}^\gamma\right) a_\gamma - b_\alpha{}^\gamma\left(\Gamma^\lambda_{\beta\gamma} a_\lambda + b_{\beta\gamma} a_3\right)$$

from which:

$$\partial_\alpha b_\beta{}^\gamma + b_\beta{}^\lambda \Gamma^\gamma_{\alpha\lambda} = \partial_\beta b_\alpha{}^\gamma + b_\alpha{}^\lambda \Gamma^\gamma_{\beta\lambda}$$

which demonstrates (3.80).

[11] Gaspare Mainardi (1800–1879) and Delfino Codazzi (1824–1873), Italian mathematicians.

The **Gauss relations** are obtained by multiplying (3.79) by a_γ:

$$\boxed{R_{\alpha\gamma\lambda\beta} = b_{\alpha\beta}b_{\lambda\gamma} - b_{\alpha\lambda}b_{\beta\gamma}} \tag{3.81}$$

Note: This result is also obtained by developing the expression of R in E_3^* obtained by (2.62), using relations (3.67) and (3.70).

Given the symmetry of b, the components of R are zero unless $\alpha \neq \lambda$ and $\beta \neq \gamma$. Therefore R has four non-zero components, such as:

$$R_{2112} = R_{1221} = -R_{1212} - R_{2121}$$

$$R_{2112} = b_{22}b_{11} - \left(b_{12}\right)^2 = \det\left(b_{\alpha\beta}\right)$$

This last relation makes it possible to connect R to the Gaussian curvature K. Indeed:

$$K = \det\left(b_\alpha{}^\beta\right) = \hat{a}^{-1}\det\left(b_{\alpha\beta}\right)$$

The **Gauss theorem** (1828) is then obtained using (2.62).

Theorem:

$$K = \frac{R_{1221}}{\hat{a}} = \frac{a_{\alpha 2}}{\hat{a}}\left(\partial_2\Gamma^\alpha_{11} - \partial_1\Gamma^\alpha_{21} + \Gamma^\alpha_{2\delta}\Gamma^\delta_{11} - \Gamma^\alpha_{1\delta}\Gamma^\delta_{21}\right) \tag{3.82}$$

or, in the symmetrical form:

$$\boxed{K = \frac{1}{2}a_{\alpha\beta}e^{\beta\lambda}e^{\gamma\mu}\left(\partial_\gamma\Gamma^\alpha_{\lambda\mu} + \Gamma^\alpha_{\gamma\delta}\Gamma^\delta_{\lambda\mu}\right) = \frac{1}{2}a^{\lambda\mu}\left(\partial_\gamma\Gamma^\alpha_{\lambda\mu} + \Gamma^\alpha_{\gamma\delta}\Gamma^\delta_{\lambda\mu}\right)} \tag{3.83}$$

This allows in particular to calculate the solid angle (§3.1.4.6). This theorem is important because it shows that the Gaussian curvature K (invariant of the curvature tensor) depends only on the metric tensor a by (3.68).

Note: Gauss's theorem can also be demonstrated by writing $\partial_\alpha\partial_\beta a_\gamma = \partial_\beta\partial_\alpha a_\gamma$. The first member is written as:

$$\partial_\alpha\partial_\beta a_\gamma = \left(\partial_\alpha\Gamma^\mu_{\beta\gamma} + \Gamma^\lambda_{\beta\gamma}\Gamma^\mu_{\alpha\lambda} - b_{\beta\gamma}b_\alpha{}^\mu\right)a_\mu + \left(\partial_\alpha b_{\beta\gamma} + \Gamma^\lambda_{\beta\gamma}b_{\alpha\lambda}\right)a_3$$

The second member is decomposed analogously and subtracted from the first. The a_3 component of the result leads again to the Mainardi–Codazzi relationship. The component on a_μ leads to the relationship:

$$\partial_\alpha \Gamma^\mu_{\beta\gamma} - \partial_\beta \Gamma^\mu_{\alpha\gamma} + \Gamma^\lambda_{\beta\gamma}\Gamma^\mu_{\alpha\lambda} - \Gamma^\lambda_{\alpha\gamma}\Gamma^\mu_{\beta\lambda} = a_{\gamma\delta}\left(b_\beta{}^\delta b_\alpha{}^\mu - b_\alpha{}^\delta b_\beta{}^\mu\right)$$

The case $\alpha = \beta$ is identically verified and it is therefore permissible, without restricting the generality, to take $\alpha = 1$ and $\beta = 2$. It becomes:

$$\partial_1 \Gamma^\mu_{2\gamma} - \partial_2 \Gamma^\mu_{1\gamma} + \Gamma^\lambda_{2\gamma}\Gamma^\mu_{1\lambda} - \Gamma^\lambda_{1\gamma}\Gamma^\mu_{2\lambda} = a_{\gamma\delta}\left(b_2{}^\delta b_1{}^\mu - b_1{}^\delta b_2{}^\mu\right)$$

Taking more specifically $\gamma = 1$ and combining the relations obtained:

$$a_{\mu 2}\left(\partial_1 \Gamma^\mu_{21} - \partial_2 \Gamma^\mu_{11} + \Gamma^\lambda_{21}\Gamma^\mu_{1\lambda} - \Gamma^\lambda_{11}\Gamma^\mu_{2\lambda}\right) = a_{\mu 2} a_{1\delta}\left(b_2{}^\delta b_1{}^\mu - b_1{}^\delta b_2{}^\mu\right)$$

When δ and μ are equal, the term in parentheses of the second member is zero and the pair (δ, μ) thus takes the values $(1,2)$ and $(2,1)$; finally:

$$a_{\mu 2}\left(\partial_1 \Gamma^\mu_{21} - \partial_2 \Gamma^\mu_{11} + \Gamma^\lambda_{21}\Gamma^\mu_{1\lambda} - \Gamma^\lambda_{11}\Gamma^\mu_{2\lambda}\right) = \left(a_{12}a_{12} - a_{11}a_{22}\right) K$$

which is identical to (3.82).

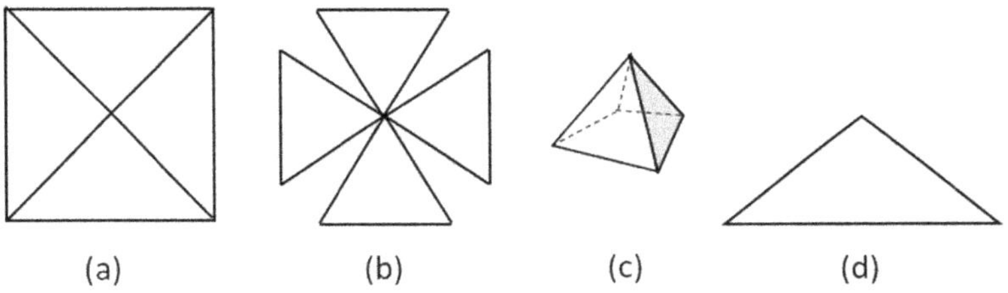

Figure 3.12

One way to illustrate Gauss's theorem summarily is as follows: a square is cut along its two diagonals to form four equal isosceles triangles (Figure 3.12a). The four triangles undergo identical deformation in their plane such that their bases are reduced (Figure 3.12b). To restore continuity (remove cuts), it is necessary to form a pyramid with the four triangles (Figure 3.12c). The transformation from (a) to (c) implies a variation in area and shape. If the bases are stretched instead of being reduced, it is not possible to create such a pyramid (Figure 3.12d).

3.2.2.3 Consequences of the Gauss theorem

If the Gaussian curvature of a surface is zero in all points, the Riemann–Christoffel tensor is zero and the surface is a Euclidean space. The surface is said to be **developable**: it has then the same metric as a plane and it can be "unrolled" on a plane without changing the distance between any two points (a cylinder or a cone can be obtained from a sheet of paper, without extension). Conversely, a developable surface cannot be transformed into a non-developable surface without the metric being modified (a sheet of paper can not be transformed into a sphere without extension).

In an isometric flexure (inextensional) where a is preserved, the Gaussian curvature is preserved. Conversely, a change of shape with modification of the Gaussian curvature cannot be done without a change in metrics

The Mainardi–Codazzi (3.80) and Gauss (3.83) equations are integrability conditions of the Weingarten (3.24) and Gauss (3.60) formulas. In other words, these are the conditions for two tensors a and b chosen *a priori* to be the first two fundamental forms of a surface (disregarding a rigid displacement).

Example: Cylinder.
All connection coefficients Γ and Gaussian curvature K are zero. As a result, the relations of Gauss and Mainardi–Codazzi are identically verified.

Exemple: Sphere.
The connection coefficients Γ were calculated in §3.2.1.7. The curvature tensor is constant on the sphere, and its covariant derivative is zero, which is also true term by term, for example:

Photo 3.2. Sphere (Paris, France)

$$\nabla_\theta b_{\theta\theta} = \partial_\theta b_{\theta\theta} - 2\Gamma^{\alpha}_{\theta\theta} b_{\alpha\theta}$$

$$= 0 - 2\Gamma^{\varphi}_{\theta\theta} b_{\varphi\theta} = 0$$

$$\nabla_\varphi b_{\theta\theta} = \partial_\varphi b_{\theta\theta} - 2\Gamma^{\alpha}_{\varphi\theta} b_{\alpha\theta} = \partial_\varphi\left(-R_0{}^2 \cos^2\varphi\right) + 2\tan\varphi\left(-R_0{}^2\cos^2\varphi\right) = 0$$

The Mainardi–Codazzi equations are thus verified.

The second member of (3.83) reduces to:

$$\frac{a_{22}}{\hat{a}}\left(\partial_\varphi \Gamma^{\varphi}_{\theta\theta} - \Gamma^{\varphi}_{\theta\theta}\Gamma^{\theta}_{\varphi\theta}\right) = \frac{R_0{}^2}{R_0{}^4\cos^2\varphi}\left[\partial_\varphi\left(\sin\varphi\ \cos\varphi\right) + \sin\varphi\ \cos\varphi\ \tan\varphi\right]$$

$$= \frac{1}{R_0{}^2} = K$$

which makes it possible to verify the Gauss equation.

3.3 DEFORMATION OF SURFACES

3.3.1 Characterisation of surface deformation

A surface is characterised by its metrics (intrinsic) and its shape (curvature tensor, extrinsic). A rigid transformation of the surface is an arbitrary composition of translations and global rotations. A rigid transformation does not, obviously, alter the two basic forms a

and b. Conversely, if a and b are kept in a transformation, this transformation is a rigid transformation:

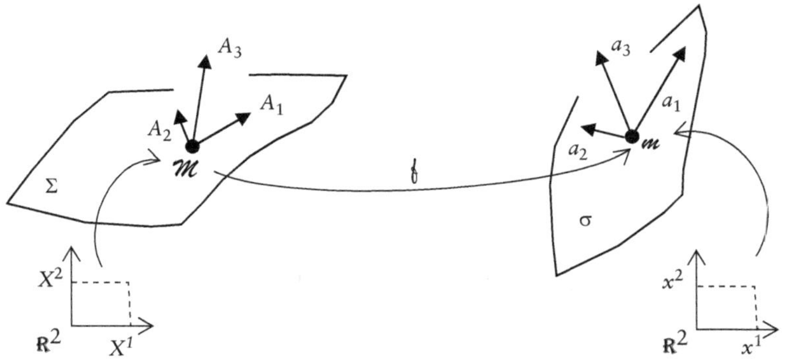

Figure 3.13

Theorem:

Let Σ and σ be two surfaces of class C^2 and $\mathfrak{f}$ a diffeomorphism of Σ on σ (Figure 3.13) such that, at every point $\mathcal{M}$ of Σ, the images by $\bar{\mathfrak{f}}$ of the two fundamental forms A and B in $\mathcal{M}$ are the fundamental forms a and b of σ at $m = \mathfrak{f}(\mathcal{M})$:

$$\overset{2}{\otimes}\bar{\mathfrak{f}}^T(a) = A \qquad ; \qquad \overset{2}{\otimes}\bar{\mathfrak{f}}^T(b) = B$$

(where $\bar{\mathfrak{f}}$ denotes the linear application tangent to $\mathfrak{f}$ and $\bar{\mathfrak{f}}^T$ its transposed).

then $\mathfrak{f}$ is a rigid transformation.

In practice, the function $\mathfrak{f}$ is determined by the expression of the parameters x^λ of σ as functions of the parameters X^α of Σ, which also makes it possible to take X^α as parameters of σ. To express x^λ in terms of X^α is a way to make a basis change in the plane T_m tangent to σ': $a'_\alpha = \dfrac{\partial m}{\partial X^\alpha}$ is the natural basis associated with coordinates X^α. So:

$$a'_\alpha = \frac{\partial x^\lambda}{\partial X^\alpha} a_\lambda$$

$\left(\dfrac{\partial x^\lambda}{\partial X^\alpha}\right)$ are the components of the tangent linear application $\bar{\mathfrak{f}}$, which is a linear application from $T_{\mathcal{M}}$ to T_m. They form the Jacobian of transformation:

$$\bar{\mathfrak{f}}(A_\alpha) = \frac{\partial x^\lambda}{\partial X^\alpha} a_\lambda = a'_\alpha$$

If $a_{\lambda\mu} = a_\lambda \cdot a_\mu$ are the components of the metric tensor in basis (a_λ) and $a'_{\alpha\beta} = a'_\alpha \cdot a'_\beta$ in basis (a'_α):

$$a'_{\alpha\beta} = \frac{\partial x^\lambda}{\partial X^\alpha}\frac{\partial x^\mu}{\partial X^\beta} a_{\lambda\mu}$$

If $\mathfrak{f}$ is a rigid transformation, the metric is preserved:

$$\forall(\alpha,\beta): \qquad \overline{\mathfrak{f}}(A_\alpha)\cdot\overline{\mathfrak{f}}(A_\beta) = A_\alpha \cdot A_\beta$$

that is:

$$a'_{\alpha\beta} = A_{\alpha\beta}$$

Similarly, the curvatures are preserved. Indeed, if the transformation is a translation U:

$$\mathscr{m} = \mathscr{M} + U$$

Then, in a cartesian coordinate system:

$$a'_\alpha = \frac{\partial \mathscr{m}}{\partial x^\alpha} = \frac{\partial \mathscr{M}}{\partial x^\alpha} = A_\alpha \text{ and } a_3 = A_3$$

then:

$$\frac{\partial^2 \mathscr{m}}{\partial X^\alpha \partial X^\beta} = \frac{\partial^2 \mathscr{M}}{\partial X^\alpha \partial X^\beta} \Rightarrow b'_{\alpha\beta} = B_{\alpha\beta}$$

Similarly, if the transformation is a rotation Ω around a point O:

$$O\mathscr{m} = \Omega \times O\mathscr{M}$$

then, in a cartesian coordinate system:

$$a'_\alpha = \Omega \times A_\alpha$$

$$a_3 = \frac{a'_1 \times a'_2}{\|a'_1 \times a'_2\|} = \frac{(\Omega \times A_1)\times(\Omega \times A_2)}{\|A_1 \times A_2\|} = \Omega \times A_3$$

so:

$$\frac{\partial^2 \mathscr{m}}{\partial X^\alpha \partial X^\beta} = \Omega \times \frac{\partial^2 \mathscr{M}}{\partial X^\alpha \partial X^\beta}$$

then:

$$b'_{\alpha\beta} = \left(\Omega \times \frac{\partial^2 \mathscr{M}}{\partial X^\alpha \partial X^\beta}\right)\cdot(\Omega \times A_3) = B_{\alpha\beta}$$

Generalisation is immediate to any rigid movement and:

$$b'_{\alpha\beta} = \frac{\partial x^\lambda}{\partial X^\alpha}\frac{\partial x^\mu}{\partial X^\beta} b_{\lambda\mu} = B_{\alpha\beta}$$

In this case, if the curvilinear coordinates used for σ are the images of those used for Σ, then A and a on the one hand, B and b on the other hand, have the same covariant components.

Proof of the theorem: The idea of the proof is to use the fact that, in the ambient Euclidean space, the rigid transformation is an isometry, characterised by a null strain tensor; the surface is "thickened" by generating a three-dimensional volume Ω using normal coordinates:

$$\mathcal{P} \in \Omega \quad \Leftrightarrow \quad \mathcal{P} = \mathcal{M} + X^3 A_3(\mathcal{M}), \qquad \mathcal{M} \in \Sigma, \qquad X^3 \in \left]a, b\right[$$

where A_3 is the vector normal to Σ at $\mathcal{M}$.

On the other hand, $\mathfrak{f}$ is extended to $\tilde{\mathfrak{f}}$ by the relation:

$$p = \tilde{\mathfrak{f}}(\mathcal{P}) = \mathfrak{f}(\mathcal{M}) + X^3 a_3$$

where a_3 is the vector normal to σ at $m = \mathfrak{f}(\mathcal{M})$.

For Ω to be unambiguously defined by the above formulation, and $\tilde{\mathfrak{f}}$ to be a diffeomorphism, the conditions (3.50) must be satisfied for Σ and σ, ie that $|a|$ or b (according to the concavity) is less than the smallest of the radii of curvature of the two surfaces. This restriction on a and b does not restrain the demonstration. By (3.43), the metric tensor in $\mathcal{P}$ is written:

$$G_{\alpha\beta} = \left(\delta_\alpha^\lambda - x^3 B_\alpha{}^\lambda\right)\left(\delta_\beta^\mu - x^3 B_\beta{}^\mu\right) A_{\lambda\mu}$$

$$G_{33} = 1 \qquad\qquad G_{3\alpha} = 0$$

If the same curvilinear coordinates (that is to say, transformed by $\mathfrak{f}$) are used in $\mathcal{M}$ and m, the components of δ, A and B are retained in the transformation, and noting g'_α the basis in p associated with parameters X_α:

$$g'_{ij} = G_{ij}$$

in any point $p = \tilde{\mathfrak{f}}(\mathcal{P})$; $\tilde{\mathfrak{f}}$ is then an isometry, which demonstrates the theorem by restriction to $\mathfrak{f}$.

3.3.2 Tensors of deformation of a surface

The above theorem is important since it makes it possible to characterise the deformation of a surface by the variations of the metric and curvature tensors. So let Σ and σ be two configurations of the same surface (during a movement, for example): in Lagrange coordinates on Σ, the variations of these two tensors are:

a) The **strain tensor** of the surface, ε:

$$\varepsilon(\mathcal{M}) = \frac{1}{2}\left[\overset{2}{\otimes}\bar{\mathfrak{f}}^T(a) - A\right] \tag{3.84}$$

either, in differential form:

$$\varepsilon(\mathcal{M}) = \varepsilon_{\alpha\beta} dX^\alpha dX^\beta = \frac{1}{2}\left(dm^2 - d\mathcal{M}^2\right) = \frac{1}{2}\left(\frac{\partial x^\lambda}{\partial X^\alpha}\frac{\partial x^\mu}{\partial X^\beta} a_{\lambda\mu} - A_{\alpha\beta}\right) dX^\alpha dX^\beta \tag{3.85}$$

b) The tensor variation of curvature of the surface, κ:

$$\kappa(\mathcal{M}) = \overset{2}{\otimes} \bar{f}^T(b) - B \tag{3.86}$$

either, in differential form:

$$\begin{aligned}\kappa(\mathcal{M}) &= \kappa_{\alpha\beta} dX^\alpha dX^\beta = -\left(dm \cdot da_3 - d\mathcal{M} \cdot dA_3\right) \\ &= \left(\frac{\partial x^\lambda}{\partial X^\alpha}\frac{\partial x^\mu}{\partial X^\beta} b_{\lambda\mu} - B_{\alpha\beta}\right) dX^\alpha dX^\beta\end{aligned} \tag{3.87}$$

Example: let an initially flat sheet of paper, of width L, wound into a cylinder of radius R_0, then $L = 2\pi R_0$. The transformation considered is expressed by (Figure 3.14):

$$\begin{cases} x \\ y \end{cases} \rightarrow \begin{cases} \theta = \dfrac{x}{R_0} \\ z = y \end{cases}$$

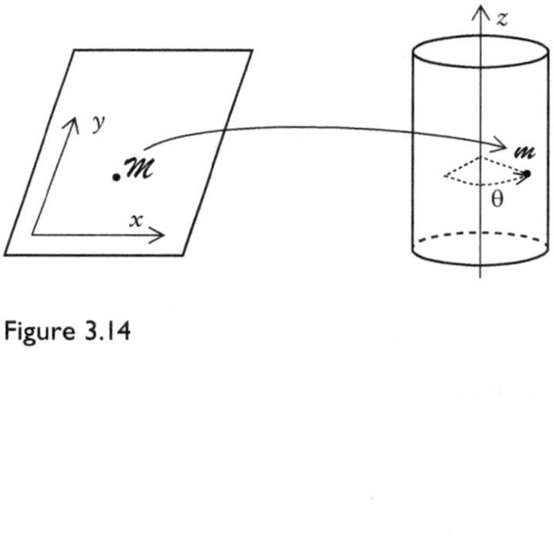

Figure 3.14

successively:

$$dm^2 = dx^2 + dy^2$$

$$dm = a_\theta d\theta + a_z dz$$

$$dm^2 = R_0{}^2 d\theta^2 + dz^2 = dx^2 + dy^2$$

$$da_3 = \frac{a_\theta}{R_0} d\theta$$

$$dm \cdot da_3 = R_0 d\theta^2 = \frac{dx^2}{R_0}$$

From which: $\varepsilon = 0$, which was obvious and:

$$\kappa_{xx} = -\frac{1}{R_0} \quad ; \qquad \kappa_{xy} = \kappa_{yy} = 0$$

In many cases encountered in practice, the same parameterisation is used for both configurations, which simplifies (3.85) and (3.87):

$$\varepsilon(\mathcal{M}) = \frac{1}{2}\left(a_{\alpha\beta} - A_{\alpha\beta}\right) dX^\alpha dX^\beta = \frac{1}{2}\left(\frac{\partial m}{\partial X^\alpha}\frac{\partial m}{\partial X^\beta} - \frac{\partial \mathcal{M}}{\partial X^\alpha}\frac{\partial \mathcal{M}}{\partial X^\beta}\right) dX^\alpha dX^\beta \tag{3.88}$$

$$\kappa(\mathcal{M}) = \left(b_{\alpha\beta} - B_{\alpha\beta}\right) dX^\alpha dX^\beta = -\left(\frac{\partial m}{\partial X^\alpha}\frac{\partial a_3}{\partial X^\beta} - \frac{\partial \mathcal{M}}{\partial X^\alpha}\frac{\partial A_3}{\partial X^\beta}\right) dX^\alpha dX^\beta \tag{3.89}$$

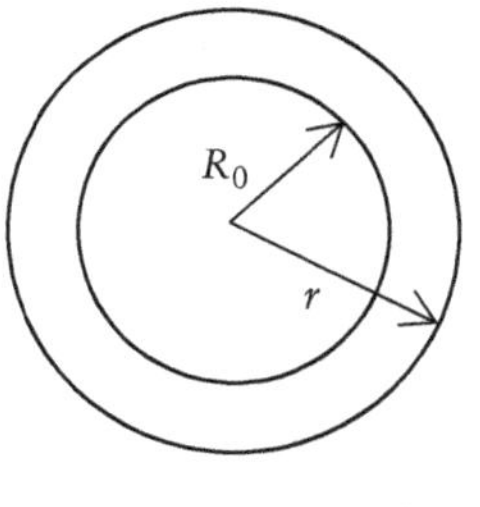

Figure 3.15

Example: let a sphere of radius R_0 inflated in a sphere of radius r (Figure 3.15):

Spherical coordinates θ, φ are used on both spheres:

$$d\mathcal{M} = A_\theta d\theta + A_\varphi d\varphi$$

$$dA_3 = \frac{A_\theta}{R_0} d\theta + \frac{A_\varphi}{R_0} d\varphi = \frac{d\mathcal{M}}{R_0}$$

from which:

$$A = {R_0}^2 \cos^2 \varphi \, d\theta^2 + {R_0}^2 d\varphi^2$$

$$B = -d\mathcal{M} \bullet dA_3 = -\left(R_0 \cos^2 \varphi \, d\theta^2 + R_0 d\varphi^2\right)$$

and:

$$a = r^2 \cos^2 \varphi \, d\theta^2 + r^2 d\varphi^2$$

$$b = -d\mathcal{m} \bullet da_3 = -\left(r \cos^2 \varphi \, d\theta^2 + r d\varphi^2\right)$$

It becomes:

$$\varepsilon = \frac{1}{2}(a - A) = \frac{1}{2}\left[\left(r^2 - {R_0}^2\right)\cos^2 \varphi \, d\theta^2 + \left(r^2 - {R_0}^2\right) d\varphi^2\right]$$

$$\kappa = b - B = -\left[\left(r - R_0\right)\cos^2 \varphi \, d\theta^2 + \left(r - R_0\right) d\varphi^2\right]$$

In physical coordinates (orthonormal coordinate system) associated with spherical coordinates:

$$\tilde{\varepsilon}_{\theta\theta} = \tilde{\varepsilon}_{\varphi\varphi} = -\frac{1}{2}\left(1 - \frac{r^2}{{R_0}^2}\right) \quad ; \quad \tilde{\varepsilon}_{\theta\varphi} = 0$$

$$\tilde{\kappa}_{\theta\theta} = \tilde{\kappa}_{\varphi\varphi} = \frac{1}{R_0}\left(1 - \frac{r}{R_0}\right) \quad ; \quad \tilde{\kappa}_{\theta\varphi} = 0$$

Note that the components of κ are not equal to $-\left(\frac{1}{r} - \frac{1}{R_0}\right)$, because $\frac{1}{r}$ is the curvature of the deformed sphere, expressed in the plane tangent to the deformed surface, whereas, in κ, it must be expressed in the plane tangent to the undeformed surface (through $\overline{\mathfrak{f}}^{T}$).

3.3.3 Expression of deformation variables as a function of displacement

3.3.3.1 General case

Let $\mathcal{M}\mathcal{m} = \xi(\mathcal{M})$, where ξ is the displacement. The field ξ is a field in $\mathbf{R}^3$ defined on Σ. ξ is decomposed on the normal A_3 and the tangent plane $T_{\mathcal{M}}$ (Figure 3.16):

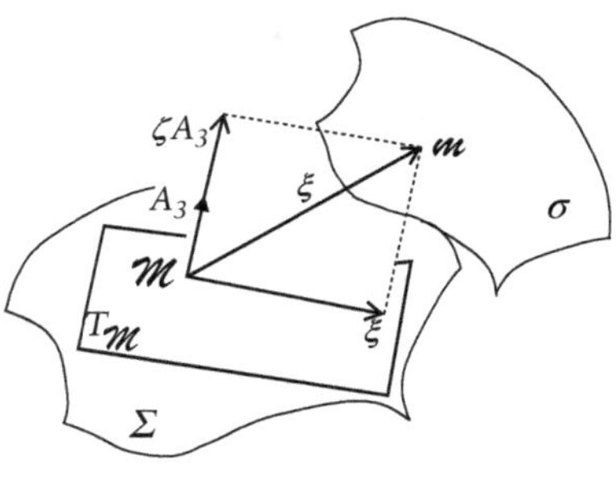

Figure 3.16

$$\boldsymbol{m} = \mathfrak{f}(\mathcal{M}) = \mathcal{M} + \xi(\mathcal{M}) = \mathcal{M} + \xi^{\beta}\left(X^{\alpha}\right)A_{\beta} + \zeta\left(X^{\alpha}\right)A_{3} \tag{3.90}$$

Let:

$$\underline{\xi} = \xi^{\beta} A_{\beta} \tag{3.91}$$

the projection of ξ on the tangent plane $T_{\mathcal{M}}$.

For a vector Y of $T_{\mathcal{M}}$:

$$\begin{aligned} \bar{\mathfrak{f}}(Y) &= Y + D_Y\xi \\ &= Y + \nabla_Y \underline{\xi} + \nabla_Y\left(\zeta A_3\right) + B\left(Y, \underline{\xi}\right) A_3 \end{aligned} \tag{3.92}$$

As there is no ambiguity on the differentiation carried out, $\nabla_Y \underline{\xi}$ is noted $\nabla_Y \xi$. ε is a two-fold covariant tensor. It is a bilinear form on $T_{\mathcal{M}}$:

$$\begin{aligned} &\forall (X,Y) \in T_{\mathcal{M}} \otimes T_{\mathcal{M}} : \\ &\varepsilon(X,Y) = \frac{1}{2}\left[\bar{\mathfrak{f}}(X) \cdot \bar{\mathfrak{f}}(Y) - X \cdot Y\right] = \frac{1}{2}\left[X \cdot D_Y\xi + Y \cdot D_X\xi + D_X\xi \cdot D_Y\xi\right] \end{aligned} \tag{3.93}$$

Then:

- the local basis (a_{α}, a_3) on σ,
- the strain tensor ε,
- the curvature variation tensor κ,

are expressed as functions of displacement. By relation (3.90), the surface σ is generated by the same parameterisation $\left(X^{\alpha}\right)$ as the reference surface Σ. Relations (3.88) and (3.89) can in particular be used.

Differentiating relation (90) or applying (92) for $Y = A_{\alpha}$, it becomes:

$$a_{\alpha} = \partial_{\alpha}\boldsymbol{m} = A_{\alpha} + \nabla_{\alpha}\xi^{\beta}\, A_{\beta} + B_{\alpha}{}^{\beta}\, \xi_{\beta}\, A_3 + \partial_{\alpha}\zeta\, A_3 - \zeta\, B_{\alpha}{}^{\beta}\, A_{\beta}$$

Noting, for further simplifications:

$$D_{\alpha}\xi = \mathcal{A}_{\alpha}{}^{\beta} A_{\beta} + \mathcal{B}_{\alpha} A_3 = \mathcal{A}_{\alpha}{}^{i} A_i \tag{3.94}$$

the above expression is rewritten:

$$\boxed{\begin{aligned} &a_{\alpha} = A_{\alpha} + \mathcal{A}_{\alpha}{}^{\beta} A_{\beta} + \mathcal{B}_{\alpha} A_3 \\ &\text{with:} \\ &\mathcal{A}_{\alpha}{}^{\beta} = \nabla_{\alpha}\xi^{\beta} - \zeta\, B_{\alpha}{}^{\beta} \\ &\mathcal{A}_{\alpha}{}^{3} = \mathcal{B}_{\alpha} = B_{\alpha}{}^{\beta}\xi_{\beta} + \partial_{\alpha}\zeta \end{aligned}} \tag{3.95}$$

With these notations, (3.92) is rewritten:

$$\bar{f}(Y) = Y + Y^{\alpha} \mathcal{A}_{\alpha}{}^{\beta} A_{\beta} + Y^{\alpha} \mathcal{B}_{\alpha} A_3 = Y + Y^{\alpha} \mathcal{A}_{\alpha}{}^{i} A_i \tag{3.96}$$

Also:

$$a_{\alpha} \cdot A_{\beta} = A_{\alpha\beta} + \mathcal{A}_{\alpha\beta}$$

$$a_{\alpha} \cdot A_3 = \mathcal{B}_{\alpha}$$

The expression of the strain tensor is obtained immediately by (3.95):

$$\boxed{\varepsilon_{\alpha\beta} = \frac{1}{2}\left(\mathcal{A}_{\alpha\beta} + \mathcal{A}_{\beta\alpha} + \mathcal{A}_{\alpha}{}^{\gamma} \mathcal{A}_{\beta\gamma} + \mathcal{B}_{\alpha} \mathcal{B}_{\beta}\right)} \tag{3.97}$$

The vector a_3 is noted:

$$a_3 = \mathcal{N}^{\alpha} A_{\alpha} + A_3 \cos\omega$$

where:

$$\mathcal{N}_{\alpha} = a_3 \cdot A_{\alpha}$$

$$\cos\omega = a_3 \cdot A_3$$

ω is the angle of rotation of the normal vector.
Otherwise:

$$\hat{A} = \det(A_{\alpha\beta}) = \|A_1 \times A_2\|^2 = (A_1, A_2, A_3)^2$$

$\sqrt{\hat{A}}\, A_3 = A_1 \times A_2$ and $\sqrt{\hat{a}}\, a_3 = a_1 \times a_2$

By replacing this last expression a_{α} by expression (3.95) and noting $\mathcal{E} = \sqrt{\hat{A}}\, \tilde{e}$ the two-fundamental form on Σ, it becomes:

$$\begin{aligned} a_1 \times a_2 = {} & A_1 \times A_2 + \left(\mathcal{A}_2{}^{\gamma} \mathcal{E}_{1\gamma} - \mathcal{A}_1{}^{\gamma} \mathcal{E}_{2\gamma}\right) A_3 + \mathcal{A}_1{}^{\beta} \mathcal{A}_2{}^{\gamma} \mathcal{E}_{\beta\gamma} A_3 \\ & + \left(\mathcal{B}_1 \mathcal{E}_{2\lambda} - \mathcal{B}_2 \mathcal{E}_{1\lambda}\right) A^{\lambda} + \left(\mathcal{B}_1 \mathcal{A}_2{}^{\gamma} - \mathcal{B}_2 \mathcal{A}_1{}^{\gamma}\right) \mathcal{E}_{\gamma\lambda} A^{\lambda} \\ = {} & \sqrt{\hat{A}}\, A_3 + \Lambda \sqrt{\hat{A}}\, A_3 - \sqrt{\hat{A}}\, \mathcal{B}_{\alpha} A^{\alpha} + \delta_{12}^{\alpha\beta}\, \mathcal{B}_{\alpha} \mathcal{A}_{\beta}{}^{\gamma} \mathcal{E}_{\gamma\lambda} A^{\lambda} \end{aligned}$$

where:

$$\Lambda = \mathrm{TR}(\mathcal{A}) + \det\left(\mathcal{A}_{\alpha}{}^{\beta}\right)$$

By multiplying this equation by A_3 and A_{α}, it finally comes:

$$
\boxed{
\begin{aligned}
&a_3 = \mathcal{N}^\alpha A_\alpha + A_3 \cos\omega \\
&\text{with:} \qquad \mathcal{N}_\alpha = \sqrt{\frac{\hat{A}}{\hat{a}}}\left(-\mathcal{B}_\alpha + \delta_{12}^{\lambda\mu}\mathcal{B}_\lambda \mathcal{A}_\mu{}^\gamma \tilde{e}_{\gamma\alpha}\right) \\
&\qquad\qquad \cos\omega = (1+\Lambda)\sqrt{\frac{\hat{A}}{\hat{a}}} \\
&\qquad\qquad \hat{a} = \det\left(A_{\alpha\beta} + 2\varepsilon_{\alpha\beta}\right) \\
&\qquad\qquad \Lambda = \mathcal{A}_\beta{}^\beta + \det\left(\mathcal{A}_\alpha{}^\beta\right)
\end{aligned}
}
\tag{3.98}
$$

then, writing that $a_\alpha \cdot a_3 = 0$:

$$
\mathcal{N}_\beta\left(\delta_\alpha^\beta + \mathcal{A}_\alpha{}^\beta\right) + \mathcal{B}_\alpha \cos\omega = 0 \tag{3.99}
$$

The curvature tensor is obtained using relation (3.25), which requires expressing $\partial_\alpha a_\beta$. From relationships (3.60) and (3.95):

$$
\begin{aligned}
\partial_\alpha a_\beta = &\left[\partial_\alpha\left(\delta_\beta^\gamma + \mathcal{A}_\beta{}^\gamma\right) + \Gamma_{\alpha\mu}^\gamma\left(\delta_\beta^\mu + \mathcal{A}_\beta{}^\mu\right) - B_\alpha{}^\gamma \mathcal{B}_\beta\right] A_\gamma \\
&+ \left[\partial_\alpha \mathcal{B}_\beta + B_{\alpha\gamma}\left(\delta_\beta^\gamma + \mathcal{A}_\beta{}^\gamma\right)\right] A_3
\end{aligned}
$$

from which:

$$
\begin{aligned}
b_{\alpha\beta} &= \partial_\alpha a_\beta \cdot \left(\mathcal{N}_\mu A^\mu + A_3 \cos\omega\right) \\
&= \left(\partial_\alpha \mathcal{A}_\beta{}^\gamma + \Gamma_{\alpha\mu}^\gamma \mathcal{A}_\beta{}^\mu + \Gamma_{\alpha\beta}^\gamma - B_\alpha{}^\gamma \mathcal{B}_\beta\right)\mathcal{N}_\gamma \\
&\quad + \left(\partial_\alpha \mathcal{B}_\beta + B_{\alpha\beta} + B_{\alpha\gamma}\mathcal{A}_\beta{}^\gamma\right)\cos\omega
\end{aligned}
\tag{3.100}
$$

and the tensor variation of curvature:

$$
\boxed{
\begin{aligned}
\kappa_{\alpha\beta} = &\left(\partial_\alpha \mathcal{A}_\beta{}^\gamma + \Gamma_{\alpha\mu}^\gamma \mathcal{A}_\beta{}^\mu + \Gamma_{\alpha\beta}^\gamma - B_\alpha{}^\gamma \mathcal{B}_\beta\right)\mathcal{N}_\gamma \\
&+ \left(\partial_\alpha \mathcal{B}_\beta + B_{\alpha\gamma}\mathcal{A}_\beta{}^\gamma\right)\cos\omega - B_{\alpha\beta}\left(1 - \cos\omega\right)
\end{aligned}
}
\tag{3.101}
$$

The components of ε and κ are expressed according to the components of $D\xi$

The curvature tensor b being symmetric, this ensures by (3.100) that the expression of κ given by (3.101) is symmetric.

3.3.3.2 Small displacement cases

In this section, the foregoing relationships are expressed in the assumption of small displacements: the three components of the displacement are then of the first order with respect to a characteristic length and the higher-order terms are neglected in the various relations.

Note: this hypothesis, which gives the same importance to all the components, is not always the most representative of the physical phenomena: indeed, in certain situations, and by analogy with the behaviour of beams, it appears that the normal component is preponderant, which leads to considering that the other components are of the second order. The linearization must then be adapted accordingly (see Chapter 8).

Relations (3.95) are unchanged, $\mathcal{A}$ and $\mathcal{B}$ being of the first order, and:

$$\boxed{\varepsilon_{\alpha\beta} = \frac{1}{2}\left(\mathcal{A}_{\alpha\beta} + \mathcal{A}_{\beta\alpha}\right) = \frac{1}{2}\left(\nabla_\alpha \xi_\beta + \nabla_\beta \xi_\alpha\right) - \zeta B_{\alpha\beta}} \tag{3.102}$$

or:

$$\varepsilon_{\alpha\beta} = \frac{1}{2}\left(D_\alpha \xi_\beta + D_\beta \xi_\alpha\right)$$

Then successively:

$$a_1 \times a_2 = \left(1 + \mathcal{A}_\alpha{}^\alpha\right)\sqrt{\hat{A}}\, A_3 - \sqrt{\hat{A}}\, \mathcal{B}_\alpha A^\alpha$$

$$\left\|a_1 \times a_2\right\|^2 = \left[\left(1 + \mathcal{A}_\alpha{}^\alpha\right)^2 + A_{\alpha\beta}\mathcal{B}^\alpha\, \mathcal{B}^\beta\right]\hat{A} \approx \left(1 + 2\mathcal{A}_\alpha{}^\alpha\right)\hat{A}$$

$$a_3 = \left(1 - \mathcal{A}_\alpha{}^\alpha\right)\left(A_3 + \mathcal{A}_\alpha{}^\alpha A_3 - \mathcal{B}_\alpha A^\alpha\right) \approx A_3 - \mathcal{B}^\alpha A_\alpha = A_3 - \mathcal{B}$$

from which:

$$\boxed{a_3 = A_3 - \left(B_{\alpha\beta}\xi^\beta + \partial_\alpha \zeta\right)A^{\alpha\gamma} A_\gamma} \tag{3.103}$$

The rotation of the normal vector $a_3 - A_3$ is orthogonal to A_3 and:

$$a_3 - A_3 = \omega \times A_3 \tag{3.104}$$

The vector $\boldsymbol{\omega}$ is in the tangent plane: $\omega = \omega^\alpha A_\alpha$. Considering both relations (3.103) and (3.104), it becomes:

$$\begin{aligned} \mathcal{B} &= A_3 \times \omega \\ \omega &= \mathcal{B} \times A_3 \end{aligned} \tag{3.105}$$

The three vectors $\mathcal{B}$, $\boldsymbol{\omega}$ and A_3 are orthogonal and $\boldsymbol{\omega}$ is entirely determined by the displacement of the surface. In small displacements: $\cos\omega \approx 1$ (small rotation) and $\mathcal{N}^\alpha \approx -\mathcal{B}^\alpha$. Where from immediately:

$$b_{\alpha\beta} = \partial_\alpha \mathcal{B}_\beta + B_{\alpha\beta} + B_{\alpha\gamma}\mathcal{A}_\beta{}^\gamma - \Gamma^\gamma_{\alpha\beta}\mathcal{B}_\gamma \tag{3.106}$$

and:

$$\boxed{\kappa_{\alpha\beta} = \nabla_\alpha \mathcal{B}_\beta + B_{\alpha\gamma}\mathcal{A}_\beta{}^\gamma} \tag{3.107}$$

that is:

$$\boxed{\kappa_{\alpha\beta} \approx \nabla_\alpha\left(\partial_\beta\zeta + B_{\beta\gamma}\xi^\gamma\right) + B_{\alpha\gamma}\left(\nabla_\beta\xi^\gamma - \zeta B_\beta{}^\gamma\right)} \tag{3.108}$$

In expression (3.108), it should be understood that ∇_α applies to the form of components: $B_\beta = \partial_\beta\zeta + B_{\beta\gamma}\xi^\gamma$, resulting in the components of a twice covariant tensor. The symmetry of (3.107) is demonstrated by expanding by (3.95) the equality $\partial_\alpha a_\beta = \partial_\beta a_\alpha$ which the component following a_3 precisely expresses the desired symmetry.

By application of (3.95), formulas (3.102), (3.103), and (3.107) above can be written in a more synthetic form using the tensor $\mathcal{A}$:

$$\begin{aligned} \varepsilon_{\alpha\beta} &= \frac{1}{2}\left(\mathcal{A}_{\alpha\beta} + \mathcal{A}_{\beta\alpha}\right) \\ a_3 &= A_3 - \mathcal{A}_{\alpha 3}A^\alpha \\ \kappa_{\alpha\beta} &= \nabla_\alpha \mathcal{A}_{\beta 3} + B_\alpha{}^\gamma \mathcal{A}_{\beta\gamma} \end{aligned} \tag{3.109}$$

Finally, by application of (3.94), $D_\alpha\xi$ can be rewritten:

$$D_\alpha\xi = \mathcal{A}_{\alpha i}A^i = \frac{1}{2}\left(\mathcal{A}_{\alpha\beta} + \mathcal{A}_{\beta\alpha}\right)A^\beta + \frac{1}{2}\left(\mathcal{A}_{\alpha\beta} - \mathcal{A}_{\beta\alpha}\right)A^\beta + \mathcal{A}_{\alpha 3}A^3$$

that is:

$$D_\alpha\xi = \varepsilon_{\alpha\beta}A^\beta + \mathfrak{a}_{\alpha\beta}A^\beta + \mathcal{A}_{\alpha 3}A^3 \tag{3.110}$$

where $\mathfrak{a}$ is the antisymmetric part of the restriction on the surface of the twice-covariant tensor $\mathcal{A}$.

3.3.4 Compatibility equations

Let two tensors ε and κ given *a priori* (that is six components, considering the symmetries). ε and κ are the tensors of strain and curvature variation of a transformation of the surface if they satisfy equations (3.97) and (3.101), which show only the three components of displacement. As a result, the six components of ε and κ cannot be independent (only three are independent) and they are linked by compatibility equations.

To establish the three compatibility equations, it should be written that the metrics and the curvature of the deformed configuration are compatible, that is, they respect the Mainardi–Codazzi and Gauss equations. As by construction the reference surface respects these equations, the compatibility equations can be written by the difference between the two configurations.

To simplify the presentation, the reference surface and the deformed surface have the same parameterisation (x^α), which is for example the case if the deformed surface is defined using the displacement (but if the deformed surface is defined from the displacement, it automatically respects the conditions of compatibility!). In that case: $a_{\alpha\beta} = A_{\alpha\beta} + 2\varepsilon_{\alpha\beta}$ and $b_{\alpha\beta} = B_{\alpha\beta} + \kappa_{\alpha\beta}$. Let g be the difference between the final and initial Gaussian curvatures: $g = k - K$, which is calculated from the development of $\det\left(b_{\bullet}^{\bullet}\right) = \det\left(a^{\beta\gamma} b_{\alpha\beta}\right)$ as a function of the initial metric and curvature, using relations (3.16) and (3.20).

In the **case of small deformations**, the variation of metric can be neglected and the development of the determinant is written:

$$\det(b_{\bullet}^{\bullet}) \approx \det(B_{\bullet}^{\bullet} + \kappa_{\bullet}^{\bullet}) = \det(B_{\bullet}^{\bullet}) + \tilde{\varepsilon}^{\alpha\lambda}\tilde{\varepsilon}_{\beta\mu} B_{\alpha}{}^{\beta} \kappa_{\lambda}{}^{\mu} + \det(\kappa_{\bullet}^{\bullet}) \tag{3.111}$$

3.3.4.1 Case where the reference surface is a plane

This particular case is important since it makes it possible to obtain information on the deformation of the plates (shells with a flat reference surface, see Chapter 7). The reference surface is a plane, its curvature is zero, as well as its Gaussian curvature. The curvature variation in the transformation is then equal to the curvature of the deformed surface. The Mainardi–Codazzi equations then give two compatibility conditions, ie, the parameterisation being unique:

$$\nabla_1 \kappa_{\alpha 2} = \nabla_2 \kappa_{\alpha 1}$$

In this expression, the differentiation is that of the deformed surface.

The Gaussian equation allows expressing the relation between ε and κ, replacing $b_{..}$ by $B_{..} + \kappa_{..}$ and $a_{..}$ by $A_{..} + 2\varepsilon_{..}$ in (3.68), (3.16), and (3.83).

In the **case of small deformations**, the variation of metric is negligible and this relation is simplified, (3.68) and (3.83) being reduced then to:

$$\Gamma_{\alpha\beta}^{\gamma} \approx \frac{1}{2} A^{\gamma\delta} \left[\partial_\alpha \left(A_{\delta\beta} + 2\varepsilon_{\delta\beta} \right) + \partial_\beta \left(A_{\alpha\delta} + 2\varepsilon_{\alpha\delta} \right) - \partial_\delta \left(A_{\alpha\beta} + 2\varepsilon_{\alpha\beta} \right) \right]$$

$$k \approx \frac{1}{2} A_{\alpha\beta} \mathcal{E}^{\beta\lambda} \mathcal{E}^{\gamma\mu} \partial_\gamma \Gamma_{\lambda\mu}^{\alpha}$$

These expressions are further simplified when using a Cartesian coordinate system in the reference plane. So, by combining the two previous relationships and taking into account the definition of the Gaussian curvature k of the deformed surface:

$$\boxed{\det\left(\kappa_{\bullet}^{\bullet}\right) = \frac{1}{2} A_{\alpha\beta} \mathcal{E}^{\beta\lambda} \mathcal{E}^{\gamma\mu} \partial_\gamma \left[A^{\alpha\delta} \left(\partial_\lambda \varepsilon_{\delta\mu} + \partial_\mu \varepsilon_{\lambda\delta} - \partial_\delta \varepsilon_{\lambda\mu} \right) \right]} \tag{3.112}$$

or, in the case of orthonormal **Cartesian coordinates** (x, y) in the plane:

$$\boxed{k = \kappa_{xx}\kappa_{yy} - \kappa_{xy}{}^2 = 2\partial_{xy}^2 \varepsilon_{xy} - \partial_{yy}^2 \varepsilon_{xx} - \partial_{xx}^2 \varepsilon_{yy}} \tag{3.113}$$

which restores the classical condition of compatibility in small deformations for a transformation in the plane, for which $\kappa = 0$.

A consequence of (3.113) is important for the understanding of the behaviour of a plate: as soon as it undergoes a flexure due to a transverse loading (the load on a floor for example), it also undergoes membrane forces caused by the change of metrics, because of the emergence of curvatures related to flexure.

Exemple: Plate in torsion

A rectangular plate whose coordinates in the plane are x and y is subjected to a transverse displacement $\xi = Cxy\,\vec{k}$.

The following results concerning the geometry of the deformed configuration can be obtained directly or by applying the results of §3.2.1.8.

$$a_x = \vec{\mathbf{i}} + Cy\vec{\mathbf{k}}$$

$$a_y = \vec{\mathbf{j}} + Cx\vec{\mathbf{k}}$$

$$a_3 = \left(\vec{\mathbf{k}} - Cy\vec{\mathbf{i}} - Cx\vec{\mathbf{j}}\right)\hat{a}^{-1/2}$$

$$\hat{a} = 1 + C^2x^2 + C^2y^2$$

$$b_{xx} = b_{yy} = 0\,;\quad b_{xy} = C\hat{a}^{-1/2}$$

$$\Gamma^{\alpha}_{xx} = \Gamma^{\alpha}_{yy} = 0\,;\ \Gamma^{x}_{xy} = C^2y\,\hat{a}^{-1}\,;\ \Gamma^{y}_{xy} = C^2x\,\hat{a}^{-1}$$

$$k = -C^2\,\hat{a}^{-1}$$

The Mainardi–Codazzi relations $\nabla_x b_{xy} = \nabla_y b_{xx}$ and $\nabla_y b_{xy} = \nabla_x b_{yy}$ and Gauss' theorem can be verified from the expressions above (which is not a surprise since the deformed surface is defined from the displacement). In the transformation from the plane configuration, the deformation variables are:

$$\varepsilon_{xx} = \frac{1}{2}C^2y^2\,;\ \varepsilon_{yy} = \frac{1}{2}C^2x^2\,;\ \ \varepsilon_{xy} = \frac{1}{2}C^2xy$$

$$\kappa_{xx} = \kappa_{yy} = 0\,;\quad \kappa_{xy} = C\hat{a}^{-1/2}$$

In the case of small transformations, Cxy is of the first order (hence C) and the strains ε are of the second order (the metric a is reduced to identity at first-order). The only non-zero variation of curvature is the torsion κ_{xy}, which is equal to C. It is then consistent to take in the deformed configuration:

$$\Gamma^{\alpha}_{xx} = \Gamma^{\alpha}_{yy} = 0\,;\ \Gamma^{x}_{xy} = C^2y\,;\ \Gamma^{y}_{xy} = C^2x$$

$$K = -C^2$$

In these approximations, the verification of the Mainardi–Codazzi and Gauss relations must be made taking into account the good orders of magnitude. In fact, if the displacement is of the first order, the curvatures are of the first order, whereas the Gaussian curvature, the connection coefficients and the strains are of the second order. As a result, the Mainardi–Codazzi relations are of the first order and with respect to this order (the Levi-Civita derivatives being null to the first order); on the other hand, the Gaussian equation is of the second order. Indeed, relation (3.113) is verified, such as relation (3.112), taking into account the expression of the deformations, here of the second order, without neglecting them. A calculation from the above approximations without consideration of the orders of magnitude does not make it possible to verify correctly the conditions of compatibility. This is consistent with the assumptions made in small strains, developed in Chapter 8.

3.3.4.2 *Case where the reference surface is developable*

A developable surface has a zero Gaussian curvature. This is the case of cylinders, very important in practice. In small strains, formulas (112) and (113) are applicable.

3.3.4.3 *Case of shallow surfaces or almost developable*

A shallow surface is a surface whose curvature is sufficiently small for the metric to be comparable to that of a plate: the lengths and angles on the surface are assimilated to those of the underlying plate (which should therefore be defined, as well as its parametrisation).

In **small strains**, the variations of curvature nevertheless remain much smaller than the curvatures (the surface does not inverse by changing the sign of the curvature). In these circumstances, by (3.111):

$$g \approx \tilde{\mathcal{E}}^{\alpha\lambda}\tilde{\mathcal{E}}_{\beta\mu}B_{\alpha}{}^{\beta}\kappa_{\lambda}{}^{\mu} \tag{3.114}$$

This expression is reduced, in the principal curvature coordinate system of the reference surface, to:

$$g = \frac{\kappa_{11}}{R_2} + \frac{\kappa_{22}}{R_1} \tag{3.115}$$

3.4 CURVES DRAWN ON A SURFACE

3.4.1 Coordinate systems associated with a curve

3.4.1.1 *Frénet–Serret coordinate system*

A curve of $E_3^\bullet$ is an application such that:

$$s \in \mathbb{R} \quad \rightarrow \quad m(s) \in E_3^\bullet$$

supposed here twice differentiable. To simplify the presentation, s is the curvilinear abscissa of the curve.

The Frénet[12]–Serret[13] coordinate system of the curve is formed of the three vectors:

a) the tangent vector (unitary) **t** such that:

$$\mathbf{t} = \frac{dm}{ds} \tag{3.116}$$

[12]Jean Frédéric Frénet, French mathematician, astronomer and meteorologist (1816–1900)
[13]Joseph Serret, French mathematician and astronomer (1819–1885).

b) the normal (unitary) vector **n** such that:

$$\frac{d\mathbf{t}}{ds} = \rho\mathbf{n} \quad (3.117)$$

and such that (**t**, **n**) is direct. ρ is the curvature of the curve.

The differentiation with respect to s of identity $\mathbf{t}^2 = 1$ shows that **t** and **n** are orthogonal.

c) the binormal vector (unitary) **b** such that:

$$\mathbf{b} = \mathbf{t} \times \mathbf{n} \quad (3.118)$$

Relations $\frac{d}{ds}(\mathbf{b}^2) = 0$ and $\frac{d\mathbf{b}}{ds} = \mathbf{t} \times \frac{d\mathbf{n}}{ds}$ show that $\frac{d\mathbf{b}}{ds}$ is orthogonal to **b** and **t**, so collinear to **n**; let:

$$\frac{d\mathbf{b}}{ds} = -\tau\,\mathbf{n} \quad (3.119)$$

τ is the **torsion** of the curve.

Previous relationships allow to be written as:

$$\frac{d\mathbf{n}}{ds} = -\rho\,\mathbf{t} + \tau\,\mathbf{b} \quad (3.120)$$

If the curve is plane, the torsion is zero. τ measures the tendency of the curve to deviate from the tangent plane (**t**, **n**).

Previous results can be rewritten:

$$\frac{d\mathbf{t}}{ds} = A\mathbf{t} = \Omega \times \mathbf{t}$$

$$\frac{d\mathbf{n}}{ds} = A\mathbf{n} = \Omega \times \mathbf{n} \quad (3.121)$$

$$\frac{d\mathbf{b}}{ds} = A\mathbf{b} = \Omega \times \mathbf{b}$$

where A denotes the antisymmetric matrix:

$$A = \begin{pmatrix} 0 & \rho & 0 \\ -\rho & 0 & \tau \\ 0 & -\tau & 0 \end{pmatrix} \quad (3.122)$$

and Ω the vector:

$$\Omega = \begin{pmatrix} \tau \\ 0 \\ \rho \end{pmatrix} \quad (3.123)$$

By linearity, the derivative along the curve of any vector X of constant components on (**t**, **n**, **b**) is written:

$$\frac{dX}{ds} = AX = \Omega \times X \tag{3.124}$$

A curve is a plane if it is inscribed in a plane (**t**, **n**) independent of the point $\mathcal{m}$ considered. It is clear that, in this case, **b** is a constant vector, the torsion τ is zero, and:

$$\frac{d\mathbf{n}}{ds} = -\rho\,\mathbf{t}$$

3.4.1.2 Darboux coordinate system

In what follows, the curve is drawn on the surface σ. It is then defined by the transformation:

$$s \in \mathbb{R} \;\rightarrow\; \mathcal{m}\left[x^\alpha(s)\right] \in \sigma$$

where $(x^\alpha) \in \mathbb{R}^2$ are the curvilinear coordinates of the surface.

It is important to note that, although **t** belongs to the plane $T_{\mathcal{m}}$ tangent to the surface, **n** is usually not normal to the surface.

Let $\mathcal{P}$ be the plane $(\mathbf{n}, a_3)$ normal to **t** and ν the unit vector of its trace on $T_{\mathcal{m}}$, such that (**t**, ν) be direct (Figure 3.17). ν is called **geodesic normal.**

The coordinate system $(\mathbf{t}, \nu, a_3)$ is orthonormal; it is the **Darboux**[14] **coordinate system.** Vectors **n**, **b**, and a_3 belong to the plane $\mathcal{P}$ normal to **t**.

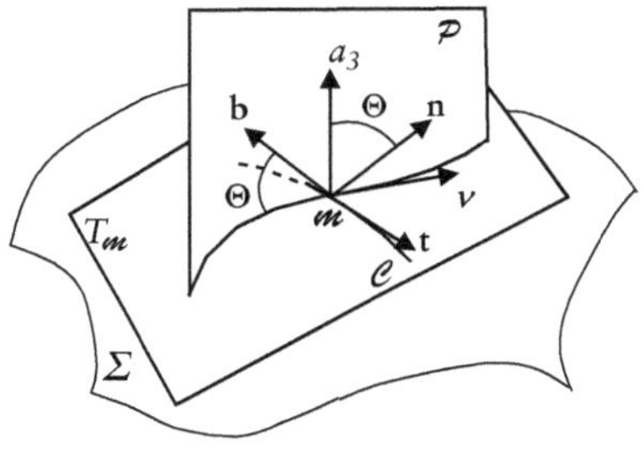

Figure 3.17

Then:

a) the **normal curvature** ρ_n, projection of $\frac{d\mathbf{t}}{ds}$ on the vector a_3 normal to the surface; it is the curvature of the projection of the curve on the plane (**t**, a_3) ;
b) the **geodesic curvature** ρ_g, projection of $\frac{d\mathbf{t}}{ds}$ on ν; it is the curvature of the projection of the curve on the plane (**t**, ν).

Finally:

$$\frac{d\mathbf{t}}{ds} = \rho_g \nu + \rho_n a_3 \tag{3.125}$$

If Θ denotes the angle $(a_3, \mathbf{n})$ in this plane (Figure 3.17), ρ_n and ρ_g verify:

$$\rho_g = \rho\ \sin\Theta$$

[14] Jean Gaston Darboux, French mathematician (1842–1917).

$$\rho_n = \rho \ \cos\Theta$$

This last relation expresses Meusnier's theorem. Differentiation formulas in the Darboux system are written:

$$\frac{d\mathbf{t}}{ds} = \rho_g \nu + \rho_n a_3 = \rho\,\mathbf{n}$$

$$\frac{d\nu}{ds} = -\rho_g \mathbf{t} + \tau_g a_3 \tag{3.126}$$

$$\frac{da_3}{ds} = -\rho_n \mathbf{t} - \tau_g \nu$$

where τ_g denotes the **geodesic torsion**. The relation:

$$\tau_g = \tau + \frac{d\Theta}{ds} \tag{3.127}$$

is obtained easily by expressing a_3 in terms of **n** and **b**, and then differentiating.

Example: helix drawn on a cylinder

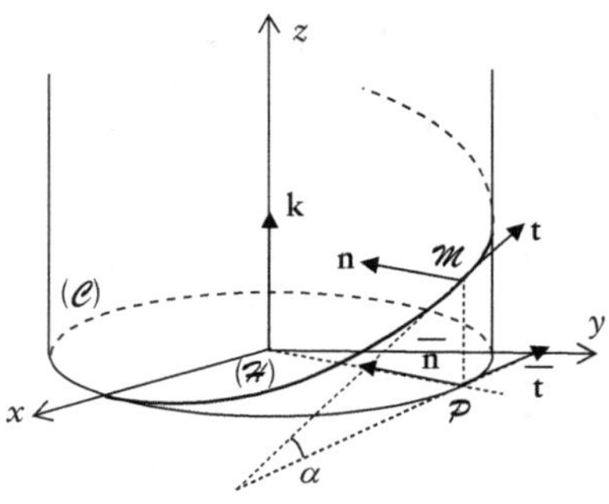

Figure 3.18

Let the cylinder with radius R_0 be parameterised by the curvilinear abscissa $\sigma = \theta\, R_0$ of the base circle ($\mathcal{C}$) in $z = 0$and the coordinate z (Figure 3.18). Let ($\mathcal{H}$) the helix drawn on the cylinder, with the equation:

$$z(\sigma) = \sigma \tan\alpha$$

with $0 \le \alpha \le \dfrac{\pi}{2}$.

Let $\mathcal{P}$ be the projection of current point $\mathcal{M}$ of ($\mathcal{H}$) on ($\mathcal{C}$). Then:

$$\mathcal{M} = \mathcal{P} + z(\sigma)\,\mathbf{k}$$

Let $\bar{\mathbf{t}}$ (resp. $\bar{\mathbf{n}}$) be the tangent (resp. the normal) to ($\mathcal{C}$) in $\mathcal{P}$. The following results are then obtained for the helix:

$$ds = \sqrt{1+\tan^2\alpha}\ d\sigma = \frac{d\sigma}{\cos\alpha}$$

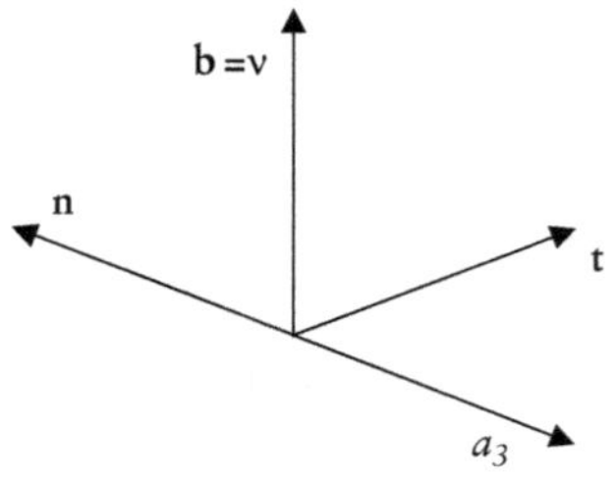

Figure 3.19

- Vectors of the Frénet system :

$$\mathbf{t} = \frac{d\mathcal{M}}{ds} = \frac{d\mathcal{M}}{d\sigma}\frac{d\sigma}{ds} = \left(\bar{\mathbf{t}} + z'\mathbf{k}\right)\cos\alpha = \bar{\mathbf{t}}\ \cos\alpha + \mathbf{k}\ \sin\alpha$$

$$\rho\mathbf{n} = \frac{d\mathbf{t}}{ds} = \frac{d\bar{\mathbf{t}}}{d\sigma}\cos^2\alpha = \frac{\bar{\mathbf{n}}}{R_0}\cos^2\alpha\ ; \quad \mathbf{n} = \bar{\mathbf{n}}$$

$$\mathbf{b}=\mathbf{t}\times\mathbf{n}=\left(\overline{\mathbf{t}}+z'\mathbf{k}\right)\times\overline{\mathbf{n}}\cos\alpha=\left(\mathbf{k}-z'\overline{\mathbf{t}}\right)\cos\alpha=\mathbf{k}\cos\alpha-\overline{\mathbf{t}}\sin\alpha$$

The angle between $\mathbf{t}$ and $\overline{\mathbf{t}}$ is constant, equal to α, as $\mathbf{t}\cdot\overline{\mathbf{t}}=\cos\alpha$.

- relation with the Darboux coordinate system:

$\mathbf{t}\times\mathbf{k}=-\mathbf{n}\cos\alpha$, then $a_3=-\mathbf{n}$

$$\begin{aligned}\nu&=a_3\times\mathbf{t}=a_3\times\left(\overline{\mathbf{t}}\cos\alpha+\mathbf{k}\sin\alpha\right)\\&=\mathbf{k}\cos\alpha-\overline{\mathbf{t}}\sin\alpha=\mathbf{b}\end{aligned}$$

The Darboux system is therefore deduced from the Frenet system by a rotation around $\mathbf{t}$, such that $\Theta=\pi$ (Figure 3.19).

- differentiation of the basis vectors:

$$\begin{aligned}\frac{d\mathbf{b}}{ds}&=-\frac{d\overline{\mathbf{t}}}{d\sigma}\sin\alpha\cos\alpha\\&=-\frac{\mathbf{n}}{R_0}\sin\alpha\cos\alpha=-\tau\,\mathbf{n}\end{aligned}$$

$$\rho=\frac{1}{R}\cos^2\alpha\,;\quad\tau=\frac{\sin\alpha\,\cos\alpha}{R}$$

The curvature and torsion of the helix are constant.

$$\frac{d\mathbf{n}}{ds}=\frac{d\overline{\mathbf{n}}}{d\sigma}\cos\alpha=-\frac{\overline{\mathbf{t}}}{R_0}\cos\alpha=-\left(\frac{\mathbf{t}}{R_0}-\frac{\sin\alpha}{R_0}\mathbf{k}\right)$$

$$\frac{d\nu}{ds}=\frac{d\mathbf{b}}{ds}=-\tau\,\mathbf{n}=\tau\,a_3\;\Rightarrow\;\begin{cases}\rho_g=0\\\tau_g=\tau\;\Rightarrow\;\dfrac{d\Theta}{ds}=0\;\left(\Theta=\pi\right)\end{cases}$$

$$\frac{d\mathbf{t}}{ds}=\rho\,\mathbf{n}=-\rho\,a_3\qquad\Rightarrow\quad\rho_n=-\rho$$

- If $\alpha=0$, $(\mathcal{H})\equiv(\mathcal{C})$, $\rho=\dfrac{1}{R_0}$ and $\tau=0$ (plan curve).
- If $\alpha=\dfrac{\pi}{2}$, $(\mathcal{H})$ is a vertical line, and $\rho=\tau=0$.

3.4.2 Link with the curvature tensor of the surface

By using the vectors a_1 and a_2 tangent to the surface, the differentiation along the curve becomes:

$$\mathbf{t} = \frac{d\boldsymbol{m}}{ds} = \frac{dx^\alpha}{ds} a_\alpha = t^\alpha a_\alpha$$
$$\rho\,\mathbf{n} = \frac{d^2\boldsymbol{m}}{ds^2} = \frac{d^2x^\alpha}{ds^2} a_\alpha + t^\alpha t^\beta \frac{\partial^2\boldsymbol{m}}{\partial x^\alpha \partial x^\beta} \tag{3.128}$$

In the Darboux coordinate system:

$$\frac{da_3}{ds} = \frac{\partial a_3}{\partial x^\alpha} t^\alpha = -b_\alpha{}^\beta t^\alpha a_\beta = -\, b(\mathbf{t})$$
$$= -\rho_n \mathbf{t} - \tau_g \nu$$

from which:

$$\rho_g = \frac{d^2x^\alpha}{ds^2} \nu_\alpha + t^\alpha t^\beta \frac{\partial^2\boldsymbol{m}}{\partial x^\alpha \partial x^\beta} \cdot \nu$$
$$\rho_n = \rho\,\mathbf{n} \cdot a_3 = \mathbf{t} \cdot b(\mathbf{t}) = b(\mathbf{t},\mathbf{t}) = b_{\alpha\beta} t^\alpha t^\beta \tag{3.129}$$
$$\tau_g = \nu \cdot b(\mathbf{t}) = b(\mathbf{t}, \nu)$$

Thus, an observer who runs along the line (direction t) would rotate with the curvature b_{tt} in the direction of his run (rotation around ν), but would also undergo a lateral deviation (torsion of axis t) due to $b_{t\nu}$

These relations also show that ρ_n and τ_g depend only on the vector tangent to the curve and on the shape of the surface. If b is expressed in the orthonormal basis ($\mathbf{t}$, ν):

$$\rho_n = b_{tt}$$
$$\tau_g = b_{t\nu}$$

Noting:

$$\mathbf{t} = t^\alpha a_\alpha$$
$$\nu = \nu^\beta a_\beta = a_3 \times \mathbf{t} = t^\alpha a_3 \times a_\alpha$$
$$= e_{\alpha\beta} t^\alpha a^\beta$$

It becomes:

$$\tau_g = b_{\alpha\beta} \nu^\alpha t^\beta = e_{\alpha\gamma} b^\gamma{}_\beta\, t^\alpha t^\beta \tag{3.130}$$

In the principal curvature coordinate system, these formulas are written:

$$\rho_n = \frac{1}{R_1} \cos^2\omega + \frac{1}{R_2} \sin^2\omega$$
$$\tau_g = \left(\frac{1}{R_2} - \frac{1}{R_1}\right) \sin\omega \cos\omega \tag{3.131}$$

where ω denotes the angle of **t** with the first curvature line.

The geodesic torsion is null and the normal curvature extremal, along the main curvature lines.

According to (3.124), for two vectors X and Y of $\mathbb{R}^3$, of constant components in the Darboux system:

$$Y \cdot \frac{dX}{ds} = Y \cdot (\Omega \times X) = (Y \times \Omega) \cdot X$$

The scalar triple product (X, Y, Ω) is worth for $X = \mathbf{t}$ and for Y:

- $Y = \nu$: $$(\mathbf{t},\ \nu,\ \Omega) = \rho_g \tag{3.132}$$
- $Y = a_3$: $$(\mathbf{t},\ a_3,\ \Omega) = \rho_n \tag{3.133}$$

In addition, the geodesic curvature is related to the covariant differentiation along the curve. Indeed:

$$\nabla_{\mathbf{t}} \mathbf{t} = \underbrace{\text{Proj}}_{T_{\mathcal{M}}} \left(\frac{d\mathbf{t}}{ds} \right) = \rho_g \nu$$

Moreover:

$$\nabla_{\mathbf{t}} \mathbf{t} = t^\alpha \left(\frac{\partial t^\beta}{\partial x^\alpha} + \Gamma^\beta_{\alpha\gamma} t^\gamma \right) a_\beta = \left(\frac{d^2 x^\beta}{ds^2} + \Gamma^\beta_{\alpha\gamma} \frac{dx^\alpha}{ds} \frac{dx^\gamma}{ds} \right) a_\beta$$

It becomes :

$$\rho_g = \nu \cdot \nabla_{\mathbf{t}} \mathbf{t} = \nu \cdot \frac{d\mathbf{t}}{ds} = \rho \nu \cdot \mathbf{n} = \left(\frac{d^2 x^\beta}{ds^2} + \Gamma^\beta_{\alpha\gamma} \frac{dx^\alpha}{ds} \frac{dx^\gamma}{ds} \right) \frac{dx^\delta}{ds} e_{\delta\beta} \tag{3.134}$$

Example: Helix drawn on a cylinder.

The example in §3.4.1.2 is considered again. The basic vectors of the tangent plane at $\mathcal{M}$ are **t** and **k**. So:

$$\rho_n = b(\mathbf{t}, \mathbf{t}) = \begin{pmatrix} \cos\alpha & \sin\alpha \end{pmatrix} \begin{pmatrix} -\dfrac{1}{R_0} & 0 \\ 0 & 0 \end{pmatrix} \begin{pmatrix} \cos\alpha \\ \sin\alpha \end{pmatrix} = -\frac{\cos^2\alpha}{R_0}$$

$$\tau_g = b(\mathbf{t}, \nu) = \begin{pmatrix} \cos\alpha & \sin\alpha \end{pmatrix} \begin{pmatrix} -\dfrac{1}{R_0} & 0 \\ 0 & 0 \end{pmatrix} \begin{pmatrix} -\sin\alpha \\ \cos\alpha \end{pmatrix} = -\frac{\cos\alpha \sin\alpha}{R_0}$$

$$\omega = \alpha$$

$$\nabla_{\mathbf{t}} \mathbf{t} = \underbrace{\text{Proj}}_{\perp a_3} \left(\frac{d\mathbf{t}}{ds} \right) = 0 \ \Rightarrow \ \rho_g = 0$$

3.4.3 Special lines on a surface

It has been established that the principal lines of curvature are characterised by: $\tau_g = 0$. Other particular lines are also of interest. The most important are presented in this paragraph:

a) A **geodesic** of the surface is a curve $\gamma(s)$ such that the field of vectors tangent to the curve is parallel along γ, that is to say:

$$\nabla_{\mathbf{t}}\mathbf{t} = 0 \tag{3.135}$$

where $\mathbf{t} = \dfrac{d\gamma(s)}{ds}$.

According to the relations established previously, a necessary and sufficient condition for a curve of the surface to be a geodesic is that its geodesic curvature is null in all points: $\rho_g = 0$ (**n** is collinear with a_3).

The definition implies that:

$$\nabla_{\mathbf{t}}\mathbf{t} = 0 \quad \Leftrightarrow \quad \left[\frac{d^2x^\alpha}{ds^2} + \Gamma^\alpha_{\beta\gamma}\frac{dx^\beta}{ds}\frac{dx^\gamma}{ds}\right] = 0 \tag{3.136}$$

These two relations are the necessary and sufficient conditions (Euler's equations) for the length to be extremal along a curved arc: the arcs of geodesics are thus the arcs of the curve drawn on a surface in which the length between two given points is extreme.

If the geodesic curvature ρ_g is null:

- ether $\rho = 0$; then the geodesic is a straight line;
- or $\sin\Theta = 0$ ($\Theta = 0$ or π), in which case:

$$\rho_n = \pm\rho \text{ and } \tau_g = \tau$$

and the plane osculator to the geodesic is normal to the surface at all points: the normal to the surface and the normal to the curve are then identical.

In view of the relations established in §3.2, $b_{\alpha\beta}$ is:

- if $\alpha = \beta$, the curvature of the geodesic tangent to a_α;
- if $\alpha \neq \beta$, the torsion of the same geodesic.

Example: The helix is a geodesic of the cylinder.

b) An **asymptotic direction** is a direction in which the curvature of the surface is zero. If $\mathbf{t} = t^\alpha a_\alpha$ is a unit vector in the asymptotic direction:

$$\rho_n = b_{\alpha\beta}t^\alpha t^\beta = 0$$

An **asymptotic line** is tangent at every point to an asymptotic direction on the surface and thus satisfies the equation:

$$b_{\alpha\beta}\frac{dx^{\alpha}}{ds}\frac{dx^{\beta}}{ds}=0 \tag{3.137}$$

which means that the normal curvature ρ_n is zero at any point, in which case:

- either $\rho = 0$: the asymptotic line is a straight line;
- or $\cos\Theta = 0$ $(\Theta = \pm\frac{\pi}{2})$, which leads to:

$\mathbf{n} = \nu$ and $\mathbf{b} = a_3$

In this case:

$$\frac{d\mathbf{b}}{ds}=\frac{da_3}{ds}=-b_{\alpha}{}^{\beta}t^{\alpha}a_{\beta}=-\tau\,\mathbf{n}=-\tau_g\nu$$

The torsion of the asymptotic line is then equal to its geodesic torsion and:

$$\tau^2 = b_{\alpha}{}^{\beta}b_{\beta\gamma}t^{\alpha}t^{\gamma} = c_{\alpha\gamma}t^{\alpha}t^{\gamma}$$

Now, according to formula (3.36) applied to the vector **t** tangent to the asymptotic line and according to (3.137):

$$c_{\alpha\gamma}t^{\alpha}t^{\gamma} + Ka_{\alpha\gamma}t^{\alpha}t^{\gamma} = 0$$

Noting that $a_{\alpha\gamma}t^{\alpha}t^{\gamma} = 1$, it becomes:

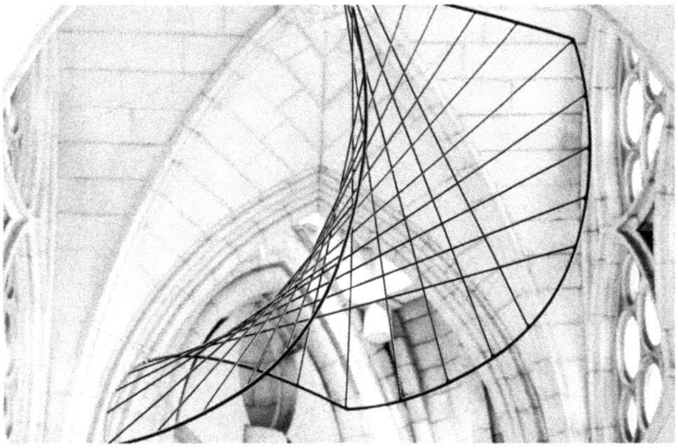

Photo 3.3. Asymptotic lines on a negative Gaussian curvature surface (Barcelona, Spain)

$$\tau^2 = -K = -\frac{1}{R_1R_2} \tag{3.138}$$

(Enneper[15] formula).

The asymptotic directions are real if and only if the Gaussian curvature is negative: this is obvious since the curvature varies then continuously between the two extreme values $\frac{1}{R_1}$ and $\frac{1}{R_2}$, which are of opposite signs.

[15]Alfred Enneper, German mathematician (1830–1885).

MAIN RESULTS

GEOMETRY OF THE SURFACE

Vectors of the basis: $a_\alpha = \dfrac{\partial \boldsymbol{m}}{\partial x^\alpha} = \partial_\alpha \boldsymbol{m}$ $\qquad a_3 = \dfrac{a_1 \times a_2}{\|a_1 \times a_2\|}$

Metric tensor: $a_{\alpha\beta} = a_\alpha \bullet a_\beta \qquad \|a_1 \times a_2\| = \sqrt{\hat{a}}$

Curvature tensor: $b_{\alpha\beta} = -a_\beta \bullet \partial_\alpha a_3 = a_3 \bullet \partial^2_{\alpha\beta} \boldsymbol{m}$

Basis of $E_3^\bullet$ in the neibourhood of the surface: $g_i = m_i{}^j a_j$ with: $m_\alpha{}^\beta = \delta^\beta_\alpha - x^3 b_\alpha{}^\beta$

$m_3{}^i = \delta^i_3$

DIFFERENTIATION ON THE SURFACE

Differentiation of a vector field on the surface: $D_X Y = \nabla_X Y + b(X,Y)\, a_3$

Differentiation of a vector of the basis: $D_\alpha a_\beta = \partial_\alpha a_\beta = \nabla_\alpha a_\beta + b_{\alpha\beta} a_3 = \Gamma^\gamma_{\alpha\beta} a_\gamma + b_{\alpha\beta} a_3$

Levi-Civita differentiation of a vector: $\nabla_\alpha Y^\beta = \partial_\alpha Y^\beta + \Gamma^\beta_{\alpha\gamma} Y^\gamma$

with: $\Gamma^\gamma_{\alpha\beta} = \dfrac{1}{2} a^{\gamma\delta}\left(\partial_\alpha a_{\delta\beta} + \partial_\beta a_{\alpha\delta} - \partial_\delta a_{\alpha\beta}\right) = a^\gamma\, \partial_\alpha a_\beta = -a_\beta\, \partial_\alpha a^\gamma$

Differentiation of a vector in E_3'' $(X \in T_{\boldsymbol{m}})$:

$$D_X U = X^\alpha \left[\left(\partial_\alpha U^\beta + \Gamma^\beta_{\alpha\gamma} U^\gamma - b_\alpha{}^\beta U^3\right) a_\beta + \left(b_{\alpha\beta} U^\beta + \partial_\alpha U^3\right) a_3\right]$$

Mainardi–Codazzi relation: $\nabla_\lambda b_{\alpha\beta} = \nabla_\beta b_{\alpha\lambda}$

Gauss theorem: $K = \dfrac{1}{2} a^{\lambda\mu}\left(\partial_\gamma \Gamma^\alpha_{\lambda\mu} + \Gamma^\alpha_{\gamma\delta}\Gamma^\delta_{\lambda\mu}\right)$

DEFORMATION OF THE SURFACE

Strain: $\varepsilon(\mathcal{M}) = \varepsilon_{\alpha\beta} dX^\alpha dX^\beta = \dfrac{1}{2}\left(d\boldsymbol{m}^2 - d\mathcal{M}^2\right)$

Curvature variation: $\kappa(\mathcal{M}) = \kappa_{\alpha\beta} dX^\alpha dX^\beta = -\left(d\boldsymbol{m} \bullet da_3 - d\mathcal{M} \bullet dA_3\right)$

In small displacements: $\varepsilon_{\alpha\beta} = \dfrac{1}{2}\left(\nabla_\alpha \xi_\beta + \nabla_\beta \xi_\alpha\right) - \zeta B_{\alpha\beta}$

$$a_3 = A_3 - \left(B_{\alpha\beta}\xi^\beta + \partial_\alpha \zeta\right) A^{\alpha\gamma} A_\gamma = A_3 + \omega \times A_3$$

$$\kappa_{\alpha\beta} \approx \nabla_\alpha\left(\partial_\beta \zeta + B_{\beta\gamma}\xi^\gamma\right) + B_{\alpha\gamma}\left(\nabla_\beta \xi^\gamma - \zeta B_\beta{}^\gamma\right)$$

Small deformation of a plane: $\kappa_{xx}\kappa_{yy} - \kappa_{xy}{}^2 = 2\partial^2_{xy}\varepsilon_{xy} - \partial^2_{yy}\varepsilon_{xx} - \partial^2_{xx}\varepsilon_{yy}$

EXERCISES

Exercise 3.1

What are the principal curvatures of a circular cone? Using the angle θ along a horizontal circle and z the vertical coordinate as parameters, calculate the metric tensor and the connection coefficients in the natural basis (Figure 3.20).

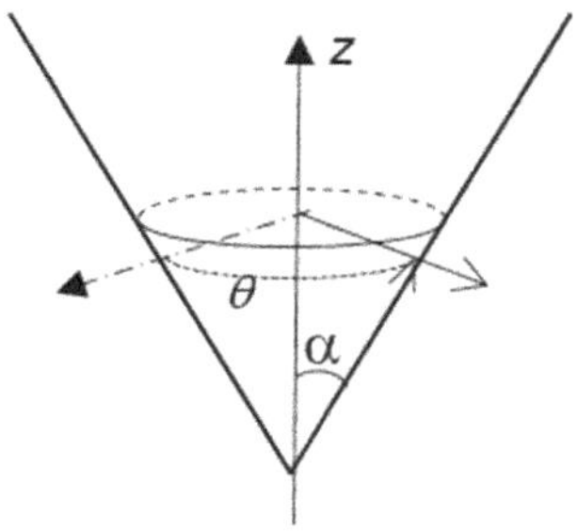

Figure 3.20

Answer:
The curvature of the generatrix is zero. The other curvature is given by Meusnier's theorem, that is, for the circumferential radius of curvature:

$$R_\theta = z\frac{\sin\alpha}{\cos^2\alpha}$$

$$a_{\theta\theta} = z^2\tan^2\alpha \ ; \ a_{zz} = \frac{1}{\cos^2\alpha}$$

$$\Gamma^z_{\theta\theta} = -z\sin^2\alpha \ ; \ \Gamma^\theta_{\theta z} = \frac{1}{z} \ ; \text{ others are zero}$$

Exercise 3.2

An ellipsoid of revolution of axis z (Figure 3.21) is parameterised by the angles θ and ϕ (figure 3.21), in this order.

Calculate, at any point M on the surface:

- The natural coordinate system associated with the curvilinear parameters.
- The metric tensor and the curvature tensor in the natural coordinate system. Verify the curvature of a line ϕ = cst.
- Riemannian connection coefficients.

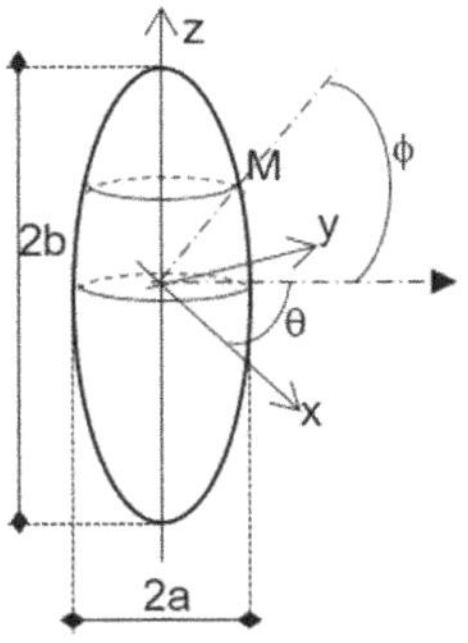

Figure 3.21

What is the physical coordinate system associated with the natural coordinate system and what are the curvatures in this system?

Calculate the derivatives, with respect to a_θ, of the vector defined at any point by its expression in the physical coordinate system:

$$V = 2\,\tilde{a}_\theta + 3\,\tilde{a}_\phi + 4\,a_3$$

Answer:
Point M and the vectors of the natural basis are expressed in the coordinate system (x, y, z):

$$M = \begin{vmatrix} a\cos\phi\cos\theta \\ a\cos\phi\sin\theta \\ b\sin\phi \end{vmatrix} \quad a_\theta = \begin{vmatrix} -a\cos\phi\sin\theta \\ a\cos\phi\cos\theta \\ 0 \end{vmatrix} \quad a_\phi = \begin{vmatrix} -a\sin\phi\cos\theta \\ -a\sin\phi\sin\theta \\ b\cos\phi \end{vmatrix} \quad a_3 = \frac{1}{c}\begin{vmatrix} b\cos\phi\cos\theta \\ b\cos\phi\sin\theta \\ a\sin\phi \end{vmatrix}$$

$$c(\phi) = \left(b^2\cos^2\phi + a^2\sin^2\phi\right)^{1/2} \quad a_{\theta\theta} = a^2\cos^2\phi = r(\phi)^2 \ ; \ a_{\phi\phi} = c^2 \ ; \ a_{\theta\phi} = 0$$

r is the radius of a circle where z (or ϕ) is constant.

In the natural coordinate system:

$$b_{\theta\theta} = -\frac{ab}{c}\cos^2\phi \;\; ; \; b_{\phi\phi} = -\frac{ab}{c} \;\; ; \; b_{\theta\phi} = 0$$

$$\Gamma^{\theta}_{\theta\phi} = -\tan\phi \;\; ; \; \Gamma^{\phi}_{\theta\theta} = \frac{a^2}{c^2}\sin\phi\cos\phi \;\; ; \; \Gamma^{\phi}_{\phi\phi} = \frac{a^2-b^2}{c^2}\sin\phi\cos\phi \;\; ; \text{ other } \Gamma \text{ values are zero.}$$

In the physical coordinate system:

$$\tilde{a}_\theta = \begin{vmatrix} -\sin\theta \\ \cos\theta \\ 0 \end{vmatrix} \quad \tilde{a}_\phi = \frac{1}{c}\begin{vmatrix} -a\sin\phi\cos\theta \\ -a\sin\phi\sin\theta \\ b\cos\phi \end{vmatrix} \qquad \tilde{b}_{\theta\theta} = -\frac{b}{ac} \;\; ; \; \tilde{b}_{\phi\phi} = -\frac{ab}{c^3}$$

$$V = \frac{2}{a\cos\phi}\, a_\theta + \frac{3}{c}\, a_\phi + 4\, a_3$$

$$D_\theta V = \partial_\theta V + \frac{2}{a\cos\phi}\Gamma^{\phi}_{\theta\theta} a_\phi + \frac{3}{c}\Gamma^{\theta}_{\theta\phi} a_\theta - 4 b_\theta{}^\theta a_\theta = \frac{2a}{c^2}\sin\phi\, a_\phi + \frac{4b - 3a\tan\phi}{ac}\, a_\theta$$

$$= \frac{2a}{c}\sin\phi\,\tilde{a}_\phi + \frac{4b\cos\phi - 3a\sin\phi}{c}\,\tilde{a}_\theta$$

Exercise 3.3

A torus is generated by the rotation of a circle around an axis z belonging to its plane (Figure 3.22).

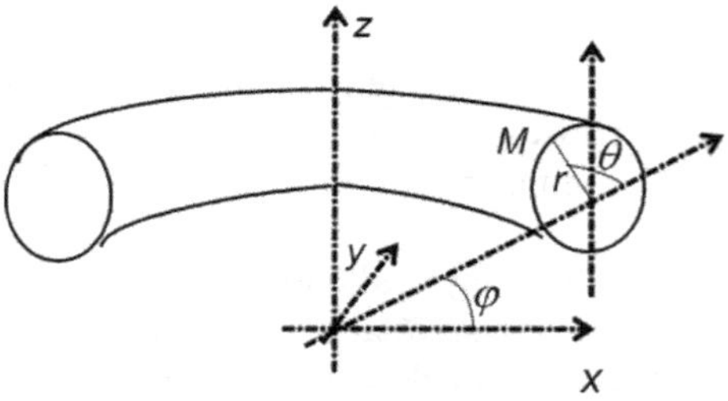

Figure 3.22

1) Calculate the natural local coordinate system. On what condition(s) is it defined? Calculate the metric and curvature tensors, then the connection coefficients in this system. What are the principal lines of curvature?
2) Verify the Mainardi–Codazzi and Gauss equations.
3) $\overline{\overline{N}}$ being a twice-contravariant tensor defined in the plane tangent to the torus, calculate $\nabla_\alpha N^{\alpha\theta}$ and $\nabla_\alpha N^{\alpha\varphi}$.
4) Calculate the orthonormal physical coordinate system associated with the natural coordinate system, then the metric and curvature tensors in this coordinate system. Interpret the principal curvatures by Meusnier's theorem. Express the components of $\overline{\overline{N}}$ in the orthonormal coordinate system as a function of those in the natural coordinate system.
5) Determine the geodesic equations. Which particular curves verify these equations?

Answer:

1) The point M and the vectors of the natural basis are expressed in the coordinate system (x, y, z) :

$$M = \begin{vmatrix} (R+r\cos\theta)\cos\varphi \\ (R+r\cos\theta)\sin\varphi \\ r\sin\theta \end{vmatrix} \quad a_\varphi = \begin{vmatrix} -(R+r\cos\theta)\sin\varphi \\ (R+r\cos\theta)\cos\varphi \\ 0 \end{vmatrix}$$

$$a_\theta = \begin{vmatrix} -r\sin\theta\cos\varphi \\ -r\sin\theta\sin\varphi \\ r\cos\theta \end{vmatrix} \quad a_3 = \begin{vmatrix} \cos\theta\cos\varphi \\ \cos\theta\sin\varphi \\ \sin\theta \end{vmatrix}$$

Condition: $R > r$

$$a_{\varphi\varphi} = (R+r\cos\theta)^2 \quad ; \quad a_{\theta\theta} = r^2 \quad ; \quad a_{\theta\varphi} = 0$$

$$b_{\varphi\varphi} = -(R+r\cos\theta)\cos\theta \quad ; \quad b_{\theta\theta} = -r \quad ; \quad b_{\theta\varphi} = 0$$

The principal lines of curvature are the circles defined by θ constant and by φ constant.

$\Gamma^{\varphi}_{\varphi\theta} = -\dfrac{r}{R+r\cos\theta}\sin\theta \quad ; \quad \Gamma^{\theta}_{\varphi\varphi} = \dfrac{R+r\cos\theta}{r}\sin\theta$; other Γ values are zero.

2) Use the results of 1) to check relationships.

3)

$$\nabla_\alpha N^{\alpha\varphi} = \partial_\alpha N^{\alpha\varphi} + \Gamma^{\alpha}_{\alpha\gamma} N^{\gamma\varphi} + \Gamma^{\varphi}_{\alpha\gamma} N^{\alpha\gamma} = \partial_\alpha N^{\alpha\varphi} - 3\frac{r\sin\theta}{R+r\cos\theta} N^{\theta\varphi}$$

$$\nabla_\alpha N^{\alpha\theta} = \partial_\alpha N^{\alpha\theta} + \Gamma^{\alpha}_{\alpha\gamma} N^{\gamma\theta} + \Gamma^{\theta}_{\alpha\gamma} N^{\alpha\gamma} = \partial_\alpha N^{\alpha\theta} - \frac{r\sin\theta}{R+r\cos\theta} N^{\theta\theta} + \frac{R+r\cos\theta}{r}\sin\theta\, N^{\varphi\varphi}$$

4) Orthonormal physical system:

$$\tilde{a}_\varphi = \begin{vmatrix} -\sin\varphi \\ \cos\varphi \\ 0 \end{vmatrix} \quad \tilde{a}_\theta = \begin{vmatrix} -\sin\theta\cos\varphi \\ -\sin\theta\sin\varphi \\ \cos\theta \end{vmatrix} \qquad \tilde{b}_{\varphi\varphi} = -\frac{\cos\theta}{R+r\cos\theta} \quad ; \quad \tilde{b}_{\theta\theta} = -\frac{1}{r}$$

Meusnier theorem: $$R_\varphi = -\frac{R+r\cos\theta}{\cos\theta}$$

$$\tilde{N}^{\varphi\varphi} = (R+r\cos\theta)^2 N^{\varphi\varphi} \quad ; \quad \tilde{N}^{\theta\varphi} = r(R+r\cos\theta) N^{\theta\varphi} \quad ; \quad \tilde{N}^{\theta\theta} = r^2 N^{\theta\theta}$$

5) Equations of geodesics. In the orthonormal physical system:

$$\frac{d^2\varphi}{ds^2} - \frac{2r\sin\theta}{R + r\cos\theta}\frac{d\theta}{ds}\frac{d\varphi}{ds} = 0$$

$$\frac{d^2\theta}{ds^2} + \frac{R + r\cos\theta}{r}\sin\theta\left(\frac{d\varphi}{ds}\right)^2 = 0$$

Particular solutions:

- Small circles φ= cst.
- Large circles $\theta = 0$ and $\theta = \pi$.

Exercise 3.4

Calculate $\nabla_\beta b_\gamma{}^\alpha$ in the principal curvature coordinate system, as a function of the radii of curvature and the connection coefficients. Express the latter as functions of the characteristic curvatures of the two principal curvature lines. Calculate $\frac{\partial}{\partial s_2}\left(\frac{1}{R_1}\right)$ function of these characteristics. Application to the torus (Figure 3.22).

Answer :

$$\nabla_\beta b_\gamma{}^\alpha = \partial_\beta\left(\frac{1}{R_\alpha}\right)\delta_\gamma^\alpha + \Gamma_{\beta\gamma}^\alpha\left(\frac{1}{R_\gamma} - \frac{1}{R_\alpha}\right)$$

(in the second member, there is no summation on α and γ).
In the principal curvature system :

$$\Gamma_{11}^1 = \Gamma_{22}^2 = 0;\ \Gamma_{11}^2 = -\Gamma_{12}^1 = \rho_{g1};\ \Gamma_{12}^2 = -\Gamma_{22}^1 = \rho_{g2}$$

where ρ_{g1} an ρ_{g2} are the geodesic curvatures of lines *1* and *2* respectively. By application of the Mainardi–Codazzi theorem:

$$\frac{\partial}{\partial s_2}\left(\frac{1}{R_1}\right) = \rho_{g1}\left(\frac{1}{R_1} - \frac{1}{R_2}\right)$$

In the case of the torus, the main lines of curvature are:

- circles defined by ϕ = constant, of radius $R + r\cos\phi$; the angle of the normal **n** with $-a_3$ is ϕ;
- the small circles defined by θ = constant, of radius r. **n** and a_3 are opposite.

$$\frac{1}{R_1} = -\frac{\cos\phi}{R + r\cos\phi} = \rho_{n1}\ ;\ \rho_{g1} = \frac{1}{R_1}\tan\phi\ ;\ \frac{1}{R_2} = -\frac{1}{r} = \rho_{n2}\ ;\ \rho_{g2} = 0$$

$$\frac{\partial}{\partial s_2}\left(\frac{1}{R_1}\right) = \frac{1}{r}\frac{d}{d\phi}\left(-\frac{\cos\phi}{R + r\cos\phi}\right) = \rho_{g1}\left(\frac{1}{R_1} - \frac{1}{R_2}\right)\ ;\ \frac{\partial}{\partial s_1}\left(\frac{1}{R_2}\right) = 0$$

Chapter 4

The shell as a surfacic solid

In accordance with the methods of the theory of structures, the equations of motion are established by a "direct" (or "natural" or "global") approach, which consists of abstracting the three-dimensional reality and assimilating the shell to its middle surface, according to the methods of the theory of structures: the evolution of the shell is described by variables defined on its middle surface, in particular the displacement, the generalised variables of deformation and stress, the shell being considered as a two-dimensional material medium immersed in $E_3^\bullet$. *The evolution equations are obtained by applying the principles of mechanics and take the form of partial differential equations of quantities defined on the surface.*

In the first section, the results of the previous chapter are supplemented by some kinematic considerations and the virtual deformations are introduced.

In the second section, the equations of motion are established either by the application of general momentum theorems or by the Principle of Virtual Powers (PVP). Both approaches are equivalent, but the equations are different; it appears that the PVP makes it easier to understand the boundary conditions, on the one hand, and to better control modelling, on the other hand. The equivalence between the two approaches is established in the fourth section.

In the third section, the problem of movement is posited, which shows the need to introduce constitutive laws. Linear elasticity is then taken as an example.

In the fifth section, it is shown that by completing the deformation variables on the surface, a richer shell model can be built.

Sections four and five explore the previous topics in further depth, and can be omitted on first reading.

4.1 KINEMATICS OF SURFACES

4.1.1 Velocity and acceleration

4.1.1.1 Parametrisation

In the approach used in this chapter, the shell is assimilated to its middle surface and the thickness does not intervene directly. Therefore, all the quantities considered are defined on the middle surface.

During a movement, the shell occupies successive positions over time[1] t. $\boldsymbol{m}$ is any point of the middle surface at time t.

[1] Time is introduced here for convenience to describe the change of configuration, even for static problems.

DOI: 10.1201/9780429440403-4

As is often the case in solid mechanics, it is convenient to describe the evolution of the shell in Lagrange coordinates, from a reference position Σ of the middle surface (or initial position); the reference position is equipped with a coordinate system associated with the curvilinear coordinates $\left(X^{\alpha}\right)$. At time t, the surface occupies the "current" position σ and the position of a point of the surface is expressed according to its initial position on Σ and time[2]:

$$\boldsymbol{m}\left(X^{\alpha},t\right) \tag{4.1}$$

In practice, it is the coordinates x^{β} of $\boldsymbol{m}$ that are expressed as a function of X^{α} and t. In this representation, the same parameters X^{α} can be used for successive configurations of the surface. This is particularly the case when the current position is determined by the displacement from Σ:

$$\boldsymbol{m} = \mathcal{M}\left(X^{\alpha}\right)+\xi\left(X^{\alpha},t\right) \tag{4.2}$$

On the contrary, the Eulerian representation abstracts from the initial position of the shell: at each instant, the configuration σ is defined by a bijection of $\mathbb{R}^2$ in $\mathbb{R}^3$ and the different physical quantities are defined on the current surface, so as functions of x^1 and x^2 only. Even when it is possible to define a reference position, it may be useful to express certain quantities ϕ (such as for example velocity or acceleration) in Euler coordinates, that is to say using the current parameters of the surface, so that these quantities are expressed in the form: $\phi\left[x^{\alpha}(t)\right]$.

4.1.1.2 Velocity

Let v be the velocity of a point of the surface in its current configuration σ. It is a field of $\mathbb{R}^3$ defined on σ:

$$v = \frac{d\boldsymbol{m}}{dt} = \frac{d\xi}{dt} = \dot{\xi} \tag{4.3}$$

Using relation (4.2), it becomes, in Lagrange coordinates:

$$v = \frac{\partial \boldsymbol{m}}{\partial t}\left(X^{\alpha},t\right)$$

In Euler coordinates, the position is defined using the parameters of the current surface, defined at each instant: $\boldsymbol{m}\left[x^{\beta}(t)\right]$. The parametrisation of the surface varies continuously over time. The velocity is then expressed using the parameters of the surface at time t, in the form:

$$v = v^i\, \mathrm{a}_i$$

[2] More generally, when a quantity is defined on the "current" surface, it is designated by a lowercase letter, whereas the same letter in upper case designates it when it is defined on the reference surface.

Since the velocity field is given on σ, it is useful to express certain quantities in Euler variables, as a function of v, notably the time derivatives of the vectors of the basis.

In the current configuration, the differentiation along a line of coordinate x^α and the differentiation with respect to time commute, results in:

$$a_\alpha = \frac{\partial \boldsymbol{m}}{\partial x^\alpha} \quad \Rightarrow \quad \dot{a}_\alpha = \frac{d}{dt}\left[\frac{\partial \boldsymbol{m}}{\partial x^\alpha}\right] = D_\alpha v \tag{4.4}$$

Yet:

$$D_\alpha v = \left(\nabla_\alpha v^\beta - b_\alpha{}^\beta v^3\right) a_\beta + \left(\partial_\alpha v^3 + b_{\alpha\beta} v^\beta\right) a_3 \tag{4.5}$$

hence the relations:

$$\begin{aligned} \dot{a}_\alpha \bullet a_\beta &= \nabla_\alpha v_\beta - b_{\alpha\beta} v^3 = v_{\alpha\beta} \\ \dot{a}_\alpha \bullet a_3 &= \partial_\alpha v^3 + b_{\alpha\beta} v^\beta = v_{\alpha 3} \end{aligned} \tag{4.6}$$

These expressions are to be compared with expressions (3.95).

On the other hand, $\dot{a}_3$ is in the tangent plane and:

$$\dot{a}_3 \bullet a_\alpha = -\,\dot{a}_\alpha \bullet a_3 = -\left(\partial_\alpha v^3 + b_{\alpha\beta} v^\beta\right) = -\,v_{\alpha 3} \tag{4.7}$$

Finally:

$$\boxed{\begin{aligned} \dot{a}_\alpha &= a^{\beta\gamma} v_{\alpha\beta} a_\gamma + v_\alpha{}^3 a_3 = v_{\alpha\beta} a^\beta + v_{\alpha 3} a^3 \\ \dot{a}_3 &= -\,a^{\alpha\beta} v_{\beta 3} a_\alpha = -\,v_{\beta 3} a^\beta \end{aligned}} \tag{4.8}$$

Note: In Lagrange variables, $\dot{a}_\alpha$ and $\dot{a}_3$ are simply obtained by differentiation of expressions (3.95) and (3.98).

$\dot{a}_3$ is linked to the angular velocity vector Ω in $\boldsymbol{m}$ through relationships:

$$\begin{aligned} \dot{a}_3 &= \Omega \times a_3 = -\,\Omega^\alpha e_\alpha{}^\lambda a_\lambda \\ \Omega &= a_3 \times \dot{a}_3 \end{aligned} \tag{4.9}$$

$\dot{a}_3$ completely determines the local rotation velocity of the middle surface. Ω also verifies the relationship:

$$\Omega^2 = \left(\dot{a}_3\right)^2 = \left(a^{\alpha\beta} v_{\beta 3} a_\alpha\right)\left(a^{\lambda\mu} v_{\mu 3} a_\lambda\right) = v_{\beta 3}\, v_{\mu 3}\, a^{\beta\mu} \tag{4.10}$$

Finally, note the following results for the metric tensor:

$$\dot{a}_{\alpha\beta} = \dot{\overline{a_\alpha \bullet a_\beta}} = \dot{a}_\alpha \bullet a_\beta + a_\alpha \bullet \dot{a}_\beta = v_{\alpha\beta} + v_{\beta\alpha} \tag{4.11}$$

$$\dot{\hat{a}} = \overline{\det\left(a_{\alpha\beta}\right)}^{\,\cdot} = \frac{\partial \hat{a}}{\partial a_{\alpha\beta}} \dot{a}_{\alpha\beta}$$

It is easy to verify that:

$$\frac{\partial \hat{a}}{\partial a_{\alpha\beta}} = \hat{a}\, a^{\alpha\beta}$$

which leads to:

$$\dot{\hat{a}} = \hat{a}\, a^{\alpha\beta} \dot{a}_{\alpha\beta} = 2\, \hat{a}\, v_{\alpha}{}^{\alpha} = 2\, \hat{a} \left(\nabla_{\alpha} v^{\alpha} - b_{\alpha}{}^{\alpha} v^{3}\right) \tag{4.12}$$

Then, differentiating the relationship: $a^{\alpha\beta} a_{\beta\gamma} = \delta^{\alpha}_{\gamma}$, it becomes:

$$\dot{a}^{\alpha\beta} a_{\beta\gamma} + a^{\alpha\beta} \dot{a}_{\beta\gamma} = 0$$

from which

$$\dot{a}^{\alpha\beta} = -\, a^{\alpha\gamma} a^{\beta\delta} \dot{a}_{\gamma\delta} \tag{4.13}$$

4.1.1.3 Strain and variation of curvature rates

The strain and variation of curvature tensors, expressed as Lagrange variables on the initial surface, were introduced in Chapter 3 (formulas (3.84) and (3.86)). Strain and variation of curvature in Euler variables can be expressed in a similar way on the current surface:

$$\begin{aligned} e(\boldsymbol{m}) &= \frac{1}{2}\left[a - \overset{2}{\otimes} \bar{f}^{T-1}(A)\right] \\ k(\boldsymbol{m}) &= b - \overset{2}{\otimes} \bar{f}^{T-1}(B) \end{aligned} \tag{4.14}$$

ε and κ are respectively the transforms of e and k by $\overset{2}{\otimes} \bar{f}^{T}$. The strain rate tensor $\dot{e}$ expressed in Euler variables is deduced from (4.11) and (4.6):

$$\boxed{\dot{e}_{\alpha\beta} = \frac{1}{2} \dot{a}_{\alpha\beta} = \frac{1}{2}\left(\nabla_{\alpha} v_{\beta} + \nabla_{\beta} v_{\alpha} - 2 b_{\alpha\beta} v^{3}\right)} \tag{4.15}$$

On the other hand:

$$b_{\alpha\beta} = -\, a_{\beta} \bullet \partial_{\alpha} a_{3} \quad \Rightarrow \quad \dot{b}_{\alpha\beta} = -\, \dot{a}_{\beta} \bullet \partial_{\alpha} a_{3} - a_{\beta} \bullet \partial_{\alpha} \dot{a}_{3}$$

That is:

$$\dot{b}_{\alpha\beta} = -\left(v_{\beta\gamma} a^{\gamma} + v_{\alpha 3} a_{3}\right) \bullet \partial_{\alpha} a_{3} - a_{\beta} \bullet \partial_{\alpha}\left(-\, v_{\gamma 3} a^{\gamma}\right)$$

Yet:

$$a_\beta \bullet \partial_\alpha \left(v_{\gamma 3} a^\gamma\right) = a_\beta \bullet \nabla_\alpha \left(v_{\gamma 3} a^\gamma\right) = \partial_\alpha v_{\beta 3} + a_\beta \bullet \left(-\Gamma^\delta_{\alpha\gamma} v_{\delta 3} a^{\gamma\lambda} a_\lambda\right)$$

$$= \partial_\alpha v_{\beta 3} - \Gamma^\delta_{\alpha\beta} v_{\delta 3} = \nabla_\alpha v_{\beta 3} = \nabla_\alpha \left(\partial_\beta v^3 + b_{\beta\gamma} v^\gamma\right)$$

The variation of curvature rate tensor of components $\dot{k}_{\alpha\beta}$ is deduced:

$$\boxed{\dot{k}_{\alpha\beta} = \dot{b}_{\alpha\beta} = \nabla_\alpha \left(\partial_\beta v^3 + b_{\beta\gamma} v^\gamma\right) + b_\alpha{}^\gamma \left(\nabla_\beta v_\gamma - b_{\beta\gamma} v^3\right)} \tag{4.16}$$

As expected, formulas (4.15) and (4.16) resemble formulas (3.102) and (3.108) giving ε and κ in small displacements.

4.1.1.4 Spin Tensor

Formulas (4.8) make it possible to write:

$$\dot{a}_i = v_{ij} a^j \tag{4.17}$$

with:

$$v_{3\alpha} = -v_{\alpha 3}$$

$$v_{33} = 0$$

The tensor v can be decomposed into its symmetrical part η and its antisymmetric part ꞩ:

$$v_{ij} = \eta_{ij} + ꞩ_{ij}$$

where:

$$\begin{cases} \eta_{\alpha\beta} = \dfrac{1}{2}\left(v_{\alpha\beta} + v_{\beta\alpha}\right) = \dfrac{1}{2}\dot{a}_{\alpha\beta} = \dot{e}_{\alpha\beta} \\ \eta_{3\alpha} = \dfrac{1}{2}\left(v_{3\alpha} + v_{\alpha 3}\right) = 0 \\ \eta_{33} = 0 \end{cases} \tag{4.18}$$

The restriction of the tensor η to the tangent plane is therefore the strain rate tensor. On the other hand:

$$\begin{cases} ꞩ_{\alpha\beta} = \dfrac{1}{2}\left(v_{\alpha\beta} - v_{\beta\alpha}\right) = \nabla_\alpha v_\beta - \nabla_\beta v_\alpha = -ꞩ_{\beta\alpha} \\ ꞩ_{\alpha 3} = v_{\alpha 3} = -ꞩ_{3\alpha} \\ ꞩ_{33} = 0 \end{cases} \tag{4.19}$$

The antisymmetric tensor $\mathfrak{s}$ is called **spin tensor.**

The velocities calculated in §4.1.1.2 above can be expressed as functions of η and $\mathfrak{s}$. In addition, by (4.12), (4.13), and (4.6):

$$\dot{a}^\alpha = \overline{a^{\alpha\beta} a_\beta} = -2a^{\alpha\gamma} a^{\beta\delta} \eta_{\gamma\delta} a_\beta + a^{\alpha\beta} \left(\eta_{\beta k} + \mathfrak{s}_{\beta k}\right) a^k$$

either, since $\eta_{\beta 3} = 0$:

$$\dot{a}^\alpha = a^{\alpha\beta} \left(\mathfrak{s}_{\beta k} - \eta_{\beta k}\right) a^k \tag{4.20}$$

Similarly:

$$\dot{a}_3 = \left(\mathfrak{s}_{3k} - \eta_{3k}\right) a^k \tag{4.21}$$

The spin tensor is of some use in establishing Cosserat's oriented surface motion equations (see §4.5).

4.1.1.5 Acceleration

Let:

$$\gamma = \gamma^i a_i = \frac{d^2 \boldsymbol{m}}{dt^2} = \frac{d^2 \xi}{dt^2} \tag{4.22}$$

be the acceleration vector of material points of the middle surface. In Lagrange variables, γ is written:

$$\gamma = \frac{\partial^2 \xi}{\partial t^2}\left(X^\alpha, t\right)$$

The expression of γ in Euler variables is obtained by differentiating v with respect to t:

$$\frac{dv}{dt} = \frac{d}{dt}\left(v^\alpha a_\alpha + v^3 a_3\right) = \dot{v}^\alpha a_\alpha + v^\alpha v_{\alpha\beta}\, a^\beta + v^\alpha v_{\alpha 3}\, a_3 + \dot{v}^3 a_3 - v^3 v_{\alpha 3}\, a^\alpha$$

from which:

$$\begin{aligned}\gamma = \left[\dot{v}^\beta + v^\alpha \left(\nabla_\alpha v^\beta - b_\alpha{}^\beta v^3\right) - v^3 \left(a^{\beta\gamma} \partial_\gamma v^3 + b_\gamma{}^\beta v^\gamma\right)\right] a_\beta \\ + \left[\dot{v}^3 + v^\alpha \left(\partial_\alpha v^3 + b_{\alpha\beta} v^\beta\right)\right] a_3\end{aligned} \tag{4.23}$$

4.1.2 Conservation laws

4.1.2.1 General form

The laws of thermomechanics can often be expressed by the conservation of certain quantities. In the framework of shell theory, the conservation laws are expressed as functions of physical quantities defined on the middle surface and representative of the state of the shell. Let Φ be a tensor function representing a physical quantity (mass, energy, quantity of movement...) and ϕ its surface density defined at each instant on the configuration σ (ϕ is thus defined on an open of $\mathbb{R}^2$):

$$\Phi = \int_\sigma \phi \, d\sigma$$

In the case where the quantity Φ is conserved over time, its time derivative is zero at any moment:

$$\frac{d\Phi}{dt} = 0$$

As a first step, the quantities are expressed in Lagrange variables. Then let Ξ the surface dilatation between the reference configuration Σ and the current configuration σ and J the Jacobian of the transformation $x^\alpha\left(X^\beta\right)$; they are connected by relationships:

$$d\sigma = \left(1+\Xi\right) d\Sigma = \hat{a}^{1/2} dx^1 dx^2 = \hat{a}^{1/2} J \; dX^1 dX^2 = \left(\frac{\hat{a}}{\hat{A}}\right)^{1/2} J \; d\Sigma$$

Hence the expression of the surface dilatation:

$$\boxed{1+\Xi = \left(\frac{\hat{a}}{\hat{A}}\right)^{1/2} J} \tag{4.24}$$

Let ϕ_0 be the expression of ϕ on Σ:

$$\phi_0 = \phi\left(t = 0\right)$$

then:

$$\Phi = \int_\Sigma \phi_0 d\Sigma = \int_\sigma \phi \, d\sigma = \int_\Sigma \phi \left(1+\Xi\right) d\Sigma$$

hence the expression of the conservation law in Lagrange variables:

$$\boxed{\phi_0 = \phi\left(1+\Xi\right) = \phi \left(\frac{\hat{a}}{\hat{A}}\right)^{1/2} J} \tag{4.25}$$

This expression is simplified if the same coordinate system X^α is used in all configurations. In this case, $\boldsymbol{m}\left(X^\alpha, t\right)$ is the expression of the position of a material point at time t, $\mathcal{M}$ its position at the initial instant and:

$$\begin{cases} \mathcal{M} = \boldsymbol{m}\left(X^\alpha, 0\right) \\ J = 1 \end{cases}$$

Then:

$$\hat{A}^{1/2} \phi_0 = \hat{a}^{1/2} \phi \tag{4.26}$$

$$1+\Xi=\left(\frac{\hat{a}}{\hat{A}}\right)^{1/2} \tag{4.27}$$

To obtain the expression of this conservation law in Euler variables, the balance of Φ is written:

$$\frac{d}{dt}\int_\sigma \phi \, d\sigma = 0$$

To calculate the derivative, it is useful to introduce a reference position (which can be arbitrary: it is an intermediate calculation that does not intervene in the result). This allows the differentiation to be expressed on an integral with fixed limits:

$$\frac{d}{dt}\int_\sigma \phi \, d\sigma = \frac{d}{dt}\int_\Sigma \phi(1+\Xi)\, d\Sigma = \int_\Sigma \frac{d}{dt}\left[\phi(1+\Xi)\right] d\Sigma = \int_\sigma \frac{d}{dt}\left[\phi(1+\Xi)\right]\frac{d\sigma}{1+\Xi}$$

Hence the expression of the conservation law:

$$\dot{\phi}+\phi\frac{\dot{\Xi}}{1+\Xi}=0 \tag{4.28}$$

This expression is not very interesting, since Ξ depends on the arbitrary reference configuration. To present it in a form independent of Σ, Ξ is expressed according to the transformation, using a unique coordinate system X^α. Differentiating (4.27):

$$\dot{\Xi}=\frac{1}{2}\frac{\dot{\hat{a}}}{\left(\hat{a}\hat{A}\right)^{1/2}}$$

Then:

$$\frac{\dot{\Xi}}{1+\Xi}=\frac{1}{2}\frac{\dot{\hat{a}}}{\hat{a}}$$

On the other hand, (4.12) allows it to be written as:

$$\frac{1}{2}\frac{\dot{\hat{a}}}{\hat{a}}=v_\alpha{}^\alpha$$

Hence the conservation law in Euler variables:

$$\boxed{\dot{\phi}+\phi\left(\nabla_\alpha v^\alpha - b_\alpha{}^\alpha v^3\right)=0} \tag{4.29}$$

4.1.2.2 Mass continuity equation

The above results are applied to mass conservation. If ρ denotes the mass per unit area:

- in Lagrange variables (using the coordinates of the reference position):

$$\hat{A}^{1/2}\rho_0=\hat{a}^{1/2}\rho \tag{4.30}$$

- in Euler variables:

$$\dot{\rho}+\rho\left(\nabla_\alpha v^\alpha - b_\alpha{}^\alpha v^3\right)=0 \tag{4.31}$$

4.1.3 Virtual Movements

4.1.3.1 Characterisation of a virtual movement

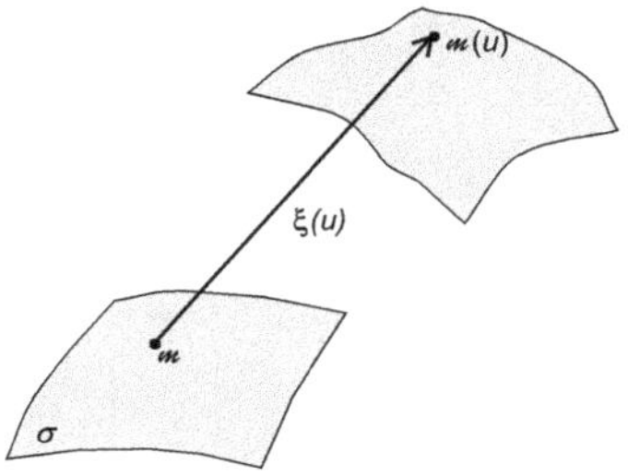

Figure 4.1

The shell is assimilated to its middle surface; the strain energy is then expressed as a function of deformation variables defined on the middle surface. It has been seen previously that for the surface alone (without thickness), the deformation is entirely defined by the surface deformation and its variation of curvature. It is a question of characterising its virtual movements, in order to evaluate the virtual power of deformation and to apply the Principle of the Virtual Powers[3].

A virtual movement of the surface σ in its current configuration is defined as a movement $\mathcal{m}\left(x^\alpha,u\right)$ from the current position for which $u=0$ (Figure 4.1). u is an evolution parameter also called **virtual time.**

The virtual movements considered here are compatible, that is, they respect the continuity of the surface and the vector normal to σ remains normal to the surface during the virtual movement. This last hypothesis is fundamental and brings a restriction that can be lifted to enrich the model (see §4.5).

The displacement between the current position taken for reference and the virtual position, for which $u \neq 0$, is:

$$\xi(u)=\overrightarrow{\mathcal{m}(0)\mathcal{m}(u)}$$

By definition, the virtual velocity is the derivative of $\xi(u)$ with respect to u at $u=0$:

$$\overset{*}{\xi}=\lim_{u\to 0}\left[\frac{1}{u}\xi(u)\right] \tag{4.32}$$

The virtual velocity field $\overset{*}{\xi}$ is therefore a vector field defined in Euler variables on the current configuration. $\overset{*}{\xi}$ breaks down on the local basis in:

$$\boxed{\overset{*}{\xi}=\overset{*}{\xi}{}^\alpha\, a_\alpha+\overset{*}{\zeta} a_3} \tag{4.33}$$

The real velocity is a particular value of virtual velocity. During the virtual transformation, the different interesting geometric entities vary virtually; their virtual velocity of variation is given below:

[3] Although "Principle of Virtual Work" is more often used, "Principle of the Virtual Powers" is used here more consistently. However, there is no practical difference.

- Vectors of the basis:

Differentiating with respect to x^α, from relation:

$$\boldsymbol{m}(u) = \boldsymbol{\mathcal{M}}(0) + \xi(u)$$

It becomes:

$$\partial_\alpha \boldsymbol{m}(u) = a_\alpha(u) = a_\alpha(0) + D_\alpha \xi(u)$$

from which:

$$\overset{*}{a}_\alpha = \lim_{u \to 0} \frac{1}{u}\left[a_\alpha(u) - a_\alpha(0)\right] = \lim_{u \to 0} \frac{1}{u}\left[D_\alpha \xi(u)\right]$$

That is to say:

$$\boxed{\overset{*}{a}_\alpha = D_\alpha \overset{*}{\xi} = \left(\nabla_\alpha \overset{*}{\xi}{}^\beta - b_\alpha{}^\beta \overset{*}{\zeta}\right) a_\beta + \left(\partial_\alpha \overset{*}{\zeta} + b_{\alpha\beta} \overset{*}{\xi}{}^\beta\right) a_3 = \overset{*}{\mathcal{A}}_\alpha{}^\beta a_\beta + \overset{*}{\mathcal{B}}_\alpha \, a_3} \tag{4.34}$$

With a relation similar to relation (4.6), which is noted, by analogy with expressions (3.101):

$$\begin{aligned} \overset{*}{\mathcal{A}}_\alpha{}^\beta &= \nabla_\alpha \overset{*}{\xi}{}^\beta - \overset{*}{\zeta} b_\alpha{}^\beta \\ \overset{*}{\mathcal{B}}_\alpha &= b_\alpha{}^\beta \overset{*}{\xi}{}^\beta + \partial_\alpha \overset{*}{\zeta} \end{aligned} \tag{4.35}$$

Note: $\overset{*}{\mathcal{A}}$ and $\overset{*}{\mathcal{B}}$ are notations and these quantities are not obtained by virtualisation of $\mathcal{A}$ and $\mathcal{B}$.

- Dual basis:

Differentiating with respect to u the relation $\left\langle a_\alpha, a^\beta \right\rangle = \delta_\alpha^\beta$, it becomes:

$$< a_\alpha, \overset{*}{a}{}^\beta > = - < \overset{*}{a}_\alpha, a^\beta > = - < D_\alpha \overset{*}{\xi}, a^\beta >$$

from which:

$$\boxed{\overset{*}{a}{}^\beta = - a^{\alpha\beta} D_\alpha \overset{*}{\xi} = - a^{\alpha\beta}\left(\nabla_\alpha \overset{*}{\xi}{}^\gamma - b_\alpha{}^\gamma \overset{*}{\zeta}\right) a_\gamma} \tag{4.36}$$

The virtual velocity of the dual form is not the contravariant representative of the virtual velocity of the corresponding vector of the basis

Relation: $\overset{*}{a}{}^\beta = a^{\alpha\beta} \overset{*}{a}_\alpha$ is not verified.

This non-equality is generalised to all tensors. For example:

$$\overset{*}{T}_\alpha{}^\beta \neq a_{\alpha\gamma}\overset{*}{T}{}^{\alpha\beta}$$

This is also the case for derivatives with respect to real-time (velocities).

- Normal vector:

Differentiating with respect to u the relation $a_\alpha \cdot a_3 = 0$:

$$a_\alpha \cdot \overset{*}{a}_3 = -\overset{*}{a}_\alpha \cdot a_3 = -D_\alpha \overset{*}{\xi} \cdot a_3$$

either, by applying relations (4.34) and (4.35):

$$a_\alpha \cdot \overset{*}{a}_3 = -\left(\partial_\alpha \overset{*}{\zeta} + b_{\alpha\beta}\overset{*}{\xi}{}^\beta\right)$$

Hence, noting that $\overset{*}{a}_3$ is orthogonal to a_3:

$$\boxed{\overset{*}{a}_3 = -\left(\partial_\alpha \overset{*}{\zeta} + b_{\alpha\beta}\overset{*}{\xi}{}^\beta\right)a^\alpha} \tag{4.37}$$

The relation (4.37) can also be written as:

$$\overset{*}{a}_3 = -\overset{*}{\mathcal{B}} \tag{4.38}$$

and is analogous to (4.8). $\overset{*}{\mathcal{A}}_{\alpha\beta}$ and $\overset{*}{\mathcal{B}}_\alpha$ play the virtual velocity the same role as $v_{\alpha\beta}$ and $v_{\alpha 3}$ for the real velocity (see (4.6)).

4.1.3.2 Virtual rotation velocity

During the virtual movement, $a_3(u)$ remains of constant length and normal to the surface (rigid body displacement). Therefore, if $\mathcal{p}$ denotes the end extremity of the vector a_3:

$$\overset{*}{\mathcal{p}} = \overset{*}{\mathcal{m}} + \overset{*}{\omega} \times a_3$$

where $\overset{*}{\omega}$ designates the virtual rotation velocity at $\mathcal{m}$, to calculate the virtual velocity of any indeformable vector rigidly connected to a_3.

But by definition:

$$\mathcal{p} = \mathcal{m} + a_3$$

which leads, by virtualisation, to the relationship:

$$\overset{*}{a}_3 = \overset{*}{\omega} \times a_3$$

a_3 and $\overset{*}{a}_3$ being orthogonal, it comes:

$$\overset{*}{\omega} = a_3 \times \overset{*}{a}_3 = \overset{*}{\mathcal{B}} \times a_3 \tag{4.39}$$

This relationship can also be obtained from (3.105).

4.1.3.3 *Virtual rates of deformation and curvature variation*

As previously seen, variations in metric and curvature are sufficient to describe the deformation of the middle surface. The associated virtual variations are the subject of the present paragraph. In the case where the thickness is considered variable during the transformation, it is necessary to introduce other deformation variables; a possible approach is presented in §4.5. The virtual transformation of the surface is expressed in Lagrange variables from the current position, from where (relations (3.88) and (3.89)):

$$\varepsilon(u) = \frac{1}{2}\left[a(u) - a(0)\right]$$

$$\kappa(u) = b(u) - b(0)$$

The virtual velocity of deformation and of variation of curvature are defined, in Euler coordinates on the current surface, by:

$$\overset{*}{e} = \lim_{u \to 0} \frac{1}{u}\left[\varepsilon(u) - \varepsilon(0)\right] = \lim_{u \to 0} \frac{1}{u}\varepsilon(u)$$

$$\overset{*}{k} = \lim_{u \to 0} \frac{1}{u}\left[\kappa(u) - \kappa(0)\right] = \lim_{u \to 0} \frac{1}{u}\kappa(u)$$

It is clear that:

$$\overset{*}{e} = \frac{1}{2}\overset{*}{a}$$

That is:

$$\overset{*}{e}_{\alpha\beta} = \frac{1}{2}\overset{*}{a}_{\alpha\beta} = \frac{1}{2}\overset{*}{\overline{a_\alpha \cdot a_\beta}} = \frac{1}{2}\left[\overset{*}{a}_\alpha \cdot a_\beta + a_\alpha \cdot \overset{*}{a}_\beta\right]$$

$$= \frac{1}{2}\left[D_\alpha \overset{*}{\xi} \cdot a_\beta + a_\alpha \cdot D_\beta \overset{*}{\xi}\right] = \frac{1}{2}\left(\overset{*}{\mathcal{A}}_{\alpha\beta} + \overset{*}{\mathcal{A}}_{\alpha\beta}\right)$$

from which, using (4.35):

$$\boxed{\overset{*}{e}_{\alpha\beta} = \frac{1}{2}\overset{*}{a}_{\alpha\beta} = \frac{1}{2}\left[\nabla_\alpha \overset{*}{\xi}_\beta + \nabla_\beta \overset{*}{\xi}_\alpha - 2b_{\alpha\beta}\overset{*}{\zeta}\right]} \tag{4.40}$$

Similarly:

$$\overset{*}{k} = \overset{*}{b}$$

Then, $\overset{*}{b}_{\alpha\beta}$ denoting the covariant components of $\overset{*}{b}$:

$$\overset{*}{b}_{\alpha\beta} = \lim_{u\to 0}\frac{1}{u}\left[-a_\beta \bullet \partial_\alpha a_3\right] = -a_\beta \bullet \partial_\alpha \overset{*}{a}_3 - \overset{*}{a}_\beta \bullet \partial_\alpha a_3$$

$$\overset{*}{b}_{\alpha\beta} = \nabla_\alpha \overset{*}{\mathcal{B}}_\beta + b_\alpha{}^\gamma \overset{*}{\mathcal{A}}_{\beta\gamma} \tag{4.41}$$

Either, by (4.35):

$$\boxed{\overset{*}{k}_{\alpha\beta} = \overset{*}{b}_{\alpha\beta} = \nabla_\alpha\left(\partial_\beta \overset{*}{\zeta} + b_{\beta\gamma}\overset{*}{\xi}{}^\gamma\right) + b_\alpha{}^\gamma\left(\nabla_\beta \overset{*}{\xi}_\gamma - b_{\beta\gamma}\overset{*}{\zeta}\right)} \tag{4.42}$$

Formulas (4.40) and (4.42) resemble formulas (3.102) and (3.108) obtained for ε and κ in the case of small displacements.

This property (already encountered in three-dimensional theory) is general. Let $\mathcal{P}(x)$ a polynomial of the components of x, which the linear part is $\mathcal{P}^l(x)$. The virtual velocity of $\mathcal{P}$ is:

$$\overset{*}{\mathcal{P}} = \lim_{u\to 0}\frac{1}{u}\left[\mathcal{P}\left(\xi(u)\right)\right] = \partial_i\mathcal{P}(0)\left(\frac{d\xi^i}{du}\right)_{u=0} = \partial_i\mathcal{P}(0)\,\overset{*}{\xi}{}^i$$

It can be established without difficulty that:

$$\partial_i\mathcal{P}(0) = \partial_i\mathcal{P}^l(0)$$

and then:

$$\overset{*}{\mathcal{P}} = \overset{*}{\mathcal{P}}{}^l \tag{4.43}$$

Formulas (4.40) and (4.42) are written in the intrinsic form:

$$\boxed{\begin{aligned} \overset{*}{e} &= \frac{1}{2}\left[D\overset{*}{\xi} + D^T\overset{*}{\xi}\right] \\ \overset{*}{k} &= -D\overset{*}{a}_3 + b\bullet D\overset{*}{\xi} \end{aligned}} \tag{4.44}$$

For future use, the formulas above are usefully supplemented by the calculation of $\overset{*}{a}{}^{\alpha\beta}$ and $\overset{*}{b}{}^{\alpha\beta}$:

Differentiating relation $a^{\alpha\beta}\,a_{\beta\gamma} = \delta^\alpha_\gamma$, it comes:

$$\overset{*}{a}{}^{\alpha\beta}\,a_{\beta\gamma} = -a^{\alpha\beta}\,\overset{*}{a}_{\beta\gamma}$$

either:

$$\overset{*}{a}{}^{\alpha\beta} = -a^{\alpha\gamma}a^{\beta\delta}\,\overset{*}{a}_{\gamma\delta} \tag{4.45}$$

Moreover, by differentiating the relation:

$$b^{\alpha\beta} = a^{\alpha\lambda} a^{\beta\mu} b_{\lambda\mu}$$

it becomes:

$$\overset{*}{b}{}^{\alpha\beta} = \overset{*}{a}{}^{\alpha\lambda} a^{\beta\mu} b_{\lambda\mu} + a^{\alpha\lambda} \overset{*}{a}{}^{\beta\mu} b_{\lambda\mu} + a^{\alpha\lambda} a^{\beta\mu} \overset{*}{b}_{\lambda\mu}$$

which leads, in view of (4.45), to:

$$\overset{*}{b}{}^{\alpha\beta} = -\left(a^{\alpha\gamma} a^{\lambda\delta} a^{\beta\mu} + a^{\alpha\lambda} a^{\beta\gamma} a^{\mu\delta}\right) b_{\lambda\mu} \overset{*}{a}_{\gamma\delta} + a^{\alpha\lambda} a^{\beta\mu} \overset{*}{b}_{\lambda\mu} \tag{4.46}$$

where it is clear that:

$$\overset{*}{b}{}^{\alpha\beta} \neq a^{\alpha\lambda} a^{\beta\mu} \overset{*}{b}_{\lambda\mu}$$

(4.45) and (4.46) confirm that the derivative of a contravariant component is not a component of the derivative.

4.1.3.4 Virtual transformation of a curve drawn on a surface

Boundary conditions are obtained from expressions written on the edge of the shell. It is useful, to obtain such expressions, to study the virtual variations of the vectors of the basis along such an edge. Let $(\mathcal{C})$ a curve drawn on σ, which is intended to later represent the edge of σ, s its curvilinear abscissa and $(\mathbf{t}, \nu, a_3)$ its Darboux coordinate system (see §3.4.1.2):

- $\mathbf{t}$ is tangent to $\mathcal{C}$. Let $\mathbf{t} = t^\alpha a_\alpha$;
- $\nu = a_3 \times \mathbf{t}$ belongs to T_{m}. This is the vector in the tangent plane to σ normal to $\mathbf{t}$.

Differentiating relation: $m(u) = \mathcal{M}(0) + \xi(u)$ at a point m belonging to the curve and by application of relation (4.34):

$$\overset{*}{\mathbf{t}} = \frac{\partial \overset{*}{\xi}}{\partial s} \tag{4.47}$$

$\overset{*}{\nu}$ is obtained by its three components in the Darboux coordinate system:

$$\begin{cases} \nu \cdot \overset{*}{\nu} = 0 \\ \overset{*}{\nu}_t = \mathbf{t} \cdot \overset{*}{\nu} = -\nu \cdot \overset{*}{\mathbf{t}} = -\nu \cdot \dfrac{\partial \overset{*}{\xi}}{\partial s} \\ \overset{*}{\nu}_3 = a_3 \cdot \overset{*}{\nu} = -\nu \cdot \overset{*}{a}_3 = \overset{*}{\mathcal{B}}_\nu = \partial_\nu \overset{*}{\zeta} + b_{\nu\beta} \overset{*}{\xi}{}^\beta \end{cases} \tag{4.48}$$

These relationships show in particular that $\overset{*}{\mathbf{t}}$ and $\overset{*}{\nu}_t$ depend only on the values that $\overset{*}{\xi}$ takes on the curve.

Similarly, $\overset{*}{a}_3$ can be characterised by its components in the Darboux coordinate system, which are easily obtained by noting that, for any vector X of T_{m}:

$$X \cdot \overset{*}{a}_3 = X^\alpha a_\alpha \cdot \overset{*}{a}_3 = -X^\alpha\, a_3 \cdot D_\alpha \overset{*}{\xi} = -a_3 \cdot D_X \overset{*}{\xi} \tag{4.49}$$

Applying (4.49) to $\mathbf{t}$ and ν:

$$\begin{cases} \nu \cdot \overset{*}{a}_3 = -\overset{*}{\mathcal{B}}_\nu = -a_3 \cdot \overset{*}{\nu} = -a_3 \cdot \dfrac{\partial \overset{*}{\xi}}{\partial \nu} \\ \mathbf{t} \cdot \overset{*}{a}_3 = -\overset{*}{\mathcal{B}}_t = -a_3 \cdot \overset{*}{\mathbf{t}} = -a_3 \cdot \dfrac{\partial \overset{*}{\xi}}{\partial s} \end{cases} \tag{4.50}$$

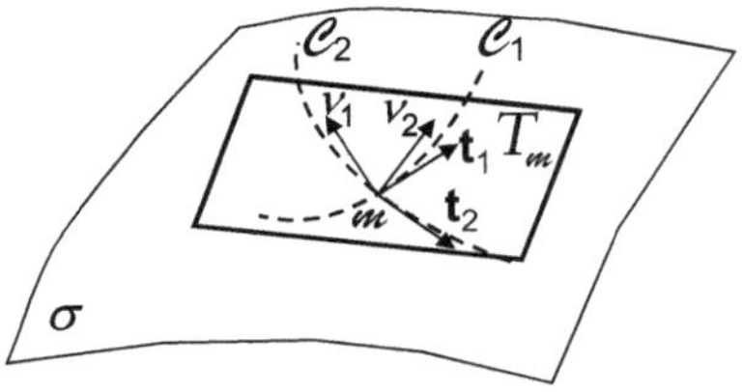

Figure 4.2

The second relation shows that the component of the $\overset{*}{a}_3$ tangent to the curve depends only on the values of $\overset{*}{\xi}$ taken on the curve. This circumstance is related to the fact that $\overset{*}{a}_3$ is compatible, that is to say, that the **normal to the surface remains normal during the virtual movement.** On the other hand, the component on ν depends in general on the values taken by $\overset{*}{\xi}$ in the surface.

The relations established previously assume that the curve is regular: if at a singular point m exists a discontinuity of the tangent and if the surface is regular, the vectors $\mathbf{t}$ and ν undergo a discontinuity at the passage of m, separating locally $\partial\sigma$ into two parts [1] and [2] (Figure 4.2).

If, for example, the two portions of the curve are orthogonal, the tangent and normal vectors are linked by the relations $\mathbf{t}_2 = \nu_1$, $\nu_2 = -\mathbf{t}_1$ (the sign depending on the orientation of the curves). $\overset{*}{a}_3$ being continuous, it is clear by (4.50) that, in the vicinity of the singular point, $\nu \cdot \overset{*}{a}_3$ on a branch depends on the values of $\overset{*}{\xi}$ on the other branch. In general, if $[\bullet]$ designates the vector discontinuity (the jump in m), (4.49) allows it to be written as, assuming that the field $\overset{*}{\xi}$ is continuously differentiable:

$$[\mathbf{t}] \cdot \overset{*}{a}_3 = -a_3 \cdot \left(D_{\mathbf{t}_2} \overset{*}{\xi} - D_{\mathbf{t}_1} \overset{*}{\xi} \right) = -a_3 \cdot D_{[\mathbf{t}]} \overset{*}{\xi} \tag{4.51}$$

The discontinuity of ν is equal to the discontinuity of $\mathbf{t}$, subject to an adequate orientation, the two fields being orthogonal in the tangent plane:

$$[\nu] \cdot \overset{*}{a}_3 = [\mathbf{t}] \cdot \overset{*}{a}_3 = -a_3 \cdot D_{[\mathbf{t}]} \overset{*}{\xi} \tag{4.52}$$

4.2 EQUATIONS OF MOTION

4.2.1 Application of general momentum theorems

4.2.1.1 Introduction

The equations of motion are obtained, in this chapter, by two methods:

- the application of the fundamental law (general momentum theorems);
- the application of the principle of virtual powers (PVP).

In both cases, the principle of mechanics is applied to the middle surface, whose behaviour is supposed to be representative of the behaviour of the shell as a whole. If the first method makes it possible to quickly obtain the local equations of motion, it presents some difficulties in the interpretation of the boundary conditions, which are more easily solved by the second method. In addition, the energetic method makes it possible to better control the approximations made in the different theories of shells.

The internal actions and generalised stresses introduced in the two methods differ, as well as the equilibrium equations and the boundary conditions obtained. The different quantities and relationships are compared in §4.4.

4.2.1.2 Nature of actions applied to a shell part

Let a shell represented by a surface σ, delimited by a closed curve $\partial\sigma$. It is a question of expressing the equilibrium of any part d of σ, d being delimited by a closed curve ∂d drawn on σ. d is subject to actions outside σ, to the liaison actions exercised by the complement $d^{\perp}$ of d in σ along ∂d and to the inertia actions.

4.2.1.2.1 External actions

In common cases, the external actions applied to the shell are:

- actions at distance, mainly due to gravitation;
- contact actions, for example due to loads directly applied on the shell or to its periphery or to pressures or thrusts due to liquids or gases, soils, etc., which are generally applied to one side of the shell;
- dimensional variations due to material shrinkage, temperature or imposed displacements;
- concentrated or distributed reactions associated with the liaisons to which the shell is subjected.

Imposed deformations and displacements are provisionally set aside, these actions are either volumic actions applied within the solid (like gravity), or surface actions applied on both sides of the shell, or actions applied at the edge of the shell. Their resultant is expressed on the middle surface and is reduced in practice to:

- surfacic forces:

$$p = p^{\alpha} a_{\alpha} + p^{3} a_{3} = p^{i} a_{i} \tag{4.53}$$

and surfacic moments:

$$c = c^i a_i \tag{4.54}$$

applied to the middle surface;

- linear forces:

$$q = q^i a_i \tag{4.55}$$

and linear moments:

$$m = m^i a_i \tag{4.56}$$

applied at the edge $\partial\sigma$ of σ.

The way to obtain these resulting actions from their local distribution is developed in §5.5, where the three-dimensional aspect is examined. It is also possible to take into account all other cases encountered in practice (concentrated loads, distributed loads on a curve…) considering that p, c, q, and m are distributions; nevertheless, to simplify the presentation, this possibility is not developed here.

4.2.1.2.2 Inertial actions

If it is a question of studying a movement and not of seeking a new state of equilibrium, it is necessary to take into account inertial actions. The middle surface is considered without thickness, only the surface density of mass ρ participates in the inertial actions, of which surface density is:

$$-\rho\frac{d^2\xi}{dt^2} = -\rho\left(\gamma^\alpha a_\alpha + \gamma^3 a_3\right) \tag{4.57}$$

where the acceleration γ is given by (4.23). This expression can be completed to take into account the influence of thickness. This aspect is addressed in §4.2.2.3, in the context of the application of the PVP, but is not considered here as it does not present any particular difficulty. The surface density of the inertial actions defined by (4.57) is added to the density of forces p defined by (4.53), so it is not necessary to make it appear in the rest of the reasoning, without this detracts from its generality.

4.2.1.2.3 Actions exercised by $d^\perp$ on d

By analogy with the Cauchy[4] Postulate of the three-dimensional theory, the action exerted by $d^\perp$ on d is assumed to be reduced to contact actions along ∂d, of the same nature (forces N and moments M per unit length of edge ∂d) as the forces q and the moments m exerted on the edge $\partial\sigma$ (Figure 4.3).

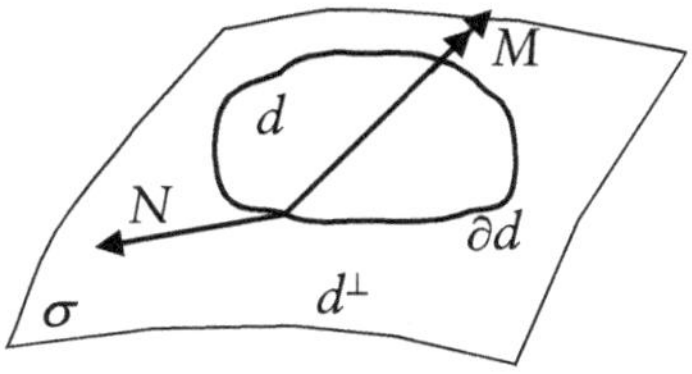

Figure 4.3

[4] Augustin-Louis Cauchy, French mathematician, engineer and physicist (1789–1857).

Thus, by postulate, these actions N and M depend only on the point considered, the time t and the orientation of the edge, that is to say, the vector ν normal to ∂d in the tangent plane. It is logical that N and M do not depend on a_3, which is fixed for a given point, and does not intervene in the local orientation of ∂d. It is also assumed that N and M vary continuously as a function of $\mathcal{m}$ and ν.

Physically, it is accepted by this postulate that the actions between d and $d^\perp$ are local actions (depending only on the points of the edge) and not on actions at a distance. There is no force of attraction or repulsion (for example magnetic) between the two portions of the shell. This hypothesis is satisfactory in the usual cases.

4.2.1.3 Expression of internal actions

It remains to demonstrate, as in the three-dimensional case, that the forces N and moments of M depend linearly on the normal vector ν.

Let two planes $(\mathcal{P}_1)$ and $(\mathcal{P}_2)$ passing through $\mathcal{m}$ and containing a_3, be orthogonal to each other (Figure 4.4). Their traces at σ are plane curves (straight sections), respectively $(\mathcal{C}_1)$ and $(\mathcal{C}_2)$.

Let the plane $(\mathcal{P})$ be parallel to a_3, at the distance h from a_3 and which orientation is defined by the angle α of its normal ν' with respect to the plane $(\mathcal{P}_1)$.

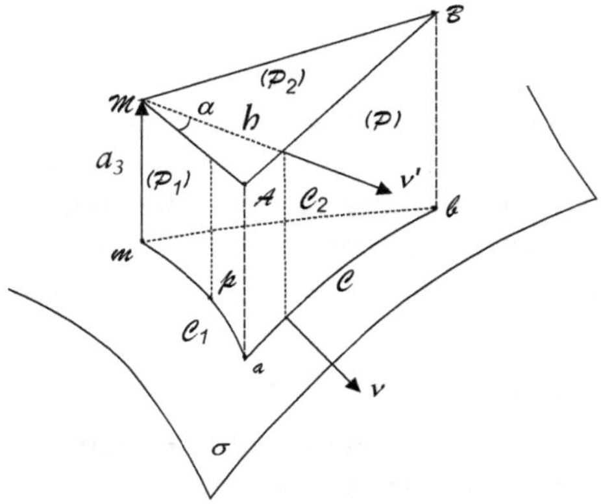

Figure 4.4

The trace of $(\mathcal{P})$ on σ is a cross-section, which cuts $(\mathcal{C}_1)$ and $(\mathcal{C}_2)$ in $\mathcal{a}$ and $\mathcal{b}$ respectively, if α is not a multiple of $\frac{\pi}{2}$. For a given angle α, $\mathcal{a}$ and $\mathcal{b}$ depend only on h and:

$$\underbrace{\lim}_{h\to 0}\mathcal{a}(h) = \underbrace{\lim}_{h\to 0}\mathcal{b}(h) = \mathcal{m}$$

The area of the curvilinear triangle $\mathcal{mab}$ considered as domain d is of the order of h^2. The triangle $\mathcal{mab}$ is the projection along a_3, on the surface, of the triangle $\mathcal{MAB}$, the intersection of the trihedron formed by the planes $(\mathcal{P}_1)$, $(\mathcal{P}_2)$ and $(\mathcal{P})$, and of a plane perpendicular to a_3. In the triangle $\mathcal{MAB}$:

$$\|\mathcal{MA}\| = \frac{h}{\cos\alpha}\ ;\ \|\mathcal{MB}\| = \frac{h}{\sin\alpha}\ ;\ \|\mathcal{AB}\| = \frac{h}{\cos\alpha\sin\alpha}$$

Also, the curvilinear abscissa along each of the sides of this triangle can be expressed as a function of a single parameter ξ, which varies from 0 to h. In addition, the curvilinear abscissa along each side of $\mathcal{MAB}$ can be used as a parameter for the corresponding side of the triangle $\mathcal{mab}$. For example, along $\mathcal{ma}$:

$$\frac{ds_1}{d\xi} = \frac{ds_1}{du_1}\frac{du_1}{d\xi} = \frac{1}{\cos\alpha}\frac{ds_1}{du_1}$$

where u_1 is the curvilinear abscissa along $\mathcal{MAB}$, s_1 the curvilinear abscissa along $\boldsymbol{mab}$. Let $\mathcal{P}$ be the current point on $\mathcal{MAB}$ and $\boldsymbol{p}$ the corresponding point on $\boldsymbol{mab}$, projection of $\mathcal{P}$ on a_3, normal to the surface. Taking, without limiting the scope of reasoning:

$$\boldsymbol{m}\mathcal{M} = a_3$$

it becomes:

$$\boldsymbol{mp} = \boldsymbol{m}\mathcal{M} + \mathcal{MP} + \boldsymbol{p}\mathcal{P} = \mathcal{MP} + (1-\zeta)\, a_3 \tag{4.58}$$

where ζ locally defines the position of the surface.

$\dfrac{d(\boldsymbol{mp})}{ds_1} = \mathbf{t}_1(\boldsymbol{p})$ is the vector tangent to $(\mathcal{C}_1)$ at the current point, and:

$$\mathcal{MP} = u_1\, \mathbf{t}_1(\boldsymbol{m})$$

Differentiating expression (4.58) with respect to u_1:

$$\frac{d(\boldsymbol{mp})}{du_1} = \frac{d(\mathcal{MP})}{du_1} - \frac{d\zeta}{du_1} a_3$$

either, for $u_1 = 0$:

$$\frac{ds_1}{du_1}(0)\, \mathbf{t}_1(\boldsymbol{m}) = \mathbf{t}_1(\boldsymbol{m}) - \frac{d\zeta}{du_1}(0)\, a_3$$

which leads to relationships:

$$\frac{ds_1}{du_1}(0) = 1 \qquad \frac{d\zeta}{du_1}(\boldsymbol{m}) = 0 \tag{4.59}$$

Similar results are obtained along the curves $(\mathcal{C}_2)$ and $(\mathcal{C})$ at the points of origin of the curvilinear abscissae.

Contact actions that are applied to d are written as:

- $(-N_2, -M_2)$ on $\boldsymbol{ma}$;
- $(-N_1, -M_1)$ on $\boldsymbol{mb}$;
- $(N(\nu, h), M(\nu, h))$ on $\boldsymbol{ab}$;

and the resulting surface screw of all the actions applied inside the triangle $\boldsymbol{mab}$ is (p, c) including the inertial actions. The balance of the forces applied to the triangle $\boldsymbol{mab}$ is written as:

$$\int_{\boldsymbol{m}}^{\boldsymbol{a}} -N_2\, ds_1 + \int_{\boldsymbol{m}}^{\boldsymbol{b}} -N_1\, ds_2 + \int_{\boldsymbol{a}}^{\boldsymbol{b}} N\, ds + \iint_d p\, d\sigma = 0$$

and can be rewritten as:

$$\int_0^h -\frac{N_2}{\cos\alpha}\frac{ds_1}{du_1} d\xi + \int_0^h -\frac{N_1}{\sin\alpha}\frac{ds_2}{du_2} d\xi + \int_0^h \frac{N}{\sin\alpha\cos\alpha}\frac{ds}{du} d\xi + \iint_d p\, d\sigma = 0$$

By dividing this relation by h and tending h towards 0, the last term, which is of the order of h^2, vanishes and it remains, given (4.59):

$$-\frac{N_2}{\cos\alpha}-\frac{N_1}{\sin\alpha}+\frac{N}{\sin\alpha\cos\alpha}=0$$

that is:

$$N = N_2\sin\alpha + N_1\cos\alpha \tag{4.60}$$

Similarly, by writing the balance of moments in relation to the point $\mathscr{m}$:

$$\int_{\mathscr{m}}^{a} -(M_2+\mathscr{mp}\times N_2)\,ds_1+\int_{\mathscr{m}}^{b} -(M_1+\mathscr{mp}\times N_1)\,ds_2$$
$$+\int_{a}^{b}(M+\mathscr{mp}\times N)\,ds+\iint_d (c+\mathscr{mp}\times p)\,d\sigma=0$$

By following a method similar to the case of the resultant and noting that all the points $\mathscr{p}$ tend towards $\mathscr{m}$, it becomes:

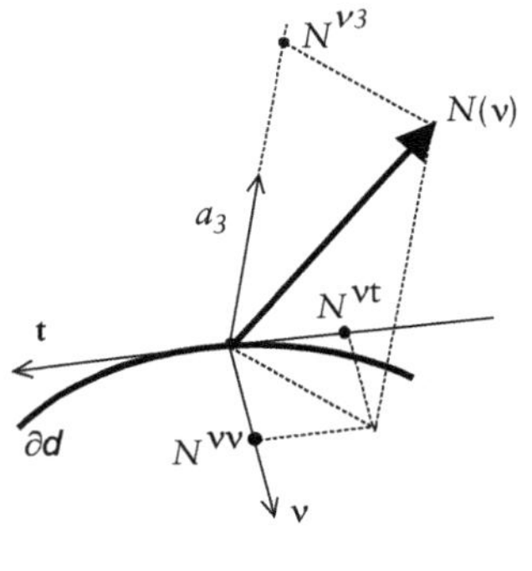

Figure 4.5

$$M = M_2\sin\alpha + M_1\cos\alpha \tag{4.61}$$

(4.60) and (4.61) show the linearity property of N and M as functions of v. Finally:

$$N(v)=N^i a_i = N^{\alpha i} v_\alpha a_i \tag{4.62}$$

Figure 4.5 shows the decomposition of $N(v)$ on the vectors $(\mathbf{t}, v, a_3)$ alongside ∂d. The decomposition of $M(v)$ is similar:

$$M(v)=M^i a_i = M^{\alpha i} v_\alpha a_i \tag{4.63}$$

Tensors N and M of components $N^{\alpha i}$ and $M^{\alpha i}$ represent applications of $T_{\mathscr{m}}$ in $\mathbb{R}^3$. They do not benefit, in principle, from any property of symmetry.

Linearity results in particular that $N(-v)=-N(v)$ and $M(-v)=-M(v)$ (action–reaction).

4.2.1.4 Expression of local equilibrium

In this paragraph, domain d is assumed to be totally included in the middle surface σ, with no point of ∂d belonging to $\partial\sigma$. Under these conditions, the equilibrium of d does not involve the reactions on the edge ∂d. The equilibrium of the shell portion d is then expressed by the following six equations (the screw of the actions external to d is zero):

$$\begin{cases} \int_{\partial d} N\, ds + \int_d p\, d\sigma = 0 \\ \int_{\partial d} \left(M + \mathcal{OP} \times N\right) ds + \int_d \left(c + \mathcal{OP} \times p\right) d\sigma = 0 \end{cases} \tag{4.64}$$

- FIRST EQUATION:

Given (4.62), it is rewritten, in terms of components:

$$\int_{\partial d} N^{\alpha i} \nu_\alpha a_i\, ds + \int_d p^i a_i\, d\sigma = 0$$

either, by applying the Stokes theorem:

$$\int_d \left[D_\alpha \left(N^{\alpha i} a_i\right) + p^i a_i \right] d\sigma = 0$$

Since the domain $d \subset \sigma$ is arbitrary, the balance of forces can be deduced:

$$\boxed{D_\alpha \left(N^{\alpha i} a_i\right) + p^i a_i = 0} \tag{4.65}$$

by developing the differentiation:

$$\boxed{\begin{array}{l} \nabla_\alpha N^{\alpha\gamma} - b_\alpha{}^\gamma N^\alpha \quad + p^\gamma = 0 \\ N^{\alpha\beta} b_{\alpha\beta} + \nabla_\alpha N^{\alpha 3} + p^3 = 0 \end{array}} \tag{4.66}$$

These three equations respectively express the balance of forces in the tangent plane and along the normal a_3. Inertial actions can be easily restored in these equations. The surface tensor of components $N^{\alpha\beta}$ is called the **tensor of in-plane** (or **membrane**) forces and $N^{\alpha 3}$ is called **out-of-plane** (or **transverse**) **shear force** in the direction α.

- SECOND EQUATION:

Noting, in order to simplify the second equation, that:

$$\int_{\partial d} \mathcal{OP} \times N\, ds = \int_{\partial d} \mathcal{OP} \times \left(N^{\alpha i} \nu_\alpha a_i\right) ds = \int_d D_\alpha \left[\mathcal{OP} \times \left(N^{\alpha i} a_i\right)\right] d\sigma$$

$$= \int_d D_\alpha \mathcal{OP} \times \left(N^{\alpha i} a_i\right) d\sigma + \int_d \mathcal{OP} \times D_\alpha \left(N^{\alpha i} a_i\right) d\sigma$$

the second equation of (4.64) is reduced, given (4.63) and (4.65), to:

$$\int_d \left[D_\alpha \left(M^{\alpha i} a_i\right) + D_\alpha \mathcal{OP} \times \left(N^{\alpha i} a_i\right) + c^i a_i \right] d\sigma = 0$$

d being any and noting that $D_\alpha \mathcal{OP} = a_\alpha$, it becomes:

$$\boxed{D_\alpha \left(M^{\alpha i} a_i\right) + a_\alpha \times \left(N^{\alpha i} a_i\right) + c^i a_i = 0} \tag{4.67}$$

then, after the development of the derivate and the cross-product:

$$\boxed{\begin{array}{l} \nabla_\alpha M^{\alpha\beta} - b_\alpha{}^\beta M^{\alpha 3} - N^{\alpha 3} e_{\alpha\gamma} a^{\gamma\beta} + c^\beta = 0 \\ M^{\alpha\beta} b_{\alpha\beta} + \nabla_\alpha M^{\alpha 3} + N^{\alpha\beta} e_{\alpha\beta} + c^3 = 0 \end{array}} \quad (4.68)$$

that is three equations expressing respectively the balance of the moments in the tangent plane and along the normal vector. The surface tensor of components $M^{\alpha\beta}$ is called the **tensor of bending moments** and the term $M^{\alpha 3}$ is the **moment of axis normal to the surface**, applied to the facet of normal a_α.

Equations of equilibrium of the shells were established for the first time, using this method, by Love[5] (1893).

4.2.1.5 Boundary conditions

Shells can be classified into three categories, from the point of view of the boundary conditions:

- **closed shells** (example: the sphere), without edge. There is no boundary condition to express. On the other hand, conditions of periodicity expressing closure can be written, most often following the coordinate lines.
- **open shells**, bounded by an edge along which boundary conditions are to be expressed.
- **semi-closed shells**, such as cylinders open along two circles perpendicular to generators (a piece of pipe), along which boundary conditions are to be expressed. Periodicity conditions must be written along the generators.

In the last two cases, the boundary conditions on the edge are of the **kinematic** type (imposed displacements or rotations) or of the **mechanical** type (imposed forces or moments). In the first type, the forces and moments at the edge are the reactions associated with displacements or imposed rotations; these reactions are then determined from the internal actions, according to the relations established below. In the second type, the boundary conditions give conditions relating to internal actions to be determined. There may nevertheless be elastic support conditions, which give linear conditions between displacements and rotations, on the one hand, and the associated reactions, on the other hand.

Where there are intermediate points or support lines, the reasoning in this chapter must be adapted to take into account the corresponding reactions within σ. In most cases, it is also possible to cut the shell into several shell elements whose edges pass through the intermediate supports, by writing conditions of kinematic continuity.

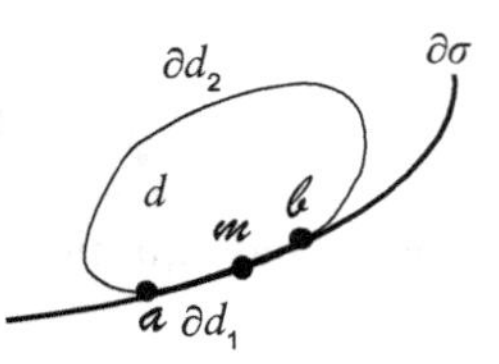

Figure 4.6

To obtain the boundary conditions at a point m of the edge ∂d, it is necessary to write the equilibrium of a domain d delimited by its edge ∂d, made up of two parts (Figure 4.6):

[5] Augustus Edward Hough Love, English mathematician (1863–1940).

- ∂d_1, part of $\partial\sigma$ containing $\mathcal{m}$, limited by two points $\mathcal{a}$ and $\mathcal{b}$;
- ∂d_2, any curve drawn on σ, limited by $\mathcal{a}$ and $\mathcal{b}$, so that there is continuity of the tangent to ∂d at these two points.

$\mathcal{O}$ is the point of calculation of the moment and $\mathcal{P}$ the current point of the surface, the equilibrium of d is written as:

$$\begin{cases} \int_{\partial d_1} q^i a_i \, ds + \int_{\partial d_2} N^{\alpha i} \nu_\alpha a_i \, ds + \int_d p^i a_i \, d\sigma = 0 \\ \int_{\partial d_1} \left[m^i a_i + \mathcal{OP} \times \left(q^i a_i \right) \right] ds + \int_{\partial d_2} \left[M^{\alpha i} \nu_\alpha a_i + \mathcal{OP} \times \left(N^{\alpha i} \nu_\alpha a_i \right) \right] ds \\ \qquad + \int_d \left(c + \mathcal{OP} \times p \right) d\sigma = 0 \end{cases}$$

and is rewritten:

$$\begin{cases} \int_{\partial d_1} \left(q^i - N^{\alpha i} \nu_\alpha \right) a_i \, ds + \int_{\partial d} N^{\alpha i} \nu_\alpha a_i \, ds + \int_d p^i a_i \, d\sigma = 0 \\ \int_{\partial d_1} \left[\left(m^i - M^{\alpha i} \nu_\alpha \right) a_i + \mathcal{OP} \times \left(q^i - N^{\alpha i} \nu_\alpha \right) a_i \right] ds \\ \qquad + \int_{\partial d} \left[M^{\alpha i} \nu_\alpha a_i + \mathcal{OP} \times \left(N^{\alpha i} \nu_\alpha a_i \right) \right] ds + \int_d \left(c + \mathcal{OP} \times p \right) d\sigma = 0 \end{cases}$$

either, taking into account equilibrium equations (4.64):

$$\begin{cases} \int_{\partial d_1} \left(q^i - N^{\alpha i} \nu_\alpha \right) a_i \, ds = 0 \\ \int_{\partial d_1} \left[\left(m^i - M^{\alpha i} \nu_\alpha \right) a_i + \mathcal{OP} \times \left(q^i - N^{\alpha i} \nu_\alpha \right) a_i \right] ds = 0 \end{cases} \tag{4.69}$$

(4.69) translates the equilibrium of a "ribbon" based on ∂d_1, subjected on one side to the actions internal to the surface, on the other to external actions at the edge.

At this point, the easiest way to obtain the boundary conditions is to write that the relations (4.69) are true regardless of the edge interval ∂d_1 considered, containing $\mathcal{m}$, which leads to expressing the equilibrium at each point of the edge. In this case, the reactions balance the internal actions applied directly on the edge:

$$\begin{cases} N^{\alpha i} \nu_\alpha = q^i \\ M^{\alpha i} \nu_\alpha = m^i \end{cases} \tag{4.70}$$

that is six conditions. However, these conditions, which result from equilibrium at a point, and which have been (historically) presented first, are not the only ones possible and they are not always compatible with the kinematic hypotheses of the shell deformation, as it is demonstrated in §4.2.2. It is established below that other boundary conditions than those given by (4.70) are possible.

Figure 4.7 shows that there is no need to respect the second relationship (4.70) to balance moments applied on a line representing the edge. Indeed, these can be balanced by reaction densities distributed along the line. There can therefore be self-equilibrium between reaction moments and force densities, the shell only playing the role of dispatcher with associated internal actions whose values are consistent with the shape of the distribution.

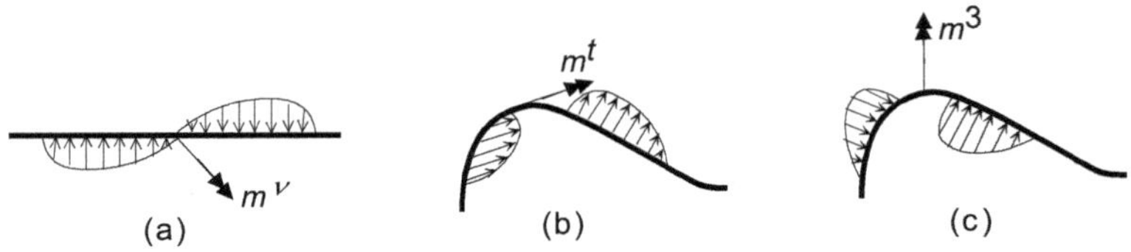

Figure 4.7

These figures show that:

- the balances presented there are not punctual; they are ensured by densities of forces distributed on the edge and are thus not translated by relations (4.70). The equilibrium of the edge is then written locally and no longer punctually.
- the role of curvature in case (b). In this case, where the represented curvature does not change sign, the sum of the reactions is not zero, and the equilibrium of the reactions alone is therefore not possible.
- the way in which the reactions are distributed along the edge depends on the behaviour of the shell itself, in particular its rigidity, which can not be demonstrated by simple equilibrium considerations on the curve.

For the rest of the reasoning, it is posed:

$$\mu(s) = \left(m^i - M^{\nu i}\right) a_i(s)$$

$$r(s) = \left(q^i - N^{\nu i}\right) a_i(s)$$

Figure 4.7 suggests that the functions $\mu(s)$ and $r(s)$ may not be zero, unlike relations (4.70); they verify equations (4.69), that is, by writing them for a segment of the edge between the abscissae 0 and s, since the segment ∂d_1 is arbitrary and calculating the moment for example at the origin point $\boldsymbol{a}$ of the curvilinear abscissa:

$$\begin{cases} \int_0^s r \, ds = 0 \\ \int_0^s \left(\mu + \boldsymbol{ap} \times r\right) ds = 0 \end{cases} \tag{4.71a}$$

where $\boldsymbol{p}$ is the current point in $\left[0, s\right]$.

Let $R(s)$ be a primitive of $r(s)$. The first equation (4.71a) then simply integrates into:

$$\left[R\right]_0^s = R(s) - R(0) = 0 \quad \forall s$$

Since the value of s is any, a different equilibrium of (4.70) is possible only if there are concentrated forces at the ends of the segment $\left[0, s\right]$ which have not been taken into account until now. It is therefore necessary to introduce concentrated forces at the ends of the segment $[0,s]$ into the equilibrium of the edge segment and replace (4.71a) with:

$$\begin{cases} V_0 + V_s + \int_0^s r\, ds = 0 \\ \boldsymbol{am} \times V_s + \int_0^s (\mu + \boldsymbol{ap} \times r)\, ds = 0 \end{cases} \tag{4.71b}$$

where $\boldsymbol{m}$ is the point with abscissa s.

The existence of these forces at the ends of the segment is easily conceivable by considering the sketches of Figure 4.7, since, for any segment of the edge, part of the reactions allowing its equilibrium is outside this segment. Nevertheless, they can not be highlighted by the surface equilibrium model used here, since they do not appear in (4.69); indeed, they are related to the behaviour of the shell in the thickness. This is more easily demonstrated by the PPV (§4.2.2); it suffices here to indicate that imposing a restriction on the displacements of the points in the thickness of the shell (and therefore of the liaisons) implies reactions distributed in the thickness, the result of which, for example in $s = 0$, is precisely the force V_0.

In the kinematics of Kirchhoff–Love (see §5.2), the liaisons prevent the distortions $\bar{\varepsilon}_{3\alpha}$, the reactions are thus transverse shear stresses $T_{3\alpha}$ which results in a force perpendicular to $\mathbf{t}$.

The second equation (4.71b) is then written:

$$\boldsymbol{am} \times V_s + \int_0^s \left[\mu + \boldsymbol{ap} \times \frac{dR}{ds}\right] ds = \boldsymbol{am} \times V_s + \int_0^s \left[\mu + \frac{d}{ds}(\boldsymbol{ap} \times R) - \mathbf{t} \times R\right] ds$$

$$= \int_0^s [\mu - \mathbf{t} \times R]\, ds = 0$$

since $R(s)$ is equal to $-V_s$, s being any. Finally:

$$\mu - \mathbf{t} \times R = 0 \qquad \forall s$$

which implies:

$$\begin{aligned} &\mu^t = 0 \\ &\mathbf{t} \times (\mu - \mathbf{t} \times R) = \mathbf{t} \times \mu + R - (R \cdot \mathbf{t})\,\mathbf{t} = 0 \end{aligned} \tag{4.72}$$

The second relationship is verified when $\mathbf{t} \times \mu + R = 0$, for which $R \cdot \mathbf{t} = 0$, either, by differentiating:

$$r + \frac{d}{ds}(\mathbf{t} \times \mu) = 0 \tag{4.73}$$

By referring this relation to the first equation (4.71b), where s is any, it becomes:

$$V_s + \mathbf{t} \times \mu(s) = 0 \quad \forall s$$

The force V_s is in a plane perpendicular to the edge and is therefore a "shear force" in the sense of the beams. s being any, this confirms that the hypothesis $R\cdot\mathbf{t}=0$ taken above does not lead to a contradiction. The above considerations are illustrated in the examples on Kirchhoff–Love shells, for example in 7.2.1.8

Taking into account differentiation relations in the Darboux coordinate system (see (3.122)):

$$\frac{d}{ds}(\mathbf{t}\times\mu)=\frac{d}{ds}\left(\mu^{\nu}a_3-\mu^3\nu\right)$$
$$=\left(\mu^3\rho_g-\mu^{\nu}\rho_n\right)\mathbf{t}-\left(\frac{d\mu^3}{ds}+\mu^{\nu}\tau_g\right)\nu+\left(\frac{d\mu^{\nu}}{ds}-\mu^3\tau_g\right)a_3$$

relations (4.73) between $r(s)$ and $\mu(s)$ are finally written:

$$\begin{aligned}
&\mu^t=0\\
&r^t+\mu^3\rho_g-\mu^{\nu}\rho_n=0\\
&r^{\nu}-\frac{d\mu^3}{ds}-\mu^{\nu}\tau_g=0\\
&r^3+\frac{d\mu^{\nu}}{ds}-\mu^3\tau_g=0
\end{aligned}\tag{4.74}$$

In the current case here μ^3 is zero:

$$\boxed{\begin{aligned}
&m^t-M^{\nu t}=0\\
&q^t-N^{\nu t}-\left(m^{\nu}-M^{\nu\nu}\right)\rho_n=0\\
&q^{\nu}-N^{\nu\nu}-\left(m^{\nu}-M^{\nu\nu}\right)\tau_g=0\\
&q^3-N^{\nu 3}+\frac{d}{ds}\left(m^{\nu}-M^{\nu\nu}\right)=0
\end{aligned}}\tag{4.75}$$

with $\rho_n=b_{tt}$, $\tau_g=b_{t\nu}$ (see §3.4.2). These expressions are **sufficient** to ensure equilibrium in a domain such as ∂d_1, however small, but not reduced to a point, and subject to a particular behaviour in the thickness of the shell. They are verified if conditions (4.70) are verified, in particular $m^{\nu}=M^{\nu\nu}$. The choice of "good" conditions, therefore, depends on the behaviour of the shell in its thickness. This aspect is developed in paragraph 4.2.2 below. It should be noted further that, according to the Saint-Venant[6] Principle, the differences between the possible boundary conditions have influence only in the vicinity of the edges.

[6] Adhémar Jean Claude Barré de Saint-Venant, French mechanician and mathematician (1797–1886).

In the case of a straight edge, the boundary conditions (4.75) are reduced to:

$$m^t = M^{vt}$$

$$q^\alpha = N^{\alpha v}$$

$$q^3 + \frac{\partial m^v}{\partial s} = N^{v3} + \frac{\partial M^{vv}}{\partial s}$$

4.2.2 Application of the principle of virtual powers

The objective is to establish the equations of evolution of the surface (local equations and boundary conditions), the different powers involved in the movement are specified below in order to apply the Principle of Virtual Powers (PVP).

4.2.2.1 Virtual power of deformation

As seen in 3.3.2, the deformation and curvature variation tensors of the middle surface completely characterise the deformation of the surface. It is therefore natural to choose e and k as deformation variables (in Euler variables), in the study of the evolution of the surface. This approach is obviously restrictive since it is possible to imagine that other terms due to the behaviour of the shell in its thickness intervene in the expression of the strain energy. Then, a more complete approach is presented in §4.5. The theory built on the approach developed in this paragraph is said **restricted** and is consistent with Kirchhoff's hypotheses presented in Chapter 5. It is developed on the assumption that, in the current state, the virtual power of deformation $\overset{*}{\mathcal{W}}_f$ is a linear function of virtual velocities $\overset{*}{e}$ and $\overset{*}{k}$ of deformation variables, considering that the strain energy is due only to the surface deformation and the variation of curvature (flexion).

This leads to the introduction of two tensor fields N and M (generalised or global stresses), such as:

$$\overset{*}{\mathcal{W}}_f = \int_\sigma \left(\mathsf{N} \bullet \overset{*}{e} + \mathsf{M} \bullet \overset{*}{k} \right) d\sigma \tag{4.76}$$

The tensors N and M are completely saturated $\overset{*}{e}$ and $\overset{*}{k}$ so the result is a scalar. Only the symmetrical parts of N and M participate in the integral, $\overset{*}{e}$ and $\overset{*}{k}$ is symmetrical. N and M are therefore assumed to be symmetrical. In order to reveal the antisymmetric parts, the expression of the virtual deformation power should be supplemented by using antisymmetric deformation variables.

The tensors N and M are respectively called the **tensor of membrane forces** and the **tensor of bending moments**: it is the mechanical interpretation in terms of forces and moments resulting from the internal actions, in connection with the tensors N and M introduced in §4.2.1, which justifies these appellations. Despite the similarities of notation and naming, the tensors N and N on the one hand, M and M on the other hand, are not identical. The relations which bind them are established in §4.4. The tensors N and M are also known as the Sanders[7] resultants.

Formulae (4.40) and (4.42) make it possible to write:

$$\mathsf{N}\bullet\overset{*}{e}=\mathsf{N}^{\alpha\beta}\left(\nabla_\alpha\overset{*}{\xi}_\beta-b_{\alpha\beta}\overset{*}{\zeta}\right)$$

$$\mathsf{M}\bullet\overset{*}{k}=\mathsf{M}^{\alpha\beta}\left(\nabla_\alpha\left(\partial_\beta\overset{*}{\zeta}+b_{\beta\gamma}\overset{*}{\xi}^\gamma\right)+b_\alpha{}^\gamma\left(\nabla_\beta\overset{*}{\xi}_\gamma-b_{\beta\gamma}\overset{*}{\zeta}\right)\right)$$

Some terms of the sum can be grouped together by posing:

$$\mathsf{L}^{\alpha\beta}=\mathsf{N}^{\alpha\beta}+b_\gamma{}^\beta\mathsf{M}^{\alpha\gamma} \tag{4.77}$$

The tensor L has no particular physical meaning. In general, it is not symmetrical; it is, therefore, necessary to pay particular attention to the order of the indices in the definition (4.77). It becomes:

$$\mathsf{N}\bullet\overset{*}{e}+\mathsf{M}\bullet\overset{*}{k}=\mathsf{L}^{\alpha\beta}\left(\nabla_\alpha\overset{*}{\xi}_\beta-b_{\alpha\beta}\overset{*}{\zeta}\right)+\mathsf{M}^{\alpha\beta}\nabla_\alpha\left(\partial_\beta\overset{*}{\zeta}+b_{\beta\gamma}\overset{*}{\xi}^\gamma\right) \tag{4.78}$$

The relation (4.78) can not be exploited directly, $\overset{*}{\xi}$ intervening also by its derivatives, which are not independent of $\overset{*}{\xi}$. For the application of PVP, it is necessary to proceed to integrations by parts, and then to apply the Stokes[8] formula for a surface:

$$\int_\sigma \nabla_\alpha X^\alpha d\sigma=\oint_{\partial\sigma}X^\alpha\nu_\alpha ds \tag{4.79}$$

where X denotes a vector of $T_{\mathcal{M}}$ (its proof is analogous to that of the formula divergence – flux in $\mathbb{R}^3$ for the differentiation D). It comes successively, also using (4.37):

- $\mathsf{L}^{\alpha\beta}\nabla_\alpha\overset{*}{\xi}_\beta=\nabla_\alpha\left(\mathsf{L}^{\alpha\beta}\overset{*}{\xi}_\beta\right)-\nabla_\alpha\mathsf{L}^{\alpha\beta}\bullet\overset{*}{\xi}_\beta$

- $\mathsf{M}^{\alpha\beta}\nabla_\alpha\left(\partial_\beta\overset{*}{\zeta}+b_{\beta\gamma}\overset{*}{\xi}^\gamma\right)=-\mathsf{M}^{\alpha\beta}\nabla_\alpha\left(a_\beta\bullet\overset{*}{a}_3\right)$

$$=-\nabla_\alpha\left(\mathsf{M}^{\alpha\beta}\,a_\beta\bullet\overset{*}{a}_3\right)+\left(a_\beta\bullet\overset{*}{a}_3\right)\nabla_\alpha\mathsf{M}^{\alpha\beta}$$

- $\left(a_\beta\bullet\overset{*}{a}_3\right)\nabla_\alpha\mathsf{M}^{\alpha\beta}=-\left(\partial_\beta\overset{*}{\zeta}+b_{\beta\gamma}\overset{*}{\xi}^\gamma\right)\nabla_\alpha\mathsf{M}^{\alpha\beta}$

$$=-\nabla_\beta\left(\overset{*}{\zeta}.\nabla_\alpha\mathsf{M}^{\alpha\beta}\right)+\overset{*}{\zeta}\nabla_\beta\nabla_\alpha\mathsf{M}^{\alpha\beta}-b_{\beta\gamma}\overset{*}{\xi}^\gamma\,\nabla_\alpha\mathsf{M}^{\alpha\beta}$$

[7] John Lyell Sanders, Jr., American mechanician (1924–1998).
[8] George Gabriel Stokes, Anglo-Irish physicist and mathematician (1819–1903).

hence, by integration on σ, and use of (4.79):

$$\overset{*}{\mathcal{W}}_f = -\int_\sigma \left[\left(\nabla_\alpha \mathsf{L}^{\alpha\beta} + b_\gamma{}^\beta \nabla_\alpha \mathsf{M}^{\alpha\gamma} \right) \overset{*}{\xi}_\beta + \left(-\nabla_\beta \nabla_\alpha \mathsf{M}^{\alpha\beta} + b_{\alpha\beta} \mathsf{L}^{\alpha\beta} \right) \overset{*}{\zeta} \right] d\sigma$$
$$+ \int_{\partial\sigma} \left[\mathsf{L}^{\alpha\beta} \overset{*}{\xi}_\beta - \nabla_\beta \mathsf{M}^{\beta\alpha} \bullet \overset{*}{\zeta} - \mathsf{M}^{\alpha\beta} \, a_\beta \bullet \overset{*}{a}_3 \right] \nu_\alpha \, ds \tag{4.80}$$

As was seen in §4.1.3.4, $\overset{*}{a}_3$ is not independent of $\overset{*}{\xi}$ along the edge. To be able to switch from PVP to local equations, which is done in §4.2.2.4, it is necessary to keep only independent virtual velocities. By (4.37):

$$a_\beta \bullet \overset{*}{a}_3 = -\overset{*}{\mathcal{B}}_\beta = -\left(\partial_\beta \overset{*}{\zeta} + b_\beta{}^\alpha \overset{*}{\xi}_\alpha \right)$$

Thus, the last term of the edge integral must be transformed to integrate $\partial_s \overset{*}{\zeta}$ that depends on the values of $\overset{*}{\zeta}$ along the edge. The edge integral is then rewritten (the edge is closed):

$$\int_{\partial\sigma} \left[\left(\mathsf{L}^{\nu\beta} + b_\gamma{}^\beta \mathsf{M}^{\nu\gamma} \right) \overset{*}{\xi}_\beta - \left(\frac{\partial \mathsf{M}^{\nu t}}{\partial s} + \nabla_\beta \mathsf{M}^{\beta\nu} \right) \overset{*}{\zeta} + \mathsf{M}^{\nu\nu} \frac{\partial \overset{*}{\zeta}}{\partial \nu} \right] ds \tag{4.81}$$

The integration by parts nevertheless assumes that $\mathsf{M}^{\nu t}$ is continuously differentiable along the edge, which implies, in particular, that ν is continuous and thus excludes the "corners" (see §4.2.2.7). It also assumes that the function $\partial_s \overset{*}{\zeta}$ is not identically zero on the considered edge segment.

4.2.2.2 Virtual power of external actions

External actions were introduced in §4.2.1.2.1.

On the edge $\partial\sigma$ of σ are applied known actions and **liaison actions**, reactions from outside on σ. These actions are represented by linear densities of forces q and moments m on the edge $\partial\sigma$, which can be decomposed either in the local coordinate system of the surface, or in the Darboux coordinate system of the edge; for example, for q:

$$q = q^\alpha a_\alpha + q^3 a_3 = q^t \mathbf{t} + q^\nu \nu + q^3 a_3 \tag{4.82}$$

The density of moments m works in the local virtual rotation $\overset{*}{\omega}$, which, by (4.39), is orthogonal to a_3. Thus, only the tangential components of the moment work in the virtual field. It is therefore not necessary to consider the normal component of m that does not intervene in this theory and:

$$m = m^\alpha a_\alpha = m^t \, \mathbf{t} + m^\nu \nu \tag{4.83}$$

On σ, the density of force p and the density of moment c are applied. The latter works in the virtual rotation of the normal and its normal component do not intervene, for the same reason as for m.

The virtual power of external forces is therefore in the form:

$$\overset{*}{\mathcal{W}}_e = \int_\sigma \left(p \bullet \overset{*}{\xi} + c \bullet \overset{*}{\omega} \right) d\sigma + \int_{\partial\sigma} \left(q \bullet \overset{*}{\xi} + m \bullet \overset{*}{\omega} \right) ds \tag{4.84}$$

By (4.37) and (4.39):

$$m \bullet \overset{*}{\omega} = \left(m^t \, \mathbf{t} + m^\nu \, \nu \right) \bullet \left(a_3 \times \overset{*}{a_3} \right)$$

$$= \left(m^t \, \mathbf{t} \wedge a_3 + m^\nu \, \nu \wedge a_3 \right) \bullet \overset{*}{a_3} = m^\nu \, \mathbf{t} \bullet \overset{*}{a_3} - m^t \, \nu \bullet \overset{*}{a_3} \tag{4.85}$$

$$= -m^\nu \, \overset{*}{\mathcal{B}}_t + m^t \, \overset{*}{\mathcal{B}}_\nu = -m^\nu \left(\frac{\partial \overset{*}{\zeta}}{\partial s} + b_t^{\ \alpha} \, \overset{*}{\xi}_\alpha \right) + m^t \left(\frac{\partial \overset{*}{\zeta}}{\partial \nu} + b_\nu^{\ \alpha} \, \overset{*}{\xi}_\alpha \right)$$

Similarly:

$$c \bullet \overset{*}{\omega} = \left(c^\gamma a_\gamma \right) \bullet \left(a_3 \times \overset{*}{a_3} \right) = c^\gamma \left(a_3 \times a_\gamma \right) \bullet \left(\partial_\beta \overset{*}{\zeta} + b_\beta^{\ \alpha} \, \overset{*}{\xi}_\alpha \right) a^\beta \tag{4.86}$$

By integrating into parts the term $\partial_\beta \overset{*}{\zeta}$ in (4.86) and the edge term $\frac{\partial \overset{*}{\zeta}}{\partial s}$ into (4.85), (4.84) becomes, for a closed edge:

$$\overset{*}{\mathcal{W}}_e = \int_\sigma \left[\left(p^\alpha + c^\gamma \, b^{\alpha\beta} \left(a_3 \times a_\gamma \right) \cdot a_\beta \right) \overset{*}{\xi}_\alpha + \left(p^3 - \nabla_\beta \left(c^\gamma \left(a_3 \times a_\gamma \right) \cdot a^\beta \right) \right) \overset{*}{\zeta} \right] d\sigma$$

$$+ \int_{\partial\sigma} \begin{bmatrix} \left(q^\alpha - m^\nu \, b_t^{\ \alpha} + m^t \, b_\nu^{\ \alpha} \right) \overset{*}{\xi}_\alpha \\ + m^t \dfrac{\partial \overset{*}{\zeta}}{\partial \nu} + \left(q^3 + c^\gamma \left(a_3 \times a_\gamma \right) \cdot \nu + \dfrac{\partial m^\nu}{\partial s} \right) \overset{*}{\zeta} \end{bmatrix} ds \tag{4.87}$$

The integration by parts supposes that m^ν is continuously differentiable along the edge, which implies, in particular, that ν is continuous and thus excludes the corners (see §4.2.2.7). It also assumes that the function $\partial_s \overset{*}{\zeta}$ is not identically zero on the considered edge segment.

4.2.2.3 Virtual power of inertial actions

When the middle surface is considered without thickness, only the mass density ρ participates in the inertia actions:

$$\overset{*}{\mathcal{W}}_j = -\int_\sigma \rho \frac{d^2\xi}{dt^2} \bullet \overset{*}{\xi} d\sigma = -\int_\sigma \rho \left(\gamma^\alpha \, \overset{*}{\xi}_\alpha + \gamma^3 \, \overset{*}{\zeta} \right) d\sigma \tag{4.88}$$

where the acceleration γ is given by (4.23).

In anticipation of subsequent developments, this expression can already be completed by supposing that the surface has a thickness and that the material points situated on a_3 remain rigidly connected to the surface during its evolution. This hypothesis, falling within the framework of the Kirchhoff–Love theory developed in the next chapter, is also a sufficient approximation for the other theories. It consists in supposing that a normal segment inscribed in the thickness undergoes a rigid displacement. In this case, mass inertia I of the material segment normal to the surface is introduced, equal to $\mathrm{I} = \rho \frac{e^3}{12}$ if the shell is homogeneous. The angular velocity vector $\dot{\theta}$ of the normal is given by the relation (4.9):

$$\dot{\theta} = a_3 \times \dot{a}_3$$

The angular acceleration is deduced:

$$\ddot{\theta} = a_3 \times \ddot{a}_3$$

and the virtual power of rotational inertia actions, using the expression (4.39) of the virtual rotational velocity:

$$\overset{*}{\mathcal{W}}{}^{r}_{j} = -\int_{\sigma} \mathrm{I}\left(a_3 \times \ddot{a}_3\right) \cdot \left(a_3 \wedge \overset{*}{a}_3 \right) d\sigma = -\int_{\sigma} \mathrm{I}\, \ddot{a}_3 \cdot \overset{*}{a}_3\, d\sigma \tag{4.89}$$

as:

$$\left(a_3 \times \ddot{a}_3\right) \cdot \left(a_3 \times \overset{*}{a}_3 \right) = \overset{*}{a}_3 \cdot \left[\left(a_3 \times \ddot{a}_3\right) \times a_3 \right] = \overset{*}{a}_3 \cdot \left[\left(a_3\right)^2 \ddot{a}_3 - \left(a_3 \cdot \ddot{a}_3\right) a_3 \right] = \overset{*}{a}_3 \cdot \ddot{a}_3$$

This expression can be calculated from (4.8), (4.11), and (4.13) by evaluating the following quantities:

$$\left(\overline{\dot{a}_3 \cdot a_\alpha}^{\,\cdot} \right) = \ddot{a}_3 \cdot a_\alpha + \dot{a}_3 \cdot \dot{a}_\alpha = \ddot{a}_3 \cdot a_\alpha - v_{\beta 3} v_\alpha{}^\beta = -\frac{dv_{\alpha 3}}{dt}$$

$$\left(\overline{\dot{a}_3 \cdot a_3}^{\,\cdot} \right) = \ddot{a}_3 \cdot a_3 + \left(\dot{a}_3\right)^2 = \ddot{a}_3 \cdot a_3 + v^{\gamma 3} v_{\gamma 3} = 0$$

from which:

$$\ddot{a}_3 = \left(-\frac{dv_{\beta 3}}{dt} a^{\alpha\beta} + v_{\beta 3} v^{\alpha\beta} \right) a_\alpha - v^{\gamma 3} v_{\gamma 3} a_3 \tag{4.90}$$

which is noted:

$$\ddot{a}_3 = \ddot{a}^{\alpha 3}\, a_\alpha + \ddot{a}^{33} a_3$$

from which:

$$\ddot{a}_3 \bullet \overset{*}{a}_3 = -\ddot{a}^{\beta 3}\left(\partial_\beta \overset{*}{\zeta} + b_\beta{}^\alpha \overset{*}{\xi}_\alpha\right) \tag{4.91}$$

Then:

$$\overset{*}{\mathcal{W}}{}_j^r = \int_\sigma \mathrm{I}\left(\ddot{a}^{\beta 3}\, b_\beta{}^\alpha\, \overset{*}{\xi}_\alpha - \nabla_\beta \ddot{a}^{\beta 3}\, \overset{*}{\zeta}\right) d\sigma + \int_{\partial\sigma} \mathrm{I}\, \ddot{a}^{\beta 3} \nu_\beta\, \overset{*}{\zeta}\, d\sigma \tag{4.92}$$

This rotational kinetic energy can be added to the translational kinetic energy given by (4.88). Nevertheless, the calculations that can be performed *a posteriori* show that, especially for thin shells, the rotational kinetic energy is negligible compared to the translational kinetic energy. Indeed, the ratio of the two energies is of the order of $(e/R)^2$, where R denotes the minimum radius of curvature of the shell.

4.2.2.4 *Application of the Principle of Virtual Powers*

The application of the PVP:

$$\forall \overset{*}{\xi}: \quad \overset{*}{\mathcal{W}}_i + \overset{*}{\mathcal{W}}_e + \overset{*}{\mathcal{W}}_j = 0 \tag{4.93}$$

with:

$$\overset{*}{\mathcal{W}}_i = -\overset{*}{\mathcal{W}}_f$$

leads to the relation, true irrespective of the virtual velocity $\overset{*}{\xi}$, where are grouped the expressions (4.80), (4.81), (4.87), (4.88), and (4.92):

$$\begin{aligned}
&\int_\sigma \begin{bmatrix} \left(\nabla_\beta \mathsf{L}^{\beta\alpha} + b_\gamma{}^\alpha \nabla_\beta \mathsf{M}^{\beta\gamma} + p^\alpha + c^\gamma b^{\alpha\beta} \mathsf{e}_{\gamma\beta} - \rho\gamma^\alpha + \mathrm{I}\, \ddot{a}^{\beta 3} b_\beta{}^\alpha\right)\overset{*}{\xi}_\alpha + \\ \left(-\nabla_\beta \nabla_\alpha \mathsf{M}^{\alpha\beta} + b_{\alpha\beta}\mathsf{L}^{\alpha\beta} + p^3 - \nabla_\beta\left(c^\alpha \mathsf{e}_\alpha{}^\beta\right) - \rho\gamma^3 - \mathrm{I}\, \nabla_\alpha \ddot{a}^{\alpha 3}\right)\overset{*}{\zeta} \end{bmatrix} d\sigma \\
&- \int_{\partial\sigma} \begin{bmatrix} \left(\mathsf{L}^{\nu\beta} + b_\gamma{}^\beta \mathsf{M}^{\nu\gamma} - q^\beta + m^\nu b_t{}^\beta - m^t b_\nu{}^\beta\right)\overset{*}{\xi}_\beta \\ -\left(\dfrac{\partial \mathsf{M}^{\nu t}}{\partial s} + \nabla_\alpha \mathsf{M}^{\alpha\nu} + q^3 + \nu_\beta c^\alpha \mathsf{e}_\alpha{}^\beta + \dfrac{\partial m^\nu}{\partial s} + \mathrm{I}\, \ddot{a}^{\nu 3}\right)\overset{*}{\zeta} \\ +\left(\mathsf{M}^{\nu\nu} - m^t\right)\dfrac{\partial \overset{*}{\zeta}}{\partial \nu} \end{bmatrix} ds = 0
\end{aligned} \tag{4.94}$$

4.2.2.5 Local equations

The local (or indefinite) equations, in Euler variables, are obtained by considering any virtual velocities, zero on the edge:

$$\forall m \in \sigma: \quad \begin{cases} \nabla_\beta L^{\beta\alpha} + b_\gamma{}^\alpha \nabla_\beta M^{\beta\gamma} + p^\alpha + c^\gamma b^{\alpha\beta} e_{\gamma\beta} - \rho\gamma^\alpha + I\ddot{a}^{\beta 3} b_\beta{}^\alpha = 0 \\ -\nabla_\beta \nabla_\alpha M^{\alpha\beta} + b_{\alpha\beta} L^{\alpha\beta} + p^3 - \nabla_\beta \left(c^\alpha e_\alpha{}^\beta \right) - \rho\gamma^3 - I\nabla_\alpha \ddot{a}^{\alpha 3} = 0 \end{cases} \tag{4.95}$$

(three equations for six unknowns, in static).

The first line gives two equations representing a balance of actions associated with $\overset{*}{\xi}_\alpha$, that is, the resultant of force densities tangent to the surface. The last relation, associated with $\overset{*}{\zeta}$, relates to the resultant of the densities of forces normal to the surface. Equivalent expressions are obtained by replacing $L^{\alpha\beta}$ by its definition (4.77).

Moreover, it is interesting to introduce the **shear force**, vector in T_m of components:

$$Q^\beta = -\nabla_\alpha M^{\alpha\beta} \tag{4.96}$$

Finally, in the usual cases, the density of moment c is zero and the rotational kinetic energy can be neglected. Under these conditions, (4.95) is rewritten:

$$\forall m \in \sigma: \quad \begin{cases} \nabla_\beta N^{\beta\alpha} - 2b_\gamma{}^\alpha Q^\gamma + M^{\beta\gamma} \nabla_\beta b_\gamma{}^\alpha + p^\alpha - \rho\gamma^\alpha = 0 \\ \nabla_\alpha Q^\alpha + b_{\alpha\beta} N^{\alpha\beta} + b_{\alpha\beta} b_\gamma{}^\beta M^{\alpha\gamma} + p^3 - \rho\gamma^3 = 0 \\ Q^\gamma + \nabla_\beta M^{\beta\gamma} = 0 \end{cases} \tag{4.97}$$

(five equations for eight unknowns, in static).

4.2.2.6 Boundary conditions

The edge integral of (4.94) is zero for virtual velocities $\overset{*}{\xi}_\alpha$, $\overset{*}{\zeta}$, and $\partial_\nu \overset{*}{\zeta}$ independent and arbitrary, which leads to boundary conditions:

$$\forall m \in \partial\sigma: \quad \begin{cases} L^{\nu\beta} + b_\gamma{}^\beta M^{\nu\gamma} - q^\beta + m^\nu b_t{}^\beta - m^t b_\nu{}^\beta = 0 \\ \dfrac{\partial M^{\nu t}}{\partial s} + \nabla_\alpha M^{\alpha\nu} + q^3 + \nu_\beta c^\alpha e_\alpha{}^\beta + \dfrac{\partial m^\nu}{\partial s} + I\,\ddot{a}^{\nu 3} = 0 \\ M^{\nu\nu} - m^t = 0 \end{cases} \tag{4.98}$$

The first line is also written as:

$$L^{\nu\beta} + b_t{}^\beta \left(M^{\nu t} + m^\nu \right) - q^\beta = 0$$

In the case where c and the rotational kinetic energy are neglected, and simplifying the first line given the last relationship, (4.98) becomes:

$$\forall \boldsymbol{m} \in \partial\sigma: \quad \begin{cases} \mathsf{N}^{\nu\beta} + b_t{}^{\beta}\left(2\mathsf{M}^{\nu t} + m^{\nu}\right) + b_{\nu}{}^{\beta}\mathsf{M}^{\nu\nu} - q^{\beta} = 0 \\ \mathsf{Q}^{\nu} - \dfrac{\partial}{\partial s}\left(\mathsf{M}^{\nu t} + m^{\nu}\right) - q^{3} = 0 \\ \mathsf{M}^{\nu\nu} - m^{t} = 0 \end{cases} \tag{4.99}$$

To write these boundary conditions, it was made a choice of direct coordinate system $(\mathbf{t}, \mathrm{v}, a_3)$. It should be noted further that m^{ν} (the moment with an axis normal to the edge) is zero in many applications.

The last relation reflects the equilibrium of the moments of axis **t**; if there is no fixed-end moment (rotation free about **t**), $\mathsf{M}^{\nu\nu}$ is zero. The first two relations of (4.98) translate the balance of the forces applied to the edge, in the plane tangent to σ. They involve the in-plane forces $\mathsf{N}^{\nu t}$ and $\mathsf{N}^{\nu\nu}$, but also the bending moments. Even when the reaction moment m^{ν} is zero, the reaction q^{t} (resp. q^{ν}) is not equal to $\mathsf{N}^{\nu t}$ (resp. $\mathsf{N}^{\nu\nu}$). The third relation reflects the balance of forces applied to the edge in the direction of the normal to σ. Again, even if m^{ν} is zero, the shear force Q^{ν} at the edge does not balance the normal reaction q^{3}. The intervention of the term $\mathsf{M}^{\nu t}$ is explained by the fact that, in the application of the PVP, and when the function $\overset{*}{\xi}$ varies along the edge (which is the case since $\overset{*}{\xi}$ is any), there is simultaneously a translation along a_3 (in which work Q^{ν} and q^{3}) and rotation around ν (in which work m^{ν} and $\mathsf{M}^{\nu t}$). A twisting moment $\mathsf{M}^{\nu t}$ of the edge is balanced by vertical reactions which the resulting moment is precisely equal to $\mathsf{M}^{\nu t}$. If $\mathsf{M}^{\nu t}$ is constant along the edge, these reactions are balanced gradually and have no contribution to q^{3}. This intervention of the moment in the balance of forces at the edge has already been reported for tensors N and M in §4.2.1.5. It appears more naturally here, as being due to the fact that the kinematics of the normal is related to the kinematics of the surface. An illustration is given in §7.1.1.3, in the case of plates.

In the model developed here, the moments work in the rotation of the normal, which leads to the results exposed above (local equations and boundary conditions), this rotation plays an important mechanical role, consistent with Kirchhoff–Love's theory exposed in Chapter 5. A more general point of view is considered in §4.5, where the movement of particles in the thickness of the shell is no longer related to the rotation of the normal, which leads in particular to boundary conditions where the reaction moments directly balance the moments of the stresses.

In practice, three situations of boundary conditions are commonly encountered:

- **clamped edge:** displacements and rotation of axis parallel to the edge are prevented:

  along $\partial\sigma$: $\xi(s) = 0$; $\partial_{\nu}\zeta(s) = 0$

  The imposed conditions are kinematic, formula (4.98) make it possible to determine the reactions (see nevertheless the last point below).

- **simply supported edge:** displacements normal to the surface are prevented along the edge, however the surface can turn around the tangent ($\partial_v\zeta \neq 0$). In addition, the reaction moment m^t is zero, the rotational movements around the tangent being free:

$$\text{along } \partial\sigma: \quad \zeta(s)=0\,;\ \mathsf{M}^{vv}=0$$

Liaisons relating to the horizontal components of the displacement can be added.

- **free edge:** in this case, translations and rotations are left free; there is no connection with the outside and all reactions are zero. If there is no given external force applied to the edge, (4.99) reduces on $\partial\sigma$ to:

$$\text{along } \partial\sigma: \begin{cases} \mathsf{N}^{v\beta} + 2b_t{}^{\beta}\mathsf{M}^{vt} = 0 \\ \mathsf{Q}^{v} - \dfrac{\partial \mathsf{M}^{vt}}{\partial s} = 0 \\ \mathsf{M}^{vv} = 0 \end{cases}$$

- **edge continuously supported or clamped:** A continuous support condition $\zeta(s)=0$ implies that the rotation with axis v: $\partial_s\zeta(s)$ is zero along the entire length of the support. In this case, a compatible virtual velocity field $\partial_s\overset{*}{\zeta}$ is zero along the considered edge segment and there is no need to proceed to the integrations leading to expressions (4.81) and (4.87). The terms $\dfrac{\partial \mathsf{M}^{vt}}{\partial s}$ and $\dfrac{\partial m^{v}}{\partial s}$ then do not intervene under the conditions (4.98) and (4.99) and the boundary conditions are reduced to the "natural" conditions where, in (4.99), the second line is reduced to:

$$\mathsf{Q}^{v} = q^{3}$$

These are the conditions (4.177) obtained in §4.5 in the framework of the general theory or the conditions (4.70) obtained by the equilibrium of the edge. Thus, the boundary conditions to be used in the restricted theory (of Kirchhoff–Love) are to be determined carefully according to the imposed relationships.

In a corner, the intersection of two supported edges, the two rotations are zero.

4.2.2.7 *Singular points of the edges*

When the edge has a discontinuity of the tangent (corner), the integrations by part of the terms $\mathsf{M}^{vt}\dfrac{\partial\overset{*}{\zeta}}{\partial s}$ (§4.2.2.1) and $m^{v}\dfrac{\partial\overset{*}{\zeta}}{\partial s}$ (§4.2.2.2) must take into account the discontinuity. Indeed, even in the hypothesis that the tensor M is perfectly continuous on the domain σ, it can not vary continuously along the edge at the discontinuity of v.

To simplify the presentation, the following assumptions are made:

- the singularity is unique in the interval under consideration;
- the surface σ is regular;
- the density of moment c is assumed to be zero, which is the current situation;
- the problem is analysed in statics.

Let m_0 the point of abscissa s_0 on the edge $\partial\sigma$, where the singularity occurs, correspond to a discontinuity of the tangent (corner). Let $\mathbf{t}_1$ (resp., $\mathbf{t}_2$) and ν_1 (resp., ν_2) be the tangent and normal vectors to the edge at point m_0. Given the regularity of the surface in the vicinity of m_0, a_3 is defined in a unique way; $\mathbf{t}_1$, $\mathbf{t}_2$, ν_1 and ν_2 are in the tangent plane. [Q] denoting the discontinuity of a quantity Q at m_0, $\left[\mathbf{t}_2-\mathbf{t}_1\right]=\left[\nu_2-\nu_1\right]$ is the common discontinuity of the tangent and of the geodesic normal at m_0.

The integrations by parts are modified to take into account this discontinuity; integrating along the edge between a point immediately after m_0 to a point immediately before m_0:

$$\int_{\partial\sigma} \mathsf{M}^{\nu t}\frac{\partial \overset{*}{\zeta}}{\partial s}\,ds=\lim_{h\to 0}\left[\mathsf{M}^{\nu t}\overset{*}{\zeta}\right]_{s_0+h}^{s_0-h}-\int_{\partial\sigma}\frac{\partial \mathsf{M}^{\nu t}}{\partial s}\overset{*}{\zeta}\,ds=-\left[\mathsf{M}^{\nu t}\right]\overset{*}{\zeta}-\int_{\partial\sigma}\frac{\partial \mathsf{M}^{\nu t}}{\partial s}\overset{*}{\zeta}\,ds \tag{4.100}$$

which shows the discontinuity of $\mathsf{M}^{\nu t}$ at m_0. The term $m^{\nu}\dfrac{\partial \overset{*}{\zeta}}{\partial s}$ is integrated in a similar way and shows the discontinuity of m^{ν}. The edge integral of (4.94) is therefore replaced by:

$$-\int_{\partial\sigma}\left[\begin{array}{l}\left(\mathsf{L}^{\nu\beta}+b_{\gamma}{}^{\beta}\mathsf{M}^{\nu\gamma}-q^{\beta}+m^{\nu}b_{t}{}^{\beta}-m^{t}b_{\nu}{}^{\beta}\right)\overset{*}{\xi}_{\beta}+\left(\mathsf{M}^{\nu\nu}-m^{t}\right)\dfrac{\partial\overset{*}{\zeta}}{\partial\nu}\\ -\left(\dfrac{\partial \mathsf{M}^{\nu t}}{\partial s}+\nabla_{\alpha}\mathsf{M}^{\alpha\nu}+q^{3}+\nu_{\beta}c^{\alpha}e_{\alpha}{}^{\beta}+\dfrac{\partial m^{\nu}}{\partial s}+\mathrm{I}\,\ddot{a}^{\nu 3}\right)\overset{*}{\zeta}\end{array}\right]ds$$
$$-\left(-\left[\mathsf{M}^{\nu t}\right]+\left[m^{\nu}\right]\right)\overset{*}{\zeta}(m_0)=0 \tag{4.101}$$

The integral on σ is understood almost everywhere except in m_0 for the term integrated by parts. By application of the PVP, the term under the integral is zero, which leads to the boundary conditions (4.98), with a correction to be made in m_0.

Having cancelled the integral term, the virtual power is reduced to $\left[\mathsf{M}^{\nu t}-m^{\nu}\right]\overset{*}{\zeta}_0$. If the corner is simply supported, it is possible that a force R **normal to the shell and concentrated in the corner** works in the direction normal to the surface. The liaison actions defined in §4.2.2.2 must then be supplemented by a concentrated force R which virtual power $R\overset{*}{\zeta}_0$ is to be added to the expression (4.101). R then verifies the relationship:

$$\boxed{R=\left[m^{\nu}-\mathsf{M}^{\nu t}\right]} \tag{4.102}$$

If the corner does not undergo any liaison, the discontinuity of $\mathsf{M}^{\nu t}$ is zero.

The existence of this concentrated reaction is due to the fact that the moments m^{ν} and $\mathsf{M}^{\nu t}$ work in the rotation of a_3, which is related to the assumptions of the restricted theory (Kirchhoff). In the general theory where the material points on the normal segment are no longer constrained to stay on a_3, this reaction does not exist.

Example: Let a plate with an edge singular at point $\mathcal{O}$ and the two orthogonal edge portions, tangent to the two Cartesian axes $\mathcal{O}x$ and $\mathcal{O}y$ in $\mathcal{O}$ (Figure 4.8). So:

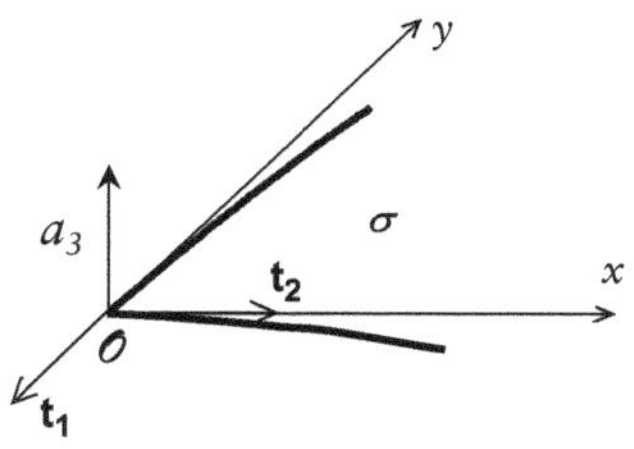

Figure 4.8

$$\mathbf{t}_1 = -\mathbf{j}\,;\quad \mathbf{t}_2 = \mathbf{i}$$

$$\nu_1 = \mathbf{i}\,;\quad \nu_2 = \mathbf{j}$$

where i and j denote the unit vectors of the axes. In these conditions:

$$\mathsf{M}^{\nu t}(1) = \mathbf{t}_1 \cdot \left(\mathsf{M} \cdot \nu_1\right) = -\,\mathsf{M}^{xy}$$

$$\mathsf{M}^{\nu t}(2) = \mathbf{t}_2 \cdot \left(\mathsf{M} \cdot \nu_2\right) = \mathsf{M}^{xy}$$

which leads to:

$$\left[\mathsf{M}^{\nu t}\right] = 2\mathsf{M}^{xy} \tag{4.103}$$

If the plate is supported on the corner and m^{ν} is zero, then the reaction is equal to $-2\mathsf{M}^{xy}$.

4.2.3 Special cases

Starting from the local equations (4.97) and the boundary conditions (4.99), three particular important cases should be considered in practice.

4.2.3.1 Membranes

Membrane theory (introduced in a simple way in Chapter 1) is a simplifying theory in which it is accepted that the tensor of bending moments is nil (in fact negligible). This implies in particular that:

$$\begin{aligned} &\mathsf{Q} = -\operatorname{div}\mathsf{M} = 0 \\ &\mathsf{L} = \mathsf{N} = N \end{aligned} \tag{4.104}$$

In membrane theory, it is not necessary to distinguish the generalised stresses N and the resultant of the internal actions N. It follows, in static:

- local equilibrium equations:

$$\forall \boldsymbol{m} \in \sigma:\quad \begin{cases} \nabla_{\beta}\mathsf{N}^{\beta\alpha} + p^{\alpha} = 0 \\ b_{\alpha\beta}\mathsf{N}^{\alpha\beta} + p^{3} = 0 \end{cases} \tag{4.105}$$

- boundary conditions:

$$\forall m \in \partial\sigma: \quad N^{\nu\beta} - q^{\beta} = 0 \tag{4.106}$$

So necessarily:

$$q^3 = 0$$

In membrane theory, there cannot be any force normally applied to the surface at the edge; the reactions are necessarily in the tangent plane.

The main interest of the membrane theory is that, in statics and subject to the compatibility of the boundary conditions, the equation (4.105) makes it possible to completely determine the tensor of the membrane (stretching) stresses: the shell is statically determined. The first two equations of (4.105), relative to the tangential equilibrium, are partial differential equations of the first order, whereas the third equation expressing the normal equilibrium contains no derivative. Moreover, the two boundary conditions (4.106) are sufficient to completely determine the components of **N**.

The membrane solution is a particular solution of equilibrium equations (4.95). Even when it is not the exact solution, it often gives interesting information on the behaviour of double curvature shells. The conditions of application of the membrane theory and usual examples are studied in Chapter 6.

4.2.3.2 Plates

When the surface σ is flat, the shell is called a plate. If it can be admitted that the curvature is zero or very small in the current position (negligible geometric changes...), then $\mathsf{N} = N$, and the local equations (4.97) are rewritten:

$$\forall m \in \sigma: \quad \begin{cases} \nabla_\beta \mathsf{N}^{\beta\alpha} + p^\alpha - \rho\gamma^\alpha = 0 \\ -\nabla_\gamma \nabla_\beta \mathsf{M}^{\beta\gamma} + p^3 - \rho\gamma^3 = 0 \end{cases} \tag{4.107}$$

with boundary conditions:

$$\forall m \in \partial\sigma: \quad \begin{cases} \mathsf{N}^{\nu\beta} - q^\beta = 0 \\ \nabla_\alpha \mathsf{M}^{\alpha\nu} + \dfrac{\partial\left(\mathsf{M}^{\nu t} + m^\nu\right)}{\partial s} + q^3 = 0 \\ \mathsf{M}^{\nu\nu} - m^t = 0 \end{cases} \tag{4.108}$$

Introducing the shear force **Q**, the second condition is written:

$$\mathsf{Q}^\nu = \frac{\partial\left(\mathsf{M}^{\nu t} + m^\nu\right)}{\partial s} + q^3 \tag{4.109}$$

Relations (4.107) and (4.108) show that stretching forces and bending moments are not coupled by equilibrium equations. On the other hand, these relations are not sufficient to completely determine M and N. If the constitutive laws are also decoupled, the equations can be solved separately to obtain M and N, which is a considerable simplification. The partial differential equations thus obtained are first order on N and second order on M. They correspond to a fully linearised plate theory.

The theory of plates is detailed in Chapter 7, where it is highlighted that the equations are not decoupled when the changes in geometry are not neglected.

4.2.3.3 *Shallow shells*

A shell with low curvature or **shallow shell** is a shell whose radii of curvature are large, in its different configurations, as compared to a characteristic dimension (for example the span ℓ):

$$\frac{\ell}{\|R_2\|} << 1$$

where R_2 is the smallest radius of curvature in absolute value. It is reasonable then to assume that curvatures vary little. An example of such a shell is a spherical dome with a large radius of curvature. The connection coefficients are of the order of the curvatures, if the basic vectors are normalised (3.127). In such a situation, the terms in $\frac{1}{R^2}$ and the variations of curvature can be neglected, the equilibrium equations (3.97) then being reduced to:

$$\forall \mathcal{m} \in \sigma: \qquad \begin{cases} \nabla_\beta \mathsf{N}^{\beta\alpha} + 2b_\gamma{}^\alpha \partial_\beta \mathsf{M}^{\beta\gamma} + p^\alpha - \rho\gamma^\alpha = 0 \\ -\nabla_\gamma \nabla_\beta \mathsf{M}^{\beta\gamma} + b_{\alpha\beta} \mathsf{N}^{\alpha\beta} + p^3 - \rho\gamma^3 = 0 \end{cases} \qquad (3.110)$$

More precise indications are given on shallow shells in Chapter 8.

In the absence of applied moment c, given the hypothesis of low curvatures, the last equation of (3.68) can be written $N^{\alpha\beta} = N^{\beta\alpha}$, it is no longer necessary to distinguish equations (3.110) from equations (3.68) related to the resultant stresses.

4.2.4 Double surface concept

Calladine[9] introduced the idea that the shell can be considered as the superposition of two surfaces:

- a surface that behaves like a membrane;
- a surface behaving only in bending.

Indeed, limiting to the case of static for simplification, equations (4.97) can be rewritten:

$$\begin{cases} \nabla_\beta \mathsf{N}^{\beta\alpha} + p^\alpha - \mathcal{p}^\alpha = 0 & \text{(a)} \\ b_{\alpha\beta} \mathsf{N}^{\alpha\beta} + p^3 - \mathcal{p}^3 = 0 & \text{(b)} \\ \nabla_\alpha \mathsf{Q}^\alpha + b_{\alpha\beta} b_\gamma{}^\beta \mathsf{M}^{\alpha\gamma} + \mathcal{p}^3 = 0 & \text{(c)} \end{cases} \qquad (4.111)$$

[9] Christopher Reuben Calladine, British engineer (1935–). See last chapter.

with:

$$\begin{cases} \mathcal{P}^\alpha = 2b_\gamma{}^\alpha \mathsf{Q}^\gamma - \mathsf{M}^{\beta\gamma}\nabla_\beta b_\gamma{}^\alpha \\ \mathcal{P}^3 = -\nabla_\alpha \mathsf{Q}^\alpha - b_{\alpha\beta} b_\gamma{}^\beta \mathsf{M}^{\alpha\gamma} \end{cases}$$

Equations (a) and (b) are the equations of a membrane, while (c) only relate to flexion.

The equality of the displacements of the two surfaces in every point provides the complementary relations compensating for the introduction of the three auxiliary variables $\mathcal{P}^\alpha$, $\mathcal{P}^3$.

In practice, this concept is interesting in the case of shallow shells. In this case, the two tangential components $\mathcal{P}^\alpha$ can be neglected, which allows a direct resolution of the membrane problem (a) + (b) and the equation (c) is reduced to the equation of a plate in bending. Then there remains only one auxiliary variable $\mathcal{P}^3$ and the compatibility of deformation of the two surfaces, which makes it possible to completely solve the problem, is expressed by equalising the transverse displacements .

4.3 PROBLEM OF MOVEMENT AND CONSTITUTIVE LAWS

4.3.1 Need to take into account the behaviour of materials

The equations of motion (local (4.97) and at the boundaries (4.99)) are not sufficient, except in static in the case of the membrane theory, to determine the generalised stresses M and N. Indeed, these are represented by six independent scalar fields, taking into account the symmetries of M and N. The local equations are three in number. There are therefore three relations missing, which must reflect the behaviour of the materials.

A **global constitutive law** of a shell connects the generalised stresses (M, N) to the deformation variables (e, k) (integrated from a three-dimensional constitutive law). In the general case, this relation depends on the history of the material and of the shell: the material could be yielded during its manufacture or loading prior to the situation considered. The constitutive laws can be written in the form:

$$\mathsf{N} = \mathcal{F}\left(\mathsf{T}, e, k\right)$$

$$\mathsf{M} = \mathcal{G}\left(\mathsf{T}, e, k\right)$$

where T denotes the temperature, $\mathcal{F}$ and $\mathcal{G}$ are two functionals. In this form, laws are not the most general, since generalised stresses can also depend on strain rates and time.

These laws result in six scalar equalities covering the components of the different tensors. The introduction of the constitutive laws allows the "closing" of the problem (remains to demonstrate the existence and possibly the uniqueness of the solutions). Indeed:

- the unknowns are 17:
  - the temperature field T;
  - the surfacic mass density ρ;
  - the displacement field ξ (3 scalar functions);
  - deformation fields (e, k) (6 scalar functions);
  - generalised stresses field (M, N) (6 scalar functions).
- there are also 17 equations:

- the equation of heat;
- the continuity equation;
- three equations of motion;
- six relations expressing the deformation fields as functions of the displacement field;
- six relations reflecting the behaviour of the material.

Mechanical boundary conditions must be supplemented by temperature or heat flow conditions at the edge $\partial\sigma$.

4.3.2 Elastic global constitutive laws

4.3.2.1 Expression of elastic constitutive laws from the thermodynamic potential

In elasticity, the thermodynamic potential of the system depends only on its state, not on the history of loading. By analogy with the three-dimensional theory or the theory of beams, a surface density of thermodynamic potential being dependent only on the temperature and deformation variables is introduced. In Euler variables, this thermodynamic potential writes: $\Phi = \int_\sigma \phi(e,k,\mathrm{T})\, d\sigma$.

By (4.76):

$$\overset{*}{\mathcal{W}}_f = \int_\sigma \left(\mathsf{N} \bullet \overset{*}{e} + \mathsf{M} \bullet \overset{*}{k} \right) d\sigma = \int_\sigma \left(\frac{\partial \phi}{\partial e} \bullet \overset{*}{e} + \frac{\partial \phi}{\partial k} \bullet \overset{*}{k} \right) d\sigma$$

which is true for any virtual deformation, hence:

$$\boxed{\mathsf{N} = \frac{\partial \phi}{\partial e} \quad ; \quad \mathsf{M} = \frac{\partial \phi}{\partial k}} \tag{4.112}$$

This presentation is symbolic: in fact, the differentiations should be written component by component: $\mathsf{N}^{\alpha\beta} = \frac{\partial \phi}{\partial e^{\alpha\beta}}$. Similar expressions can be obtained in Lagrange variables by expressing as a function of ε and κ.

In small deformations, the Lagrange variables are used to express the thermodynamic potential as a function of τ, temperature variation, ε and κ (deformation fields in Lagrange variables, from a reference position Σ):

$$\Phi = \int_\sigma \phi(\varepsilon, \kappa, \tau)\, d\sigma \tag{4.113}$$

In the same way as above, it comes:

$$\boxed{\mathsf{N} = \frac{\partial \phi}{\partial \varepsilon} \quad ; \quad \mathsf{M} = \frac{\partial \phi}{\partial \kappa}} \tag{4.114}$$

4.3.2.2 Linearly elastic shells

To obtain a linearly thermoelastic constitutive law in small deformations, the potential ϕ is developed to the second order, from a reference state $(\mathcal{E}_0)$, as a function of the deformation variables ε, κ defined in Lagrange variables from $(\mathcal{E}_0)$ and of the temperature variation τ:

$$\phi(\tau,\varepsilon,\kappa) = -s_0\tau + \mathsf{N}_0 \bullet \varepsilon + \mathsf{M}_0 \bullet \kappa + \frac{1}{2}\left(A\tau^2 + 2\tau B \bullet \varepsilon + 2\tau B' \bullet \kappa + \varepsilon \bullet \Lambda \bullet \varepsilon + \kappa \bullet \mathrm{H} \bullet \kappa + 2\varepsilon \bullet \mathrm{K} \bullet \kappa\right) \tag{4.115}$$

where: s_0 is the entropy at the state $(\mathcal{E}_0)$;
N_0 and M_0 are tensors of stretching and bending prestresses in the state $(\mathcal{E}_0)$;
A is a coefficient of thermal expansion;
B and B' are tensors containing the coefficients of thermoelasticity;
Λ, H and K are tensors containing the elasticity coefficients.

This formulation assumes that the temperature is constant in the thickness of the shell, which is consistent with the surface approach, but is insufficient in some cases: an approach taking into account the temperature gradients (see for example §7.1.3.3) or a three-dimensional approach is then necessary.

The expression (4.115) of the potential makes it possible to obtain the constitutive law, by (4.114):

$$\begin{aligned} \mathsf{N} &= \mathsf{N}_0 + B\tau + \Lambda \bullet \varepsilon + \mathrm{K} \bullet \kappa \\ \mathsf{M} &= \mathsf{M}_0 + B'\tau + \mathrm{H} \bullet \kappa + \mathrm{K} \bullet \varepsilon \end{aligned} \tag{4.116}$$

Four-fold contravariant tensors Λ, H and K benefit from the symmetries of ε and κ, for example:

$$\mathrm{K}^{\alpha\beta\gamma\delta} = \mathrm{K}^{\alpha\beta\delta\gamma} = \mathrm{K}^{\beta\alpha\gamma\delta} = \mathrm{K}^{\beta\alpha\delta\gamma} \tag{4.117}$$

The various coefficients depend on the materials used, but also on the constitution of the shell (variations of the mechanical properties in the thickness, as in the case of composites). They can be determined by direct measurements of an element of the shell, but this is not always feasible under acceptable economic conditions. Predictive behaviour models are obtained from three-dimensional laws, the shell is considered a three-dimensional solid. In practice, the surface constitutive laws are deduced from the three-dimensional constitutive laws by means of certain kinematic assumptions of behaviour in the thickness of the shell (see Chapter 5).

4.4 COMPARISON OF THE EQUATIONS OBTAINED BY THE EQUILIBRIUM EQUATIONS AND BY THE PVP

4.4.1 Comparison of local equations

The expression of the shell strain energy, in view of the application of the Principle of Virtual Powers, leads to introduce generalised stresses associated with the deformations. It is interesting to relate these generalised stresses to the resultants of the internal actions introduced in §4.2.1. In three-dimensional theory, the stress tensor involved in the expression of the strain energy can be directly interpreted as representing the actions internal to a solid. In the case of shells, the comparison is less direct and a calculation effort is necessary to compare the equations of movement obtained by the PVP and those obtained by applying the fundamental momentum law to a piece of shell. This comparison is the subject of this paragraph.

Relationships (4.66) and (4.68) provide six scalar equations, whereas relations (4.97), to which they can be compared, provide only five equations. But the tensors **N** and **M**, generalised stresses associated with e and k, are symmetrical, which gives additional relations.

The densities of external moments c applied to the surface, the internal moments $M^{\alpha 3}$ of axis a_3 and the rotational inertia actions are not considered here, which corresponds to the current situation, to simplify the presentation. Translational inertia actions are "hidden" in external actions because they are of the same form. Actions that have not been taken into account can be restored without difficulty. Equations (4.66) and (4.68) are thus reduced to:

$$\begin{array}{|l|}\hline \nabla_\alpha N^{\alpha\gamma} - b_\alpha{}^\gamma N^{\alpha 3} + p^\gamma = 0 \\ N^{\alpha\beta} b_{\alpha\beta} + \nabla_\alpha N^{\alpha 3} + p^3 = 0 \\ \nabla_\alpha M^{\alpha\beta} - N^{\alpha 3} e_{\alpha\gamma} a^{\gamma\beta} = 0 \\ M^{\alpha\beta} b_{\alpha\beta} + N^{\alpha\beta} e_{\alpha\beta} = 0 \\ \hline \end{array} \tag{4.118}$$

These relations should be compared to relations (4.97), which are reduced to:

$$\begin{array}{|l|}\hline \nabla_\beta \mathsf{N}^{\beta\alpha} - 2b_\gamma{}^\alpha \mathsf{Q}^\gamma + \mathsf{M}^{\beta\gamma} \nabla_\beta b_\gamma{}^\alpha + p^\alpha = 0 \\ \nabla_\alpha \mathsf{Q}^\alpha + b_{\alpha\beta} \mathsf{N}^{\alpha\beta} + b_{\alpha\beta} b_\gamma{}^\beta \mathsf{M}^{\alpha\gamma} + p^3 = 0 \\ \mathsf{Q}^\gamma + \nabla_\beta \mathsf{M}^{\beta\gamma} = 0 \\ \hline \end{array} \tag{4.119}$$

Despite the similarity of equations (4.118) and (4.119), they cannot be compared term by term: there is no coincidence between the tensors **N** and N on the one hand, **M** and M on the other hand.

Let a facet of normal ν and $H(\nu)$ the vector of $T_{\mathcal{M}}$ obtained by rotation of $\frac{\pi}{2}$ of $M(\nu)$ (Figure 4.9):

$$H(\nu) = a_3 \times M(\nu) = a_3 \times \left(M^{\alpha\gamma} \nu_\alpha a_\gamma \right) = M^{\alpha\gamma} \nu_\alpha \, e_{\gamma\delta} \, a^{\delta\beta} a_\beta \tag{4.120}$$

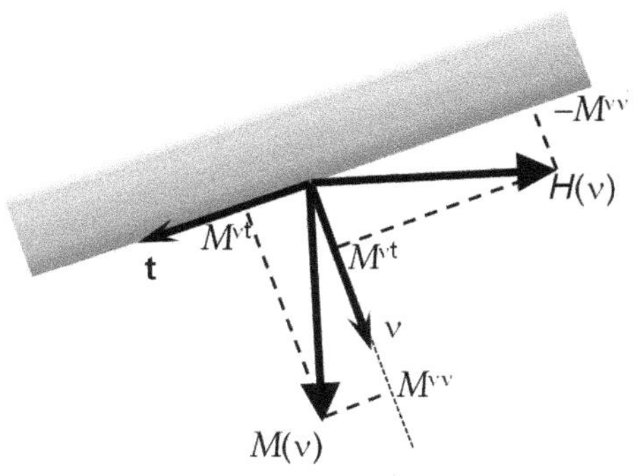

Figure 4.9

Then let H the tensor on $T_{\mathcal{M}}$, of components:

$$H^{\alpha\beta} = M^{\alpha\gamma} e_{\gamma\delta} \, a^{\delta\beta} \tag{4.121}$$

such that:

$$H(\nu) = H^{\alpha\beta} \nu_\alpha a_\beta$$

According to relation (3.17):

$$H^{\alpha\beta} e_{\lambda\beta} = M^{\alpha\gamma} e_{\gamma\delta} \, a^{\delta\beta} e_{\lambda\beta}$$
$$= M^{\alpha\gamma} a_{\gamma\lambda}$$

hence the relationship inverse of (4.121):

$$M^{\alpha\mu} = H^{\alpha\beta} a^{\lambda\mu} \,\mathrm{e}_{\lambda\beta} \tag{4.122}$$

Finally, let:

$$\mathcal{M}^{\alpha\beta} = \frac{1}{2}\left(H^{\alpha\beta} + H^{\beta\alpha}\right) \tag{4.123}$$

and:

$$\mathcal{A}^{\alpha\beta} = \frac{1}{2}\left(H^{\alpha\beta} - H^{\beta\alpha}\right)$$

the symmetrical and antisymmetrical parts of H.

The tensor with components:

$$\mathcal{N}^{\alpha\beta} = N^{\alpha\beta} - b_\lambda{}^\beta H^{\lambda\alpha} \tag{4.124}$$

is symmetrical:

$$N^{\alpha\beta} - b_\lambda{}^\beta H^{\lambda\alpha} = N^{\beta\alpha} - b_\lambda{}^\alpha H^{\lambda\beta} \tag{4.125}$$

Indeed, its product totally contracted with the antisymmetric tensor e:

$$\begin{aligned}\mathcal{N}^{\alpha\beta}\mathrm{e}_{\alpha\beta} &= N^{\alpha\beta}\mathrm{e}_{\alpha\beta} - b_\lambda{}^\beta H^{\lambda\alpha}\mathrm{e}_{\alpha\beta} = N^{\alpha\beta}\mathrm{e}_{\alpha\beta} - b_\lambda{}^\beta \mathrm{e}_{\alpha\beta}\,\mathrm{e}_{\gamma\delta} a^{\delta\alpha} M^{\lambda\gamma}\\ &= N^{\alpha\beta}\mathrm{e}_{\alpha\beta} + b_{\alpha\beta} M^{\alpha\beta}\end{aligned}$$

is null according to the last relation (4.118), which demonstrates symmetry.

The third line of (4.118) is rewritten:

$$\mathrm{e}_{\beta\delta}\nabla_\alpha M^{\alpha\beta} = N^{\alpha 3}\mathrm{e}_{\alpha\gamma}\,\mathrm{e}_{\beta\delta} a^{\gamma\beta} = -\,a_{\alpha\delta} N^{\alpha 3}$$

where it comes from, with (4.122):

$$N^{\lambda 3} = -\,\mathrm{e}_{\beta\delta} a^{\lambda\delta}\,\nabla_\alpha M^{\alpha\beta} = -\,\nabla_\alpha H^{\alpha\lambda} - a^{\lambda\delta}\mathrm{e}_{\beta\delta} H^{\alpha\mu}\,\nabla_\alpha\left(a^{\gamma\beta}\mathrm{e}_{\gamma\mu}\right)$$

Now, according to (3.17), (3.72), and (3.74):

$$\nabla_\alpha\left(a^{\gamma\beta}\mathrm{e}_{\gamma\mu}\,\mathrm{e}_{\beta\delta}\right) = \nabla_\alpha a_{\mu\delta} = 0$$

$$\nabla_\alpha a^{\gamma\beta} = 0$$

Finally :

$$N^{\lambda 3} = -\nabla_\alpha H^{\alpha\lambda} \tag{4.126}$$

So, according to (4.123), (4.124), and (4.126):

$$\begin{aligned}\nabla_\alpha N^{\alpha\gamma} - b_\alpha{}^\gamma N^{\alpha 3} &= \nabla_\alpha\left(\mathcal{N}^{\alpha\gamma} + b_\mu{}^\gamma H^{\mu\alpha}\right) + b_\alpha{}^\gamma \nabla_\mu H^{\mu\alpha} \\ &= \nabla_\alpha \mathcal{N}^{\alpha\gamma} + H^{\mu\alpha}\nabla_\alpha b_\mu{}^\gamma + 2b_\mu{}^\gamma \nabla_\alpha \mathcal{M}^{\alpha\mu}\end{aligned} \tag{4.127}$$

Now, by (4.123):

$$H^{\mu\alpha}\ \nabla_\alpha b_\mu{}^\gamma = \mathcal{M}^{\alpha\mu}\ \nabla_\alpha b_\mu{}^\gamma + \mathcal{A}^{\alpha\mu}\ \nabla_\alpha b_\mu{}^\gamma$$

The second term of the second member is zero because it is the totally contracted product of an antisymmetric tensor and a symmetric tensor. Indeed, according to the Mainardi–Codazzi equations:

$$\nabla_\alpha b_\mu{}^\gamma = \nabla_\mu b_\alpha{}^\gamma$$

By posing then:

$$\mathcal{Q}^\lambda = -\nabla_\alpha \mathcal{M}^{\alpha\lambda} \tag{4.128}$$

which is not equal to $N^{\lambda 3}$, because of the antisymmetric part of H, (4.127) is rewritten:

$$\nabla_\alpha N^{\alpha\gamma} - b_\alpha{}^\gamma N^{\alpha 3} = \nabla_\alpha \mathcal{N}^{\alpha\gamma} + \mathcal{M}^{\alpha\mu}\ \nabla_\alpha b_\mu{}^\gamma - 2b_\mu{}^\gamma \mathcal{Q}^\mu$$

Hence the equivalence of the first relations of (4.118) and (4.119), with $\mathcal{N} \equiv \mathsf{N}$, $\mathcal{M} \equiv \mathsf{M}$ and $\mathcal{Q} \equiv \mathrm{Q}$.

Finally, the totally contracted product $c \bullet \mathcal{A}$ is zero (where c denotes the third symmetrical fundamental form, see §3.1.3.6):

$$N^{\alpha\beta} b_{\alpha\beta} = \left(\mathcal{N}^{\alpha\beta} + b_\lambda{}^\beta H^{\lambda\alpha}\right) b_{\alpha\beta} = \mathcal{N}^{\alpha\beta} b_{\alpha\beta} + b_\lambda{}^\beta b_{\alpha\beta} \mathcal{M}^{\lambda\alpha}$$

To finally demonstrate the equivalence of the second lines of (4.118) and (4.119), it must be shown that:

$$\nabla_\alpha N^{\alpha 3} = \nabla_\alpha \mathcal{Q}^\alpha$$

which implies:

$$\nabla_\alpha \nabla_\beta \mathcal{A}^{\alpha\beta} = 0$$

This last relation can be demonstrated by developing the double differentiation, and by using the antisymmetry of $\mathcal{A}$, which product totally contracted with a symmetrical quantity is null:

$$\nabla_\alpha \nabla_\beta \mathcal{A}^{\alpha\beta} = \partial_\alpha \left(\partial_\beta \mathcal{A}^{\alpha\beta} + \Gamma^\alpha_{\beta\lambda} \mathcal{A}^{\lambda\beta} + \Gamma^\beta_{\beta\lambda} \mathcal{A}^{\alpha\lambda} \right)$$
$$+ \Gamma^\alpha_{\alpha\mu} \left(\partial_\beta \mathcal{A}^{\mu\beta} + \Gamma^\mu_{\beta\lambda} \mathcal{A}^{\lambda\beta} + \Gamma^\beta_{\beta\lambda} \mathcal{A}^{\mu\lambda} \right)$$
$$- \Gamma^\gamma_{\alpha\beta} \left(\partial_\gamma \mathcal{A}^{\alpha\beta} + \Gamma^\alpha_{\alpha\lambda} \mathcal{A}^{\lambda\beta} + \Gamma^\beta_{\gamma\lambda} \mathcal{A}^{\alpha\lambda} \right)$$
$$+ \Gamma^\beta_{\alpha\gamma} \left(\partial_\beta \mathcal{A}^{\alpha\gamma} + \Gamma^\alpha_{\beta\lambda} \mathcal{A}^{\lambda\gamma} + \Gamma^\gamma_{\beta\lambda} \mathcal{A}^{\alpha\lambda} \right)$$

and taking into account the equality obtained by differentiating (3.75):

$$\partial_\alpha \Gamma^\beta_{\beta\lambda} = \partial_\lambda \Gamma^\beta_{\beta\alpha}$$

The following points are particularly noteworthy:

- only the symmetrical part $\mathcal{M}$ of the tensor H intervenes in equilibrium equations.
- the moment $M(\mathrm{v})$ applied to the facet is orthogonal to $H(\mathrm{v})$, associated with the tensor of moments.

Strictly speaking, the tensors N, M, and Q are identical to the tensors $\mathcal{N}$, $\mathcal{M}$, and $\mathcal{Q}$ respectively only to a self-equilibrated field. However, this self-balanced field being arbitrary, there is no physical objection to formally equating these tensors (which is done in the following).

In conclusion:

$$\forall \mathrm{v} \in T_{\mathcal{M}} : H(\mathrm{v}) = a_3 \times M(\mathrm{v})$$

$$\begin{aligned} \mathsf{N}^{\alpha\beta} &= N^{\alpha\beta} - b_\lambda{}^\beta H^{\lambda\alpha} \\ \mathsf{M}^{\alpha\beta} &= \frac{1}{2}\left(H^{\alpha\beta} + H^{\beta\alpha}\right) \\ \mathsf{Q}^\beta &= N^{\beta 3} + \frac{1}{2}\nabla_\alpha\left(H^{\alpha\beta} - H^{\beta\alpha}\right) \end{aligned} \tag{4.129}$$

In the orthonormal coordinate system $\left(\mathbf{t}, \mathrm{v}, a_3\right)$:

$$\begin{aligned} &H^{t\mathrm{v}} = M^{tt} \qquad\qquad H^{\mathrm{v}t} = -M^{\mathrm{vv}} \\ &\mathsf{M}^{tt} = H^{tt} = -M^{t\mathrm{v}} \qquad \mathsf{M}^{\mathrm{vv}} = H^{\mathrm{vv}} = M^{\mathrm{v}t} \\ &\mathsf{M}^{t\mathrm{v}} = \mathsf{M}^{\mathrm{v}t} = \frac{1}{2}\left(M^{tt} - M^{\mathrm{vv}}\right) \end{aligned} \tag{4.130}$$

$$\begin{aligned} \mathsf{N}^{\mathrm{vv}} &= N^{\mathrm{vv}} - b_t{}^{\mathrm{v}} H^{t\mathrm{v}} - b_{\mathrm{v}}{}^{\mathrm{v}} H^{\mathrm{vv}} = N^{\mathrm{vv}} - b_t{}^{\mathrm{v}} M^{tt} - b_{\mathrm{v}}{}^{\mathrm{v}} M^{\mathrm{v}t} \\ \mathsf{N}^{tt} &= N^{tt} - b_t{}^t H^{tt} - b_{\mathrm{v}}{}^t H^{\mathrm{v}t} = N^{tt} + b_t{}^t M^{t\mathrm{v}} + b_{\mathrm{v}}{}^t M^{\mathrm{vv}} \\ \mathsf{N}^{t\mathrm{v}} &= N^{t\mathrm{v}} - b_t{}^{\mathrm{v}} H^{tt} - b_{\mathrm{v}}{}^{\mathrm{v}} H^{\mathrm{v}t} = N^{t\mathrm{v}} + b_t{}^{\mathrm{v}} M^{t\mathrm{v}} + b_{\mathrm{v}}{}^{\mathrm{v}} M^{\mathrm{vv}} \\ &= N^{\mathrm{v}t} - b_t{}^t H^{t\mathrm{v}} - b_{\mathrm{v}}{}^t H^{\mathrm{vv}} = N^{\mathrm{v}t} - b_t{}^t M^{tt} - b_{\mathrm{v}}{}^t M^{\mathrm{v}t} \end{aligned} \tag{4.131}$$

$$\mathsf{Q}^{t} = N^{t3} - \frac{1}{2}\partial_{v}\left(M^{tt} + M^{vv}\right)$$
$$\mathsf{Q}^{v} = N^{v3} + \frac{1}{2}\partial_{t}\left(M^{tt} + M^{vv}\right) \quad (4.132)$$

4.4.2 Example: sphere under internal pressure

If the material is homogeneous and isotropic, considering the spherical symmetry, the tensors N and M are homogeneous and isotropic:

$$N = n\, a \qquad M = m\, a$$

$N^{\alpha 3}$ is then zero and the equations (4.118) are reduced to:

$$-2\frac{n}{R} + p = 0$$
$$-2\frac{m}{R} \quad = 0$$

where R is the radius of the deformed sphere. The equilibrium is a membrane one. Starting now from relations (4.119) with the assumption that the tensors N and M are homogeneous and isotropic, by spherical symmetry:

$$\mathsf{N} = \mathsf{n}\, a \qquad \mathsf{M} = \mathsf{m}\, a$$

It becomes:

$$-2\frac{\mathsf{n}}{R} + 2\frac{\mathsf{m}}{R^2} + p = 0$$

Finally, note that $M(\mathrm{v}) = m\,\mathrm{v}$ and $M(\mathbf{t}) = m\,\mathbf{t}$, the tensor H is equal to:

$$H^{vt} = -\,H^{tv} = -\,m = 0$$
$$H^{vv} = H^{tt} = 0$$

H is zero and, therefore, $\mathcal{M}$ and $\mathcal{Q}$ are zero. Finally:

$$\mathcal{N}^{vv} = \mathcal{N}^{tt} = n$$
$$\mathcal{N}^{vt} = \mathcal{N}^{tv} = \frac{m}{R} = 0$$

$\mathcal{N}$ and N being regarded as identical, then necessarily $\mathsf{m} = 0$ and the equations are identical:

$$\mathsf{n} = n = \frac{pR}{2}$$
$$\mathsf{m} = m = 0$$

4.4.3 Boundary conditions

By limiting to the case where $c = 0$, the boundary conditions applied to N and M, the resultants of internal actions in the restricted theory, are obtained by introducing relations (4.130) to (4.132) in the boundary conditions (4.99) relating to the generalised stresses **N** and **M**. It becomes:

$$\boxed{\begin{aligned} & N^{\nu t} + \left(m^{\nu} - M^{\nu\nu}\right)\rho_n - q^t = 0 \\ & N^{\nu\nu} + \left(m^{\nu} - M^{\nu\nu}\right)\tau_g - q^{\nu} = 0 \\ & m^t - M^{\nu t} = 0 \\ & N^{\nu 3} - \frac{d}{ds}\left(m^{\nu} - M^{\nu\nu}\right) - q^3 = 0 \end{aligned}} \tag{4.133}$$

analogous to relationships (4.75).

4.5 MECHANICS OF COSSERAT-ORIENTED SURFACES

4.5.1 Kinematics

4.5.1.1 Principle

In the approach described in §4.2.2, the strain energy considered is solely due to the deformation of the middle surface. This is consistent with the Kirchhoff–Love theory exposed in Chapter 5. In this theory, the material points initially located on the normal to the middle surface remain on this normal during evolution. This hypothesis amounts to neglecting the distortions $\bar{\varepsilon}_{\alpha 3}$ due to the shear force. On the other hand, the normal segment is supposed to remain rigid to express the tangent components of the local strain tensor $\bar{\varepsilon}_{\alpha\beta}$: the variations of thickness of the shell during the movement are neglected. It thus appears that the shell strain energy is reduced, in Kirchhoff's theory, to the strain energy of its middle surface: Kirchhoff's theory is also called a **restricted theory**.

The choice of the restricted theory does not make it possible to account for neglected terms in the expression of the strain energy, in particular those due to the shear force distortion and the variation in thickness of the shell. To take into account the strain energy associated with such phenomena, additional deformation variables should be considered.

In this case, the representation of the middle surface by its single position at any moment is insufficient, because it has been shown in §3.3.1 that ε and κ completely characterises, by itself, the deformation of the surface. It is therefore necessary to complete the description of the kinematics of the surface by quantities adding to its geometric position.

In 1909, E. and F. Cosserat[10] developed the theory of oriented mediums, where vector fields are introduced that are not constrained to follow rotations of the normal

[10] Eugène-Maurice-Pierre Cosserat, French mathematician and astronomer (1866–1931); François Cosserat, French engineer and mathematician (1852–1914).

vector. A surface with such vector fields is called a **Cosserat surface.** For the practical needs of the development of shell theory, a single field of such vectors (called directors) is useful.

The application to the shells of the theory of oriented mediums, synthesising various previous works, notably by Green[11] (1965), has been formalised by Naghdi[12] (1972). The approach used in these works is the application of momentum laws. Below, only the energy approach is developed.

4.5.1.2 Parameterisation of the Cosserat surface

A Cosserat surface is defined in parametric form by the pair consisting of the geometric position of any point $\boldsymbol{m}$ of the surface and a field of directors $\boldsymbol{d}$ defined on the surface, a pair called the surface **configuration:**

$$\sigma:(x_1,x_2)\in\mathbf{R}^2\to\begin{cases}\boldsymbol{m}(x_1,x_2)\in\sigma\subset\mathbf{R}^3\\ \boldsymbol{d}(x_1,x_2)\in\{\mathbf{R}^3\}_{\boldsymbol{m}}\end{cases}\tag{4.134}$$

During an evolution, in the Lagrangian description, the surface takes successive configurations, starting from an initial configuration Σ defined by:

$$\Sigma:(X_1,X_2)\in\mathbf{R}^2\to\begin{cases}\mathcal{M}(X_1,X_2)\in\Sigma\subset\mathbf{R}^3\\ \mathcal{D}(X_1,X_2)\in\{\mathbf{R}^3\}_{\mathcal{M}}\end{cases}\tag{4.135}$$

To simplify the presentation, the **same parameterisation**, in Lagrange variables, is used for all the successive configurations, in which case the configuration at time t is written:

$$\sigma(t):(X_1,X_2)\in\mathbf{R}^2\to\begin{cases}\boldsymbol{m}(X_1,X_2,t)\in\sigma\subset\mathbf{R}^3\\ \boldsymbol{d}(X_1,X_2,t)\in\{\mathbf{R}^3\}_{\boldsymbol{m}}\end{cases}\tag{4.136}$$

with:

$$\begin{cases}\mathcal{M}(X_1,X_2)=\boldsymbol{m}(X_1,X_2,0)\\ \mathcal{D}(X_1,X_2)=\boldsymbol{d}(X_1,X_2,0)\end{cases}\tag{4.137}$$

It is this simplified representation (which does not restrict the generality of the reasoning) that is adopted below.

In practice, the vector $\boldsymbol{d}$ is intended to represent the position of the material points located on the normal in the reference configuration: its variations in length are related to variations in the thickness of the shell, and its rotations to shear force distortions. Without detracting from the generality, $\mathcal{D}$ can be put in the form:

$$\mathcal{D}=\bar{D}A_3\tag{4.138}$$

[11] Albert Edward Green, British mathematician (1912–1999).
[12] Paul Mansour Naghdi, American (born Iranian) engineer (1924–1994).

In general, it is sufficient to take $\bar{D} = -1$ (which is consistent with the notations of Chapter 5, as demonstrated below), but there may be some interest in showing the variations in the thickness of the shell as a function of X^α, to which case:

$$\bar{D} = -\frac{e}{e_0} \tag{4.139}$$

e_0 being a reference thickness. If the thickness of the shell is constant in the reference state, then:

$$\bar{D} = -1 \; ; \; \mathcal{D} = -A_3$$

In the context of the restricted theory, $\boldsymbol{d}$ is equal to $\bar{d}\, a_3$, $\bar{d}$ being independent of time $\left(\bar{d} = \bar{D}\right)$, and then there is no need to consider the field of directors, since a_3 is entirely defined from the position $\boldsymbol{m}\left(X_1, X_2, t\right)$ of the surface.

4.5.1.3 Kinematics of the director

The relationships established in §4.1.1 concerning the kinematics of the surface remain applicable. They must be supplemented by relations concerning the evolution of the director.

In Euler variables:

$$\boldsymbol{d} = d^i\, a_i = d_i\, a^i \tag{4.140}$$

By differentiation with respect to time:

$$\dot{\boldsymbol{d}} = \dot{d}_i a^i + d_i \dot{a}^i$$

either, by (4.20) and (4.21):

$$\dot{\boldsymbol{d}} = \left[\dot{d}_k + d^i\left(\mathrm{s}_{ik} - \eta_{ik}\right)\right] a^k \tag{4.141}$$

On the other hand, differentiating $\boldsymbol{d}$ with respect to the curvilinear coordinates x^α:

$$D_\alpha \boldsymbol{d} = \partial_\alpha d_i \bullet a^i + d_i\, D_\alpha a^i$$

by (3.63):

$$D_\alpha \boldsymbol{d} = \left(\partial_\alpha d_\gamma - \Gamma^\beta_{\alpha\gamma} d_\beta - b_{\alpha\gamma} d_3\right) a^\gamma + \left(\partial_\alpha d_3 + b_\alpha{}^\beta d_\beta\right) a^3$$

which is noted:

$$D_\alpha \boldsymbol{d} = \mathrm{p}_{\alpha i} a^i = \mathrm{p}_\alpha{}^i a_i$$

$$\text{with:} \quad \mathrm{p}_{\alpha\beta} = \nabla_\alpha d_\beta - b_{\alpha\beta} d_3 \tag{4.142}$$

$$\mathrm{p}_{\alpha 3} = \partial_\alpha d_3 + b_\alpha{}^\beta d_\beta$$

The derivative of $\dot{\mathbf{d}}$ with respect to x^{α} is deduced:

$$D_{\alpha}\dot{\mathbf{d}} = \dot{\rho}_{\alpha i}\, a^{i} + \rho_{\alpha i}\, \dot{a}^{i} = \left[\, \dot{\rho}_{\alpha k} + \rho_{\alpha}{}^{i}\left(s_{ik} - \eta_{ik}\right)\right] a^{k} \tag{4.143}$$

4.5.1.4 Deformation and rate of deformation

The strain energy is related to the movement of macroscopic material particles. In the framework of the theory presented here, part of this energy is due to the variations of the metric of the surface (extensions, variations of angles), that is the stretching energy, another part representing the bending energy is linked to the variations of grad $\mathbf{d}$ ($\mathbf{d}$ bears the material particles in the thickness and no longer a_3: grad $\mathbf{d}$ somehow represents a curvature related to the material points), the last part representing both the shear force distortion energy and the energy related to the thickness variation is due to the variation of the director.

As a result, the deformation variables (Lagrangian) are:

$$\begin{cases} \varepsilon_{\alpha\beta} = \dfrac{1}{2}\left(a_{\alpha\beta} - A_{\alpha\beta}\right) \\ \rho_{\alpha i} = \rho_{\alpha i} - \rho_{\alpha i} \\ \vartheta_{i} = \mathbf{d}_{i} - \mathcal{D}_{i} \end{cases} \tag{4.144}$$

and the strain rates expressed in Euler variables:

$$\begin{cases} \dot{e}_{\alpha\beta} = \dfrac{1}{2}\dot{a}_{\alpha\beta} \\ \dot{r}_{\alpha i} = \dot{\rho}_{\alpha i} \\ \dot{y}_{i} = \dot{\mathbf{d}}_{i} \end{cases} \tag{4.145}$$

In the context of the restricted theory, the director $\mathbf{d}$ is obliged to follow the relation which relates it to the normal:

$$\mathbf{d} = -\, a_3$$

and then:

$$\begin{cases} \mathbf{d}_{\alpha} = \mathcal{D}_{\alpha} = 0 \\ \mathbf{d}_{3} = \mathcal{D}_{3} = -1 \end{cases} \quad \Rightarrow \quad \begin{cases} \rho_{\alpha\beta} = b_{\alpha\beta} \\ \rho_{\alpha 3} = 0 \end{cases} \tag{4.146}$$

In this case, ρ is reduced to the curvature variation tensor κ and ϑ is zero.

The choice of the deformation variables (4.144) is only sensible if they are zero in a rigid displacement. In such a case, the director is rigidly bound to the surface and undergoes the same rotation, so that such a movement can be written:

$$\begin{cases} \mathcal{M}m = \mathcal{M}_0 m_0 + Q(\mathcal{M}_0\mathcal{M}) \\ d = Q(\mathcal{D}) \end{cases} \tag{4.147}$$

where Q is a rotation operator, represented by an antisymmetric matrix $[Q]$ such that $[Q][Q]^T = [Q]^T[Q] = [I]$ and $\det([Q]) = 1$.

It is clear that the metric is then preserved (see §3.3.1).

Vectors of the basis undergo the same rotation:

$$a_i = Q(A_i)$$

By linearity of Q:

$$d = Q(\mathcal{D}) = \mathcal{D}^i\, Q(A_i) = \mathcal{D}^i\, a_i = \mathcal{D}^i\, a_{ij}a^j$$

but the metric being conserved $\mathcal{D}^i\, a_{ij} = \mathcal{D}_j$, then:

$$d = \mathcal{D}_i\, a^i$$

The covariant components of d and $\mathcal{D}$ are equal, and therefore ϑ is zero. The same reasoning is applied to the vector $D_\alpha d$, which implies with (4.142) that the components $\rho_{\alpha i}$ are kept. Therefore tensor ρ is zero.

4.5.1.5 Expression of deformation variables by displacement and director

The transformation defined by (4.136) and (4.137) can be written as:

$$\begin{cases} m = \mathcal{M} + \xi \\ d = \mathcal{D} + \delta \end{cases} \tag{4.148}$$

where:

- ξ is the displacement field;
- δ is the field of director variation.

Now the question is to express ε, ρ and ϑ as functions of ξ and δ. Only the case of small displacements is examined. At first, $\mathcal{D}$ is considered as any.

It must first be noted that ϑ is not equal to δ, because in the expression (4.144), d_i is expressed in the base a^i, as well as $a_{\alpha i}$ and $\rho_{\alpha i}$. Intrinsically:

$$\vartheta = \mathfrak{f}^{-T}\left[b(d)\right] - b(\mathcal{D}) \tag{4.149}$$

while $\delta = \mathfrak{f}^{-1}(d) - \mathcal{D}$. In the restricted theory, ϑ is zero (whereas δ is not). ϑ is a good deformation variable, since it allows us to measure the gap between the restricted theory (Kirchhoff) and the more general theory exposed here.

$a_\alpha, a_{\alpha\beta}, \varepsilon_{\alpha\beta}, b_{\alpha\beta}$ and $\kappa_{\alpha\beta}$ are obtained by relations (3.95), (3.102), (3.106), and (3.108), which remain applicable here. Recalling that $a_\alpha = A_\alpha + D_\alpha \xi$, a development limited to first-order leads to:

$$\boldsymbol{d}_\alpha = \boldsymbol{d} \bullet a_\alpha \approx \mathcal{D} \bullet A_\alpha + \mathcal{D} \bullet D_\alpha \xi + \delta \bullet A_\alpha \approx \mathcal{D}_\alpha + \delta_\alpha + \mathcal{A}_\alpha{}^\beta \mathcal{D}_\beta + \mathcal{B}_\alpha \mathcal{D}_3 \tag{4.150}$$

that is:

$$\boxed{\vartheta_\alpha \approx \delta_\alpha + \left(\nabla_\alpha \xi^\beta - \zeta B_\alpha{}^\beta\right) \mathcal{D}_\beta + \left(B_\alpha{}^\beta \xi_\beta + \partial_\alpha \zeta\right) \mathcal{D}_3} \tag{4.151}$$

Noting on the other hand, by (3.103), that: $a_3 = A_3 - \mathcal{B}^\beta A_\beta$, it becomes:

$$\boldsymbol{d}_3 = \boldsymbol{d} \bullet a_3 = \mathcal{D} \bullet A_3 + \delta \bullet A_3 + \mathcal{D} \bullet \left(-\mathcal{B}^\beta A_\beta\right) = \mathcal{D}_3 + \delta_3 - \mathcal{B}^\beta \mathcal{D}_\beta \tag{4.152}$$

that is:

$$\boxed{\vartheta_3 = \delta_3 - \left(B_\beta{}^\gamma \xi_\gamma + \partial_\beta \zeta\right) \mathcal{D}^\beta} \tag{4.153}$$

Given the definition (3.94), by posing $\mathcal{A}_{33} = 0$ and $\mathcal{A}_{3\alpha} = -\mathcal{A}_{\alpha 3} = -\mathcal{B}_\alpha$, formulas (4.151) and (4.153) are rewritten, in the simplified form:

$$\vartheta_i \approx \delta_i + \mathcal{A}_{ij} \mathcal{D}^j \tag{4.154}$$

Finally, from (4.142):

$$\begin{aligned} \mathfrak{p}_{\alpha\beta} &= D_\alpha \boldsymbol{d} \bullet a_\beta \approx D_\alpha \mathcal{D} \bullet A_\beta + D_\alpha \delta \bullet A_\beta + D_\alpha \mathcal{D} \bullet D_\beta \xi \\ &\approx \mathcal{P}_{\alpha\beta} + \nabla_\alpha \delta_\beta - B_{\alpha\beta} \delta_3 + \left(\mathcal{P}_\alpha{}^i A_i\right)\left(\mathcal{A}_\beta{}^j A_j\right) \\ &\approx \mathcal{P}_{\alpha\beta} + \nabla_\alpha \delta_\beta - B_{\alpha\beta} \delta_3 + \mathcal{B}_\beta \mathcal{P}_{\alpha 3} + \mathcal{A}_\beta{}^\gamma \mathcal{P}_{\alpha\gamma} \end{aligned} \tag{4.155}$$

(where $\mathcal{P}$ is fully defined in the initial configuration), resulting in:

$$\rho_{\alpha\beta} \approx \nabla_\alpha \delta_\beta - B_{\alpha\beta} \delta_3 + \mathcal{A}_\beta{}^j \mathcal{P}_{\alpha j} \tag{4.156}$$

that is:

$$\boxed{\rho_{\alpha\beta} \approx \nabla_\alpha \delta_\beta - B_{\alpha\beta} \delta_3 + \left(\nabla_\beta \xi^\gamma - \zeta B_\beta{}^\gamma\right) \mathcal{P}_{\alpha\gamma} + \left(B_\beta{}^\gamma \xi_\gamma + \partial_\beta \zeta\right) \mathcal{P}_{\alpha 3}} \tag{4.157}$$

and:

$$\begin{aligned} \mathfrak{p}_{\alpha 3} &= D_\alpha \boldsymbol{d} \bullet a_3 \approx D_\alpha \mathcal{D} \bullet A_3 + D_\alpha \delta \bullet A_3 + D_\alpha \mathcal{D} \bullet \left(-\mathcal{B}^\beta A_\beta\right) \\ &\approx \mathcal{P}_{\alpha 3} + \partial_\alpha \delta_3 + B_\alpha{}^\beta \delta_\beta - \mathcal{B}^\beta \mathcal{P}_{\alpha\beta} \end{aligned} \tag{4.158}$$

that is:

$$\boxed{\rho_{\alpha 3} \approx \partial_\alpha \delta_3 + B_\alpha{}^\beta \delta_\beta - \left(B_\beta{}^\gamma \xi_\gamma + \partial_\beta \zeta\right) \mathcal{P}_\alpha{}^\beta} \tag{4.159}$$

or:

$$\rho_{\alpha,3} \approx \partial_\alpha \delta_3 + B_\alpha{}^\beta \delta_\beta + \mathcal{A}_3{}^j \mathcal{P}_{\alpha j} \tag{4.160}$$

Since general formulas have been established, it is interesting to adopt simplification (4.138), in which case the above relations are reduced to:

$$\boxed{\begin{aligned} &\vartheta_\alpha \approx \delta_\alpha + \left(B_\alpha{}^\beta \xi_\beta + \partial_\alpha \zeta\right) \bar{\mathrm{D}} \\ &\vartheta_3 \approx \delta_3 \end{aligned}} \tag{4.161}$$

$$\rho_{\alpha\beta} \approx \nabla_\alpha \delta_\beta - B_{\alpha\beta}\delta_3 - \left(\nabla_\beta \xi^\gamma - \zeta B_\beta{}^\gamma\right) B_{\alpha\gamma} \bar{\mathrm{D}} + \left(B_\beta{}^\gamma \xi_\gamma + \partial_\beta \zeta\right) \partial_\alpha \bar{\mathrm{D}}$$

$$\rho_{\alpha 3} \approx \partial_\alpha \delta_3 + B_\alpha{}^\beta \delta_\beta + \left(B_\beta{}^\gamma \xi_\gamma + \partial_\beta \zeta\right) B_\alpha{}^\beta \bar{\mathrm{D}}$$

These last two expressions can be transformed by (4.161). Noting that:

$$\mathcal{B}_\beta \, \partial_\alpha \bar{D} = \nabla_\alpha \left(\mathcal{B}_\beta \bar{D}\right) - \bar{D} \nabla_\alpha \mathcal{B}_\beta$$

and:

$$\nabla_\alpha \left(\mathcal{B}_\beta \bar{D}\right) = \nabla_\alpha \left(\vartheta_\beta - \delta_\beta\right)$$

it becomes, using (3.107):

$$\boxed{\begin{aligned} &\rho_{\alpha\beta} \approx \nabla_\alpha \vartheta_\beta - B_{\alpha\beta} \vartheta_3 - \bar{\mathrm{D}} \kappa_{\alpha\beta} \\ &\rho_{\alpha 3} \approx \partial_\alpha \vartheta_3 + B_\alpha{}^\beta \vartheta_\beta \end{aligned}} \tag{4.162}$$

In the context of the restricted theory: $\vartheta = 0, \rho_{\alpha 3} = 0$ and $\rho_{\alpha\beta} = \kappa_{\alpha\beta}$. So $\delta_3 = 0$ and $\delta_\alpha = B_\alpha{}^\beta \xi_\beta + \partial_\alpha \zeta$, analogous to relationship (3.103).

It is interesting to bring up the symmetric and antisymmetric parts of $\rho_{\alpha\beta}$:

$$\boxed{\begin{aligned} &\rho_{\alpha\beta} = {}^s\rho_{\alpha\beta} + {}^a\rho_{\alpha\beta} \\ &{}^s\rho_{\alpha\beta} = \frac{1}{2}\left(\nabla_\alpha \vartheta_\beta + \nabla_\beta \vartheta_\alpha\right) - B_{\alpha\beta} \vartheta_3 - \bar{\mathrm{D}} \kappa_{\alpha\beta} \\ &{}^a\rho_{\alpha\beta} = \frac{1}{2}\left(\nabla_\alpha \vartheta_\beta - \nabla_\beta \vartheta_\alpha\right) \end{aligned}} \tag{4.163}$$

${}^s\rho$ can be written:

$${}^s\rho(\vartheta, \xi) = \varepsilon(\vartheta) - \bar{D} \kappa(\xi) \tag{4.164}$$

Subsequently, and without restricting the generality of the reasoning, it is made choice of $\bar{D} = -1$ which is the case of shells of constant thickness. It is clear in this hypothesis that ϑ is equal to the difference between the vector δ, director displacement and the rotation of the vector a_3. ϑ characterises the director's distortion and length variation.

4.5.1.6 Virtual movements

In the current position of the Cosserat surface, a virtual movement is constituted by any change of position of the surface and by any variation of the director field, independently of one another. Such a movement is therefore characterised by independent virtual velocity $\overset{*}{\xi}$ and director virtual velocity $\overset{*}{\delta}$. Under these conditions, the virtual deformation rates are $\overset{*}{y}$, the virtual velocity of ϑ, $\overset{*}{e}\left(\overset{*}{\xi}\right)$ defined by (4.40), and $\overset{*}{r}\left(\overset{*}{\xi},\overset{*}{y}\right)$, the virtual velocity of ρ, with:

$$\begin{aligned}\overset{*}{r}_{\alpha\beta}\left(\overset{*}{\xi},\overset{*}{y}\right) &= \nabla_\alpha \overset{*}{y}_\beta - b_{\alpha\beta}\overset{*}{y}_3 + \overset{*}{k}_{\alpha\beta}\left(\overset{*}{\xi}\right) \\ &= \overset{*}{e}_{\alpha\beta}\left(\overset{*}{y}\right) + \overset{*}{k}_{\alpha\beta}\left(\overset{*}{\xi}\right) + \frac{1}{2}\left(\nabla_\alpha \overset{*}{y}_\beta - \nabla_\beta \overset{*}{y}_\alpha\right)\end{aligned} \quad (4.165)$$

$$\overset{*}{r}_{\alpha 3} = \partial_\alpha \overset{*}{y}_3 + b_\alpha{}^\beta \overset{*}{y}_\beta \quad (4.166)$$

$\overset{*}{k}\left(\overset{*}{\xi}\right)$ being defined by (4.42).

$\overset{*}{y}$ is obtained from relationships (4.161):

$$\begin{aligned}\overset{*}{y}_\alpha &= \overset{*}{\delta}_\alpha - \left(b_\alpha{}^\beta \overset{*}{\xi}_\beta + \partial_\alpha \overset{*}{\zeta}\right) \\ \overset{*}{y}_3 &= \overset{*}{\delta}_3\end{aligned} \quad (4.167)$$

4.5.2 Equations of movement

4.5.2.1 Virtual power of deformation

Following the method used in §4.2.2 to establish the equations of motion by applying the PVP, it is postulated that the virtual power of deformation $\overset{*}{\mathcal{W}}_f$ is a linear function of the virtual velocities of the deformation variables:

$$\overset{*}{\mathcal{W}}_f = \int_\sigma \left(\mathsf{Q}\bullet\overset{*}{y} + \mathsf{N}\bullet\overset{*}{e}\left(\overset{*}{\xi}\right) + \mathsf{M}\bullet\overset{*}{r}\right) d\sigma \quad (4.168)$$

Q is a vector of $\mathbb{R}^3_{m}$. N and M are two-fold contravariant tensors. N is a symmetrical tensor on T_{m}, M is a tensor of $T_{m} \otimes \mathbb{R}^3_{m}$, not necessarily symmetrical (as $\overset{*}{r}$).

The last term can be rewritten to account for (4.165) in the form:

$$\mathsf{M}\bullet\overset{*}{r}=\mathsf{M}^{\alpha\beta}\left[\nabla_\alpha\overset{*}{y}_\beta-b_{\alpha\beta}\overset{*}{y}_3+k_{\alpha\beta}\left(\overset{*}{\xi}\right)\right]+\mathsf{M}^{\alpha 3}\overset{*}{r}_{\alpha 3}\tag{4.169}$$

The expression (4.168) must be transformed by integrations by parts in order to take into account the relations existing between $\overset{*}{\xi}$ and $\overset{*}{\delta}$ on the one hand, their derivatives on the other hand.

The transformation of the term $\int_\sigma\left(\mathsf{N}\bullet e\left(\overset{*}{\xi}\right)+\mathsf{M}\bullet k\left(\overset{*}{\xi}\right)\right)d\sigma$ was carried out in §4.2.2.1 and appears in relations (4.80) and (4.81). Other terms are transformed in a similar way:

$$\int_\sigma \mathsf{M}^{\alpha\beta}\left(\nabla_\alpha\overset{*}{y}_\beta-b_{\alpha\beta}\overset{*}{y}_3\right)d\sigma=\int_{\partial\sigma}\mathsf{M}^{\alpha\beta}\,\nu_\alpha\overset{*}{y}_\beta\,ds-\int_\sigma\left(\nabla_\alpha\mathsf{M}^{\alpha\beta}\,\overset{*}{y}_\beta+\mathsf{M}^{\alpha\beta}\,b_{\alpha\beta}\overset{*}{y}_3\right)d\sigma$$

$$\int_\sigma \mathsf{M}^{\alpha 3}\,\overset{*}{r}_{\alpha 3}\,d\sigma=\int_\sigma \mathsf{M}^{\alpha 3}\left(\partial_\alpha\overset{*}{y}_3+b_\alpha{}^\beta\overset{*}{y}_\beta\right)d\sigma$$

$$=\int_{\partial\sigma}\mathsf{M}^{\alpha 3}\,\nu_\alpha\overset{*}{y}_3\,ds-\int_\sigma\left(\nabla_\alpha\mathsf{M}^{\alpha 3}\overset{*}{y}_3-\mathsf{M}^{\alpha 3}\,b_\alpha{}^\beta\overset{*}{y}_\beta\right)d\sigma$$

Taking into account the transformations above, $\overset{*}{\mathcal{W}}_f$ is rewritten, assuming a regular edge (without a corner):

$$\begin{aligned}\overset{*}{\mathcal{W}}_f=&-\int_\sigma\left[\begin{array}{l}\left(\nabla_\alpha\mathsf{L}^{\alpha\beta}+b_\gamma{}^\beta\nabla_\alpha\mathsf{M}^{\alpha\gamma}\right)\overset{*}{\xi}_\beta+\left(-\nabla_\beta\nabla_\alpha\mathsf{M}^{\alpha\beta}+b_{\alpha\beta}\mathsf{L}^{\alpha\beta}\right)\overset{*}{\zeta}\\+\left(\nabla_\alpha\mathsf{M}^{\alpha\beta}-\mathsf{M}^{\alpha 3}\,b_\alpha{}^\beta-\mathsf{Q}^\beta\right)\overset{*}{y}_\beta+\left(\mathsf{M}^{\alpha\beta}\,b_{\alpha\beta}+\nabla_\pm\mathsf{M}^{\alpha 3}-\mathsf{Q}^3\right)\overset{*}{y}_3\end{array}\right]d\sigma\\&+\int_{\partial\sigma}\left[\left(\mathsf{L}^{\nu\beta}+b_\gamma{}^\beta\mathsf{M}^{\nu\gamma}\right)\overset{*}{\xi}_\beta-\left(\frac{\partial\mathsf{M}^{\nu t}}{\partial s}+\nabla_\alpha\mathsf{M}^{\alpha\nu}\right)\overset{*}{\zeta}+\mathsf{M}^{\nu\nu}\frac{\partial\overset{*}{\zeta}}{\partial\nu}+\mathsf{M}^{\nu i}\overset{*}{y}_i\right]ds\end{aligned}\tag{4.170}$$

Where $\mathsf{L}^{\alpha\beta}$ is defined by (4.77) and $\overset{*}{y}$ by relations (4.167). This expression needs to be transformed again, only to show virtual velocities $\overset{*}{\xi}$ and $\overset{*}{\delta}$. Now, in the expression of $\overset{*}{y}_\beta$ intervenes the derivative $\partial_\beta\overset{*}{\zeta}$. Integration by parts of the integral on σ leads to:

$$\overset{*}{\mathcal{W}}_f=-\int_\sigma\left[\begin{array}{l}\left(\nabla_\alpha\mathsf{L}^{\alpha\beta}+b_\gamma{}^\beta\nabla_\alpha\mathsf{M}^{\alpha\gamma}\right)\overset{*}{\xi}_\beta+\left(-\nabla_\beta\nabla_\alpha\mathsf{M}^{\alpha\beta}+b_{\alpha\beta}\mathsf{L}^{\alpha\beta}\right)\overset{*}{\zeta}\\-\left(\nabla_\alpha\mathsf{M}^{\alpha\gamma}-\mathsf{M}^{\alpha 3}\,b_\alpha{}^\gamma-\mathsf{Q}^\gamma\right)b_\gamma{}^\beta\,\overset{*}{\xi}_\beta\\+\nabla_\beta\left(\nabla_\alpha\mathsf{M}^{\alpha\beta}-\mathsf{M}^{\alpha 3}\,b_\alpha{}^\beta-\mathsf{Q}^\beta\right)\overset{*}{\zeta}\\+\left(\nabla_\alpha\mathsf{M}^{\alpha\beta}-\mathsf{M}^{\alpha 3}\,b_\alpha{}^\beta-\mathsf{Q}^\beta\right)\overset{*}{\delta}_\beta+\left(\mathsf{M}^{\alpha\beta}\,b_{\alpha\beta}+\nabla_\alpha\mathsf{M}^{\alpha 3}-\mathsf{Q}^3\right)\overset{*}{\delta}_3\end{array}\right]d\sigma$$

$$
+\int_{\partial\sigma}\left[\begin{array}{l}\left(\mathsf{L}^{\nu\beta}+b_\gamma{}^\beta\mathsf{M}^{\nu\gamma}\right)\overset{*}{\xi}_\beta-\left(\dfrac{\partial\mathsf{M}^{\nu t}}{\partial s}+\nabla_\alpha\mathsf{M}^{\alpha\nu}\right)\overset{*}{\zeta}\\[2ex]+\left(\nabla_\alpha\mathsf{M}^{\alpha\nu}-\mathsf{M}^{\alpha 3}\,b_\alpha{}^\nu-\mathsf{Q}^\nu\right)\overset{*}{\zeta}+\mathsf{M}^{\nu\nu}\dfrac{\partial\overset{*}{\zeta}}{\partial\nu}+\mathsf{M}^{\nu i}\overset{*}{y}_i\end{array}\right]ds
$$

A new transformation is needed, to express $\overset{*}{y}_i$ along the edge. After integration by parts of the last term of the expression:

$$
\mathsf{M}^{\nu i}\overset{*}{y}_i=\mathsf{M}^{\nu i}\overset{*}{\delta}_i-\mathsf{M}^{\nu\alpha}b_\alpha{}^\beta\overset{*}{\xi}_\beta-\mathsf{M}^{\nu\nu}\partial_\nu\overset{*}{\zeta}-\mathsf{M}^{\nu t}\partial_s\overset{*}{\zeta}
$$

taking into account that the edge is closed and regular and by grouping the terms adequately, the edge integral is rewritten:

$$
\int_{\partial\sigma}\left[\begin{array}{l}\left(\mathsf{L}^{\nu\beta}+b_\gamma{}^\beta\mathsf{M}^{\nu\gamma}-b_\gamma{}^\beta\mathsf{M}^{\nu\gamma}\right)\overset{*}{\xi}_\beta\\[2ex]-\left(\dfrac{\partial\mathsf{M}^{\nu t}}{\partial s}+\nabla_\alpha\mathsf{M}^{\alpha\nu}-\nabla_\alpha\mathsf{M}^{\alpha\nu}+\mathsf{M}^{\alpha 3}\,b_\alpha{}^\nu+\mathsf{Q}^\nu-\partial_s\mathsf{M}^{\nu t}\right)\overset{*}{\zeta}+\mathsf{M}^{\nu i}\overset{*}{\delta}_i\end{array}\right]ds
$$

hence the expression of the virtual power of deformation:

$$
\overset{*}{\mathcal{W}}_f=-\int_\sigma\left[\begin{array}{l}\left[\nabla_\alpha\mathsf{L}^{\alpha\beta}+b_\gamma{}^\beta\nabla_\alpha\mathsf{M}^{\alpha\gamma}-\left(\nabla_\alpha\mathsf{M}^{\alpha\gamma}-b_\alpha{}^\gamma\mathsf{M}^{\alpha 3}-\mathsf{Q}^\gamma\right)b_\gamma{}^\beta\right]\overset{*}{\xi}_\beta\\[2ex]+\left[-\nabla_\beta\nabla_\alpha\mathsf{M}^{\alpha\beta}+b_{\alpha\beta}\mathsf{L}^{\alpha\beta}+\nabla_\beta\left(\nabla_\alpha\mathsf{M}^{\alpha\beta}-b_\alpha{}^\beta\mathsf{M}^{\alpha 3}-\mathsf{Q}^\beta\right)\right]\overset{*}{\zeta}\\[2ex]+\left(\nabla_\alpha\mathsf{M}^{\alpha\beta}-b_\alpha{}^\beta\mathsf{M}^{\alpha 3}-\mathsf{Q}^\beta\right)\overset{*}{\delta}_\beta+\left(b_{\alpha\beta}\mathsf{M}^{\alpha\beta}+\nabla_\alpha\mathsf{M}^{\alpha 3}-\mathsf{Q}^3\right)\overset{*}{\delta}_3\end{array}\right]d\sigma \tag{4.171}
$$

$$
+\int_{\partial\sigma}\left[\mathsf{L}^{\nu\beta}\overset{*}{\xi}_\beta-\left(\mathsf{M}^{\alpha 3}\,b_\alpha{}^\nu+\mathsf{Q}^\nu\right)\overset{*}{\zeta}+\mathsf{M}^{\nu i}\overset{*}{\delta}_i\right]ds
$$

4.5.2.2 Virtual power of external actions

External actions are introduced according to the model used:

$$
\overset{*}{\mathcal{W}}_e=\int_\sigma\left(p\cdot\overset{*}{\xi}+c\cdot\overset{*}{\delta}\right)d\sigma+\int_{\partial\sigma}\left(q\cdot\overset{*}{\xi}+m\cdot\overset{*}{\delta}\right)ds \tag{4.172}
$$

$\overset{*}{\delta}$ being dimensionless, c is homogeneous at a surface density of moments, m at a lineic moment. In this expression, $(\overset{*}{\delta}_\alpha)$, which represents the rotation of material points in the thickness, is substituted for the rotation of the normal used in (4.84). $\overset{*}{\delta}_3$ representing the variation in thickness of the shell over time, c^3 and m^3 are densities of forces distributed in the thickness and directed according to the normal, which tend to modify this thickness.

Note: the densities c and m thus introduced are not exactly identical to the densities c and m introduced in (4.84). Indeed, in restricted theory, for example by (4.39):

$$c \cdot \overset{*}{\delta} = -c \cdot \overset{*}{a}_3 = c \cdot \left(a_3 \times \overset{*}{\omega} \right)$$

instead of $c \cdot \overset{*}{\omega}$: the vector c was turned from $\frac{\pi}{2}$ in the tangent plane.

4.5.2.3 *Virtual power of inertial actions*

The velocity of the points in the thickness is related to the velocity of the director ($\dot{d}_\alpha$ for the rotation, $\dot{d}_3$ for the elongation with respect to the surface). It is then logical to take (in agreement with the model):

$$\overset{*}{\mathcal{W}}_j = -\int_\sigma \left(\rho\gamma \cdot \overset{*}{\xi} + \mathrm{I}\, \ddot{d}^\alpha\, \overset{*}{\delta}_\alpha + \mathrm{J}\, \ddot{d}^3\, \overset{*}{\delta}_3 \right) d\sigma \qquad (4.173)$$

where I is the mass inertia of the normal segment with respect to an axis of the tangent plane. J is the mass inertia related to the variation of thickness of the shell. The corresponding term can usually be neglected.

4.5.2.4 *Local equations*

By application of the PVP: $\overset{*}{\mathcal{W}}_f = \overset{*}{\mathcal{W}}_e + \overset{*}{\mathcal{W}}_j$ and considering the expressions (4.171), (4.172), and (4.173), the fields $\overset{*}{\xi}$ and $\overset{*}{\delta}$ being any, the local equations are obtained in Euler variables:

$$\forall m \in \sigma:$$

$$\begin{cases} \nabla_\beta \mathrm{L}^{\beta\alpha} + b_\gamma{}^\alpha \nabla_\beta \mathrm{M}^{\beta\gamma} - \left(\nabla_\beta \mathrm{M}^{\beta\gamma} - b_\beta{}^\gamma \mathrm{M}^{\beta 3} - \mathrm{Q}^\gamma \right) b_\gamma{}^\alpha + p^\alpha - \rho\gamma^\alpha = 0 \\ -\nabla_\beta \nabla_\gamma \mathrm{M}^{\gamma\beta} + b_{\gamma\beta} \mathrm{L}^{\gamma\beta} + \nabla_\beta \left(\nabla_\gamma \mathrm{M}^{\gamma\beta} - b_\gamma{}^\beta \mathrm{M}^{\gamma 3} - \mathrm{Q}^\beta \right) + p^3 - \rho\gamma^3 = 0 \\ \nabla_\beta \mathrm{M}^{\beta\alpha} - b_\beta{}^\alpha \mathrm{M}^{\beta 3} - \mathrm{Q}^\alpha + c^\alpha - \mathrm{I}\, \ddot{d}^\alpha = 0 \\ b_{\gamma\beta} \mathrm{M}^{\gamma\beta} + \nabla_\gamma \mathrm{M}^{\gamma 3} - \mathrm{Q}^3 + c^3 - \mathrm{J}\, \ddot{d}^3 = 0 \end{cases} \qquad (4.174)$$

either six equations for 10 unknowns in static.

The first two lines of (4.174) resemble equations (4.95) and have the same meaning. The third line expresses the equilibrium with respect to the moments tending to turn all the points situated on the normal. It completes the definition of shear force (4.96) to which it is reduced in the context of the restricted theory, with the sign depending, however, on the way Q has been defined by (4.96). The last relation expresses the equilibrium vis-à-vis the movements of variation of the thickness; it allows the expression of a relationship between $\mathsf{M}^{\alpha 3}$ and Q^3. By eliminating the shear force of the first two lines of (4.174) using the third line, it becomes:

$$\begin{cases} \nabla_\beta \mathsf{L}^{\beta\alpha} + b_\gamma{}^\alpha \nabla_\beta \mathsf{M}^{\beta\gamma} + b_\beta{}^\alpha c^\beta - \mathrm{I}\, b_\beta{}^\alpha \ddot{d}^\beta + p^\alpha - \rho\gamma^\alpha = 0 \\ b_{\gamma\beta} \mathsf{L}^{\gamma\beta} - \nabla_\beta \nabla_\gamma \mathsf{M}^{\gamma\beta} - \nabla_\beta c^\beta + \nabla_\beta\left(\mathrm{I}\, \ddot{d}^\beta\right) + p^3 - \rho\gamma^3 = 0 \end{cases} \tag{4.175}$$

which are exactly the same as relations (4.95).

The shear force Q is here introduced as a generalised stress associated with the deformation ϑ. In the framework of the restricted theory, it is a Lagrange multiplier vector associated with the relation $\vartheta = 0$. It is also interesting to note the similarity of the third and fourth lines of (4.174) with equations (4.68) reflecting the balance of moments.

In common cases where c is zero and the kinetic energy of rotation is negligible, the equilibrium equations are reduced to:

$$\begin{cases} \nabla_\beta \mathsf{L}^{\beta\alpha} + b_\gamma{}^\alpha \nabla_\beta \mathsf{M}^{\beta\gamma} + p^\alpha - \rho\gamma^\alpha = 0 \\ b_{\gamma\beta} \mathsf{L}^{\gamma\beta} - \nabla_\beta \nabla_\gamma \mathsf{M}^{\gamma\beta} + p^3 - \rho\gamma^3 = 0 \\ \mathsf{Q}^\alpha = \nabla_\beta \mathsf{M}^{\beta\alpha} - b_\beta{}^\alpha \mathsf{M}^{\beta 3} \\ \mathsf{Q}^3 = b_{\gamma\beta} \mathsf{M}^{\gamma\beta} + \nabla_\gamma \mathsf{M} \end{cases} \tag{4.176}$$

4.5.2.5 Boundary conditions

Similarly, the boundary conditions are written:

$$\boxed{\begin{array}{l} \forall \boldsymbol{m} \in \partial\sigma: \\ \begin{cases} \mathsf{L}^{\nu\alpha} - q^\alpha = 0 \\ b_\beta{}^\nu \mathsf{M}^{\beta 3} + \mathsf{Q}^\nu + q^3 = 0 \\ \mathsf{M}^{\nu i} - m^i = 0 \end{cases} \end{array}} \tag{4.177}$$

either six relations, which are comparable to relations (4.98) or (4.99) obtained in the restricted theory.

Here, the moments applied to the edge are directly balanced by the tensor M of moments, contrary to the case of the restricted theory, as it was explained in §4.2.2.6. No condition relates to Q^3, which is logical, since it is a force applied in the direction of the normal, associated with the variation of thickness and which is therefore independent of ν.

4.5.2.6 Conclusion

Taking into account deformation variables related to the evolution of the director, in addition to the deformation variables ε and κ considered in restricted theory, made it possible to clarify and extend the equilibrium equations, without modifying their form. The generalised stresses introduced, associated with the new deformation variables, can be introduced in restricted theory in the form of Lagrange multipliers when these deformation variables are kept null.

The complete resolution of the movement problem requires the introduction of constitutive laws covering all the deformation variables and associated generalised stresses.

Photo 4.1: The masonry vaults are a very ancient traditional construction type (Jerusalem, Israel)

MAIN RESULTS

Kinematics

Acceleration in Euler variables

$$\gamma = \left[\dot{v}^\beta + v^\alpha\left(\nabla_\alpha v^\beta - b_\alpha{}^\beta v^3\right) - v^3\left(a^{\beta\gamma}\partial_\gamma v^3 + b_\gamma{}^\beta v^\gamma\right)\right] a_\beta$$
$$+\left[\dot{v}^3 + v^\alpha\left(\partial_\alpha v^3 + b_{\alpha\beta}v^\beta\right)\right] a_3$$

Conservation of mass $\dot{\rho} + \rho\left(\nabla_\alpha v^\alpha - b_\alpha{}^\alpha v^3\right) = 0$

Virtual velocities

$$\overset{*}{a}_\alpha = D_\alpha \overset{*}{\xi} = \left(\nabla_\alpha \overset{*}{\xi}{}^\beta - b_\alpha{}^\beta \overset{*}{\zeta}\right) a_\beta + \left(\partial_\alpha \overset{*}{\zeta} + b_{\alpha\beta}\overset{*}{\xi}{}^\beta\right) a_3 \qquad \overset{*}{a}_3 = -\left(\partial_\alpha \overset{*}{\zeta} + b_{\alpha\beta}\overset{*}{\xi}{}^\beta\right) a^\alpha$$

$$\overset{*}{\omega} = a_3 \times \overset{*}{a}_3$$

$$\overset{*}{e}_{\alpha\beta} = \frac{1}{2}\left[\nabla_\alpha \overset{*}{\xi}_\beta + \nabla_\beta \overset{*}{\xi}_\alpha - 2b_{\alpha\beta}\overset{*}{\zeta}\right] \qquad \overset{*}{k}_{\alpha\beta} = \nabla_\alpha\left(\partial_\beta \overset{*}{\zeta} + b_{\beta\gamma}\overset{*}{\xi}{}^\gamma\right) + b_\alpha{}^\gamma\left(\nabla_\beta \overset{*}{\xi}_\gamma - b_{\beta\gamma}\overset{*}{\zeta}\right)$$

Equations of movement (restricted theory)

on generalised stresses (symmetric N and M):

$\forall \boldsymbol{m} \in \sigma$:

Local equations

$$\begin{cases} \nabla_\beta \mathrm{N}^{\beta\alpha} - 2b_\gamma{}^\alpha \mathrm{Q}^\gamma + \mathrm{M}^{\beta\gamma}\nabla_\beta b_\gamma{}^\alpha + p^\alpha - \rho\gamma^\alpha = 0 \\ \nabla_\alpha \mathrm{Q}^\alpha + b_{\alpha\beta}\mathrm{N}^{\alpha\beta} + b_{\alpha\beta}b_\gamma{}^\beta \mathrm{M}^{\alpha\gamma} + p^3 - \rho\gamma^3 = 0 \\ \mathrm{Q}^\gamma + \nabla_\beta \mathrm{M}^{\beta\gamma} = 0 \end{cases}$$

$\forall \boldsymbol{m} \in \partial\sigma$:

Boundary conditions

$$\begin{cases} \mathrm{N}^{\nu\beta} + b_t{}^\beta\left(2\mathrm{M}^{\nu\tau} + m^\nu\right) + b_\nu{}^\beta \mathrm{M}^{\nu\nu} - q^\beta = 0 \\ \mathrm{Q}^\nu - \dfrac{\partial}{\partial s}\left(\mathrm{M}^{\nu t} + m^\nu\right) - q^3 = 0 \\ \mathrm{M}^{\nu\nu} - m^t = 0 \end{cases}$$

Constitutive law of linear elasticity

Thermodynamic potential: $\Phi = \int_\sigma \phi\left(e, k, \mathrm{T}\right) d\sigma$

Constitutive law: $\mathrm{N} = \dfrac{\partial\phi}{\partial e} \quad ; \quad \mathrm{M} = \dfrac{\partial\phi}{\partial k}$

Linear elasticity:

$$\mathrm{N} = \mathrm{N}_0 + B\tau + \Lambda \bullet \varepsilon + \mathrm{K} \bullet \kappa$$
$$\mathrm{M} = \mathrm{M}_0 + B'\tau + \mathrm{H} \bullet \kappa + \mathrm{K} \bullet \varepsilon$$

EXERCISES

Exercise 4.1

A drum skin has the shape of a circular disc and consists of a homogeneous and isotropic elastic membrane in which flexural stiffness is negligible. The membrane is clamped on its periphery and initially tensioned isotropically by a membrane tension N_0 (identical in all directions). The variations in stretching forces during the skin movements remain low compared to N_0, which should be justified. It is further noted that, under the effect of a moderate transverse loading, the displacements in the plane of the membrane are negligible compared to the transverse displacement, which can nevertheless be considered small.

What is the deformation of the drum skin under the effect of a pressure difference p between the inside and the outside of the drum?

Calculate the first natural pulsation of the drum.

Answer:
The elastic constitutive law of the membrane is written (cf. (4.116)): $\mathsf{N} = \mathsf{N}_0 a + \Lambda \bullet \varepsilon$. The behaviour being isotropic, the tensor Λ depends only on two mechanical properties, as in 3D.

After (2.97), $\varepsilon_{\alpha\beta} \approx \dfrac{1}{2}\partial_\alpha\zeta\,\partial_\beta\zeta$, by keeping only the preponderant terms, which justifies neglecting the term of elasticity compared to N_0 when the tension is high, which is the case of the drums.

Given the assumptions, (4.76) is reduced to:

$$\overset{*}{\mathcal{W}}_f = \int_\sigma \mathsf{N}\bullet \overset{*}{e}\, d\sigma \approx \int_\sigma \mathsf{N}_0 a \bullet \left(-b_{\alpha\beta}\,\overset{*}{\zeta}\right) d\sigma = -\int_\sigma \mathsf{N}_0\, b_\alpha{}^\alpha\, \overset{*}{\zeta}\, d\sigma$$

Considering (4.84) and (4.88), the application of the PVP leads to the equation of transverse movement, where, according to usage, notation ζ is replaced by w:

$$\mathsf{N}_0\, b_\alpha{}^\alpha + p - \rho\frac{d^2 w}{dt^2} = 0$$

Under the static effect of a pressure difference p, the equation allows us to conclude that $b_\alpha{}^\alpha = \dfrac{1}{R_1} + \dfrac{1}{R_2}$ is constant. The deformation does not depend on the direction and is therefore axisymmetric. As a result, the two radii are constant; consequently, the generatrices are circles and the deformed shape is therefore a sphere of radius $R = \dfrac{2\mathsf{N}_0}{p}$.

The first mode of the drum is axisymmetric and w only depends on r, the distance to the centre. The two radii of curvature are, in the physical coordinate system: $b_{11} = w''$ and $b_{22} = \dfrac{w'}{r}$. The equation of motion is therefore written as:

$$\mathsf{N}_0\left(w'' + \frac{w'}{r}\right) - \rho\frac{d^2 w}{dt^2} = 0$$

By looking for a vibration mode of form $w(r,t) = W(r)\sin\omega t$, there comes the equation in W:

$$W'' r^2 + W' r + \frac{\rho\omega^2}{N_0} W r^2 = 0$$

which solutions are Bessel's functions J_0:

$$W(r) = \sum_{k=0}^{\infty} \frac{(-1)^k}{(k!)^2} (\lambda r)^{2k} \text{ with } \lambda^2 = \omega^2 \Big/ \left(\frac{N_0}{\rho}\right)$$

The first zero of J_0 is obtained for about 2,405, which therefore leads to the first pulsation: $\omega_1 \approx \frac{2{,}405}{R}\sqrt{\frac{N_0}{\rho}}$.

Exercise 4.2

Write the local equations (4.119) relating to the generalised stresses in statics in the principal curvature coordinate system. Deduce their expression for a cylindrical shell and compare them in this case with the local equations (4.118) relating to the stress resultants; check the equivalence of the two systems of local equations.

Answer:
The principal curvature coordinate system (a_1; a_2) is orthonormal; the associated radii of curvature are R_1 and R_2 respectively. $\nabla_\beta b_\gamma{}^\alpha$ was calculated in Exercise 3.4. There comes the local equations relating to the generalised stresses:

$$\begin{cases} \nabla_\beta N^{\beta 1} - 2\dfrac{Q^1}{R_1} + M^{\beta 1}\partial_\beta\left(\dfrac{1}{R_1}\right) + M^{\beta 2}\Gamma^1_{\beta 2}\left(\dfrac{1}{R_2} - \dfrac{1}{R_1}\right) + p^1 = 0 \\ \nabla_\beta N^{\beta 2} - 2\dfrac{Q^2}{R_2} + M^{\beta 2}\partial_\beta\left(\dfrac{1}{R_2}\right) + M^{\beta 1}\Gamma^1_{\beta 1}\left(\dfrac{1}{R_1} - \dfrac{1}{R_2}\right) + p^2 = 0 \\ \nabla_\beta Q^\beta + \dfrac{N^{11}}{R_1} + \dfrac{N^{22}}{R_2} + \dfrac{M^{11}}{R_1^2} + \dfrac{M^{22}}{R_2^2} + p^3 = 0 \\ Q^\gamma + \nabla_\beta M^{\beta\gamma} = 0 \end{cases}$$

If the shell is cylindrical, $a_1 = a_s$ and $a_2 = a_z$ being the vectors of the principal basis, $1/R_2$ is zero and R_1 is equal to $-R$, where R is the radius of the circle generating the cylinder. So:

$$\begin{cases} \dfrac{1}{R}\partial_\theta N^{11} + \partial_z N^{21} + 2\dfrac{Q^1}{R} + p^1 = 0 \quad ; \quad \dfrac{1}{R}\partial_\theta N^{12} + \partial_z N^{22} + p^2 = 0 \\ \dfrac{1}{R}\partial_\theta Q^1 + \partial_z Q^2 - \dfrac{N^{11}}{R} + \dfrac{M^{11}}{R^2} + p^3 = 0 \\ Q^1 + \dfrac{1}{R}\partial_\theta M^{11} + \partial_z M^{21} = 0 \quad ; \quad Q^2 + \dfrac{1}{R}\partial_\theta M^{12} + \partial_z M^{22} = 0 \end{cases} \qquad \text{(i)}$$

The local equations on the resultants of stresses reduce to:

$$\left\{\begin{array}{l} \dfrac{1}{R}\partial_\theta N^{11} + \partial_z N^{21} + \dfrac{N^{13}}{R} + p^1 = 0 \;\; ; \;\; \dfrac{1}{R}\partial_\theta N^{12} + \partial_z N^{22} + p^2 = 0 \\ -\dfrac{N^{11}}{R} + \dfrac{1}{R}\partial_\theta N^{13} + \partial_z N^{23} + p^3 = 0 \\ \dfrac{1}{R}\partial_\theta M^{11} + \partial_z M^{21} + N^{23} = 0 \;\; ; \;\; \dfrac{1}{R}\partial_\theta M^{12} + \partial_z M^{22} - N^{13} = 0 \\ -\dfrac{M^{11}}{R} + N^{12} - N^{21} = 0 \end{array}\right. \qquad \text{(ii)}$$

The last equation of (ii) implies that $\mathsf{N}^{12} = \mathsf{N}^{21}$, given the relationships below.
Using (4.121), (4.123), (4.124), and (4.128):

$$H^{11} = -M^{12} \;\; ; \;\; H^{12} = M^{11} \;\; ; \;\; H^{21} = -M^{22} \;\; ; \;\; H^{22} = M^{21}$$

$$\mathsf{M}^{11} = -M^{12} \;\; ; \;\; \mathsf{M}^{12} = \mathsf{M}^{21} = \frac{1}{2}\left(M^{11} - M^{22}\right) \;\; ; \;\; \mathsf{M}^{22} = M^{21}$$

$$\mathsf{N}^{11} = N^{11} - \frac{M^{12}}{R} \;\; ; \;\; \mathsf{N}^{12} = N^{12} \;\; ; \;\; \mathsf{N}^{21} = N^{21} + \frac{M^{11}}{R} \;\; ; \;\; \mathsf{N}^{22} = N^{22}$$

$$\mathsf{Q}^1 = \frac{1}{R}\partial_\theta M^{12} - \frac{1}{2}\partial_z\left(M^{11} - M^{22}\right) \;\; ; \;\; \mathsf{Q}^2 = -\frac{1}{2R}\partial_\theta\left(M^{11} - M^{22}\right) - \partial_z M^{21}$$

Equation (ii) is obtained by replacing the generalised stresses in equation (i) using these relations.

Exercise 4.3

The subject is to study the equilibrium of a circular membrane, filled with a heavy fluid (density γ). In its initial configuration, the membrane is a flat disk of radius R. In the final configuration, the theme of the study, the membrane is completely filled: the surface of the fluid then reaches the level of support of the membrane (Figure 4.10).

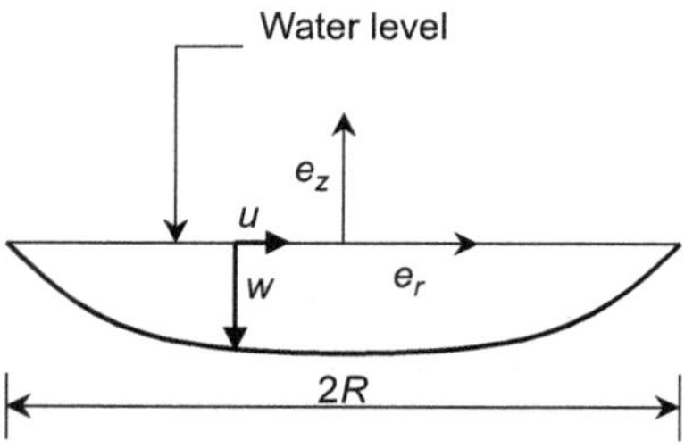

Figure 4.10

Calculate the tensor of in-plane strains function of the displacement, decomposed on the orthonormal cylindrical basis: $\xi = u(r)\, e_r + w(r)\, e_z$, which cannot be considered "small".

Calculate the virtual power of internal actions and the virtual power of pressure $p = -\gamma z$. Deduce the equilibrium equations, as well as the boundary conditions. Comment on the equations obtained (without trying to solve them), if the membrane has a linear elastic behaviour.

Answer:

$$\tilde{\varepsilon}_{rr} = u' + \frac{1}{2}\left(u'^2 + w'^2\right) \qquad \tilde{\varepsilon}_{\theta\theta} = \frac{u}{r} + \frac{1}{2}\frac{u^2}{r^2}$$

The virtual power of deformation is expressed in the deformed configuration σ by introducing generalised stresses (Euler variables), then transformed to be expressed on the reference configuration Σ.

$$\overset{*}{\mathcal{W}}_{def} = \int_\sigma \left(\mathsf{N}^{11}\,\overset{*}{\tilde{\varepsilon}}_{rr} + \mathsf{N}^{22}\,\overset{*}{\tilde{\varepsilon}}_{\theta\theta} \right) d\sigma$$

$$= 2\pi \int_0^R \left\{ \begin{array}{c} \mathsf{N}^{11}\left[\overset{*}{u}'(1+u') + \overset{*}{w}'w'\right] \\ + \mathsf{N}^{22}\dfrac{\overset{*}{u}}{r}\left(1+\dfrac{u}{r}\right) \end{array} \right\} \left(1+\frac{u}{r}\right)\sqrt{(1+u')^2 + w'^2}\, r\, dr$$

The virtual power of the water pressure is expressed on the deformed configuration, the pressure being normal to the deformed membrane.

$$\overset{*}{\mathcal{W}}_e = \int_\sigma p a_3 \cdot \overset{*}{\xi}\, d\sigma = 2\pi \int_0^R \gamma w \left(1+\frac{u}{r}\right)\left[w'\overset{*}{u} - (1+u')\overset{*}{w}\right] r\, dr$$

Equilibrium equations obtained by applying the PVP:

$$\left\langle \begin{array}{l} \dfrac{d}{dr}\left[\mathsf{N}^{11}(1+u')\left(1+\dfrac{u}{r}\right)\sqrt{(1+u')^2 + w'^2}\; r\right] \\ \qquad -\mathsf{N}^{22}\left(1+\dfrac{u}{r}\right)^2 \sqrt{(1+u')^2 + w'^2} + \gamma w w'\left(1+\dfrac{u}{r}\right) r = 0 \end{array} \right\rangle$$

$$\left\langle \frac{d}{dr}\left[\mathsf{N}^{11} w'\left(1+\frac{u}{r}\right)\sqrt{(1+u')^2 + w'^2}\; r\right] - \gamma w\left(1+\frac{u}{r}\right)(1+u') r = 0 \right\rangle$$

with the conditions of fixity of the outer edge with radius R: $u(R) = w(R) = 0$.

The reactions along the outer edge are obtained by the application of PVP:

$$\mathrm{R}_r = \mathsf{N}^{11} R(1+u')\left(1+\frac{u}{R}\right)\sqrt{(1+u')^2 + w'^2} \qquad \mathrm{R}_z = \mathsf{N}^{11} R w'\left(1+\frac{u}{R}\right)\sqrt{(1+u')^2 + w'^2}$$

If the membrane has an elastic linear behaviour, the membrane forces are expressed linearly as functions of the strains, therefore of u and w and their derivatives. The two equilibrium equations are then expressed as functions of the two components of the displacement to be determined. They are very nonlinear and a solution can be searched numerically, for example by using a discretisation by finite differences.

Chapter 5

Shell as a three-dimensional solid

In this chapter, the "semi-global" point of view adopted in the previous chapter, where mechanical quantities are defined on the middle surface, is related to the "local" point of view of three-dimensional solid mechanics. The introduction of the screw resultant of the three-dimensional stresses in the thickness of the shell makes it possible to obtain the equilibrium equations by integration of the local equations. These equations are identical to those obtained in the previous chapter.

It is then possible to relate the local deformations to the deformation variables of the middle surface by adopting hypotheses on the kinematic behaviour of the shell in its thickness. The Kirchhoff–Love hypotheses are the most used; they include neglecting the distortions due to shear force. These can be taken into account thanks to the Reissner–Mindlin assumptions.

The usual global constitutive laws are then established in one or the other family of hypotheses. This allows, in particular, to express of the tensor of the three-dimensional stresses from the generalised stresses defined on the middle surface.

The consequences of the hypothesis of thin shells are examined in each paragraph, which makes it possible to introduce gradual simplifications very useful in applications.

Sections five and six respectively, deal with the establishment of equilibrium equations from the local equilibrium and with supplements on the global constitutive laws, are in-depth studies which may not be addressed during a first reading.

5.1 STRESSES AND EQUILIBRIUM EQUATIONS

5.1.1 Parameter setting and metric

5.1.1.1 Normal coordinates

The shell is actually a three-dimensional solid having a thickness, the material is distributed on both sides of the average surface σ. A part of the shell occupies a volume ω which relies on a subdomain d of σ (the edge $\partial\omega$ of ω is generated by the circulation of the vector normal to σ along the edge ∂d of d). In a given configuration, the normal coordinates $\left(x^\alpha, x^3 = x_3\right)$ defined on σ (§3.1.4) are used on the whole area occupied by the shell[1].

[1] The conditions of uniqueness of the normal coordinates are naturally satisfied in a real shell.

DOI: 10.1201/9780429440403-5

Isomorphism $\mathfrak{m}$ of components:

$$\begin{cases} \mathfrak{m}_\alpha{}^\beta = a_\alpha{}^\beta - x^3 b_\alpha{}^\beta \\ \mathfrak{m}_3{}^\beta = 0 \\ \mathfrak{m}_3{}^3 = 1 \end{cases}$$

allows defining at any point in the thickness of the shell the natural basis of $\mathbb{R}^3$ such that:

$$g_i = \mathfrak{m}_i{}^j a_j \tag{5.1}$$

In the principle curvature coordinate system (Figure 5.1), $\mathfrak{m}$ is diagonal and:

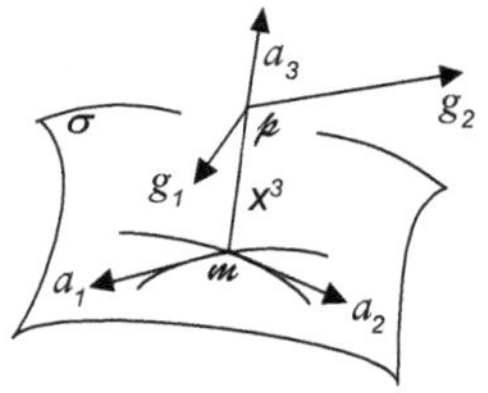

$$\begin{aligned} g_1 &= \left(1 - \frac{x^3}{R_1}\right) a_1 \\ g_2 &= \left(1 - \frac{x^3}{R_2}\right) a_2 \end{aligned} \tag{5.2}$$

Figure 5.1

The basis (g_i) is then orthogonal, and each vector g_i is parallel to the vector a_i. It is often convenient to use the associated orthonormal basis $(\tilde{g}_i)$, simply translated from the principal basis of curvature (a_i), which is done by many authors.

5.1.1.2 Metric in normal coordinates

The metric tensor at any point p of the shell has components:

$$g_{ij} = \mathfrak{m}_i{}^k \mathfrak{m}_j{}^l a_{kl} \tag{5.3}$$

with:

$$\begin{cases} g_{\alpha\beta} = a_{\alpha\beta} - 2x^3 b_{\alpha\beta} + \left(x^3\right)^2 c_{\alpha\beta} \\ g_{\alpha 3} = 0 \\ g_{33} = 1 \end{cases} \tag{5.4}$$

The metric of the shell at a point $p\left(x^i\right)$ in the thickness is therefore entirely determined by the three fundamental forms of the middle surface σ. In the coordinate system (g_i) **not normalised** associated with the principle curvature coordinate system of the middle surface:

$$g_{\alpha\beta} = \begin{cases} \left(1 - \dfrac{x^3}{R_\alpha}\right)^2 & \text{if } \alpha = \beta \\ 0 & \text{if } \alpha \neq \beta \end{cases} \tag{5.5}$$

5.1.1.3 Volumic element

All quantities defined in $\pmb{p}$ are expressed in the local coordinate system (g_i). Volume integrals are reduced to surface integrals on σ by calculating the volume element $dV(\pmb{p})$ in the system (a_i):

$$dV(\pmb{p}) = \hat{g}\, d\sigma\, dx^3 = \hat{m}\, \hat{a}\, dx^1\, dx^2\, dx^3 \tag{5.6}$$

where $\hat{m}$ is the determinant of endomorphism m:

$$\hat{m} = \det\left(m_i{}^j\right) = \det\left(m_\alpha{}^\beta\right) = \det\left(a_\alpha{}^\beta - x^3 b_\alpha{}^\beta\right) \tag{5.7}$$

$\hat{m}$ depends on the point $\pmb{p}$ considered, more precisely x^3, the metric and the curvature in $\pmb{m}$. In the natural coordinate system (g_i) associated with the principal coordinate system of curvature of the middle surface:

$$\hat{m} = \left(1 - \frac{x^3}{R_1}\right)\left(1 - \frac{x^3}{R_2}\right) \tag{5.8}$$

5.1.2 Kinematics

Let $\bar{\xi}$ the field of displacement of any point of the shell located at the distance x^3 of the middle surface, $\bar{\xi}^i$ its components in the basis (g_i), and $\bar{U}^i$ its components in the associated basis (a_i):

$$\bar{\xi} = \bar{\xi}^i g_i = \bar{U}^i a_i \text{ with } \bar{\xi}^3 = \bar{U}^3$$

Applying the relations (3.63) and with $a^i = m_j{}^i g^j$ resulting from (3.43), it becomes, by differentiation in $E_3^\bullet$:

$$D_\beta \bar{\xi}_\alpha = m_\alpha{}^\gamma \left(\nabla_\beta \bar{U}_\gamma - b_{\alpha\gamma} \bar{U}^3\right)$$

$$D_3 \bar{\xi}_\alpha = m_\alpha{}^\gamma\, \partial_3 \bar{U}_\gamma$$

$$D_\alpha \bar{\xi}_3 = \partial_\alpha \bar{U}_3 + b_\alpha{}^\gamma \bar{U}_\gamma$$

This makes it possible to express the components of the linearised strain tensor on the basis (g_i) according to the components $\bar{U}^i$:

$$\varepsilon_{\alpha\beta} = \frac{1}{2}\left(D_\alpha \bar{\xi}_\beta + D_\beta \bar{\xi}_\alpha\right) = \frac{1}{2}\left[m_\alpha{}^\gamma \left(\nabla_\beta \bar{U}_\gamma - b_{\alpha\gamma}\bar{U}^3\right) + m_\beta{}^\gamma \left(\nabla_\alpha \bar{U}_\gamma - b_{\alpha\gamma}\bar{U}^3\right)\right]$$

$$\varepsilon_{\alpha 3} = \frac{1}{2}\left(D_\alpha \bar{\xi}_3 + D_3 \bar{\xi}_\alpha\right) = \frac{1}{2}\left[\partial_\alpha \bar{U}_3 + b_\alpha{}^\gamma \bar{U}_\gamma + m_\alpha{}^\gamma\, \partial_3 \bar{U}_\gamma\right]$$

$$\varepsilon_{33} = \partial_3 \bar{\xi}_3 = \partial_3 \bar{U}_3$$

5.1.3 Screw resultant of local stresses

The objective is to establish the relations between the three-dimensional stress tensor $\overline{\overline{T}}$ in Euler variables existing in the thickness of the shell and the internal actions represented by the tensors N and M and the vector V as resultants of stresses introduced in the previous chapter.

If the shell is thin enough that the metric outside the middle surface is equivalent to that of the middle surface (the corresponding assumptions are detailed in §5.1.5), the resultants of the stresses in the thickness on a facet of normal a_β are simply a force per unit length of components $\int_{-e/2}^{e/2} T^{\alpha\beta}\, dx^3$ and $\int_{-e/2}^{e/2} T^{3\beta}\, dx^3$ and a moment per unit length of components $-\int_{-e/2}^{e/2} T^{\alpha\beta}\, x^3\, dx^3$.

But, to treat this question in all generality, it is necessary to take into account the variation of metric in the thickness of the shell. This is the goal pursued by introducing the two new tensors $\bar{N}$ and $\bar{M}$ defined by the integrals given by (5.9) along the normal to the surface:

$$\begin{cases} \bar{N}^{\alpha\beta} = \int_{-e/2}^{e/2} T^{\alpha\gamma}\, \mathfrak{m}_\gamma{}^\beta \hat{\mathfrak{m}}\, dx^3 \\ \bar{M}^{\alpha\beta} = -\int_{-e/2}^{e/2} T^{\alpha\gamma}\, \mathfrak{m}_\gamma{}^\beta \hat{\mathfrak{m}}\, x^3\, dx^3 \end{cases} \tag{5.9}$$

As well as the vector $\bar{Q}$ défined by:

$$\bar{Q}^\beta = \int_{-e/2}^{e/2} T^{3\beta}\, \hat{\mathfrak{m}}\, dx^3 \tag{5.10}$$

It is shown below that these are the right quantities to express the resultants of stresses.

The tensors $\bar{N}$, $\bar{M}$ and the vector $\bar{Q}$ are defined fields on the middle surface σ. As well as the three-dimensional stresses T and the generalised stresses, they are defined (in Euler variables) on the current configuration, that is to say without reference to a possible initial configuration. There is no hypothesis on the evolution of the shell.

In the case where the stresses are expressed in the orthonormal physical coordinate system $(\tilde{g}_i)$ associated with the principal curvature coordinate system, the relations (5.9) and (5.10) are written, taking into account relations (5.2), (5.5), and (5.8):

$$\begin{aligned} \bar{N}^{11} &= \int_{-e/2}^{e/2} T^{11}\left(1-\frac{x^3}{R_2}\right) dx^3 & \bar{N}^{12} &= \int_{-e/2}^{e/2} T^{12}\left(1-\frac{x^3}{R_2}\right) dx^3 \\ \bar{N}^{21} &= \int_{-e/2}^{e/2} T^{21}\left(1-\frac{x^3}{R_1}\right) dx^3 & \bar{N}^{22} &= \int_{-e/2}^{e/2} T^{22}\left(1-\frac{x^3}{R_1}\right) dx^3 \end{aligned} \tag{5.11}$$

$$\begin{aligned} \bar{M}^{11} &= -\int_{-e/2}^{e/2} T^{11}\left(1-\frac{x^3}{R_2}\right) x^3\, dx^3 & \bar{M}^{12} &= -\int_{-e/2}^{e/2} T^{12}\left(1-\frac{x^3}{R_2}\right) x^3\, dx^3 \\ \bar{M}^{21} &= -\int_{-e/2}^{e/2} T^{21}\left(1-\frac{x^3}{R_1}\right) x^3\, dx^3 & \bar{M}^{22} &= -\int_{-e/2}^{e/2} T^{22}\left(1-\frac{x^3}{R_1}\right) x^3\, dx^3 \end{aligned} \tag{5.12}$$

$$\bar{Q}^1 = \int_{-e/2}^{e/2} T^{31}\left(1 - \frac{x^3}{R_2}\right) dx^3$$
$$\bar{Q}^2 = \int_{-e/2}^{e/2} T^{32}\left(1 - \frac{x^3}{R_1}\right) dx^3 \tag{5.13}$$

These relationships are very often used to establish equations of motion following the method described below. It clearly appears that $\bar{N}^{12}$ and $\bar{N}^{21}$, on the one hand, $\bar{M}^{12}$ and $\bar{M}^{21}$ on the other hand, are not equal, despite the symmetry of the stress tensor $\bar{\bar{T}}$. Tensors $\bar{N}$ and $\bar{M}$ are not symmetrical.

Demonstration. It is a question of connecting $\bar{N}$, $\bar{M}$, and $\bar{Q}$ to the screw resulting from the stresses applied to a facet inscribed in the shell. The stresses are expressed in the local coordinate system (g_i). It should be noted first of all that stresses T^{33} does not intervene in the reasoning that follows.

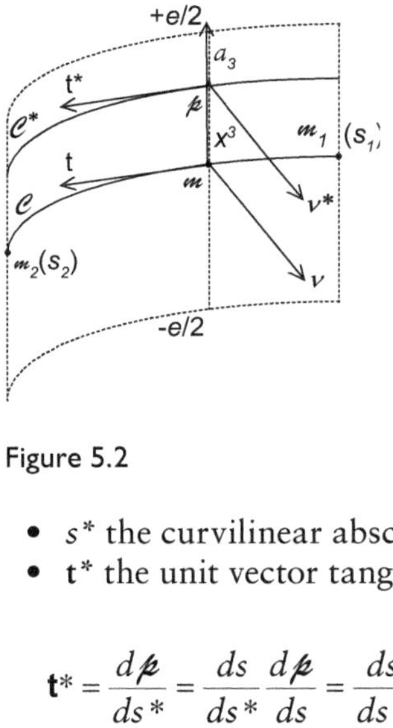

Figure 5.2

Let a curve $(\mathcal{C})$ be drawn on the surface σ. The set of normal segments carried by a_3 based on $(\mathcal{C})$ generates a curved surface inscribed in the thickness of the shell (Figure 5.2). s denoting the curvilinear abscissa along $(\mathcal{C})$, let $\mathcal{F}$ be the facet, part of the curved surface defined above, limited by two values s_1 and s_2 of s.

Let $\mathbf{t} = t^\alpha a_\alpha = \dfrac{dm}{ds}$ the vector tangent to $(\mathcal{C})$ in m and ν be the vector normal to the facet at this point: $\nu = a_3 \times \mathbf{t}$. Let $(\mathcal{C}^*)$ be the curve parallel to $(\mathcal{C})$, a set of points p such that $p = m + x^3 a_3$, where $m \in (\mathcal{C})$ and x^3 is constant.

Then:

- s^* the curvilinear abscissa along $(\mathcal{C}^*)$;
- $\mathbf{t}^*$ the unit vector tangent to $(\mathcal{C}^*)$:

$$\mathbf{t}^* = \frac{dp}{ds^*} = \frac{ds}{ds^*}\frac{dp}{ds} = \frac{ds}{ds^*}\mathrm{m}(\mathbf{t})$$

- ν^* the unit vector normal to the facet in p :

$$\nu^* = a_3 \times \mathbf{t}^*$$

But on the other hand:

$$\nu = \nu^\alpha a_\alpha = a_3 \times \mathbf{t} = t^\alpha\, \mathrm{e}_{\alpha\lambda}(m)\, a^{\lambda\mu}\, a_\mu$$

$$\nu^* = \nu^{*\alpha}\, g_\alpha = a_3 \times \mathbf{t}^* = a_3 \times \left(\frac{ds}{ds^*} t^\alpha g_\alpha\right) = \frac{ds}{ds^*} t^\alpha\, \mathrm{e}_{\alpha\lambda}(p)\, g^{\lambda\mu}\, g_\mu$$

from which:

$$\nu_\beta = t^\alpha \, \mathrm{e}_{\alpha\beta}(\boldsymbol{m}) = t^\alpha \, \hat{a}^{1/2} \, \tilde{\mathrm{e}}_{\alpha\beta}$$

$$\nu^*_\beta = \frac{ds}{ds^*} t^\alpha \, \mathrm{e}_{\alpha\beta}(\boldsymbol{p}) = \frac{ds}{ds^*} t^\alpha \, \hat{g}^{1/2} \, \tilde{\mathrm{e}}_{\alpha\beta}$$

It becomes:

$$\nu^*_\beta = \frac{ds}{ds^*} \hat{\mathrm{m}} \, \nu_\beta \tag{5.14}$$

because:

$$\left(\frac{\hat{g}}{\hat{a}}\right)^{1/2} = \hat{\mathrm{m}}$$

The geometry and parameterisation of the facet are thus fully defined.

The stress vector at any point $\boldsymbol{p}$ of $\mathcal{F}$ is expressed by:

$$\mathbf{T} = \overline{\overline{T}} \bullet \nu^* = T^{\alpha i} \bullet \nu^*_\alpha \, g_i$$

$\overline{\overline{T}}$ and ν^* being expressed in the system g_i, the resultant $\mathcal{R}$ of $\mathbf{T}$ on $\mathcal{F}$ is:

$$\mathcal{R} = \int_{\mathcal{F}} \overline{\overline{T}} \bullet \nu^* \, dS(\boldsymbol{p}) = \int_{\mathcal{F}} T^{\alpha i} \bullet \nu^*_\alpha \, g_i \, ds^* \, dx^3$$

either by (5.14):

$$\mathcal{R} = \int_{s_1}^{s_2} \nu_\alpha \left(\int_{-e/2}^{e/2} T^{\alpha i} \, \mathrm{m}_i^{\,j} \, a_j \, \hat{\mathrm{m}} \, dx^3 \right) ds$$

or again, according to (5.9) and (5.10):

$$\mathcal{R} = \int_{s_1}^{s_2} \left(\bar{\mathrm{N}}^{\alpha\beta} a_\beta + \bar{Q}^\alpha a_3 \right) \nu_\alpha \, ds \tag{5.15}$$

The density of resultant forces (resultant of stresses in the thickness, along the normal a_3) is therefore the vector:

$$\boxed{\frac{d\mathcal{R}}{ds} = \bar{\mathrm{N}}^{\alpha\beta} \nu_\alpha a_\beta + \bar{Q}^\alpha \nu_\alpha a_3} \tag{5.16}$$

Let $\boldsymbol{m}_1$ be the point of $(\mathcal{C})$ with abscissa s_1 and $\mathcal{M}$ the moment of T with respect to $\boldsymbol{m}_1$:

$$\mathcal{M} = \int_{\mathcal{F}} \boldsymbol{m}_1\boldsymbol{p} \times \left(\overline{\overline{T}} \bullet \nu^* \right) dS(\boldsymbol{p}) = \int_{\mathcal{F}} \left(\boldsymbol{m}_1\boldsymbol{m} + x^3 a_3 \right) \times \left(T^{\alpha i} \bullet \nu^*_\alpha \right) g_i \, ds^* \, dx^3$$

$$\mathcal{M}=\int_{s_1}^{s_2} \nu_\alpha \boldsymbol{m}_1\boldsymbol{m}\times\left(\int_{-e/2}^{e/2} T^{\alpha i}\ \mathrm{m}_i^{\ j}\ a_j\ \hat{\mathrm{m}}\ dx^3\right)ds$$

$$+\int_{s_1}^{s_2} \nu_\alpha a_3\times a_\beta\left(\int_{-e/2}^{e/2} T^{\alpha\gamma}\ \mathrm{m}_\gamma^{\ \beta}\ \hat{\mathrm{m}}\ x^3\ dx^3\right)ds$$

$$\mathcal{M}=\int_{s_1}^{s_2} \nu_\alpha \boldsymbol{m}_1\boldsymbol{m}\times\frac{d\mathcal{R}}{ds}ds+\int_{s_1}^{s_2} \bar{M}^{\alpha\beta}\nu_\alpha a_\beta\times a_3\ ds$$

The density of resultant moment (resultant moment of the stresses in the thickness, along a_3 with respect to $\boldsymbol{m}$) is therefore:

$$\frac{\mathrm{d}\mathcal{M}}{ds}=\bar{M}^{\alpha\beta}\nu_\alpha a_\beta\times a_3 \tag{5.17}$$

Expressing $\bar{M}$ in the orthonormal system ($\mathbf{t}$, ν), (5.17) is written:

$$\frac{\mathrm{d}\mathcal{M}}{ds}=\bar{M}^{\nu\nu}\mathbf{t}-\bar{M}\ \nu \tag{5.18}$$

$\bar{M}^{\nu\nu}$ is the bending moment applied to the facet in $\boldsymbol{m}$, $-\bar{M}^{\nu t}$ is called the torsion moment (it tends to "twist" the facet).

The expression of the generalised stresses function to the three-dimensional stresses was obtained by Love[2] (1893), in the principal curvature coordinate system, and by Zerna[3] (1949), in any coordinate system.

5.1.4 Obtaining equilibrium equations from three-dimensional equations

Thus established, the direct relations between the screw resulting from the stresses on a facet $\mathcal{F}$ and the tensors$\bar{N}$, $\bar{M}$, and $\bar{Q}$, it is possible to establish equations of equilibrium of a shell by integrating the local equations of the three-dimensional theory:

$$\operatorname{div}\bar{\bar{T}}+\bar{\rho}\left(F-\gamma\right)=0$$

in the thickness of the shell (Novozhilov[4], 1943, in the principal curvature system, and Naghdi, 1963, in any coordinate system). The calculations, detailed in §5.5, lead to the following equations, where some terms have been considered null:

$$\begin{aligned}
&\nabla_\alpha \bar{N}^{\alpha\beta}-b_\alpha^{\ \beta}\bar{Q}^\alpha+p^\beta-\rho\gamma^\beta=0\\
&\nabla_\alpha \bar{Q}^\alpha+b_{\alpha\beta}\bar{N}^{\alpha\beta}+p^3-\rho\gamma^3=0\\
&\nabla_\alpha \bar{M}^{\alpha\beta}+\bar{Q}^\beta+c^\beta=0\\
&\bar{N}^{\alpha\beta}-b_\gamma^{\ \beta}\bar{M}^{\gamma\alpha}=\bar{N}^{\beta\alpha}-b_\gamma^{\ \alpha}\bar{M}^{\gamma\beta}
\end{aligned} \tag{5.19}$$

[2] Augustus Edward Hough Love, British mathematician (1863–1940).
[3] Wolfgang Zerna, German Engineer (1916–2005).
[4] Valentin Valentinovich Novozhilov, Russian engineer (1910–1987).

In fact, the relations of the last line of (5.19) are identically verified. Indeed, from the definitions (5.9):

$$\bar{N}^{\alpha\beta} - b_\gamma{}^\beta \bar{M}^{\gamma\alpha} = \int_{-e/2}^{e/2} \left(a_\gamma{}^\alpha \mathfrak{m}_\delta{}^\beta + x^3 b_\gamma{}^\beta \mathfrak{m}_\delta{}^\alpha \right) T^{\gamma\delta} \hat{\mathfrak{m}}\, dx^3$$

$$= \int_{-e/2}^{e/2} \left[a_\gamma{}^\alpha a_\delta{}^\beta - x^3 \left(a_\gamma{}^\alpha b_\delta{}^\beta + a_\delta{}^\alpha b_\gamma{}^\beta \right) + \left(x^3 \right)^2 b_\gamma{}^\beta b_\delta{}^\alpha \right] T^{\gamma\delta} \hat{\mathfrak{m}}\, dx^3$$

The resulting relation is symmetrical in α and β, taking into account the symmetry of the tensor $\bar{\bar{T}}$. This property is even easier to demonstrate in the main curvature coordinate system, using relations (5.11) and (5.12). In reality, this symmetry relation is only interesting when $\alpha = \beta$, in which case:

$$\bar{N}^{12} - b_\gamma{}^2 \bar{M} \quad = \bar{N}^{21} - b_\gamma{}^1 \bar{M}^{\gamma 2}$$

There are therefore only five independent equations in relations (5.19). These equations are almost identical in static to equations (4.66) and (4.68) obtained by writing the overall equilibrium of the shell. They differ in the fact that the tensor $\bar{M}$ is identical to the tensor H introduced by (4.120) and not to the tensor M: $\bar{M}(\mathrm{v})$ and $M(\mathrm{v})$ are orthogonal in $T_{\mathfrak{m}}$. H satisfies equations (4.125) and (4.126), identical to the third and fourth lines of (5.19). Thus, the tensors N, H, and Q of §4.4.1 are respectively equivalent to $\bar{N}$, $\bar{M}$, and $\bar{Q}$, which thus correctly define the resultants of stresses. Due to the equivalence demonstrated in §4.4.1, equations (5.19) are also equivalent to equations (4.97) obtained by PVP. The relations between these different tensors are obtained again in Section 5.2 in the particular case of the Kirchhoff[5]–Love hypotheses.

5.1.5 Case of thin shells

5.1.5.1 Hypothesis of thin shells

/ H4 / A shell is thin when its thickness in any point is small vis-à-vis its characteristic dimensions and its radii of curvature:

$$\frac{e}{\ell} << 1 \text{ and } \frac{e}{|R_1|} << 1$$

where R_1 is the smallest radius of curvature in absolute value; ℓ is for example the smallest span of the shell (for an unclosed shell). In the case of a plate, the first inequality is the only one to retain.

From a physical point of view, the hypothesis of thin shells must be related to the kinematic hypotheses set out in the following paragraphs.

From a mathematical point of view, the hypothesis of thin shells is expressed by retaining in the different expressions only the terms of the first order in x^3. It is also known as "first approximation theory".

[5] Gustav Robert Kirchhoff, German physicist (1824–1887).

5.1.5.2 Consequences on the resultant of stresses in the thickness

The most extreme simplification consists in considering that the endomorphism $\mathbb{m}$ is reduced to the identity: the shell is sufficiently thin so that the change of metric in the thickness can be neglected. In this case, expressions (5.9) and (5.10) are simplified in:

$$\boxed{\begin{aligned} \bar{N}^{\alpha\beta} &= \int_{-e/2}^{e/2} T^{\alpha\beta}\, dx^3 \\ \bar{M}^{\alpha\beta} &= -\int_{-e/2}^{e/2} T^{\alpha\beta}\, x^3\, dx^3 \end{aligned}} \tag{5.20}$$

$$\boxed{\bar{Q}^{\beta} = \int_{-e/2}^{e/2} T^{3\beta}\, dx^3} \tag{5.21}$$

In this hypothesis, tensors $\bar{N}$ and $\bar{M}$ are symmetrical. But the last equilibrium equation (5.19) is written in the main curvature coordinates for the component (*1,2*):

$$\bar{N}^{12} - \bar{N}^{21} + \frac{\bar{M}^{12}}{R_1} - \frac{\bar{M}^{21}}{R_2} = 0 \tag{5.22}$$

The approximation given by (5.20) therefore violates the equilibrium condition (5.22), due to the symmetry of $\bar{N}$ and $\bar{M}$. The approximation is however acceptable in the following cases:

- the shell is spherical: $R_1 = R_2$;
- the shell is a plate: $b = 0$;
- the situations where N and M are diagonal simultaneously, for example, axisymmetric loaded shells.

If these conditions are not fulfilled, the approximation (5.20) is acceptable if $\dfrac{\bar{M}^{12}}{R_1} - \dfrac{\bar{M}^{21}}{R_2}$ is negligible. This is the case with shallow shells or when the twisting moments are small vis-à-vis the bending moments.

In cases where it is not desirable to make this approximation because the above conditions are not met, it is best to keep the exact relationships (5.9) and (5.10), which are most often expressed in the principal curvature coordinates by (5.11), (5.12), and (5.13). Koiter[6] (1960), Novozhilov (1964), and Naghdi (1963) among other authors have particularly discussed the degree of approximation made by the hypothesis of thin shells. Other approximations are shown in §5.6.2 for the establishment of constitutive laws for elastic thick shells.

5.2 THEORY OF KIRCHHOFF–LOVE SHELLS

5.2.1 Kirchhoff kinematic assumptions

The notions of the middle surface and of the normal vector are purely geometrical. Material points on the middle surface in a given configuration do not necessarily remain

[6] Warner Tjardus Koiter, Dutch engineer (1914–1997).

on the middle surface in a mechanical transformation. Similarly, all the material points in coincidence with a line segment normal to σ deform over time and generally remains neither straight nor normal to the surface.

The local behaviour of the shell can therefore only be approached by the three-dimensional theory. In order to retain the benefits of two-dimensional modelling, some simplifying assumptions are needed to appreciate the local behaviour of the shell. These hypotheses find their justification, in validity domains to be specified, only by the good adequacy of the results obtained with respect to the experimental observations or, in certain particular situations, with respect to the results of the three-dimensional theory.

Among these hypotheses, Kirchhoff's hypotheses are the simplest, and give satisfactory results in most cases, especially when the shell is sufficiently thin. They generalise in the case of shells the Bernoulli-Euler hypotheses for the beams. They were used for the first time by Kirchhoff (1850) to establish the theory of plate bending, then taken up by Love (1888) to establish a linear theory of shells.

/ H1 / The material points on the middle surface in a given configuration remain on the middle surface during the transformation.

This first hypothesis is fundamental to establishing a theory of shells, so it is generally accepted.

/ H2 / (Kirchhoff) The set of material points initially situated on the normal to Σ remain a segment of the line that is a rigid body and constantly normal to σ during the transformation.

This second hypothesis (Kirchhoff) actually comprises two hypotheses: the indeformability of the normal segment and the hypothesis of normality.

These two hypotheses make it possible to express the transformation at any point $\mathcal{P}\left(X^\alpha, X^3 = x^3\right)$ of the shell, in Lagrange coordinates (Figure 5.3):

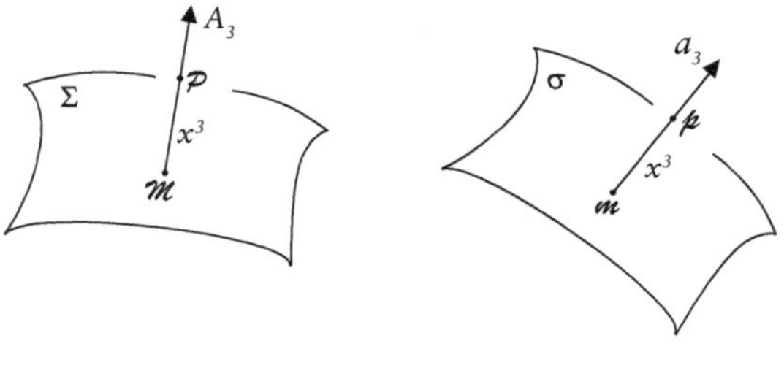

Figure 5.3

INITIAL CONFIGURATION Σ **ACTUAL CONFIGURATION σ**

$$\begin{array}{lcl} \mathcal{M} \in \Sigma & & m \in \sigma, \quad m = f(\mathcal{M}) \\ & \longrightarrow & \\ \mathcal{P} \notin \Sigma, \quad \mathcal{P} = \mathcal{M} + x^3 A_3 & & p = m + x^3 a_3 \end{array} \tag{5.23}$$

The first line expresses hypothesis H1, $\mathcal{M}$ and m being two geometric positions of the same material point permanently located on the middle surface. The second line expresses hypothesis H2:

- the normal remains straight and normal;
- the normal does not lengthen (the thickness remains constant over time).

The displacement of any point $\mathcal{P}$[7] can therefore be expressed as a function of the displacement of point $\mathcal{M}$, projection of $\mathcal{P}$ on Σ:

$$\bar{\xi}(\mathcal{P}) = \xi(\mathcal{M}) + x^3(a_3 - A_3) \tag{5.24}$$

which shows that the study of the evolution of the shell is brought back to that of the middle surface.

In small displacements, (5.24) is written, by introducing the rotation vector ω of the normal:

$$\bar{\xi}(\mathcal{P}) = \xi(\mathcal{M}) + x^3\omega \times A_3 \tag{5.25}$$

that is to say:

$$\begin{aligned} \bar{\xi}^\alpha(\mathcal{P}) &= \xi^\alpha - x^3\,\omega^\lambda\,\varepsilon_\lambda{}^\alpha \\ \bar{\xi}^3(\mathcal{P}) &= \zeta \end{aligned} \tag{5.26}$$

Finally, in the thickness of the shell, $\mathcal{M}$ is the endomorphism:

$$\mathcal{M} = A - x^3 B$$

being the value of the endomorphism m on the reference configuration, x^3 being constant during the transformation.

5.2.2 Consequences of the kinematics of the transformation

5.2.2.1 Velocity and acceleration

The velocity is obtained by differentiating the displacement:

$$\bar{V}(p) = V(m) + x^3\dot{a}_3 \tag{5.27}$$

and is expressed entirely according to the kinematics of the middle surface thanks to the relations of §4.1.1.2:

$$\bar{V}^i(p) = v^i - x^3\left(\partial_\alpha v^3 + b_{\alpha\beta}v^\beta\right)a^{\alpha i} \tag{5.28}$$

[7] Some quantities defined on the middle surface and in the ambient space have close physical meanings and the same symbol is assigned to them. To differentiate them, the symbol is upperlined for the quantity defined in the ambient space. This is for example the case for strains.

The velocities of m and p in the direction of the normal are identical. It becomes:

$$\bar{V}^i(p) = v^\alpha a_\alpha + v^3 a_3 \tag{5.29}$$

with:

$$\bar{V}_\alpha(p) = v_\alpha - x^3\left(\partial_\alpha v^3 + b_{\alpha\beta} v^\beta\right)$$

$$\bar{V}_3(p) = v_3$$

where v^i are the components of $V(m)$ in the basis a_i.

As $\dot{a}_3 = \Omega \times a_3$, previous relationships can also be written:

$$\bar{V}(p) = V(m) + x^3 \Omega \times a_3 \tag{5.30}$$

where Ω is the angular velocity vector of the normal.

Acceleration is obtained by differentiating (5.27):

$$\bar{\gamma}(p) = \gamma(m) + x^3 \ddot{a}_3 \tag{5.31}$$

$\gamma(m)$ is determined by relation (4.23) and is noted:

$$\gamma(m) = \gamma = \gamma^\alpha a_\alpha + \gamma^3 a_3 \tag{5.32}$$

$\ddot{a}_3$ is calculated from the relations of §4.1.1.2.:

$$\overset{\cdot}{\overbrace{\left(\dot{a}_3 \bullet a_\alpha\right)}} = \ddot{a}_3 \bullet a_\alpha + \dot{a}_3 \bullet \dot{a}_\alpha = \ddot{a}_3 \bullet a_\alpha - v_{\lambda 3}\, a^{\lambda\beta} a_\beta \bullet \dot{a}_\alpha = -\frac{dv_{\alpha 3}}{dt}$$

$$\overset{\cdot}{\overbrace{\left(\dot{a}_3 \bullet a_3\right)}} = \ddot{a}_3 \bullet a_3 + \left(\dot{a}_3\right)^2 = \ddot{a}_3 \bullet a_3 - a^{\lambda\mu} v_{\lambda 3} v_{\mu 3} = 0$$

from where:

$$\ddot{a}_3 \bullet a_\alpha = -\frac{dv_{\alpha 3}}{dt} + v_{\lambda 3}\, a^{\lambda\beta} a_\beta \bullet \dot{a}_\alpha$$

$$\ddot{a}_3 \bullet a_3 = a^{\lambda\mu}\, v_{\lambda 3}\, v_{\mu 3}$$

It finally becomes:

$$\bar{\gamma}(p) = \left[\gamma^\alpha + x^3 a^{\alpha\beta}\left(v_{\beta\lambda}\, v_{\mu 3}\, a^{\lambda\mu} - \frac{dv_{\alpha 3}}{dt}\right)\right] a_\alpha + \left[\gamma^3 + x^3 v_{\lambda 3}\, v_{\mu 3}\, a^{\lambda\mu}\right] a_3 \tag{5.33}$$

5.2.2.2 Deformations

Kirchhoff's kinematic hypotheses make it possible to calculate the three-dimensional strain tensor $\bar{\varepsilon}$ in Lagrangian variables at any point $\mathcal{P}$ of the shell:

- $\bar{\varepsilon}_{\alpha 3}(\mathcal{P}) = 0$: the normal vector remains normal;
- $\bar{\varepsilon}_{33}(\mathcal{P}) = 0$: the normal vector does not elongate;

$$\bar{\varepsilon}_{\alpha\beta}(\mathcal{P}) = \frac{1}{2}\left(g_{ij}\frac{\partial x^i}{\partial X^\alpha}\frac{\partial x^j}{\partial X^\beta} - G_{\alpha\beta} \right)$$

In Kirchhoff–Love theory, the layers parallel to the middle surface are in plane strain

where the strain tensor $\bar{\varepsilon}$ in the Lagrange variable is expressed in the local coordinate system G_i.

Tensors g and G, taking into account Kirchhoff's hypotheses, have analogous formulations given by (5.1). Noting that $\frac{\partial x^3}{\partial X^\alpha} = 0$, it becomes:

$$\bar{\varepsilon}_{\alpha\beta}(\mathcal{P}) = \frac{1}{2}\left(a_{\lambda\mu}\frac{\partial x^\lambda}{\partial X^\alpha}\frac{\partial x^\mu}{\partial X^\beta} - A_{\alpha\beta} \right) - x^3\left(b_{\lambda\mu}\frac{\partial x^\lambda}{\partial X^\alpha}\frac{\partial x^\mu}{\partial X^\beta} - B_{\alpha\beta} \right) + \frac{\left(x^3\right)^2}{2}\left(c_{\lambda\mu}\frac{\partial x^\lambda}{\partial X^\alpha}\frac{\partial x^\mu}{\partial X^\beta} - C_{\alpha\beta} \right) \tag{5.34}$$

which is rewritten:

$$\bar{\varepsilon}_{\alpha\beta}(\mathcal{P}) = \varepsilon_{\alpha\beta} - x^3\kappa_{\alpha\beta} + \left(x^3\right)^2\tau_{\alpha\beta} \tag{5.35}$$

where ε and κ are the tensors of deformation and of variation of curvature of the middle surface Σ and τ the tensor of variation of the third fundamental form on Σ, which components are:

$$\tau_{\alpha\beta}(\mathcal{P}) = \frac{1}{2}\left(c_{\lambda\mu}\frac{\partial x^\lambda}{\partial X^\alpha}\frac{\partial x^\mu}{\partial X^\beta} - C_{\alpha\beta} \right) \tag{5.36}$$

In intrinsic notations:

$$\bar{\varepsilon}(\mathcal{P}) = \varepsilon - x^3\kappa + \left(x^3\right)^2\tau \tag{5.37}$$

$\bar{\varepsilon}(\mathcal{P})$ can be expressed as a function of displacement by replacing ε, κ and τ by their expressions using relations (3.95) to (3.101). This is not detailed here in the general case. Only the case of small displacements is examined below. The same parameterisation is used on the reference surface and the actual surface, then, according to relations (3.95) and (3.103):

$$a_\alpha = \left(\delta_\alpha{}^\beta + \mathcal{A}_\alpha{}^\beta\right)A_\beta + \mathcal{B}_\alpha A_3$$

$$a_3 = A_3 - \mathcal{B}_\alpha A^\alpha$$

Differentiating relation:

$$\mathscr{p} = \mathscr{m} + x^3 a_3$$

it becomes:

$$g_\alpha = a_\alpha - x^3\left[\left(B_\alpha{}^\gamma + \nabla_\alpha \mathcal{B}^\gamma\right)A_\gamma + \mathcal{B}^\gamma B_{\alpha\gamma} A_3\right]$$

hence the deformation:

$$\bar{\varepsilon}_{\alpha\beta} = \frac{1}{2}\left[\begin{array}{c}\left(\delta_\alpha{}^\gamma + \mathcal{A}_\alpha{}^\gamma - x^3\nabla_\alpha \mathcal{B}^\gamma - x^3 B_\alpha{}^\gamma\right)\left(\delta_\beta{}^\delta + \mathcal{A}_\beta{}^\delta - x^3\nabla_\beta \mathcal{B}^\delta - x^3 B_\beta{}^\delta\right)A_{\gamma\delta} \\ + \mathcal{B}^\gamma \mathcal{M}_{\alpha\gamma}\mathcal{B}^\delta \mathcal{M}_{\beta\delta} - \mathcal{M}_{\alpha\gamma}\mathcal{M}_\beta{}^\gamma\end{array}\right]$$

By retaining only the terms of displacement of the first order:

$$\bar{\varepsilon}_{\alpha\beta} \approx \frac{1}{2}\left\{\begin{array}{c}A_{\alpha\beta} + \mathcal{A}_{\alpha\beta} + \mathcal{A}_{\beta\alpha} - x^3\left(B_{\alpha\beta} + B_{\beta\alpha} + \nabla_\alpha \mathcal{B}_\beta + \nabla_\beta \mathcal{B}_\alpha\right) \\ - x^3\left[B_\alpha{}^\gamma\left(\mathcal{A}_{\beta\gamma} - x^3\nabla_\beta \mathcal{B}_\gamma\right) + B_\beta{}^\gamma\left(\mathcal{A}_{\alpha\gamma} - x^3\nabla_\alpha \mathcal{B}_\gamma\right)\right] \\ + \left(x^3\right)^2 B_\alpha{}^\gamma B_{\beta\gamma} - \mathcal{M}_{\alpha\gamma}\mathcal{M}_\beta{}^\gamma\end{array}\right\}$$

then, reordering and simplifying:

$$\bar{\varepsilon}_{\alpha\beta} \approx \frac{1}{2}\left[\begin{array}{c}\mathcal{M}_\alpha{}^\gamma \mathcal{A}_{\beta\gamma} + \mathcal{M}_\beta{}^\gamma \mathcal{A}_{\alpha\gamma} - x^3\left(\nabla_\alpha \mathcal{B}_\beta + \nabla_\beta \mathcal{B}_\alpha\right) \\ + \left(x^3\right)^2\left(B_\alpha{}^\gamma \nabla_\beta \mathcal{B}_\gamma + B_\beta{}^\gamma \nabla_\alpha \mathcal{B}_\gamma\right)\end{array}\right] \tag{5.38}$$

which gives at the same time the expression of $\tau_{\alpha\beta}$, term factor of $\left(x^3\right)^2$, as a function of the (small) displacement.

A new simplification leads to:

$$\bar{\varepsilon}_{\alpha\beta} \approx \frac{1}{2}\left[\mathcal{M}_\alpha{}^\gamma\left(\mathcal{A}_{\beta\gamma} - x^3\nabla_\beta \mathcal{B}_\gamma\right) + \mathcal{M}_\beta{}^\gamma\left(\mathcal{A}_{\alpha\gamma} - x^3\nabla_\alpha \mathcal{B}_\gamma\right)\right] \tag{5.39}$$

By developing and reordering this last equality, the local strain tensor is also expressed as a function of the deformation variables of the middle surface:

$$\begin{aligned}\bar{\varepsilon}_{\alpha\beta}(\mathcal{P}) \approx & \left[\delta_\alpha{}^\lambda \delta_\beta{}^\mu - \left(x^3\right)^2 B_\alpha{}^\lambda B_\beta{}^\mu\right]\varepsilon_{\lambda\mu} \\ & - x^3\left[\delta_\alpha{}^\lambda \delta_\beta{}^\mu - \frac{x^3}{2}\left(B_\alpha{}^\lambda \delta_\beta{}^\mu + \delta_\alpha{}^\lambda B_\beta{}^\mu\right)\right]\kappa_{\lambda\mu}\end{aligned} \tag{5.40}$$

The term of the second degree in x^3 contains the expression of $\tau_{\alpha\beta}$ as a function of the displacement, that is, in small displacements:

$$\tau_{\alpha\beta} \approx -B_\alpha{}^\lambda B_\beta{}^\mu \varepsilon_{\lambda\mu} + \frac{1}{2}\left(B_\alpha{}^\lambda \delta_\beta{}^\mu + \delta_\alpha{}^\lambda B_\beta{}^\mu\right)\kappa_{\lambda\mu}$$

where $\varepsilon_{\lambda\mu}$ and $\kappa_{\lambda\mu}$ are respectively given by (3.102) and (3.108), either, replacing by the components of the displacement and then simplifying with the help of Mainardi–Codazzi relations:

$$\tau_{\alpha\beta} \approx \frac{1}{2}\left[\begin{array}{c} B_\alpha{}^\gamma \nabla_\beta \nabla_\gamma \zeta + B_\beta{}^\gamma \nabla_\alpha \nabla_\gamma \zeta + \left(B_\alpha{}^\lambda \nabla_\beta B_{\lambda\gamma} + B_\beta{}^\lambda \nabla_\alpha B_{\lambda\gamma}\right)\xi^\gamma \\ + C_{\alpha\gamma}\nabla_\beta \xi^\gamma + C_{\beta\gamma}\nabla_\alpha \xi^\gamma \end{array}\right]$$

which is symmetrical.

In the orthonormal coordinate system associated with the principal curvature system, relations (5.40) are written:

$$\begin{aligned}
\overline{\varepsilon}_{11} &\approx \left(1-\frac{x^3}{R_1}\right)^{-1}\left[\left(1+\frac{x^3}{R_1}\right)\varepsilon_{11} - x^3\kappa_{11}\right] \\
\overline{\varepsilon}_{22} &\approx \left(1-\frac{x^3}{R_2}\right)^{-1}\left[\left(1+\frac{x^3}{R_2}\right)\varepsilon_{22} - x^3\kappa_{22}\right] \\
\overline{\varepsilon}_{12} &\approx \left(1-\frac{x^3}{R_1}\right)^{-1}\left(1-\frac{x^3}{R_2}\right)^{-1}\left[\left(1-\frac{\left(x^3\right)^2}{R_1R_2}\right)\varepsilon_{12} - x^3\left(1-\frac{x^3}{2}\left(\frac{1}{R_1}+\frac{1}{R_2}\right)\right)\kappa_{12}\right]
\end{aligned} \qquad (5.41)$$

5.2.2.3 *Virtual deformation*

Let us consider virtual transformations respecting Kirchhoff's hypotheses. The rate of virtual deformation at any point of the shell can be calculated by the virtualisation of relation (5.40). The expression obtained is established directly below from the general relationship (5.37). Taking as a reference configuration for the virtual transformation the current configuration, according to (5.24):

$$\xi(\mathscr{p},u) = \xi(\mathscr{m},u) + x^3\left[a_3(u) - a_3(0)\right]$$

The virtual strain rate in $\mathscr{p}$ is then written, according to (5.37):

$$\overset{*}{\overline{e}}(\mathscr{p}) = \overset{*}{e} - x^3\,\overset{*}{k} + \frac{\left(x^3\right)^2}{2}\lim_{u\to 0}\frac{1}{u}\left[\tau(u) - \tau(0)\right]$$

$\overset{*}{e}$ is the virtual deformation rate and $\overset{*}{k}$ the virtual rate of variation of curvature of the surface; $\overset{*}{e}$ and $\overset{*}{k}$ are given by relations (4.40) and (4.42). On the other hand:

$$\lim_{u\to 0}\frac{1}{u}\left[\tau(u) - \tau(0)\right] = \overset{*}{c}$$

But:

$$\overset{*}{c}_{\alpha\beta} = \overset{*}{\overbrace{\left(b_\alpha{}^\lambda b_{\beta\lambda}\right)}} = \overset{*}{b}_\alpha{}^\lambda\, b_{\beta\lambda} + b_\alpha{}^\lambda\, \overset{*}{b}_{\beta\lambda}$$

and:

$$\overset{*}{b}_\alpha{}^\lambda = \overset{*}{\overbrace{b_{\alpha\mu} a^{\mu\lambda}}} = \overset{*}{b}_{\alpha\mu}\, a^{\mu\lambda} + b_{\alpha\mu}\, \overset{*}{a}{}^{\mu\lambda}$$

Expressing $\overset{*}{a}{}^{\mu\lambda}$ by relationship (4.45), it becomes:

$$\overset{*}{c}_{\alpha\beta} = a^{\mu\lambda} b_{\beta\lambda}\, \overset{*}{b}_{\alpha\mu} + b_\alpha{}^\lambda\, \overset{*}{b}_{\beta\lambda} - a^{\mu\gamma} a^{\lambda\delta} b_{\alpha\mu} b_{\beta\lambda}\, \overset{*}{a}_{\gamma\delta}$$

By grouping the different terms:

$$\overset{*}{\bar{e}}_{\alpha\beta}(\not{p}) = \left[\delta_\alpha{}^\lambda \delta_\beta{}^\mu - \left(x^3\right)^2 b_\alpha{}^\lambda b_\beta{}^\mu\right] \overset{*}{e}_{\lambda\mu}$$
$$- x^3 \left[\delta_\alpha{}^\lambda \delta_\beta{}^\mu - \frac{x^3}{2}\left(b_\alpha{}^\lambda \delta_\beta{}^\mu + \delta_\alpha{}^\lambda b_\beta{}^\mu\right)\right] \overset{*}{k}_{\lambda\mu} \tag{5.42}$$

On the other hand:

$$\overset{*}{\bar{e}}_{i3}(\not{p}) = 0 \tag{5.43}$$

In conclusion, the virtual deformation at any point of the shell is expressed as a function of the virtual deformation $\overset{*}{e}$ and the virtual variation of curvature $\overset{*}{k}$ of the middle surface. $\overset{*}{e}$ and $\overset{*}{k}$, therefore $\overset{*}{\bar{e}}(\not{p})$, are expressed as functions of the virtual velocity field using relationships (4.40), (4.42), (5.52), and (5.43).

Because of Kirchhoff's hypotheses, by which the motion of the normal is entirely determined by the motion of the middle surface, the virtual deformations in the thickness depend only on the virtual deformations of the middle surface. There is therefore perfect compatibility between Kirchhoff's hypotheses and the deformation hypotheses of the restricted theory, used in §4.2.2 to establish equations of motion.

5.2.3 Continuity equation and conservation equations

5.2.3.1 Density and mass per unit area

Let $\bar{\rho}$ the density of the shell, depending on the point $\not{p}$. A surfacic field ρ (mass per unit area or surfacic mass), defined on σ, is associated with the volumic field $\bar{\rho}$.

Let ω be a domain cut into the shell by the set of normals resting on the edge ∂d of area d of σ (Figure 5.4).

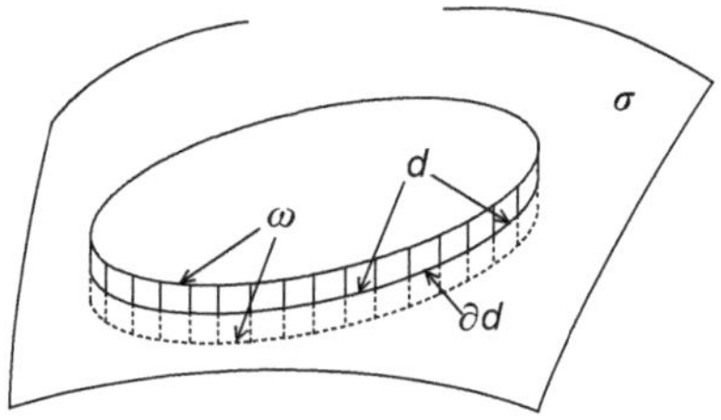

Figure 5.4

By definition of ρ, the mass m of d is written:

$$m = \int_d \rho \, d\sigma = \int_\omega \bar{\rho} \, dV \tag{5.44}$$

This equality is rewritten:

$$m = \int_d \rho \, \hat{a}^{1/2} dx^1 dx^2 = \int_\omega \bar{\rho} \, \hat{g}^{1/2} dx^1 dx^2 dx^3$$

It is verified regardless of the domain d, which implies the relation:

$$\rho \, \hat{a}^{1/2} = \int_{-e/2}^{e/2} \bar{\rho} \, \hat{g}^{1/2} dx^3 \tag{5.45}$$

In addition, in the normal coordinate system:

$$\hat{g}^{1/2} = \hat{\mathrm{m}} \, \hat{\mathrm{a}}^{1/2}$$

which allows it to be rewritten (5.45):

$$\boxed{\rho = \int_{-e/2}^{e/2} \bar{\rho} \, \hat{\mathrm{m}} \, dx^3} \tag{5.46}$$

5.2.3.2 Continuity equation

It is natural to follow a given mass over time. Now, taking into account Kirchhoff's hypothesis, a set of material points situated in the thickness of the shell remains delimited over time by a domain generated by the normals at the middle surface. By using, in Lagrange coordinates, a single coordinate system X^i for all configurations, the continuity equation (4.30) shows that the quantity $\rho \, \hat{a}^{1/2}$, therefore $\int_{-e/2}^{e/2} \bar{\rho} \, \hat{g}^{1/2} dx^3$, remains constant during the evolution of the shell.

Let Θ be the volume dilation in $\mathcal{P}$ from the reference configuration. By analogy with formula (4.24) and noting that Kirchhoff's hypotheses imply that the Jacobian of the transformation $x^i\left(X^j\right)$ is the same as the Jacobian of the transformation $x^\alpha\left(X^\beta\right)$, J, it becomes:

$$1 + \Theta = J\left(\frac{\hat{g}}{\hat{G}}\right)^{1/2} = \left(1 + \Xi\right)\frac{\hat{\mathrm{m}}}{\hat{\mathcal{M}}} \tag{5.47}$$

The three-dimensional continuity equation is therefore written as:

$$\bar{\rho}_0 = \bar{\rho}\left(1 + \Theta\right) = \bar{\rho}\left(1 + \Xi\right)\frac{\hat{\mathrm{m}}}{\hat{\mathcal{M}}} \tag{5.48}$$

either:

$$\boxed{\bar{\rho}_0 \, \hat{\mathcal{M}} = \bar{\rho} \, \hat{\mathrm{m}} \left(1 + \Xi\right)} \tag{5.49}$$

This relation reveals the effect of the change of curvature on the variations of density, in the thickness of the shell.

The expression of the continuity equation in Euler variables is obtained by deriving expression (5.49) with respect to time:

$$\frac{d}{dt}\left[\overline{\rho}\,\hat{\mathfrak{m}}\left(1+\Xi\right)\right]=0$$

which can be rewritten as:

$$\dot{\overline{\rho}}+\overline{\rho}\left(\frac{\dot{\hat{\mathfrak{m}}}}{\hat{\mathfrak{m}}}+\frac{\dot{\Xi}}{1+\Xi}\right)=0 \tag{5.50}$$

The last term has already been calculated in §4.1.2.1. To calculate the derivative of $\hat{\mathfrak{m}}$, it is useful to introduce the inverse endomorphism ℓ of $\mathfrak{m}$. So:

$$\dot{\hat{\mathfrak{m}}}=\frac{\partial\hat{\mathfrak{m}}}{\partial\mathfrak{m}_\alpha{}^\beta}\dot{\mathfrak{m}}_\alpha{}^\beta=\hat{\mathfrak{m}}\,\ell^\alpha{}_\beta\,\dot{\mathfrak{m}}_\alpha{}^\beta=\hat{\mathfrak{m}}\,\mathrm{TR}\left(\mathfrak{m}^{-1}.\dot{\mathfrak{m}}\right)$$

On the other hand:

$$\dot{\mathfrak{m}}_\alpha{}^\beta=-x^3\,\overbrace{b_{\alpha\lambda}a^{\lambda\beta}}^{\cdot}=-x^3\left(\dot{b}_{\alpha\lambda}a^{\lambda\beta}+b_{\alpha\lambda}\dot{a}^{\lambda\beta}\right)$$

and:

$$\dot{a}^{\lambda\beta}=-\,a^{\mu\beta}a^{\lambda\gamma}\dot{a}_{\gamma\mu}$$

(relationship obtained by writing that $\overbrace{a^{\lambda\gamma}a_{\gamma\mu}}^{\cdot}=0$). Finally, the continuity equation is written:

$$\boxed{\dot{\overline{\rho}}+\overline{\rho}\left(v_\alpha{}^\alpha-x^3\ell^{\alpha\beta}\dot{b}_{\alpha\beta}+x^3\ell^{\alpha\beta}b_\alpha{}^\gamma\dot{a}_{\gamma\beta}\right)=0} \tag{5.51}$$

where $v_{\alpha\beta}$, $\dot{a}_{\gamma\beta}$, and $\dot{b}_{\alpha\beta}$ are respectively given by relations (4.6), (4.11), and (4.16).

5.2.3.3 Conservation equations

Relationships (5.49) to (5.51) can be generalised by substituting for $\overline{\rho}$ any physical quantity represented by a volumic field $\overline{\phi}$ which quantity on a domain ω:

$$\Phi=\int_\omega\overline{\phi}\,dV$$

remains constant over time. A surface density ϕ can on the other hand be associated to $\overline{\phi}$ with relations similar to (5.45) and (5.46).

5.2.4 Equilibrium equations

5.2.4.1 Expressions of deformation energy

Relationships (5.42) and (5.43) make it possible to express the deformation of virtual power in three-dimensional theory as a function of $\overset{*}{e}$ and $\overset{*}{k}$. Indeed, $\overline{\overline{T}}$ being the tensor of local stresses in Euler variables:

$$\forall d, \forall \overset{*}{\bar{e}}, \quad \overset{*}{\mathcal{W}}_f = \int_\omega T^{ij}(p)\overset{*}{\bar{e}}_{ij}(p)\, dV(p) = \int_\omega T^{\alpha\beta}(p)\overset{*}{\bar{e}}_{\alpha\beta}(p)\, dV(p)$$

or:

$$\forall d, \forall \overset{*}{e}, \forall \overset{*}{k}$$

$$\overset{*}{\mathcal{W}}_f = \int_\omega T^{\alpha\beta}(p)\left[\delta_\alpha{}^\lambda\delta_\beta{}^\mu - \left(x^3\right)^2 b_\alpha{}^\lambda b_\beta{}^\mu\right]\overset{*}{e}_{\lambda\mu}\, dV(p)$$
$$+\int_\omega -x^3\, T^{\alpha\beta}(p)\left[\delta_\alpha{}^\lambda\delta_\beta{}^\mu - \frac{x^3}{2}\left(b_\alpha{}^\lambda\delta_\beta{}^\mu + \delta_\alpha{}^\lambda b_\beta{}^\mu\right)\right]\overset{*}{k}_{\lambda\mu}\, dV(p) \tag{5.52}$$

The virtual power of deformation is then written:

$$\overset{*}{\mathcal{W}}_f = \int_d \left\{\begin{array}{l} \displaystyle\int_{-e/2}^{+e/2} T^{\alpha\beta}(p)\left[\delta_\alpha{}^\lambda\delta_\beta{}^\mu - \left(x^3\right)^2 b_\alpha{}^\lambda b_\beta{}^\mu\right]\hat{m}\, dx^3 \bullet \overset{*}{e}_{\lambda\mu} \\ \displaystyle -\int_{-e/2}^{+e/2} x^3\, T^{\alpha\beta}(p)\left[\begin{array}{l}\delta_\alpha{}^\lambda\delta_\beta{}^\mu \\ -\dfrac{x^3}{2}\left(b_\alpha{}^\lambda\delta_\beta{}^\mu + \delta_\alpha{}^\lambda b_\beta{}^\mu\right)\end{array}\right]\hat{m}\, dx^3 \bullet \overset{*}{k}_{\lambda\mu} \end{array}\right\} d\sigma \tag{5.53}$$

The KL (Kirchhoff–Love) hypotheses thus made it possible to express the virtual power of deformation as a function of the deformation variables of the middle surface $\overset{*}{e}$ and $\overset{*}{k}$ only. This result is consistent with the hypotheses of the previous chapter concerning the strain energy in the restricted theory.

By assimilating the (virtual) power of three-dimensional deformation to its (restricted) expression as a function of $\overset{*}{e}$ and $\overset{*}{k}$, certain phenomena which can only have a representation in three-dimensional theory and can not be modelled only by the deformation variables e and k, are not taken into account: distortions due to shear force, variations in shell thickness, cracks, nonlinear temperature distribution…

5.2.4.2 Equilibrium equations

The expression of the virtual power of deformation is, in Kirchhoff's hypotheses, formally identical to the expression (4.76) and, likewise, the virtual powers of the external actions and of the inertial actions that can be assimilated to their global formulations given respectively by (4.87), (4.88), and (4.92), the equilibrium equations are given, in the framework of Kirchhoff's hypotheses, by (4.95) for the local equations and by (4.98) for the boundary conditions.

5.2.4.3 Relationships between generalised stresses and resultant stresses

As in the case of the beams, the objective is to connect the generalised stresses (of Sanders) N and M or the tensors resulting from the internal actions N and M introduced in the previous chapter to the tensor $\overline{\overline{T}}$ of the local stresses existing in the thickness of the shell.

Expressions (4.76) and (5.53) of the deformation virtual power can be considered formally equal, which makes it possible to deduce the expression of the generalised stresses associated with the deformations as a function of local stresses, the virtual velocity considered to be arbitrary (over σ):

$$\begin{aligned} \mathsf{N}^{\lambda\mu} &= \int_{-e/2}^{+e/2} T^{\alpha\beta}(\not{p})\left[\delta_\alpha{}^\lambda\delta_\beta{}^\mu-\left(x^3\right)^2 b_\alpha{}^\lambda b_\beta{}^\mu\right]\hat{\mathfrak{m}}\,dx^3 \\ \mathsf{M}^{\lambda\mu} &= -\int_{-e/2}^{+e/2} x^3\, T^{\alpha\beta}(\not{p})\left[\delta_\alpha{}^\lambda\delta_\beta{}^\mu-\frac{x^3}{2}\left(b_\alpha{}^\lambda\delta_\beta{}^\mu+\delta_\alpha{}^\lambda b_\beta{}^\mu\right)\right]\hat{\mathfrak{m}}\,dx^3 \end{aligned} \tag{5.54}$$

It is now a question of establishing, in the framework of the KL hypotheses, the relations between the generalised stresses N and M and the screw resulting from the local stresses applied to a facet, defined by relations (5.9) and (5.10). Noting that:

$$\begin{aligned} \delta_\lambda{}^\alpha\mathfrak{m}_\gamma{}^\beta+\delta_\gamma{}^\beta\mathfrak{m}_\lambda{}^\alpha &= \delta_\lambda{}^\alpha\left(\delta_\gamma{}^\beta-x^3b_\gamma{}^\beta\right)+\delta_\gamma{}^\beta\left(\delta_\lambda{}^\alpha-x^3b_\lambda{}^\alpha\right) \\ &= 2\left[\delta_\lambda{}^\alpha\delta_\gamma{}^\beta-\frac{x^3}{2}\left(\delta_\lambda{}^\alpha b_\gamma{}^\beta+b_\lambda{}^\alpha\delta_\gamma{}^\beta\right)\right] \end{aligned}$$

and:

$$\begin{aligned} \delta_\lambda{}^\alpha\delta_\gamma{}^\beta-\left(x^3\right)^2 b_\lambda{}^\alpha b_\gamma{}^\beta &= \delta_\lambda{}^\alpha\left(\delta_\gamma{}^\beta-x^3b_\gamma{}^\beta\right)+x^3b_\gamma{}^\beta\left(\delta_\lambda{}^\alpha-x^3b_\lambda{}^\alpha\right) \\ &= \delta_\lambda{}^\alpha\mathfrak{m}_\gamma{}^\beta+x^3b_\gamma{}^\beta\mathfrak{m}_\gamma{}^\alpha \end{aligned}$$

Relations (5.50) and (5.11) make it possible to obtain the following relations between the different tensors:

$$\begin{cases} \mathsf{N}^{\alpha\beta}=\bar{N}^{\alpha\beta}-b_\lambda{}^\beta\bar{M}^{\lambda\alpha} \\ \mathsf{M}^{\alpha\beta}=\dfrac{1}{2}\left(\bar{M}^{\alpha\beta}+\bar{M}^{\beta\alpha}\right) \end{cases} \tag{5.55}$$

Relations (5.11) and (5.12) show that $\bar{N}$ and $\bar{M}$ are not symmetrical in general. The symmetry of N and M (by definition) is respected by relations (5.55). These relations make it possible to find, in the context of the KL hypotheses and taking into account the remarks made in §5.1.3, the relations (4.129) established in the general case. Also:

$$\mathsf{L}^{\alpha\beta}=\bar{N}^{\alpha\beta}+\frac{1}{2}b_\gamma{}^\beta\left(\bar{M}^{\alpha\gamma}-\bar{M}^{\gamma\alpha}\right)$$

5.2.5 Case of thin shells

5.2.5.1 Consequences for deformations

Applying the results of §5.2.3 to the case of thin shells, it becomes:

$$\boxed{\bar{\varepsilon}(\not{p}) = \varepsilon - x^3\kappa} \tag{5.56}$$

The local deformation involves only the variations of metric and curvature of the middle surface. Similarly, the virtual deformation is written as:

$$\boxed{\overset{*}{\bar{e}}(\not{p}) = \overset{*}{e} - x^3\overset{*}{k}} \tag{5.57}$$

5.2.5.2 Consequences for generalised stresses

Expressions (5.54) are simplified by neglecting the metric variations in the thickness and keeping only the terms of the first order in x^3. Comparing them to relations (5.20):

$$\boxed{\begin{aligned} \bar{N}^{\alpha\beta} &= \int_{-e/2}^{+e/2} T^{\alpha\beta}(\not{p})\, dx^3 = \mathsf{N}^{\alpha\beta} \\ \bar{M}^{\alpha\beta} &= -\int_{-e/2}^{+e/2} x^3\, T^{\alpha\beta}(\not{p})\, dx^3 = \mathsf{M}^{\alpha\beta} \end{aligned}} \tag{5.58}$$

Therefore there is no need, in the case of thin shells, to distinguish between generalised stresses and the resultant of stresses.

5.3 THEORY OF REISSNER–MINDLIN SHELLS

5.3.1 Reissner–Mindlin hypotheses

The consideration of shear deformations was initiated by Reissner[8] (1945) and continued by Mindlin[9] (1951) for elastic plates. Compared to the KL assumptions, the normal segment stiffness assumption is maintained, but it does not remain normal to the middle surface during the transformation.

The hypothesis / H2 / is replaced by the hypothesis / H2'/:

/H2'/ The set of material points initially located on the normal to Σ is submitted to a rigid displacement composed of the translation of the point of the middle surface and a rotation not linked to the rotation of the normal vector.

[8] Max Erich (Eric) Reissner, German-American engineer and mathematician (1913–1996).
[9] Raymond David Mindlin, American engineer (1906–1987).

This hypothesis makes it possible to express the transformation of any point $\mathcal{P}(X^\alpha, X^3 = x^3)$ of the shell into a point p, by denoting **d** the unit vector carrying mp (Figure 5.5):

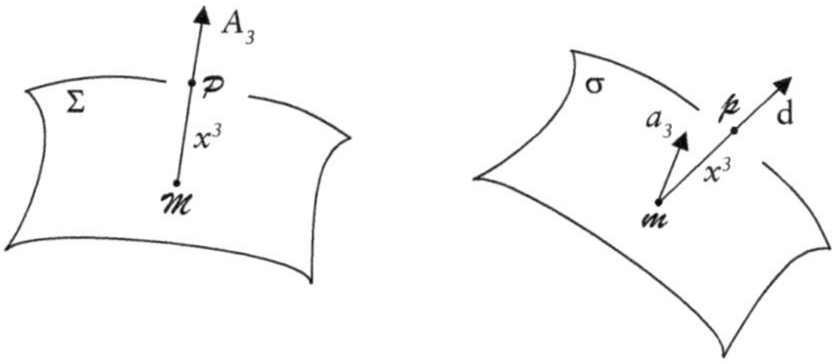

Figure 5.5

INITIAL CONFIGURATION Σ ACTUAL CONFIGURATION σ

$$\begin{array}{lcl} \mathcal{M} \in \Sigma & & m \in \sigma, \quad m = f(\mathcal{M}) \\ \mathcal{P} \notin \Sigma, \ \mathcal{P} = \mathcal{M} + x^3 A_3 & \longrightarrow & p = m + x^3 \mathbf{d} \end{array} \tag{5.59}$$

where the second line translates the hypothesis H2'.

The vector **d** being unitary is defined by two components (two angles for example). More generally, the three components of **d** are linked by a relation, for example in the basis (A_i):

$$d^i d^j A_{ij} = 1 \tag{5.60}$$

The displacement of the point $\mathcal{P}$ is thus written:

$$\mathcal{P}p = \bar{\xi}(\mathcal{P}) = \xi(\mathcal{M}) + x^3(\mathbf{d} - A_3) \tag{5.61}$$

In small displacements, the rotation of the material segment $\mathcal{MP}$ is characterised by a vector field θ belonging to T_m at every point. Thus, the displacement of a point $\mathcal{P}$ located on the normal in the reference position is written:

$$\mathcal{P}p = \bar{\xi}(\mathcal{P}) = \xi(\mathcal{M}) + \theta \times \mathcal{MP} \tag{5.62}$$

hence, in the basis A_i:

$$\begin{aligned} \bar{\xi}^\alpha(\mathcal{P}) &= \xi^\alpha - x^3 \theta^\gamma \varepsilon_{\gamma\beta} A^{\beta\alpha} \\ \bar{\xi}^3(\mathcal{P}) &= \zeta \end{aligned} \tag{5.63}$$

and the expression of **d**:

$$\mathbf{d} = A_3 + \theta \times A_3 \tag{5.64}$$

either, in components in the basis A_i:

$$d^{\beta} = -\theta^{\gamma}\varepsilon_{\gamma}{}^{\beta}$$
$$d^3 = 1 \qquad (5.65)$$

In the model used here, x^3 remains constant and the segment carrying the material points initially located on the normal remains straight and of constant length, but does not necessarily remain normal to the middle surface. The evolution of the shell is then completely described by five fields defined on the middle surface: the three components of the displacement of the middle surface, and two components of the vector **d** (or θ in small displacements).

Compared to the Kirchhoff–Love theory, the only difference is the release of two degrees of rotation to introduce distortions $\overline{\varepsilon}_{\alpha 3}$ as specified below. It should be noted that this theory is not as complete as the theory in §4-4.5, since here the variation of the thickness (always measured normally on the average surface) is related to the rotation θ and does not intervene at the first order. In addition, the variation of the distortion in the thickness is not in conformity with that obtained by the three-dimensional elastic constitutive law.

In Kirchhoff–Love theory, the expression (5.59) is reduced to (5.23) where **d** = a_3, in small displacements, by comparing expressions (5.25) and (5.62):

$$\omega = \theta$$

$\boldsymbol{\omega}$ being the rotation vector of the normal defined by (3.110).

It is necessary in the presentation above to introduce a reference configuration, which favours the Lagrange coordinates. Nevertheless, the same hypothesis for the kinematic evolution can be taken for any configuration of the shell, the material points being on the normal at a given instant occupying previously a segment not normal to the surface. This hypothesis makes it possible to obtain the expression of the quantities in Euler variables.

5.3.2 Consequences for kinematics

5.3.2.1 Velocity and acceleration

The velocity is obtained in Euler variables by writing that the segment bearing the material points situated on the normal at time t is a rigid body. If Ω denotes the angular velocity vector of a_3:

$$\overline{v}(\mathscr{p}) = v(\mathscr{m}) + x^3\Omega \times a_3$$

A similar expression is obtained by differentiating (5.62), expressed for a point $\mathscr{p}$ located at time t on the normal. Indeed, writing this relation for the displacement of $\mathscr{p}$ between the instants t and $t + h$:

$$\overline{\xi}\left[\mathscr{P}(t), h\right] = \xi\left[\mathscr{M}(t), h\right] + x^3\theta(t+h) \times a_3(t)$$

hence, differentiating with respect to h, in $h = 0$:

$$v(p) = v(m) + x^3 \dot{\theta} \times a_3 \tag{5.66}$$

$\dot{\theta} = \Omega$ is here the angular velocity vector of a_3. It is important to emphasise that this expression is obtained, in Euler variables, for a point p located at time t on the normal.

The velocity can also be obtained by differentiating (5.62) along the trajectory, which leads to the expression of the velocity of a point situated initially on the normal, in Lagrange variables:

$$V(p) = v(m) + x^3 \dot{\theta} \times A_3 \tag{5.67}$$

which therefore depends only on the velocities of the parameters $\xi(t)$ and $\theta(t)$ of the trajectory, defined on the middle reference surface. There is no identity between relations (5.66) and (5.67), which relate to different points and where the rotational velocities $\dot{\theta}$ are not necessarily the same.

The acceleration of a point p located on the normal at time t is obtained by differentiating (5.66):

$$\gamma(p) = \gamma(m) + x^3\, \ddot{\theta} \times a_3 + x^3\, \dot{\theta} \times \left(\dot{\theta} \times a_3\right) \tag{5.68}$$

5.3.2.2 Continuity equation

The mass of a portion of the shell is expressed by the relation:

$$m = \int_d \rho\, \hat{a}^{1/2} dx^1 dx^2 = \int_\omega \bar{\rho}\, \hat{g}^{1/2} dx^1 dx^2 dx^3$$

In the Reissner–Mindlin theory, the geometric quantities on the deformed position are expressed as a function of the same quantities on the reference position. Then, the position of the point p located initially in $\mathcal{P}$ on the normal to Σ is written as:

$$p\left(x^i\right) = m\left(x^1, x^2\right) + x^3\, \mathbf{d}\left(x^1, x^2\right) \tag{5.69}$$

The natural basis in p associated with this parametrisation is:

$$\begin{aligned} g_\alpha &= a_\alpha + x^3\, D_\alpha \mathbf{d} \\ g_3 &= \mathbf{d} \end{aligned} \tag{5.70}$$

where $\mathbf{d}$ verifies (5.60) and $D_\alpha \mathbf{d}$ is obtained by (4.142) and verifies the relation, obtained in differentiating the identity $\mathbf{d}^2 = 1$:

$$\mathbf{d} \cdot D_\alpha \mathbf{d} = 0 \tag{5.71}$$

It comes in particular:

$$g_1 \times g_2 = \left(a_1 + x^3 \partial_1 \mathbf{d}\right) \times \left(a_2 + x^3 \partial_2 \mathbf{d}\right)$$

$$= a_1 \times a_2 + x^3 \left(a_1 \times \partial_2 \mathbf{d} - a_2 \times \partial_1 \mathbf{d}\right) + \left(x^3\right)^2 \left(\partial_1 \mathbf{d} \times \partial_2 \mathbf{d}\right)$$

$$= \left(A_1 + \mathcal{A}_1{}^{\beta} A_\beta + \mathcal{B}_1 A_3\right) \times \left(A_2 + \mathcal{A}_2{}^{\gamma} A_\gamma + \mathcal{B}_2 A_3\right)$$

$$+ x^3 \left\{ \begin{array}{l} \left[A_1 + \mathcal{A}_1{}^{\beta} A_\beta + \mathcal{B}_1 A_3\right] \times \left[\begin{array}{l} \left(\nabla_2 d^\gamma - B_2{}^\gamma d_3\right) A_\gamma \\ + \left(\nabla_2 d_3 + B_2{}^\gamma d_\gamma\right) A_3 \end{array} \right] \\ - \left[A_2 + \mathcal{A}_2{}^{\beta} A_\beta + \mathcal{B}_2 A_3\right] \times \left[\begin{array}{l} \left(\nabla_1 d^\gamma - B_1{}^\gamma d_3\right) A_\gamma \\ + \left(\nabla_1 d_3 + B_1{}^\gamma d_\gamma\right) A_3 \end{array} \right] \end{array} \right\}$$

$$+ \left(x^3\right)^2 \left\{ \left[\begin{array}{l} \left(\nabla_1 d^\gamma - B_1{}^\gamma d_3\right) A_\gamma \\ + \left(\nabla_1 d_3 + B_1{}^\gamma d_\gamma\right) A_3 \end{array} \right] \times \left[\begin{array}{l} \left(\nabla_2 d^\delta - B_2{}^\delta d_3\right) A_\delta \\ + \left(\nabla_2 d_3 + B_2{}^\delta d_\delta\right) A_3 \end{array} \right] \right\}$$

$$g_3 = \mathbf{d} = d^i a_i = d^\alpha \left(A_\alpha + \mathcal{A}_\alpha{}^{\beta} A_\beta + \mathcal{B}_\alpha A_3\right) + d^3 \left(\mathcal{N}^\alpha A_\alpha + A_3 \cos\omega\right)$$

The mixed product between these quantities makes it possible to calculate the determinant $\sqrt{\hat{g}}$ intervening in the continuity equation.

5.3.2.3 Deformations

The purely surfacic approach adopted in §4.5 made it possible to highlight the deformation variables characterising the deformation of the Cosserat surface. The three-dimensional approach developed below makes it possible to highlight the same deformation variables, by an approach similar to that followed in the context of KL assumptions.

For the calculation of the local strain tensor, the reference surface and the current surface are related to the same parameterisation (x^i). g_3 is not orthogonal to vectors g_α. In expressing the differentiation, it becomes:

$$g_\alpha = a_\alpha + x^3 \mathfrak{p}_\alpha{}^j A_j \tag{5.72}$$

with (4.142):

$$\begin{aligned} \mathfrak{p}_{\alpha\beta} &= \nabla_\alpha d_\beta - B_{\alpha\beta} d_3 \\ \mathfrak{p}_{\alpha 3} &= \partial_\alpha d_3 + B_\alpha{}^\beta d_\beta \end{aligned} \tag{5.73}$$

In these relations, **d** is expressed in the basis A_i. The expressions of $\mathfrak{p}_{\alpha\beta}$ and $\mathfrak{p}_{\alpha 3}$ resemble those of $\mathcal{A}$ and $\mathcal{B}$ given by (3.95), replacing the displacement ξ by the vector **d**. In the assumptions KL, $\mathfrak{p}_{\alpha\beta} = -B_{\alpha\beta}$ and $\mathfrak{p}_{\alpha 3} = 0$; $\mathfrak{p}$ is a curvature associated with material points.

This established the components of the local strain tensor in the basis (G_i) are deduced from the differences of dot products:

$$\begin{aligned}
\bar{\varepsilon}_{33} &= \frac{1}{2}\left(g_{33} - G_{33}\right) = 0 \\
\bar{\varepsilon}_{3\alpha} &= \frac{1}{2} g_{3\alpha} = \frac{1}{2}\mathbf{d} \cdot \left(a_\alpha + x^3 \partial_\alpha \mathbf{d}\right) = \frac{1}{2}\mathbf{d} \cdot a_\alpha \\
\bar{\varepsilon}_{\alpha\beta} &= \frac{1}{2}\left(g_{\alpha\beta} - G_{\alpha\beta}\right) = \frac{1}{2}\left[\left(a_\alpha + x^3 \mathfrak{p}_\alpha{}^k A_k\right) \cdot \left(a_\beta + x^3 \mathfrak{p}_\beta{}^l A_l\right) - G_{\alpha\beta}\right] \\
&= \varepsilon_{\alpha\beta} + x^3\left(\rho_{\alpha\beta} + \rho_{\beta\alpha}\right) + \left(x^3\right)^2\left(\mathfrak{p}_\alpha{}^k \mathfrak{p}_{\beta k} - B_\alpha{}^\gamma B_{\beta\gamma}\right)
\end{aligned} \tag{5.74}$$

which introduced the new tensor:

$$\rho_{\alpha\beta} = \frac{1}{2}\left(\mathfrak{p}_\alpha{}^k A_k \cdot a_\beta + B_{\alpha\beta}\right) = \frac{1}{2}\left(\mathfrak{p}_{\alpha\beta} + \mathfrak{p}_\alpha{}^\gamma \mathcal{A}_{\beta\gamma} + \mathfrak{p}_\alpha{}^3 \mathcal{B}_\beta + B_{\alpha\beta}\right) \tag{5.75}$$

This tensor, nonsymmetrical in general, is reduced (except the sign) to the variation of curvature $\kappa_{\alpha\beta}$ in Kirchhoff–Love theory. It is related to the variation of rotation of the material segment carried by **d**, so it can be qualified as **material variation of curvature.** ρ has already been introduced in §4.5.1.5 as a deformation variable of the surface. Thus, (5.74), $\bar{\varepsilon}_{\alpha\beta}$ admits an expression similar to that obtained in the KL hypotheses. Reissner–Mindlin's hypotheses simply introduced the distortions $\bar{\varepsilon}_{\alpha 3}$.

These expressions can be rewritten in the basis (A_i), depending on the displacement and the rotation represented by **d**:

$$\begin{aligned}
\bar{\varepsilon}_{3\alpha} &= \frac{1}{2}\left(d^i a_i\right) \cdot \left[\left(\delta_\alpha^\gamma + \mathcal{A}_\alpha{}^\gamma\right) A_\gamma + \mathcal{B}_\alpha A_3\right] = \frac{1}{2}\left[\left(\delta_\alpha^\gamma + \mathcal{A}_\alpha{}^\gamma\right) d_\gamma + \mathcal{B}_\alpha d_3\right] \\
&= \frac{1}{2}\left[d^\beta\left(A_{\alpha\beta} + \nabla_\alpha \xi_\beta - \zeta B_{\alpha\beta}\right) + d_3\left(B_\alpha{}^\beta \xi_\beta + \partial_\alpha \zeta\right)\right]
\end{aligned} \tag{5.76}$$

$$\bar{\varepsilon}_{\alpha\beta} = \frac{1}{2}\left\{
\begin{array}{l}
2\varepsilon_{\alpha\beta} + x^3 \left[\begin{array}{l}
\left(\nabla_\alpha d^\lambda - B_\alpha{}^\lambda d^3\right)\left(\delta_\beta^\mu + \mathcal{A}_\beta{}^\mu\right) A_{\lambda\mu} \\
+\left(\nabla_\beta d^\lambda - B_\beta{}^\lambda d^3\right)\left(\delta_\alpha^\mu + \mathcal{A}_\alpha{}^\mu\right) A_{\lambda\mu} \\
+2B_{\alpha\beta} + \mathcal{B}_\alpha\left(B_{\beta\gamma} d^\gamma + \partial_\beta d^3\right) + \mathcal{B}_\beta\left(B_{\alpha\gamma} d^\gamma + \partial_\alpha d^3\right)
\end{array}\right] \\
+\left(x^3\right)^2 \left[\begin{array}{l}
\left(\nabla_\alpha d^\lambda - B_\alpha{}^\lambda d^3\right)\left(\nabla_\beta d^\mu - B_\beta{}^\mu d^3\right) A_{\lambda\mu} \\
+\left(B_{\beta\gamma} d^\gamma + \partial_\beta d^3\right)\left(B_{\alpha\gamma} d^\gamma + \partial_\alpha d^3\right) \\
-B_\alpha{}^\gamma B_{\beta\gamma}
\end{array}\right]
\end{array}\right\}$$

where $\mathcal{A}$ and $\mathcal{B}$ are defined by (3.95) as functions of the displacement, and $\varepsilon_{\alpha\beta}$ denotes the deformation of the middle surface, which is expressed by (III-97) as a function of the displacement.

5.3.2.4 Deformation in small displacements

In small displacements, these expressions are reduced to:

$$\begin{aligned}
&\bar{\varepsilon}_{33} = 0 \\
&\bar{\varepsilon}_{3\alpha} = \frac{1}{2}\left(d_\alpha + B_\alpha{}^\beta \xi_\beta + \partial_\alpha \zeta\right) = \frac{1}{2}\left(d_\alpha + \mathcal{B}_\alpha\right) \\
&\bar{\varepsilon}_{\alpha\beta} = \frac{1}{2}\left\{
\begin{array}{l}
\left(\nabla_\alpha \xi_\beta + \nabla_\beta \xi_\alpha - 2\zeta B_{\alpha\beta}\right) + x^3 \left[\begin{array}{r} \nabla_\alpha d_\beta + \nabla_\beta d_\alpha \\ - B_\alpha{}^\lambda\left(\nabla_\beta \xi_\lambda - \zeta B_{\beta\lambda}\right) \\ - B_\beta{}^\lambda\left(\nabla_\alpha \xi_\lambda - \zeta B_{\alpha\lambda}\right) \end{array}\right] \\
\qquad - \left(x^3\right)^2 \left[B_{\beta\gamma} \nabla_\alpha d^\gamma + B_{\alpha\gamma} \nabla_\beta d^\gamma\right]
\end{array}
\right\} \\
&\quad = \frac{1}{2}\left\{\mathcal{M}_\alpha{}^\lambda\left(\nabla_\beta \xi_\lambda - \zeta B_{\beta\lambda} + x^3 \nabla_\beta d_\lambda\right) + \mathcal{M}_\beta{}^\lambda\left(\nabla_\alpha \xi_\lambda - \zeta B_{\alpha\lambda} + x^3 \nabla_\alpha d_\lambda\right)\right\} \\
&\quad = \frac{1}{2}\left\{\mathcal{M}_\alpha{}^\lambda\left(\mathcal{A}_{\beta\lambda} + x^3 \nabla_\beta d_\lambda\right) + \mathcal{M}_\beta{}^\lambda\left(\mathcal{A}_{\alpha\lambda} + x^3 \nabla_\alpha d_\lambda\right)\right\}
\end{aligned} \tag{5.77}$$

where the components d^i can be expressed as functions of θ by (5.65). The last expression is similar to the relation (5.39) established in the case of KL assumptions. In Kirchhoff–Love theory, the distortions $\bar{\varepsilon}_{3\alpha}$ are zero and the vector d is the normal vector, related to displacement by relations:

$$d_\alpha + B_\alpha{}^\beta \xi_\beta + \partial_\alpha \zeta = d_\alpha + \mathcal{B}_\alpha = 0 \tag{5.78}$$

The term in x^3 in the first expression of $\bar{\varepsilon}_{\alpha\beta}$ is then reduced to the variation of curvature $\kappa_{\alpha\beta}$ (taking into account the symmetry of the latter) and the expression (5.77) of $\bar{\varepsilon}_{\alpha\beta}$ is reduced to (5.39).

The expression (5.77) of $\bar{\varepsilon}_{\alpha\beta}$ can reorder to reveal the deformation variables of the middle surface:

$$\begin{aligned}
\bar{\varepsilon}_{\alpha\beta}(\mathcal{P}) = &\left[\delta_\alpha{}^\lambda \delta_\beta{}^\mu - \left(x^3\right)^2 B_\alpha{}^\lambda B_\beta{}^\mu\right] \varepsilon_{\lambda\mu} \\
&+ \frac{x^3}{2}\left[\delta_\alpha{}^\lambda\left(\delta_\beta{}^\mu - x^3 B_\beta{}^\mu\right) + \delta_\beta{}^\lambda\left(\delta_\alpha{}^\mu - x^3 B_\alpha{}^\mu\right)\right] \rho_{\lambda\mu}
\end{aligned} \tag{5.79}$$

with $\rho_{\alpha\beta}$ calculated in small displacements:

$$\rho_{\alpha\beta} = \nabla_\alpha d_\beta - B_\alpha{}^\gamma\left(\nabla_\beta \xi_\gamma - \zeta\, B_{\beta\gamma}\right) \tag{5.80}$$

The deformations of the surface parallel to the middle surface, therefore, depend only on the tensors ε and ρ of the middle surface.

Note: The three-dimensional strain tensor can also be calculated directly from the displacement. For example, in small displacements:

$$\bar{\varepsilon}_{\alpha 3}=\frac{1}{2}\left(D_\alpha \bar{\xi}_3+D_3\bar{\xi}_\alpha\right)=\frac{1}{2}\left(\partial_\alpha\zeta-2\bar{\Gamma}^i_{\alpha 3}\bar{\xi}_i+\partial_3\bar{\xi}_\alpha\right)$$

The different quantities are expressed in the basis (G_i). With the help of (3.67), and noting that:

$$\bar{\xi}=\mathcal{M}_\alpha{}^\beta\left(\xi_\beta-x^3\theta^\gamma\varepsilon_{\gamma\beta}\right)G^\alpha+\zeta G_3$$

the relation (5.77) is then easily demonstrated.

In the hypothesis of thin shells, expressions (5.76) and (5.77) are simplified by neglecting the terms in $(x^3)^2$.

It should be remembered that the above relationships giving the strain components are expressed in a basis that is not orthonormal. **In the orthonormal coordinate system associated with the principal curvature system,** $\bar{\varepsilon}$ is rewritten, in small displacements:

$$\bar{\varepsilon}_{13}=\frac{1}{2}\left(1-\frac{x^3}{R_1}\right)^{-1}\left(\frac{\xi^1}{R_1}+\partial_1\zeta+\theta^2\right)$$

$$\bar{\varepsilon}_{23}=\frac{1}{2}\left(1-\frac{x^3}{R_2}\right)^{-1}\left(\frac{\xi^2}{R_2}+\partial_2\zeta-\theta^1\right)$$

$$\bar{\varepsilon}_{11}=\left(1-\frac{x^3}{R_1}\right)^{-1}\left(\nabla_1\xi_1-\frac{\zeta}{R_1}+x^3\,\nabla_1\theta_2\right)=\left(1-\frac{x^3}{R_1}\right)^{-1}\left[\left(1+\frac{x^3}{R_1}\right)\varepsilon_{11}+x^3\rho_{11}\right] \tag{5.81}$$

$$\bar{\varepsilon}_{22}=\left(1-\frac{x^3}{R_2}\right)^{-1}\left(\nabla_2\xi_2-\frac{\zeta}{R_2}-x^3\,\nabla_2\theta_1\right)=\left(1-\frac{x^3}{R_2}\right)^{-1}\left[\left(1+\frac{x^3}{R_2}\right)\varepsilon_{22}+x^3\rho_{22}\right]$$

$$\bar{\varepsilon}_{12}=\frac{1}{2}\left[\left(1-\frac{x^3}{R_2}\right)^{-1}\left(\nabla_2\xi_1+x^3\,\nabla_2\theta_2\right)+\left(1-\frac{x^3}{R_1}\right)^{-1}\left(\nabla_1\xi_2-x^3\,\nabla_1\theta_1\right)\right]$$

$$=\left(1-\frac{x^3}{R_1}\right)^{-1}\left(1-\frac{x^3}{R_2}\right)^{-1}\left\{\left(1-\frac{\left(x^3\right)^2}{R_1R_2}\right)\varepsilon_{12}+\frac{x^3}{2}\left[\left(1-\frac{x^3}{R_2}\right)\rho_{12}+\left(1-\frac{x^3}{R_1}\right)\rho_{21}\right]\right\}$$

where:

$$\varepsilon_{11} = \nabla_1 \xi_1 - \frac{\zeta}{R_1}$$

$$\varepsilon_{22} = \nabla_2 \xi_2 - \frac{\zeta}{R_2} \tag{5.82}$$

$$\varepsilon_{12} = \frac{1}{2}\left(\nabla_2 \xi_1 + \nabla_1 \xi_2\right)$$

are the components of the middle surface strain tensor and:

$$\rho_{11} = -\frac{1}{R_1}\left(\nabla_1 \xi_1 - \frac{\zeta}{R_1}\right) + \nabla_1 \theta_2$$

$$\rho_{22} = -\frac{1}{R_2}\left(\nabla_2 \xi_2 - \frac{\zeta}{R_2}\right) - \nabla_2 \theta_1 \tag{5.83}$$

$$\rho_{12} = -\frac{1}{R_1}\nabla_2 \xi_1 - \nabla_1 \theta_1$$

$$\rho_{21} = -\frac{1}{R_2}\nabla_1 \xi_2 + \nabla_2 \theta_2$$

are the components of the tensor representing the material curvature variation associated with the material segment initially carried by A_3.

In these expressions, the differentiations are made with respect to the curvilinear abscissae along the main lines of curvature.

If the shell can be considered thin, $\mathcal{M}_\alpha{}^\beta$ is reduced to $\delta_\alpha{}^\beta$ and $\overline{\varepsilon}_{\alpha 3}$ is constant in thickness. Moreover, in Kirchhoff–Love theory, $\overline{\varepsilon}_{\alpha 3}$ is zero and:

$$\theta^1 = \frac{\xi^2}{R_2} + \partial_2 \zeta$$

$$\theta^2 = -\left(\frac{\xi^1}{R_1} + \partial_1 \zeta\right)$$

Example: In a coordinate system (x, y) of the surface such that $\sqrt{\hat{a}} = 1$ and the connection coefficients are zero (it is for example possible on a cylinder or a plane), expressions (5.77) are reduced, in the hypothesis of thin shells, to:

$$\overline{\varepsilon}_{xx} = \varepsilon_{xx} + x^3 \partial_x \theta^y$$

$$\overline{\varepsilon}_{yy} = \varepsilon_{yy} - x^3 \partial_y \theta^x$$

$$\bar{\varepsilon}_{xy} = \varepsilon_{xy} + \frac{1}{2} x^3 \left(\partial_y \theta^y - \partial_x \theta^x \right)$$

$$\bar{\varepsilon}_{3x} = \frac{1}{2} \left(\partial_x \zeta + \theta^y + B_x{}^\beta \xi_\beta \right) \tag{5.84}$$

$$\bar{\varepsilon}_{3y} = \frac{1}{2} \left(\partial_y \zeta - \theta^x + B_y{}^\beta \xi_\beta \right)$$

5.3.2.5 Virtual deformations

The virtual deformations expressed in Euler variables are obtained from relations (5.77) assuming that, in all successive configurations, the Reissner–Mindlin kinematics is respected:

$$\overset{*}{\bar{e}}_{3\alpha} = \frac{1}{2} \left(-\overset{*}{\theta}{}^\gamma e_{\gamma\alpha} + b_\alpha{}^\beta \overset{*}{\xi}_\beta + \partial_\alpha \overset{*}{\zeta} \right) = \frac{1}{2} \left(\overset{*}{d}_\alpha + \overset{*}{\mathfrak{B}}_\alpha \right) = \overset{*}{e}_{3\alpha}$$

$$\overset{*}{\bar{e}}_{\alpha\beta} = \frac{1}{2} \begin{bmatrix} \mathfrak{m}_\alpha{}^\lambda \left(\nabla_\beta \overset{*}{\xi}_\lambda - \overset{*}{\zeta} b_{\beta\lambda} \right) + \mathfrak{m}_\beta{}^\lambda \left(\nabla_\alpha \overset{*}{\xi}_\lambda - \overset{*}{\zeta} b_{\alpha\lambda} \right) \\ - x^3 e_{\lambda\gamma} \left(\mathfrak{m}_\beta{}^\gamma \nabla_\alpha \overset{*}{\theta}{}^\lambda + \mathfrak{m}_\alpha{}^\gamma \nabla_\beta \overset{*}{\theta}{}^\lambda \right) \end{bmatrix} \tag{5.85}$$

The expression of $\overset{*}{\bar{e}}_{\alpha\beta}$ can also be obtained as a function of $\overset{*}{e}$ and $\overset{*}{r}$ (virtual variation of the material curvature tensor) defined on the middle surface, by virtualisation of relation (5.79):

$$\overset{*}{\bar{e}}_{\alpha\beta} = \left[\delta_\alpha{}^\lambda \delta_\beta{}^\mu - \left(x^3 \right)^2 b_\alpha{}^\lambda b_\beta{}^\mu \right] \overset{*}{e}_{\lambda\mu}$$
$$+ \frac{x^3}{2} \left[\delta_\alpha{}^\lambda \left(\delta_\beta{}^\mu - x^3 b_\beta{}^\mu \right) + \delta_\beta{}^\lambda \left(\delta_\alpha{}^\mu - x^3 b_\alpha{}^\mu \right) \right] \overset{*}{r}_{\lambda\mu} \tag{5.86}$$

with:

$$\overset{*}{r}_{\alpha\beta} = \nabla_\alpha \overset{*}{d}_\beta - b_\alpha^\gamma \left(\nabla_\beta \overset{*}{\xi}_\gamma - \overset{*}{\zeta} b_{\beta\gamma} \right) \tag{5.87}$$

to be compared to (4.165) and $\overset{*}{e}$ given by (4.40).

5.3.3 Application of PVP

5.3.3.1 Virtual power of deformation

As in the case of the Kirchhoff–Love theory, the virtual deformation power can be evaluated in two ways:

- either directly according to the variables of deformation of the middle surface,
- or from its expression in three-dimensional theory.

The first approach is based on expressions (5.86) and (5.87), which show that the local virtual strain tensor $\overset{*}{\bar{e}}_{ij}$ depends only on $\overset{*}{e}_{\alpha 3}$, $\overset{*}{e}$ and $\overset{*}{r}$, deformation variables defined on the middle surface. Under these conditions, the deformation virtual power of the shell is expressed by introducing generalised stresses:

$$\overset{*}{\mathcal{W}}_f = \int_\sigma \left[\mathsf{N} \bullet \overset{*}{e} + \mathsf{M} \bullet \overset{*}{r} + \mathsf{Q}^\alpha \overset{*}{e}_{\alpha 3} \right] d\sigma \tag{5.88}$$

where the tensor M is not necessarily symmetrical and the strain energy is supplemented by the distortion energy. This is the approach used in §4.5. In the hypothesis of thin shells, only the symmetrical part of $\overset{*}{r}$ intervenes in (5.86) and M can be taken as symmetrical.

The deformation virtual power can also be obtained from its expression in the three-dimensional theory (for a volume ω based on σ):

$$\begin{aligned}
\overset{*}{\mathcal{W}}_f &= \int_\omega T^{ij}(p)\overset{*}{\bar{e}}_{ij}(p) dV(p) = \int_\omega T^{\alpha\beta}(p)\overset{*}{\bar{e}}_{\alpha\beta}(p)\, dV(p) \\
&\quad + 2\int_\omega T^{3\beta}(p)\overset{*}{\bar{e}}_{3\beta}(p)\, dV(p) \\
&= \int_\omega T^{\alpha\beta}(p) \begin{bmatrix} \mathfrak{m}_\alpha{}^\lambda \left(\nabla_\beta \overset{*}{\xi}_\lambda - \overset{*}{\zeta} b_{\beta\lambda} \right) \\ -\dfrac{1}{2} x^3 \mathfrak{e}_{\lambda\gamma} \left(\mathfrak{m}_\beta{}^\gamma \nabla_\alpha \overset{*}{\theta}{}^\lambda + \mathfrak{m}_\alpha{}^\gamma \nabla_\beta \overset{*}{\theta}{}^\lambda \right) \end{bmatrix} dV(p) \\
&\quad + \int_\omega T^{3\beta}(p) \left(\partial_\alpha \overset{*}{\zeta} - \overset{*}{\theta}{}^\gamma \mathfrak{e}_{\gamma\alpha} + b_\alpha{}^\beta \overset{*}{\xi}_\beta \right) dV(p)
\end{aligned} \tag{5.89}$$

Either, by introducing the resultants (5.9) and (5.10):

$$\overset{*}{\mathcal{W}}_f = \int_\sigma \bar{N}^{\alpha\beta} \left(\nabla_\beta \overset{*}{\xi}_\alpha - \overset{*}{\zeta} b_{\beta\alpha} \right) + \bar{M}^{\alpha\beta} \mathfrak{e}_{\lambda\beta} \nabla_\alpha \overset{*}{\theta}{}^\lambda + \bar{Q}^\alpha \left(\partial_\alpha \overset{*}{\zeta} - \overset{*}{\theta}{}^\gamma \mathfrak{e}_{\gamma\alpha} + b_\alpha{}^\lambda \overset{*}{\xi}_\lambda \right) d\sigma \tag{5.90}$$

In the formula (5.90), the virtual distortion $\overset{*}{\theta}$ is not related to the virtual velocity $\overset{*}{\xi}$ and the shear force $\bar{Q}^\alpha$ is working during the virtual movement. Virtual power (5.90) is expressed only in terms of quantities defined on the middle surface.

It is then necessary to proceed to integrations by parts, necessary to reveal only independent virtual velocities. It becomes:

$$\begin{aligned}
\overset{*}{\mathcal{W}}_f &= \int_{\partial\sigma} \left[\bar{N}^{\nu\beta} \overset{*}{\xi}_\beta + \bar{M}^{\nu\beta} \mathfrak{e}_{\lambda\beta} \overset{*}{\theta}{}^\lambda + \bar{Q}^\nu \overset{*}{\zeta} \right] ds \\
&\quad + \int_\sigma \begin{bmatrix} \left(-\nabla_\alpha \bar{N}^{\alpha\beta} + b_\alpha{}^\beta \bar{Q}^\alpha \right) \overset{*}{\xi}_\beta - \left(\bar{N}^{\alpha\beta} b_{\alpha\beta} + \nabla_\alpha \bar{Q}^\alpha \right) \overset{*}{\zeta} \\ - \left(\mathfrak{e}_{\lambda\beta} \nabla_\alpha \bar{M}^{\alpha\beta} + \mathfrak{e}_{\lambda\alpha} \bar{Q}^\alpha \right) \overset{*}{\theta}{}^\lambda \end{bmatrix} d\sigma
\end{aligned} \tag{5.91}$$

5.3.3.2 *Virtual power of external actions*

External actions are defined in §4.2.2.2. Their virtual power is deduced from (4.84), taking $\overset{*}{\omega} = \overset{*}{\theta}$ (the torques work in the rotation of the material segment):

$$\overset{*}{\mathcal{W}}_e = \int_\sigma \left(\mathbf{p} \cdot \overset{*}{\xi} + \mathbf{c} \cdot \overset{*}{\theta} \right) d\sigma + \int_\sigma \left(\mathbf{q} \cdot \overset{*}{\xi} + \mathbf{m} \cdot \overset{*}{\theta} \right) ds \tag{5.92}$$

5.3.3.3 *Equilibrium equations*

Starting from the expression (5.88) of the deformation virtual power, equilibrium equations are established as in §4.2.2 and §4.5.2. They relate to the generalised stresses **N**, **M**, and **Q**. In fact, this is a special case of that envisaged in §4.5, the vector d considered here being a particular director vector, of constant length, which leads to $\mathsf{M}^{\alpha 3} = 0$; the relation (5.88) can, in particular, be compared with (4.168). To compare the two systems of equilibrium equations thus obtained, it should be noted that the moments **c** and **m** introduced in the two cases are not quite identical (see remark of §4.5.2.2). The local equations and the boundary conditions are therefore deduced directly from (4.174) and (4.177), in which c_α is to be replaced by $-\,e_{\alpha\beta}c^\beta$ and m_α par $-\,e_{\alpha\beta}m^\beta$.

In the case of the statics, these equilibrium equations relating to the generalised stresses are written:

$$\boxed{\begin{aligned} &\nabla_\beta \mathsf{L}^{\beta\alpha} + b_\gamma{}^\alpha \nabla_\beta \mathsf{M}^{\beta\gamma} + p^\alpha = 0 \\ &-\nabla_\beta \nabla_\alpha \mathsf{M}^{\alpha\beta} + b_{\alpha\beta}\mathsf{L}^{\alpha\beta} + p^3 = 0 \\ &\mathsf{Q}^\alpha = \nabla_\beta \mathsf{M}^{\beta\alpha} - e^\alpha{}_\beta c^\beta \end{aligned}} \tag{5.93}$$

with the boundary conditions:

$$\boxed{\begin{aligned} &\mathsf{L}^{\nu\beta} - q^\beta = 0 \\ &\mathsf{Q}^\nu + q^3 = 0 \\ &\mathsf{M}^{\nu\alpha} + e^\alpha{}_\beta\, m^\beta = 0 \end{aligned}} \tag{5.94}$$

In contrast to the Kirchhoff–Love theory, the shear force has emerged here as a generalised stress associated with distortion $\overline{\varepsilon}_{3\alpha}$.

Now starting from the expression (5.91) for the deformation virtual power, and not considering the inertial actions, the total virtual power is written as:

$$\begin{aligned} \overset{*}{\mathcal{W}}_f = &\int_{\partial\sigma} \left[\left(\bar{N}^{\nu\beta} - q^\beta \right) \overset{*}{\xi}_\beta + \left(\bar{Q}^\nu - q^3 \right) \overset{*}{\zeta} + \left(\bar{M}^{\nu\beta} e_{\lambda\beta} - m_\lambda \right) \overset{*}{\theta}{}^\lambda \right] ds \\ &+ \int_\sigma \left[\begin{aligned} &\left(-\nabla_\alpha \bar{N}^{\alpha\beta} + b_\alpha{}^\beta \bar{Q}^\alpha - p^\beta \right) \overset{*}{\xi}_\beta - \left(\bar{N}^{\alpha\beta} b_{\alpha\beta} + \nabla_\alpha \bar{Q}^\alpha + p^3 \right) \overset{*}{\zeta} \\ &- \left(e_{\lambda\beta} \left(\nabla_\alpha \bar{M}^{\alpha\beta} + \bar{Q}^\beta \right) + c_\lambda \right) \overset{*}{\theta}{}^\lambda \end{aligned} \right] d\sigma \end{aligned} \tag{5.95}$$

from where are obtained:

- Local equations:

$$\boxed{\begin{aligned}&\nabla_\alpha \bar{N}^{\alpha\beta} - b_\alpha{}^\beta \bar{Q}^\alpha + p^\beta = 0\\&\bar{N}^{\alpha\beta} b_{\alpha\beta} + \nabla_\alpha \bar{Q}^\alpha + p^3 = 0\\&e_{\lambda\beta}\left(\nabla_\alpha \bar{M}^{\alpha\beta} + \bar{Q}^\beta\right) + c_\lambda = 0\end{aligned}} \tag{5.96}$$

- Boundary conditions:

$$\boxed{\begin{aligned}&\bar{N}^{\nu\beta} - q^\beta = 0\\&\bar{Q}^\nu - q^3 = 0\\&\bar{M}^{\nu\beta} e_{\lambda\beta} - m_\lambda = 0\end{aligned}} \tag{5.97}$$

The equilibrium equations thus obtained are obviously identical to equations (5.19), in which the relations of the last line are identically verified. The boundary conditions (5.95) confirm that, in Reissner–Mindlin's theory, the moments balance themselves directly on the edge and do not interfere in the equilibrium of the normal reaction.

The two sets of relationships are then immediately comparable:

$$\begin{aligned}&\bar{N}^{\alpha\beta} = \mathsf{L}^{\alpha\beta}\\&\bar{M}^{\alpha\beta} = \mathsf{M}^{\alpha\beta}\\&\bar{Q}^\alpha = -\mathsf{Q}^\alpha\end{aligned} \tag{5.98}$$

and it is recalled that in this case, $\mathsf{M}^{\alpha\beta}$ is not symmetrical.

5.4 SHELL CONSTITUTIVE LAWS

5.4.1 Hypotheses of plane stresses

5.4.1.1 Stress T^{33}

Tridimensional stresses are represented by a symetric twice contravariant tensor $\bar{\bar{T}}$ in Euler variables[10].

In a shell, the stresses ($T^{\alpha\beta}$) are usually predominant (see example below). This leads to the hypothesis of plane stresses, where T^{33} is neglected.

For example, let a thin spherical balloon under internal pressure p; in membrane theory, the stresses in the plane tangent to σ are:

$$T^{\alpha\beta} = \frac{pR}{2e} a^{\alpha\beta}$$

[10]Notations T and e are retained here for stresses and deformations in Euler variables. The constitutive laws are most often expressed in Lagrange variables (see §5.2).

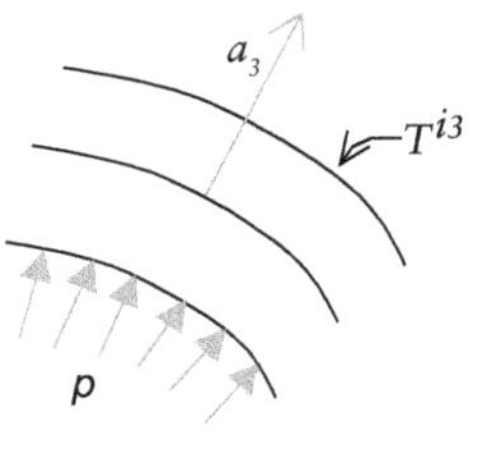

Figure 5.6

The pinch stress T^{33} varies between $-p$ and 0 and the shears $T^{\alpha 3}$ are zero on both faces (Figure 5.6). The ratio between the stresses $T^{\alpha\beta}$ and the stress T^{33} is of the order of $\frac{R}{e}$, which justifies neglecting T^{33}. This result has also been obtained in three-dimensional elasticity, in §1.2.2.2, in the case of a cylinder.

A similar result is obtained for bending plates (see Chapter 7). Nevertheless, when there is flexure, the shear stresses $T^{\alpha 3}$ balance the shear force – cf. (5.10) - and are not negligible in thickness.

In general, the stress T^{33} is of the order of magnitude of the external loads, whereas the stresses $T^{\alpha\beta}$ are in the ratio $\frac{R}{e}$ with them. It is therefore natural to neglect T^{33} vis-à-vis $T^{\alpha\beta}$.

/H3/

$$\boxed{T^{33} = 0} \tag{5.99}$$

If the shear stresses $T^{\alpha 3}$ do not intervene in the expression of the deformations $\bar{e}_{\alpha\beta}$ by the constitutive law, this one is then a law of plane stresses for $T^{\alpha\beta}$.

Hypothesis /H3/ differs from the assumption of indeformability of the normal. Indeed, for an isotropic elastic body, hypothesis /H3/ implies that:

$$e_{33} = -\frac{\nu}{1-\nu} e^{\alpha}{}_{\alpha}$$

which is contrary to the assumption of indeformability of the normal $e_{33} = 0$. This is the hypothesis of the plane stresses that best reflects the physical reality, as regards the rigidity of the shell. It is very widely used in practice. Both hypotheses are tantamount to neglecting the term $T^{33} e_{33}$ in the expression of the strain energy of the shell and are therefore only apparently contradictory.

The relations established in §5.2 and 5.3 are based either on Kirchhoff's assumptions or on Reissner–Mindlin's assumptions. In both cases, the variation in the thickness of the shell during its transformation is neglected. These are the assumptions that are used to express deformations. On the other hand, hypothesis /H3/ is used to express the stresses:

$$\bar{\bar{T}} = \begin{pmatrix} \left(T^{\alpha\beta}\right) & & T^{13} \\ & & T^{23} \\ T^{13} & T^{23} & 0 \end{pmatrix} \tag{5.100}$$

Note: The assumption of plane stresses is generally acceptable for homogeneous materials, where the strain energy corresponding to the variation of thickness is linked to the Poisson effect and of an order of magnitude much lower than that due to the stresses $T^{\alpha\beta}$. It can nevertheless be in default.

For example, in the case of a sandwich material where the core is made of a very flexible material (honeycomb type). The application of pressure on one face of the shell leads to a compression of the core, so to energy related to the variation in thickness which may not be negligible. Also, when the sandwich shell is curved, a bending moment creates normal forces in the skins and compression of the core is necessary to ensure equilibrium (Figure 5.7), which again leads to energy related to thickness variation.

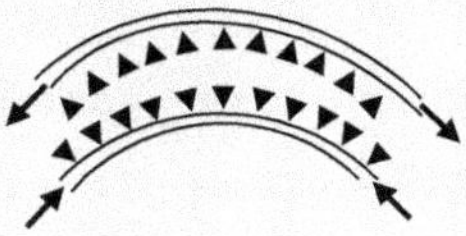

Figure 5.7

Certain models (see §4.5) allow taking into account the energy related to the variation of the thickness.

5.4.1.2 Case of Kirchhoff hypothesis

Moreover, in the context of KL hypotheses (hypothesis H2), it is natural to neglect the strain energy associated with shears $T^{\alpha 3}$, the associated distortions being neglected:

$$T^{3i} \overset{*}{e}_{3i} \approx 0$$

Thus, only the terms $T^{\alpha\beta}$ relating to the tangent plane are retained in the expression of the strain energy, whereas the stresses $T^{\alpha 3}$ are not zero since they balance the shear force.

5.4.2 Linearly elastic shells

5.4.2.1 Expression of the local law in-plane stresses

The local constitutive law connects the stress tensor σ expressed in Lagrange variables and the three-dimensional strain tensor $\bar{\varepsilon}$.

Contrary to the direct approach used in the previous chapter, the constitutive law of the shell is established from the constitutive law of the constitutive material. This approach is only possible if kinematic assumptions are made, in addition to the hypothesis of plane stresses (see §5.4.2.3). First, it is a matter of establishing the consequences of the hypothesis of plane stresses, consequences which are independent of the kinematic hypotheses considered later.

Only the case of a shell made of a linearly thermoelastic material, in small displacements, is examined in this paragraph. Because of the hypothesis of small displacements, the points of view of Lagrange and Euler are confused. The fact of confusing the two points of view makes it possible to use the relations (5.9) with σ in Lagrange variables, instead of T. The case where the hypothesis of small displacements is not retained can be treated analogously, either in keeping the Lagrangian presentation (but it is then necessary to express the equilibrium equations and the boundary conditions in Lagrange variables), either by making the constitutive laws relate to quantities expressed in Euler variables. The cases of small transformations where the hypothesis of small displacements is not retained are examined in Chapters 7 and 8.

Since a reference state $\mathcal{E}_0$ has been defined, the thermodynamic volumetric density $\overline{\phi}$, linearised at the second order, is expressed at any point of the shell as a function of the normal thermodynamic variables, namely:

- temperature variation $\tau(\mathcal{P})$,
- strain tensor $\overline{\varepsilon}(\mathcal{P})$,

which are fields of $\mathbb{R}^3$. According to the three-dimensional theory:

$$\overline{\rho}_0\,\overline{\phi}(\mathcal{P}) = -\overline{\rho}_0\, s_0\, \tau + \sigma_0 \cdot \overline{\varepsilon} + \frac{1}{2}\left[a\,\Theta^2 + 2\Theta\,\overline{\alpha}\cdot\overline{\varepsilon} + \overline{\varepsilon}\cdot\overline{\Lambda}\cdot\overline{\varepsilon}\right] \tag{5.101}$$

where:

$\overline{\rho}_0$ is the density (volumic mass) at point $\mathcal{P}$ in the state $\mathcal{E}_0$;
s_0 is the entropy density at point $\mathcal{P}$ in the state $\mathcal{E}_0$;
σ_0 is the prestress field at point $\mathcal{P}$ in the state $\mathcal{E}_0$;

$a = -\dfrac{c_0}{\mathsf{T}_0}$, where:

c_0 is the specific heat capacity, at point $\mathcal{P}$ in the state $\mathcal{E}_0$,
T_0 is the temperature at point $\mathcal{P}$ in the state $\mathcal{E}_0$;
$\overline{\alpha}$ is the thermoelastic coupling tensor at point $\mathcal{P}$ in the state $\mathcal{E}_0$ (2 times contravariant);
$\overline{\Lambda}$ is the elasticity tensor at point $\mathcal{P}$ in the state $\mathcal{E}_0$ (4 times contravariant).

Given the symmetry of $\overline{\varepsilon}(\mathcal{P})$, $\overline{\alpha}$ is symmetrical, and $\overline{\Lambda}$ verifies the symmetry relations:

$$\overline{\Lambda}^{ijkl} = \overline{\Lambda}^{jikl} = \overline{\Lambda}^{klij} \tag{5.102}$$

and therefore contains at most 21 independent coefficients.

The local constitutive law is given by:

$$\sigma^{ij}(\mathcal{P}) = \overline{\rho}_0 \frac{\partial \overline{\phi}}{\partial \overline{\varepsilon}_{ij}} = \sigma_0^{\,ij}(\mathcal{P}) + \tau(\mathcal{P})\cdot\overline{\alpha}^{ij}(\mathcal{P}) + \overline{\Lambda}^{ijkl}(\mathcal{P})\,\overline{\varepsilon}_{kl}(\mathcal{P}) \tag{5.103}$$

In the usual situation where $\sigma_0^{\,33}$ is neglected, the hypothesis of plane stresses leads to the relation:

$$\sigma^{33} = \overline{\Lambda}^{33kl}\,\overline{\varepsilon}_{kl} + \tau\,\overline{\alpha}^{33} = 0$$

which allows expressing $\overline{\varepsilon}_{33}$ according to $\overline{\varepsilon}_{\alpha\beta}$. It finally becomes:

$$\boxed{\sigma^{\alpha\beta} = \sigma_0^{\,\alpha\beta} + \tau\left(\overline{\alpha}^{\alpha\beta} - \frac{\overline{\Lambda}^{33\alpha\beta}\,\overline{\alpha}^{33}}{\overline{\Lambda}^{3333}}\right) + \left(\overline{\Lambda}^{\alpha\beta\gamma\delta} - \frac{\overline{\Lambda}^{33\alpha\beta}\,\overline{\Lambda}^{33\gamma\delta}}{\overline{\Lambda}^{3333}}\right)\overline{\varepsilon}_{\gamma\delta}} \tag{5.104}$$

where all the quantities depend on point $\mathcal{P}$. It has been assumed, to obtain (5.104), that the distortions $\varepsilon_{3\alpha}$ are null (hypotheses KL) or that the couplings between the stresses $\sigma_{\alpha\beta}$ and σ_{33} on the one hand and $\varepsilon_{3\alpha}$ on the other hand are negligible.

5.4.2.2 *Case of homogeneous and isotropic material*

In the particular case of a homogeneous and isotropic material, the material is also thermally isotropic and $\bar{\alpha}^{ij} = \bar{\alpha} G^{ij}$. Then:

$$\sigma^{ij} = \sigma_0{}^{ij} + \bar{\alpha}\,\tau\, G^{ij} + \frac{E}{1+\nu}\left(\frac{\nu}{1-2\nu}\bar{\varepsilon}_k^k\, G^{ij} + \bar{\varepsilon}^{ij}\right) \tag{5.105}$$

In consideration of a strain field with zero stresses, $\bar{\alpha}$ is expressed as a function of the coefficient of linear expansion λ_l:

$$\bar{\alpha} = -\frac{\lambda_\ell\, E}{1-2\nu} \tag{5.106}$$

Since the vector A_3 is unitary and normal to the plane formed by the two other vectors, it becomes, assuming that $\sigma_0{}^{33} = 0$:

$$\sigma^{33} = \bar{\alpha}\,\tau + \frac{E}{1+\nu}\left(\frac{\nu}{1-2\nu}\left(\bar{\varepsilon}_\gamma^\gamma + \bar{\varepsilon}_{33}\right) + \bar{\varepsilon}_{33}\right) = 0$$

from where:

$$\bar{\varepsilon}_{33} = -\frac{(1-2\nu)(1+\nu)}{(1-\nu)\,E}\bar{\alpha}\,\tau - \frac{\nu}{1-\nu}\bar{\varepsilon}_\gamma^\gamma \tag{5.107}$$

and the elastic constitutive law:

$$\sigma^{\alpha\beta} = \sigma_0{}^{\alpha\beta} + \alpha\,\tau\, G^{\alpha\beta} + \frac{E}{1-\nu^2}\left((1-\nu)\bar{\varepsilon}^{\alpha\beta} + \nu\,\bar{\varepsilon}_\gamma^\gamma\, G^{\alpha\beta}\right) \tag{5.108}$$

(law of plane stresses), where:

$$\alpha = \frac{1-2\nu}{1-\nu}\bar{\alpha} = -\frac{\lambda_\ell\, E}{1-\nu} \tag{5.109}$$

5.4.2.3 *Obtaining the shell constitutive law*

The constitutive law connecting $\bar{N}$ and $\bar{M}$ to the global deformation variables can be obtained by integrating the local constitutive law (5.104) using relations (5.9). (5.104) contains the components of the tensors $\bar{\alpha}$ and $\bar{\Lambda}$ expressed in the reference coordinate system (G_α), which depends on x^3. Similarly, $\bar{\varepsilon}(\mathcal{P})$ can be expressed in terms of x^3, if kinematic assumptions are made. The integration according to the variable x^3 is generally carried out from the development in series of $\sigma^{\alpha\beta}(\mathcal{P})$ as a function of x^3, $\mathfrak{m}_{\alpha\beta}$ being taken as equal to its expression $A_{\alpha\beta} - x^3\, B_{\alpha\beta}$ on the undistorted shell, because of the hypothesis of small displacements (cf. §5.4.5). This integration is simplified in the framework of the hypothesis of thin shells

Similarly, the law connecting a "mean" distortion γ_α characteristic of deformations $\bar{\varepsilon}_{\alpha 3}$ due to shear forces $\bar{Q}^\alpha$ can be obtained by integrating a local law by relation (5.10). This also makes it necessary to express the variation of the deformation $\bar{\varepsilon}_{\alpha 3}$ as a function of x^3, which could in principle be done by means of adequate kinematic hypotheses. In reality, the kinematic models generally used are not rich enough to allow the correct representation of the variation of distortion as a function of x^3 and the law is then obtained by energy considerations (see §5.4.4.2 below).

The constitutive laws are explained in Sections 5.4.3 and 5.4.4, in the case of thin shells and in Section 5.4.5, under less restrictive assumptions.

5.4.3 Elastic thin shells in Love theory

5.4.3.1 Hypotheses of Love theory

In this Section, it is a question of establishing the shell constitutive laws relating to the normal forces and the bending moments.

The following assumptions are considered throughout §5.4.3:

- hypothesis of thin shells,
- Kirchhoff–Love's assumptions,
- hypothesis of plane stresses,
- hypothesis of small displacements (vis-à-vis the thickness of the shell),
- the material is linearly elastic.

This set of hypotheses is the basis of the **Love theory**, also called the **first approximation theory**.

From a physical point of view, the KL hypotheses are consistent with the hypothesis of thin shells. Indeed, when the shell can not be considered thin, generally appear distortions $\bar{\varepsilon}_{\alpha 3}$ incompatible with KL assumptions (unless such a thick shell undergoes purely membrane forces or pure bending). Similarly, the assumption of plane stresses is better verified that the shell is thinner.

5.4.3.2 Linearly elastic solid

The hypothesis of thin shells leads to writing[11]:

$$g_{\alpha\beta} = a_{\alpha\beta} \;\; ; \;\; g^{\alpha\beta} = a^{\alpha\beta}$$
$$m_\alpha{}^\beta = \ell_\alpha{}^\beta = \delta_\alpha^\beta \qquad (5.110)$$

from which, in particular:

$$\bar{\varepsilon}_{\alpha\beta}(\mathcal{P}) = \varepsilon_{\alpha\beta} - x^3 \kappa_{\alpha\beta} \qquad (5.111)$$

[11]Due to the hypothesis of small displacements, these different quantities are confused with their Lagrangian counterparts, respectively G, A, $\mathcal{M}$. The notation in lowercase letters is preserved for these tensors in the following.

The local law is rewritten after having expressed that $\sigma^{33} = 0$:

$$\sigma^{\alpha\beta}(\mathcal{P}) = \sigma_0{}^{\alpha\beta}(\mathcal{P}) + \bar{A}^{\alpha\beta}(\mathcal{P})\,\tau(\mathcal{P}) + \bar{L}^{\alpha\beta\gamma\delta}(\mathcal{P})\,\bar{\varepsilon}_{\gamma\delta}(\mathcal{P}) \tag{5.112}$$

(law of plane stresses) and, by integration:

$$\begin{aligned}
N^{\alpha\beta} &= N_0{}^{\alpha\beta} + \int_{-e/2}^{e/2} \bar{A}^{\alpha\beta}(\mathcal{P}) \cdot \tau(\mathcal{P})\,dx^3 + \varepsilon_{\gamma\delta}\int_{-e/2}^{e/2} \bar{L}^{\alpha\beta\gamma\delta}(\mathcal{P})\,dx^3 \\
&\quad - \kappa_{\gamma\delta}\int_{-e/2}^{e/2} \bar{L}^{\alpha\beta\gamma\delta}(\mathcal{P})\,x^3\,dx^3 \\
M^{\alpha\beta} &= M_0{}^{\alpha\beta} + \int_{-e/2}^{e/2} \bar{A}^{\alpha\beta}(\mathcal{P}).\,\tau(\mathcal{P})\,x^3 dx^3 - \varepsilon_{\gamma\delta}\int_{-e/2}^{e/2} \bar{L}^{\alpha\beta\gamma\delta}(\mathcal{P})\;x^3\,dx^3\;\bar{\varepsilon}_{\alpha 3} \\
&\quad + \kappa_{\gamma\delta}\int_{-e/2}^{e/2} \bar{L}^{\alpha\beta\gamma\delta}(\mathcal{P})\left(x^3\right)^2 dx^3
\end{aligned} \tag{5.113}$$

which shows a symmetrical coupling. The purely elastic part can be written in the form:

$$\begin{aligned}
N^{\alpha\beta} &= K^{\alpha\beta\gamma\delta}\varepsilon_{\gamma\delta} - H^{\alpha\beta\gamma\delta}\kappa_{\gamma\delta} \\
M^{\alpha\beta} &= -\,H^{\alpha\beta\gamma\delta}\varepsilon_{\gamma\delta} + D^{\alpha\beta\gamma\delta}\kappa_{\gamma\delta}
\end{aligned} \tag{5.114}$$

If the elastic properties are symmetrical with respect to the middle surface (case of homogeneous materials and most composites): $L^{\alpha\beta\gamma\delta}\left(x^3\right) = L^{\alpha\beta\gamma\delta}\left(-x^3\right)$, the coupling term is zero. When this is not the case, it may be interesting to take as the middle surface, no longer the surface located mid-thickness, but, where possible, the surface situated at a distance such that the coupling term is null (called neutral surface), then:

$$\int \bar{L}^{\alpha\beta\gamma\delta}(\mathcal{P})\,dx^3 = H^{\alpha\beta\gamma\delta} = 0$$

This is obviously only possible if this dimension is defined in a unique way, that is to say, whatever α, β, γ, or δ. This method is only of limited interest in general but becomes very interesting in the case of plates, where equilibrium equations are decoupled in linearised theory.

The constitutive laws of rotationally symmetrical and orthotropic shells are developed in §5.6.1.

5.4.3.3 Homogeneous and isotropic solid

In the case of isotropic solids both thermally and mechanically:

$$\bar{A}^{\alpha\beta} = \alpha\, a^{\alpha\beta}$$

$$
\begin{aligned}
N^{\alpha\beta} &= \int_{-e/2}^{e/2} \sigma^{\alpha\beta} dx^3 = \int_{-e/2}^{e/2} \sigma_0{}^{\alpha\beta} dx^3 + \int_{-e/2}^{e/2} \alpha\,\tau\, a^{\alpha\beta} dx^3 \\
&+ \frac{E}{1-\nu^2}\int_{-e/2}^{e/2}\left[(1-\nu)\left(\varepsilon^{\alpha\beta} - x^3\kappa^{\alpha\beta}\right) + \nu\left(\varepsilon^{\gamma}{}_{\gamma} - x^3\kappa^{\gamma}{}_{\gamma}\right) a^{\alpha\beta}\right] dx^3 \\
M^{\alpha\beta} &= -\int_{-e/2}^{e/2} \sigma^{\alpha\beta}\, x^3\, dx^3 = -\int_{-e/2}^{e/2} \sigma_0{}^{\alpha\beta}\, x^3\, dx^3 - \int_{-e/2}^{e/2} \alpha\,\tau\, a^{\alpha\beta}\, x^3\, dx^3 \\
&- \frac{E}{1-\nu^2}\int_{-e/2}^{e/2}\left[(1-\nu)\left(\varepsilon^{\alpha\beta} - x^3\kappa^{\alpha\beta}\right) + \nu\left(\varepsilon^{\gamma}{}_{\gamma} - x^3\kappa^{\gamma}{}_{\gamma}\right) a^{\alpha\beta}\right] x^3\, dx^3
\end{aligned}
\tag{5.115}
$$

and the properties E, ν, and α are constant in thickness due to the homogeneity assumption. Relations (5.113) become: (5.115) Assuming that the temperature varies linearly in thickness:

$$
\tau(\mathcal{P}) = \Theta(\mathcal{M}) + \frac{\Delta\Theta(\mathcal{M})}{e} x^3 \tag{5.116}
$$

Noting that:

$$
\int_{-e/2}^{e/2} x^3\, dx^3 = 0
$$

and posing:

$$
\mathrm{I} = \int_{-e/2}^{e/2} \left(x^3\right)^2 dx^3 = \frac{Ee^3}{12} \tag{5.117}
$$

the shell constitutive laws are finally written:

$$
\boxed{
\begin{aligned}
N^{\alpha\beta} &= N_0{}^{\alpha\beta} + \alpha e\,\Theta\, a^{\alpha\beta} + \frac{Ee}{1-\nu^2}\left[(1-\nu)\varepsilon^{\alpha\beta} + \nu\,\varepsilon^{\gamma}{}_{\gamma}\, a^{\alpha\beta}\right] \\
M^{\alpha\beta} &= M_0{}^{\alpha\beta} + \alpha \mathrm{I}\,\frac{\Delta\Theta}{e}\, a^{\alpha\beta} + \frac{Ee^3}{12\left(1-\nu^2\right)}\left[(1-\nu)\kappa^{\alpha\beta} + \nu\,\kappa^{\gamma}{}_{\gamma}\, a^{\alpha\beta}\right]
\end{aligned}
}
\tag{5.118}
$$

The effects of surface deformation and curvature variation are decoupled. It appears in these laws "global" elastic stiffness:

$$
\boxed{K = \frac{Ee}{1-\nu^2} \;;\; D = \frac{Ee^3}{12\left(1-\nu^2\right)}} \tag{5.119}
$$

- K is the rigidity of the shell to the extension;
- D is the stiffness of the shell to bending.

These two geometric characteristics of the shell are to be compared with the area and the inertia of the section of a beam of unit width and height equal to e, and show the effect of Poisson's ratio, which does not exist in the beam model.

The average temperature variations generate normal forces, while "temperature gradients" (temperature difference between the two faces of the shell, reduced to the thickness) create bending moments. In general, by continuing the development of $\tau(\mathcal{P})$as a function of x^3, even terms occur in N and odd terms in M.

The elastic strain potential associated with relationships (5.118) can be written as:

$$\mathcal{F} = \frac{K}{2}\int_{\Sigma}\left[(1-\nu)\,\varepsilon^{\alpha\beta}\varepsilon_{\alpha\beta} + \nu\left(\varepsilon_{\gamma}{}^{\gamma}\right)^2\right]dS + \frac{D}{2}\int_{\Sigma}\left[(1-\nu)\,\kappa^{\alpha\beta}\kappa_{\alpha\beta} + \nu\left(\kappa_{\gamma}{}^{\gamma}\right)^2\right]dS \tag{5.120}$$

The two terms correspond respectively to the extension energy and to the bending energy, the two energies being decoupled in the case considered.

Example: Comparison with three-dimensional theory.
It is a question of verifying that the law of plane stresses adopted better reflects the rigidity of the shell. Let a cylinder be subjected to internal pressure. If the ends are fixed longitudinally ($\varepsilon_{zz} = 0$), the three-dimensional solution (see §1.2.2.2) makes it possible to obtain the normal displacement at every point:

$$w(r) = \frac{(1+\nu)\,p}{E}\,\frac{a^2}{b^2 - a^2}\left[(1-2\nu)\,r + \frac{b^2}{r}\right]$$

If the shell is very thin, e can be neglected in front of R and the displacement at mid-thickness is then approximately:

$$w \approx \frac{pR^2}{Ee}\left(1-\nu^2\right)$$

In a thin shell model such as $\varepsilon_{zz} = 0$ and $\tilde{\varepsilon}_{00} = \dfrac{w}{R}$, the membrane balance is such that $\tilde{N}^{00} = pR$ and the same displacement is obtained with the assumption of plane stresses. On the other hand, preserving the three-dimensional law (5.105) and omitting the hypothesis of plane stresses, $\tilde{\sigma}^{00} = \dfrac{pR}{e}$ is written as a function of the deformation, expressing Kirchhoff's hypothesis $\varepsilon_{rr} = 0$:

$$\tilde{\sigma}^{00} = \frac{E}{1+\nu}\,\frac{1-\nu}{1-2\nu}\,\tilde{\varepsilon}^{00}$$

This would lead to:

$$w = \frac{pR^2}{Ee}\,\frac{(1+\nu)(1-2\nu)}{1-\nu}$$

which is lower than the previous value because $1-2\nu < (1-\nu)^2$. Indeed, this last value of w is obtained with $\varepsilon_{rr} = 0$ (Kirchhoff hypothesis), which stiffens the shell artificially. ■

By keeping only the terms of pure elasticity, the constitutive law reverses:

$$\begin{aligned} \varepsilon &= \frac{1}{Ee}\left[(1+\nu)\,N - \nu\,N^{\alpha}{}_{\alpha}\,a\right] \\ \kappa &= \frac{12}{Ee^3}\left[(1+\nu)\,M - \nu\,M^{\alpha}{}_{\alpha}\,a\right] \end{aligned} \tag{5.121}$$

and:

$$\varepsilon^{\alpha}{}_{\alpha} = \frac{1-\nu}{Ee} N^{\alpha}{}_{\alpha}$$
$$\kappa^{\alpha}{}_{\alpha} = \frac{12(1-\nu)}{Ee^3} M^{\alpha}{}_{\alpha} \tag{5.122}$$

Local strains are obtained by (5.56):

$$\overline{\varepsilon} = \varepsilon - x^3 \kappa$$

from which:

$$\overline{\varepsilon}^{\alpha}_{\alpha} = \frac{1-\nu}{Ee}\left(N^{\alpha}{}_{\alpha} - \frac{12\, x^3}{e^2} M^{\alpha}{}_{\alpha} \right)$$

and the local stresses by (5.108):

$$\boxed{\sigma^{\alpha\beta} = \frac{1}{e} N^{\alpha\beta} - \frac{12\, x^3}{e^3} M^{\alpha\beta}} \tag{5.123}$$

a formula similar to that obtained for beams.

5.4.3.4 *Shear stress due to shear force in a homogeneous and isotropic shell*

An expression of the shear stresses is obtained by proceeding in a manner analogous to that which makes it possible to obtain the Jouravsky theorem for the beams: the demonstration is made in static, neglecting the mass forces F and the surface forces Φ. The local equilibrium equations are then written, in projection on the tangent plane (cf (5.139), considering the hypothesis of thin shells):

$$\nabla_{\alpha}\sigma^{\alpha\beta} - B_{\alpha}{}^{\alpha}\sigma^{3\alpha} - 2B_{\alpha}{}^{\beta}\sigma^{3\alpha} + \partial_3\sigma^{3\beta} = 0$$

$\sigma^{3\beta}$ being of the order of $\frac{Q^{\beta}}{e}$, the two terms containing the curvature in the equilibrium equation are negligible. By (123) and considering equilibrium equations (5.19):

$$\nabla_{\alpha}\sigma^{\alpha\beta} = \frac{1}{e}\nabla_{\alpha}N^{\alpha\beta} - \frac{12\, x^3}{e^3}\nabla_{\alpha}M^{\alpha\beta} = \left(\frac{1}{e} B_{\alpha}{}^{\beta} + \frac{12\, x^3}{e^3} \delta_{\alpha}{}^{\beta} \right) Q^{\alpha}$$

In this last expression, the term $\frac{1}{e} B_{\alpha}{}^{\beta}$ is negligible infront of the other terms, given the hypothesis of thin shells. It stays:

$$\frac{12\, x^3}{e^3} Q^{\beta} + \partial_3\sigma^{3\beta} = 0$$

Shear stresses are assumed to be zero on both sides, it comes by integration:

$$\boxed{\sigma^{3\beta} = -\frac{6Q^\beta}{e^3}\left(\left(x^3\right)^2 - \frac{e^2}{4}\right)} \tag{5.124}$$

The shear distribution is parabolic, as in a rectangular beam section. Notably, the maximum shear is $\frac{3}{2}\frac{Q^\beta}{e}$, 50% higher than the average shear $\frac{Q^\beta}{e}$.

5.4.4 Thin elastic shells in Reissner–Mindlin theory

5.4.4.1 Membrane stresses

In the case of thin shells, the expression (5.74) of the strains $\overline{\varepsilon}_{\alpha\beta}$ reduces to:

$$\overline{\varepsilon}_{\alpha\beta} = \varepsilon_{\alpha\beta} - x^3\kappa_{\alpha\beta} \tag{5.125}$$

noting:

$$\kappa_{\alpha\beta} = -\frac{1}{2}\left[\nabla_\alpha d_\beta + \nabla_\beta d_\alpha - B_\alpha{}^\lambda\left(\nabla_\beta \xi_\lambda - \zeta B_{\beta\lambda}\right) - B_\beta{}^\lambda\left(\nabla_\alpha \xi_\lambda - \zeta\, B_{\alpha\lambda}\right)\right] \tag{5.126}$$

(symmetrical part of ρ). $\kappa_{\alpha\beta}$ represents the variation of curvature associated with the material segment carried by **d**. Physically, this quantity is quite analogous to the variation of curvature of the surface in Kirchhoff–Love theory and it is thus the generalised deformation intervening in the constitutive laws. Starting from (5.125), the constitutive laws obtained in the various cases in Love's theory in the preceding paragraph can be applied here, taking the expression (5.126) of κ.

In the case of homogeneous and isotropic linear elastic material, the stresses $\sigma^{\alpha\beta}$ are linearly distributed in the thickness and their expressions given by (5.123).

5.4.4.2 Taking into account the distortion energy

This paragraph is limited to the case of homogeneous, isotropic linear elasticity. In the Reissner–Mindlin theory, the strain energy must be supplemented by the distortion energy. However, in linear elasticity, stresses and strains are linked by the relation:

$$\sigma^{3\alpha} = 2G\,\varepsilon^{3\alpha}$$

where $\varepsilon_{3\alpha}$ is given in small displacements by (5.77). So, by (5.10) and considering the hypothesis of thin shells, by which $m_\alpha{}^\beta = \delta_\alpha{}^\beta$, by simple integration:

$$Q^\alpha = G\,e\,\gamma^\alpha \tag{5.127}$$

where the distorsion γ^α is constant in the thickness:

$$\gamma_\alpha = d_\alpha + B_\alpha{}^\beta\,\xi_\beta + \partial_\alpha\zeta \tag{5.128}$$

It is clear that the assumptions of Reissner–Mindlin and thin shells lead to a constant distortion, thus a constant shear $\sigma^{\alpha 3}$ in the elastic case. This result is obviously not realistic, since, on the one hand, it generally violates the boundary conditions on both sides of the shell and, on the other hand, it is in contradiction with the parabolic distribution of the shear stresses obtained in §5.4.3.3 from simple equilibrium considerations, taking into account the linear distribution of normal stresses. This artificially stiffens the shell.

In the hypothesis of parabolic distribution of shear stresses given by (5.124):

$$\sigma^{\alpha 3} = \frac{6Q^\alpha}{e}\left(\frac{1}{4} - \left(\frac{x^3}{e}\right)^2\right) \tag{5.129}$$

which can be adopted here, the stresses $\sigma^{\alpha\beta}$ varying linearly in the thickness, and which satisfies the equilibrium (5.10) in the thickness:

$$Q^\alpha = \int_{-e/2}^{+e/2} \sigma^{\alpha 3} dx^3 \tag{5.130}$$

A maximum value of the distortion is obtained on the middle surface, where it is worth:

$$\gamma'^\alpha = 2\,\varepsilon^{3\alpha} = \frac{3}{2}\frac{Q^\alpha}{Ge} \tag{5.131}$$

It is clear that the value of γ^α given by (5.128) corresponds to the average of the distortions corresponding to the parabolic distribution of the shear stresses.

One way to obtain a better assessment of shear stiffness is to correct shear stiffness Ge by substituting a reduced stiffness kGe. k is less than 1 to reflect greater flexibility than that given by RM assumptions: for a given shear force, the distortion is greater. By (5.130), a lower limit value of k is 2/3. This leads to a distortion (called **Mindlin distortion**), more representative of the average distortion, given by:

$$\gamma'^\alpha = \frac{Q^\alpha}{kGe} \tag{5.132}$$

To obtain a value of reduced stiffness closer to physical reality, a method is to calculate the distortion energy $\mathcal{F}_d$ due to the shear force from the shear stresses (complementary energy[12]), thus from the shear force, and then connect it to the distortions. $\mathcal{F}_d$ is calculated from local stresses on volume ω based on Σ:

$$\mathcal{F}_d = \frac{1}{2}\int_\omega 2 \times \frac{1}{2G}\sigma^{\alpha 3}\sigma_{\alpha 3}\, dV$$

So, introducing the distribution (5.129):

$$\mathcal{F}_d = \frac{1}{2}\int_\Sigma\int_{-e/2}^{+e/2} \frac{72}{2Ge^2} Q^\alpha Q_\alpha \left(\frac{1}{4} - \left(\frac{x^3}{e}\right)^2\right)^2 dx^3 d\sigma = \frac{1}{2}\frac{6}{5Ge}\int_\Sigma Q^\alpha Q_\alpha d\sigma \tag{5.133}$$

[12]For convenience, the strain energy and the complementary energy are confused as they are equal in linear elasticity.

from which it emerges that the coefficient k is equal to 5/6. This value of the coefficient k is therefore related to the parabolic distribution of the shear stresses in the thickness and it corresponds to a distortion energy evaluated from the three-dimensional theory. It confirms that the shell is more flexible with respect to shear than the RM hypothesis suggests. It better respects the experimental results in homogeneous materials.

Other approaches have been developed to evaluate the shear force correction, notably the comparison with three-dimensional elastic solutions, which show that the coefficient k depends on the Poisson's ratio.

In composite materials, shear distribution is obtained from the stress distribution $\sigma^{\alpha\beta}$ in the thickness. They obviously lead to different values of reduced stiffness (see §5.6.3).

The introduction of this corrected stiffness makes it possible to express the shear force distortion energy either as a function of the shear force or the average distortion γ^α given by (5.128):

$$\mathcal{F}_d = \frac{1}{2}\frac{1}{kGe}\int_\Sigma Q^\alpha Q_\alpha d\sigma = \frac{1}{2}kGe\int_\Sigma \gamma^\alpha \gamma_\alpha d\sigma \tag{5.134}$$

5.4.4.3 Elastic potential of Reissner-Mindlin thin shells

Taking into account the results of the two preceding paragraphs, the elastic potential of a thin shell linearly elastic, homogeneous and isotropic is written:

$$\boxed{\begin{aligned}\mathcal{F} &= \frac{K}{2}\int_\Sigma \left[(1-\nu)\,\varepsilon^{\alpha\beta}\varepsilon_{\alpha\beta} + \nu\left(\varepsilon_\gamma{}^\gamma\right)^2\right] dS \\ &+ \frac{D}{2}\int_\Sigma \left[(1-\nu)\,\kappa^{\alpha\beta}\kappa_{\alpha\beta} + \nu\left(\kappa_\gamma{}^\gamma\right)^2\right] dS + \frac{1}{2}kGe\int_\Sigma \gamma^\alpha\gamma_\alpha\, dS\end{aligned}} \tag{5.135}$$

where γ is given by (5.128) and κ by (5.126). k can be taken as 5/6 for a homogeneous and isotropic shell. A more precise value has been proposed by Wittrick[13]: $k = \dfrac{5}{6-\nu}$.

5.5 SHELL EQUILIBRIUM EQUATIONS OBTAINED FROM LOCAL EQUILIBRIUM EQUATIONS

Equations of equilibrium of the shell were obtained directly by characterising the shell by its middle surface, either by the General Theorems (§4.2.1), or by the Principle of the Virtual Powers (§4.2.2). Relations (5.9) and (5.10) allowed the definition of adequate resultants of stresses, which could be related to the tensors N and M representing globally the internal actions, introduced in §4.2.1, then to the Sanders generalised stresses **N** and **M**.

It is also possible to obtain the equilibrium equations of the shell by integrating the local equations of the three-dimensional theory. For this, only the relations (5.9) and (5.10) defining the resultants of the three-dimensional stresses are used, without any kinematic hypothesis being necessary (except, however, to detail the expression of the acceleration).

The notations are those of §5.1.

[13]William Henry Wittrick, British mathematician and mechanician (1922–1986).

5.5.1 Expression of the local equilibrium equations

The local equations are expressed in the local coordinate system g_i associated with normal coordinates:

$$\left[D_i T^{ij} + \bar{\rho}\left(\bar{F}^i - \bar{\gamma}^i\right) g_j\right] = 0 \tag{5.136}$$

where : $(\bar{F}^i)$ are the external massic forces applied within the shell (most often, the acceleration of gravity);

$(\bar{\gamma}^i)$ are the components of acceleration at every point. These components are given by relation (5.33) in the context of KL assumptions.

The usual differentiation formulas of a contravariant two-fold tensor of $\mathbb{R}^3$ make it possible to write, given the symmetry of $\bar{\bar{\mathsf{T}}}$:

$$\begin{aligned}
D_i T^{i\beta} &= \underbrace{\partial_\alpha T^{\alpha\beta} + \bar{\Gamma}^\alpha_{\alpha\gamma} T^{\gamma\beta} + \bar{\Gamma}^\alpha_{\alpha 3} T^{3\beta} + \bar{\Gamma}^\beta_{\alpha\gamma} T^{\alpha\gamma} + \bar{\Gamma}^\beta_{\alpha 3} T^{\alpha 3}}_{D_\alpha T^{\alpha\beta}} \\
&\quad \underbrace{+ \partial_3 T^{3\beta} + \bar{\Gamma}^3_{3\alpha} T^{\alpha\beta} + \bar{\Gamma}^3_{33} T^{3\beta} + \bar{\Gamma}^\beta_{3\alpha} T^{3\alpha} + \bar{\Gamma}^\beta_{33} T^{33}}_{D_3 T^{3\beta}} \\
D_i T^{i3} &= \underbrace{\partial_\alpha T^{\alpha 3} + \bar{\Gamma}^\alpha_{\alpha\gamma} T^{\gamma 3} + \bar{\Gamma}^\alpha_{\alpha 3} T^{33} + \bar{\Gamma}^3_{\alpha\gamma} T^{\alpha\gamma} + \bar{\Gamma}^3_{\alpha 3} T^{\alpha 3}}_{D_\alpha T^{\alpha 3}} \\
&\quad \underbrace{+ \partial_3 T^{33} + 2\bar{\Gamma}^3_{3\alpha} T^{\alpha 3} + 2\bar{\Gamma}^3_{33} T^{33}}_{D_3 T^{33}}
\end{aligned} \tag{5.137}$$

The expression of the coefficients $\bar{\Gamma}$ in the coordinate system g_j has been obtained in §3.2.1.3, which makes it possible to rewrite the three-dimensional equations of equilibrium, in Euler variables, in the form:

$$\begin{aligned}
&\left[\begin{array}{l}
\partial_\alpha T^{\alpha\beta} + \left(\ell^\alpha{}_\lambda \nabla_\alpha m_\gamma{}^\lambda + \bar{\Gamma}^\alpha_{\alpha\gamma}\right) T^{\gamma\beta} - b_\alpha{}^\gamma \ell^\alpha{}_\gamma T^{3\beta} - 2 b_\alpha{}^\gamma \ell^\beta{}_\gamma T^{3\alpha} \\
\quad + \left(\ell^\beta{}_\lambda \nabla_\alpha m_\gamma{}^\lambda + \bar{\Gamma}^\beta_{\alpha\gamma}\right) T^{\alpha\gamma} + \partial_3 T^{3\beta} + \bar{\rho}\left(\bar{F}^\beta - \bar{\gamma}^\beta\right)
\end{array}\right] m_\beta{}^\mu a_\mu \\
&\quad + \left[\begin{array}{l}
\partial_\alpha T^{\alpha 3} + \left(\ell^\alpha{}_\lambda \nabla_\alpha m_\gamma{}^\lambda + \bar{\Gamma}^\alpha_{\alpha\gamma}\right) T^{\gamma 3} - b_\alpha{}^\gamma \ell^\alpha{}_\gamma T^{33} \\
\quad + m_\gamma{}^\delta b_{\alpha\delta} T^{\alpha\gamma} + \partial_3 T^{33} + \bar{\rho}\left(\bar{F}^3 - \bar{\gamma}^3\right)
\end{array}\right] a_3 = 0
\end{aligned} \tag{5.138}$$

The three equilibrium equations thus obtained, applicable at any point in the thickness of the shell, are expressed in the local coordinate system of the surface, as a function of the parameters (x^α) of the surface and of x^3. Moreover, taking into account the Mainardi–Codazzi equations:

$$\nabla_\alpha m_\gamma{}^\lambda = \nabla_\gamma m_\alpha{}^\lambda$$

a rearrangement of terms makes it possible to write:

$$\begin{aligned}
&\left[\begin{array}{l} \nabla_\alpha T^{\alpha\beta} + \ell^\alpha{}_\lambda \nabla_\gamma \mathrm{m}_\alpha{}^\lambda T^{\gamma\beta} + \ell^\beta{}_\lambda \nabla_\alpha \mathrm{m}_\gamma{}^\lambda T^{\alpha\gamma} - b_\alpha{}^\gamma \ell^\alpha{}_\gamma T^{3\beta} \\ \qquad -2 b_\alpha{}^\gamma \ell^\beta{}_\gamma T^{3\alpha} + \partial_3 T^{3\beta} + \bar{\rho}\left(\bar{F}^\beta - \bar{\gamma}^\beta\right) \end{array}\right] \mathrm{m}_\beta{}^\mu = 0 \\
&\nabla_\alpha T^{\alpha 3} + \ell^\alpha{}_\lambda \nabla_\gamma \mathrm{m}_\alpha{}^\lambda T^{\gamma 3} - b_\alpha{}^\gamma \ell^\alpha{}_\gamma T^{33} + \mathrm{m}_\gamma{}^\delta b_{\alpha\delta} T^{\alpha\gamma} + \partial_3 T^{33} + \bar{\rho}\left(\bar{F}^3 - \bar{\gamma}^3\right) = 0
\end{aligned} \tag{5.139}$$

5.5.2 Integration method

The adopted method consists in integrating the local equations (5.139) in the thickness, to reveal the elements of the screw resultant of the stresses. More precisely, to take into account the variation $\hat{g}$ of the area formed on the basis vectors (g_α) as a function of x^3, the local equations are multiplied by $\hat{\mathrm{m}}$ before integrating into the thickness. This amounts to an integration with respect to a "volume element" based on a_1 and a_2 and generated by a_3 (Figure 5.8). This element is subject, besides the stresses :

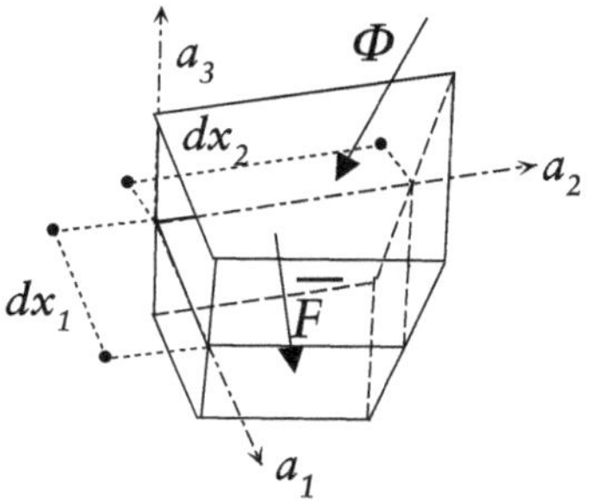

Figure 5.8

- on the one hand, to the forces of inertia $-\bar{\rho}\,\bar{\gamma}$,
- on the other hand, to external forces:
  . massic forces $\bar{F}$ inside the volume,
  . surface forces Φ that apply on both faces $(x^3 = a)$ and $(x^3 = b)$ of the element.

5.5.3 Resultant of stresses

5.5.3.1 Normal component

Given the relationships:

$$\begin{cases} \nabla_\alpha \hat{\mathrm{m}} = \hat{\mathrm{m}}\, \ell^\gamma{}_\delta \nabla_\alpha \mathrm{m}_\gamma{}^\delta = \hat{\mathrm{m}} \mathrm{TR}\left(\mathrm{m}^{-1} \nabla_\alpha \mathrm{m}\right) \\ \partial_3 \hat{\mathrm{m}} = -\, \hat{\mathrm{m}}\; \mathrm{b}_\alpha{}^\gamma \ell^\alpha{}_\gamma \end{cases}$$

the integration of the third component of the local equation leads to:

$$\int_a^b \left[\hat{\mathrm{m}}\, \nabla_\alpha T^{\alpha 3} + T^{\alpha 3}\, \nabla_\alpha \hat{\mathrm{m}} + T^{33}\, \partial_3 \hat{\mathrm{m}} + \hat{\mathrm{m}}\, \mathrm{m}_\gamma{}^\lambda\, b_{\alpha\lambda} T^{\alpha\gamma} + \hat{\mathrm{m}}\, \partial_3 T^{33} + \bar{\rho} \hat{\mathrm{m}}\left(\bar{F}^3 - \bar{\gamma}^3\right)\right] dx^3 = 0$$

that is to say:

$$\int_a^b \left[\nabla_\alpha\left(\hat{\mathrm{m}}\, T^{\alpha 3}\right) + \hat{\mathrm{m}}\, \mathrm{m}_\gamma{}^\lambda\, b_{\alpha\lambda} T^{\alpha\gamma} + \partial_3\left(\hat{\mathrm{m}}\, T^{33}\right) + \bar{\rho} \hat{\mathrm{m}}\left(\bar{F}^3 - \bar{\gamma}^3\right)\right] dx^3 = 0 \tag{5.140}$$

- Calculation of $\nabla_\alpha \bar{Q}^\alpha$:

By definition:

$$\bar{Q}^\alpha = \int_a^b T^{\alpha 3}\,\hat{m}\,dx^3$$

Let $Y^\alpha\left(x^i\right)$ be a primitive of $T^{\alpha 3}\,\hat{m}$ with respect to x^3; for example:

$$Y^\alpha = \int_0^{x^3} \left[T^{\alpha 3}\,\hat{m}\right]\left(x^\beta,\xi\right) d\xi$$

It becomes:

$$\bar{Q}^\alpha = Y^\alpha\left[x^\gamma, b\left(x^\gamma\right)\right] - Y^\alpha\left[x^\gamma, a\left(x^\gamma\right)\right]$$

from which:

$$\begin{aligned}\nabla_\beta \bar{Q}^\alpha &= \partial_\beta \bar{Q}^\alpha + \Gamma^\alpha_{\beta\lambda}\bar{Q}^\lambda \\ &= \partial_\beta Y^\alpha\left[x^\gamma, b\left(x^\gamma\right)\right] - \partial_\beta Y^\alpha\left[x^\gamma, a\left(x^\gamma\right)\right] \\ &\quad + \frac{\partial Y^\alpha}{\partial x^3}\left[x^\gamma, b\left(x^\gamma\right)\right]\partial_\beta b - \frac{\partial Y^\alpha}{\partial x^3}\left[x^\gamma, a\left(x^\gamma\right)\right]\partial_\beta a + \Gamma^\alpha_{\beta\lambda}\bar{Q}^\lambda\end{aligned}$$

But:

$$\partial_\beta Y^\alpha = \partial_\beta\left(\int_0^{x^3} T^{\alpha 3}\,\hat{m}\,d\xi\right) = \int_0^{x^3} \partial_\beta\left(T^{\alpha 3}\,\hat{m}\right) d\xi$$

Finally:

$$\nabla_\alpha \bar{Q}^\alpha = \int_a^b \nabla_\alpha\left(T^{\alpha 3}\,\hat{m}\right) dx^3 + \partial_\alpha b\left[T^{\alpha 3}\,\hat{m}\right]_{x^3=b} - \partial_\alpha a\left[T^{\alpha 3}\,\hat{m}\right]_{x^3=a} \qquad (5.141)$$

- Surface forces $\boldsymbol{\Phi}$.

Noting that:

$$\int_a^b \partial_3\left(T^{33}\,\hat{m}\right) dx^3 = \left[T^{33}\,\hat{m}\right]_{x^3=a}^{x^3=b}$$

it appears in $x^3 = b$ a contribution of the stresses T^{i3} coming from the terms in square brackets. The vector $\mathbf{n}$ of components $\left(-\frac{\partial b}{\partial x^1};\quad -\frac{\partial b}{\partial x^2};\quad 1\right)$ is the vector normal to the surface defined by $x^3 = b\left(x^1, x^2\right)$.

Then:

$$-T^{\alpha 3}\hat{m}\frac{\partial \mathscr{b}}{\partial x^{\alpha}}+\hat{m}\ T^{33}=\hat{m}\left(\overline{\overline{T}}\cdot\mathbf{n}\right)a_3 \tag{5.142}$$

Now: $\overline{\overline{T}}\cdot\mathbf{n}=\Phi$, where Φ is the density of force applied to the surface $x^3=\mathscr{b}$. The multiplication by $\hat{m}$ expresses the area of the "surface element" in $x^3=\mathscr{b}$ with respect to the middle surface. $\left[\hat{m}\,\Phi\,a_3\right]_{x^3=\mathscr{a}}^{x^3=\mathscr{b}}$ is therefore the projection on a_3 of the surfacic densities of the forces applied to the two faces of the shell.

- External force p^3 :

The integral $\int_{\mathscr{a}}^{\mathscr{b}}\bar{\rho}\,\hat{m}\,\bar{F}^3dx^3$ represents the resultant projected on a_3 of the density of forces applied within the shell. Let then:

$$p^3=\left[\hat{m}\Phi a_3\right]_{x^3=\mathscr{a}}^{x^3=\mathscr{b}}+\int_{\mathscr{a}}^{\mathscr{b}}\bar{\rho}\,\hat{m}\,\bar{F}^3dx^3 \tag{5.143}$$

p^3 is the resultant on a_3, in the volumic element considered, of external forces applied inside or on the faces of the shell.

- Acceleration:

$-\int_{\mathscr{a}}^{\mathscr{b}}\bar{\rho}\,\hat{m}\,\bar{\gamma}^3dx^3$ represents the resultant on a_3 of the inertial actions. In the hypothesis KL, and according to the relation (5.33):

$$-\int_{\mathscr{a}}^{\mathscr{b}}\bar{\rho}\,\hat{m}\,\bar{\gamma}^3dx^3=-\int_{\mathscr{a}}^{\mathscr{b}}\bar{\rho}\,\hat{m}\left(\gamma^3+x^3v_{\lambda 3}v_{\mu 3}a^{\lambda\mu}\right)dx^3$$

$$=-\gamma^3\int_{\mathscr{a}}^{\mathscr{b}}\bar{\rho}\,\hat{m}\,dx^3-v_{\lambda 3}v_{\mu 3}a^{\lambda\mu}\int_{\mathscr{a}}^{\mathscr{b}}\bar{\rho}\,\hat{m}\,x^3dx^3$$

But:

$$\int_{\mathscr{a}}^{\mathscr{b}}\bar{\rho}\,\hat{m}\,x^3dx^3=\rho\,\overline{\mathscr{mg}}$$

where $\mathscr{g}$ refers to the centre of gravity of the normal segment. In most cases, this term is nil or negligible. On the other hand, according to (4.10):

$$\Omega^2=v_{\lambda 3}v_{\mu 3}a^{\lambda\mu}$$

where Ω is the angular velocity vector of the normal segment. It becomes:

$$\int_{a}^{b} \bar{\rho}\, \hat{m}\, \bar{\gamma}^3 dx^3 = \rho\gamma^3 - \rho\Omega^2 \overline{mg}$$

By neglecting, in common cases, the term in Ω^2, the normal component of the shell local equations is written:

$$\boxed{\nabla_\alpha \bar{Q}^\alpha + b_{\alpha\beta}\bar{N}^{\alpha\beta} + p^3 - \rho\gamma^3 = 0} \tag{5.144}$$

5.5.3.2 *Tangential components*

After multiplication by $\hat{m}$, the integration of the tangential components of the local equations leads to:

$$\int_{a}^{b} \begin{bmatrix} \hat{m}\, m_\beta{}^\mu \nabla_\alpha T^{\alpha\beta} + T^{\alpha\beta} m_\beta{}^\mu \nabla_\alpha \hat{m} + T^{\alpha\gamma}\hat{m}\nabla_\alpha m_\gamma{}^\mu + m_\beta{}^\mu\, \partial_3 \hat{m}\, T^{3\beta} \\ + \hat{m} T^{3\beta} \partial_3 m_\beta{}^\mu - \hat{m}\, b_\alpha{}^\mu T^{3\alpha} + \partial_3 T^{3\alpha}\, m_\alpha{}^\mu\, \hat{m} + \bar{\rho}\, \hat{m}\, m_\beta{}^\mu \left(\bar{F}^\beta - \bar{\gamma}^\beta\right) \end{bmatrix} dx^3 = 0$$

That is to say:

$$\begin{aligned} &\int_{a}^{b} \nabla_\alpha \left(\hat{m}\, m_\beta{}^\mu T^{\alpha\beta}\right) dx^3 + \int_{a}^{b} \partial_3 \left(\hat{m}\, m_\beta{}^\mu T^{3\beta}\right) dx^3 \\ &\qquad - b_\alpha{}^\mu \int_{a}^{b} \hat{m} T^{3\alpha} dx^3 + \int_{a}^{b} \bar{\rho}\hat{m}\, m_\beta{}^\mu \left(\bar{F}^\beta - \bar{\gamma}^\beta\right) dx^3 = 0 \end{aligned} \tag{5.145}$$

- Calculation of $\nabla_\alpha \bar{N}^{\alpha\beta}$:

By definition:

$$\bar{N}^{\alpha\beta} = \int_{a}^{b} T^{\alpha\gamma} m_\gamma{}^\beta \hat{m}\, dx^3$$

By proceeding in a similar way to what has been done to $\nabla_\alpha \bar{Q}^\alpha$:

$$\begin{aligned} \nabla_\alpha \bar{N}^{\alpha\mu} = &\int_{a}^{b} \nabla_\alpha \left(\hat{m}\, m_\beta{}^\mu T^{\alpha\beta}\right) dx^3 \\ &+ \partial_\alpha b \left[T^{\alpha\gamma} m_\gamma{}^\mu \hat{m}\right]_{x^3=b} - \partial_\alpha a \left[T^{\alpha\gamma} m_\gamma{}^\mu \hat{m}\right]_{x^3=a} \end{aligned} \tag{5.146}$$

- Surfacic forces:

$$\int_{a}^{b} \partial_3 \left(\hat{m}\, m_\beta{}^\mu T^{3\beta}\right) dx^3 = \left[\hat{m}\, m_\beta{}^\mu T^{3\beta}\right]_{x^3=a}^{x^3=b}$$

The set of terms between brackets is written on the face $x^3 = \mathscr{b}$ of the shell:

$$\ell^{\mu} = -\partial_{\alpha}\mathscr{b}\left[T^{\alpha\gamma}\mathfrak{m}_{\gamma}{}^{\mu}\hat{\mathfrak{m}}\right]_{x^3=\mathscr{b}} + \left[\hat{\mathfrak{m}}\,\mathfrak{m}_{\beta}{}^{\mu}T^{3\beta}\right]_{x^3=\mathscr{b}} = \left[T^{i\gamma}\mathfrak{m}_{\gamma}{}^{\mu}\hat{\mathfrak{m}}\,\mathrm{n}_i\right]_{x^3=\mathscr{b}}$$

that is to say:

$$\ell = \ell^{\mu}a_{\mu} = \hat{\mathfrak{m}}\left(T^{i\gamma}\,\mathrm{n}_i\right)g_{\gamma} = \hat{\mathfrak{m}}\,\Phi^{\gamma}\,g_{\gamma} = \hat{\mathfrak{m}}\,\Phi \tag{5.147}$$

This is the tangential resultant of surface actions, the factor $\hat{\mathfrak{m}}$ reduces the local metric to that of the middle surface.

- External force p^{α} :

The integral $\int_{a}^{\mathscr{b}} \bar{\rho}\hat{\mathfrak{m}}\,\mathfrak{m}_{\beta}{}^{\mu}\bar{F}^{\beta}dx^3$ is the resultant following a_{μ} of the density of the forces applied within the shell. Then:

$$p^{\mu} = \ell^{\mu} + \int_{a}^{\mathscr{b}} \bar{\rho}\hat{\mathfrak{m}}\,\mathfrak{m}_{\beta}{}^{\mu}\bar{F}^{\beta}dx^3 = 0 \tag{5.148}$$

p^{μ} is the resultant along a_{μ} of the set of external forces: forces applied tangentially to both faces of the shell and forces applied within it, parallel to the middle surface.

- Acceleration:

$-\int_{a}^{\mathscr{b}} \bar{\rho}\hat{\mathfrak{m}}\,\mathfrak{m}_{\beta}{}^{\mu}\bar{\gamma}^{\beta}dx^3$ is the resultant along a_{μ} of inertia actions. In the assumptions KL and according to relation (33), where $\bar{\gamma}$ is expressed in the sytem a_i:

$$\begin{aligned}\int_{a}^{\mathscr{b}} \bar{\rho}\hat{\mathfrak{m}}\,\mathfrak{m}_{\beta}{}^{\mu}\bar{\gamma}^{\beta}dx^3 &= \int_{a}^{\mathscr{b}} \bar{\rho}\hat{\mathfrak{m}}\,\mathfrak{m}_{\beta}{}^{\mu}\left[\gamma^{\beta} + x^3a^{\beta\delta}\left(v_{\delta\lambda}v_{\gamma 3}a^{\lambda\gamma} - \dot{v}_{\delta 3}\right)\right]dx^3 \\ &= \rho\left(\delta_{\beta}{}^{\mu} - b_{\beta}{}^{\mu}\overline{\mathscr{mg}}\right)\gamma^{\beta} + \left(\rho a^{\mu\beta} - \mathrm{I}b^{\mu\beta}\right)\left(v_{\beta\lambda}v_{\gamma 3}a^{\lambda\gamma} - \dot{v}_{\beta 3}\right)\overline{\mathscr{mg}}\end{aligned}$$

where:

$$\mathrm{I} = \int_{a}^{\mathscr{b}} \bar{\rho}\hat{\mathfrak{m}}\left(x^3\right)^2 dx^3$$

denotes the mass inertia of the normal segment and:

$$v_{\beta\lambda}v_{\gamma 3}a^{\lambda\gamma} - \dot{v}_{\beta 3} = \ddot{a}_3 \bullet a_{\beta} = \left[\dot{\Omega}\times a_3 + \overbrace{\Omega\times(\Omega\times a_3)}^{-\Omega^2 a_3}\right]\bullet a_{\beta} = \left(\dot{\Omega}\times a_3\right)\bullet a_{\beta}$$

In a homogeneous shell, $\overline{\mathcal{mg}}$ is zero and $\mathcal{a} = -\mathcal{b}$.

- By preserving the only term $-\rho\gamma^{\mu}$, which is preponderant, the tangential components of the shell equations are written:

$$\boxed{\nabla_{\alpha}\bar{N}^{\alpha\mu} - b_{\alpha}{}^{\mu}\bar{Q}^{\alpha} + p^{\mu} - \rho\gamma^{\mu} = 0} \tag{5.149}$$

5.5.3.3 Resultant moment

To obtain the overall equilibrium of the normal segment with respect to the moment, each of the tangential components of the local equations is multiplied by x^3 before integrating. It comes:

$$\int_{\mathcal{a}}^{\mathcal{b}} \nabla_{\alpha}\left(\hat{m}\, m_{\beta}{}^{\mu}T^{\alpha\beta}\right)x^3dx^3 + \int_{\mathcal{a}}^{\mathcal{b}} \partial_3\left(\hat{m}\, m_{\beta}{}^{\mu}T^{3\beta}\right)x^3dx^3$$

$$- b_{\alpha}{}^{\mu}\int_{\mathcal{a}}^{\mathcal{b}} \hat{m}T^{3\alpha}x^3dx^3 + \int_{\mathcal{a}}^{\mathcal{b}} \bar{\rho}\,\hat{m}\, m_{\beta}{}^{\mu}\left(\bar{F}^{\beta} - \bar{\gamma}^{\beta}\right)x^3dx^3 = 0$$

By definition:

$$\bar{M}^{\alpha\beta} = -\int_{\mathcal{a}}^{\mathcal{b}} T^{\alpha\gamma}m_{\gamma}{}^{\beta}\hat{m}\, x^3\, dx^3 \nabla_{\alpha}\bar{M}^{\alpha\mu} = -\int_{\mathcal{a}}^{\mathcal{b}} \nabla_{\alpha}\left(\hat{m}\, m_{\beta}{}^{\mu}T^{\alpha\beta}\right)x^3dx^3$$

$$-\partial_{\alpha}\mathcal{b}\left[T^{\alpha\gamma}m_{\gamma}{}^{\mu}\hat{m}\, x^3\right]_{x^3=\mathcal{b}} + \partial_{\alpha}\mathcal{a}\left[T^{\alpha\gamma}m_{\gamma}{}^{\mu}\hat{m}\, x^3\right]_{x^3=\mathcal{a}}$$

Then:

$$\nabla_{\alpha}\bar{M}^{\alpha\mu} = -\int_{\mathcal{a}}^{\mathcal{b}} \nabla_{\alpha}\left(\hat{m}\, m_{\beta}{}^{\mu}T^{\alpha\beta}\right)x^3dx^3$$

$$-\partial_{\alpha}\mathcal{b}\left[T^{\alpha\gamma}m_{\gamma}{}^{\mu}\hat{m}\, x^3\right]_{x^3=\mathcal{b}} + \partial_{\alpha}\mathcal{a}\left[T^{\alpha\gamma}m_{\gamma}{}^{\mu}\hat{m}\, x^3\right]_{x^3=\mathcal{a}}$$

- Moment of surface actions:

$$\int_{\mathcal{a}}^{\mathcal{b}} \partial_3\left(\hat{m}\, m_{\beta}{}^{\mu}T^{3\beta}\right)x^3dx^3 = \left[\hat{m}\, m_{\beta}{}^{\mu}T^{3\beta}x^3\right]_{\mathcal{a}}^{\mathcal{b}} - \int_{\mathcal{a}}^{\mathcal{b}} \hat{m}\, m_{\beta}{}^{\mu}T^{3\beta}dx^3$$

In view of the interpretation given in the preceding paragraph, the terms in square brackets are grouped together and express the moment of the actions exerted tangentially on the surface parallel to a_{μ}, for example, on the face $x^3 = \mathcal{b}$:

$$r^{\mu} = -\mathcal{b}\,\ell^{\mu}$$

- Resultant moment of external actions:

$-\int_{\mathcal{a}}^{\mathcal{b}} \bar{\rho}\hat{m}\, m_{\beta}{}^{\mu}\, \bar{F}^{\beta}\, x^3dx^3$ is the moment relative to the middle surface of massic actions parallel to the middle surface. Then:

$$c^{\mu} = r^{\mu} - \int_{a}^{b} \bar{\rho}\hat{m}\, m_{\beta}{}^{\mu}\, \bar{F}^{\beta}\, x^3 dx^3$$

is the moment of the set of external actions directed according to a_μ, with respect to the middle surface.

- In KL assumptions, the moment of inertia actions is worth:

$$\int_{a}^{b} \bar{\rho}\hat{m}\, m_{\beta}{}^{\mu}\, \bar{\gamma}^{\beta}\, x^3 dx^3 = \left(\delta_{\beta}{}^{\mu} \overbrace{\int_{a}^{b} \bar{\rho}\hat{m}\, x^3 dx^3}^{\rho\, mg} - \mathrm{I}\, b_{\beta}{}^{\mu} \right) \bar{\gamma}^{\beta}$$

$$-\left(v_{\beta\lambda} v_{\gamma 3} a^{\lambda\gamma} - \dot{v}_{\beta 3}\right)\left(\mathrm{I} a^{\mu\beta} - b^{\mu\beta} \int_{a}^{b} \bar{\rho}\hat{m} \left(x^3\right)^3 dx^3 \right)$$

The predominant term is $-\mathrm{I}\, \ddot{a}_3 \cdot a^{\mu}$. It represents the moment of the inertial actions directed according to a_μ, with respect to the middle surface. This moment is most often neglected as explained in Chapter 4.

- By neglecting the moment resulting from the inertia actions and noting on the other hand that:

$$b_{\alpha}{}^{\mu} \int_{a}^{b} \hat{m} T^{3\alpha} x^3 dx^3 + \int_{a}^{b} \hat{m}\, m_{\beta}{}^{\mu} T^{3\beta} dx^3 = \int_{a}^{b} \hat{m}\, \delta_{\beta}{}^{\mu} T^{3\beta} dx^3 = \bar{Q}^{\mu}$$

it comes:

$$\boxed{\nabla_{\alpha} \bar{M}^{\alpha\mu} + \bar{Q}^{\mu} + c^{\mu} = 0} \qquad (5.150)$$

5.5.3.4 Symetries

The tensor of the three-dimensional stresses is symmetrical; in particular, for its restriction to planes parallel to T_m:

$$\forall(\gamma,\delta): \qquad T^{\gamma\delta} = T^{\delta\gamma}$$

To show the resultants of stresses, this equality is multiplied by $m_{\gamma}{}^{\alpha}\, m_{\delta}{}^{\beta}\, \hat{m}$ and integrated into the thickness:

$$\int_{a}^{b} T^{\gamma\delta} m_{\gamma}{}^{\alpha}\, m_{\delta}{}^{\beta}\, \hat{m}\, dx^3 = \int_{a}^{b} T^{\delta\gamma} m_{\gamma}{}^{\alpha}\, m_{\delta}{}^{\beta}\, \hat{m}\, dx^3 \qquad (5.151)$$

That is to say:

$$\begin{aligned}\delta_\gamma{}^\alpha \int_{a}^{b} T^{\gamma\delta}\, \mathfrak{m}_\delta{}^\beta\, \hat{\mathfrak{m}}\, dx^3 - b_\gamma{}^\alpha \int_{a}^{b} T^{\gamma\delta}\, \mathfrak{m}_\delta{}^\beta\, \hat{\mathfrak{m}}\, x^3 dx^3 \\ = \delta_\delta{}^\beta \int_{a}^{b} T^{\delta\gamma}\, \mathfrak{m}_\gamma{}^\alpha\, \hat{\mathfrak{m}}\, dx^3 - b_\delta{}^\beta \int_{a}^{b} T^{\delta\gamma} \mathfrak{m}_\gamma{}^\alpha\, \hat{\mathfrak{m}}\, x^3 dx^3\end{aligned} \tag{5.152}$$

It comes:

$$\bar{N}^{\alpha\beta} + b_\gamma{}^\alpha \bar{M}^{\gamma\beta} = \bar{N}^{\beta\alpha} + b_\delta{}^\beta \bar{M}^{\delta\alpha} \tag{5.153}$$

which, given the relationships (5.54), (4.116) and the remarks made in §5.1.3, leads to:

$$\boxed{\mathsf{N}^{\alpha\beta} = \bar{N}^{\alpha\beta} - b_\delta{}^\beta \bar{M}^{\delta\alpha} = \bar{N}^{\beta\alpha} - b_\gamma{}^\alpha \bar{M}^{\gamma\beta} = \mathsf{N}^{\beta\alpha}} \tag{5.154}$$

The symmetry of the tensor N can therefore also be considered as a consequence of the symmetry of the stress tensor.

The symmetry of the tensor M is natural considering (4.116).

5.6 COMPLEMENTS ON SHELL CONSTITUTIVE LAWS

5.6.1 Thin elastic shells in Love theory

5.6.1.1 Materials with symmetry of revolution

This is the case of a linearly elastic material having a symmetry of revolution in which the axis is normal to the middle surface: the physical properties of the material are the same in all directions perpendicular to the axis. This is the case of certain composite materials consisting of successive layers, each layer being homogeneous and isotropic.

By limiting the following to purely elastic terms, the constitutive law of such materials is written:

$$\begin{cases} \bar{\varepsilon}_{\alpha\beta} = \dfrac{1}{E}\left[(1+\nu)\sigma_{\alpha\beta} - \nu\sigma^\gamma{}_\gamma a_{\alpha\beta}\right] - \dfrac{\nu'}{E'}\sigma_{33}\, a_{\alpha\beta} \\ \bar{\varepsilon}_{33} = \dfrac{1}{E'}\left(\sigma_{33} - \nu'\sigma^\gamma{}_\gamma\right) \\ \bar{\varepsilon}_{\alpha 3} = \dfrac{1}{2G'}\sigma_{\alpha 3} \end{cases}$$

which depends on five coefficients:

- two moduli of elasticity E and E';
- two Poisson ratios ν and ν';
- a shear modulus G'.

In the hypothesis of plane stresses, G' does not intervene.

The constitutive law in-plane stresses is obtained while expressing that $\sigma^{33} = 0$:

$$\sigma^{\alpha\beta} = \frac{E}{1-\nu^2}\left((1-\nu)\,\bar{\varepsilon}^{\alpha\beta} + \nu\,\bar{\varepsilon}^{\gamma}_{\gamma}\,a^{\alpha\beta}\right) \tag{5.155}$$

which is analogous to the law of isotropic solids. Indeed, σ^{33} being zero, ν' does not intervene in this expression. But, in this case, E and ν depend on x^3. Only the normal strain is written according to different coefficients:

$$\bar{\varepsilon}_{33} = -\frac{\nu'}{E'}\sigma^{\gamma}{}_{\gamma} \tag{5.156}$$

The integration of the law (5.155) leads to:

$$\begin{aligned}
N^{\alpha\beta} &= \varepsilon^{\alpha\beta}\int_{-e/2}^{e/2}\frac{E}{1+\nu}dx^3 + \varepsilon^{\gamma}{}_{\gamma}\,a^{\alpha\beta}\int_{-e/2}^{e/2}\frac{E\nu}{1-\nu^2}dx^3 \\
&\quad - \kappa^{\alpha\beta}\int_{-e/2}^{e/2}\frac{E}{1+\nu}x^3dx^3 - \kappa^{\gamma}{}_{\gamma}\,a^{\alpha\beta}\int_{-e/2}^{e/2}\frac{E\nu}{1-\nu^2}x^3dx^3 \\
M^{\alpha\beta} &= -\varepsilon^{\alpha\beta}\int_{-e/2}^{e/2}\frac{E}{1+\nu}x^3dx^3 - \varepsilon^{\gamma}{}_{\gamma}\,a^{\alpha\beta}\int_{-e/2}^{e/2}\frac{E\nu}{1-\nu^2}x^3dx^3 \\
&\quad + \kappa^{\alpha\beta}\int_{-e/2}^{e/2}\frac{E}{1+\nu}\left(x^3\right)^2dx^3 + \kappa^{\gamma}{}_{\gamma}\,a^{\alpha\beta}\int_{-e/2}^{e/2}\frac{E\nu}{1-\nu^2}\left(x^3\right)^2dx^3
\end{aligned} \tag{5.157}$$

Noting:

$$\begin{aligned}
K_1 &= \int_{-e/2}^{e/2}\frac{E}{1+\nu}dx^3 & K_2 &= \int_{-e/2}^{e/2}\frac{E\nu}{1-\nu^2}dx^3 \\
H_1 &= -\int_{-e/2}^{e/2}\frac{E}{1+\nu}x^3dx^3 & H_2 &= -\int_{-e/2}^{e/2}\frac{E\nu}{1-\nu^2}x^3dx^3 \\
D_1 &= \int_{-e/2}^{e/2}\frac{E}{1+\nu}\left(x^3\right)^2dx^3 & D_2 &= \int_{-e/2}^{e/2}\frac{E\nu}{1-\nu^2}\left(x^3\right)^2dx^3
\end{aligned} \tag{5.158}$$

the constitutive law of such shells is written:

$$\boxed{\begin{aligned}
N^{\alpha\beta} &= K_1\,\varepsilon^{\alpha\beta} + K_2\,\varepsilon^{\gamma}{}_{\gamma}\,a^{\alpha\beta} + H_1\,\kappa^{\alpha\beta} + H_2\,\kappa^{\gamma}{}_{\gamma}\,a^{\alpha\beta} \\
M^{\alpha\beta} &= H_1\,\varepsilon^{\alpha\beta} + H_2\,\varepsilon^{\gamma}{}_{\gamma}\,a^{\alpha\beta} + D_1\,\kappa^{\alpha\beta} + D_2\,\kappa^{\gamma}{}_{\gamma}\,a^{\alpha\beta}
\end{aligned}} \tag{5.159}$$

The fact that the mechanical properties depend on x^3 leads to a coupling between extension and flexion. It is then possible to choose a middle surface not located mid-thickness, such that there is no coupling (see §5.4.3.2). The law (5.159) is reduced to (5.118) when the mechanical properties are constant in thickness.

5.6.1.2 Orthotropic material

This is a situation a little more general than the previous one. An orthotropic material has three planes of symmetry of its mechanical properties; these three planes are orthogonal. This situation is quite common in composite materials, where some layers are made of woven threads (the weft stiffness is different from the warp stiffness).

Shells with orthogonal stiffener networks of different stiffnesses can also be assimilated to orthotropic shells (this situation is detailed in Chapter 7, in the case of plates). Shells of planes or boats are also often made in this way. In practice, the normal to (Σ) plays a particular role since the shell consists of superimposed "layers" parallel to (Σ), each layer being orthotropic, and the case considered here is where two of the plane of symmetry are normal to (Σ).

In the preceding hypotheses, the constitutive law is written in the basis (G_i), limiting itself to purely elastic terms:

$$\begin{pmatrix} \sigma_{11} \\ \sigma_{22} \\ \sigma_{33} \\ \sigma_{12} \\ \sigma_{13} \\ \sigma_{23} \end{pmatrix} = \begin{pmatrix} \lambda_{1111} & \lambda_{1122} & \lambda_{1133} & 0 & 0 & 0 \\ \lambda_{1122} & \lambda_{2222} & \lambda_{2233} & 0 & 0 & 0 \\ \lambda_{1133} & \lambda_{2233} & \lambda_{3333} & 0 & 0 & 0 \\ 0 & 0 & 0 & \lambda_{1212} & 0 & 0 \\ 0 & 0 & 0 & 0 & \lambda_{1313} & 0 \\ 0 & 0 & 0 & 0 & 0 & \lambda_{2323} \end{pmatrix} \begin{pmatrix} \overline{\varepsilon}_{11} \\ \overline{\varepsilon}_{22} \\ \overline{\varepsilon}_{33} \\ \overline{\varepsilon}_{12} \\ \overline{\varepsilon}_{13} \\ \overline{\varepsilon}_{23} \end{pmatrix} \tag{5.160}$$

where the matrix of elasticity coefficients is symmetrical. The hypothesis of plane stresses leads in particular to the relation (in an orthonormal coordinate system (G_α)):

$$\overline{\varepsilon}_{33} = -\frac{1}{\lambda_{3333}}\left(\lambda_{1133}\,\overline{\varepsilon}_{11} + \lambda_{2233}\,\overline{\varepsilon}_{22}\right)$$

which makes it possible to write the local constitutive law in-plane stresses:

$$\begin{cases} \sigma_{11} = \left(\lambda_{1111} - \dfrac{(\lambda_{1133})^2}{\lambda_{3333}}\right)\overline{\varepsilon}_{11} + \left(\lambda_{1122} - \dfrac{\lambda_{1133}\,\lambda_{2233}}{\lambda_{3333}}\right)\overline{\varepsilon}_{22} \\ \sigma_{22} = \left(\lambda_{2222} - \dfrac{(\lambda_{2233})^2}{\lambda_{3333}}\right)\overline{\varepsilon}_{22} + \left(\lambda_{1122} - \dfrac{\lambda_{1133}\,\lambda_{2233}}{\lambda_{3333}}\right)\overline{\varepsilon}_{11} \\ \sigma_{12} = \lambda_{1212}\,\overline{\varepsilon}_{12} \end{cases} \tag{5.161}$$

or:

$$\begin{cases} \sigma_{11} = \Lambda_1\,\overline{\varepsilon}_{11} + \Lambda'\,\overline{\varepsilon}_{22} \\ \sigma_{22} = \Lambda_2\,\overline{\varepsilon}_{22} + \Lambda'\,\overline{\varepsilon}_{11} \\ \sigma_{12} = \Lambda_{12}\,\overline{\varepsilon}_{12} \end{cases} \tag{5.162}$$

There are therefore four coefficients of elasticity.

The integration of these relations makes it possible to obtain the generalised stresses:

$$\left\{\begin{aligned}
N^{11} &= \varepsilon^{11}\int_{-e/2}^{+e/2}\Lambda_1\,dx^3 + \varepsilon^{22}\int_{-e/2}^{+e/2}\Lambda'\,dx^3 \\
&\qquad -\kappa^{11}\int_{-e/2}^{+e/2}\Lambda_1\,x^3dx^3 - \kappa^{22}\int_{-e/2}^{+e/2}\Lambda'\,x^3dx^3 \\
N^{22} &= \varepsilon^{22}\int_{-e/2}^{+e/2}\Lambda_2\,dx^3 + \varepsilon^{11}\int_{-e/2}^{+e/2}\Lambda'\,dx^3 \\
&\qquad -\kappa^{22}\int_{-e/2}^{+e/2}\Lambda_2\,x^3dx^3 - \kappa^{11}\int_{-e/2}^{+e/2}\Lambda'\,x^3dx^3 \\
N^{12} &= \varepsilon^{12}\int_{-e/2}^{+e/2}\Lambda_{12}\,dx^3 - \kappa^{12}\int_{-e/2}^{+e/2}\Lambda_{12}\,x^3dx^3
\end{aligned}\right. \tag{5.163}$$

$$\left\{\begin{aligned}
M^{11} &= -\varepsilon^{11}\int_{-e/2}^{+e/2}\Lambda_1\,x^3dx^3 - \varepsilon^{22}\int_{-e/2}^{+e/2}\Lambda'\,x^3dx^3 \\
&\qquad +\kappa^{11}\int_{-e/2}^{+e/2}\Lambda_1\left(x^3\right)^2dx^3 + \kappa^{22}\int_{-e/2}^{+e/2}\Lambda'\left(x^3\right)^2dx^3 \\
M^{22} &= -\varepsilon^{22}\int_{-e/2}^{+e/2}\Lambda_2\,x^3dx^3 - \varepsilon^{11}\int_{-e/2}^{+e/2}\Lambda'\,x^3dx^3 \\
&\qquad +\kappa^{22}\int_{-e/2}^{+e/2}\Lambda_2\left(x^3\right)^2dx^3 + \kappa^{11}\int_{-e/2}^{+e/2}\Lambda'\left(x^3\right)^2dx^3 \\
M^{12} &= -\varepsilon^{12}\int_{-e/2}^{+e/2}\Lambda_{12}\,x^3dx^3 + \kappa^{12}\int_{-e/2}^{+e/2}\Lambda_{12}\left(x^3\right)^2dx^3
\end{aligned}\right.$$

either, by introducing the 12 coefficients:

$$K_1 = \int_{-e/2}^{+e/2}\Lambda_1\,dx^3\ ;\quad K_2 = \int_{-e/2}^{+e/2}\Lambda_2\,dx^3\ ;$$
$$K_{12} = \int_{-e/2}^{+e/2}\Lambda_{12}\,dx^3\ ;\quad K' = \int_{-e/2}^{+e/2}\Lambda'\,dx^3$$

$$H_{11} = -\int_{-e/2}^{+e/2}\Lambda_1\,x^3dx^3\ ;\quad H_{22} = -\int_{-e/2}^{+e/2}\Lambda_2\,x^3dx^3\ ;$$
$$H_{12} = -\int_{-e/2}^{+e/2}\Lambda_{12}\,x^3dx^3\ ;\quad H' = -\int_{-e/2}^{+e/2}\Lambda'\,x^3dx^3 \tag{5.164}$$

$$D_{11}=\int_{-e/2}^{+e/2}\Lambda_1\left(x^3\right)^2 dx^3\ ;\quad D_{22}=\int_{-e/2}^{+e/2}\Lambda_2\left(x^3\right)^2 dx^3\ ;$$

$$D_{12}=\int_{-e/2}^{+e/2}\Lambda_{12}\left(x^3\right)^2 dx^3\ ;\quad D'=\int_{-e/2}^{+e/2}\Lambda'\left(x^3\right)^2 dx^3$$

comes the constitutive law of the shell:

$$\begin{cases} N^{11}=K_{11}\,\varepsilon^{11}+K'\,\varepsilon^{22}+H_{11}\,\kappa^{11}+H'\,\kappa^{22} \\ N^{22}=K'\,\varepsilon^{11}+K_{22}\,\varepsilon^{22}+H'\,\kappa^{11}+H_{22}\,\kappa^{22} \\ N^{12}=K_{12}\,\varepsilon^{12}+H_{12}\,\kappa^{12} \end{cases}$$
$$\begin{cases} M^{11}=H_{11}\,\varepsilon^{11}+H'\,\varepsilon^{22}+D_{11}\,\kappa^{11}+D'\,\kappa^{22} \\ M^{22}=H'\,\varepsilon^{11}+H_{22}\,\varepsilon^{22}+D'\,\kappa^{11}+D_{22}\,\kappa^{22} \\ M^{12}=H_{12}\,\varepsilon^{12}+D_{12}\,\kappa^{12} \end{cases} \qquad (5.165)$$

In the very particular case where the material is homogeneous in the direction of the thickness, the following relationships are verified for all the indices:

$$H_{\bullet\bullet}=0$$

$$K_{\bullet\bullet}=\Lambda_{\bullet\bullet}\,e$$

$$D_{\bullet\bullet}=\Lambda_{\bullet\bullet}\,\mathrm{I}$$

and the 12 coefficients determined by (5.164) are reduced to the thickness e, the inertia I and the four coefficients $\Lambda_{\bullet\bullet}$, that is, six coefficients.

In the case of ribbed shells, if the ribs have the same rigidity in both orthogonal directions, the normal to the shell is an axis of revolution. It is then possible to choose the position of the neutral surface so that membrane effects and bending effects are decoupled in the laws (5.130); H_1 is then zero and H_2 can be neglected in the case of ribbed shells. If the rigidities of the ribs are different, it is not possible to define the same neutral surface in both directions and the laws are coupled in at least one direction.

5.6.2 Thick elastic shells

5.6.2.1 General method

The general method has been described in §5.4.2.3. In the case of a thick shell, it is no longer possible to neglect the terms in x^3 in the integration of (5.9) and (5.10).

For example, the isothermal evolution of a homogeneous and isotropic elastic solid, which constitutive law is given by (5.118), then:

$$\bar{N}^{\alpha\beta}=\frac{E}{1-\nu^2}\int_{-e/2}^{e/2}\left[(1-\nu)\bar{\varepsilon}^{\alpha\gamma}+\nu\,\bar{\varepsilon}_\delta^\delta\,g^{\alpha\gamma}\right]m_\gamma{}^\beta\,\hat{m}\,dx^3$$
$$\bar{M}^{\alpha\beta}=-\frac{E}{1-\nu^2}\int_{-e/2}^{e/2}\left[(1-\nu)\bar{\varepsilon}^{\alpha\gamma}+\nu\,\bar{\varepsilon}_\delta^\delta\,g^{\alpha\gamma}\right]m_\gamma{}^\beta\,\hat{m}\,x^3\,dx^3 \qquad (5.166)$$

with:

$$\overline{\varepsilon}^{\alpha\gamma}(\mathcal{P}) = g^{\alpha\lambda}\, g^{\gamma\mu}\, \overline{\varepsilon}_{\lambda\mu}(\mathcal{P}) \tag{5.167}$$

$m_\alpha{}^\beta$ and g_α can be assimilated to their initial values $\mathcal{M}_\alpha{}^\beta$ and G_α in small perturbations.

To be able to develop the calculation, it is necessary to specify the expression of $\overline{\varepsilon}$ as a function of x^3. In the context of Kirchhoff's hypotheses:

$$\overline{\varepsilon}^{\alpha\beta}(\mathcal{P}) = \ell^\alpha{}_\delta\, \ell^\lambda{}_\phi\, \ell^\beta{}_\gamma\, \ell^\mu{}_\nu\, a^{\delta\phi}\, a^{\gamma\nu}\left[\varepsilon_{\lambda\mu} - x^3\kappa_{\lambda\mu} + \left(x^3\right)^2 \tau_{\lambda\mu}\right]$$

where ℓ is the endomorphism inverse of the endomorphism m, which is expressed as a function of x^3.

Note: similar expressions are obtained with Reissner-Mindlin's hypotheses, starting from relations (5.74).

The expressions obtained can then be developed in a series of power of x^3 :

$$\begin{cases} \overline{\varepsilon}^{\alpha\gamma} m_\gamma{}^\beta\, \hat{m} = \sum\limits_{n=0}^{\infty} C_n^{\alpha\beta}\left(x^3\right)^n \\ \overline{\varepsilon}_\delta^\delta\, g^{\alpha\gamma}\, m_\gamma{}^\beta\, \hat{m} = \sum\limits_{n=0}^{\infty} D_n^{\alpha\beta}\left(x^3\right)^n \end{cases} \tag{5.168}$$

For example, relying on the relationship (40):

$$\overline{\varepsilon}^{\alpha\gamma} m_\gamma{}^\beta\, \hat{m} = \left(\varepsilon^{\alpha\beta} - x^3\kappa^{\alpha\beta}\right) + C_1 x^3\left(\varepsilon^{\alpha\beta} - x^3\kappa^{\alpha\beta}\right) + \left(x^3\right)^2 \tau^{\alpha\beta} + \ldots$$

where C_1 is a constant not developed here.

It is clear that the integration introduces a coupling, that is to say, the surfacic deformation influences the bending moment and the curvature influences the membrane force. N and M can then be deduced from $\overline{N}$ and $\overline{M}$ by relations (5.55). Obviously, in this hypothesis, there is no constitutive law dealing with shear force, but a similar path can be followed in the context of Reissner-Mindlin's hypotheses. N and M being thus obtained as functions of ε and κ, thus of the displacement, can be reported in the equilibrium equations to obtain three equilibrium equations relating to the three components of the displacement, with the associated boundary conditions. In particular, the method described above was developed by Vlasov[14] in the main curvature coordinate system (see following paragraphs).

The surfacic density of thermodynamic potential is obtained by integrating $\overline{\phi}$ into the thickness by relation (5.18):

$$\boxed{\rho_0\, \phi = \int_{-e/2}^{e/2} \overline{\rho}_0\, \overline{\phi}\, \hat{m}\, dx^3} \tag{5.169}$$

[14] Vasily Zacharovich Vlasov, Russian engineer (1906–1958).

5.6.2.2 Shell constitutive law in the main curvature coordinate system

In particular, the method described in 5.6.2.1 was developed by Vlasov in the principal curvature coordinate system. In the orthonormal coordinate system associated with the principal curvature coordinate system, the covariant, mixed and contravariant components of the different tensors are equal. Starting from expressions (5.41) of the local deformation tensor, the stresses are written using the constitutive law (5.108):

$$\sigma_{11} = \frac{E}{1-\nu^2}\left[\frac{\left(1+\frac{x^3}{R_1}\right)\varepsilon_{11} - x^3\kappa_{11}}{1-\frac{x^3}{R_1}} + \nu\frac{\left(1+\frac{x^3}{R_2}\right)\varepsilon_{22} - x^3\kappa_{22}}{1-\frac{x^3}{R_2}}\right]$$

$$\sigma_{22} = \frac{E}{1-\nu^2}\left[\frac{\left(1+\frac{x^3}{R_2}\right)\varepsilon_{22} - x^3\kappa_{22}}{1-\frac{x^3}{R_2}} + \nu\frac{\left(1+\frac{x^3}{R_1}\right)\varepsilon_{11} - x^3\kappa_{11}}{1-\frac{x^3}{R_1}}\right] \tag{5.170}$$

$$\sigma_{12} = \frac{E}{1+\nu}\left[\frac{\left(1-\frac{\left(x^3\right)^2}{R_1R_2}\right)\varepsilon_{12} - x^3\left[1-\frac{x^3}{2}\left(\frac{1}{R_1}+\frac{1}{R_2}\right)\right]\kappa_{12}}{\left(1-\frac{x^3}{R_1}\right)\left(1-\frac{x^3}{R_2}\right)}\right]$$

Reporting in (5.11) and (5.12), it becomes:

$$\bar{N}_{11} = \frac{E}{1-\nu^2}\int_{-e/2}^{+e/2}\left\{\frac{1-\frac{x^3}{R_2}}{1-\frac{x^3}{R_1}}\left[\left(1+\frac{x^3}{R_1}\right)\varepsilon_{11} - x^3\kappa_{11}\right] + \nu\left[\left(1+\frac{x^3}{R_2}\right)\varepsilon_{22} - x^3\kappa_{22}\right]\right\}dx^3$$

$$\bar{N}_{22} = \frac{E}{1-\nu^2}\int_{-e/2}^{+e/2}\left\{\frac{1-\frac{x^3}{R_1}}{1-\frac{x^3}{R_2}}\left[\left(1+\frac{x^3}{R_2}\right)\varepsilon_{22} - x^3\kappa_{22}\right] + \nu\left[\left(1+\frac{x^3}{R_1}\right)\varepsilon_{11} - x^3\kappa_{11}\right]\right\}dx^3 \tag{5.171}$$

$$\bar{N}_{12} = \frac{E}{1+\nu}\int_{-e/2}^{+e/2}\left[\frac{\left(1-\frac{\left(x^3\right)^2}{R_1R_2}\right)\varepsilon_{12} - x^3\left[1-\frac{x^3}{2}\left(\frac{1}{R_1}+\frac{1}{R_2}\right)\right]\kappa_{12}}{1-\frac{x^3}{R_1}}\right]dx^3$$

$$\bar{N}_{21} = \frac{E}{1+\nu}\int_{-e/2}^{+e/2}\left[\frac{\left(1-\frac{\left(x^3\right)^2}{R_1R_2}\right)\varepsilon_{12} - x^3\left[1-\frac{x^3}{2}\left(\frac{1}{R_1}+\frac{1}{R_2}\right)\right]\kappa_{12}}{1-\frac{x^3}{R_2}}\right]dx^3$$

$$\bar{M}_{11} = -\frac{E}{1-\nu^2}\int_{-e/2}^{+e/2}\left\{\begin{array}{l}\frac{1-\frac{x^3}{R_2}}{1-\frac{x^3}{R_1}}\left[\left(1+\frac{x^3}{R_1}\right)\varepsilon_{11} - x^3\kappa_{11}\right] \\ \qquad +\nu\left[\left(1+\frac{x^3}{R_2}\right)\varepsilon_{22} - x^3\kappa_{22}\right]\end{array}\right\}x^3dx^3$$

$$\bar{M}_{22} = -\frac{E}{1-\nu^2}\int_{-e/2}^{+e/2}\left\{\begin{array}{l}\frac{1-\frac{x^3}{R_1}}{1-\frac{x^3}{R_2}}\left[\left(1+\frac{x^3}{R_2}\right)\varepsilon_{22} - x^3\kappa_{22}\right] \\ \qquad +\nu\left[\left(1+\frac{x^3}{R_1}\right)\varepsilon_{11} - x^3\kappa_{11}\right]\end{array}\right\}x^3dx^3$$

$$\bar{M}_{12} = -\frac{E}{1+\nu}\int_{-e/2}^{+e/2}\left[\frac{\left(1-\frac{\left(x^3\right)^2}{R_1R_2}\right)\varepsilon_{12} - x^3\left[1-\frac{x^3}{2}\left(\frac{1}{R_1}+\frac{1}{R_2}\right)\right]\kappa_{12}}{1-\frac{x^3}{R_1}}\right]x^3dx^3$$

$$\bar{M}_{21} = -\frac{E}{1+\nu}\int_{-e/2}^{+e/2}\left[\frac{\left(1-\frac{\left(x^3\right)^2}{R_1R_2}\right)\varepsilon_{12} - x^3\left[1-\frac{x^3}{2}\left(\frac{1}{R_1}+\frac{1}{R_2}\right)\right]\kappa_{12}}{1-\frac{x^3}{R_2}}\right]x^3dx^3$$

5.6.2.3 Flügge–Lüre–Byrne method

In this method, $\mathcal{L}_\alpha{}^\beta$ is developed to second order in x^3, for example:

$$\left(1-\frac{x^3}{R_1}\right)^{-1} \approx 1+\frac{x^3}{R_1}+\left(\frac{x^3}{R_1}\right)^2$$

By reporting to expressions (5.171), neglecting the terms of order greater than $(x^3)^2$ under the integrals, then integrating, the shell constitutive laws are written:

$$\bar{N}_{11} = K\left\{\varepsilon_{11} + \nu\varepsilon_{22} - \frac{e^2}{12}\left(\frac{1}{R_1} - \frac{1}{R_2}\right)\left[\kappa_{11} - \frac{2\varepsilon_{11}}{R_1}\right]\right\}$$

$$\bar{N}_{22} = K\left\{\varepsilon_{22} + \nu\varepsilon_{11} - \frac{e^2}{12}\left(\frac{1}{R_2} - \frac{1}{R_1}\right)\left[\kappa_{22} - \frac{2\varepsilon_{22}}{R_2}\right]\right\}$$

$$\bar{N}_{12} = K(1-\nu)\left\{\varepsilon_{12} - \frac{e^2}{12}\left(\frac{1}{R_1} - \frac{1}{R_2}\right)\left[\frac{1}{2}\kappa_{12} - \frac{\varepsilon_{12}}{R_1}\right]\right\}$$

$$\bar{N}_{21} = K(1-\nu)\left\{\varepsilon_{12} - \frac{e^2}{12}\left(\frac{1}{R_2} - \frac{1}{R_1}\right)\left[\frac{1}{2}\kappa_{12} - \frac{\varepsilon_{12}}{R_2}\right]\right\}$$

$$\bar{M}_{11} = D\left[\kappa_{11} + \nu\,\kappa_{22} - \left(\frac{2}{R_1} - \frac{1}{R_2}\right)\varepsilon_{11} - \nu\frac{\varepsilon_{22}}{R_2}\right]$$

$$\bar{M}_{22} = D\left[\kappa_{22} + \nu\,\kappa_{11} - \left(\frac{2}{R_2} - \frac{1}{R_1}\right)\varepsilon_{22} - \nu\frac{\varepsilon_{11}}{R_1}\right]$$

$$\bar{M}_{12} = D(1-\nu)\left[\kappa_{12} - \frac{\varepsilon_{12}}{R_1}\right] \qquad (5.172)$$

$$\bar{M}_{21} = D(1-\nu)\left[\kappa_{12} - \frac{\varepsilon_{12}}{R_2}\right]$$

These expressions can be reported in the equilibrium equations to form Flügge[15]–Lüre[16]–Byrne equilibrium equations. Relations (5.171) and (5.172) show that $\bar{N}$ and $\bar{M}$ are not symmetrical. The last equilibrium relation (5.22) is then verified, which removes the inconsistency introduced by the hypothesis of thin shells. Moreover, the deformations due to a rigid displacement are well zero in this method.

5.6.3 Shells of composite materials

5.6.3.1 Multilayer materials

Composite shells often consist of layers of different materials, each layer being parallel to the middle surface and made of an isotropic material (Figure 5.9). A shell thus formed has a particular orthotropic behaviour, since it benefits from the symmetry of revolution around the normal[17].

Figure 5.9

[15] Wilhelm Flügge, German engineer and mathematician (1904–1990).

[16] Anatoliy Isakovich Lure, Russian engineer and mathematician (1901–1980).

[17] When a layer consists of a fabric, it is not, in general, isotropic because weft and warp do not have the same stiffness. The behaviour is then othotropic without symmetry of revolution. The general case of orthotopic shells is developed in §5.6.1.2, but the more specific case considered here is important in industrial practice.

To simplify the presentation, only the case of plates is considered; in particular, the metric G in the thickness is assimilated to the metric A of the middle plane. The thickness of each layer is considered to be sufficiently thin so that the local stresses σ^{ij} can be considered constant in each layer; if this is not the case, the thick layers can be subdivided into thin layers of the same properties. Only the terms of elasticity are considered.

Each layer i is characterised by its thickness e_i, its Young modulus E_i and its Poisson's ratio ν_i. z_i denotes the ordinate of layer i according to a_3 and n is the number of layers (Figure 5.10). Leaving aside the terms of prestressing, relations (5.113) are written with the local law of plane stresses (108):

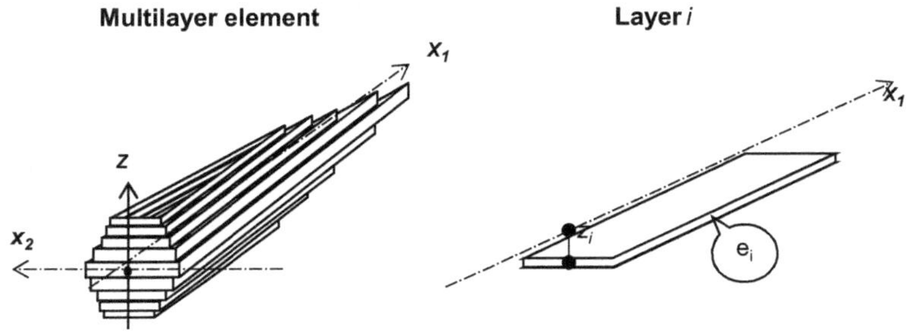

Figure 5.10

$$
\begin{aligned}
N^{\alpha\beta} &= \sum_{i=1}^{n} \sigma_i^{\alpha\beta} e_i = \sum_{i=1}^{n} \frac{E_i e_i}{1-\nu_i^2}\left[(1-\nu_i)\bar{\varepsilon}^{\alpha\beta} + \nu_i\, \bar{\varepsilon}_\gamma^\gamma\, G^{\alpha\beta}\right] \\
&= \sum_{i=1}^{n} \frac{E_i e_i}{1-\nu_i^2}\left[(1-\nu_i)\left(\varepsilon^{\alpha\beta} - z_i \kappa^{\alpha\beta}\right) + \nu_i\left(\varepsilon^{\gamma}{}_{\gamma} - z_i \kappa^{\gamma}{}_{\gamma}\right) G^{\alpha\beta}\right] \\
M^{\alpha\beta} &= -\sum_{i=1}^{n} \sigma_i^{\alpha\beta} e_i z_i = -\sum_{i=1}^{n} \frac{E_i e_i z_i}{1-\nu_i^2}\left[(1-\nu_i)\bar{\varepsilon}^{\alpha\beta} + \nu_i\, \bar{\varepsilon}_\gamma^\gamma\, G^{\alpha\beta}\right] \\
&= -\sum_{i=1}^{n} \frac{E_i e_i}{1-\nu_i^2}\left[(1-\nu_i)\left(z_i \varepsilon^{\alpha\beta} - z_i^2 \kappa^{\alpha\beta}\right) + \nu_i\left(z_i \varepsilon^{\gamma}{}_{\gamma} - z_i^2 \kappa^{\gamma}{}_{\gamma}\right) G^{\alpha\beta}\right]
\end{aligned}
\tag{5.173}
$$

These expressions show the following coefficients of elasticity:

$$
\begin{aligned}
K &= \sum_{i=1}^{n} \frac{E_i e_i}{1-\nu_i^2}; \quad K' = \sum_{i=1}^{n} \frac{\nu_i E_i e_i}{1-\nu_i^2} \\
H &= \sum_{i=1}^{n} \frac{E_i e_i z_i}{1-\nu_i^2}; \quad H' = \sum_{i=1}^{n} \frac{\nu_i E_i e_i z_i}{1-\nu_i^2} \\
D &= \sum_{i=1}^{n} \frac{E_i e_i z_i^2}{1-\nu_i^2}; \quad D' = \sum_{i=1}^{n} \frac{\nu_i E_i e_i z_i^2}{1-\nu_i^2}
\end{aligned}
\tag{5.174}
$$

In an orthonormal system, the constitutive laws thus obtained are written:

$$\begin{aligned} N^{11} &= K\varepsilon^{11} + K'\varepsilon^{22} - H\kappa^{11} - H'\kappa^{22} \\ N^{22} &= K\varepsilon^{22} + K'\varepsilon^{11} - H\kappa^{22} - H'\kappa^{11} \\ N^{12} &= (K - K')\varepsilon^{12} - (H - H')\kappa^{12} \end{aligned} \tag{5.175}$$

$$\begin{aligned} M^{11} &= D\kappa^{11} + D'\kappa^{22} - H\varepsilon^{11} - H'\varepsilon^{22} \\ M^{22} &= D\kappa^{22} + D'\kappa^{11} - H\varepsilon^{22} - H'\varepsilon^{11} \\ M^{12} &= (D - D')\kappa^{12} - (H - H')\varepsilon^{12} \end{aligned} \tag{5.176}$$

The shape of these constitutive laws is identical to that obtained in the general case of rotationally symmetrical shells (§5.6.1.1). If the stack is symmetrical about mid-thickness, H and H' are zero. In asymmetrical cases, it is possible to choose the position of the middle surface such that H is zero; this position corresponds to the barycentre of the membrane rigidities of the layers. In this case, H' is not necessarily zero, but is small enough to be neglected, if the dissymmetry of the Poisson coefficients is not too strong. In conclusion and with these assumptions, the laws are decoupled, the hypothesis is taken in what follows. Thus, in an orthonormal system, the constitutive laws and the inverse laws are written:

$$\begin{aligned} N^{11} &= K\varepsilon^{11} + K'\varepsilon^{22} & M^{11} &= D\kappa^{11} + D'\kappa^{22} \\ N^{22} &= K\varepsilon^{22} + K'\varepsilon^{11} & M^{22} &= D\kappa^{22} + D'\kappa^{11} \\ N^{12} &= (K - K')\varepsilon^{12} & M^{12} &= (D - D')\kappa^{12} \end{aligned} \tag{5.177}$$

$$\begin{aligned} \varepsilon^{11} &= \frac{1}{K^2 - K'^2}\left(KN^{11} - K'N^{22}\right) & \kappa^{11} &= \frac{1}{D^2 - D'^2}\left(DM^{11} - D'M^{22}\right) \\ \varepsilon^{22} &= \frac{1}{K^2 - K'^2}\left(KN^{22} - K'N^{11}\right) & \kappa^{22} &= \frac{1}{D^2 - D'^2}\left(DM^{22} - D'M^{11}\right) \\ \varepsilon^{12} &= \frac{1}{K^2 - K'^2}(K + K')N^{12} & \kappa^{12} &= \frac{1}{D^2 - D'^2}(D + D')M^{12} \end{aligned} \tag{5.178}$$

either, in any coordinate system:

$$\begin{aligned} N^{\alpha\beta} &= (K - K')\varepsilon^{\alpha\beta} + K'\varepsilon^{\gamma}{}_{\gamma}A^{\alpha\beta} \\ M^{\alpha\beta} &= (D - D')\kappa^{\alpha\beta} + D'\kappa^{\gamma}{}_{\gamma}A^{\alpha\beta} \end{aligned} \tag{5.179}$$

$$\begin{aligned} \varepsilon^{\alpha\beta} &= \frac{1}{K^2 - K'^2}\left[(K + K')N^{\alpha\beta} - K'N^{\gamma}{}_{\gamma}A^{\alpha\beta}\right] \\ \kappa^{\alpha\beta} &= \frac{1}{D^2 - D'^2}\left[(D + D')M^{\alpha\beta} - D'M^{\gamma}{}_{\gamma}A^{\alpha\beta}\right] \end{aligned} \tag{5.180}$$

which makes it possible to calculate the stresses in a layer i. Noting that:

$$\varepsilon^{\gamma}{}_{\gamma} = \frac{1}{K^2 - K'^2}(K - K')N^{\gamma}{}_{\gamma}\,; \quad \kappa^{\gamma}{}_{\gamma} = \frac{1}{D^2 - D'^2}(D - D')M^{\gamma}{}_{\gamma}$$

it becomes by (5.108) and (5.180):

$$\begin{aligned}\sigma_i^{\alpha\beta} &= \frac{E_i}{1-\nu_i^2}\left[(1-\nu_i)\left(\varepsilon^{\alpha\beta} - z_i\kappa^{\alpha\beta}\right) + \nu_i\left(\varepsilon^{\gamma}{}_{\gamma} - z_i\kappa^{\gamma}{}_{\gamma}\right)G^{\alpha\beta}\right] \\ &= \frac{E_i}{1-\nu_i^2}\left[\left(\begin{aligned}&\frac{1}{K^2-K'^2}\left[(1-\nu_i)(K+K')N^{\alpha\beta} + (\nu_i K - K')N^{\gamma}{}_{\gamma}A^{\alpha\beta}\right] \\ &-z_i\frac{1}{D^2-D'^2}\left[(1-\nu_i)(D+D')M^{\alpha\beta} + (\nu_i D - D')M^{\gamma}{}_{\gamma}A^{\alpha\beta}\right]\end{aligned}\right)\right]\end{aligned} \tag{5.181}$$

To establish the constitutive law relating to the shear force, only flexion is considered (without membrane forces). The shear stress in a layer is obtained by discretising the local equations:

$$\frac{\partial\sigma^{\alpha 3}}{\partial z} = -\nabla_{\beta}\sigma^{\beta\alpha}$$

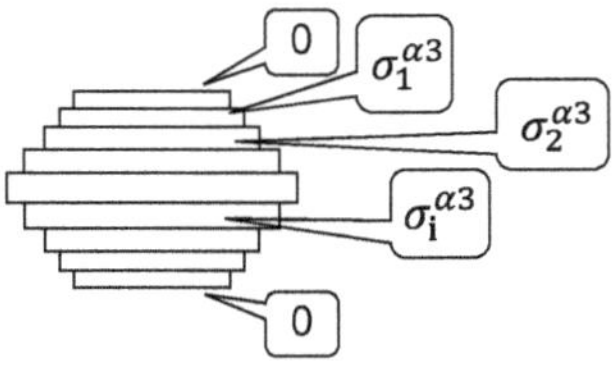

Figure 5.11

Considering that the shear is zero on the upper face (Figure 5.11):

$$\sigma_i^{\alpha 3} = -\sum_{j\geq i}^{n} e_j \nabla_{\beta}\sigma_j^{\beta\alpha} \tag{5.182}$$

The calculation of $\nabla_{\beta}\sigma_j^{\beta\alpha}$:

$$\nabla_{\beta}\sigma_j^{\beta\alpha} = z_i\frac{E_i}{1-\nu_i^2}\frac{1}{D^2-D'^2}\left[(1-\nu_i)(D+D')Q^{\alpha} - (\nu_i D - D')\mathrm{grad}(\mathrm{TR}(M))^{\alpha}\right] \tag{5.183}$$

leads to the expression of the shear stresses knowing M and Q. The moment tensor-dependent term is zero when the Poisson's ratio is constant in the thickness and generally negligible. In this case, the shear stresses are reduced to:

$$\sigma_i^{\alpha 3} = -\frac{Q^{\alpha}}{D-D'}\sum_{j\geq i}^{n}\frac{E_i e_j z_j}{1+\nu_j} \tag{5.184}$$

The shear is also zero on the lower face, that is to say:

$$\sum_{j=1}^{n}\frac{E_i e_j z_j}{1+\nu_j} = 0 \tag{5.185}$$

which is actually verified, H and H' being zero.

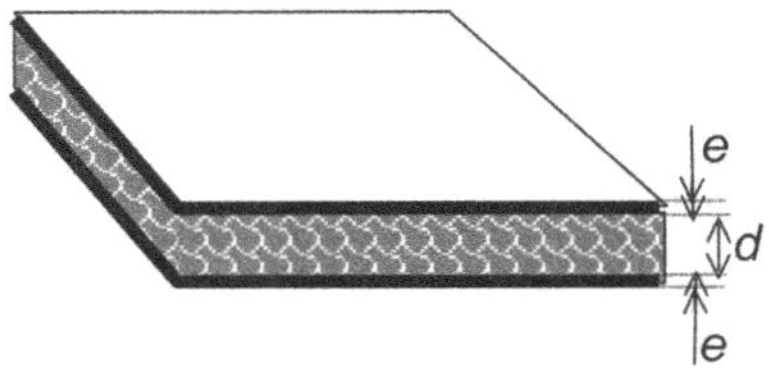

Figure 5.12

5.6.3.2 Sandwich shell

A sandwich shell is formed of two layers of a fairly rigid material (eg metal sheet or plywood wood) thin, called skins, separated by a "spacer" material of low rigidity (eg foam or honeycomb), called heart or web (figure 5.12). The web can not be considered thin and its membrane rigidity is negligible. With regard to normal forces and moments, a sandwich is comparable to a multilayer, with only two layers apart, the two skins of Young's modulus E and Poisson's ratio ν. The corresponding properties are given by (5.174):

$$K = \frac{2Ee}{1-\nu^2}\,;\quad K' = \nu K\,;\quad D = \frac{Eed^2}{2\left(1-\nu^2\right)}\,;\quad D' = \nu D\,;\quad H = H' = 0 \tag{5.186}$$

In skins, (5.181) is then reduced to:

$$\sigma^{\alpha\beta} = \frac{1}{e}\left(\frac{N^{\alpha\beta}}{2} \mp \frac{M^{\alpha\beta}}{d}\right) \tag{5.187}$$

For example, for a single unidirectional bending with axial force $\left(N^{xx}, M^{xx}\right)$ in the x direction, the stress in the tense skin is reduced to:

$$\sigma^{xx} = \left(\frac{N^{xx}}{2e} + \frac{M^{xx}}{ed}\right) \tag{5.188}$$

what the equilibrium allows to calculate easily. Considering only the flexion, the force in one of the skins is equal to $\frac{M^{xx}}{d}$ and the shear stress between the skin and the core is therefore equal to $\sigma^{x3} = \frac{Q^x}{d}$, that is to say, the mean shear stress, given the geometric approximations. As a result, the shear stress varies little in the core and can therefore be considered constant.

The shear stress strain energy is evaluated from the shear stress distribution, which is physically more accurate than if it were evaluated from the Reissner-Mindlin distortion (which would give high stress in the skins): given the very high relative stiffness of the skins and their proximity to the intrados and extrados, most of the energy of shear deformation is in the core.

G_a denoting the shear modulus of the core, the shear force strain energy density is written:

$$\mathcal{F}_c = \frac{1}{2}\int_0^d \frac{\sigma^{\alpha 3}\sigma_{\alpha 3}}{G_a}\,dx^3 = \frac{1}{2}\frac{Q^\alpha Q_\alpha}{G_a d} = \frac{1}{2} G_a d\,\gamma^\alpha \gamma_\alpha \tag{5.189}$$

and must be taken into account when calculating displacements: the KL hypotheses would not give a correct result. As a result, the reduced stiffness is equal to $G_a d$, which justifies the expression of the energy as a function of the distortion, here almost constant in the core.

MAIN RESULTS

Resultant of stresses

$$\bar{N}^{\alpha\beta} = \int_{-e/2}^{e/2} T^{\alpha\gamma} \mathfrak{m}_\gamma{}^\beta \hat{\mathfrak{m}}\, dx^3$$

$$\bar{Q}^\beta = \int_{-e/2}^{e/2} T^{3\beta}\, \hat{\mathfrak{m}}\, dx^3$$

$$\bar{M}^{\alpha\beta} = -\int_{-e/2}^{e/2} T^{\alpha\gamma} \mathfrak{m}_\gamma{}^\beta \hat{\mathfrak{m}}\, x^3\, dx^3$$

DEFORMATIONS

Deformation in Kirchhoff–Love theory

$$\bar{\varepsilon}(\mathcal{P}) = \varepsilon - x^3 \kappa + \left(x^3\right)^2 \tau$$

Deformation in Reissner–Mindlin theory

$$\bar{\varepsilon}_{3\alpha} = \frac{1}{2}\left(d_\alpha + B_\alpha{}^\beta \xi_\beta + \partial_\alpha \zeta\right) \text{ (small displacements)}$$

$$\bar{\varepsilon}_{\alpha\beta} = \frac{1}{2}\left\{\mathcal{M}_\alpha{}^\lambda\left(\nabla_\beta \xi_\lambda - \zeta B_{\beta\lambda} + x^3 \nabla_\beta d_\lambda\right) + \mathcal{M}_\beta{}^\lambda\left(\nabla_\alpha \xi_\lambda - \zeta B_{\alpha\lambda} + x^3 \nabla_\alpha d_\lambda\right)\right\}$$

HOMOGENEOUS AND ISOTROPIC LINEAR ELASTIC THIN SHELL

Shell constitutive law (Kirchhoff–Love)

$$N^{\alpha\beta} = N_0{}^{\alpha\beta} + \alpha \mathrm{e}\, \Theta\, a^{\alpha\beta} + \frac{\mathrm{Ee}}{1-\nu^2}\left[(1-\nu)\, \varepsilon^{\alpha\beta} + \nu\, \varepsilon^\gamma{}_\gamma\, a^{\alpha\beta}\right]$$

$$M^{\alpha\beta} = M_0{}^{\alpha\beta} + \alpha \mathrm{I} \frac{\Delta\Theta}{\mathrm{e}} a^{\alpha\beta} + \frac{\mathrm{Ee}^3}{12\left(1-\nu^2\right)}\left[(1-\nu)\, \kappa^{\alpha\beta} + \nu\, \kappa^\gamma{}_\gamma\, a^{\alpha\beta}\right]$$

Inverse law for elastic terms (Kirchhoff–Love)

$$\varepsilon = \frac{1}{Ee}\left[(1+\nu)\, N - \nu\, N^\alpha{}_\alpha\, a\right]$$

$$\kappa = \frac{12}{Ee^3}\left[(1+\nu)\, M - \nu\, M^\alpha{}_\alpha\, a\right]$$

Elastic potential (Reissner–Mindlin)

$$\mathcal{F} = \frac{K}{2}\int_\Sigma \left[(1-\nu)\varepsilon^{\alpha\beta}\varepsilon_{\alpha\beta} + \nu\left(\varepsilon_\gamma{}^\gamma\right)^2\right] dS + \frac{D}{2}\int_\Sigma \left[(1-\nu)\kappa^{\alpha\beta}\kappa_{\alpha\beta} + \nu\left(\kappa_\gamma{}^\gamma\right)^2\right] dS$$

$$+ \frac{1}{2} kGe \int_\Sigma \gamma^\alpha \gamma_\alpha dS$$

EXERCISES

Exercise 5.1

Check that the equilibrium equation (5.22) is respected with the expressions:

- (5.171) for thick shells,
- (5.172) in the Flügge–Lüre–Byrne approximation.

Answer:
It suffices to transfer the expressions given in (5.22) to show that the equilibrium is respected in both cases, which is not generally the case with the thin shell hypothesis.

Exercise 5.2

The purpose of this exercise is to compare the approximations discussed in this chapter with a three-dimensional solution. To do this, the case of the cylinder subjected to internal pressure is considered, in the case where the bottoms are fixed in the direction of the axis, that is in-plane deformation (see §1.2.2.2). To simplify the calculations, the Poisson's ratio is taken equal to 0.

1. The mean radius of the cylinder is R and its thickness e. Using the three-dimensional solution, express the radial displacement $w = u(R)$, the orthoradial deformation ε_{11} and the variation of curvature κ_{11} of the mean circle of radius R. Calculate the tensors of membrane forces and bending moments as functions of w and develop them to the second order of e/R.
2. Compare i) the orthoradial deformation at any point in the thickness of the cylinder by assimilating it to a thick shell with ii) its three-dimensional expression.
3. Calculate the tensors of membrane forces and bending moments in the hypotheses below:
   - thick shell;
   - Flügge–Lüre–Byrne;
   - membrane theory;

and compare them with the tensors obtained in question 1.

Answer:
1. In-plane strain, the axial strain ε_{22} and the variation of curvature κ_{22} in the direction of the axis are zero.

$$w = \frac{p}{E}\frac{a^2}{b^2 - a^2}\left(R + \frac{b^2}{R}\right) = \frac{\mathcal{P}R}{E}\left[1 + \left(1 + \frac{e}{2R}\right)^2\right] \quad \text{with } \mathcal{P} = p\frac{a^2}{b^2 - a^2}$$

$$\varepsilon_{11} = \frac{w}{R} \; ; \; \kappa_{11} = -\frac{w}{R^2}$$

$$N^{11}=\mathscr{p}\frac{2\,\mathrm{Re}}{\mathscr{a}}\approx Ee\frac{w}{R}\left(1+\frac{1}{8}\frac{e^2}{R^2}\right)\;;\;N^{22}=2\nu E\mathscr{p}=0$$

$$M^{11}=\mathscr{p}\left(-eR\frac{\mathscr{b}}{\mathscr{a}}+\mathscr{b}^2\mathrm{Ln}\frac{\mathscr{b}}{\mathscr{a}}\right)\approx-\frac{Ee^3}{12}\frac{w}{R^2}\left(1-\frac{e}{R}\right)\;;\;M^{22}=\nu\mathscr{p}\frac{e^2}{4}=0$$

2. $\tilde{\varepsilon}_{11}=\dfrac{\mathscr{p}}{E}\left(1+\dfrac{\mathscr{b}^2}{Rr}\right)$ to compare with the three-dimensional solution $\tilde{\varepsilon}_{11}=\dfrac{\mathscr{p}}{E}\left(1+\dfrac{\mathscr{b}^2}{r^2}\right)$, where r is the polar coordinate of the point in the thickness.
3. Expressions (5.171) for a thick shell:

$$N^{11}=ER\frac{w}{R}\mathrm{Ln}\frac{\mathscr{b}}{\mathscr{a}}\approx Ee\frac{w}{R}\left(1+\frac{1}{12}\frac{e^2}{R^2}\right)\;;\;N^{22}=Ee\nu\frac{w}{R}=0$$

$$M^{11}=-E\frac{w}{R}R\left(e-R\,\mathrm{Ln}\frac{\mathscr{b}}{\mathscr{a}}\right)\approx-\frac{Ee^3}{12}\frac{w}{R^2}\left(1+3\frac{e}{R}\right)\;;\;M^{22}=0$$

Expressions (5.172) F–L–B approximation:

$$N^{11}=Ee\frac{w}{R}\left(1+\frac{1}{12}\frac{e^2}{R^2}\right)\;;\;N^{22}=0$$

$$M^{11}=-\frac{Ee^3}{12}\frac{w}{R^2}\;;\;M^{22}=0$$

In both cases, the main terms are the same as for the three-dimensional solution, only the higher order corrections in e/R differ.

In membrane theory:

$$N^{11}=Ee\frac{w}{R}\;;\;N^{22}=Ee\nu\frac{w}{R}=0\;;\;M^{11}=M^{22}=0$$

Exercise 5.3

Establish the Flügge–Lure–Byrne equilibrium equations in the case of a circular cylinder in bending under the effect of a loading $p=p^z a_z+p^3 a_3$. The quantities are expressed in the principal curvature basis (a_z, a_s) where $s = R\theta$. The deformations are calculated from the small displacement $\xi=u(z,s)a_z+v(z,s)a_s+w(z,s)a_3$. Comment on the equations obtained.

Answer:

- First step: expression of the deformations according to the components of the displacement.

$$(3.102) \Rightarrow \quad \varepsilon_{zz} = \frac{\partial u}{\partial z}, \quad \varepsilon_{zs} = \frac{1}{2}\left(\frac{\partial u}{\partial s} + \frac{\partial v}{\partial z}\right), \quad \varepsilon_{ss} = \frac{\partial v}{\partial s} - \frac{w}{R} \qquad (3.108) \Rightarrow \quad \kappa_{zz} = \frac{\partial^2 w}{\partial z^2}, \quad \kappa_{zs} = \frac{\partial^2 w}{\partial z\,\partial s} + \frac{1}{R}\frac{\partial v}{\partial z}, \quad \kappa_{ss} = \frac{\partial^2 w}{\partial s^2} - \frac{w}{R^2} + \frac{2}{R}\frac{\partial v}{\partial s} \qquad \text{(i)}$$

- Second step: expression of the stress resultants by the Flügge–Lure–Byrne relations (5.172):

$$\begin{aligned}
\bar{N}_{zz} &= K\left[\varepsilon_{zz} + \nu\varepsilon_{ss} + \frac{e^2}{12R}\kappa_{zz}\right] & \bar{M}_{zz} &= D\left[\kappa_{zz} + \nu\kappa_{ss} + \frac{1}{R}\left(\varepsilon_{zz} - \nu\varepsilon_{ss}\right)\right]\\
\bar{N}_{zs} &= K(1-\nu)\left[\varepsilon_{zs} + \frac{e^2}{12R}\frac{\kappa_{zs}}{2}\right] & \bar{M}_{zs} &= D(1-\nu)\kappa_{zs}\\
\bar{N}_{sz} &= K(1-\nu)\left[\varepsilon_{zs} - \frac{e^2}{12R}\left(\frac{\kappa_{zs}}{2} - \frac{\varepsilon_{zs}}{R}\right)\right] & \bar{M}_{sz} &= D(1-\nu)\left[\kappa_{zs} - \frac{\varepsilon_{zs}}{R}\right]\\
\bar{N}_{ss} &= K\left[\varepsilon_{ss} + \nu\varepsilon_{zz} - \frac{e^2}{12R}\left(\kappa_{ss} - \frac{2\varepsilon_{ss}}{R}\right)\right] & \bar{M}_{ss} &= D\left[\kappa_{ss} + \nu\kappa_{zz} - 2\frac{\varepsilon_{ss}}{R}\right]
\end{aligned} \qquad \text{(ii)}$$

- Third step: expression of the equilibrium equations (5.19) statically on the cylinder, with $p^s = 0$. The shear force is expressed as a function of the moments and removed from the equations:

$$\begin{aligned}
&\partial_z\bar{N}_{zz} + \partial_s\bar{N}_{sz} + p^z = 0\\
&\partial_z\left(\bar{N}_{zs} + \frac{\bar{M}_{zs}}{R}\right) + \partial_s\left(\bar{N}_{ss} + \frac{\bar{M}_{ss}}{R}\right) = 0\\
&\partial^2_{z^2}\bar{M}_{zz} + \partial^2_{zs}\left(\bar{M}_{zs} + \bar{M}_{sz}\right) + \partial^2_{s^2}\bar{M}_{ss} - \frac{\bar{N}_{ss}}{R} - p^3 = 0
\end{aligned} \qquad \text{(iii)}$$

The sought equations are obtained by transfering (i) in (ii), then the result in (iii). They relate to the two components of displacement and their derivatives.

$$\Rightarrow \quad \lambda R\partial^3_{z^3}w-\frac{1}{2}(1-\nu)\lambda R\partial^3_{s^2z}w+\partial^2_{z^2}u+\frac{1}{2}(1-\nu)(1+\lambda)\partial^2_{s^2}u+\frac{1}{2}(1+\nu)\partial^2_{zs}v-\nu\frac{\partial_z w}{R}+\frac{p^z}{K}=0$$

$$\Rightarrow \quad \frac{1}{2}(1-\nu)\left(\partial^2_{zs}u+\partial^2_{z^2}v\right)+(1-\nu)\lambda R\frac{3}{2}\partial^3_{z^2s}w+(1-\nu)\lambda\frac{3}{2}\partial^2_{z^2}v+\partial^2_{s^2}v-\frac{\partial_s w}{R}+\nu\,\partial^2_{zs}u+\lambda R\nu\partial^3_{z^2s}w=0$$

$$\begin{aligned}\Rightarrow \quad &\partial^4_{z^4}w+\nu\,\partial^4_{z^2s^2}w-\nu\frac{\partial^2_{z^2}w}{R^2}+\nu\frac{2}{R}\partial^3_{z^2s}v+\frac{1}{R}\partial^3_{z^3}u-\nu\frac{1}{R}\partial^3_{z^2s}v+\nu\frac{1}{R}\frac{\partial^2_{z^2}w}{R}\\
&+2(1-\nu)\partial^4_{z^2s^2}w+2(1-\nu)\frac{1}{R}\partial^3_{z^2s}v-\frac{1}{2R}(1-\nu)\left(\partial^3_{s^2z}u+\partial^3_{z^2s}v\right)\\
&+\partial^4_{s^4}w-\frac{\partial^2_{s^2}w}{R^2}+\frac{2}{R}\partial^3_{s^3}v+\nu\,\partial^4_{z^2s^2}w-\frac{2}{R}\partial^3_{s^3}v+\frac{2}{R}\frac{\partial^2_{s^2}w}{R}-\frac{1}{\lambda R^3}\partial_s v+\frac{1}{\lambda R^3}\frac{w}{R}-\frac{1}{\lambda R^3}\nu\,\partial_z u\\
&+\frac{1}{R^2}\partial^2_{s^2}w-\frac{1}{R^2}\frac{w}{R^2}+\frac{1}{R^2}\frac{2}{R}\partial_s v-\frac{2}{R^3}\partial_s v+\frac{2}{R^3}\frac{w}{R}-\frac{p^3}{D}=0\end{aligned}$$

with $\lambda=\dfrac{e^2}{12R^2}$.

Despite the precision provided by the kinematic assumption, the differential equations remain linear.

Chapter 6

Equilibrium of membrane shells

The methods of membrane theory are discussed in more detail in this chapter, where application examples are also proposed.

The equations and laws to be implemented in the framework of the membrane theory are recalled in §1, which delimits the scope of this theory. Some general methods of resolution are then presented.

Classical analytical solutions are given for cylindrical shells (§ 2), for shells of revolution (§ 3), for helical shells (§ 5) and for shells of any shape that can be defined by an explicit function (§ 4). In addition to the formal interest presented by these solutions, they make it possible to highlight the limits of the membrane theory and to specify to what extent they can constitute an acceptable approximation. They highlight in particular the role of curvature in the distribution of stresses.

6.1 THE CLASSICAL MEMBRANE THEORY

6.1.1 Assumptions and evolution equations

6.1.1.1 What is a membrane theory?

Some shells have a flexural stiffness low enough that the bending energy can be neglected in front of the membrane (stretching) energy. The **membrane theory** is a simplification of the shell theory where the bending moments are neglected and where the equilibrium of the shell is ensured only by the membrane (stretching) forces.

The tent is a good example of such a situation. It also shows that if the flexural rigidity is very low, the shell is not stable when compressed. That's why the camper must tense his tent in all directions to keep its shape under the wind, without scalloping.

But this example gives an image that is too restrictive. In fact, the membrane theory is also used for shells with a certain flexural rigidity and consequently able to work also in compression; it is then necessary to see under which conditions the bending moments can be effectively neglected. This chapter shows that the solutions obtained by the membrane theory are of interest in many cases and they allow us to specify its limits.

DOI: 10.1201/9780429440403-6

6.1.1.2 Equilibrium equations

Thus, the membrane theory is a simplified shell theory, where it is admitted that the tensor of bending moments, and consequently the shear forces, are null:

$$M = 0$$
$$Q = -\operatorname{div} M = 0 \qquad (6.1)$$

By the relations (4.129), it is clear that N and N are coincident and symmetrical: it is not necessary to distinguish between generalised stresses and resultant stresses.

Under these conditions, the local static equilibrium equations of the shell are written:

$$\boxed{\begin{aligned}\nabla_\alpha N^{\alpha\beta} + p^\beta &= 0\\ b_{\alpha\beta} N^{\alpha\beta} + p^3 &= 0\end{aligned}} \qquad (6.2)$$

where the first line corresponds to the first two equilibrium equations (tangential equilibrium); they are partial differential equations that depend on the first fundamental form a, while the third equation (normal equilibrium) has no derivative and is expressed as a function of the second fundamental form b.

The first two equations correspond to an equilibrium in the tangent plane (2D space), whereas the last equation corresponds to the transverse equilibrium and depends on the shape of the shell, and therefore its definition in 3D space

The associated boundary conditions are written:

$$\boxed{N^{\alpha\beta} \nu_\beta - q^\alpha = 0} \qquad (6.3)$$

with the following notations already introduced:

- N denotes the tensor of membrane forces;
- ∇ denotes the covariant derivation on the middle surface;
- b is the curvature tensor of the middle surface;
- p^β (resp., p^3) is the tangent (resp. normal) components of the force density applied to the middle surface of the membrane;
- ν is the normal vector at the edge inscribed in the plane tangent to the middle surface (Figure 6.1);
- q^β is the components in the tangent plane of the reactions at the edge (the normal reaction is necessarily zero in membrane theory).

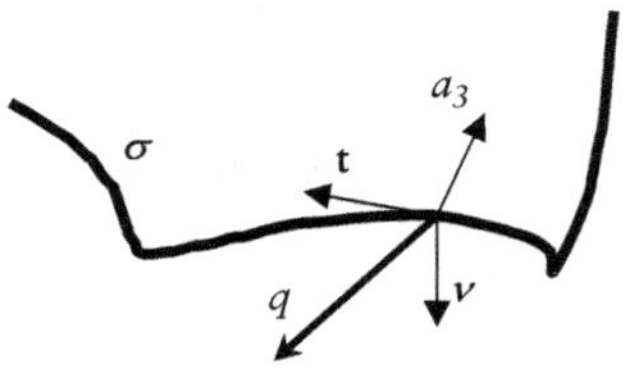

Figure 6.1

Remark: The equilibrium equations can be directly established within the framework of the membrane theory, following the method of §4.2.1. The main steps of this method are recalled below:

Any domain d is inscribed in the middle surface of the shell.

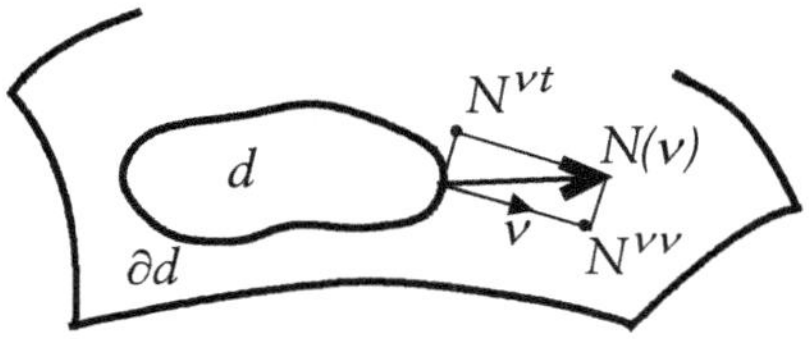

Figure 6.2

Along the edge ∂d of d, the reactions are represented by a lineic density of forces N inscribed in the plane tangent to the surface (Figure 6.2).

N depends linearly on the vector ν normal to the surface at the edge:

$$N = N^{\alpha\beta}\nu_\beta a_\alpha$$

The equilibrium of d is written as:

- For the resultant force:

$$\int_{\partial d} N^{\alpha\beta}\nu_\beta a_\alpha \, ds + \int_d p^i a_i \, d\sigma = 0$$

that is to say:

$$\int_d \left[D_\beta \left(N^{\alpha\beta} a_\alpha \right) + p^i a_i \right] d\sigma = 0$$

The term under the integral is null, the domain d being any, which, after development, leads to the shell local equations (6.2).

- For the resultant moment:

$$\int_{\partial d} OP \wedge \left(N^{\alpha\beta}\nu_\beta a_\alpha \right) ds + \int_d OP \wedge \left(p^i a_i \right) d\sigma = 0$$

that is to say:

$$\int_d \left[D_\beta \left(OP \wedge N^{\alpha\beta} a_\alpha \right) + OP \wedge \left(p^i a_i \right) \right] d\sigma = 0$$

Developing and noting that $D_\beta OP = a_\beta$, then using the equation of the resultant force, it becomes:

$$N^{\alpha\beta} a_\beta \wedge a_\alpha = 0$$

which demonstrates that N is symmetrical.

When **the geometry changes are neglected**, b, ∇ and ν are expressed on the reference configuration and are therefore known, and in this case only the membrane forces are unknown. In static, the Equation (6.2) and the boundary conditions (6.3) are then sufficient to determine the tensor of the membrane forces N: **the shell is statically determined.** In this case, it is not necessary to have a constitutive law to determine N. The determination of local (3D) stresses simply requires a distribution hypothesis in the thickness.

The local equations (6.2) can be supplemented by the terms of inertia $-\rho\gamma$ to study the movements, but it is then necessary to introduce a constitutive law to solve them, the displacement field intervening in the equations.

The equilibrium equations recalled above are formulated in Euler variables (on the deformed configuration, which is unknown) and do not require the introduction of a reference state. When the displacements can not be considered small enough to neglect the geometry changes, it is necessary to take into account the constitutive law.

Nevertheless, taking into account large displacements faces some difficulties: let, for example, a balloon in a state of spherical equilibrium under internal pressure. So:

$$N = \frac{pR_0}{2} a$$

R_0, the radius of the sphere, and a, the metric tensor, depend on the deformed geometry of the sphere. But R_0, for example, depends on the material of the balloon and is not known *a priori*. It is therefore necessary to define a reference state of the balloon (which one, by the way?), To study the (large) displacement at equilibrium and carry out a complete analysis taking into account the constitutive law.

In the case where kinematic conditions are imposed at the boundaries, it is necessary to involve the displacement, therefore the constitutive laws. In practice, the membrane solution does not allow most often meet all conditions of kinematic continuity, including those at the boundaries. Thus, the classical solutions studied in this chapter assume a rigid shell and relate only to the membrane forces, including at the boundaries.

Note: The system of Equation (6.2) is a first-order system with two unknowns, once eliminated N^{22}. Therefore, there can be only two independent boundary conditions.

6.1.1.3 Membrane forces / local stress relationships

The following assumptions are adopted throughout the chapter:

- Kirchhoff–Love assumptions;
- hypothesis of thin shells with moderate curvature.

Let $\overline{\overline{T}}$ be the tensor of local stresses; the assumptions lead to:

$$\boxed{N^{\alpha\beta} = e\, T^{\alpha\beta}} \tag{6.4}$$

where e is the thickness of the shell. T^{i3} components are zero (or negligible).

6.1.1.4 Deformations

In membrane theory, when it is necessary to calculate the deformations, for example, to obtain the displacements, the strain variable to be considered is the tensor of surface deformation of the middle surface, of components $\varepsilon_{\alpha\beta}$.

During a change of equilibrium, the curvature tensor may vary, which introduces, by the constitutive law, bending moments. The hypotheses of the membrane theory are tantamount to neglecting the energy associated with flexion with respect to extensional energy. The curvature variation tensor is therefore not taken into consideration, even if it is not exactly zero (see §6.1.2.1).

For structures with a certain bending stiffness, the bending energy may no longer be negligible for "non-small" displacements (orders of magnitude are specified in Chapter 8).

It thus appears that the membrane theory is of practical interest for these structures only in the **hypothesis of small displacements**, which is adopted below.

In small displacements, the geometric entities involved in the equilibrium equations (∇, b, ν) are expressed in the known reference position, N remaining then the only unknown. The strain tensor, constant in the thickness of the shell according to the KL assumptions, is expressed in the linearised form (formula 3.108):

$$\boxed{\varepsilon_{\alpha\beta} = \frac{1}{2}\left(\nabla_\alpha \xi_\beta + \nabla_\beta \xi_\alpha\right) - \zeta B_{\alpha\beta}} \tag{6.5}$$

$\bar{\varepsilon}$ being independent of x^3, it is not necessary in this chapter to refer to x^3: only intervenes ε, deformation of the middle surface.

When the deformations can be determined by a constitutive law from the membrane forces, the relations (6.5) make it possible to calculate the displacement fields. Since the total number of unknowns (membrane forces and displacement) is equal to the number of equations, there is no need to introduce compatibility conditions, which are automatically verified.

6.1.1.5 Homogeneous and isotropic linear elastic behaviour

Assuming the homogeneous thermoelastic and isotropic linear material and in the hypothesis of plane stresses, according to (5.118), the membrane constitutive law is written:

$$\boxed{N = N_0 + \alpha e \Theta\, a + \frac{Ee}{1-\nu^2}\left[\nu\, \varepsilon^\alpha{}_\alpha\, a + (1-\nu)\,\varepsilon\right]} \tag{6.6}$$

The difference in temperature with respect to the reference state only intervenes by its average value in the thickness of the membrane.

By restricting itself to the purely elastic part, the inverse law is written:

$$\boxed{\varepsilon = \frac{1}{Ee}\left[(1+\nu)\,N - \nu\, N^\alpha{}_\alpha\, a\right]} \tag{6.7}$$

This inverse law is used most often because it makes it possible to go back to the displacement after having determined the membrane forces. The associated elastic potential is reduced to membrane energy:

$$\boxed{\mathcal{F} = \frac{1}{2}\frac{Ee}{1-\nu^2}\int_\Sigma \left[\nu\left(\varepsilon^\alpha{}_\alpha\right)^2 + (1-\nu)\,\varepsilon_{\alpha\beta}\varepsilon^{\alpha\beta}\right] dS} \tag{6.8}$$

Structures with negligible flexural stiffness (canvases and similar, cable networks…) have negligible bending energy in all configurations and are constantly in membrane equilibrium. These membranes are generally under tension, so as not to undergo compression in service; indeed, they would be quickly unstable because of their lack of flexural rigidity. They undergo great displacements during their installation involving pretensions, then small displacements in service. Their rigidity in service depends largely on the pretensions and not only their extensional rigidity.

6.1.1.6 General linearly elastic behaviour

By reporting the hypothesis $M^{\alpha\beta} = 0$ in the constitutive laws of elasticity (5.113), it appears that:

$$-\varepsilon_{\gamma\delta} \int_{-e/2}^{e/2} \overline{L}^{\alpha\beta\gamma\delta}\ x^3\ dx^3 = 0$$

$$\kappa_{\gamma\delta} \int_{-e/2}^{e/2} \overline{L}^{\alpha\beta\gamma\delta} \left(x^3\right)^2\ dx^3 = 0$$

Deformations $\varepsilon_{\gamma\delta}$ can not be zero in general, because of the existence of membrane forces, and:

$$\int_{-e/2}^{e/2} \overline{L}^{\alpha\beta\gamma\delta}\ x^3\ dx^3 = 0$$

The material must therefore be symmetrical in thickness with respect to the middle surface.

The second equality can be achieved in two different ways:

- either the bending stiffness is zero or negligible:

$$\int_{-e/2}^{e/2} \overline{L}^{\alpha\beta\gamma\delta} \left(x^3\right)^2\ dx^3 \approx 0$$

This condition is well verified if the shell is very thin, which results in the condition $\dfrac{e}{|R|} << 1$ (see §6.1.2.2).

- or the variation of curvature is zero or negligible: $\kappa_{\alpha\beta} \approx 0$, which is only respected if conditions relating to external actions and boundary conditions are met (§6.1.2.3).

In both cases, the shell constitutive law is reduced to:

$$N^{\alpha\beta} = \varepsilon_{\gamma\delta} \int_{-e/2}^{e/2} \overline{L}^{\alpha\beta\gamma\delta}\ dx^3$$

and the deformations are uniform in thickness. The membrane state is intrinsic: it does not depend on shape changes.

6.1.2 Validity of the membrane theory

6.1.2.1 Preponderance of membrane energy

Example: Let a sphere of radius R_0 (> 0) be subjected to an internal pressure p (Figure 6.3). By symmetry, the sphere undergoes a normal displacement w with respect to the reference position (unloaded sphere). Normal is oriented outward.

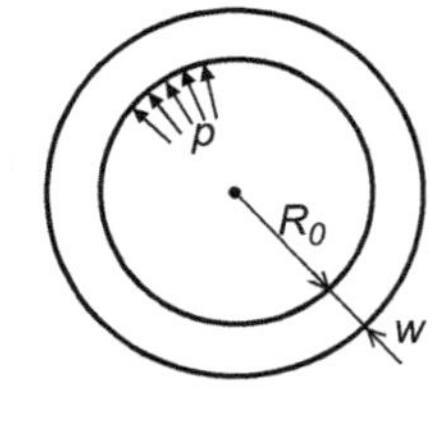

Figure 6.3

Assuming small displacements and in the orthonormal coordinate system associated with spherical coordinates (θ, ϕ):

$$\begin{cases} \tilde{\varepsilon}_{\theta\theta} = \tilde{\varepsilon}_{\varphi\varphi} \approx \dfrac{w}{R_0} \\ \tilde{\kappa}_{\theta\theta} = \tilde{\kappa}_{\varphi\varphi} = -\dfrac{w}{R_0{}^2} \end{cases}$$

Taking into account the bending energy, the strain energy is written as:

- membrane energy:

$$\mathcal{T}_m = \frac{1}{2}\frac{Ee}{1-\nu^2}\int_\Sigma \left[\nu\left(\varepsilon^\alpha{}_\alpha\right)^2 + (1-\nu)\,\varepsilon_{\alpha\beta}\varepsilon^{\alpha\beta}\right]dS = \frac{1}{2}\frac{Ee}{1-\nu^2}2S\frac{w^2}{R_0{}^2}(1+\nu)$$

- bending energy:

$$\mathcal{T}_f = \frac{1}{2}\frac{Ee^3}{12\left(1-\nu^2\right)}\int_\Sigma \left[\nu\left(\kappa^\alpha{}_\alpha\right)^2 + (1-\nu)\,\kappa_{\alpha\beta}\,\kappa^{\alpha\beta}\right]dS = \frac{1}{2}\frac{Ee^3}{12\left(1-\nu^2\right)}2S\frac{w^2}{R_0{}^4}(1+\nu)$$

that is to say:

$$\mathcal{T} = \mathcal{T}_m + \mathcal{T}_f == \frac{1}{2}\frac{Ee}{1-\nu}2S\frac{w^2}{R_0{}^2}\left(1+\frac{e^2}{12R_0{}^2}\right) = \frac{1}{2}Cw^2$$

The Principle of Virtual Powers is written as:

$$Cw\overset{*}{w} = pS\overset{*}{w}$$

hence the solution:

$$w = \frac{pS}{C} = \frac{1-\nu}{Ee\left(1+\dfrac{e^2}{12R_0{}^2}\right)}\frac{pR_0{}^2}{2}$$

The ratio of the two types of energies is:

$$\frac{\mathcal{T}_f}{\mathcal{T}_m} = \frac{e^2}{12R_0{}^2}$$

which shows that the bending energy is negligible compared to the membrane energy in the case of a thin shell and justifies *a posteriori* the use of the membrane theory.

The order of the approximation thus consented to the membrane forces in this approach is specified by the relations:

$$\tilde{N}_{\theta\theta} = \tilde{N}_{\phi\phi} = \frac{pR_0}{2\left(1+\dfrac{e^2}{12R_0{}^2}\right)} \approx \frac{pR_0}{2}\left(1-\frac{e^2}{12R_0{}^2}\right)$$

Note: Membrane forces only assure the equilibrium (without bending moments) in Euler variables. The values obtained above are in Lagrange variables, as it is necessary to specify a change of equilibrium to calculate the variation of strain energy.

6.1.2.2 Connection of two shells with different curvatures

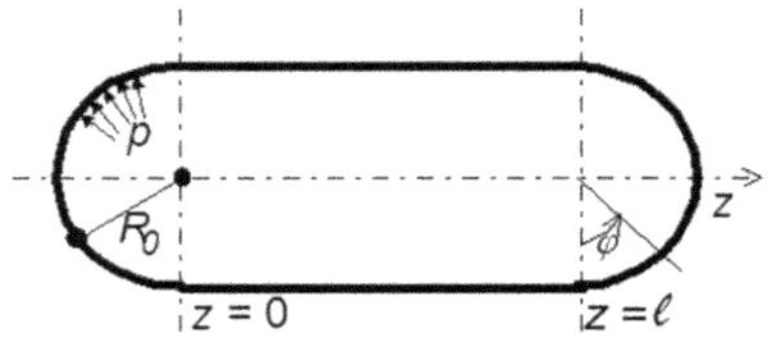

Figure 6.4

Example: Let a tank consisting of a cylindrical shell closed at its ends by two hemispherical bottoms. R_0 (> 0) is the radius, common to the sphere and the cylinder. This tank contains a gas at pressure p.

θ designates the angle of revolution around the axis of the reservoir, z the ordinate along the generatrices of the cylinder and φ the latitude on the hemispheres (Figure 6.4).

Using the orthonormal coordinate systems associated with the cylindrical and spherical coordinates, the membrane equilibrium state is written (§1.1.2.):

- In the cylinder:

$$\tilde{N}^{\theta\theta} = pR_0 \ ; \ \tilde{N}^{zz} = \frac{pR_0}{2} \ ; \ \tilde{N}^{z\theta} = 0$$

- In a hemisphere:

$$\tilde{N}^{\theta\theta} = \tilde{N}^{\varphi\varphi} = \frac{pR_0}{2} \ ; \ \tilde{N}^{\varphi\theta} = 0.$$

This membrane state is determined for the complete tank, the continuity being ensured at the interface between cylinder and hemisphere:

$$\tilde{N}^{zz} = \tilde{N}^{\varphi\varphi} \ ; \ \tilde{N}^{z\theta} = \tilde{N}^{\varphi\theta}$$

Nevertheless, it is not enough that the stress field is statically admissible[1], it is necessary to ensure the compatibility of displacements (**kinematically admissible** displacement field).

The material is linearly elastic and verifies (6.7). In the cylinder, the displacement has no more than two components: w normal and v on z:

$$\xi = w(z)\, e_r + v(z)\, e_z$$

The transformation is expressed by:

$$\mathcal{M} = R_0\, e_r + z e_z \ \rightarrow \ \mathcal{m} = (R_0 + w)\, e_r + (z + v)\, e_z$$

1 "Statically admissible" means that the stress field satisfies all conditions of equilibrium (local and boundary conditions).

hence the strain tensor:

$$\varepsilon = \frac{1}{2}\left(d\mathcal{m}^2 - d\mathcal{M}^2\right) = \frac{1}{2}\left[\left(w'^2 + (1+v')^2\right)dz^2 + (R_0+w)^2\, d\theta^2 - dz^2 - R_0^{\,2}d\theta^2\right]$$

and, in the associated physical coordinate system, in small displacements:

$$\tilde{\varepsilon}_{zz} = v' \ ; \ \tilde{\varepsilon}_{\theta\theta} = \frac{w}{R_0} \ ; \ \tilde{\varepsilon}_{\theta z} = 0$$

On the other hand, by (6.7):

$$\tilde{\varepsilon}_{zz} = \frac{(1-2\nu)\,pR_0}{2Ee} \ ; \ \tilde{\varepsilon}_{\theta\theta} = \frac{(2-\nu)\,pR_0}{2Ee}$$

from where the displacements:

$$w = \frac{(2-\nu)\,pR_0^{\,2}}{2Ee} \ ; \ v = \frac{(1-2\nu)\,pR_0 z}{2Ee} + C$$

By fixing $v = 0$ in $z = 0$, it comes, if the cylinder has a length ℓ:

$$v(\ell) = \frac{(1-2\nu)\,pR_0\ell}{2Ee}$$

This is the lengthening of the cylinder due to the bottom effect.

In the sphere, the displacement has a normal component w and a component v following the meridian:

$$\xi = w(\varphi)\, e_r + v(\varphi)\, \tilde{e}_\varphi$$

where $\tilde{e}_\varphi$ denotes the unit vector tangent to the meridian.

Here, the rigid displacement $v(\ell)$ at the abscissa $z = \ell$ has been eliminated for the hemisphere in $\varphi = 0$.

Transformation can be expressed by:

$$\mathcal{M} = R_0 e_r \ \rightarrow \ \mathcal{m} = (R_0 + w)\, e_r + \frac{v}{R_0} e_\varphi$$

where e_φ denotes the vector of the natural basis associated with the latitude, in spherical coordinates. By differentiation of $\mathcal{m}$:

$$d\mathcal{m} = w' e_r\, d\varphi + \left(\frac{v'}{R_0} + \frac{R_0 + w}{R_0}\right) e_\varphi d\varphi + \left(\frac{R_0 + w}{R_0} - \frac{v}{R_0}\tan\varphi\right) e_\theta d\theta$$

hence the deformation tensor:

$$\varepsilon = \frac{1}{2}\left[\begin{array}{l} w'^2 d\varphi^2 + \left(\dfrac{v'}{R_0} + \dfrac{R_0+w}{R_0}\right)^2 R_0^{\,2} d\varphi^2 \\ \qquad + \left(\dfrac{R_0+w}{R_0} - \dfrac{v}{R_0}\mathrm{tg}\varphi\right)^2 R_0^{\,2}\cos^2\varphi\, d\theta^2 - R_0^{\,2} d\varphi^2 - R_0^{\,2}\cos^2\varphi\, d\theta^2 \end{array}\right]$$

or, in the associated orthonormal coordinate system, in small displacements:

$$\tilde{\varepsilon}_{\varphi\varphi} = \frac{v' + w}{R_0} \; ; \; \tilde{\varepsilon}_{\theta\theta} = \frac{w}{R_0} - \frac{v}{R_0}\tan\varphi$$

On the other hand, using (6.7):

$$\tilde{\varepsilon}_{\varphi\varphi} = \tilde{\varepsilon}_{\theta\theta} = \frac{(1-\nu)\,pR_0}{2Ee}$$

From which:

$$v' + v\tan\varphi = 0$$

The solution of the equation in v is: $v = A\cos\varphi$. At $\varphi = 0$ (interface with the cylinder), v is continuous, therefore zero considering the convention in $z = \ell$; so $A = 0$, hence:

$$w = \frac{(1-\nu)\,pR_0^{\,2}}{2Ee}$$

In conclusion, the longitudinal displacements v are due to the extension of the cylinder alone under the effect of pressure on the bottoms $\frac{pR_0}{2}$. The continuity of this displacement is ensured at the passage between the cylinder and hemisphere.

On the other hand, the normal displacement w is discontinuous at the interface between the two geometrical shapes: the state of membrane equilibrium is thus physically not possible if the thickness of the shell is constant; there is then necessarily a flexure at the connection between the cylinder and hemisphere due to the discontinuity of the curvatures in the longitudinal direction, discontinuity appearing in membrane theory. Nevertheless, there is a solution to ensure the continuity of w at the right of the interface: it consists in having different thicknesses between the sphere: e_s and the cylinder: e_c. These thicknesses would then be in the ratio:

$$\frac{e_c}{e_s} = \frac{2-\nu}{1-\nu}$$

For a metal tank, this ratio is equal to 2.4, a little higher than the ratio of the maximum stresses in the two parts of the tank; this gives an almost homogeneous resistance.

6.1.2.3 Conclusions

For the membrane equilibrium state to be the exact solution for the equilibrium of the shell, it is necessary that:

- on the one hand, the external forces applied to the shell are compatible with equilibrium Equation (6.2) and boundary conditions (6.3), which are limited in number given the order of the equations: no moment applied to the surface, no non-tangent reaction to the surface along the edge.
- on the other hand, the compatibility of the deformations must be ensured between the different parts of the shell and with the outside. Incompatibilities may be due to:

kinematic boundary conditions (clamping conditions, for example): the membrane equilibrium is only compatible with some of the conditions relating to displacements, at the limits; it is not possible to impose displacements normal to the surface, ie conditions of the type $\zeta = 0$ or $\frac{\partial \zeta}{\partial v} = 0$.

- discontinuities in the normal component of applied forces or concentrated forces;
- discontinuities of thickness;
- discontinuities of the curvature tensor (example of §6.1.2.3).

External forces concentrated at points or linearly can be balanced in membrane theory only if there is a discontinuity of the curvature (Figure 6.5).

The model of the membrane theory gives an acceptable solution only if the variations of thickness, curvature and forces applied on the shell are weak.

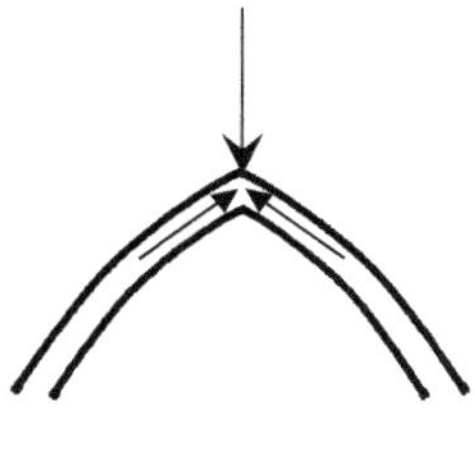

Figure 6.5

Membrane forces are entirely determined, the membrane theory imposes rather strong restrictions on the displacement fields and thus leaves little freedom for the expression of kinematic liaisons.

Imposing non-compatible kinematic constraints causes bending in the shell.

These rather restrictive conditions might suggest that membrane theory is of little practical interest. This is not so because the membrane equilibrium state remains a good approximation in many situations, "far enough" from the singularities inducing bending in the shell (the latter aspect is addressed in Chapter 8). It determines the general distribution of membrane forces, highlighting in particular the role of the curvature.

This theory is also of historical interest in that it has long made it possible to dimension the shells without the need to use the calculation tools necessary for the study of flexion. Even today, it keeps its place as a tool for overall design and pre-dimensioning. Moreover, the membrane theory can provide in some cases a particular solution of equilibrium equations, useful in the search for a complete solution (see Chapter 8).

6.1.3 General methods of resolution

6.1.3.1 Direct methods

The very nature of membrane equilibrium equations leads to an attempt to solve them directly to obtain the distribution of membrane forces. The displacement method is not generally used because it requires the implementation of a constitutive law useless for the determination of stresses. The displacements are obtained *a posteriori* (cf examples of §6.1.2).

It is interesting to express equilibrium equations in an orthonormal coordinate system associated with orthogonal coordinate lines ($\mathcal{C}_1$) and ($\mathcal{C}_2$).

Let s_1 (resp. s_2) be the curvilinear abscissa, $\mathbf{t}_1$ (resp., $\mathbf{t}_2$) the vector tangent to ($\mathcal{C}_1$) [resp. ($\mathcal{C}_2$)] in m, such as (Figure 6.6):

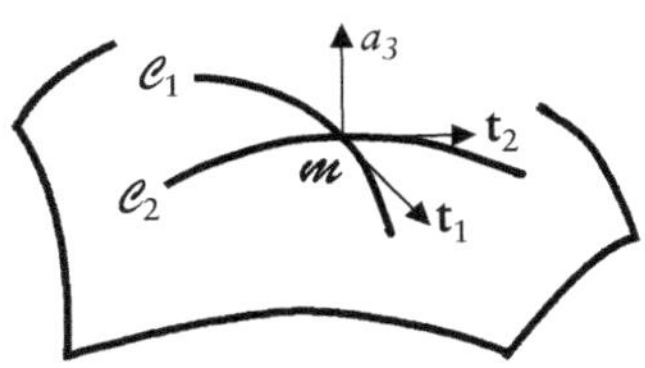

Figure 6.6

$$a_3 = \mathbf{t}_1 \times \mathbf{t}_2$$

($\mathbf{t}_1$, $\mathbf{t}_2$, a_3) is Darboux's trihedron of ($\mathcal{C}_1$) in m, ($\mathbf{t}_2$, $-\mathbf{t}_1$, a_3) that of ($\mathcal{C}_2$).

Finally, let us consider the geodesic curvatures $\frac{1}{r_1}$ of ($\mathcal{C}_1$) and $\frac{1}{r_2}$ of ($\mathcal{C}_2$) in m. Given the derivation relations (3.126) in the Darboux coordinate system, the connection coefficients in the coordinate system ($\mathbf{t}_1$, $\mathbf{t}_2$, a_3) are easily obtained:

$$\Gamma^2_{11} = -\Gamma^1_{12} = \mathbf{t}_2 \bullet \frac{d\mathbf{t}_1}{ds_1} = \mathbf{t}_2 \bullet \frac{d^2 m}{ds_1^2} = \frac{1}{r_1}$$

$$\Gamma^1_{22} = -\Gamma^2_{21} = -\mathbf{t}_1 \bullet \frac{d\mathbf{t}_2}{ds_2} = -t_1 \bullet \frac{d^2 m}{ds_2^2} = -\frac{1}{r_2}$$

other Γ being zero.

In this coordinate system, the first two equilibrium equations are written:

$$\begin{aligned} &\partial_1 N^{11} + \partial_2 N^{12} + \frac{1}{r_2}\left(N^{11} - N^{22}\right) - \frac{3}{r_1} N^{12} + p^1 = 0 \\ &\partial_1 N^{12} + \partial_2 N^{22} + \frac{1}{r_1}\left(N^{11} - N^{22}\right) + \frac{3}{r_2} N^{12} + p^2 = 0 \end{aligned} \tag{6.9}$$

If the coordinate lines form orthogonal geodetic networks, $\frac{1}{r_1}$ and $\frac{1}{r_2}$ are zero, Equation (6.9) simply rewrite:

$$\partial_\alpha N^{\alpha\beta} + p^\beta = 0 \qquad \beta \in [1, 2] \tag{6.10}$$

Equation (6.10) is for example applicable to cylinders.

The third equation (6.2) deals directly with the $N^{\alpha\beta}$ and not their derivatives.

In the case where the coordinate lines are orthogonal, b_{12} is equal to their geodesic torsion (unitary unit vectors).

In the principal curvature coordinate system:

$$\boxed{\frac{N^{11}}{R_1} + \frac{N^{22}}{R_2} + p^3 = 0} \tag{6.11}$$

This relation comes in addition to the relations (6.9) and reflects the normal equilibrium at the surface of a shell portion delimited by four lines of curvature (Figure 6.7).

The contribution of each term is easily understood on an arc portion (Figure 6.8), where it appears that the normal component at point A due to N applied at B is equal to:

$$N \sin\alpha \approx \frac{N}{R} \Delta s$$

The tangential forces N^{12} and N^{21} do not intervene in this projection.

A practical consequence is that to tension a membrane (stretched fabric) in all directions by applying tractions on the edges, it is necessary that its shape is hyperbolic.

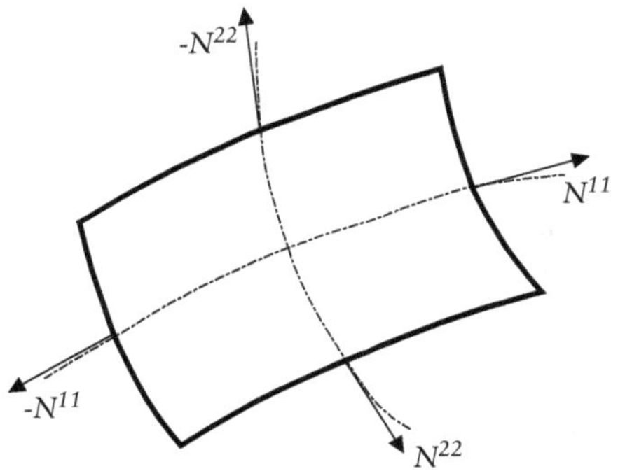

Figure 6.7

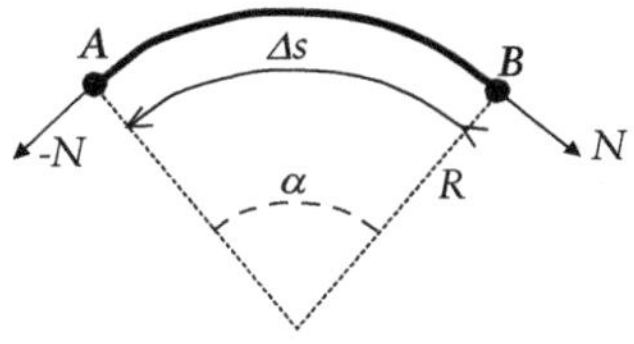

Figure 6.8

The particular situation where $p^3 = 0$ makes it possible to highlight one aspect of the importance of the local shape of the shell: if the radii of curvature are of the same sign (elliptic point), the forces N^{11} and N^{22} are of contrary signs. In the case where the point is hyperbolic, they are of the same sign.

The relation (6.11) gives very important information, especially when the symmetry conditions give the main curvatures a particular role. In some cases (sphere or cylinder under pressure, for example), it allows us to directly obtain the desired result.

In other cases, associated with an overall balance of a shell portion avoiding writing the first two equilibrium equations, allowing us to obtain the forces in the principal curvature coordinate system.

Example: In the case of the spherical dome subjected to its own weight, examined in §1.2.4.2, the writing of the vertical equilibrium of the spherical dome made it possible to obtain:

$$\tilde{N}^{\theta\varphi} = 0$$

$$\tilde{N}^{\varphi\varphi} = -\frac{pR_0}{1+\cos\varphi}$$

(in the principal curvature coordinate system, orthonormed).

By (6.11), it becomes:

$$\frac{\tilde{N}^{\theta\theta}}{R_0} + \frac{\tilde{N}^{\varphi\varphi}}{R_0} + p\cos\varphi = 0$$

From which:

$$\tilde{N}^{\theta\theta} = pR_0\left(\frac{1}{1+\cos\varphi} - \cos\varphi\right)$$

The dome is isotropically compressed at its summit, the compression is worth $-\frac{pR_0}{2}$. It remains compressed at every point in the direction of φ, according to which the compression increases with φ (the self-weight follows the line from the top to the springs). In the case of a hemisphere (semicircular vault), it reaches $-pR_0$ on the diametral plane. In the horizontal direction, the compression decreases from the top and the dome is stretched in this direction from the 51° 50' azimuth. This explains the collapse of masonry vaults which opening angle exceeds this limit.

The asymptotic lines of the surface, when they exist ($K \leq 0$), can also be used as coordinate lines. In this case, the third equation makes it possible to directly determine the shear:

$$2\, b_{12}\, N^{12} + p^3 = 0$$

b_{12} is then equal to the torsion of the asymptotic line and it comes, by application of the Enneper formula (3.139):

$$N^{12} = -\frac{p^3}{2}\sqrt{-R_1\, R_2} \tag{6.12}$$

Examples of application of this relation are given in §6.4.2

6.1.3.2 Case of Euclidean surfaces: stress function

By analogy with the Airy[2] function used in plane elasticity, a stress function Ψ can be introduced such that, in cases where (p^β) derives from a potential Φ such that $p_\beta = -\nabla_\beta \Phi$:

$$N^{\alpha\beta} = e^{\alpha\lambda} e^{\beta\mu} \nabla_\lambda \mathrm{grad}\Psi_\mu + \Phi\, a^{\alpha\beta} \tag{6.13}$$

The symmetry of N is respected if $\nabla_1 \nabla_2 \Psi = \nabla_2 \nabla_1 \Psi$, that is, **if the surface is Euclidean** (see §2.2.7). The two tangential equilibrium equations are then written:

$$\nabla_\alpha N^{\alpha\beta} + p^\beta = e^{\alpha\lambda} e^{\beta\mu} \nabla_\alpha \nabla_\lambda \nabla_\mu \Psi = 0$$

two components of which are:

$$\hat{a}^{-1}\left(\nabla_1 \nabla_2 \nabla_2 \Psi - \nabla_2 \nabla_1 \nabla_2 \Psi\right)$$

$$\hat{a}^{-1}\left(\nabla_2 \nabla_1 \nabla_1 \Psi - \nabla_1 \nabla_2 \nabla_1 \Psi\right)$$

and are obviously zero if ∇_1 and ∇_2 switch.

Equally, it is possible to introduce another Airy function ψ' such that:

$$N^{\alpha\beta} = \Delta\Psi'\, a^{\alpha\beta} - \nabla^\alpha \nabla^\beta \Psi' + \Phi\, a^{\alpha\beta} \tag{6.14}$$

where $\nabla^\beta \Psi'$ are the components of grad Ψ' and $\nabla^\alpha = a^{\alpha\gamma} \nabla_\gamma$. Then:

$$\nabla_\alpha N^{\alpha\beta} + p^\beta = \nabla^\beta\left(\nabla_\gamma \nabla^\gamma \Psi'\right) - \nabla_\alpha\left(\nabla^\alpha \nabla^\beta \Psi'\right) = 0$$

Again, this expression is null if the surface is Euclidean, that is to say, if its Gaussian curvature is zero ($K = 0$).

The two tangential equilibrium equations are then satisfied and the normal equilibrium is written:

[2] George Biddell Airy, British mathematician, astronomer and physicist (1801 – 1892).

- in the case of the function ψ:

$$b_{\alpha\beta}e^{\alpha\lambda}e^{\beta\mu}\nabla_{\lambda}\text{grad}\,\Psi_{\mu}+\Phi\text{TR}(b)+p^{3}=0$$

linear equation with partial derivatives of the second order, in ψ.

- in the case of the function ψ':

$$\text{TR}(b)(\Delta\Psi'+\Phi)-b_{\alpha\beta}\nabla^{\alpha}\nabla^{\beta}\Psi'+p^{3}=0$$

The method of the stress function can be used for all Euclidean surfaces (cylinders, PH, conoids, etc.) or, as an approximate solution, for non-Euclidean surfaces where at least one of the main curvatures is small; this is particularly the case of shallow shells.

Photo 6.1 - Concrete roof made of cylinders and conoids (Quebec-Canada)

6.2 CYLINDRICAL SHELLS

Cylinders are used in a large number of practical situations (pipes, pressure vessels, roofs, etc.). A cylinder is obtained by translation of a line (the **generatrix**) on a plane curve (the **director**). The circular cylinder is a very common case where the director is a circular arc and the generatrix is normal to the director's plane. Given the particular importance of quasi-cylinders in practice, a specific paragraph is devoted to them.

6.2.1 Consequences of geometry

6.2.1.1 Parameterisation, natural basis, fundamental forms

The cylinder is parameterised by the abscissa z along the generatrices and the curvilinear abscissa s along the director. It is developable.

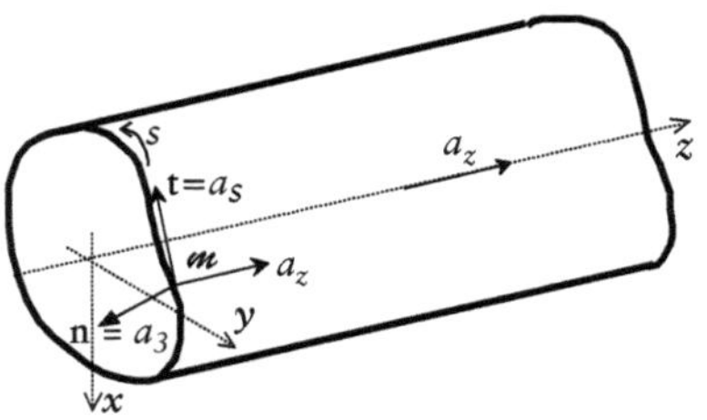

Figure 6.9

The natural basis then consists of the vectors (Figure 6.9):

$$a_z = \frac{\partial m}{\partial z} = e_3\ ;\ a_s = \frac{\partial m}{\partial s} = \mathbf{t}\ ;$$

$$a_3 = a_z \times a_s$$

and is orthonormal.

Relationships:

$$\begin{cases} \partial_z a_z = \partial_s a_z = \partial_z a_s = 0 \\ \partial_s a_s = \dfrac{\mathbf{n}}{R} = \dfrac{a_3}{R} \end{cases}$$

(where R is the radius of the cross section, **t** its tangent and **n** its normal) show that the connection coefficients $\Gamma^{\gamma}_{\alpha\beta}$ are zero and that $b_{ss} = \dfrac{1}{R}$. On the cylinder, the Levi-Civita derivative ∇_α is coincident with the partial derivative ∂_α. The basis thus constituted is an orthonormal Cartesian base, principal of curvature.

Note: with the conventions taken in Figure 6.9, **n** and a_3 are confounded and directed to the centre of curvature of the cross-section. The radius R of the curve and the radius of curvature of the surface are then equal and positive. It is obviously possible to take s as the first parameter, which changes the orientation of the normal vector and leads to take two opposite radius.

6.2.1.2 Equilibrium equations

The orthonormal coordinate system (a_z, a_s) is used. In this coordinate system, the local equations of equilibrium address the physical components of the tensor of the membrane forces:

$$\boxed{\begin{array}{l} \partial_z N^{11} + \partial_s N^{12} + p^1 = 0 \\ \partial_z N^{12} + \partial_s N^{22} + p^2 = 0 \\ \dfrac{N^{22}}{R} + p^3 = 0 \end{array}} \tag{6.15}$$

where the index *1* (or *2*) indicates the component following a_z (resp. a_s).

The last equation depends only on the normal "circumferential" force N^{22}, which is obtained directly:

$$N^{22} = -p^3\, R \tag{6.16}$$

N^{12}, then N^{11}, are obtained by successively reporting in the second, then in the first equation:

$$\begin{cases} N^{12} = \int_0^z \left[-p^2 + \frac{\partial}{\partial s}\left(R\,p^3\right) \right] dz + k_1(s) \\ N^{11} = \int_0^z \left\{ -p^1 + \int_0^u \frac{\partial}{\partial s}\left[p^2 - \frac{\partial}{\partial s}\left(R\,p^3\right) \right] du \right\} dz + k_2(s) \end{cases} \tag{6.17}$$

This system can also be reduced to the only equation:

$$\partial^2_{z^2} N^{11} + \partial^2_{s^2}\left(R\,p^3\right) - \partial_s p^2 + \partial_z p^1 = 0 \tag{6.18}$$

which integrates by double quadrature, the external actions being given. The double integration in z gives rise to two arbitrary functions $k_1(s)$ and $k_2(s)$, determined by the boundary conditions in s (or the periodicity conditions).

6.2.1.3 Deformation

Calculations are made under the assumption of small displacements. The displacement is decomposed in the local coordinate system in the form:

$$\xi = u\,a_z + v\,a_s + w\,a_3$$

where were introduced the usual notations:

$$\xi^1 = u; \quad \xi^2 = v; \quad \zeta = w$$

By application of the relation (6.5), it comes the linearised deformation tensor:

$$\boxed{\begin{aligned} \varepsilon_{11} &= \partial_z u \\ \varepsilon_{22} &= \partial_s v - \frac{w}{R} \\ \varepsilon_{12} &= \frac{1}{2}\left(\partial_s u + \partial_z v\right) \end{aligned}} \tag{6.19}$$

6.2.1.4 Linearly elastic behaviour

In the case of the linearly elastic, homogeneous and isotropic material, taking into account (6.19), the isothermal constitutive law without prestressing (6.6) is reduced to the relations:

$$\begin{cases} N^{11} = K\left(\varepsilon_{11} + \nu\,\varepsilon_{22}\right) = K\left[\partial_z u + \nu\left(\partial_s v - \frac{w}{R}\right)\right] \\ N^{22} = K\left(\varepsilon_{22} + \nu\,\varepsilon_{11}\right) = K\left(\partial_s v - \frac{w}{R} + \nu\,\partial_z u\right) \\ N^{12} = K(1-\nu)\varepsilon_{12} = \frac{K}{2}(1-\nu)\left(\partial_s u + \partial_z v\right) \end{cases} \tag{6.20}$$

By placing these expressions in the equilibrium equations, the equations obtained relate solely to displacement (equations similar to the Navier equations in 3D elasticity):

$$\begin{cases} \dfrac{1+\nu}{2}\partial^2_{sz}v+\dfrac{1-\nu}{2}\partial^2_{s^2}u+\partial^2_{z^2}u-\nu\,\partial_z\left(\dfrac{w}{R}\right)+\dfrac{p^1}{K}=0 \\ \partial^2_{s^2}v-\partial_s\left(\dfrac{w}{R}\right)+\dfrac{1+\nu}{2}\partial^2_{sz}u+\dfrac{1-\nu}{2}\partial^2_{z^2}v+\dfrac{p^2}{K}=0 \\ \partial_s v-\dfrac{w}{R}+\nu\,\partial_z u+\dfrac{p^3R}{K}=0 \end{cases} \qquad (6.21)$$

which can be solved in some cases. Most often, there is a need to express boundary conditions for the forces, through constitutive laws.

6.2.1.5 Case of circular cylinders

In this particular case, which is very important in practice, the polar angle θ and the curvilinear abscissa s are linked by the relation:

$$s=\theta\,R_0$$

where R_0 is the radius of the circle, positive and constant. θ can also be used as a second parameter, with the associated vector as vector a_θ instead of the normed vector a_s, the two vectors being linked by:

$$a_\theta=\frac{\partial \mathcal{m}}{\partial\theta}=R_0\,a_s$$

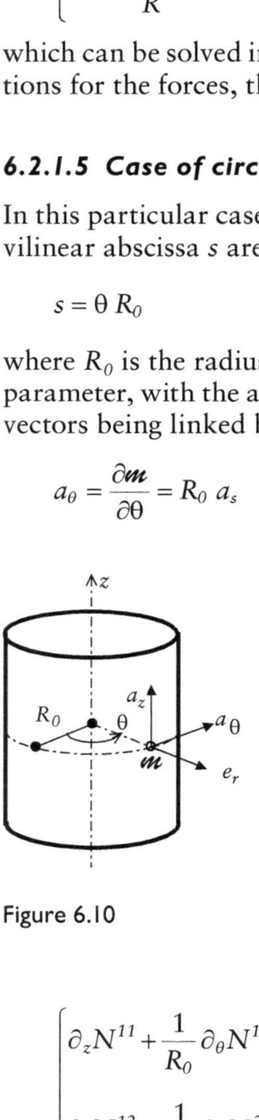

Figure 6.10

Partial derivatives with respect to s and θ are linked by the relation:

$$\frac{\partial}{\partial s}=\frac{1}{R_0}\frac{\partial}{\partial\theta}$$

The vector a_3 normal to the cylinder directed towards the centre of curvature is opposite to the radial vector e_r of the cylindrical coordinates (Figure 6.10). It is obtained by $a_3=a_z\times a_s$. The radius of curvature is positive, as in the general case.

In basis (a_θ, a_z), the first two fundamental forms are written:

$$a_{\bullet\bullet}=\begin{pmatrix}R_0^{\,2} & 0\\ 0 & 1\end{pmatrix} \qquad b_{\bullet\bullet}=\begin{pmatrix}R_0 & 0\\ 0 & 0\end{pmatrix}$$

The equilibrium equations (15) are rewritten:

$$\begin{cases} \partial_z N^{11}+\dfrac{1}{R_0}\partial_\theta N^{12}+p^1=0 \\ \partial_z N^{12}+\dfrac{1}{R_0}\partial_\theta N^{22}+p^2=0 \\ N^{22}+p^3R_0=0 \end{cases} \qquad (6.22)$$

where the membrane forces and the loading are expressed in the orthonormal coordinate system (a_z, a_s), thus in physical components.

Example : Cylindrical tank filled with liquid.

A cylindrical reservoir of constant thickness is filled to the brim with a liquid of density ρ. The vertical z-axis is directed upwards and its origin is the bottom of the tank (Figure 6.11).

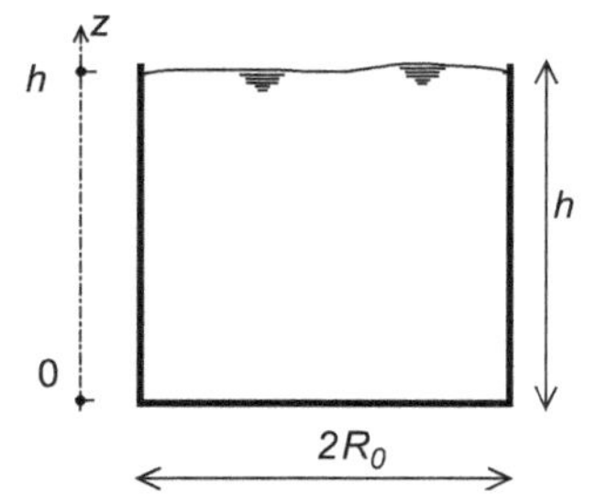

Figure 6.11

Here:

$$p_1 = p_2 = 0$$

$$p_3 = -\rho g (h - z)$$

Membrane forces $N^{\alpha\beta}$ can easily be determined:

$$N^{22} = \rho g (h - z) R_0$$

$$N^{12} = N^{11} = 0$$

Temporarily abstracting from this result, the solution can be sought from (21). Given the symmetry of revolution, v is zero, and all the derivatives with respect to s. Under these conditions equations (21) are reduced to:

$$\begin{cases} \partial^2_{z^2} u - \nu \, \partial_z \left(\dfrac{w}{R_0} \right) = 0 \\ -\dfrac{w}{R_0} + \nu \, \partial_z u + \dfrac{p^3 R_0}{K} = 0 \end{cases}$$

The first equation integrates in:

$$\partial_z u - \nu \frac{w}{R_0} = \text{cst}$$

N^{11} being zero at the top of the tank, the constant at the second member is zero and therefore $N^{11} = 0$ over the entire height. By eliminating $\partial_z u$, w is written:

$$w = -\frac{\rho g R_0{}^2}{Ee} (h - z)$$

Then:

$$N^{22} = K \left(-\frac{w}{R_0} + \nu \, \partial_z u \right) = -Ee \frac{w}{R_0} = \rho g R_0 (h - z)$$

then, by reporting the expression of w in $\partial_z u$, integrating and determining the unique integration constant by the condition $u(0) = 0$, it comes:

Photo 6.2 - Cylindrical tank

$$u = -\nu \frac{\rho g R_0}{Ee} z \left(h - \frac{z}{2} \right)$$

which is entirely due to the Poisson effect. In this problem, two conditions have been expressed, one relating to the vertical normal force N^{11}, in $z = h$, the other relating to the vertical displacement u, in $z = 0$.

w is wholly determinate and no additional conditions relating to w or $\frac{\partial w}{\partial z}$ (for example, clamping in the lower slab) can be imposed, except to violate the membrane equilibrium conditions. In practice, since a tank is always closed at its base, a localised flexion develops at the bottom (see § 8.-2.3.1).

6.2.2 Some classic examples of problems with cylindrical shells

6.2.2.1 Vault in half circular cylinder

A fairly common form of hall roof consists of juxtaposed cylinder portions to cross large spans with relatively small thicknesses. The section of such cylinders is usually circular or parabolic (see § 2.2.4).

A barrel vault consists of a half-cylinder cut along a horizontal diametral plane. Its span is 2ℓ and rests at both ends on tympana rigid in their plane, but very flexible transversely to their plane. The vault is subject to its own weight, its thickness e being constant. $\varpi = \rho g e$ is the constant weight per unit area (Figure 6.12).

The angle θ is zero along a vertical axis and the origin of z is taken in the middle of the vault. Equilibrium equations are reduced to:

$$\begin{cases} \partial_z N^{11} + \dfrac{1}{R_0} \partial_\theta N^{12} = 0 \\ \partial_z N^{12} + \dfrac{1}{R_0} \partial_\theta N^{22} + \varpi \sin\theta = 0 \\ N^{22} + \varpi R_0 \cos\theta = 0 \end{cases}$$

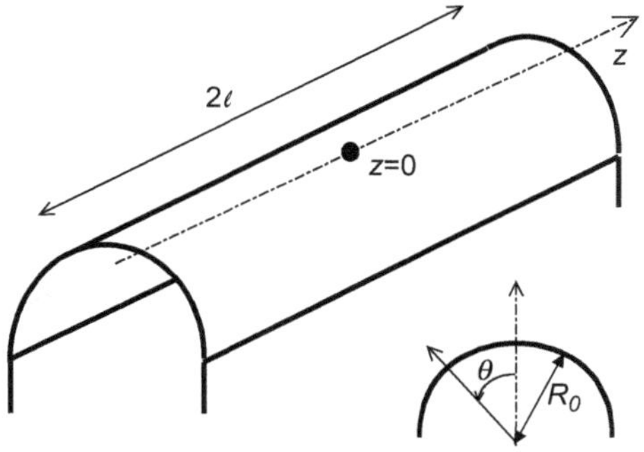

Figure 6.12

and admit the general solution:

$$N^{11} = \frac{1}{R_0}\left(\varpi z^2 \cos\theta - z f'(\theta)\right) + g(\theta)$$

$$N^{12} = -2\varpi z \sin\theta + f(\theta)$$

$$N^{22} = -\varpi R_0 \cos\theta$$

By symmetry of the vault and of the loading, N^{12} is antisymmetric in z, so the function $f(\theta)$ is null. In addition, considering the condition of flexibility of the tympana in the z direction, $N^{11}(\pm\ell) = 0$, which allows to completely determine the solution:

$$N^{11} = \frac{\varpi \cos\theta}{R_0}\left(z^2 - \frac{\ell^2}{4}\right)$$

$$N^{12} = -2\varpi z \sin\theta$$

$$N^{22} = -\varpi R_0 \cos\theta$$

The shear vector applied at a point of a vertical section is N^{12} **t**. Its resultant for θ varying from $-\pi/2$ to $\pi/2$ is worth $\pi\varpi z\,\mathbf{i}$, the shear force due to the weight.

Along the horizontal edge, N^{12} and N^{22} should be zero (free edge condition), but in the above membrane solution N^{12} is not zero. In order for the membrane solution to become established, it is necessary to have a "shear beam" along the edge. Even under these conditions, the solution can only be achieved if the displacements are kinematically admissible, which must be verified. The longitudinal force N^{11} is always a compression, whereas the vault would be stretched in the lower part if it were considered as a beam crossing the same span; this state of compression is possible thanks to the shear beams arranged along the horizontal edges.

In considering the reverse constitutive law, it comes:

$$\varepsilon_{11} = \partial_z u = \frac{1}{Ee}\left(N^{11} - \nu N^{22}\right) = \frac{1}{Ee}\frac{\varpi \cos\theta}{R_0}\left(z^2 - \frac{\ell^2}{4} + \nu R_0^{\,2}\right)$$

Taking into account zero longitudinal displacement in the middle: $u(z=0)=0$, by integration:

$$u = \frac{1}{Ee}\frac{\varpi z \cos\theta}{R_0}\left(\frac{z^2}{3} - \frac{\ell^2}{4} + \nu R_0^2\right)$$

then:

$$\varepsilon_{12} = \frac{1}{2}\left(\partial_s u + \partial_z v\right) = \frac{N^{12}}{K(1-\nu)} = -\frac{2\varpi}{K(1-\nu)} z \sin\theta$$

hence:

$$\partial_z v = \frac{\varpi \sin\theta}{Ee}\frac{z}{R_0^2}\left(\frac{z^2}{3} - \frac{\ell^2}{4} - (4+3\nu)R_0^2\right)$$

Integrating:

$$v = \frac{\varpi \sin\theta}{2Ee}\frac{z^2}{R_0^2}\left(\frac{z^2}{6} - \frac{\ell^2}{4} - (4+3\nu)R_0^2\right) + h(\theta)$$

v is zero in $z = \pm\ell$, which makes it possible to completely determine the function h. Finally:

$$v = \frac{\varpi \sin\theta}{2EeR_0^2}\left[\frac{1}{6}\left(z^4 - \frac{\ell^4}{16}\right) - \left(\frac{\ell^2}{4} + (4+3\nu)R_0^2\right)\left(z^2 - \frac{\ell^2}{4}\right)\right]$$

The transverse displacement can then be determined in $\theta = \pm\frac{\pi}{2}$ by:

$$\varepsilon_{22} = \partial_s v - \frac{w}{R} = \frac{1}{Ee}\left(N^{22} - \nu N^{11}\right) = -\frac{1}{Ee}\frac{\varpi \cos\theta}{R_0}\left(R_0^2 + \nu\left(z^2 - \frac{\ell^2}{4}\right)\right)$$

From where:

$$w = \frac{\varpi \cos\theta}{EeR_0^2}\left[\frac{1}{12}\left(z^4 - \frac{\ell^4}{16}\right) + R_0^4 - \frac{1}{2}\left(\frac{\ell^2}{4} + (4+\nu)R_0^2\right)\left(z^2 - \frac{\ell^2}{4}\right)\right]$$

w is zero along the edge beam, but not on the tympanum.

In conclusion, the conditions for achieving membrane equilibrium are difficult to achieve. On the tympanum, only the displacement v can be zero. On the horizontal edges, there is no compatibility between the displacements of the vault and the displacements of the border beam; in fact, the shear applied to the beam is linear in z, as the applied moment (the lever arm is constant if the section of the beam is constant) and the curvature. On the contrary, the membrane solution gives a quadratic curvature. Finally, the membrane solution is unrealistic and it is therefore necessary to take into account the flexion, cf. § 8.2.3.3.

6.2.2.2 Pipe under pressure

Let be a horizontal cylindrical pipe containing a fluid under pressure (Figure 6.13). The fluid is at rest and the pressure it exerts on the pipe is expressed by:

$$p = p_0 - \rho g R_0 \cos\theta$$

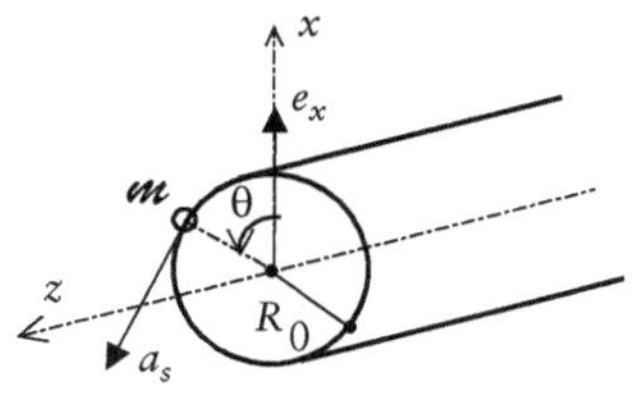

Figure 6.13

where ρg is the density of the liquid.

The resolution of equilibrium equations gives successively:

$$\begin{cases} N^{22} = p_0 R_0 - \rho g R_0{}^2 \cos\theta \\ N^{12} = -\rho g R_0 z \sin\theta + f(\theta) \\ N^{11} = \rho g \dfrac{z^2}{2} \cos\theta + g(\theta) - \dfrac{z}{R_0} f'(\theta) \end{cases}$$

where $f(\theta)$ and $g(\theta)$ are two functions to be determined.

If, for example, the pipe is supported at both ends $z = \pm \ell$ by cradles that prevent any movement following z (clamping allowing movements in the plane of the section), by symmetry of geometry and loading, N^{12} is zero in the middle section of the pipe section, in $z = 0$. Then: $f(\theta) = 0$.

The solution is entirely determined by the boundary conditions that are written:

$$\begin{cases} u(-\ell) = 0 \\ u(\ell) = u(-\ell) + \displaystyle\int_{-\ell}^{+\ell} \varepsilon_{11} dz \end{cases}$$

If the pipe consists of a linearly elastic, homogeneous and isotropic material:

$$\int_{-\ell}^{+\ell} \varepsilon_{11} dz = \frac{1}{Ee} \int_{-\ell}^{+\ell} \left(N^{11} - \nu\, N^{22}\right) dz = 0$$

either, replacing forces with their value:

$$\int_{-\ell}^{+\ell} \left(\rho g \frac{z^2}{2} \cos\theta + g(\theta) - \nu\, p_0 R_0 + \nu\, \rho g R_0{}^2 \cos\theta \right) dz = 0$$

which allows to determine $g(\theta)$:

$$g(\theta) = \nu\, p_0 R_0 - \left(\frac{\ell^2}{6} + \nu\, R_0{}^2 \right) \rho g \cos\theta$$

then N^{11}:

$$N^{11} = \nu\, p_0 R_0 - \left(\frac{\ell^2}{6} - \frac{z^2}{2} + \nu\, R_0{}^2 \right) \rho g \cos\theta$$

The other components, v and w, of displacement are determined by relations (7) and (20). The calculation shows that w can not be zero in the end sections. The membrane theory can not provide an exact solution in case the ends are clamped; it is then necessary to take into account the flexion (cf § 8.2.3.1).

It is interesting to compare the solution thus obtained with what is possible to obtain from the beam theory. It should first be noted that the latter can not give any information on circumferential stress N^{22}. By calling $q = \rho g \pi R_0^2$ the linear weight of the fluid contained in the pipe, the bending moment and the shear force of the beam clamped at its two ends $z = \pm \ell$ are:

$$\text{in } z = \pm\ell: \begin{cases} V = q\ell \\ M = -\dfrac{q\ell^2}{3} \end{cases} ; \quad \text{in } z = 0: \begin{cases} V = 0 \\ M = \dfrac{q\ell^2}{6} \end{cases}$$

In a thin tube:

- geometric inertia: $I = \pi R_0^3 e$
- static moment: $m = 2\int_0^\theta xeR_0 \, d\theta = 2eR_0^2 \sin\theta$

From where the membrane forces at the ends $z = \pm \ell$, given by the beam theory:

$$N^{11} = e\,\sigma^{11} = -\frac{Mex}{I} = \frac{q\ell^2}{3\pi R_0^2}\cos\theta = \rho g\frac{\ell^2}{3}\cos\theta$$

$$N^{12} = e\,\sigma^{12} = -\frac{Vm}{2I} = \frac{q\ell}{\pi R_0}\sin\theta = \rho g R_0 \ell \sin\theta$$

The results obtained for N^{12} are identical in both theories. The preponderant term in N^{11} is the same; it is nevertheless corrected, in membrane theory, by the term — νN^{22} due to the Poisson effect, because of the two-dimensional nature of the shell and the boundary conditions relating to displacement.

The same comparison made assuming that the pipe would simply be supported at both ends shows that N^{11} and N^{12} obtained in the shell model and in the beam model are identical, the shell model providing N^{22} in addition. This result was not obtained in the case of the semi-cylindrical vault of § 6.2.2.1.

6.2.2.3 Non-axisymmetric loading on a cylindrical tank

A tank consists of a circular cylindrical skirt of height h. This tank is subjected to the wind, which generates a pressure considered as symmetrical with respect to the direction of the wind and constant according to the elevation z (Figure 6.14).

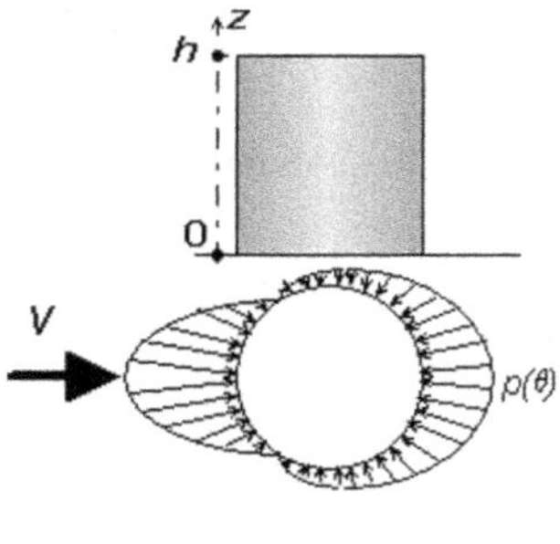

Figure 6.14

By choosing its axis of symmetry as the origin of the angles, this pressure can be represented by an even and periodic function decomposed in Fourier series, in the form: $p(\theta)=\sum_{n=0}^{\infty} p_n \cos n\theta$ where:

$$\begin{cases} p_0 = \dfrac{1}{\pi}\displaystyle\int_0^{\pi} p(\theta)\, d\theta \\ p_n = \dfrac{2}{\pi}\displaystyle\int_0^{\pi} p(\theta)\cos n\theta\, d\theta \end{cases}$$

Under the effect of a pressure mode $p^3 = p_n \cos n\theta$, the solution of equilibrium equations (19) is written:

$$\begin{cases} N^{22} = p_n R_0 \cos n\theta \\ N^{12} = p_n\, n\, z \sin n\theta + f_n(\theta) \\ N^{11} = -p_n\, n^2 \dfrac{z^2}{2R_0} \cos n\theta - \dfrac{z}{R_0} f_n'(\theta) + g_n(\theta) \end{cases}$$

The edge $z = h$ is free, which makes it possible to write:

$$\forall\theta: \begin{cases} N^{12}(z=h) = p_n\, n\, h \sin n\theta + f_n(\theta) = 0 \\ N^{11}(z=h) = -p_n\, n^2 \dfrac{h^2}{2R_0} \cos n\theta - \dfrac{h}{R_0} f_n'(\theta) + g_n(\theta) = 0 \end{cases}$$

from where:

$$\begin{cases} f_n(\theta) = -p_n\, n\, h \sin n\theta \\ g_n(\theta) = -p_n\, n^2 \dfrac{h^2}{2R_0} \cos n\theta \end{cases}$$

The solution is finally obtained by superposition of the different modes:

$$\begin{cases} N^{11} = -\displaystyle\sum_{n=0}^{\infty} p_n\, n^2 \dfrac{(z-h)^2}{2R_0} \cos n\theta \\ N^{12} = \displaystyle\sum_{n=0}^{\infty} p_n\, n\,(z-h) \sin n\theta \\ N^{22} = \displaystyle\sum_{n=0}^{\infty} p_n R_0 \cos n\theta \end{cases}$$

This solution is expressed independently of the boundary conditions in $z = 0$, which makes it possible to obtain reactions that are compatible with the membrane theory.

The calculation of displacements is also made mode by mode. For the mode of order n, the deformations are deduced from the membrane forces by the constuitutive law (7):

$$\begin{cases} \varepsilon_{11} = -\dfrac{p_n}{Ee}\left[\nu R_0 + n^2\dfrac{(z-h)^2}{2R_0}\right]\cos n\theta \\ \varepsilon_{12} = \dfrac{p_n}{Ee}(1+\nu)(z-h)\, n \sin n\theta \\ \varepsilon_{22} = \dfrac{p_n}{Ee}\left[R_0 + \nu n^2\dfrac{(z-h)^2}{2R_0}\right]\cos n\theta \end{cases}$$

then displacements as $\xi = ua_1 + va_2 + wa_3$ are obtained by (20); successively:

$$\partial_z u = -\frac{p_n}{Ee}\left[\nu R_0 + n^2\frac{(z-h)^2}{2R_0}\right]\cos n\theta$$

From where:

$$u = -\frac{p_n}{Ee}\left[\nu R_0 z + n^2\frac{(z-h)^3 + h^3}{6R_0}\right]\cos n\theta$$

where u, vertical displacement, is assumed to be zero in $z = 0$ (tank placed on the ground). Then:

$$\partial_z v = \frac{p_n}{Ee}\left[2n(1+\nu)(z-h) - \nu nz - n^3\frac{(z-h)^3 + h^3}{6R_0^2}\right]\sin n\theta$$

hence, assuming zero tangential displacement in $z = 0$:

$$v = \frac{p_n}{Ee}\left\{n(1+\nu)\left[(z-h)^2 - h^2\right] - \nu n\frac{z^2}{2} - n^3\frac{(z-h)^4 - h^4 + 4h^3 z}{24R_0^2}\right\}\sin n\theta$$

Finally:

$$w = -\frac{p_n R_0^2}{Ee}\left\{\begin{matrix} 1 + \nu n^2\dfrac{(z-h)^2}{2R_0^2} - n^2(1+\nu)\left[\dfrac{(z-h)^2 - h^2}{R_0^2}\right] \\ + \nu n^2\dfrac{z^2}{2R_0^2} + n^4\dfrac{(z-h)^4 - h^4 + 4h^3 z}{24R_0^4} \end{matrix}\right\}\cos n\theta$$

This last expression imposes the normal displacements at the bottom:

$$w(0) = -\frac{p_n R_0^2}{Ee}\left\{1 + \nu n^2 \frac{h^2}{2R_0^2}\right\}\cos n\theta$$

$$\partial_z w(0) = -\frac{p_n}{Ee}(2+\nu)\, n^2 h \cos n\theta$$

It is not possible to simultaneously cancel out the v and w components at the bottom: the membrane theory is not sufficient in case of clamping conditions.

For any pressure p, the conditions $u(\theta) = v(\theta) = 0$ at the bottom still give the same conditions for each of the modes, by Fourier series decomposition. The displacements are thus obtained by superposition of the solutions obtained for the modes separately.

6.2.2.4 Parabolic vault

A roof element is formed of a cylinder with horizontal generatrix which cross section in a vertical plane is a parabola arc of vertical axis. The two ends, distant from 2ℓ, rest on tympanums, which are stiff in their plane, but allow out-of-plane displacements (along the z-axis). The vault is supported on two generatrices located on the same horizontal plane. The axes and the curvilinear abscissa along the cross-section are oriented as shown in figure 6.15.

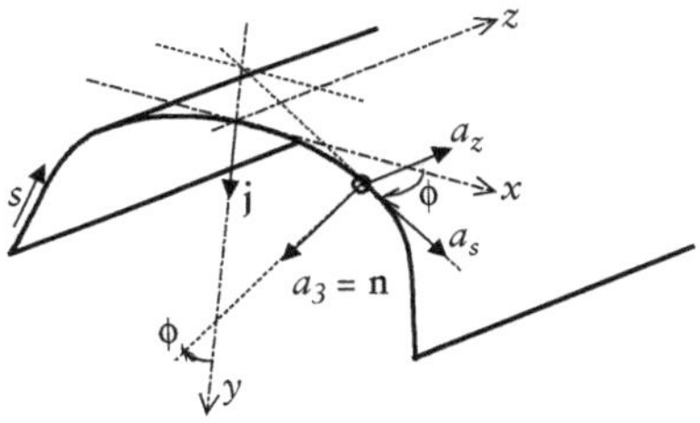

Figure 6.15

Let ϕ the angle of the tangent a_s to the parabola with the x axis:

$$\frac{dx}{ds} = \cos\phi\,; \quad \frac{dy}{ds} = \sin\phi$$

ϕ is also the angle of the normal **n** to the parabola with the y axis. $a_3 = a_z \times a_s$ is coincident with **n**.

The equation of the parabola is written:

$$x^2 = 2R_0 y$$

where R_0 is the radius of curvature at the top of the vault ($R_0 > 0$).
Noting:

$$\frac{x}{R_0} = \frac{dy}{dx} = \tan\phi$$

the parameterisation of the surface according to ϕ and z is written:

$$\mathcal{m}\begin{cases} R_0 \tan\phi \\ \dfrac{R_0}{2}\tan^2\phi \\ z \end{cases}$$

hence the natural basis:

$$a_z\,; \qquad a_\varphi = \begin{cases} \dfrac{R_0}{\cos^2\phi} \\ \dfrac{R_0\,\tan\phi}{\cos^2\phi} \\ 0 \end{cases}; \qquad a_3 = \begin{cases} -\sin\phi \\ \cos\phi \\ 0 \end{cases}$$

s being the curvilinear abscissa along the parabola, it comes:

$$R = \frac{ds}{d\phi} = \frac{R_0}{\cos^3\phi}$$

The load cases studied are the self weight and the snow load, for which the forces applied following z are zero: $p^1 = 0$.

By eliminating N^{22}, which is directly determined by the third equilibrium equation (6.16), the second equation (6.15) is rewritten:

$$-\frac{1}{R}\partial_\phi\left(R\,p^3\right) + \partial_z N^{12} + p^2 = 0$$

It makes it possible to determine N^{12}, the value of which is then reported in the first equation:

$$-\frac{1}{R}\partial_\phi\left(R\,p^3\right) + \partial_z N^{12} + p^2 = 0$$

► The self weight **p**, of linear density p_0 constant when the thickness is constant, is uniform with respect to the curvilinear abscissa:

$$\frac{d\mathbf{p}}{ds} = p_0\,\mathbf{j}$$

and breaks down in the local coordinate system in:

$$p^2 = p_0 \sin\phi$$

$$p^3 = p_0 \cos\phi$$

It comes:

$$N^{22} = -\frac{p_0 R_0}{\cos^2\phi}$$

$$N^{12} = p_0\, z \sin\phi + f(\phi)$$

By symmetry, N^{12} is zero in the middle section. Taking $z = 0$ in this section, then necessarily $f(0) = 0$. The second equation then integrates into:

$$N^{11} = -p_0 \frac{z^2}{2R_0} \cos^4 \phi + g(\phi)$$

At both ends, the movements along the z axis are free and $N^{11}(\pm\ell) = 0$.

From where finally the solution:

$$N^{11} = p_0 \frac{\ell^2 - z^2}{2R_0} \cos^4 \phi$$

$$N^{12} = p_0 z \sin \phi$$

$$N^{22} = -\frac{p_0 R_0}{\cos^2 \phi}$$

It should be noted that the stress N^{11} (along the generatrices) is a traction in the whole of the shell, which is not obvious *a priori*. There is in fact a general bending of the shell which results in normal stresses N^{11} and shear N^{12} as in a beam. This is related to the fact that the resultant of the self-weight increases when ϕ increases. In practice, the influence of boundary conditions on an edge $\phi = \pm\phi_0$ should be taken into account, the parabola being necessarily finite.

It is clear that if this edge is free, the solution can not be a membrane one, since N^{22} and N^{12} determined by the solution above can not be canceled out there. As in the case of the circular vault, the solution adopted in practice consists in setting up a rim beam (stiffener) which makes it possible to balance the membrane forces, that is to say a shear and the weight of the vault. This does not guarantee the exact realisation of the membrane theory, compatibility of movements between the shell and the stiffener is not necessarily ensured.

Such a vault is often directly supported on the ground via a sufficiently rigid footing, in which case the balance of the membrane forces on the horizontal edges is easier.

Photo 6.3 - Cylinder roof with edge stiffener (Quebec-Canada)

► The snow load is usually characterised by a load density uniformly distributed on a horizontal plane:

$$\frac{d\mathrm{p}}{dx} = p_n\,\mathrm{j}$$

But:

$$\frac{d\mathrm{p}}{dx} = \frac{d\mathrm{p}}{ds}\frac{ds}{dx} = \frac{1}{\cos\phi}\frac{d\mathrm{p}}{ds}$$

Then:

$$\frac{d\mathrm{p}}{ds} = p_n \cos\phi\,\mathrm{j} = p_n \cos\phi\left(a_s \sin\phi + a_3 \cos\phi\right)$$

By reporting the components of the load:

$$\begin{cases} p^2 = p_n \cos\phi \sin\phi \\ p^3 = p_n \cos^2\phi \end{cases}$$

in the equations, the solution is written:

$$N^{22} = -\frac{p_n R_0}{\cos\phi}$$

$$N^{12} = N^{11} = 0$$

Indeed, the parabolic arc is funicular of the uniform charge density.

In conclusion, for the snow load, the parabolic vault behaves like a series of side-by-side arches, each arch being in funicular equilibrium. In the case of self weight, this funicular equilibrium is not possible and the shell resists in membrane equilibrium with forces in both directions. Nevertheless, if there is a free edge in $\phi = \pm\,\phi_0$, the membrane equilibrium is not respected; an edge stiffener (or a direct support on the ground) makes it possible to respect it approximately, a local bending being inevitable in the shell taking into account the incompatibility of deformation between shell and stiffener.

6.2.3 Quasi-cylindrical shell

6.2.3.1 Equilibrium equations of the quasi-cylindrical shell

A quasi-cylindrical shell is such that the generatrices are slightly curved, instead of being straight. It may be a barrel if the curvature of the generatrix is positive (the surface is elliptical, Figure 6.16a) or a diabolo if it is negative (hyperbolic surface, Figure 6.16b).

The case considered here is that of an axisymmetrical surface, with a constant curvature $\frac{1}{R'}$ of the generatrices.

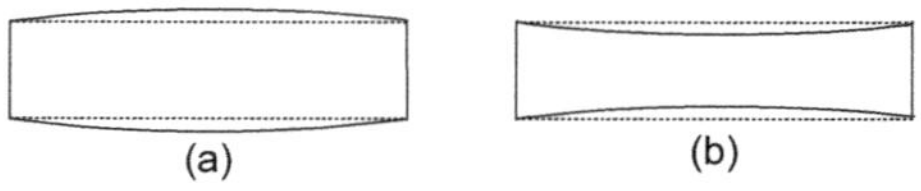

Figure 6.16

Photo 6.4 - Rum barrels (Martinique-France)

In the case where $R' >> R_0$, the surface thus defined is quasi-developable and the description of the metric of the cylinder may be assumed preserved. Only the third equilibrium equation is modified to take into account the curvature of the generatrix.

Thus, in the case where the loading is purely transverse and constant:

$$\begin{aligned} &\partial_z N^{11} + \frac{1}{R_0}\partial_\theta N^{12} = 0 \\ &\partial_z N^{12} + \frac{1}{R_0}\partial_\theta N^{22} = 0 \\ &\frac{N^{11}}{R'} + \frac{N^{22}}{R_0} - p = 0 \end{aligned} \tag{6.23}$$

By elimination, there comes the partial differential equation for the longitudinal force N^{11}:

$$\partial^2_{z^2} N^{11} + \frac{1}{R'R_0}\partial^2_{\theta^2} N^{11} = 0 \tag{6.24}$$

The solution of this equation depends on the sign of R'. In all cases, the solution is periodic in θ. N^{11} is determined by the boundary conditions. Only N^{22} depends on the transverse load. If no conditions are imposed on them at the boundaries, N^{11} and N^{12} are zero and $N^{22} = pR_0$. This simple result is due to the approximation on the metric. Thus, non-zero values of N^{11} and N^{12} can only come from boundary conditions. In the following, p is taken as zero and the effect of the boundary conditions is analysed, the solutions of the two types of actions being superposable.

Then, a behaviour of the shell different from that of the cylinder can be demonstrated by applying two opposite densities of forces $q(\theta)$, directed according to a_z, at both ends of the shell.

In the case of a circular cylinder, N^{22} is zero by (16) and N^{12} by (15). N^{11} does not depend on z. So, in this case, the forces applied longitudinally to the ends are transmitted directly along the generatrices.

6.2.3.2 Case of the barrel

In this case $R' > 0$. The loading $q(\theta)$ being periodic, it can be developed in Fourier series and the study can be limited to loading $q_n \cos n\theta$. It is logical to look for solutions of the form $f_n(z)\cos n\theta$, which, by referring to (24), leads to the equation $f_n'' - \frac{n^2}{c^2} f_n = 0$, by posing $c^2 = R_0 R'$, directly related to the Gauss curvature. Finally, the solution for N^{11} is written in the general form:

$$N^{11} = \left[A_n \,\text{ch}\left(n\frac{z}{c}\right) + B_n \,\text{sh}\left(n\frac{z}{c}\right)\right] \cos n\theta$$

The shell being symmetrical as well as the loading, the origin of z is taken in the median plane and the boundary conditions are expressed in $\pm \ell$: $N^{11}(\pm\ell) = q(\theta)$, which finally leads to the solution:

$$N^{11} = q_n \frac{\text{ch}\left(n\frac{z}{c}\right)}{\text{ch}\left(n\frac{\ell}{c}\right)} \cos n\theta$$

The other membrane forces corresponding to this solution are obtained by using again equilibrium equations and noting that, by symmetry, N^{12} is zero in the median plane. It comes:

$$N^{22} = -q_n \frac{R}{R'} \frac{\text{ch}\left(n\frac{z}{c}\right)}{\text{ch}\left(n\frac{\ell}{c}\right)} \cos n\theta \; ; \qquad N^{12} = -c q_n \frac{1}{R'} \frac{\text{sh}\left(n\frac{z}{c}\right)}{\text{ch}\left(n\frac{\ell}{c}\right)} \sin n\theta$$

Two remarks are needed in view of the result:

- N^{12} being not zero at the ends, the membrane solution is only possible if a shear is applied in reaction in $z = \pm\ell$.
- In the median plane, N^{22} is not zero, but conversely N^{11} is lower than at the ends. The barrel shape of the shell makes it possible to spread the forces by distributing them between the two directions.

6.2.3.3 Case of the diabolo

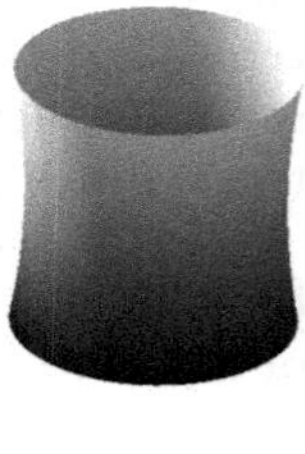

In this case, $R' < 0$. Equation (24) is hyperbolic and its general solution is written (noting $c^2 = -R_0 R'$):

$$N^{11} = f(z + c\theta) + g(z - c\theta)$$

It is simpler here to take the origin of the z-axis at one end. The boundary conditions are then written:

$$N^{11}(z=0)=f(c\theta)+g(-c\theta)=q(\theta)$$

$$N^{11}(z=2\ell)=f(2\ell+c\theta)+g(2\ell-c\theta)=q(\theta)$$

Here, nothing stands in the way of giving an independent condition on N^{12} at the ends, for example a zero shear condition, which results in:

$$N^{12}(z=0)=N^{12}(z=2\ell)=0$$

The function N^{12} (of θ) being zero along the edges, this leads, by the equilibrium equations (23):

$$\partial_z N^{11}(z=0)=f'(c\theta)+g'(-c\theta)=0$$

$$\partial_z N^{11}(z=2\ell)=f'(2\ell+c\theta)+g'(2\ell-c\theta)=0$$

The first condition gives, by integration in θ: $f(c\theta)-g(-c\theta)=\text{cst}=K$.

Adding with the expression of N^{11} at the edge, it comes $f(c\theta)=\frac{1}{2}\left[q(\theta)+K\right]$ and $g(-c\theta)=\frac{1}{2}\left[q(\theta)-K\right]$. The constant K does not play a role and can be null. Finally:

$$N^{11}=\frac{1}{2}\left[q\left(\frac{z}{c}+\theta\right)+q\left(-\frac{z}{c}+\theta\right)\right]$$

By eliminating N^{11} in the equilibrium equations (6.23), it appears that N^{12} satisfies the same equation (6.24) as N^{11} and thus admits a solution of the same form. Indeed, the first and third equilibrium equations lead to:

$$\partial_z N^{12}=\frac{1}{R'}\partial_\theta N^{11} \Rightarrow N^{12}=\frac{c}{2R'}\left[q\left(\frac{z}{c}+\theta\right)-q\left(-\frac{z}{c}+\theta\right)\right]-h(\theta)$$

The term in square brackets is zero in $z=0$ and therefore $h(\theta)=0$. N^{22} is determined by the third equilibrium equation; it has the same sign as N^{11}.

Three remarks are needed in view of the result:

- The density of forces $q(\theta)$ applied along the edge propagates without dispersion along the characteristic lines $z+c\theta=\text{cst}$ and $z-c\theta=\text{cst}$.
- Unlike the barrel case, the shear condition can be freely selected at the end $z=0$.
- The membrane solution is only possible if the conditions at the other end are respected. q being periodic with period 2π, the conditions are respected if $\frac{2\ell}{c}=2n\pi$, n being any integer. The condition of validity of the membrane theory relates to the geometry of the shell. This is because the force $q(\theta)$ applied to one end is transmitted

along the characteristic lines. If the geometric condition is not respected, a membrane solution is still possible, but it is necessary to adapt the conditions at the end $z = 2\ell$ (or following the characteristic lines) to make them compatible with the membrane solution, entirely determined by the conditions in $z = 0$.

In conclusion of this study on quasi-cylindrical shells, it clearly appears that a slight longitudinal curvature substantially modifies the distribution of forces and that the sign of this curvature plays an essential role, the right cylinder being a borderline case between elliptical shell and hyperbolic shell. This also highlights the important role that geometrical imperfections, inevitable during the construction of the shell, can play.

Photo 6.5 - Cylindrical tanks in an industrial installation (Loiret-France)

6.3 SHELLS OF REVOLUTION

The middle surface of a shell of revolution is generated by rotation of a curve called generatrix around an axis. Many shells have this geometry, for example: shell of air-cooler, dome. The circular cylinder is a special case of surface of revolution.

6.3.1 Consequences of geometry

6.3.1.1 Parameterisation, natural basis

The middle surface is of revolution around the z axis and is parameterised by:

- s curvilinear abscissa along the generatrix;
- θ polar angle in the plane xy.

The associated natural basis is (Figure 6.17) :

$$\begin{cases} a_s = \dfrac{\partial \mathcal{m}}{\partial s} \\[2ex] a_\theta = \dfrac{\partial \mathcal{m}}{\partial \theta} \end{cases}$$

a_s is the unit tangent to the generatrix, a_θ is tangent to the intersection circle of the surface and of the plane parallel to the plane $\mathcal{O}xy$ passing through m. a_3 is defined as $\dfrac{a_s \times a_\theta}{\left\| a_s \times a_\theta \right\|}$.

The metric tensor is worth in this basis:

$$a_{\bullet\bullet} = \begin{pmatrix} 1 & 0 \\ 0 & r^2 \end{pmatrix}$$

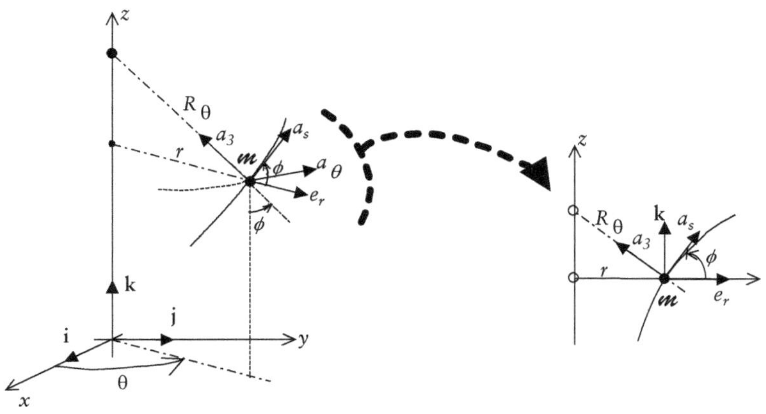

Figure 6.17

where r is the distance from the point m to the z axis and is usually not a radius of curvature of the surface.

The generatrix is most often described by a function $r(z)$ and the coordinates of the current point can be expressed by:

$$\begin{cases} x = r(z)\cos\theta \\ y = r(z)\sin\theta \\ z \end{cases}$$

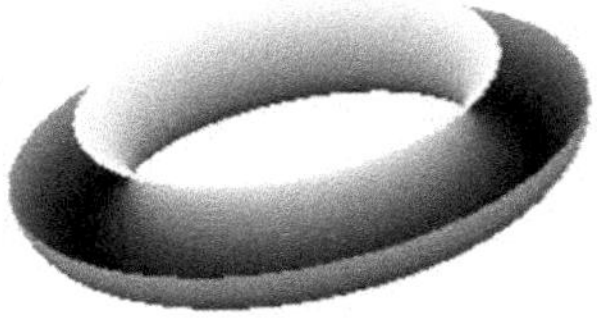

In the parameterisation (z, θ), the basic vectors are:

$$a_z = \begin{cases} r'\cos\theta \\ r'\sin\theta \\ 1 \end{cases} \quad a_\theta = \begin{cases} -r\sin\theta \\ r\cos\theta \\ 0 \end{cases} \quad a_3 = -\frac{1}{\sqrt{1+r'^2}} \begin{cases} \cos\theta \\ \sin\theta \\ -r' \end{cases}$$

with:

$$a_z = a_s \frac{ds}{dz} = a_s\sqrt{1+r'^2}$$

Let also be the orthonormal basis (a_1, a_2) associated with (a_s, a_θ), such that:

$$a_1 = a_s = \frac{a_z}{\sqrt{1+r'^2}} \quad ; \quad a_2 = \frac{a_\theta}{r}$$

In the orthonormal basis, the curvature b_{11} is determined by the curvature of the generatrix ($\mathbf{n}$ vector normal to the generatrix is confounded with a_3: $\mathbf{n} = a_3$). Given the orientation of a_3, b_{22} is positive and b_{11} is positive with the radius of the generator if r'' is negative (Figure 6.17).

$$b_{11} = \frac{1}{R_s} = -\frac{r''}{\left(1+r'^2\right)^{3/2}}$$

Curvature b_{22} is obtained by Meusnier's theorem:

$$b_{22} = \frac{1}{R_\theta} = \frac{\sin\phi}{r}\,;\ b_{\theta\theta} = r\sin\phi$$

where ϕ is the angle (e_r, a_s), which verifies:

$$\cos\phi = \frac{dr}{ds} = \frac{r'}{\sqrt{1+r'^2}}\,;\quad \sin\phi = \frac{dz}{ds} = \frac{1}{\sqrt{1+r'^2}} \tag{6.25}$$

ϕ is another common parameter of the generatrix and:

$$a_\phi = \frac{ds}{d\phi} a_s = R_s a_s$$

Then:

$$r' = \cot\phi\ ; \qquad a_3 = \begin{cases} -\cos\theta\sin\phi \\ -\sin\theta\sin\phi \\ \cos\phi \end{cases}$$

In the fixed orthonormal coordinate system (i, j, k), the vectors of the basis are decomposed into:

$$\begin{cases} a_1 = \left(\mathbf{i}\cos\theta + \mathbf{j}\sin\theta\right)\cos\phi + \mathbf{k}\sin\phi \\ a_2 = -\,\mathbf{i}\sin\theta + \mathbf{j}\cos\theta \\ a_3 = -\left(\mathbf{i}\cos\theta + \mathbf{j}\sin\theta\right)\sin\phi + \mathbf{k}\cos\phi \end{cases}$$

The Riemannian connection coefficients in the parameterisation (s, θ) are obtained by derivation of the vectors of the basis:

$$\frac{\partial a_s}{\partial s} = \frac{a_3}{R_s} \qquad \Rightarrow \qquad \Gamma^s_{ss} = \Gamma^\theta_{ss} = 0$$

$$\left.\begin{aligned} \frac{\partial a_s}{\partial \theta} &= \frac{\partial}{\partial \theta}\left(e_r\cos\phi + e_z\sin\phi\right) = \frac{\partial e_r}{\partial \theta}\cos\phi = \frac{a_\theta}{r}\cos\phi \\ \frac{\partial a_\theta}{\partial s} &= \frac{dr}{ds}\frac{\partial}{\partial r}\left(r\, a_2\right) = \frac{a_\theta}{r}\cos\phi \end{aligned}\right\} \Rightarrow \begin{cases} \Gamma^s_{s\theta} = 0 \\ \Gamma^\theta_{s\theta} = \dfrac{1}{r}\cos\phi \end{cases}$$

$$\frac{\partial a_\theta}{\partial \theta} = -r\, e_r = r \sin\phi\, a_3 - r\cos\phi\, a_s \quad \Rightarrow \quad \begin{cases} \Gamma^s_{\theta\theta} = -r\cos\phi \\ \Gamma^\theta_{\theta\theta} = 0 \end{cases}$$

Note that the Mainardi-Codazzi relation (3.80) is written $R_s \cos\phi = \partial_\phi r$, obvious relation according to (6.25).

6.3.1.2 Equilibrium equations

Local equations of equilibrium are written:

In the coordinate system (a_s, a_θ), according to (2):

$$\begin{cases} \partial_s N^{ss} + \partial_\theta N^{\theta s} + \dfrac{\cos\phi}{r} N^{ss} - r\cos\phi\, N^{\theta\theta} + p^s = 0 \\ \partial_s N^{s\theta} + \partial_\theta N^{\theta\theta} + \dfrac{3\cos\phi}{r} N^{s\theta} + p^\theta = 0 \\ \dfrac{N^{ss}}{R_s} + N^{\theta\theta}\, r\sin\phi + p^3 = 0 \end{cases} \tag{6.26}$$

In the associated reference coordinate system (a_1, a_2), according to (25):

$$\begin{cases} \partial_s N^{11} + \dfrac{1}{r}\partial_\theta N^{12} + \dfrac{\cos\phi}{r}\left(N^{11} - N^{22}\right) + p^1 = 0 \\ r\,\partial_s\left(\dfrac{N^{12}}{r}\right) + \dfrac{1}{r}\partial_\theta N^{22} + \dfrac{3\cos\phi}{r} N^{12} + p^2 = 0 \\ \dfrac{N^{11}}{R_s} + \dfrac{N^{22}\sin\phi}{r} + p^3 = 0 \end{cases} \tag{6.27}$$

which can also be written:

$$\boxed{\begin{array}{l} \partial_s\left(r\, N^{11}\right) + \partial_\theta N^{12} - N^{22}\cos\phi + r\, p^1 = 0 \\ \partial_s\left(r\, N^{12}\right) + \partial_\theta N^{22} + N^{12}\cos\phi + r\, p^2 = 0 \\ \dfrac{N^{11}}{R_s} + \dfrac{N^{22}\sin\phi}{r} + p^3 = 0 \end{array}} \tag{6.28}$$

The same result is obtained by applying the relations (9), in which $1/r_1 = 0$ (geodesic curvature of the meridian) and $r_2 = \dfrac{r}{\cos\phi}$.

6.3.1.3 Deformation

The calculations are made under the assumption of small displacements. The displacement is decomposed in the local coordinate system (with the classical notation):

$$\xi = \xi^s a_s + \xi^\theta a_\theta + \zeta\, a_3 = u\, a_1 + v\, a_2 + w\, a_3$$

By application of (5), it comes, in the coordinate system (a_s, a_θ):

$$\begin{aligned}
\varepsilon_{ss} &= \partial_s \xi_s - \frac{\zeta}{R_s} \\
\varepsilon_{s\theta} &= \frac{1}{2}\left(\partial_s \xi_\theta + \partial_\theta \xi_s\right) - \frac{\cos\phi}{r}\xi_\theta \\
\varepsilon_{\theta\theta} &= \partial_\theta \xi_\theta + \xi_s\, r\cos\phi - \zeta r \sin\phi
\end{aligned} \tag{6.29}$$

Considering relations : $\xi_s = u$; $\xi_\theta = r\,v$; $\zeta = w$, the strain tensor is written in the coordinate system (a_1, a_2):

$$\boxed{\begin{aligned}
\varepsilon_{11} &= \partial_s u - \frac{w}{R_s} \\
\varepsilon_{12} &= \frac{1}{2}\left(-\frac{v}{r}\cos\phi + \partial_s v + \frac{1}{r}\partial_\theta u\right) \\
\varepsilon_{22} &= \frac{1}{r}\partial_\theta v + \frac{u}{r}\cos\phi - \frac{w}{r}\sin\phi
\end{aligned}} \tag{6.30}$$

Note: $u \cos\phi - w \sin\phi$ is the displacement in the direction of e_r

6.3.1.4 Problems with symmetry of revolution

In this paragraph, the loading respects the symmetry of revolution, which implies that the different quantities do not depend on θ, hence:

- Equilibrium equations:

$$\begin{cases}
\dfrac{d}{ds}\left(r\,N^{11}\right) - N^{22}\cos\phi + r\,p^{1} = 0 \\
\dfrac{d}{ds}\left(r\,N^{12}\right) + N^{12}\cos\phi + r\,p^{2} = 0 \\
\dfrac{N^{11}}{R_s} + \dfrac{N^{22}\sin\phi}{r} + p^{3} = 0
\end{cases} \tag{6.31}$$

- Strain tensor:

$$\begin{aligned}
\varepsilon_{11} &= \frac{du}{ds} - \frac{w}{R_s} \\
\varepsilon_{12} &= \frac{1}{2}\left(\frac{dv}{ds} - \frac{v}{r}\cos\phi\right) \\
\varepsilon_{22} &= \frac{u}{r}\cos\phi - \frac{w}{r}\sin\phi
\end{aligned} \tag{6.32}$$

The equilibrium equations are solved simply by putting in the first equation the expression of N^{22} obtained by the third equation:

$$N^{22} = -\left(p^3 + \frac{N^{11}}{R_s}\right)\frac{r}{\sin\phi} \tag{6.33}$$

(the case where $\sin\phi = 0$ in every point corresponds to the case of the cylinder, treated previously). It comes:

$$\frac{d}{ds}\left(r\,N^{11}\right) + \left(p^3 + \frac{N^{11}}{R_s}\right) r\cot\phi + r\,p^1 = 0$$

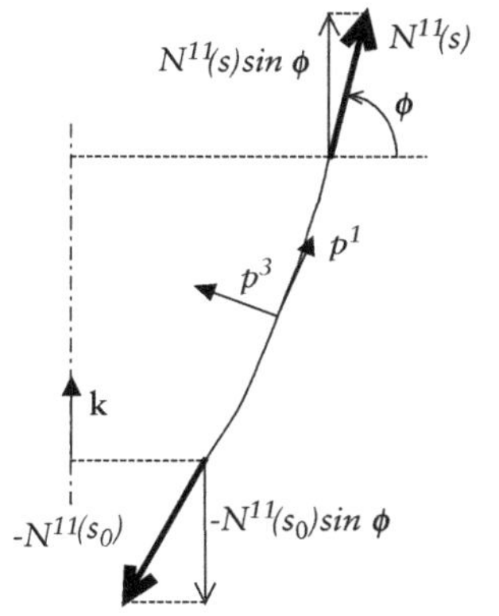

Figure 6.18

This differential equation in N^{11} is written, given the relation:

$$\frac{d(\sin\phi)}{ds} = \frac{\cos\phi}{R_s}$$

under the form:

$$\frac{d}{ds}\left(r\,N^{11}\sin\varphi\right) + r\left(p^1\sin\varphi + p^3\cos\varphi\right) = 0$$

and integrates in:

$$\left[r\,N^{11}\sin\phi\right]_{s_0}^{s} = -\int_{s_0}^{s} r\left(p^1\sin\phi + p^3\cos\phi\right)ds \tag{6.34}$$

and then makes it possible to determine N^{22} by (33).

This equality, multiplied by 2π, expresses the equilibrium, in the direction of **k**, of a shell portion delimited by the two planes perpendicular to **k**, between the abscissas s_0 and s (Figure 6.18).

The second equation of equilibrium is rewritten:

$$\frac{d}{ds}\left(r^2\,N^{12}\right) + r^2\,p^2 = 0$$

and integrates in:

$$\left[r^2\,N^{12}\right]_{s_0}^{s} = -\int_{s_0}^{s} p^2 r^2 ds \tag{6.35}$$

This equality, multiplied by 2π, expresses the equilibrium of a shell portion delimited by s_0 and s, with respect to the overall torsion around **k**.

In conclusion, the local equations are easily interpreted by equilibria of a shell slice limited by planes orthogonal to the axis of symmetry, vis-a-vis the force in the direction of this axis and the torsion around this axis.

The constitutive laws make it possible to calculate the deformations from the membrane forces determined above. The tensor ε being then given, the relations (30) make it possible to calculate the displacement.

The second relationship is written:

$$\frac{2\,\varepsilon_{12}}{\cos\phi} = \frac{dv}{dr} - \frac{v}{r}$$

and integrates in:

$$v = r\left(C_1 + \int \frac{2\,\varepsilon_{12}}{r\cos\phi}\,dr\right) \tag{6.36}$$

where C_1 is a constant determined by the boundary conditions.

An important special case is that where no load is applied in the orthoradial direction: $p^2 = q^2 = 0$. In this case: $N^{12} = 0$, which implies that $\varepsilon_{12} = 0$ in linear, homogeneous and isotropic elasticity. As a result: $v = 0$, except an overall rotation of axis **k**.

On the other hand, w can be eliminated from the other two equations (32):

$$w = R_s\left(\varepsilon_{11} + \frac{du}{ds}\right) = u\cot\phi - \frac{r\,\varepsilon_{22}}{\sin\phi} \tag{6.37}$$

hence, noting that $\frac{ds}{d\phi} = R_s$, the differential equation on u:

$$\frac{du}{d\phi}\sin\phi - u\cos\phi = -\left(R_s\sin\phi\,\varepsilon_{11} + r\,\varepsilon_{22}\right)$$

which integrates in:

$$u = \sin\phi\left(C_2 - \int \frac{R_s\sin\phi\,\varepsilon_{11} + r\,\varepsilon_{22}}{\sin^2\phi}\,d\phi\right) \tag{6.38}$$

where C_2 is a constant determined by the boundary conditions. w is then determined by (37).

6.3.1.5 Problems without symmetry of revolution

In the case of any loading inducing no symmetry of revolution, it remains the property of periodicity of the loading and the forces. Equations (28) are rewritten, eliminating N^{22} and taking as parameter ϕ rather than the curvilinear abscissa s:

$$\frac{1}{R_s}\partial_\phi\left(r\,N^{11}\right) + \partial_\theta N^{12} + \left(\frac{p^3 r}{\sin\phi} + \frac{rN^{11}}{R_s\sin\phi}\right)\cos\phi + r\,p^1 = 0$$

$$\partial_\phi\left(r\,N^{12}\right) - R_s\partial_\theta\left(\frac{p^3 r}{\sin\phi} + \frac{rN^{11}}{R_s\sin\phi}\right) + N^{12}\partial_\phi r + R_s r\,p^2 = 0$$

These equations can be simplified by introducing the auxiliary unknowns:

$$\Lambda = N^{11} r \sin\phi$$

$$M = N^{12} r^2$$

which allows to rewrite them:

$$\frac{r}{R_s}\partial_\phi\left(\frac{\Lambda}{\sin\phi}\right)+\partial_\theta\left(\frac{M}{r}\right)+\left(\frac{p^3 r^2}{\sin\phi}+\frac{r\Lambda}{R_s\sin^2\phi}\right)\cos\phi + r^2\, p^1 = 0$$

$$\partial_\phi\left(\frac{M}{r}\right) - R_s\partial_\theta\left(\frac{p^3 r}{\sin\phi}+\frac{\Lambda}{R_s\sin^2\phi}\right)+\frac{M}{r^2}\partial_\phi r + R_s r\, p^2 = 0$$

By expressing r as a function of R_θ, the elimination of M between the two equations leads to the partial differential equation for Λ:

$$\mathcal{D}\Lambda + \mathcal{P} = 0 \tag{6.39}$$

where, noting that:

$$\partial_\phi\left(\frac{M}{r}\right)+\frac{M}{r^2}\partial_\phi r = \frac{1}{r}\partial_\phi M$$

$$\partial_\phi\left[\frac{R_\theta{}^2\sin^2\phi}{R_s}\partial_\phi\left(\frac{\Lambda}{\sin\phi}\right)\right]+\partial_\phi\left(\frac{R_\theta{}^2\Lambda\cos\phi}{R_s}\right)=\partial_\phi\left[\frac{R_\theta{}^2\sin\phi}{R_s}\partial_\phi\Lambda\right]$$

$\mathcal{D}$ is the differential operator:

$$\mathcal{D}\Lambda = \frac{R_\theta \partial^2{}_{\theta^2}\Lambda}{\sin\phi}+\partial_\phi\left[\frac{R_\theta{}^2\sin\phi}{R_s}\partial_\phi\Lambda\right] \tag{6.40}$$

and where:

$$\begin{aligned}\mathcal{P} = {} & \partial_\theta\left(R_s R_\theta{}^2\sin^2\phi\, p^2\right)+\partial_\phi\left(R_\theta{}^3\sin^3\phi\, p^1\right)+\partial_\phi\left(p^3 R_\theta{}^3\sin^2\phi\cos\phi\right)\\ & + R_\theta{}^2\sin\phi\, R_s\partial^2{}_{\theta^2}p^3\end{aligned} \tag{6.41}$$

The solution Λ is necessarily periodic. The problem can therefore be solved by decomposing the loading and the forces in Fourier series. This leads to studying the elementary solution $\Lambda = \lambda_n(\phi)\cos n\theta$, for which:

$$\mathcal{D}\left[\lambda_n(\phi)\cos n\theta\right]=\left\{-n^2\frac{R_\theta\lambda_n}{\sin\phi}+\frac{d}{d\phi}\left[\frac{R_\theta{}^2\sin\phi}{R_s}\frac{d}{d\phi}(\lambda_n)\right]\right\}\cos n\theta$$

The loading corresponding to this elementary solution is:

$$p_n^1 \cos n\theta\,; \quad p_n^2 \sin n\theta\,; \quad p_n^3 \cos n\theta$$

and:

$$\mathcal{P} = \left[\begin{array}{c} np_n^2 R_s R_\theta{}^2 \sin^2\phi + \dfrac{d}{d\phi}\left(p_n^1 R_\theta{}^3 \sin^3\phi\right) \\ \\ + \dfrac{d}{d\phi}\left(p_n^3 R_\theta{}^3 \sin^2\phi \cos\phi\right) - n^2 p_n^3 R_\theta{}^2 R_s \sin\phi \end{array} \right] \cos n\theta$$

which leads to the differential equation for $\lambda_n(\phi)$:

$$\boxed{\begin{array}{l} -n^2 \dfrac{R_\theta \lambda_n}{\sin\phi} + \dfrac{d}{d\phi}\left[\dfrac{R_\theta{}^2 \sin\phi}{R_s}\dfrac{d\lambda_n}{d\phi}\right] + np_n^2 R_s R_\theta{}^2 \sin^2\phi \\ + \dfrac{d}{d\phi}\left(p_n^1 R_\theta{}^3 \sin^3\phi\right) + \dfrac{d}{d\phi}\left(p_n^3 R_\theta{}^3 \sin^2\phi \cos\phi\right) - n^2 p_n^3 R_\theta{}^2 R_s \sin\phi = 0 \end{array}} \quad (6.42)$$

An example of application of this method of resolution is given in § 6.3.2.4.

6.3.2 Some classic examples of problems involving shells of revolution

6.3.2.1 Spherical shells

6.3.2.1.1 Spherical domes under axisymmetric loading

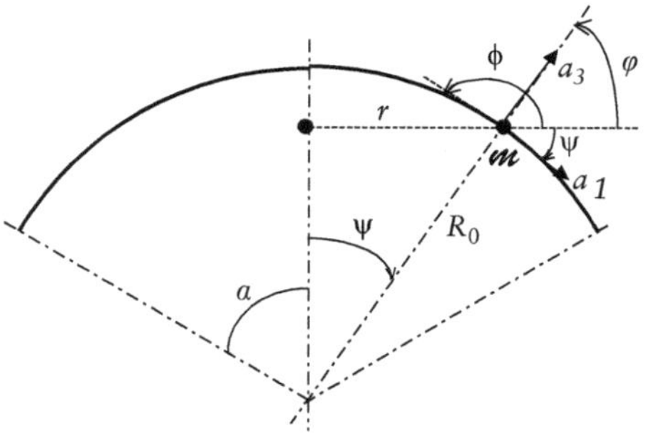

Figure 6.19

In the case of spherical domes, the parameters generally used are the angles θ and ψ, angle of the radius vector with the vertical axis of revolution. The radius of the sphere is $R_0 > 0$. It should be noted that the angle ϕ differs from the angle φ (latitude) of the spherical coordinates (Figure 6.19).

ψ is related to the angle ϕ by the relation:

$$\phi = \pi - \psi$$

ψ is limited to the opening angle α. It increases from the vertex towards the spring, so that n and a_3 are directed upwards and the radii of curvature are equal to $-R_0$.

In addition $r = R_0 \sin\psi$. The vector a_1 is the unit tangent vector such that:

$$a_1 = \frac{1}{R_0}\frac{\partial \mathcal{m}}{\partial \psi}$$

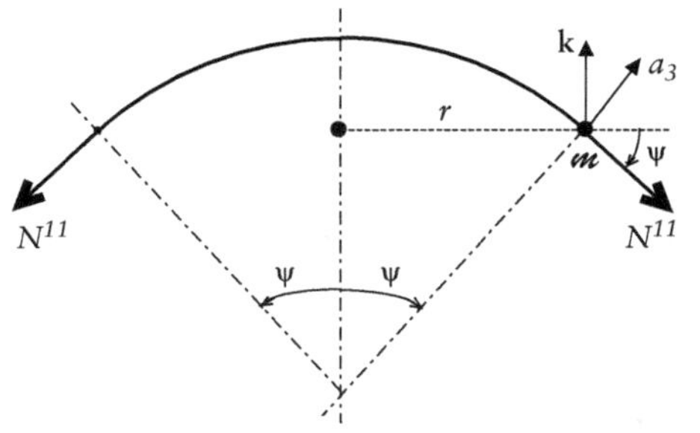

Figure 6.20

For a loading with symmetry of revolution, the membrane forces are obtained by directly writing the vertical equilibrium of a shell portion limited by a coordinate line $\psi = \text{cst}$ (Figure 6.20). Under the effect of the self-weight $q = \rho g$ (constant value per unit area, if the thickness is constant), the solution already obtained in § 1.3.1 is found with the help of (34):

$$2\pi r N^{11} \sin\psi = -\int_0^{\psi} q\, 2\pi R_0^2 \sin\psi\, d\psi = -2\pi R_0^2 q\left(1-\cos\psi\right)$$

The normal component of loading is:

$$p^3 = \left(-q\,\mathbf{k}\right)\bullet a_3 = -q\cos\psi$$

N^{22} is then obtained by the third equation of equilibrium (the radius of curvature is $-R_0$):

$$N^{11} + N^{22} = p^3 R_0$$

Hence the solution:

$$\begin{cases} N^{11} = -qR_0\dfrac{1-\cos\psi}{\sin^2\psi} = -\dfrac{qR_0}{1+\cos\psi} \\ N^{22} = qR_0\left(\dfrac{1}{1+\cos\psi} - \cos\psi\right) \end{cases}$$

N^{11} is always a compression. The circumferential force N^{22} is a compression at the top of the dome and up to the angle $\psi_0 = 51°\ 50'$; it is a tension beyond.

By expressing the deformations on the one hand according to the membrane forces by the constitutive law and on the other hand according to displacements, it comes:

$$\varepsilon_{11} = \frac{1}{Ee}\left(N^{11} - \nu N^{22}\right) = \frac{du}{R_0 d\psi} + \frac{w}{R_0}$$

$$\varepsilon_{22} = \frac{1}{Ee}\left(N^{22} - \nu N^{11}\right) = \frac{u}{R_0}\cot\psi + \frac{w}{R_0}$$

Subtracting these two expressions:

$$\frac{du}{R_0 d\psi} - \frac{u}{R_0}\cot\psi = \sin\psi \frac{d}{R_0 d\psi}\left(\frac{u}{\sin\psi}\right) = \frac{1+\nu}{Ee}\left(N^{11} - N^{22}\right) \tag{6.43}$$

Then:

$$w = \frac{R_0}{Ee}\left(N^{22} - \nu\, N^{11}\right) - u \cot \psi \tag{6.44}$$

(43) and (44) make it possible to calculate the displacements of the spherical shells submitted to axisymmetric loading. A single boundary condition completely determines u and w, which is a limitation of the membrane theory.

By reporting the previous expressions of the forces in (43), it comes:

$$\sin \psi \frac{d}{R_0 d\psi}\left(\frac{u}{\sin \psi}\right) = \frac{qR_0}{Ee}(1+\nu)\frac{-2+\cos \psi + \cos^2 \psi}{1+\cos \psi}$$

From where, by integration:

$$u(\psi) = \frac{qR_0^{\,2}(1+\nu)}{Ee}\sin \psi\left(C_1 + \int_0^{\psi} \frac{-2+\cos \psi + \cos^2 \psi}{(1+\cos \psi)\sin \psi}\, d\psi\right)$$

$$= -\frac{qR_0^{\,2}(1+\nu)}{Ee}\sin \psi\left(-C_2 + \ln\left(\frac{1}{1+\cos \psi}\right) + \frac{1}{1+\cos \psi}\right)$$

If for example it is a semicircular vault supported in $\psi = \pm\frac{\pi}{2}$, it comes:

$$u(\psi) = -\frac{qR_0^{\,2}(1+\nu)}{Ee}\sin \psi\left[\ln\left(\frac{1}{1+\cos \psi}\right) - \frac{\cos \psi}{1+\cos \psi}\right]$$

Then w is entirely determined by (44) without it being possible to impose an additional condition on springs without violating the membrane equilibrium:

$$w(\psi) = \frac{qR_0^{\,2}}{Ee}\left\{\left(\frac{1+\nu-\cos \psi - \cos^2 \psi}{1+\cos \psi}\right) + (1+\nu)\cos \psi\left[\ln\left(\frac{1}{1+\cos \psi}\right) - \frac{\cos \psi}{1+\cos \psi}\right]\right\}$$

In $\psi = \pm\frac{\pi}{2}$, w is worth $\frac{qR_0^{\,2}(1+\nu)}{Ee}$, which is generally incompatible with the boundary conditions at the spring. A flexion therefore appears at springs. The deflection at the key is worth:

$$w(0) = -\frac{qR_0^{\,2}}{Ee}\left[1 + (1+\nu)\ln 2\right]$$

Under the effect of a snow load p_N, considered constant per unit of horizontal area:

$$2\pi\, r\, N^{11} \sin\psi = -\, \pi\, r^2 p_N$$

and the normal component of the load is, reduced to the unit area of the shell:

$$p^3 = -p_N \frac{dr}{R_0 d\psi} \cos\psi = -p_N \cos^2\psi$$

Finally:

$$\begin{cases} N^{11} = -\dfrac{p_N R_0}{2} \\ N^{22} = p_N R_0 \left(\dfrac{1}{2} - \cos^2\psi\right) \\ \quad = -\dfrac{p_N R_0}{2} \cos 2\psi \end{cases}$$

The circumferential force changes sign in $\psi = \frac{\pi}{4}$.

Domes with a ridge opening (Figure 6.21) can be treated by the same method. Such an opening is covered by a skylight that transmits a vertical linear load P.

Under the effect of this load and the self weight, the vertical balance of a portion of dome is written:

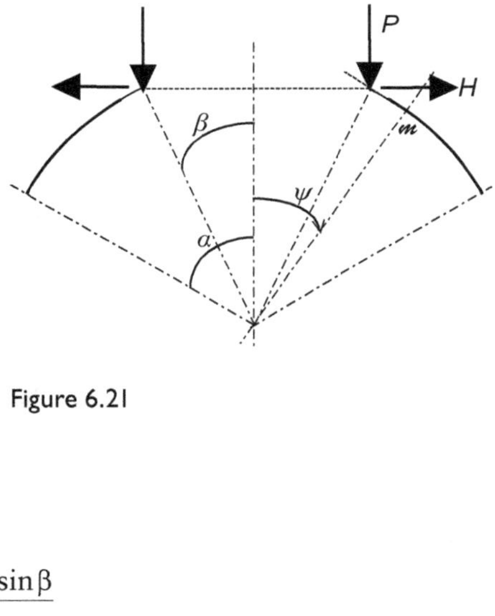

Figure 6.21

$$2\pi\, r\, N^{11} \sin\psi = -2\pi \int_{\beta}^{\psi} q R_0^2 \sin\psi\, d\psi - 2\pi P R_0 \sin\beta$$

From where:

$$\begin{cases} N^{11} = -\left(q R_0 \dfrac{\cos\beta - \cos\psi}{\sin^2\psi} + \dfrac{P \sin\beta}{\sin^2\psi}\right) \\ N^{22} = -\, q R_0 \cos\psi \left(1 + \dfrac{1}{\sin^2\psi}\right) + \dfrac{q R_0 \cos\beta + P \sin\beta}{\sin^2\psi} \end{cases}$$

At the top edge, N^{11} is worth:

$$N^{11} = -\frac{P}{\sin\beta}$$

The effort applied to the edge must be tangent to the shell so that the membrane solution is acceptable. It is necessary, to maintain the membrane equilibrium, to have a ring beam (in compression) at the edge of the hole, able to balance a horizontal linear reaction:

$$H = P \cot\beta$$

which resultant with P well balances N^{11}.

Photo 6.6 - Spherical dome with skylight (Lisbon-Portugal)

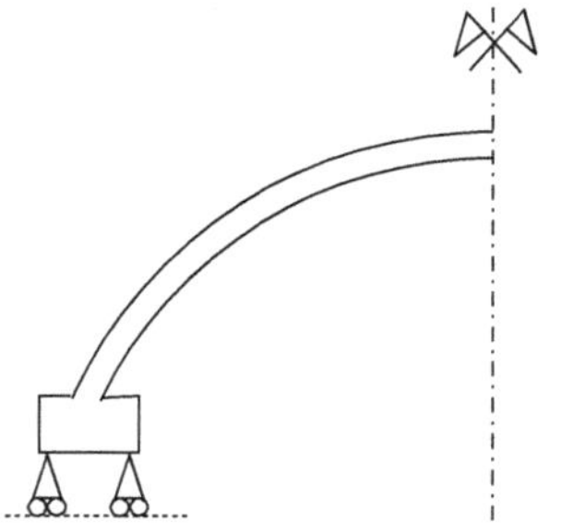

Figure 6.22

In the same way, a ring beam is arranged in the lower part of the dome to allow its support, whatever the shape of the shell (Figure 6.22): the support reaction of the shell, which is tangent to it, is decomposed into a vertical component transmitted in the columns and a horizontal component balanced by the ring beam.

The shell is generally stiffer than the ring beam, so it is clear that the compatibility of the deformations is not assured, in membrane theory, between the shell and the ring beam: there is a bending in the shell, which remains localised in the neighborhood of the ring beam.

6.3.2.1.2 Spherical dome under antisymmetric loading

The wind induces a pressure not uniformly distributed on a dome. As a first approximation (Figure 6.23 (a) and (b)), this pressure can be represented by the expression (but this is not sufficient for a complete resolution, see § 2.2.3):

$$p^3 = -\, q \sin\psi \cos\theta$$

The equilibrium equations (28) are particularised in:

$$-\,\partial_\psi\left(N^{11}\sin\psi\right) + \partial_\theta N^{12} + N^{22}\cos\psi = 0$$

$$-\,\partial_\psi\left(N^{12}\sin\psi\right) + \partial_\theta N^{22} - N^{12}\cos\psi = 0$$

$$N^{11} + N^{22} = -\, qR_0 \sin\psi \cos\theta$$

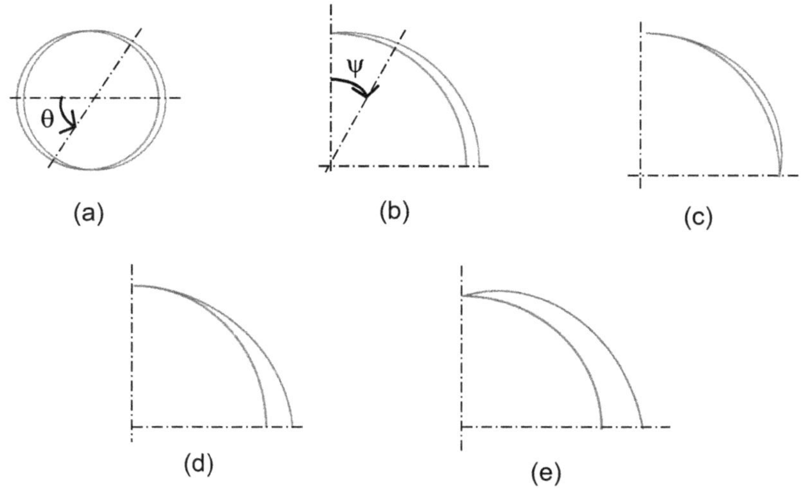

Figure 6.23 (a) pressure distribution according to θ; (b) according to ψ; (c) N^{11}; (d) N^{12}; (e) N^{22} according to ψ

The shape of these equations suggests looking for a solution in the form:

$$N^{11} = n^{11}\cos\theta\,;\ \ N^{22} = n^{22}\cos\theta\,;\ \ N^{12} = n^{12}\sin\theta$$

After elimination of n^{22}, the equations are reduced to the differential system:

$$-\partial_\psi\left(n^{11}\sin\psi\right) - n^{11}\cos\psi + n^{12} = qR_0\sin\psi\cos\psi$$

$$n^{11} - \partial_\psi\left(n^{12}\sin\psi\right) - n^{12}\cos\psi = -qR_0\sin\psi$$

As in the case of axisymmetric loadings, it is possible to simplify the search for the solution by writing an overall equilibrium of a portion of dome limited by the angle ψ. The result of the pressure on the portion is worth:

$$\iint_d p^3 a_3 dS = -q\int_0^\psi \left\{\int_0^{2\pi} \cos\theta\left[\left(\mathbf{i}\cos\theta + \mathbf{j}\sin\theta\right)\sin\psi + \mathbf{k}\cos\psi\right]d\theta\right\} R_0^2\sin^2\psi d\psi$$

$$= -\frac{\pi R_0^2 q}{3}\left[-\cos\psi\sin^2\psi + 2\left(1-\cos\psi\right)\right]\mathbf{i}$$

and its resulting moment with respect to the center of the sphere is null.

Along the horizontal cut, the resultant of membrane forces is:

$$\int_0^{2\pi}\left(N^{11}a_s + N^{12}a_2\right) r\, d\theta = \int_0^{2\pi}\left[\begin{array}{l} n^{11}\cos\theta\left[\left(\mathbf{i}\cos\theta + \mathbf{j}\sin\theta\right)\cos\psi - \mathbf{k}\sin\psi\right] \\ +n^{12}\sin\theta\left(-\mathbf{i}\sin\theta + \mathbf{j}\cos\theta\right)\end{array}\right] R_0\sin\psi\, d\theta$$

$$= \pi R_0\sin\psi\left(-n^{12} + n^{11}\cos\psi\right)\mathbf{i}$$

and their resulting moment with respect to the centre:

$$\int_0^{2\pi} \left(N^{11} a_s + N^{12} a_2\right) \wedge \left(R_0 a_3\right) r \, d\theta = \int_0^{2\pi} \left(-N^{11} a_2 + N^{12} a_s\right) R_0^2 \sin\psi \; d\theta$$

$$= \int_0^{2\pi} \begin{bmatrix} -n^{11} \cos\theta(-\mathbf{i}\sin\theta + \mathbf{j}\cos\theta) \\ +n^{12} \sin\theta\left[(\mathbf{i}\cos\theta + \mathbf{j}\sin\theta)\cos\psi - \mathbf{k}\sin\psi\right] \end{bmatrix} R_0^2 \sin\psi \; d\theta$$

$$= \pi R_0^2 \sin\psi\left(-n^{11} + n^{12}\cos\psi\right)\mathrm{j}$$

which results in $n^{11} = n^{12}\cos\psi$ and allows to calculate the complete solution:

$$N^{11} = -\frac{R_0 q \cos\psi}{3\sin^3\psi}\left[-\cos\psi \sin^2\psi + 2(1-\cos\psi)\right]\cos\theta$$

$$N^{12} = -\frac{R_0 q}{3\sin^3\psi}\left[-\cos\psi \sin^2\psi + 2(1-\cos\psi)\right]\sin\theta$$

$$N^{22} = \frac{R_0 q}{3\sin^3\psi}\left[-\cos^2\psi \sin^2\psi + 2\left(\cos\psi - \cos^2\psi\right) - 3\sin^4\psi\right]\cos\theta$$

The distribution of forces as a function of ψ is given in figure 6.23 (c) to (e), in $\theta = 0$ for N^{11} and N^{22} and in $\theta = \pi/2$ for N^{12}.

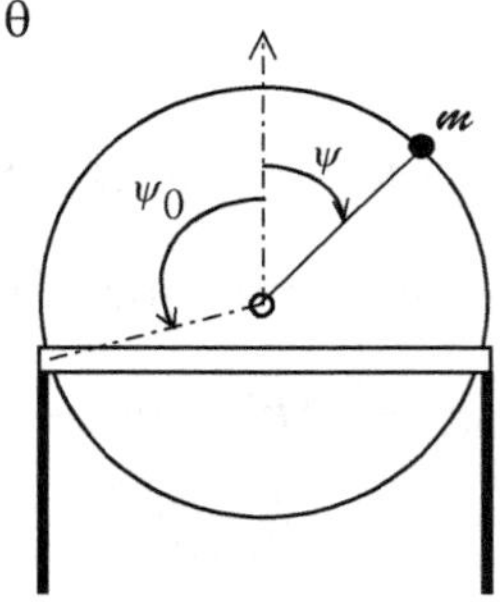

Figure 6.24

6.3.2.1.3 Spherical tank

Let a spherical tank containing a fluid under pressure, at rest. This tank rests on a continuous ring beam, at the elevation defined by the angle ψ_0 (Figure 6.24). The pressure is given by:

$$p^3 = p_0 + \gamma R_0\left(1 - \cos\psi\right)$$

where γ is the specific weight of the liquid.

As previously, N^{11} is obtained by balancing the spherical portion defined by the angle ψ, in vertical projection:

$$2\pi r N^{11} \sin\psi = 2\pi R_0^2 \int_0^{\psi} \left[p_0 + \gamma R_0\left(1-\cos\psi\right)\right]\cos\psi \sin\psi d\psi$$

$$= \frac{\pi R_0^2}{2} p_0\left(1-\cos 2\psi\right) + 2\pi R_0^3 \gamma\left[\frac{1}{6} - \frac{1}{2}\left(1 - \frac{2}{3}\cos\psi\right)\cos^2\psi\right]$$

From where:

$$N^{11} = \frac{p_0 R_0}{2} + \frac{1}{6}\gamma R_0^2 \frac{(1-\cos\psi)(1+2\cos\psi)}{1+\cos\psi}$$

Then, by the third equation of equilibrium:

$$N^{22} = p_0 R_0 + \gamma R_0^2 (1 - \cos\psi) - N^{11}$$

or:

$$N^{22} = \frac{p_0 R_0}{2} + \frac{1}{6}\gamma R_0^2 \frac{(1 - \cos\psi)(5 + 4\cos\psi)}{1 + \cos\psi}$$

At the top of the tank:

$$N^{11}(\psi = 0) = N^{22}(\psi = 0) = \frac{p_0 R_0}{2}$$

These relations are only valid for the upper part ($\psi < \psi_0$), since the reaction of the ring beam must then be taken into account. Membrane forces can be obtained in the lower part ($\psi > \psi_0$) by writing the equilibrium of the spherical portion between ψ_0 and $\psi = \pi$, which, all calculations made, leads to:

$$\begin{cases} N^{11} = \dfrac{p_0 R_0}{2} + \dfrac{1}{6}\gamma R_0^2 \left(5 + \dfrac{2\cos^2\psi}{1 - \cos\psi}\right) \\ N^{22} = \dfrac{p_0 R_0}{2} + \dfrac{1}{6}\gamma R_0^2 \left(1 - 6\cos\psi - \dfrac{2\cos^2\psi}{1 - \cos\psi}\right) \end{cases}$$

At the bottom of the tank:

$$N^{11}(\psi = \pi) = N^{22}(\psi = \pi) = \frac{p_0 R_0}{2} + \gamma R_0^2$$

Note: the value of ψ_0 does not influence the expressions of the membrane forces, but only gives the limit of application of the two pairs of expressions.

The membrane forces include a part due to the pressure p_0 at the top of the tank, giving the conventional solution under constant pressure (uniform traction equal to $p_0 R_0/2$) and a part related to the weight of the fluid contained in the tank. This second part creates a traction in both directions (ψ_0 being in practice greater than $\pi/2$). This weight equal to $\frac{4}{3}\pi\gamma R_0^3$ is necessarily balanced by the vertical component of the ring beam reaction, ie, per unit of length:

$$q_v = \frac{\frac{4}{3}\pi\gamma R_0^3}{2\pi R_0 \sin\psi_0} = \frac{2}{3}\frac{\gamma R_0^2}{\sin\psi_0}$$

At the level of the ring beam ($\psi = \psi_0$), the membrane forces are discontinuous and the discontinuities are worth:

$$\left[N^{11}\right] = N^{11}\left(\psi_0^{+}\right) - N^{11}\left(\psi_0^{-}\right) = \frac{2}{3}\frac{\gamma R_0^2}{\sin^2\psi_0} = -\left[N^{22}\right]$$

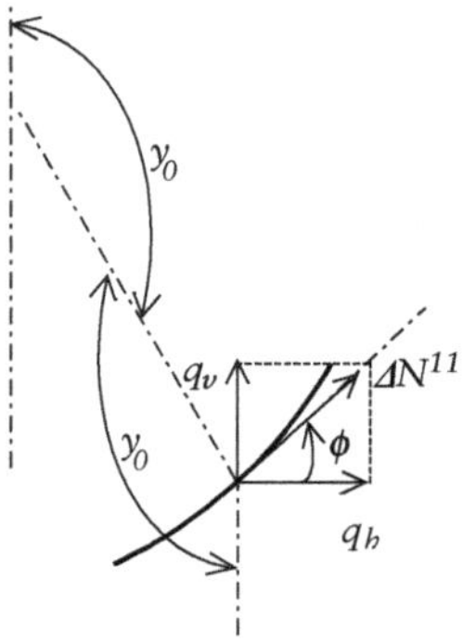

Figure 6.25

The discontinuity of N^{11} is produced by a density of linear force $\frac{2\gamma R_0^2}{3\sin^2\psi_0}$ along the ring beam which vertical projection is equal to the vertical reaction due to weight. This means that the ring beam also exerts on the sphere a density of horizontal force $\frac{2\gamma R_0^2 \cot\psi_0}{3\sin\psi_0}$ so that the resultant of the linear forces applied to the sphere is tangent to it and equal to $[N^{11}]$. For a value of ψ_0 greater than $\pi/2$, the horizontal reaction q_h is directed outwards (Figure 6.25), which may seem contrary to the first intuition. However, this is understood by considering that the bottom of the tank is in traction stronger than that due to the upper part, under the effect of the weight of the fluid, and therefore pulls on the ring beam in the tangential direction. A sufficiently rigid ring beam working as a compressed ring, arranged at the level of the support of the sphere, makes it possible to provide such a horizontal support.

Note: the constant pressure p_0 creates isotropic traction and does not intervene in the reaction.

In reality, the reactions applied by the ring beam on the sphere depend on the relative stiffness and the nature of the contact (friction...). As a result, the membrane solution is generally not respected in the vicinity of the ring beam, where bending is generated.

Moreover, the discontinuity of N^{11} induces an opposite discontinuity of N^{22}, by the third equilibrium equation. Such discontinuities lead, as seen in § 1.2.3, to discontinuities in the deformations and it is not possible to ensure the compatibility of displacements.

Indeed, in the upper part, the tangential displacement is given by (43):

$$\sin\psi\frac{d}{R\,d\psi}\left(\frac{u}{\sin\psi}\right) = -\frac{1}{3}\gamma R_0^2\frac{1+\nu}{Ee}\frac{(1-\cos\psi)(2+\cos\psi)}{1+\cos\psi}$$

similar to what was obtained for the dome under self weight. By integration, it comes:

$$u(\psi) = -\frac{1}{3}\gamma R_0^3\frac{1+\nu}{Ee}\left[\ln\left(\frac{1}{1+\cos\psi}\right) + \frac{1}{1+\cos\psi} - C_1\right]\sin\psi$$

Similarly, for the lower part:

$$\sin\psi\frac{d}{R\,d\psi}\left(\frac{u}{\sin\psi}\right) = \frac{1}{3}\gamma R_0^2\frac{1+\nu}{Ee}\left[2 + 3\cos\psi + \frac{2\cos^2\psi}{1-\cos\psi}\right]$$

From where:

$$u(\psi) = \frac{1}{3}\gamma R_0^3\frac{1+\nu}{Ee}\left[\ln\left(\frac{1}{1-\cos\psi}\right) + \frac{1}{1-\cos\psi} - C_2\right]\sin\psi$$

As expected, u does not depend on p_0. u is zero at the top and bottom of the tank. The conditions of support in $\psi = \psi_0$ make it possible to calculate the constants C_1 and C_2, while ensuring the continuity of u.

- If there is no slip between the sphere and its support, then $u(\psi_0) = 0$, it follows:
  - in the upper part $(\psi \leq \psi_0)$:

$$u(\psi) = -\frac{1}{3}\gamma R_0^3 \frac{1+\nu}{Ee}\left[\ln\left(\frac{1+\cos\psi_0}{1+\cos\psi}\right) + \frac{1}{1+\cos\psi} - \frac{1}{1+\cos\psi_0}\right]\sin\psi$$

and by (44):

$$w = \frac{p_0 R_0^2}{2Ee}(1-\nu) + \frac{\gamma R_0^3}{3Ee}\left\{\begin{matrix}\dfrac{(1-\cos\psi)}{1+\cos\psi}\left[\dfrac{5-\nu}{2} + (2-\nu)\cos\psi\right] \\ +(1+\nu)\left[\ln\left(\dfrac{1+\cos\psi_0}{1+\cos\psi}\right) + \dfrac{1}{1+\cos\psi} - \dfrac{1}{1+\cos\psi_0}\right]\cos\psi\end{matrix}\right\}$$

  - in the lower part $(\psi \geq \psi_0)$:

$$u(\psi) = \frac{1}{3}\gamma R_0^3 \frac{1+\nu}{Ee}\left[\ln\left(\frac{1-\cos\psi_0}{1-\cos\psi}\right) + \frac{1}{1-\cos\psi} - \frac{1}{1-\cos\psi_0}\right]\sin\psi$$

$$w = \frac{p_0 R_0^2}{2Ee}(1-\nu) + \frac{\gamma R_0^3}{3Ee}\left[\begin{matrix}\dfrac{1-5\nu}{2} - 3\cos\psi - (1+\nu)\dfrac{\cos^2\psi}{1-\cos\psi} \\ -(1+\nu)\left[\ln\left(\dfrac{1-\cos\psi_0}{1-\cos\psi}\right) + \dfrac{1}{1-\cos\psi} - \dfrac{1}{1-\cos\psi_0}\right]\cos\psi\end{matrix}\right]$$

The radial displacement contains a uniform swelling due to pressure p_0.

- Other kinematic conditions can be written in $\psi = \psi_0$, for example to express a possible slip. They must respect the continuity of u, which gives a condition between C_1 and C_2. In all cases, w is then discontinuous at the right of support, which is not possible. Indeed, by (44):

$$[w] = -\frac{R_0}{Ee}(1+\nu)\left[N^{11}\right] = -\frac{2}{3}\frac{1+\nu}{Ee}\frac{\gamma R_0^3}{\sin^2\psi_0}$$

The membrane theory is in default in the vicinity of the support, where it is necessary to take into account a localised flexion.

6.3.2.2 *Parabolic domes*

These are axisymmetrical domes with vertical axis which generatrix is a parabola (Figure 6.26). Parameterisation of the parabola has been given in § 2.2.4. In particular, if

R_0 denotes the radius of curvature of the parabola at the vertex, its radius at any point in the orthoradial direction is:

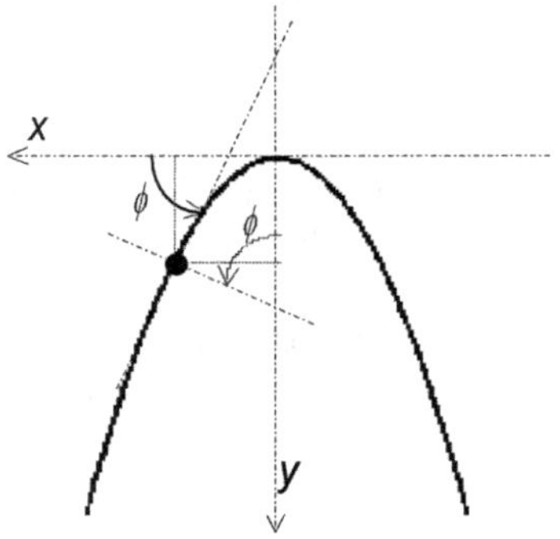

Figure 6.26

$$R_\theta = \frac{x}{\sin\phi} = \frac{R_0}{\cos\phi}$$

The membrane forces are expressed in the orthonormal coordinate system associated with (a_ϕ, a_θ). The loadings considered here are vertical and, consequently, $p^2 = 0$ and $N^{12} = 0$.

► For loading due to the self weight (assumed constant): $p^1 = p_0 \sin\phi$; $p^3 = p_0 \cos\phi$. N^{11} is obtained by writing the equilibrium of a portion of the dome delimited by a given value of ϕ:

$$2\pi R_0 \tan\phi \times N^{11} \sin\phi = -p_0 \int_0^{2\pi} d\theta \int_0^{\phi} \frac{R_0^2 \sin\phi}{\cos^4\phi} d\phi$$

$$= -\frac{2\pi R_0^2 p_0}{3}\left(\frac{1}{\cos^3\varphi} - 1\right)$$

from where N^{11} :

$$N^{11} = -\frac{p_0 R_0}{3\cos^2\phi \sin^2\phi}\left(1-\cos^3\phi\right) \xrightarrow[\phi\to 0]{} -\frac{p_0 R_0}{2}$$

N^{22} is obtained by replacing N^{11} in the third equilibrium equation:

$$N^{11}\frac{\cos^3\phi}{R_0} + N^{22}\frac{\sin\phi}{R_0 \tan\phi} + p^3 = 0 \quad \Rightarrow \quad N^{22} = -\frac{p^3 R_0}{\cos\phi} - N^{11}\cos^2\phi$$

From where:

$$N^{22} = -p^3 R_0\left(1 - \frac{1}{3}\frac{1-\cos^3\phi}{1-\cos^2\phi}\right) = -\frac{p^3 R_0}{3}\left(2 - \frac{\cos^2\phi}{1+\cos\phi}\right) \xrightarrow[\phi\to 0]{} -\frac{p_0 R_0}{2}$$

N^{22} is always negative.

► For loading due to snow: $p^1 = p_n \cos\phi \sin\phi$; $p^3 = p_n \cos^2\phi$. N^{11} is obtained by writing the equilibrium for a given value of ϕ, the resultant of the snow being given by the disk area in horizontal projection of the parabolic surface portion:

$$2\pi R_0 \tan\phi \times N^{11} \sin\phi = -p_n \pi R_0^2 \tan^2\phi$$

From where N^{11}, then N^{22} :

$$N^{11} = N^{22} = -\frac{p_n R_0}{2\cos\phi}$$

The parabolic dome remains compressed in both directions, in the two cases of load considered, whatever the value of ϕ, contrary to the case of the spherical dome. This gives it an advantage in terms of resistance which explains its wide use in masonry domes.

6.3.2.3 Conical shells

The cone is parameterised by the angle α and the abscissa s along a generatrix. The curvature of the generatrix is zero. The other radius of curvature is $s \tan\alpha$.

Note: The surface being developable, a stress function can be used.

Example : conical tank.

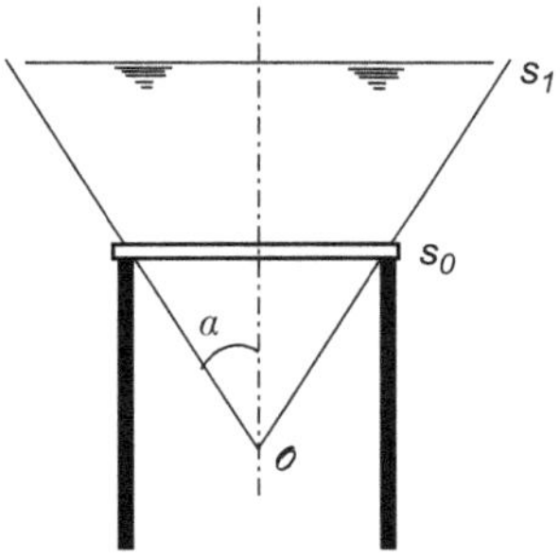

Figure 6.27

A tank consists of a conical shell with an opening angle α (Figure 6.27). It is based on a subframe located at the abscissa s_0. It contains a fluid at rest, which is flush with the top of the cone, at elevation h.

The pressure exerted by the fluid is:

$$p^3 = \gamma(s_1 - s)\cos\alpha$$

N^{22} comes from the third equation of equilibrium:

$$N^{22} = p^3 s \tan\alpha = \gamma\, s\,(s_1 - s)\sin\alpha$$

N^{11} is then obtained by the first equation, noting that $\phi = \frac{\pi}{2} - \alpha$ and $r = s\sin\alpha$:

Photo 6.7 - Water tower (Lot-France)

$$\frac{d}{ds}\left(sN^{11}\right) - N^{22} = 0$$

equation that integrates in:

$$N^{11} = \gamma\, s\left(\frac{s_1}{2} - \frac{s}{3}\right)\sin\alpha + \frac{C}{s}$$

The constant C is determined by the boundary condition at $s = s_1$, where N^{11} is zero. Finally:

$$N^{11} = \gamma\left[s\left(\frac{s_1}{2} - \frac{s}{3}\right) - \frac{s_1^{\,3}}{6s}\right]\sin\alpha$$

This solution is only valid for $s > s_0$. In $s = s_0$, the reaction of the subframe introduces a discontinuity.

For $s < s_0$, the expression of N^{22} is not modified and N^{11} is obtained in the same way. The constant C is then determined by the reaction, uniformly distributed around the circumference, which vertical resultant is opposed to the weight $\gamma\frac{\pi}{3}s_1^{\,3}\cos\alpha\,\sin^2\alpha$ of the fluid. Then, in the lower part of the tank:

$$N^{11} = \gamma\left[s\left(\frac{s_1}{2} - \frac{s}{3}\right) + \frac{s_1^{\,3}}{6}\left(\frac{1}{s_0} - \frac{1}{s}\right)\right]\sin\alpha$$

N^{11} is a compression in the upper part, but may be a tension in the lower part, depending on s_0. N^{11} tends to infinity at the point $s = 0$, where the membrane solution is not sufficient.

The reaction can not be purely vertical and there is a horizontal component to which the subframe must resist in ring. The horizontal force per unit length applied to the subframe is $\frac{1}{6}\gamma\frac{s_1^{\,3}}{s_0}\sin^2\alpha$.

In the vicinity of the support, the discontinuity of N^{11} creates a discontinuity of deformation and, here again, the membrane solution can not suffice.

Example : Umbrella under self weight.

A conical umbrella (Figure 6.28) is supported by a central mast and subjected to its self weight ρg, which components in the local coordinate system of the shell are:

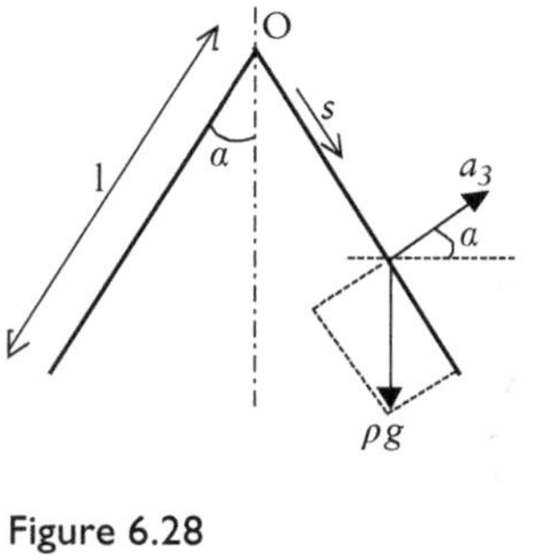

Figure 6.28

$$p_1 = \rho g\cos\alpha$$

$$p_3 = -\rho g\sin\alpha$$

The third equation of equilibrium leads to:

$$N^{22} = -\rho g\; s\,\sin\alpha\,\tan\alpha < 0$$

and the first equation is rewritten:

$$\frac{d}{ds}\left(s\,N^{11}\right) - N^{22} + \rho g\; s\;\cos\alpha = 0$$

By replacing N^{22} with the value obtained, integrating and noting that $N^{11}(\ell) = 0$, it comes:

$$N^{11} = \rho g \frac{\ell^2 - s^2}{2s\;\cos\alpha} > 0$$

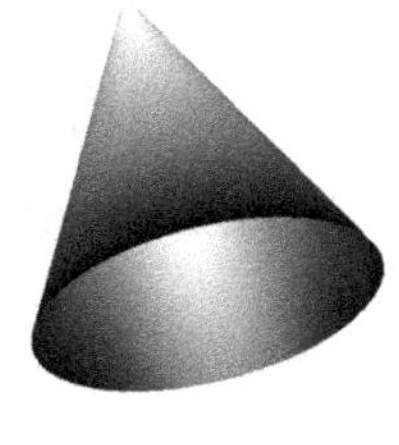

When s tends to 0, N^{11} tends to infinity and so there is local bending.
N^{11} is always a traction and N^{22} a compression.

6.3.2.4 Hyperboloid of revolution

Let be a cooling tower which middle surface is a hyperboloid of revolution (Figure 6.29).

The equation of the generatrix is:

$$\frac{r^2}{\alpha^2} - \frac{z^2}{\beta^2} = 1 \qquad (6.45)$$

The generatrix can also be defined in parametric form depending on the angle ϕ:

$$\begin{cases} r = \dfrac{\alpha^2}{C}\sin\phi \\ z = \dfrac{\beta^2}{C}\cos\phi \end{cases} \qquad (6.46)$$

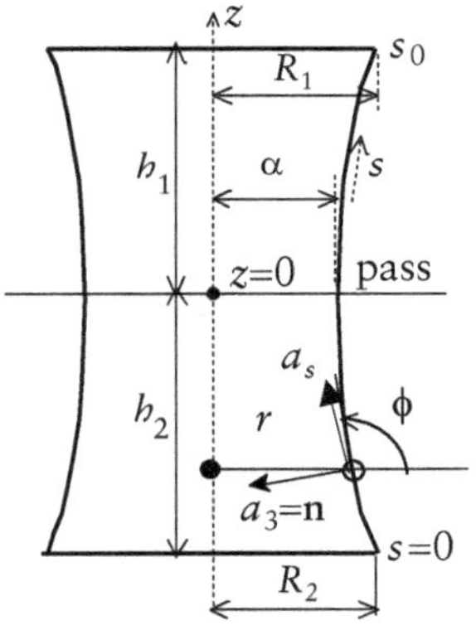

Figure 6.29

with:

$$c^2 = \alpha^2 \sin^2\phi - \beta^2 \cos^2\phi \qquad (6.47)$$

Indeed, with these expressions:

$$\begin{cases} \dfrac{dr}{d\phi} = \dfrac{\alpha^2}{C}\cos\phi - \dfrac{\alpha^2}{C^2}\sin\phi\dfrac{dC}{d\phi} = -\dfrac{\alpha^2\beta^2}{C^3}\cos\phi \\ \dfrac{dz}{d\phi} = -\dfrac{\beta^2}{C}\sin\phi - \dfrac{\beta^2}{C^2}\cos\phi\dfrac{dC}{d\phi} = -\dfrac{\alpha^2\beta^2}{C^3}\sin\phi \end{cases}$$

which check relationships (25) well:

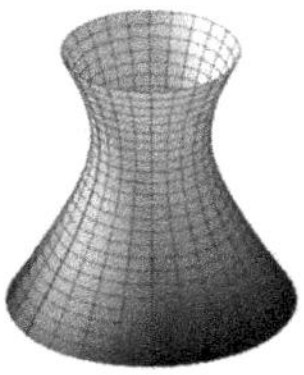

$$\begin{cases} \dfrac{dr}{ds} = \cos\phi \\ \dfrac{dz}{ds} = \sin\phi \end{cases} \quad ; \quad \frac{dr}{d\phi} = R_s\cos\phi$$

z is zero at the pass, where $\phi = \dfrac{\pi}{2}, \dfrac{dr}{dz} = 0$ and $r = \alpha$.

α and β are easily connected by (43) to the radii of the hyperboloid at the extreme elevations h_1 and h_2, defined from the pass:

$$\begin{cases} \alpha^2 = \dfrac{R_2^2 h_1^2 - R_1^2 h_2^2}{h_1^2 - h_2^2} \\ \beta^2 = \dfrac{R_2^2 h_1^2 - R_1^2 h_2^2}{R_1^2 - R_2^2} \end{cases}$$

which implies that the three quantities $h_1^2 - h_2^2$, $R_1^2 - R_2^2$ and $R_2^2 h_1^2 - R_1^2 h_2^2$ have the same sign.

Finally:

$$R_s = \frac{ds}{d\phi} = -\frac{\alpha^2\beta^2}{C^3}$$

is the radius of curvature of the generatrix, the other radius of curvature is worth $\dfrac{\alpha^2}{C}$. The curvature of the generatrix is directed outwards.

► Under the effect of the self weight of surface density: $p = \rho\, g$ the loading is written:

$$\begin{cases} p^1 = -p \sin\phi \\ p^3 = -p \cos\phi \end{cases}$$

Noting that N^{11} is zero at the top of the tower, where $s = s_0$, equation (34) becomes:

$$r\, N^{11} \sin\phi = \int_{s0}^{s} pr\left(\sin^2\phi + \cos^2\phi\right) ds$$

or:

$$\frac{\alpha^2}{C} N^{11} \sin^2\phi = \int_{\phi 0}^{\phi} -p\frac{\alpha^2}{C}\sin\phi \frac{\alpha^2\beta^2}{C^3} d\phi$$

ϕ_0 being the value of ϕ corresponding to abscissa s_0.

It comes:

$$N^{11} = -\frac{\alpha^2\beta^2 C}{\sin^2\phi}\int_{\phi 0}^{\phi} \frac{p}{C^4} \sin\phi\, d\phi$$

Photo 6.8 – Cooling tower (Loiret-France)

p depends in general of ϕ, the thickness being variable. The integration makes it possible to determine N^{11}, which remains negative. N^{22} is then obtained by equation (33). The circumferential force is positive at the top of the tower, where N^{11} is zero and cos ϕ is positive: the points are overhanging with respect to the pass and tend to deviate from the axis under the effect of the weight.

Under the pass, cos ϕ becomes negative and there is an elevation under the pass where N^{22} vanishes, then becomes negative when approaching the ground: the tower is compressed under the effect of its self weight, in both directions, in the lower part.

If p is constant, the forces can be calculated explicitly. A new parameter t of the generatrix is introduced by the relation:

$$t = \frac{\sqrt{\alpha^2+\beta^2}}{\alpha}\cos\phi$$

and it comes successively:

$$\sin^2\phi = 1-\frac{\alpha^2}{\alpha^2+\beta^2}t^2\ ;\quad \sin\phi\, d\phi = -\frac{\alpha}{\sqrt{\alpha^2+\beta^2}}dt$$

$$C^2 = \alpha^2\left(1-t^2\right)$$

$$r = \frac{\alpha^2}{\sqrt{\alpha^2+\beta^2}}\frac{\sqrt{1-t^2+\dfrac{\beta^2}{\alpha^2}}}{\sqrt{1-t^2}}\ ;\qquad z = \frac{\beta^2}{\sqrt{\alpha^2+\beta^2}}\frac{t}{\sqrt{1-t^2}}$$

$$\int_{\phi_0}^{\phi}\frac{\sin\phi}{C^4}d\phi = -\frac{\alpha}{\sqrt{\alpha^2+\beta^2}}\frac{1}{\alpha^4}\int_{t_0}^{t}\frac{1}{\left(1-t^2\right)^2}dt$$

which makes it possible to determine the membrane forces:

$$N^{11} = p\frac{\beta^2}{4\sqrt{\alpha^2+\beta^2}}\frac{\sqrt{1-t^2}}{1-\dfrac{\alpha^2}{\alpha^2+\beta^2}t^2}\left[f(t)-f(t_0)\right]$$

$$N^{22} = \frac{r}{\sin\phi}\left(p\cos\phi-\frac{N^{11}}{R_s}\right) = p\frac{\alpha^2}{\sqrt{\alpha^2+\beta^2}}\frac{t}{\sqrt{1-t^2}}+\frac{\alpha^2}{\beta^2}\left(1-t^2\right)N^{11}$$

with:

$$f(t) = \ln\frac{t+1}{t-1}+\frac{2t}{1-t^2}$$

The values of ϕ_0 and t_0 are determined from the values of r and z at the top of the shell:

$$\frac{R_1}{h_1} = \frac{\alpha^2}{\beta^2}\tan\phi_0 = \frac{\alpha^2}{\beta^2}\frac{\sqrt{1-t_0{}^2+\dfrac{\beta^2}{\alpha^2}}}{t_0}$$

► Under the effect of uniform wind over the entire height, the shell is subjected to pressure:

$$p^3 = -\sum_{n=1}^{\infty}p_n\cos n\theta$$

which effects are studied term by term. Let be the load case:

$$\begin{cases} p^1 = p^2 = 0 \\ p^3 = -p_n \cos n\theta \end{cases}$$

Given the shape of equilibrium equations, it makes sense to look for a solution in the form of:

$$\begin{cases} N^{11} = F_n(\phi)\cos n\theta \\ N^{22} = G_n(\phi)\cos n\theta \\ N^{12} = H_n(\phi)\sin n\theta \end{cases}$$

The functions F, G, H satisfy, according to (28), the system of equations:

$$\begin{cases} \dfrac{1}{R_s}\dfrac{d}{d\phi}(rF_n) + nH_n - G_n\cos\phi = 0 \\ \dfrac{1}{R_s}\dfrac{d}{d\phi}(rH_n) - nG_n + H_n\cos\phi = 0 \\ \dfrac{F_n}{R_s} + \dfrac{G_n\sin\phi}{r} = p_n \end{cases}$$

which solution can be determined numerically. The solution of the initial problem is obtained by superposition of the solutions of the modes of order n.

For any shell of revolution, an analytical approach is possible from equation (42), in which load data of the problem are reported:

$$-n^2\frac{R_\theta\lambda_n}{\sin\phi} + \frac{d}{d\phi}\left[\frac{R_\theta^2\sin\phi}{R_s}\frac{d\lambda_n}{d\phi}\right] - \frac{d}{d\phi}\left(p_n R_\theta^3\sin^2\phi\cos\phi\right) + n^2 p_n R_\theta^2 R_s\sin\phi = 0$$

The analytical solution of this equation for any mode n requires mathematical developments which trace can be found in [Gould] and which are not reproduced here. However, for the mode $n = 1$, the differential equation in ϕ is in a more easily accessible form, by posing $p_1 = p$ and $\lambda_1 = \lambda$:

$$-\frac{R_\theta\lambda}{\sin\phi} + \frac{d}{d\phi}\left[\frac{R_\theta^2\sin\phi}{R_s}\frac{d\lambda}{d\phi}\right] - \frac{d}{d\phi}\left(pR_\theta^3\sin^2\phi\cos\phi\right) + pR_\theta^2R_s\sin\phi = 0$$

At this stage, an auxiliary function ψ is introduced, such that:

$$\Psi(\phi) = \lambda(\phi)R_\theta\sin\phi$$

which makes it possible to transform the equation of mode 1 by writing successively:

$$\frac{d\Psi}{d\phi} = R_\theta\sin\phi\frac{d\lambda}{d\phi} + \lambda R_s\cos\phi$$

$$\frac{d}{d\phi}\left(\frac{R_\theta^2}{R_s}\sin\phi\frac{d\lambda}{d\phi}\right) = R_\theta \sin\phi\frac{d}{d\phi}\left(\frac{R_\theta}{R_s}\frac{d\lambda}{d\phi}\right) + R_\theta\cos\phi\frac{d\lambda}{d\phi}$$

$$-\frac{R_\theta\lambda}{\sin\phi} + \frac{d}{d\phi}\left(\frac{R_\theta^2}{R_s}\sin\phi\frac{d\lambda}{d\phi}\right) = R_\theta\sin\phi\frac{d}{d\phi}\left(\frac{1}{R_s\sin\phi}\frac{d\Psi}{d\phi}\right)$$

It comes:

$$\frac{d}{d\phi}\left(\frac{1}{R_s\sin\phi}\frac{d\Psi}{d\phi}\right) = \frac{1}{R_\theta\sin\phi}\left[\frac{d}{d\phi}\left(pR_\theta^3\sin^2\phi\cos\phi\right) - pR_\theta^2R_s\sin\phi\right] \tag{6.48}$$

which resolves by double quadrature. This expression can be used for any axisymmetric geometry. In the case of the hyperboloid of revolution, it can be solved numerically according to the real geometry.

6.3.3 Cylindrical tanks with bottom under pressure

6.3.3.1 General Relations

The tanks considered (Figure 6.30) are such that at the pole O their tangent plane is perpendicular to the axis of symmetry (no discontinuity of slope). The tank is under pressure p, so $p^3 = -p$.

At point O, where the abscissa is taken for origin, r is zero and equation (34) makes it possible to obtain:

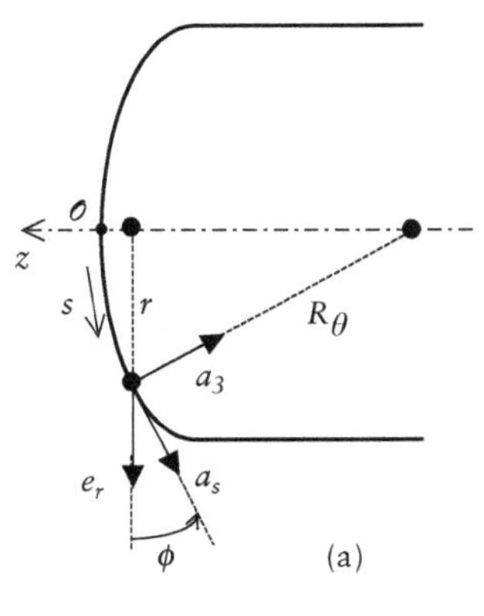

Figure 6.30

$$N^{11} = \frac{p}{r\sin\phi}\int_0^s r\cos\phi\,ds = \frac{p}{r\sin\phi}\int_0^s rdr = \frac{pr}{2\sin\phi} = \frac{pR_\theta}{2} \tag{6.49}$$

then:

$$N^{22} = \frac{1}{2}pR_\theta\left(2 - \frac{R_\theta}{R_s}\right) \tag{6.50}$$

N^{11} is always a traction. On the other hand, N^{22} can be traction or compression. If $\frac{R_\theta}{R_s} > 2$, N^{22} is a compression, which can be the cause of instability. When the tank is cylindrical, the point O is at infinity and therefore $R_\theta = R_O$, $b_{ss} = 0$, then $N^{11} = \frac{pR_O}{2}$, $N^{22} = pR_O$.

6.3.3.2 Spherical bottom

When $R_\theta = R_s = R_O$ (spherical tank): $N^{11} = N^{22} = \frac{pR_O}{2}$. The case of the hemispherical bottom (or dome) has been analysed in § 1.2.3.

If the tank is composed of a cylinder and bottoms in the form of spherical bottoms, such that there is discontinuity of the slopes, there can not be continuity of N^{11} (Figure 6.31). A stiffening beam makes it possible to balance the discontinuity thus created and to respect

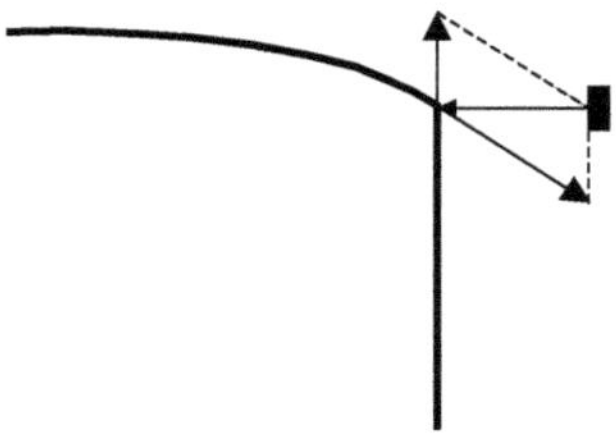

Figure 6.31

at best the membrane solution in the two parts of the shell (not considering kinematic continuity).

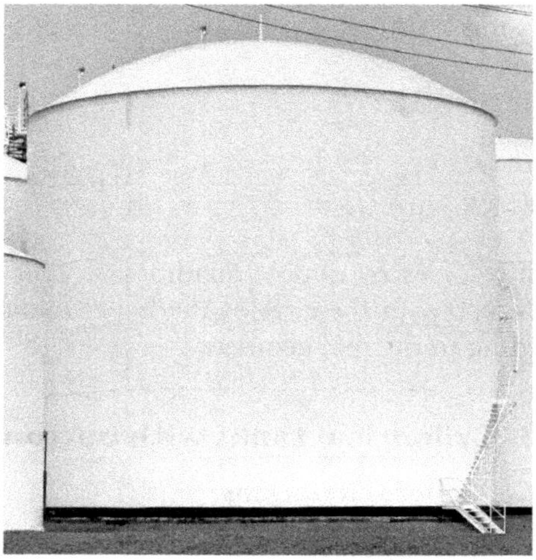

Photo 6.9. Spherical bottom tank with stiffener

6.3.3.3 Ellipsoidal bottom

The parameterisation of the ellipsoid is established according to a method similar to that used for the hyperboloid of revolution:

$$\begin{cases} r = \dfrac{\alpha^2}{C}\sin\phi \\ z = -\dfrac{\beta^2}{C}\cos\phi \end{cases}$$

with:

$$C^2 = \alpha^2\sin^2\phi + \beta^2\cos^2\phi$$

Indeed, with these expressions, it comes:

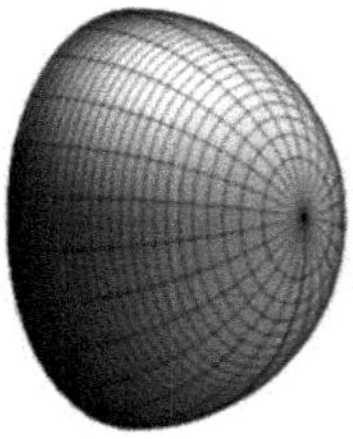

$$\begin{cases} \dfrac{dr}{d\phi} = \dfrac{\alpha^2}{C}\cos\phi - \dfrac{\alpha^2}{C^2}\sin\phi\dfrac{dC}{d\phi} = \dfrac{\alpha^2\beta^2}{C^3}\cos\phi \\ \dfrac{dz}{d\phi} = \dfrac{\beta^2}{C}\sin\phi + \dfrac{\beta^2}{C^2}\cos\phi\dfrac{dC}{d\phi} = \dfrac{\alpha^2\beta^2}{C^3}\sin\phi \end{cases}$$

which verify relationships (25), with:

$$R_s = \frac{ds}{d\phi} = \frac{\alpha^2\beta^2}{C^3}$$

Otherwise:

$$R_\theta = \frac{\alpha^2}{C}$$

The tank is closed by a half-ellipsoid of revolution. Here $\alpha = R_0$ and $\beta = h$.

The results of § 6.3.3.1 apply. It comes, through relationships (39) and (40):

$$N^{11} = \frac{pR_\theta}{2} = \frac{pR_0^{\,2}}{2\sqrt{R^2\sin^2\phi + h^2\cos^2\phi}}$$

$$N^{22} = \frac{pR_\theta}{2}\left(2 - \frac{R_\theta}{R_s}\right) = \frac{pR_\theta}{2C}\left(2 - \frac{C^2}{h^2}\right)$$

that is to say:

$$N^{22} = -p\frac{R^2}{2h^2}\frac{h^2 + \left(h^2 - R^2\right)\sin^2\phi}{\sqrt{R^2\sin^2\phi + h^2\cos^2\phi}}$$

For $h = R_0$, the results obtained give again the forces of the spherical case.

- Where $\phi = 0$, $N^{11} = N^{22} = \dfrac{pR_0^{\,2}}{2h}$.
- Where $\phi = \dfrac{\pi}{2}$, that is to say at the interface with the cylinder:

$$N^{11} = \frac{pR_0}{2}; \qquad N^{22} = \frac{pR_0}{2}\left(2 - \frac{R_0^{\,2}}{h^2}\right)$$

and the radial displacement w is worth:

$$w = R_0\varepsilon_{22} = \frac{R_0}{Ee}\left(N^{22} - \nu N^{11}\right) = \frac{pR_0^{\,2}}{2Ee}\left(2 - \nu - \frac{R_0^{\,2}}{h^2}\right)$$

In the usual cases where the ellipse is flattened ($h < R_0$), this value of w is farther away from the corresponding value of the cylinder than in the case of the hemisphere (with the same thickness). It is necessary to have large values of h so that the values of w of the ellipsoid and the cylinder come closer.

Otherwise:

$$\frac{R_\theta}{R_s} = \frac{R_0^{\,2}\sin^2\phi + h^2\cos^2\phi}{h^2}$$

This quantity is greater than 2 if h is less than $R_0/\sqrt{2}$, in which case circumferential compressions appear in the ellipsoidal bottom and there is then a risk of instability.

The results obtained above are also applicable to a completely ellipsoidal tank.

6.3.3.4 Cylindrical tank with torispherical bottom

Such a tank is a special case of the pressure vessel considered in § 3.3.1. It consists of a cylindrical body connected to the spherical bottom by a portion of torus (Figure 6.32 (a)).

The torus is the surface defining the envelope of a buoy or an inner tube. It is generated by the rotation of a circle of radius a around axis z. The distance between the center of the circle and the axis is $A > a$. is the angle that generates the circle radius a (Figure 6.32 (b)).

The main radii of curvature are a and $R_2 = a + \dfrac{A}{\cos\phi}$ (according to Meusnier's theorem).

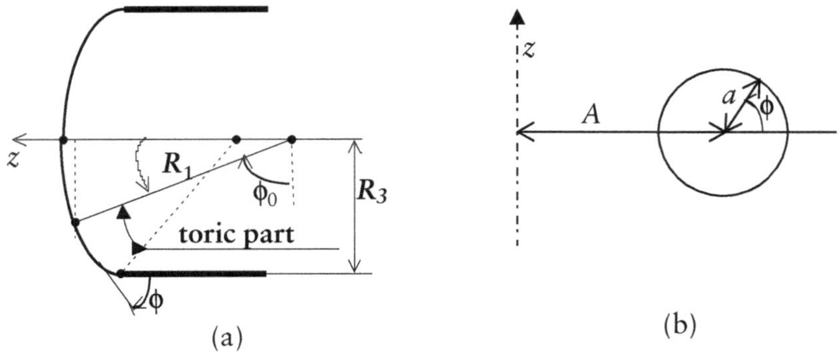

Figure 6.32

To ensure a fairly regular transition between the sphere and the torus on the one hand, between the torus and the cylinder on the other hand, the geometry is defined by ensuring the continuity of the curvature. ϕ varies from 0 to ϕ_0 (0 is the value ensuring continuity with the cylinder). φ is the angle on the sphere of radius R_1, ranging from 0 at the pole to φ_1 in continuity with the torus. φ_1, ϕ_0, A and a are determined to ensure continuities, φ_1 given:

$$A + a = R_3 \; ; \; A + a\cos\phi_0 = R_1 \sin\varphi_1 \; ; \; R_1 = a + \frac{A}{\cos\phi_0}$$

The last two relationships imply that $\sin\varphi_1 = \cos\phi_0$, which is identically verified since $\phi_0 - \varphi_1 = \pi/2$ (Figure 6.32 (a)). This also ensures the continuity of the tangent in ϕ_0. There remain three parameters to be determined for two conditions to respect, leaving a choice of design.

The membrane solution in the sphere and the cylinder is known, it remains to determine it in the toric part. At an angle ϕ between 0 and ϕ_0, N^{11} is determined from the resultant of the pressure on the bottom:

$$2\pi\left(A + a\cos\phi\right) \times N^{11}\cos\phi = \pi p\left(A + a\cos\phi\right)^2$$

From where:

$$N^{11} = \frac{p\left(A + a\cos\phi\right)}{2\cos\phi}$$

Photo 6.10 - Spherical bottom tank with toric connection

which is also obtained by (49). N^{11} is continuous at the interfaces with the cylinder and the sphere.

N^{22} is obtained by (50):

$$N^{22} = \frac{p(A + a\cos\phi)}{2\cos\phi}\left(2 - \frac{A + a\cos\phi}{a\cos\phi}\right)$$

N^{22} is not continuous at the interfaces: the solution can not be purely membrane. In addition, N^{22} is always a compression, so there is a risk of instability in the orthoradial direction.

6.4 SHELLS DEFINED BY AN EXPLICIT FUNCTION

6.4.1 General characteristics of shells defined by an explicit function

6.4.1.1 *Parametrisation; natural basis*

The shells considered here are of any form, but their formulation is nevertheless not the most general; their middle surface is defined by the function (Figure 6.33):

$$z = f(x, y) \tag{6.51}$$

x, y, z being the coordinates of a point m of the middle surface in an orthonormal Cartesian coordinate system. At each point m corresponds its projection m' 'in the plane (x, y). Such a representation of the middle surface makes it possible to approach the calculation of shells of quite different shapes. It allows in particular to represent most of the covers.

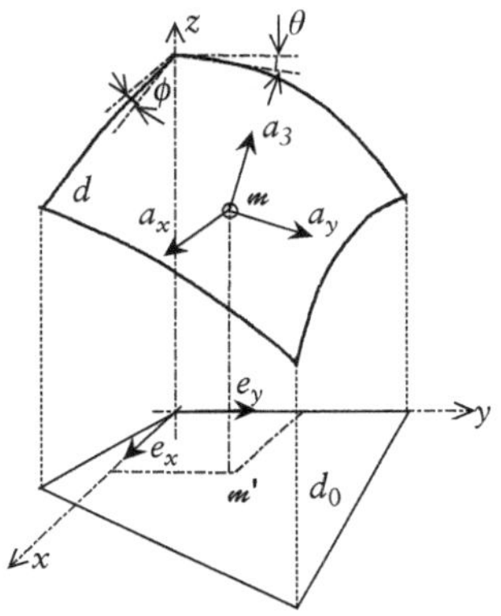

Figure 6.33

Since the parameterisation of the surface consists of the x and y coordinates, the local basis of the surface has been determined in § 3.-2.1.8 as well as the metric and curvature tensors.

6.4.1.2 *Equilibrium equations*

In the local coordinate system of the surface, equilibrium equations are written:

$$\left(\nabla_\alpha N^{\alpha\beta} + p^\beta\right) a_\beta + \left(b_{\alpha\beta} N^{\alpha\beta} + p^3\right) a_3 = 0 \tag{6.52}$$

By multiplying for example by e_x:

$$\nabla_\alpha N^{\alpha x} + p^x - \frac{p}{r}\left(b_{\alpha\beta} N^{\alpha\beta} + p^3\right) = 0 \tag{6.53}$$

then developing:

$$\partial_x N^{xx} + \partial_y N^{yx} + \frac{1}{r} N^{xx} \partial_x r + \frac{1}{r} N^{yx} \partial_y r + p^x - \frac{p}{r} p^3 = 0$$

or:

$$\partial_x\left(rN^{xx}\right)+\partial_y\left(rN^{yx}\right)+r\left(p^x-\frac{p}{r}p^3\right)=0 \tag{6.54}$$

An analogous expression is obtained by multiplying (52) by e_y. Noting:

$$\begin{aligned}\overline{N}^{xx} &= rN^{xx}\\ \overline{N}^{yy} &= rN^{yy}\\ \overline{N}^{xy} &= rN^{xy}\end{aligned} \tag{6.55}$$

the equations (52) are reduced to:

$$\begin{aligned}\frac{\partial\overline{N}^{xx}}{\partial x}+\frac{\partial\overline{N}^{xy}}{\partial y}+X=0\\ \frac{\partial\overline{N}^{yx}}{\partial x}+\frac{\partial\overline{N}^{yy}}{\partial y}+Y=0\end{aligned} \tag{6.56}$$

with:

$$\begin{aligned}X&=r\left(p^x-\frac{p}{r}p^3\right)\\ Y&=r\left(p^y-\frac{q}{r}p^3\right)\end{aligned} \tag{6.57}$$

Equations (56) express the equations of equilibrium in projection on the (x, y) plane, starting from the equilibrium of a domain d inscribed in the shell:

$$\int_d\left[\left(\nabla_\alpha N^{\alpha\beta}+p^\beta\right)a_\beta+\left(b_{\alpha\beta}N^{\alpha\beta}+p^3\right)a_3\right]d\sigma=0$$

by bringing the integral on d to an integral on the domain d_0, projection of d on (x, y). X and Y are the horizontal components of the external action in projection on the plane (x, y); the resultant of the external actions is written, r designating the area formed on the vectors a_x and a_y:

$$\int_d\left(p^x a_x+p^y a_y+p^3 a_3\right)d\sigma=\int_{d_0}\left(Xe_x+Ye_y+Ze_z\right)dxdy$$

with $d\sigma = r\,dx\,dy$.

Now writing the normal equilibrium of the shell:

$$b_{\alpha\beta}N^{\alpha\beta}+p^3=0$$

that is to say:

$$\frac{1}{r}\left(N^{xx}\frac{\partial^2 f}{\partial x^2}+2N^{xy}\frac{\partial^2 f}{\partial x\partial y}+N^{yy}\frac{\partial^2 f}{\partial y^2}\right)+p^3=0 \qquad (6.58)$$

and noting that the vertical component of the external action is worth, in projection on the plane (x, y):

$$Z=r\left(p^\beta a_\beta+p^3 a_3\right)\cdot e_z=r\left(p\,p^x+q\,p^y\right)+p^3$$

either, by reintroducing (57):

$$p^3 r^2=Z-pX-qY \qquad (6.59)$$

the last equation is written according to $\overline{N}$:

$$\overline{N}^{xx}\frac{\partial^2 f}{\partial x^2}+2\overline{N}^{xy}\frac{\partial^2 f}{\partial x\partial y}+\overline{N}^{yy}\frac{\partial^2 f}{\partial y^2}+Z-pX-qY=0 \qquad (6.60)$$

Equations (56) and (60) for pseudo-forces $\overline{N}$ are simpler than the initial equations (52).

6.4.1.3 Stress function

The interest of the transformation carried out in the preceding paragraph is that equations (56) are similar to the equilibrium equations of the two-dimensional media, for which a stress function can quite often be found. It is therefore not necessary for the surface to be ruled for such a function to exist in the (x, y) plane, which leads to introducing the Pucher[3] function $\Phi(x,y)$ such that:

$$\overline{N}^{xx}=\frac{\partial^2\Phi}{\partial y^2}-\int X dx\,;\quad \overline{N}^{yy}=\frac{\partial^2\Phi}{\partial x^2}-\int Y dy\,;\quad \overline{N}^{xy}=-\frac{\partial^2\Phi}{\partial x\partial y} \qquad (6.61)$$

for which equations (56) are verified. Equation (60) is rewritten then:

$$\boxed{\frac{\partial^2\Phi}{\partial y^2}\frac{\partial^2 f}{\partial x^2}-2\frac{\partial^2\Phi}{\partial x\partial y}\frac{\partial^2 f}{\partial x\partial y}+\frac{\partial^2\Phi}{\partial x^2}\frac{\partial^2 f}{\partial y^2}+Z-pX-qY-\frac{\partial^2 f}{\partial x^2}\int X dx-\frac{\partial^2 f}{\partial y^2}\int Y dy=0} \qquad (6.62)$$

Thus, the problem of the equilibrium of a membrane of any form explicitly defined by a function $f(x, y)$ is reduced to a partial differential equation of the second order on the stress function Φ. The boundary conditions relate to the pseudo-forces $\overline{N}$, thus on the second derivatives of Φ, by (61).

6.4.2 Translation shells

A translation shell is obtained by translating a plane curve onto another plane curve. In common cases (hyperbolic paraboloid, conoid), such a shell can be defined by an explicit function. Nevertheless, the cylinders and the cones are also translation surfaces.

[3] Adolf Pucher, Austrian engineer and professor (1902-1968).

6.4.2.1 Shells in the form of hyperbolic paraboloid

A hyperbolic paraboloid (HP) is generated by translating a convex parabola onto a concave parabola, thereby forming an anticlastic surface with negative Gaussian curvature. In the case where the two parabolas are in orthogonal planes (x, z) and (y, z), the equation of such a surface (Figure 6.34) is written:

$$z = \alpha\frac{x^2}{a^2} - \beta\frac{y^2}{b^2} \tag{6.63}$$

where α and β are positive and a and b are the lengths defined in Figure 6.34. In the parameterisation (x, y), the curvature tensor is written:

$$b_{\bullet\bullet} = \frac{1}{r}\begin{pmatrix} \frac{2\alpha}{a^2} & 0 \\ 0 & -\frac{2\beta}{b^2} \end{pmatrix}$$

with:

$$r = \sqrt{1 + \left(\frac{2\alpha x}{a^2}\right)^2 + \left(\frac{2\beta x}{b^2}\right)^2}$$

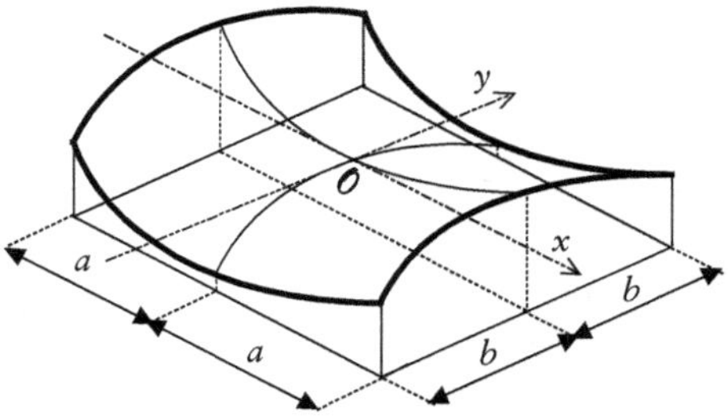

Figure 6.34

In every point, there are two asymptotic directions according to which the curvature is zero (§ 3.4.3). Equation (3.-137) for determining asymptotic lines is written:

$$\frac{2\alpha}{a^2}\left(\frac{dx}{ds}\right)^2 - \frac{2\beta}{b^2}\left(\frac{dy}{ds}\right)^2 = 0$$

The asymptotic lines are the projections on the surface of the lines of the planes (x, y) of equations:

$$y = \frac{b}{a}\sqrt{\frac{\alpha}{\beta}}x + C_1 \quad \text{and} \quad y = -\frac{b}{a}\sqrt{\frac{\alpha}{\beta}}x + C_2$$

where C_1 and C_2 are constants. These two lines are orthogonal if $\frac{\alpha}{a^2} = \frac{\beta}{b^2}$. The asymptotic lines are therefore the lines of the surface parameterised in x with equations:

$$\begin{cases} x \\ y = \frac{b}{a}\sqrt{\frac{\alpha}{\beta}}x + C_1 \\ z = -\frac{2C_1\sqrt{\alpha\beta}}{ab}x - \frac{\beta C_1^2}{b^2} \end{cases} \quad \text{and} \quad \begin{cases} x \\ y = -\frac{b}{a}\sqrt{\frac{\alpha}{\beta}}x + C_2 \\ z = \frac{2C_2\sqrt{\alpha\beta}}{ab}x - \frac{\beta C_2^2}{b^2} \end{cases}$$

This leads to another definition of the surface: a hyperbolic paraboloid (HP) is generated by two families of lines (generatrices) parallel to two distinct planes containing z. In its simplest form (Figure 6.35), if both planes are orthogonal, that is (x, z) and (y, z), the equation of such a surface is (a and b denote lengths not necessarily equal to the previous ones and (x, y) are axes different from the previous ones):

$$z = \frac{H}{ab}xy = \lambda xy \tag{6.64}$$

Then:

$$p = \lambda y \; ; \; q = \lambda x$$

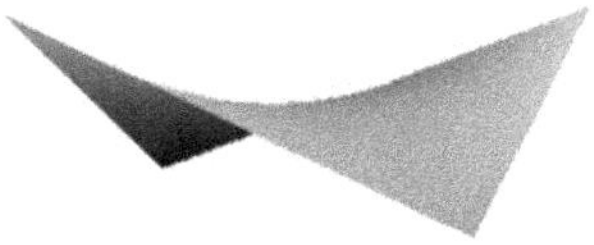

$$\frac{\partial^2 f}{\partial x^2} = 0 \; ; \; \frac{\partial^2 f}{\partial y^2} = 0 \; ; \; \frac{\partial^2 f}{\partial x \partial y} = \lambda$$

$$r^2 = 1 + \lambda^2 \left(x^2 + y^2 \right)$$

λ is the **torsion** of the surface. The curvatures in the x and y directions are zero since the traces of the normal planes in these directions are straight lines. The principal directions of curvature are at 45 ° with respect to the x and y directions and the principal curvatures are $\pm \frac{\lambda}{r}$.

Such a shell, which can be easily formed, is often used as a reinforced concrete cover element, which can be composed of several pieces of HP juxtaposed along straight edges.

Example : Hyperbolic paraboloid subjected to its self weight.

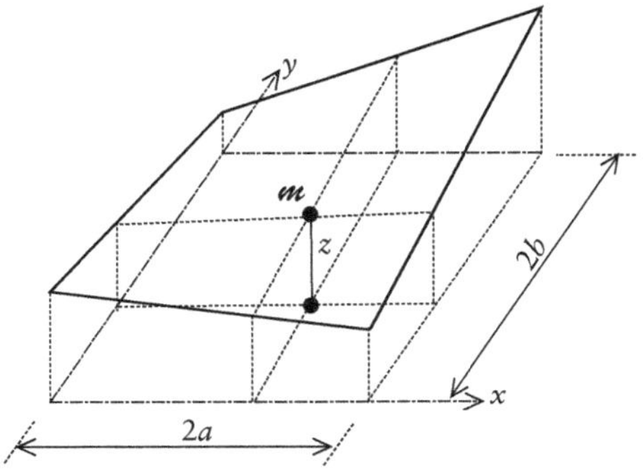

Figure 6.35

In this case, the load is written:

$$X = Y = 0 \; ; \; Z = -\rho g r$$

The equation determining the stress function becomes:

$$-2\lambda \frac{\partial^2 \Phi}{\partial x \partial y} - \rho g r = 0$$

or:

$$-\frac{\partial^2 \Phi}{\partial x \partial y} = \overline{N}^{xy} = \frac{\rho g}{2\lambda} \sqrt{1 + \lambda^2 \left(x^2 + y^2 \right)}$$

then, by (56):

$$\frac{\partial \overline{N}^{xx}}{\partial x} = -\frac{\partial \overline{N}^{xy}}{\partial y} = -\frac{1}{2} \frac{\rho g \lambda y}{\sqrt{1 + \lambda^2 \left(x^2 + y^2 \right)}}$$

and, integrating:

$$\overline{N}^{xx} = -\frac{\rho g y}{2} \ln \left[\lambda x + \sqrt{1 + \lambda^2 \left(x^2 + y^2 \right)} \right] + h(y)$$

Similarly:

$$\overline{N}^{yy} = -\frac{\rho g x}{2}\ln\left[\lambda y + \sqrt{1+\lambda^2\left(x^2+y^2\right)}\right] + g(x)$$

The functions $h(y)$ and $g(x)$ are determined by the boundary conditions. For example, if $N^{xx}(x=0)=0$ and $N^{yy}(y=0)=0$:

$$\overline{N}^{xx} = -\frac{\rho g y}{2}\ln\left[\frac{\lambda x + \sqrt{1+\lambda^2\left(x^2+y^2\right)}}{\sqrt{1+\lambda^2 y^2}}\right]$$

$$\overline{N}^{yy} = -\frac{\rho g x}{2}\ln\left[\frac{\lambda y + \sqrt{1+\lambda^2\left(x^2+y^2\right)}}{\sqrt{1+\lambda^2 y^2}}\right]$$

Actual membrane forces are obtained by (55), in particular:

$$N^{xy} = \frac{\rho g}{2\lambda}$$

It is recalled that these forces are expressed in a non-orthonormal coordinate system.

Note that for this shell having asymptotic lines in the x and y directions, N^{xy} is obtained directly by (12):

$$2b_{xy}N^{xy} = 2\frac{\lambda}{r}N^{xy} = -p^3 = -\frac{Z}{r^2} = \frac{\rho g}{r}$$

6.4.2.2 *Conoid-shaped vaults*

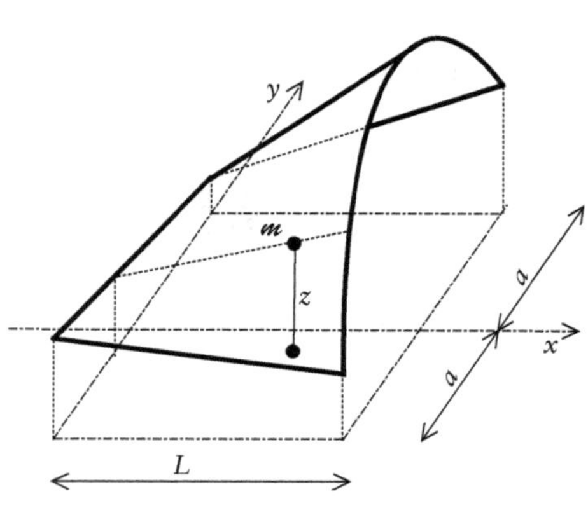

Figure 6.36

A conoid is generated by a set of straight generatrices resting on two directrices in two parallel planes (Figure 6.36). A directrix is a planar curve of equation $z = f(y)$, the other directrix is a line parallel to the plane of the previous plane curve.

The projections of the generatrices on the plane (x, y) are all parallel to a fixed direction concurrent with the plans of the directrices. If these projections are orthogonal to the planes of the directrices, the equation of the conoid is written:

$$z = f(y)\frac{x}{L} \tag{6.65}$$

The hyperbolic paraboloid is a special case of conoid where $f(y)$ is linear.

If the directrix curve is a parabola of axis z, the conoid is a parabolic conoid of equation:

$$z = \frac{H}{a^2 L}\left(a^2 - y^2\right) x \tag{6.66}$$

In this case, by posing $\lambda = \dfrac{H}{a^2 L}$:

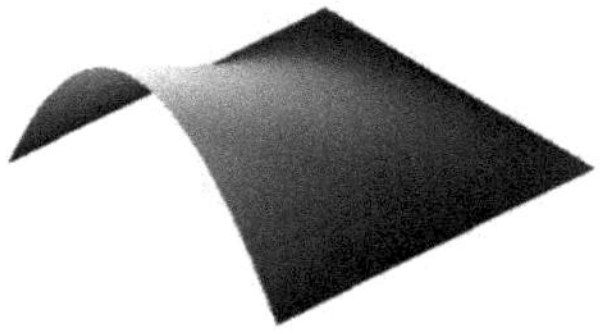

$$p = \lambda\left(a^2 - y^2\right); \qquad q = -2\lambda xy$$

$$\frac{\partial^2 f}{\partial x^2} = 0; \qquad \frac{\partial^2 f}{\partial y^2} = -2\lambda x; \qquad \frac{\partial^2 f}{\partial x \partial y} = -2\lambda y$$

$$r^2 = 1 + \left[\lambda\left(a^2 - y^2\right)\right]^2 + \left[2\lambda xy\right]^2$$

If the shell is shallow ($H << L$ and $H << a$):

$$r \approx 1 + \frac{\lambda^2}{2}\left[\left(a^2 - y^2\right)^2 + 4x^2 y^2\right]$$

Example : Straight parabolic conoid subjected to its own weight.
In this case, the load is written:

$$X = Y = 0; \quad Z = -\rho g r$$

The equation determining the stress function is written:

$$4\lambda y \frac{\partial^2 \Phi}{\partial x \partial y} - 2\lambda x \frac{\partial^2 \Phi}{\partial x^2} - \rho g r = 0$$

By choosing $y = 0$ in the plane of symmetry, the shear N^{xy} is zero in this plane if the loading is symmetrical, then:

$$\frac{\partial^2 \Phi}{\partial x \partial y}(y = 0) = 0$$

If, along the parabolic directrix, a shear beam is put in place, only the normal force N^{xx} is zero:

$$\frac{\partial^2 \Phi}{\partial y^2} = 0 \quad \text{in} \quad x = L$$

Along the straight directrix ($x = 0$) and edge generatrices ($y = \pm\, a$), the conoid is assumed to be supported to resist reactions due to N^{xx} and N^{xy}.

In the case where the shell is very shallow, the function Φ can be sought as the sum of two functions $\Phi_1(x)$ and $\Phi_2(x, y)$ such that:

$$-2\lambda x \frac{\partial^2 \Phi_1}{\partial x^2} - \frac{1}{2}\rho g\left(2 + \lambda^2 a^4\right) = C$$

where are gathered all the terms not dependent on y and:

$$4\lambda y\frac{\partial^2\Phi_2}{\partial x\partial y}-2\lambda x\frac{\partial^2\Phi_2}{\partial x^2}-\frac{1}{2}\rho g\lambda^2\left(y^4-2a^2y^2+4x^2y^2\right)=C$$

The constant C that appears at the second member of the two equations is null to finally respect the boundary conditions.

The first equation then admits for solution:

$$\Phi_1(x)=-\mu x\left(\ln x-A\right)+B$$

with:

$$\mu=\frac{1}{4}\frac{\rho g}{\lambda}\left(2+\lambda^2a^4\right)$$

A particular solution of the second equation can be sought in the form of a double series:

$$\Phi_2(x,y)=\sum_m\sum_n A_{mn}\left(x^m-L^m\right)y^n$$

which checks the boundary conditions. Any function $h(y)$ can be added to Φ_2. By reporting to the equation:

$$\sum_m\sum_n A_{mn}\left[4mn-2m(m-1)\right]x^{m-1}y^n=\frac{1}{2}\rho g\lambda\left(y^4-2a^2y^2+4x^2y^2\right)=0$$

By identifying the terms in pairs, only the terms A_{12}, A_{14} and A_{32} are not null and finally:

$$\Phi_2=\rho g\lambda y^2(x-L)\left(\frac{y^2}{32}-\frac{a^2}{8}+\frac{x^2+Lx+L^2}{6}\right)+h(y)$$

Photo 6.11 - Roof in the shape of a conoid (Barcelona-Spain)

The function $h(y)$ can be set to zero to respect the condition in $x = L$.
After summation of Φ_1 and Φ_2 and application of relations (61):

$$\overline{N}^{xx} = \rho g \lambda \left[\frac{3}{8}(x-L)y^2 - \frac{a^2(x-L)}{4} + \frac{1}{3}\left(x^3 - L^3\right) \right]$$

$$\overline{N}^{yy} = -\frac{\rho g}{4\lambda x}\left(2 + \lambda^2 a^4\right) + \rho g \lambda x y^2$$

$$\overline{N}^{xy} = -\rho g \lambda \left(\frac{y^3}{8} - \frac{a^2 y}{4} + x^2 y \right)$$

which verify the boundary conditions. Actual forces are obtained by (55).

6.5 HELICAL SHELLS

A helical shell has a middle surface consisting of an helicoid of equation:

$$\begin{cases} x = r\cos\theta \\ y = r\sin\theta \\ z = a\,\theta \end{cases}$$

r and θ designating the two parameters of the surface (Figure 6.37). The characteristics of the surface in this parametrisation are:

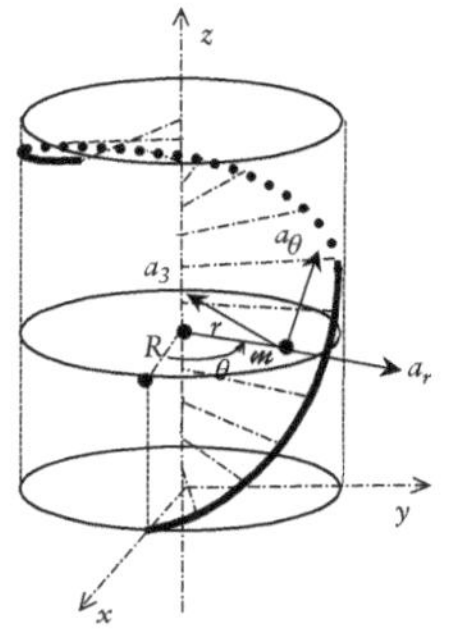

Figure 6.37

$$a_r = \begin{cases} \cos\theta \\ \sin\theta \\ 0 \end{cases} \qquad a_\theta = \begin{cases} -r\sin\theta \\ r\cos\theta \\ a \end{cases}$$

$$a_3 = \frac{1}{\sqrt{a^2 + r^2}} \begin{cases} a\sin\theta \\ -a\cos\theta \\ r \end{cases}$$

The vector a_r is the horizontal radial vector and the vector a_θ is the vector tangent to the helix defined by r = cst. Both vectors are orthogonal.

$$a_{rr} = 1 \qquad a_{\theta\theta} = a^2 + r^2 \qquad a_{r\theta} = 0$$

$$b_{rr} = 0 \qquad b_{\theta\theta} = 0 \qquad b_{r\theta} = -\frac{a}{\sqrt{a^2 + r^2}}$$

$$\Gamma^r_{\theta\theta} = -r \qquad \Gamma^\theta_{\theta r} = \frac{r}{a^2 + r^2}$$

Horizontal helices (r = cst) and generatrices (θ = cst) are asymptotic lines. Note that the vector a_3 is horizontal in $r = 0$; the shell is almost vertical near the centre.

At a point on the surface, the angle α of the normal vector with the vertical direction is defined by $\cos\alpha = \frac{r}{\sqrt{a^2+r^2}}$.

It is also interesting working in the associated orthonormal coordinate system where:

$$a_1 = a_r \qquad a_2 = \frac{a_\theta}{\sqrt{a^2+r^2}}$$

$$b_{12} = -\frac{a}{a^2+r^2}$$

Then:

$$N^{11} = N^{rr} \qquad N^{12} = N^{r\theta}\sqrt{a^2+r^2} \qquad N^{22} = N^{\theta\theta}\left(a^2+r^2\right)$$

$$p^1 = p^r \qquad p^2 = p^\theta\sqrt{a^2+r^2}$$

The equilibrium equations are written in the coordinate system (a_θ, a_r):

$$\partial_r N^{rr} + \partial_\theta N^{r\theta} - rN^{\theta\theta} + \frac{r}{a^2+r^2}N^{rr} + p^r = 0$$

$$\partial_r N^{r\theta} + \partial_\theta N^{\theta\theta} + 3\frac{r}{a^2+r^2}N^{r\theta} + p^\theta = 0$$

$$-\frac{2a}{\sqrt{a^2+r^2}}N^{r\theta} = p^3$$

and in the coordinate system (a_1, a_2):

$$\partial_r\left(N^{11}\sqrt{a^2+r^2}\right) + \partial_\theta N^{12} - \frac{r}{\sqrt{a^2+r^2}}N^{22} + p^1\sqrt{a^2+r^2} = 0$$

$$\partial_r\left[N^{12}\left(a^2+r^2\right)\right] + \partial_\theta N^{22}\cdot\sqrt{a^2+r^2} + p^2\left(a^2+r^2\right) = 0$$

$$-\frac{2a}{a^2+r^2}N^{12} = p^3$$

Example : Spiral staircase subjected to its self weight

The soffit of a spiral staircase consists of a helical surface. It rests in $r = R$ on a stringer (border beam) in the form of a helix. If p is the average surfacic weight (soffit + steps) of the staircase:

$$p^1 = 0 \qquad p^2 = -\frac{a}{\sqrt{a^2+r^2}}p \qquad p^3 = -\frac{r}{\sqrt{a^2+r^2}}p$$

then:

$$N^{12} = p\frac{r}{2a}\sqrt{a^2+r^2}$$

By including N^{12} in the second equation of equilibrium:

$$\partial_\theta N^{22}\cdot\sqrt{a^2+r^2} = -\partial_r\left[N^{12}\left(a^2+r^2\right)\right]-p^2\left(a^2+r^2\right) = -p\frac{1}{2a}\left(4r^2-a^2\right)\sqrt{a^2+r^2}$$

It comes:

$$N^{22} = -p\frac{1}{2a}\left(4r^2-a^2\right)\theta + h(r)$$

and then in the first equation of equilibrium:

$$\partial_r\left(N^{11}\sqrt{a^2+r^2}\right) = \underbrace{-\partial_\theta N^{12}}_{0} + \frac{r}{\sqrt{a^2+r^2}}N^{22} - \underbrace{p^1\sqrt{a^2+r^2}}_{0}$$

$$= \left[-p\frac{1}{2a}\left(4r^2-a^2\right)\theta + h(r)\right]\frac{r}{\sqrt{a^2+r^2}}$$

which integrates in:

$$N^{11} = p\frac{1}{6a}\left(11a^2-4r^2\right)\theta + k(r) + \frac{g(\theta)}{\sqrt{a^2+r^2}}$$

The functions $h(r)$ and $k(r)$ are linked by the expression:

$$k(r) = \frac{1}{\sqrt{a^2+r^2}}\int_0^r \frac{r\,h(r)}{\sqrt{a^2+r^2}}dr = h(r)-h(0)-\frac{1}{\sqrt{a^2+r^2}}\int_0^r h'(r)\sqrt{a^2+r^2}\,dr$$

The three functions $g(\theta)$, $h(r)$ and $k(r)$ are determined by the boundary conditions. If the soffit has a free edge in $r = 0$, $N^{11}(r = 0)$ must be zero for any value of θ, which implies that:

$$g(\theta) = -\frac{11\,pa^2}{6}\theta$$

Then:

$$N^{11} = p\frac{1}{6a}\left[11a^2\left(1-\frac{a}{\sqrt{a^2+r^2}}\right)-4r^2\right]\theta + \frac{1}{\sqrt{a^2+r^2}}\int_0^r \frac{r\,h(r)}{\sqrt{a^2+r^2}}dr$$

At the top and bottom of the stairs, it is necessary to have a shear beam to respect the membrane solution, N^{12} not being zero for any value of r. If the soffit is free at the top (in $\theta = \theta_0$) vis-à-vis the normal force N^{22}:

$$N^{22}(\theta = \theta_0) = -p\frac{1}{2a}\left(4r^2 - a^2\right)\theta_0 + h(r) = 0$$

which determines the function $h(r)$, then the membrane forces:

$$N^{11} = p\frac{1}{6a}\left[11\,a^2\left(1 - \frac{a}{\sqrt{a^2 + r^2}}\right) - 4r^2\right](\theta - \theta_0)$$

$$N^{22} = -p\frac{1}{2a}\left(4r^2 - a^2\right)(\theta - \theta_0)$$

The reactions on the stringer are then determined by the values of N^{11} and N^{12} in $r = R$.

Photo 6.12 - Spiral staircase (Paris-France)

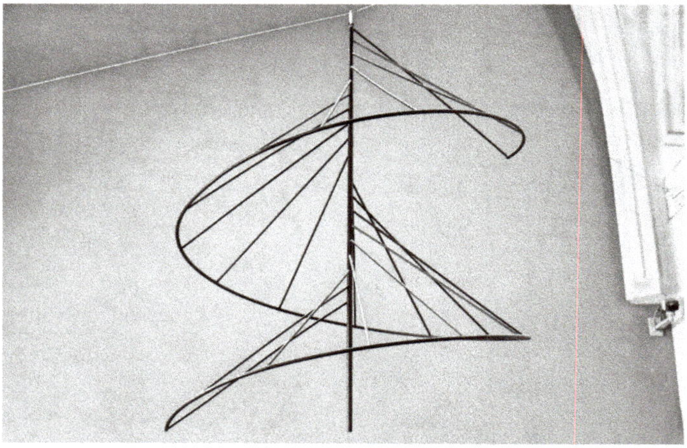

Photo 6.13 - Ruled surface (Barcelona-Spain)

MAIN RESULTS

EQUATIONS OF MOVEMENT

Local shell equations $\nabla_\alpha N^{\alpha\beta} + p^\beta = 0$

$$b_{\alpha\beta} N^{\alpha\beta} + p^3 = 0 \qquad \text{or} \qquad \frac{N^{11}}{R_1} + \frac{N^{22}}{R_2} + p^3 = 0$$

Boundary conditions $N^{\alpha\beta}\nu_\beta - q^\alpha = 0$

RELATIONSHIP BETWEEN 3D STRESSES AND MEMBRANE FORCES

$N^{\alpha\beta} = e\,T^{\alpha\beta}$

DEFORMATIONS

$$\varepsilon_{\alpha\beta} = \frac{1}{2}\left(\nabla_\alpha \xi_\beta + \nabla_\beta \xi_\alpha\right) - \zeta B_{\alpha\beta}$$

LINEAR ELASTIC BEHAVIOUR

$$N = \frac{Ee}{1-\nu^2}\left[\nu\, \varepsilon^\alpha{}_\alpha\, a + (1-\nu)\,\varepsilon\right]$$

$$\varepsilon = \frac{1}{Ee}\left[(1+\nu)\,N - \nu\, N^\alpha{}_\alpha\, a\right]$$

$$\mathcal{F} = \frac{1}{2}\frac{Ee}{1-\nu^2}\int_\Sigma \left[\nu\left(\varepsilon^\alpha{}_\alpha\right)^2 + (1-\nu)\,\varepsilon_{\alpha\beta}\varepsilon^{\alpha\beta}\right] dS$$

MEMBRANE FORCES IN AXISYMMETRIC TANKS UNDER PRESSURE

In the direction of the generatrices: $N^{11} = \dfrac{pR_\theta}{2}$

In the orthoradial direction: $N^{22} = \dfrac{1}{2} pR_\theta\left(2 - \dfrac{R_\theta}{R_s}\right)$

EXERCISES

Exercise 6.1

A conical roof supports a skylight at abscissa $s = s_0$. This brings a vertical linear load P along the support of the skylight (Figure 6.38).

Calculate the membrane forces in the roof under the effect of P and the constant self-weight ρg of the roof. What practical condition(s) must be implemented for the membrane solution to be acceptable?

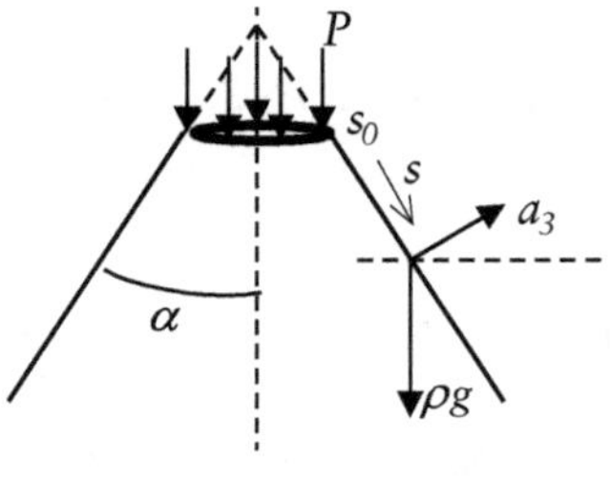

Figure 6.38

Same question under the effect of the wind exerting on the roof a pressure that can be put under the same form as in the § 2.2.3.

Answer:

$$N^{11} = \rho g \frac{s_0^{\,2} - s^2}{2s \cos\alpha} - \frac{s_0 P}{s \cos\alpha}$$

$$N^{22} = -\rho g \; s \sin\alpha \tan\alpha$$

$$N^{12} = 0$$

Only N^{11} is affected by P. So that the membrane equilibrium can be achieved, it is advisable to add at the top and at the base of the skylight a ring beam offering a horizontal reaction ensuring a total reaction tangent to the membrane (see § 3.2.1.1).

The study of stresses under the effect of the wind can be reduced to that of one of the terms of the sum: $p^3 = -p_n \cos n\theta$, for which:

$$N^{11} = -\frac{p_n s}{2}\left[\left(1 - \frac{s_0^{\,2}}{s^2}\right)\tan\alpha - \frac{n^2}{3\sin\alpha\cos\alpha}\left(1 - \frac{3s_0^{\,2}}{s^2} + \frac{2s_0^{\,3}}{s^3}\right)\right]\cos n\theta$$

$$N^{22} = -p_n \; s \tan\alpha \cos n\theta$$

$$N^{12} = -\frac{n\, p_n s}{3\cos\alpha}\left(1 - \frac{s_0^{\,3}}{s^3}\right)\sin n\theta$$

Exercise 6.2

A dome is axisymmetric around the z axis, generated by an arc of a circle not centered on z, as shown in Figure 6.39 (so-called "ogival" dome).

The dome parameters consist of angles ψ and θ. The geometric data are R_0 the radius of the circle, a the eccentricity of its centre and ψ_n the angle at the beginning of the arch.

Determine the membrane state under self-weight; what happens at the top of the vault?

Note: it is useful to consider the angle β of the current radius with the vertical.

Answer: The forces are expressed as functions of the angle β connected to ψ by the relation:

$$\cos\beta\left(\sin\beta - \frac{a}{R_0}\right) = \cos\psi\sin\psi$$

- β is indeed equal to ψ when $a = 0$.

$$N^{11} = -\rho g R_0 \frac{R_0(\cos\beta_0 - \cos\beta) - a(\beta - \beta_0)}{(R_0 \sin\beta - a)\sin\beta}$$

$$N^{22} = \rho g R_0 \left[\frac{R_0(\cos\beta_0 - \cos\beta) - a(\beta - \beta_0)}{(R_0 \sin\beta - a)\sin\beta} - \cos\beta\right]$$

These expressions are reduced to the forces found in § 6.3.2.1.1 when $a = 0$.

At the top, that is, when $\beta \to \beta_0$:

$$N^{11} \approx -\rho g R_0 \frac{\beta - \beta_0}{2 \sin\beta_0} \underset{\beta \to \beta_0}{\to} 0$$

$$N^{22} \underset{\beta \to \beta_0}{\to} -\rho g R_0 \cos\beta = -\rho g \sqrt{R_0{}^2 - a^2}$$

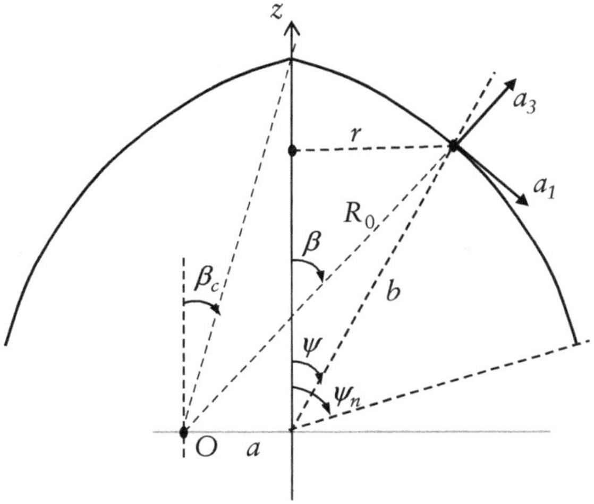

Figure 6.39

Exercise 6.3

A toric-shaped buoy is set as shown in Fig. 6.40.

Note : the geometry of the torus was the subject of Exercise 3.3.

1) Write the membrane equilibrium equations in the natural coordinate system, then in the associated orthonormal physical coordinate system.
2) The buoy is subjected to a constant internal pressure. Simplify the equations and determine the state of membrane stress. Calculate the variation in diameter along the horizontal and vertical axes, and the average horizontal displacement (in $\varphi = \pm\frac{\pi}{2}$).

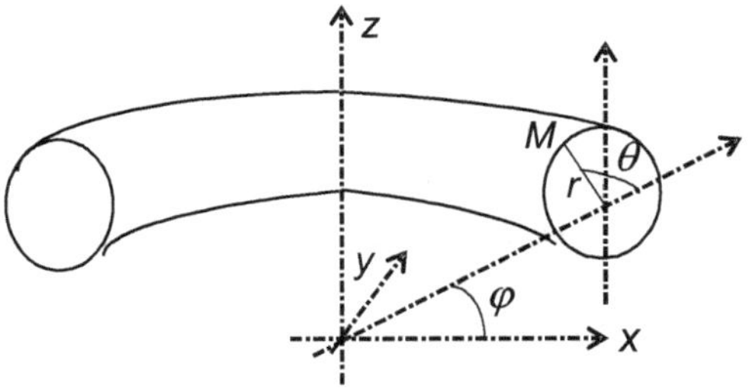

Figure 6.40

Answer :

1) In the natural coordinate system associated with the parameters:

$$\partial_\theta N^{\theta\theta} + \partial_\varphi N^{\theta\varphi} - 3\frac{r \sin\varphi}{R + r\cos\varphi} N^{\theta\varphi} + p^\theta = 0$$

$$\partial_\theta N^{\theta\varphi} + \partial_\varphi N^{\varphi\varphi} - \frac{r \sin\varphi}{R + r\cos\varphi} N^{\varphi\varphi} + \frac{R + r\cos\varphi}{r} \sin\varphi \, N^{\theta\theta} + p^\theta = 0$$

$$-(R + r\cos\phi)\cos\phi \, N^{\theta\theta} - r\, N^{\varphi\varphi} + p^3 = 0$$

In the associated orthonormal physical coordinate system:

$$\partial_\theta N^{11} + \frac{R + r\cos\varphi}{r}\partial_\varphi N^{12} - 2N^{12}\sin\varphi + p^1(R + r\cos\phi) = 0$$

$$\partial_\theta N^{12} + \frac{R + r\cos\varphi}{r}\partial_\varphi N^{22} + \left(N^{11} - N^{22}\right)\sin\varphi + p^2(R + r\cos\phi) = 0$$

$$r\cos\phi\, N^{11} + (R + r\cos\phi)N^{22} = p^3 r(R + r\cos\phi)$$

2) $p^1 = p^2 = 0$; $p^3 = p$. The problem is axisymmetric. N^{12} is zero.

$$\frac{R + r\cos\varphi}{r}\partial_\varphi N^{22} + \left(N^{11} - N^{22}\right)\sin\varphi = 0$$

$$r\cos\phi\, N^{11} + (R + r\cos\phi)N^{22} = p\, r(R + r\cos\phi)$$

The integration constants are determined by global equilibria, cf. exercise 1.3.

$$N^{11} = \frac{pr}{2}$$

$$N^{22} = pr\frac{2R + r\cos\phi}{2(R + r\cos\phi)}$$

Exercise 6.4

A hemispherical dome is subjected to non-axisymmetric loading or boundary conditions.

1) Determine the membrane solution of the homogeneous equilibrium equations (i.e. of the unloaded dome) of the form:

$$N^{11} = s_n(\psi)\cos(n\theta)$$

$$N^{12} = t_n(\psi)\sin(n\theta)$$

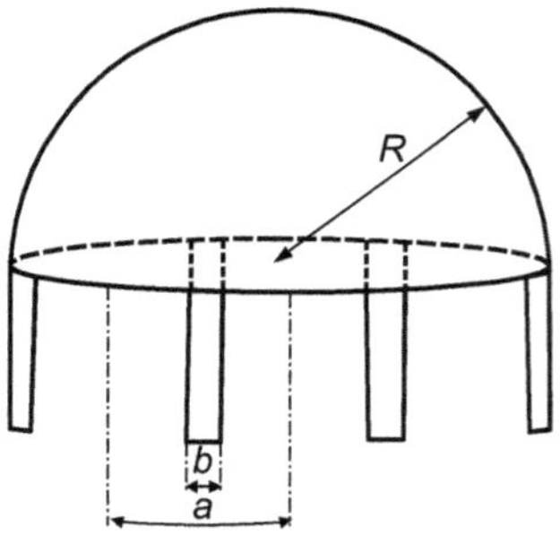

Figure 6.41

Note: Solutions can be built from the functions $\frac{\left[\tan\left(\psi/2\right)\right]^n}{\cos^2\psi}$ et $\frac{\left[\cot\left(\psi/2\right)\right]^n}{\cos^2\psi}$.

2) Determine the membrane forces of the dome supported on a number p of columns of width b under its own weight (Fig. 6.41). The columns are regularly spaced; distance

between axes is a. What becomes the solution when the columns can be considered as points ($b = 0$)?

Note: The actions and reactions to which the dome is subjected can be decomposed into Fourier series.

Answer:

1) The homogeneous equilibrium equations can be written by specialising the equations (28), taking into account the form of the desired solutions:

$$\frac{d}{d\psi}\left(\sin\psi s_n \right) + \mathrm{n}\, t_n - s_n \cos\psi = 0$$

$$\frac{d}{d\psi}\left(\sin\psi\, t_n \right) + \mathrm{n}\, s_n - t_n \cos\psi = 0$$

s_n being finite at the top of the dome, the general solution of the homogeneous equations is of the form, for n > 1 :

$$N_n^{11} = -N_n^{22} = \lambda_n \frac{\left[\tan\left(\psi/2\right)\right]^{\mathrm{n}}}{\cos^2\psi} \cos(\mathrm{n}\theta)$$

$$N_n^{12} = -\lambda_n \frac{\left[\tan\left(\psi/2\right)\right]^{\mathrm{n}}}{\cos^2\psi} \sin(\mathrm{n}\theta)$$

2) The membrane solution under the effect of the self-weight for a dome uniformly supported on its edge was determined in § 3.2.1.1, with $\alpha = \pi/2$. In the case of the hemispherical dome, $N_0^{11} = -qR_0$. To simulate a support on p columns, a uniform reaction equal to qR_0 balanced by the reactions on columns, each equal to qR_0/p.

The membrane equilibrium equations corresponding to this set of reactions are homogeneous. This "reactions" load case is added to the membrane solution of § 3.2.1.1. The reaction $r(t)$ along the edge is made up of a piecewise constant function, equal to $-qR_0$ all around, adding to it qR_0/pb on the width b of the column (which supposes that the column is thin enough so that its reaction can be distributed). The function $r(t)$ is periodic with period $a = \dfrac{2\pi R_0}{p}$ (not $2\pi R_0$), it is decomposed into Fourier series. Taking into account the parity of the function r between columns, the odd terms of the decomposition are zero:

$$r(\theta) = qR_0 \sum_{n=1}^{\infty} \frac{2}{\mathrm{n}\lambda} \sin(\mathrm{n}\lambda) \cos(\mathrm{n}p\theta) \text{ with } \lambda = \frac{\pi b}{a} = \frac{p\, b}{2R_0}$$

Taking into account the solution of the equation of the homogeneous equation found in 1), the final solution is found by adding this homogeneous solution to the axisymmetric solution of § 3.2.1.1, noting that n must be replaced by np in the Fourier decomposition:

$$N^{11} = -qR_0 \left[\frac{1}{1+\cos\psi} - \sum_{n=1}^{\infty} \frac{2}{n\lambda} \sin(n\lambda) \left(\frac{\tan^{np} \frac{\psi}{2}}{\sin\psi} \right)^2 \cos(np\theta) \right]$$

$$N^{22} = qR_0 \left[\frac{1}{1+\cos\psi} - \cos\psi + \sum_{n=1}^{\infty} \frac{2}{n\lambda} \sin(n\lambda) \left(\frac{\tan^{np} \frac{\psi}{2}}{\sin\psi} \right)^2 \cos(np\theta) \right]$$

$$N^{12} = qR_0 \sum_{n=1}^{\infty} \frac{2}{n\lambda} \sin(n\lambda) \left(\frac{\tan^{np} \frac{\psi}{2}}{\sin\psi} \right)^2 \sin(np\theta)$$

In $\psi = \pi/2$, the value of N^{11} associated with the reaction $r(\theta)$ alone is given in Fig. 6.42 (for a half-perimeter, $p = 6$ and $b/R_0 = 0{,}1$) and is indeed as expected (series calculated with 500 terms). The shear is not zero; to ensure the membrane solution, it is necessary to have a shear beam.

If $b = 0$, the reaction is concentrated on the columns and is worth along the edge:

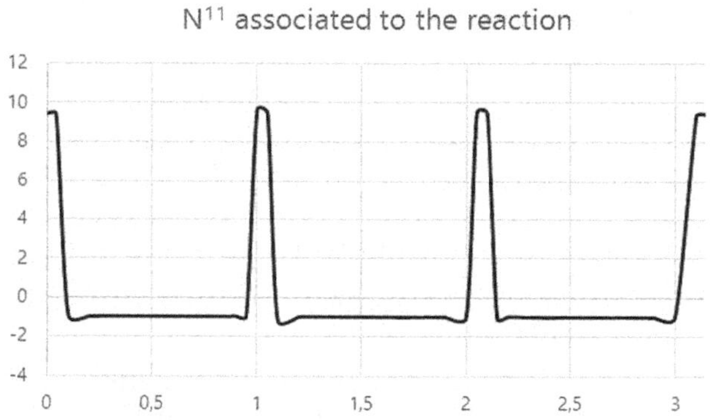

Figure 6.42

$$N^{11} = -qR_0 \left[\frac{1}{1+\cos\psi} - \sum_{n=1}^{\infty} 2\cos(np\theta) \right]$$

$$N^{12} = 2qR_0 \sum_{n=1}^{\infty} \sin(np\theta)$$

Above the columns, N^{11} tends to infinity and the shear is zero.

Chapter 7

Plates in flexion

The purpose of this chapter is the study of elastic thin plates in small transformations.

In the first section, classical equations of flexion are studied in the case where the influence of membrane forces on bending moments is negligible. The main notations and the most important properties established in the general framework of shell theory are recalled here in the narrower framework of Kirchhoff's plate theory. Special attention is paid to the shape of equilibrium equations and the distribution of stresses. The flexural equations are established in the case of isotropic and orthotropic plates.

In the second section, classical examples are studied, where analytical solutions have been established in particular for circular and rectangular plates.

The case of bending taking into account the membrane forces is examined in the third paragraph, considering, in particular, the two usual approximations: either the membrane forces are mainly due to the external loading, or they are generated by the bending because of the variation of Gaussian curvature. This allows approaching, in a practical, way the approximations described in the next chapter in a more general context.

Finally, the last section deals with the Reissner–Mindlin theory, which takes into account deformations due to shear force.

7.1 FLEXION OF PLATES: CLASSICAL THEORY

7.1.1 Equilibrium equations

7.1.1.1 Recalling the hypotheses

The purpose of this chapter is the study of elastic thin plates. The assumptions adopted in Section 7.1 and 7.2 are as follows:

- /*H1*/ **Hypothesis of thin plates.** A plate can be considered thin if the ratio of its thickness to its smallest span remains well below 1, for example, limited to 1/4 or 1/5.
- /*H2*/ **Hypothesis of plane stresses** (see § 5.4.1). Acceptable hypothesis out of the vicinity of concentrated forces and as long as the theory of thin plates remains valid, that is to say for a thickness substantially smaller than the smallest span.
- /H3/ **Kirchhoff's hypotheses.** Assumptions are consistent with the hypothesis of thin plates. It is nevertheless necessary to limit more strictly the thickness/span ratio, for example, limited to 1/20 for a homogeneous and isotropic plate. For higher ratios or for non-isotropic plates with very different layers of stiffness, the Reissner–Mindlin hypotheses should be considered; indications are given in §7.4.

DOI: 10.1201/9780429440403-7

- /H4/ **Hypothesis of small displacements.** A plate is a plane in a reference configuration. This hypothesis plays a vital role; it remains acceptable as long as the transverse displacement remains much smaller than the thickness. If this hypothesis can not be respected, the coupling between the membrane forces and the bending moments must be taken into account; indications are given in §7.3.

The equations established in this section are obtained by complete linearisation of the equations of motion. The case where geometry changes can no longer be neglected is discussed in §7.3.

7.1.1.2 Equilibrium equations and boundary conditions

Since the geometry changes are neglected during the motion, the equilibrium equations are obtained from the PVP equations (4.97) by writing that the "current" curvature tensor b is zero:

$$\boxed{\begin{aligned} &D_\alpha N^{\alpha\beta} + p^\beta - \rho\gamma^\beta = 0 \\ &D_\beta D_\alpha M^{\alpha\beta} - p^3 + \rho\gamma^3 = 0 \end{aligned}} \tag{7.1}$$

∇ reducing to the differentiation D in the plane.

Due to the hypothesis of thin shells, it is not necessary to distinguish the generalised stresses and the resultant of stresses. The equations (4.66) and (4.68) obtained by the general theorems are also reduced to (7.1) in the hypotheses taken here.

These equations are rewritten in intrinsic form:

$$\boxed{\begin{aligned} &\text{div } N + \left(p^\beta - \rho\gamma^\beta\right) a_\beta = 0 \\ &\text{div div } M - p^3 + \rho\gamma^3 = 0 \end{aligned}} \tag{7.2}$$

div designating the divergence operator in the middle plane.

In an orthonormal cartesian coordinate system of the plane, the equations in Equation (7.1) are written:

$$\boxed{\begin{aligned} &\frac{\partial N^{xx}}{\partial x} + \frac{\partial N^{xy}}{\partial y} + p^x - \rho\frac{\partial^2 \xi^x}{\partial t^2} = 0 \\ &\frac{\partial N^{xy}}{\partial x} + \frac{\partial N^{yy}}{\partial y} + p^y - \rho\frac{\partial^2 \xi^y}{\partial t^2} = 0 \\ &\frac{\partial^2 M^{xx}}{\partial x^2} + 2\frac{\partial^2 M^{xy}}{\partial x \partial y} + \frac{\partial^2 M^{yy}}{\partial y^2} - p^3 + \rho\frac{\partial^2 \zeta}{\partial t^2} = 0 \end{aligned}} \tag{7.3}$$

where the acceleration γ is expressed as a function of displacement ξ, taking into account the hypothesis of small displacements.

The first two equations can also be obtained by integrating the first two local (3D) equations, as shown in §5.5. They only contain membrane forces.

The third equation, which contains only the bending moments, can be rewritten in a similar way to (4.97), by introducing the shear vector of components Q^α :

$$\begin{cases} Q^\alpha + D_\beta M^{\beta\alpha} = 0 \\ D_\alpha Q^\alpha + p^3 - \rho\gamma^3 = 0 \end{cases} \tag{7.4}$$

either, in an orthonormal Cartesian coordinate system:

$$\begin{aligned} Q^x &= -\left(\partial_x M^{xx} + \partial_y M^{xy}\right) \\ Q^y &= -\left(\partial_x M^{xy} + \partial_y M^{yy}\right) \\ \partial_x Q^x &+ \partial_y Q^y + p^3 - \rho\gamma^3 = 0 \end{aligned} \tag{7.5}$$

These relationships can be written intrinsically:

$$\boxed{\begin{aligned} &Q + \text{div } M = 0 \\ &\text{div } Q + p^3 - \rho\gamma^3 = 0 \end{aligned}} \tag{7.6}$$

Boundary conditions are deduced from (4.99):

$$\boxed{\begin{aligned} &N^{\alpha\beta}\nu_\alpha = q^\beta \\ &Q^\nu - \frac{\partial}{\partial s}\left(m^\nu + M^{\nu t}\right) = q^3 \\ &M^{\nu\nu} = m^t \end{aligned}} \tag{7.7}$$

where **t** and **ν** designate the tangent and the normal to the edge, located in the plane of the plate, and where $Q^\nu = Q^\alpha \nu_\alpha = Q \cdot \nu$. The second line of (7.7) is Kirchhoff's condition. When the rotation of the normal segment around **ν** is blocked, which is in principle the case if the edge is continuously supported or clamped (see §4.2.2.6), it must be replaced by the Poisson condition $Q^\nu = q^3$.

Again, the curvature of the plate is approximately zero, the membrane forces and the bending moments do not appear in the same boundary conditions:

- The first line contains two conditions reflecting the equilibrium in the plane of the plate.
- The third condition reflects the balance normally to the plate, along the edge.
- The last condition is the balance with respect to bending, around the vector tangent to the edge.

In small displacements, there is no coupling between bending and membrane behaviours because of equilibrium equations and boundary conditions.

The introduction of the shear force Q makes it possible to interpret the third condition at the boundary (7.7), as a balance normal to the plate on its edge. This equilibrium involves the shear force $Q^{\nu} = Q^{\alpha}\nu_{\alpha}$ applied to the facet tangent to the edge and the normal reaction q^3. In a beam, the shear force directly balances the normal reaction.

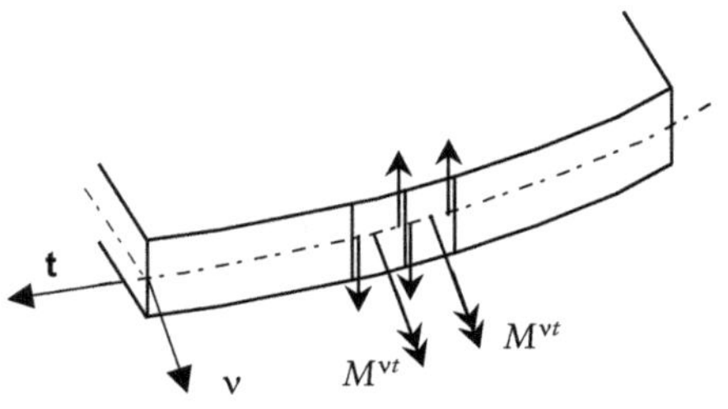

Figure 7.1

In the case of a plate, this equilibrium is disturbed by the variations of the torsion moment applied to the facet; in fact, let be two facets of the same area situated side by side, to which is applied a torsion moment $M^{\nu t}$. On each facet, the torsion moment can be balanced by two reactions normal to the plate, equal and opposite. If the moments of torsion are equal, these reactions are cancelled two by two (Figure 7.1). If the moment $M^{\nu t}$ varies along the edge, its variation participates in the reaction q^3, cf. §4.2.1.5 (Kelvin[1] and Tait[2], 1883).

7.1.1.3 Link with local stresses

The middle plane is provided with any coordinate system (x^1, x^2) or Cartesian coordinates (x, y) in a fixed orthonormal coordinate system $(\mathbf{i}, \mathbf{j})$. This system is completed by a unit vector $\mathbf{k}$ normal to the plane, associated with the coordinate z. The order of the coordinates is such that $(\mathbf{i}, \mathbf{j}, \mathbf{k})$ is direct; $\mathbf{k}$ is merged with the vector a_3 and is a constant vector.

The displacements being small, the stresses are written in Lagrange variables and noted σ.

In the coordinate system (x, y, z), taking into account the hypothesis of thin plates, by application of relations (5.20), it is noted:

$$\begin{array}{ll} N^{xx} = -\int_{-e/2}^{+e/2} \sigma^{xx} dz & M^{xx} = -\int_{-e/2}^{+e/2} \sigma^{xx} z dz \\ N^{yy} = -\int_{-e/2}^{+e/2} \sigma^{yy} dz & M^{yy} = -\int_{-e/2}^{+e/2} \sigma^{yy} z dz \\ N^{xy} = -\int_{-e/2}^{+e/2} \sigma^{xy} dz & M^{xy} = -\int_{-e/2}^{+e/2} \sigma^{xy} z dz \end{array} \tag{7.8}$$

and, using (5.21):

$$\begin{array}{l} \bar{Q}^{x} = \int_{-e/2}^{+e/2} \sigma^{xz} dz \\ \bar{Q}^{y} = \int_{-e/2}^{+e/2} \sigma^{yz} dz \end{array} \tag{7.9}$$

Let a facet of normal ν contained in the middle plane and $\mathbf{t}$ the vector of the plane contained in the facet, such that $(\mathbf{t}, \nu)$ is direct, according to the convention of §5.1.3. Let φ

[1] William Thomson, Lord Kelvin, British physicist (1824–1907).
[2] Peter Guthrie Tait, Scottish mathematical physicist (1831–1901).

be the angle of ν with the basis vector **i**. The stress vector T_ν applied locally to this facet is written according to the stress tensor expressed in the coordinate system (x, y, z):

$$T_\nu = \sigma \cdot \nu = \begin{pmatrix} \sigma^{xx}\cos\varphi + \sigma^{xy}\sin\varphi \\ \sigma^{xy}\cos\varphi + \sigma^{yy}\sin\varphi \\ \sigma^{xz}\cos\varphi + \sigma^{yz}\sin\varphi \end{pmatrix} \tag{7.10}$$

which components in the coordinate system (**t**, ν, **k**) are respectively:

$$\begin{cases} \sigma^{\nu\nu} = \nu \cdot T_\nu = \sigma^{xx}\cos^2\varphi + 2\sigma^{xy}\cos\varphi\sin\varphi + \sigma^{yy}\sin^2\varphi \\ \sigma^{\nu t} = t \cdot T_\nu = \left(\sigma^{xx} - \sigma^{yy}\right)\cos\varphi\sin\varphi + \sigma^{xy}\left(\sin^2\varphi - \cos^2\varphi\right) \\ \sigma^{\nu z} = k \cdot T_\nu = \sigma^{xz}\cos\varphi + \sigma^{yz}\sin\varphi \end{cases} \tag{7.11}$$

The resulting screw (in the thickness) of the stress vector T_ν has components:

- a resultant $N(\nu)$ in the plane:

$$N(\nu) = N^{\nu\nu}\nu + N^{\nu t}\mathbf{t} \tag{7.12}$$

with:

$$\boxed{\begin{aligned} N^{\nu\nu} &= \int_{-e/2}^{+e/2} \sigma^{\nu\nu}\,dz \\ N^{\nu t} &= \int_{-e/2}^{+e/2} \sigma^{\nu t}\,dz \end{aligned}} \tag{7.13}$$

By integration of relations (7.11), it becomes:

$$\boxed{\begin{cases} N^{\nu\nu} = N^{xx}\cos^2\varphi + 2N^{xy}\cos\varphi\sin\varphi + N^{yy}\sin^2\varphi = \nu \cdot (N \cdot \nu) \\ N^{\nu t} = \left(N^{xx} - N^{yy}\right)\cos\varphi\sin\varphi + N^{xy}\left(\sin^2\varphi - \cos^2\varphi\right) = \mathbf{t} \cdot (N \cdot \nu) \end{cases}} \tag{7.14}$$

- a resultant $Q(\nu)$ normal to the plane:

$$\boxed{Q^\nu = \int_{-e/2}^{+e/2} \sigma^{\nu z}\,dz} \tag{7.15}$$

and, by integration of (7.11):

$$\boxed{Q^\nu = Q^x\cos\varphi + Q^y\sin\varphi = Q \cdot \nu = Q(\nu)} \tag{7.16}$$

- a moment $\mathcal{M}(\nu)$ such that (Figure 7.2):

$$\mathcal{M}(\nu) = \int_{-e/2}^{+e/2} (z\,\mathbf{k}) \times T_\nu\,dz = \mathcal{M}^\nu + \mathcal{M}^t \tag{7.17}$$

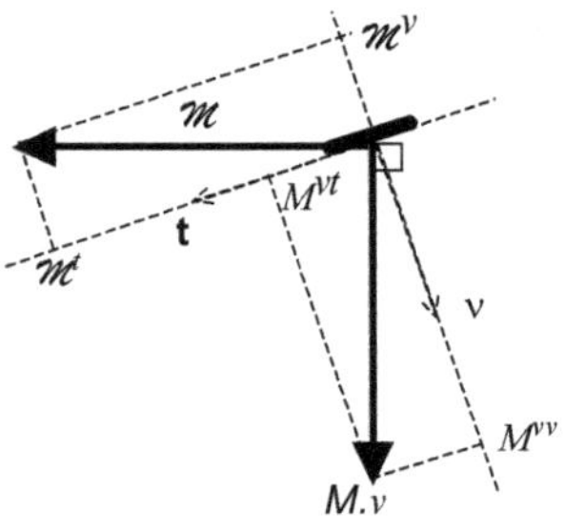

Figure 7.2

$$\begin{aligned}\mathcal{M}^t &= \int_{-e/2}^{+e/2} (z\,\mathbf{k}) \times (\sigma^{vv} v)\,dz \\ &= \left(\int_{-e/2}^{+e/2} \sigma^{vv} z\,dz\right)\mathbf{t} = M^{vv}\mathbf{t} \\ \mathcal{M}^v &= \int_{-e/2}^{+e/2} (z\,\mathbf{k}) \times (\sigma^{vt}\mathbf{t})\,dz \\ &= \left(\int_{-e/2}^{+e/2} -\sigma^{vt} z\,dz\right) v = -M^{vt} v\end{aligned} \tag{7.18}$$

That is to say:

$$\boxed{\mathcal{M}(v) = -M^{vt} v + M^{vv}\mathbf{t}} \tag{7.19}$$

with, in accordance with (5.20):

$$\boxed{\begin{aligned} M^{vv} &= -\int_{-e/2}^{+e/2} \sigma^{vv} z\,dz \\ M^{vt} &= -\int_{-e/2}^{+e/2} \sigma^{vt} z\,dz \end{aligned}} \tag{7.20}$$

either, by integration of relations (7.11):

$$\boxed{\begin{cases} M^{vv} = M^{xx}\cos^2\varphi + 2M^{xy}\cos\varphi\,\sin\varphi + M^{yy}\sin^2\varphi = v \cdot (M \cdot v) \\ M^{vt} = \left(M^{xx} - M^{yy}\right)\cos\varphi\,\sin\varphi + M^{xy}\left(\sin^2\varphi - \cos^2\varphi\right) = \mathbf{t} \cdot (M \cdot v) \end{cases}} \tag{7.21}$$

The result (7.19) specifies, in the case of the hypotheses adopted here, a general result obtained in §4.4.1, the tensor H here being symmetrical and merged with the tensor M. M^{vv} is the bending moment and M^{vt} the torsion moment applied to the facet. M^{vv} is positive if the lower face ($z \le 0$) is in tension. The resulting moment $\mathcal{M}(v)$ is therefore orthogonal to $M \cdot v$, as was shown in §4.4.1.

Note: in the (**i**, **j**, **k**) coordinate system, pay attention to the sign of the moments applied to the facets normal to **i** and **j**. For the facet of normal **i**, the tangent vector **t** is opposite to **j** and the relation (7.19) is written:

$$\mathcal{M}(\mathbf{i}) = M^{xy}\mathbf{i} - M^{xx}\mathbf{j}$$

The components of the resultant of the stress vector in the thickness are:

- N^{xx} normal to the facet, along **i**,
- N^{xy} tangent to the facet, along **j**,
- Q^x tangent to the facet, along **k**.

The resulting moment has components (Figure 7.3):

- $-M^{xx}$, moment along **j** (bending moment),
- M^{xy}, moment along **i** (torsion moment).

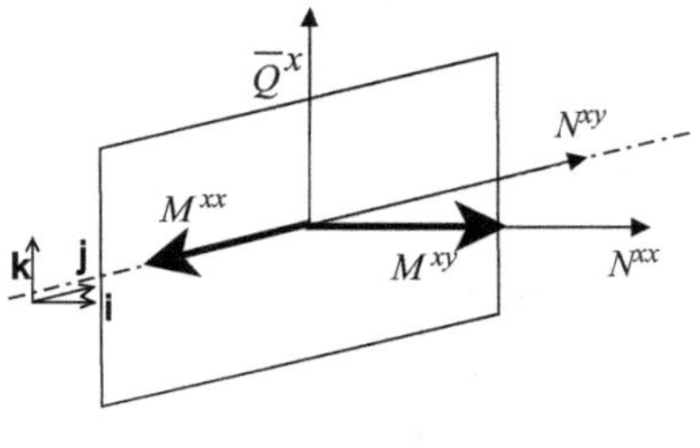

Figure 7.3

On the other hand, for the facet of normal **j**, the tangent vector is **i** and the formula (7.19) applies directly.

7.1.1.4 Extreme values of stresses

It is interesting for some applications to determine the extreme values of the shear force and the bending moment applied to a facet when the angle φ of the facet varies.

With regard to the shear force, the maximum value of Q^{v} given by (16) is obtained for $\tan\varphi = Q^{y}/Q^{x}$; it becomes:

$$\left|Q^{v}\right|_{\max} = \left\|Q\right\| = \sqrt{\left(Q^{x}\right)^{2} + \left(Q^{y}\right)^{2}} \tag{7.22}$$

It is the length of vector Q.

The bending moment M^{vv} applied to a facet is determined by (7.21). This value is extremal when v is an eigenvector of the tensor M and the corresponding moment M^{vv} is an eigenvalue of M (the torsion moment M^{vt} is then zero). This property is easily interpreted in Mohr's plane (M^{vv}, M^{vt}), cf. Figure 7.4. It becomes:

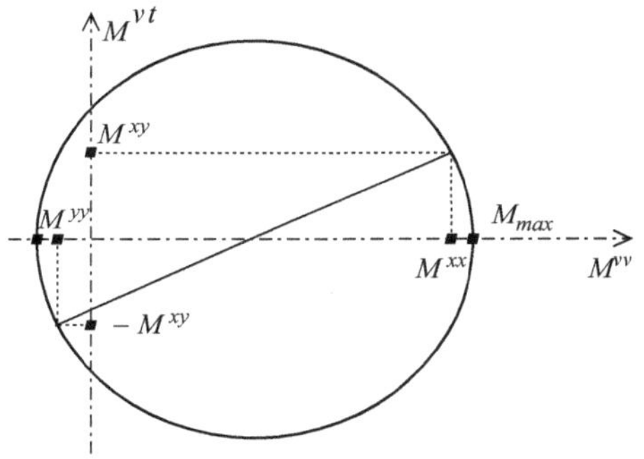

Figure 7.4

$$M^{\max}_{\min} = \frac{1}{2}\left[M^{xx} + M^{yy} \pm \sqrt{\left(M^{xx} - M^{yy}\right)^{2} + \left(2M^{xy}\right)^{2}}\right] \tag{7.23}$$

The eigen directions are the principal directions of bending.

7.1.2 Deformations

7.1.2.1 Displacement

The displacement of a point $\mathcal{M}$ of the middle plane is written:

$$\xi = \xi^{\alpha}\left(x^{1}, x^{2}\right) A_{\alpha} + w\left(x^{1}, x^{2}\right)\mathbf{k} \tag{7.24}$$

in any coordinates of the plane. In an orthonormal Cartesian coordinate system, introducing the usual notation:

$$\xi = u(x,y)\,\mathbf{i} + v(x,y)\,\mathbf{j} + w(x,y)\,\mathbf{k} \tag{7.25}$$

The transverse displacement w is assumed small with respect to the thickness, which is necessary to neglect the geometry changes in the expression of the equilibrium equations.

As in the case of beams, bending is preponderant, u and v are small in front of w and they are supposed to be second order. These assumptions (of Donnell) are specified in Chapter 8 in the more general case of shells in flexion.

7.1.2.2 Deformations

Given the above hypothesis on displacements (so-called "second-order theory"), the deformation tensor of the middle plane is of the second order and, neglecting terms of the fourth order:

$$\boxed{\varepsilon_{\alpha\beta} = \frac{1}{2}\left(D_\alpha \xi_\beta + D_\beta \xi_\alpha + \partial_\alpha w\, \partial_\beta w\right)} \tag{7.26}$$

in Cartesian coordinates:

$$\begin{aligned}
\varepsilon_{xx} &= \frac{\partial u}{\partial x} + \frac{1}{2}\left(\frac{\partial w}{\partial x}\right)^2 \\
\varepsilon_{yy} &= \frac{\partial v}{\partial y} + \frac{1}{2}\left(\frac{\partial w}{\partial y}\right)^2 \\
\varepsilon_{xy} &= \frac{1}{2}\left(\frac{\partial u}{\partial y} + \frac{\partial v}{\partial x} + \frac{\partial w}{\partial x}\frac{\partial w}{\partial y}\right)
\end{aligned} \tag{7.27}$$

Only the linear terms are retained in (7.26) and (7.27) in linearised theory (called "first order"). The following quantities are linearised and considered of the first order.

The curvature variation tensor is reduced to the curvature tensor of the plate in its deformed position. In this position, taking as parameters the parameters (x^α) of the middle plane, it comes successively:

- tangent vectors:

$$a_\alpha = \frac{\partial}{\partial x^\alpha}\left(\mathcal{M} + \xi^\gamma A_\gamma + w\,\mathbf{k}\right) = A_\alpha + \partial_\alpha \xi^\gamma \cdot A_\gamma + \xi^\gamma D_\alpha A_\gamma + \partial_\alpha w \cdot \mathbf{k}$$

- normal vector a_3:

$$\left\| a_1 \times a_2 \right\|\, a_3 = a_1 \times a_2 \approx \left(1 + \operatorname{div}\underline{\xi}\right) A_1 \times A_2 + \left(e_2{}^\mu \partial_1 w - e_1{}^\mu \partial_2 w\right) A_\mu$$

either, neglecting second-order terms:

$$a_3 = A_3 - \left(B_{\alpha\beta}\xi^\beta + \partial_\alpha \zeta\right) A^{\alpha\gamma} A_\gamma \tag{7.28}$$

- curvatures:

$$b_{\alpha\beta} = a_3 \cdot \partial_\beta a_\alpha \approx \partial^2_{\alpha\beta} w - \Gamma^\gamma_{\alpha\beta}\, \partial_\gamma w$$

neglecting second-order terms, then:

$$\kappa_{\alpha\beta} = b_{\alpha\beta} \approx \partial^2_{\alpha\beta} w - \Gamma^\gamma_{\alpha\beta}\, \partial_\gamma w \tag{7.29}$$

The deformation at any point of the thickness is then given, in the hypothesis of thin shells and in Kirchhoff–Love theory, by:

$$\bar{\varepsilon}_{\alpha\beta} = \varepsilon_{\alpha\beta} - z\, b_{\alpha\beta} \tag{7.30}$$

7.1.3 Linear elastic plates

7.1.3.1 Homogeneous and isotropic plate

In this case (see §5.4.3.3), the plate constitutive law is written:

$$N = K\left[(1-\nu)\,\varepsilon + \nu\ \mathrm{TR}\varepsilon\, G\right] \tag{7.31}$$

$$M = D\left[(1-\nu)\,b + \nu\ \mathrm{TR}b\, G\right] \tag{7.32}$$

where G denotes the metric tensor of the middle plane.

Membrane forces and bending moments appear separately in equilibrium equations and constitutive laws. The problem of bending can therefore be solved independently of membrane forces, which is the scope of the rest of this paragraph.

The constitutive law of flexion is written in an orthonormal Cartesian coordinate system:

$$\boxed{\begin{aligned} M_{xx} &= D\left(b_{xx} + \nu\, b_{yy}\right) = D\left(\frac{\partial^2 w}{\partial x^2} + \nu\frac{\partial^2 w}{\partial y^2}\right)\\ M_{yy} &= D\left(b_{yy} + \nu\, b_{xx}\right) = D\left(\frac{\partial^2 w}{\partial y^2} + \nu\frac{\partial^2 w}{\partial x^2}\right)\\ M_{xy} &= D(1-\nu)\, b_{xy} = D(1-\nu)\frac{\partial^2 w}{\partial x \partial y}\end{aligned}} \tag{7.33}$$

The shear forces are obtained by the relation (7.6):

$$\boxed{\begin{aligned} Q_x &= -D\frac{\partial}{\partial x}\left(\frac{\partial^2 w}{\partial x^2} + \frac{\partial^2 w}{\partial y^2}\right)\\ Q_y &= -D\frac{\partial}{\partial y}\left(\frac{\partial^2 w}{\partial x^2} + \frac{\partial^2 w}{\partial y^2}\right)\end{aligned}} \tag{7.34}$$

either, in intrinsic form:

$$\boxed{Q = -D\ \mathrm{grad}\Delta w} \tag{7.35}$$

Q depends on the Poisson's ratio only by the stiffness D:

$$D = \frac{Ee^3}{12\left(1-\nu^2\right)} \tag{7.36}$$

The equation of flexion of the elastic thin plate is obtained by reporting the bending constitutive law in the third equilibrium equation (7.2):

$$D\left(\frac{\partial^4 w}{\partial x^4}+2\frac{\partial^4 w}{\partial x^2\partial y^2}+\frac{\partial^4 w}{\partial y^4}\right)=p-\rho\frac{\partial^2 w}{\partial t^2} \tag{7.37}$$

where p^3 is noted p. This equation is written in the intrinsic form:

$$\boxed{D\,\Delta\Delta w = p-\rho\frac{\partial^2 w}{\partial t^2}} \tag{7.38}$$

(Sophie Germain[3] equation, 1811), where appears the double Laplacian of displacement w. To this local equation, it is necessary to add boundary conditions at the edges, which are of two types:

- kinematic conditions:
  - supported edge : $w = 0$,
  - clamped edge: $\frac{\partial w}{\partial \nu} = 0$ (ν being normal to the edge).
- the mechanical conditions: these are conditions (7.7) where the forces are expressed as functions of the displacement. For example, for a point on the edge perpendicular to **i**:
  - If the edge is articulated, the rotation around the tangent is free and:

$$D\left(\frac{\partial^2 w}{\partial x^2}+\nu\frac{\partial^2 w}{\partial y^2}\right)=0 \tag{7.39}$$

  - If the edge is not supported (no reaction):

$$-D\frac{\partial}{\partial x}\left[\frac{\partial^2 w}{\partial x^2}+(2-\nu)\frac{\partial^2 w}{\partial y^2}\right]+\frac{\partial m^x}{\partial y}=0 \tag{7.40}$$

In the case of a supported and articulated rectilinear edge, these conditions are reduced to (the equation of the edge being x = cst) :

- $w = 0$ (zero displacement),
- $\frac{\partial^2 w}{\partial x^2}=0$ (zero moment), (7.40)

where it is expressed that the curvature along the edge is zero: $\frac{\partial^2 w}{\partial y^2}=0$.

The resolution of Equation (7.38), associated with the boundary conditions, makes it possible to obtain the displacements, then the bending moments and the shear forces by (7.32) and (7.35). The reactions on the edge are then obtained by (7.40).

Noting $\mathbf{M}$ = TR(M), (7.33) allows the equation to be written as:

$$\mathbf{M}=D(1+\nu)\Delta w \tag{7.42}$$

[3] Sophie Germain, French mathematician, physicist and philosopher (1776–1831).

Equation (7.38) can then be replaced, statically, by the system of equations:

$$\begin{aligned} D\,\Delta w &= \frac{\mathsf{M}}{1+\nu} \\ (1+\nu)\Delta\mathsf{M} &= p \end{aligned} \tag{7.43}$$

On a simply supported edge, **M** is zero. This formulation may allow for simpler resolution in certain situations.

The expression of the flexural elastic energy is obtained from (5.120):

$$\mathcal{F}_f = \frac{1}{2} D \iint_{\mathcal{P}} \left[\left(\frac{\partial^2 w}{\partial x^2}\right)^2 + \left(\frac{\partial^2 w}{\partial y^2}\right)^2 + 2\nu \frac{\partial^2 w}{\partial x^2}\frac{\partial^2 w}{\partial y^2} + 2(1-\nu)\left(\frac{\partial^2 w}{\partial x \partial y}\right)^2 \right] dx\, dy \tag{7.44}$$

In this expression, the distortion energy due to shear is neglected, according to Kirchhoff's hypotheses.

In the case of a solid plate, the local stresses due to the bending moments are deduced from Hooke's law. In plane stresses:

$$\sigma^{xx}(\mathcal{P}) = \frac{1}{1-\nu^2}\left[\bar{\varepsilon}_{xx}(\mathcal{P}) + \nu\,\bar{\varepsilon}_{yy}(\mathcal{P})\right] = -\frac{E\,z}{1-\nu^2}\frac{M^{xx}}{D} = -\frac{12}{e^3} z\, M^{xx}$$

More generally:

$$\sigma^{\alpha\beta}(\mathcal{P}) = -\frac{12}{e^3} z\, M^{\alpha\beta} \tag{7.45}$$

The shear stress due to the shear force can be calculated by a method similar to that which makes it possible to obtain the Jouravski relation in beam theory: let be a zone of the plate not subject to external loading, in static mode. The first local equation is then reduced to:

$$\frac{\partial \sigma^{xx}}{\partial x} + \frac{\partial \sigma^{xy}}{\partial y} + \frac{\partial \sigma^{xz}}{\partial z} = 0$$

By relations (7.45), then expressing the first equation of equilibrium (7.4), it becomes:

$$\frac{\partial \sigma^{xz}}{\partial z} = \frac{12\,z}{e^3}\left(\frac{\partial M^{xx}}{\partial x} + \frac{\partial M^{xy}}{\partial y}\right) = -\frac{12\,z}{e^3} Q^x$$

By integrating this relationship between a level z and the upper surface of the plate, where σ^{xz} is zero:

$$\sigma^{xz} = \frac{12\,Q^x}{e^3}\int_z^{+e/2} \eta\, d\eta = \frac{6Q^x}{e}\left[\frac{1}{4} - \left(\frac{z}{2}\right)^2\right]$$

A similar result is obtained for σ^{yz} and for a facet of normal ν:

$$\sigma^{\nu z} = \frac{6Q^{\nu}}{e}\left[\frac{1}{4} - \left(\frac{z}{2}\right)^2\right] \tag{7.46}$$

The distribution of shear stresses is parabolic and the maximum value is obtained on the middle plane where:

$$\sigma^{\nu z}(z=0) = \frac{3}{2}\frac{Q^{\nu}}{e} \tag{7.47}$$

7.1.3.2 Anticlastic surface

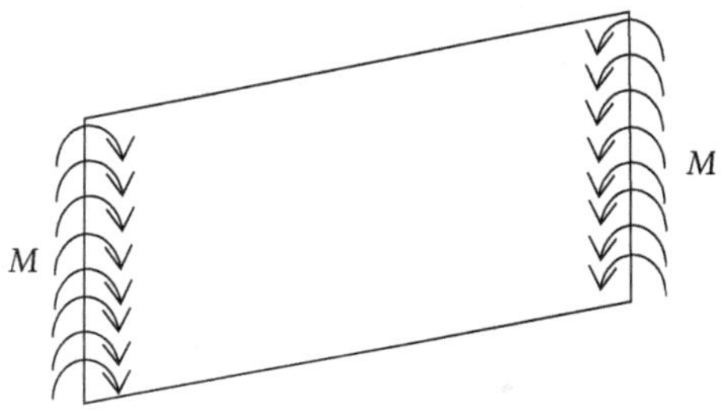

Figure 7.5

A free homogeneous rectangular plate is subjected to a uniform torque density M on two opposite sides (Figure 7.5).

The tensor of the bending moments is constant:

$$M^{xx} = M$$

$$M^{yy} = 0$$

$$M^{xy} = \text{cst}$$

which verifies the equilibrium equation and the boundary conditions.

The plate takes the form of an anticlastic surface where:

$$b_{xx} = \frac{12M}{e^3} \quad ; \quad b_{yy} = -\nu\, b_{xx} \tag{7.48}$$

The principal curvatures are of opposite signs; indeed, a compressed (or tense) plane following principal stress expands (resp., contracts) in the perpendicular direction. This creates a curvature of the opposite sign in the direction perpendicular to the main bending, proportional to the Poisson's ratio.

On the other hand, each line parallel to the x-axis behaves like a beam, the different lines taking a circular shape with a single radius, while being offset with respect to each other, according to z. By integration of relations (7.48), displacement is written:

$$w = \frac{6M}{e^3}\left(x^2 - \nu\, y^2\right) + axy + bx + cy + d$$

The origin of the x and y axes is taken from the centre of the plate. By eliminating one translation and two rotations (with respect to the x and y axes) and noting that b_{xy} is zero in the centre of the plate, w is an even function of x and y and:

$$w = \frac{6M}{e^3}\left(x^2 - \nu\, y^2\right)$$

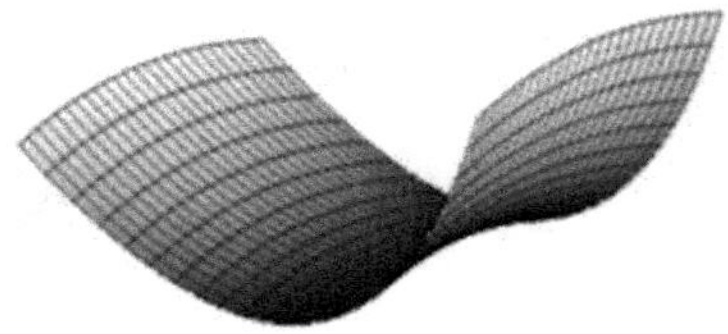

It follows that b_{xy} and the moment of torsion M^{xy} are null in all points (it is a consequence of the symmetry of the problem).

On the contrary, in a "cylindrical bending" plate (see §1.2.4.1): $b_{yy} = b_{xy} = 0$ and $M^{xx} = D\, b_{xx}$; the so-called **transversal** moment M^{yy} is connected to the **longitudinal** moment M^{xx} by $M^{yy} = \nu\, M^{xx}$: the natural b_{yy} curvature naturally existing by Poisson effect when the plate is free is hindered here, making appear a moment in reaction.

The preceding example shows that the Poisson's ratio intervenes or not in the expression of the deformations or the bending moments, but that the two fields can not be simultaneously independent of ν. These properties can be specified as follows.

Let be equation (7.38) in static; the function $\bar{w} = D\, w$ verifies the relationship:

$$\Delta\Delta\, \bar{w} = p \tag{7.49}$$

independent of the Poisson's ratio.

When the plate is subjected only to conditions of the kinematic type, for example:

$$\bar{w} = 0 \;\; ; \;\; \frac{\partial \bar{w}}{\partial x} = 0$$

or assimilated (hinges):

$$\frac{\partial^2 \bar{w}}{\partial x^2} = 0$$

which is the case if all the edges are either articulated or clamped, the solution function does not depend on the Poisson's ratio. Then, the displacement w, the shear forces, the reactions and the pulsations of the plate depends on ν only via the rigidity D. On the other hand, the bending moments depend directly on the Poisson's ratio. In the case where there is at least one free edge or elastic links, the boundary conditions explicitly contain the Poisson's ratio and the solution therefore depends directly on ν.

7.1.3.3 Stresses due to a thermal gradient in a plate clamped on its edge

The plate has any shape and its edge is clamped throughout its length. It is subjected to a thermal gradient $\frac{\Delta\Theta}{e}$, the average temperature rise Θ (measured at the level of the middle surface) being zero.

A displacement field such that $w = 0$ at any point on the plate is kinematically admissible. This field is solution if it is also statically admissible, that is to say, if it verifies the equation of equilibrium (7.6), taking into account the constitutive law. In any temperature distribution situation, it is not sufficient to check equation (7.38) because it does not take into account the terms of thermal origin.

The equation (5.118) allows the expression of the moments, for a zero displacement:

$$M^{\alpha\beta} = \alpha\, \mathrm{I} \frac{\Delta\Theta}{e} a^{\alpha\beta} = -\lambda_\ell D(1+\nu)\frac{\Delta\Theta}{e} a^{\alpha\beta} \tag{7.50}$$

λ_ℓ denoting the coefficient of linear thermal expansion, cf. (5.109).

The tensor M being homogeneous, its divergence is zero: $Q \equiv 0$ and (7.6) is verified. This is therefore the elastic solution to the problem. In this particular situation, equation (7.38) is verified.

A more intuitive way of finding this result consists in noting that, when the plate is not subjected to any connections on its edge, it takes the form of a spherical cap, the curvature of which is $-\lambda_\ell \frac{\Delta\Theta}{e}$ (the elongation of a fibre parallel to the middle plane is equal to $-\lambda_\ell \frac{\Delta\Theta}{e} z$)

As the plate is in fact clamped on the edge, fixed-end moments $M^{\nu\nu}$ appear there, which prevent the rotation of the plate along its edge. A homogeneous and spherical moment tensor ($m\ a$) gives a spherical elastic deformation such that, by (7.32), the curvature is $\frac{m}{D(1+\nu)}$. The rotation effectively vanishes at any point on the edge if the "elastic" curvature is opposite to the curvature of thermal origin, hence the result.

7.1.3.4 Orthotropic plate

The constitutive law of an orthotropic shell has been established in §5.6.1.2.

Only the case of simple bending without membrane forces is considered here. The constitutive law is written, depending on the displacement:

$$\begin{cases} M_{xx} = D_x \dfrac{\partial^2 w}{\partial x^2} + D' \dfrac{\partial^2 w}{\partial y^2} \\[2ex] M_{yy} = D' \dfrac{\partial^2 w}{\partial x^2} + D_y \dfrac{\partial^2 w}{\partial y^2} \\[2ex] M_{xy} = D_{xy} \dfrac{\partial^2 w}{\partial x \partial y} \end{cases} \tag{7.51}$$

where in particular:

$$\begin{cases} Q^x = -\dfrac{\partial}{\partial x}\left[D_x \dfrac{\partial^2 w}{\partial x^2} + (D' + D_{xy}) \dfrac{\partial^2 w}{\partial y^2} \right] \\[2ex] Q^y = -\dfrac{\partial}{\partial y}\left[D_y \dfrac{\partial^2 w}{\partial y^2} + (D' + D_{xy}) \dfrac{\partial^2 w}{\partial x^2} \right] \end{cases} \tag{7.52}$$

The equation of simple bending (in static) is obtained by introducing these expressions in the second equation of equilibrium (7.2):

$$\boxed{D_x \frac{\partial^4 w}{\partial x^4} + 2(D' + D_{xy}) \frac{\partial^4 w}{\partial x^2 \partial y^2} + D_y \frac{\partial^4 w}{\partial y^4} = p} \tag{7.53}$$

(Huber[4] Equation – 1921)

The coefficients D' and D_{xy} intervene indistinctly in the coefficient $H = D' + D_{xy}$. w therefore depends only on three coefficients.

By comparison between the constitutive laws (7.33) and (7.51), it appears that the isotropic case is a special case where:

$$D_x = D_y = D\,;\ D' = D\nu\,;\ D_{xy} = D(1-\nu)$$

Equation (7.38) is then obtained by reporting these particular values in equation (7.53).

[4] Tytus Maksymilian Huber, Polish mechanical engineer (1872–1950).

7.2 CLASSICAL EXAMPLES

7.2.1 Circular plates

7.2.1.1 General results

The case of the circular plates is usefully treated in polar coordinates (r, θ). To obtain the equation of bending (7.38), it is necessary to express the Laplacian in polar coordinates. In any coordinates (see §2.2.6):

$$\Delta w = \text{div grad } w = \frac{1}{\sqrt{\hat{a}}}\partial_\alpha\left(a^{\alpha\beta}\sqrt{\hat{a}}\ \partial_\beta w\right)$$

From which, in polar coordinates:

$$\Delta w = \frac{1}{r}\partial_r\left(r\ \partial_r w\right) + \frac{1}{r^2}\partial^2_{\theta^2} w$$

which leads to the expression of Germain's equation in polar coordinates:

$$\frac{1}{r}\partial_r\left\{r\partial_r\left[\frac{1}{r}\partial_r\left(r\ \partial_r w\right) + \frac{1}{r^2}\partial^2_{\theta^2} w\right]\right\} + \frac{1}{r^2}\partial^2_{\theta^2}\left[\frac{1}{r}\partial_r\left(r\ \partial_r w\right) + \frac{1}{r^2}\partial^2_{\theta^2} w\right] = \frac{p}{D} \quad (7.54)$$

The bending moments and shear forces are respectively obtained by relations (7.32) and (7.35). In the orthonormal coordinate system (a_1, a_2), associated with the polar coordinates of the middle plane, such as $a_1 = a_r$ and, $a_2 = \frac{a_\theta}{r}$ it becomes:

- the curvature tensor, after (7.29):

$$b_{11} = \partial^2_{r^2} w\ ;\ b_{22} = \frac{1}{r^2}\partial^2_{\theta^2} w + \frac{1}{r}\partial_r w\ ;\ b_{12} = \frac{1}{r}\partial^2_{r\theta} w - \frac{1}{r^2}\partial_\theta w \quad (7.55)$$

- by the isotropic elastic constitutive law, the bending moments:

$$\begin{cases} M_{11} = D\left[\partial^2_{r^2} w + \frac{\nu}{r^2}\left(\partial^2_{\theta^2} w + r\ \partial_r w\right)\right] \\ M_{22} = D\left[\frac{1}{r^2}\left(\partial^2_{\theta^2} w + r\ \partial_r w\right) + \nu\ \partial^2_{r^2} w\right] \\ M_{12} = \frac{D}{r^2}(1-\nu)\left(r\ \partial^2_{r\theta} w - \partial_\theta w\right) \end{cases} \quad (7.56)$$

- the shear forces:

$$\begin{cases} Q_1 = -D\frac{\partial}{\partial r}\left[\frac{1}{r}\partial_r\left(r\ \partial_r w\right) + \frac{1}{r^2}\partial^2_{\theta^2} w\right] \\ Q_2 = -\frac{D}{r}\frac{\partial}{\partial\theta}\left[\frac{1}{r}\partial_r\left(r\ \partial_r w\right) + \frac{1}{r^2}\partial^2_{\theta^2} w\right] \end{cases} \quad (7.57)$$

- and the reaction on a circular edge:

$$\begin{aligned} q^3 &= Q^1 - \frac{\partial\left(m^1 + M^{12}\right)}{r\partial\theta} \\ &= -D\frac{\partial}{\partial r}\left[\frac{1}{r}\partial_r\left(r\,\partial_r w\right) + \frac{1}{r^2}\partial^2_{\theta^2} w\right] - \frac{D}{r^3}(1-\nu)\left(r\partial^3_{r\theta^2} w - \partial^2_{\theta^2} w\right) - \frac{\partial m^1}{r\partial\theta} \end{aligned} \tag{7.58}$$

7.2.1.2 Case of axisymmetric bending

In axisymmetric bending, displacements and forces are independent of θ. Equilibrium equations are written:

$$\begin{aligned} &\text{div div } M = \frac{1}{r}\frac{d}{dr}\left[r\frac{dM^{11}}{dr} + M^{11} - M^{22}\right] = p \\ &Q^1 = -\text{div } M^r = -\left[\frac{dM^{11}}{dr} + \frac{1}{r}\left(M^{11} - M^{22}\right)\right] \end{aligned} \tag{7.59}$$

The results previously established are reduced to:

$$b_{11} = w'' \;;\; b_{22} = \frac{w'}{r} \;;\; b_{12} = 0 \tag{7.60}$$

$$\begin{cases} M_{11} = D\left(w'' + \nu\dfrac{w'}{r}\right) \\ M_{22} = D\left(\dfrac{w'}{r} + \nu w''\right) \\ M_{12} = 0 \end{cases} \tag{7.61}$$

$$\begin{cases} Q^1 = -D\dfrac{d}{dr}(\Delta w) \\ Q^2 = 0 \end{cases} \tag{7.62}$$

$$q^3 = Q^1 = -D\frac{d}{dr}\left[\frac{1}{r}\frac{d}{dr}\left(r\frac{dw}{dr}\right)\right]$$

In static, Germain's equation is reduced to:

$$\frac{1}{r}\frac{d}{dr}\left\{r\frac{d}{dr}\left[\frac{1}{r}\frac{d}{dr}\left(r\frac{dw}{dr}\right)\right]\right\} = \frac{p}{D} \tag{7.63}$$

which general solution is, in the case where p is constant:

$$w(r)=\frac{pr^4}{64D}+A\left(\frac{r^2}{4}\ln r-\frac{r^2}{4}\right)+Br^2+C\ln r+F \tag{7.64}$$

A, B, C, and F are constants to be determined by the boundary conditions. These are expressed from the following relations obtained by differentiation of (7.64):

$$w'(r)=\frac{pr^3}{16D}+A\left(\frac{r}{2}\ln r-\frac{r}{4}\right)+2Br+\frac{C}{r} \tag{7.65}$$

$$w''(r)=\frac{3pr^2}{16D}+A\left(\frac{1}{2}\ln r+\frac{1}{4}\right)+2B-\frac{C}{r^2} \tag{7.66}$$

$$\Delta w(r)=\frac{1}{r}\frac{d}{dr}\left(r\frac{dw}{dr}\right)=\frac{pr^2}{4D}+A\ln r+4B \tag{7.67}$$

$$M^{11}(r)=D\left[\frac{pr^2}{16D}(3+\nu)+\frac{A}{2}(1+\nu)\ln r+\frac{A}{4}(1-\nu)+2B(1+\nu)-\frac{C}{r^2}(1-\nu)\right]$$

$$M^{22}(r)=D\left[\frac{pr^2}{16D}(1+3\nu)+\frac{A}{2}(1+\nu)\ln r-\frac{A}{4}(1-\nu)+2B(1+\nu)+\frac{C}{r^2}(1-\nu)\right] \tag{7.68}$$

$$Q^1(r)=-D\frac{d}{dr}(\Delta w)=-\frac{pr}{2}-A\frac{D}{r}$$

The first term of the solution (7.64) is a particular solution of (7.63), thus balancing the load. In the case where p is not constant, this first term must be modified by substituting for it a particular solution adapted to the loading; the relations (7.65) to (7.68) are modified accordingly.

Finally, always in axisymmetric bending, the flexural deformation energy in a ring bounded by radii R' and R is written:

$$\mathcal{F}=\frac{1}{2}2\pi D\int_{R'}^{R}\left[(w'')^2+\left(\frac{w'}{r}\right)^2+2\nu\frac{w'w''}{r}\right]r\,dr \tag{7.69}$$

7.2.1.3 Free plate subject to a moment on the edge

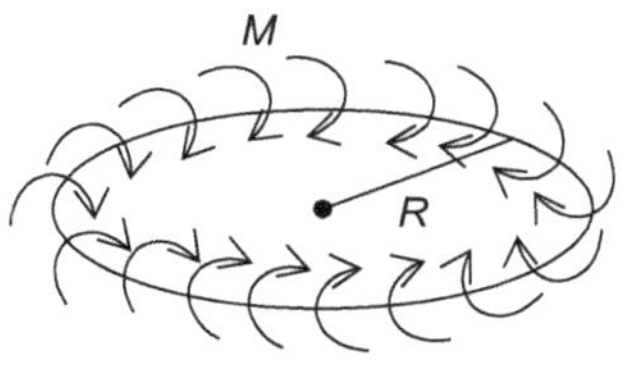

Figure 7.6

An isotropic solid plate is subjected to a uniform moment on its outer edge (Figure 7.6). Any translation is included in the solution, then it is possible to fix arbitrarily $w(0) = 0$, which entails: $C = F = 0$.

The moment is finite in the centre: $A = 0$. The condition in $r = R$ gives:

$$M^{11}(r=R)=D\left[w''+\nu\frac{w'}{r}\right]_{r=R}=2DB(1+\nu)=M$$

from which:

$$w=\frac{Mr^2}{2D(1+\nu)}$$

then:

$$\begin{cases}M^{11}=M^{22}=M\\Q^1=0\end{cases}$$

obvious results by equilibrium equations.

The same plate is pierced by a centered circular hole with radius *R'* (Figure 7.7). The integration constant *F* is still zero in this case; boundary conditions become:

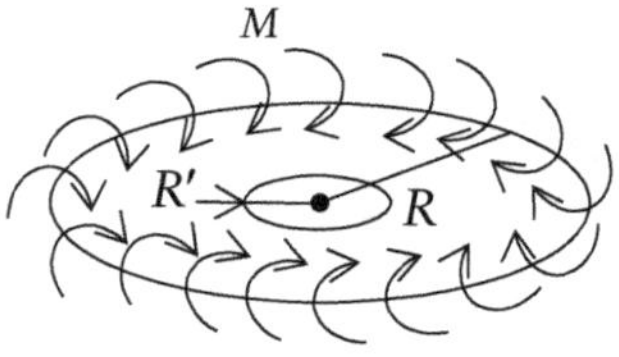

Figure 7.7

$$\text{in } r=R:\ \begin{cases}M^{11}=M\\Q^1=0\end{cases};\quad \text{in } r=R':\ \begin{cases}M^{11}=0\\Q^1=0\end{cases}$$

Now:

$$Q_1=-D\frac{d}{dr}(\Delta w)=-\frac{AD}{r}$$

therefore $A = 0$. It becomes:

$$\begin{cases}M^{11}(R)=2B(1+\nu)-\dfrac{C}{R^2}(1-\nu)=M\\M^{11}(R')=2B(1+\nu)-\dfrac{C}{R'^2}(1-\nu)=0\end{cases}$$

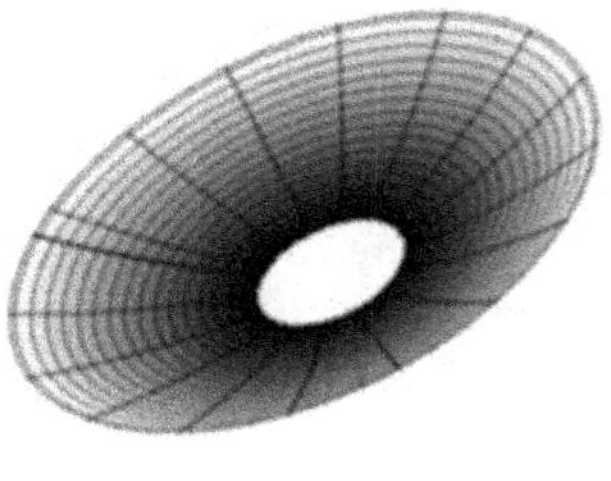

hence, a constant being undetermined:

$$w(r)=\frac{R^2R'^2}{R^2-R'^2}\frac{M}{D}\left[\frac{1}{2(1+\nu)}\frac{r^2}{R'^2}+\frac{1}{1-\nu}\ln\frac{r}{R'}\right]$$

and finally:

$$\begin{cases}M^{11}=M\dfrac{R^2}{R^2-R'^2}\left(1-\dfrac{R'^2}{r^2}\right)\\M^{22}=M\dfrac{R^2}{R^2-R'^2}\left(1+\dfrac{R'^2}{r^2}\right)\end{cases}$$

At the edge of the hole (Figure 7.8):

$$M^{22} = 2M\frac{R^2}{R^2 - R'^2} = 2M\frac{1}{1-\lambda^2}$$

with $\lambda = \frac{R'}{R}$.

The ratio $\frac{M^{22}}{M}$ expresses the supplementary stresses due to the presence of the hole. It is at least 2. It is therefore necessary to strengthen a slab around an opening.

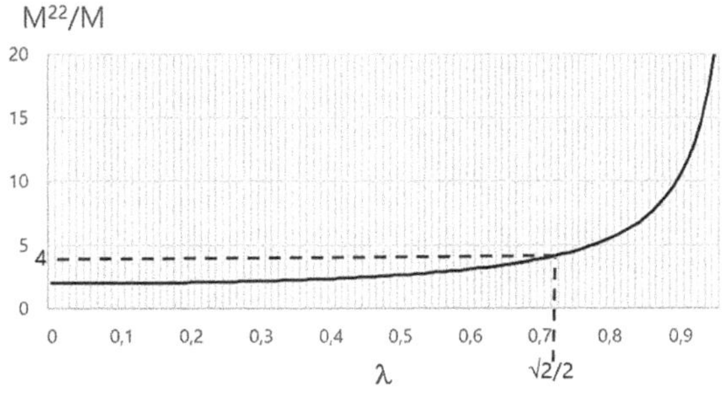

Figure 7.8

7.2.1.4 *Uniformly loaded clamped plate*

A solid circular plate, clamped on its outer edge, is subjected to a uniform load p. The transverse displacement is finite in the centre: $C = 0$. The shear force, summed on a circle of radius r, equilibrates, along the perimeter $2\pi r$, the load $\pi r^2 p$ applied to a disc of radius r. It is therefore equal to $-{pr}/{2}$, which allows us to deduce that A is zero, compared with (7.68).

The boundary conditions in $r = R$ are then written:

$$w(R) = \frac{pR^4}{64D} + BR^2 + F = 0$$

$$w'(R) = \frac{4pR^3}{64D} + 2BR = 0$$

hence the expression of the deflection:

$$w(r) = \frac{pR^4}{64D}\left[\left(\frac{r}{R}\right)^4 - 2\left(\frac{r}{R}\right)^2 + 1\right]$$

then the bending moments:

$$\begin{cases} M^{11} = \frac{pR^2}{16}\left[(3+\nu)\left(\frac{r}{R}\right)^2 - (1+\nu)\right] \\ M^{22} = \frac{pR^2}{16}\left[(1+3\nu)\left(\frac{r}{R}\right)^2 - (1+\nu)\right] \end{cases}$$

and the shear forces:

$$Q^1 = -\frac{pr}{2}; \; Q^2 = 0$$

Note that here the reaction on the periphery of a disk is equal to the shear force $-\dfrac{pR}{2}$, since M^{12} is zero.

M^{11} and M^{22} are equal and negative in the centre where they are worth $-\dfrac{pR^2}{16}(1+\nu)$; they are positive in $r = R$ where:

$$\begin{cases} M^{11}(R) = \dfrac{pR^2}{8} \\[2ex] M^{22}(R) = \nu \dfrac{pR^2}{8} \end{cases}$$

Indeed, at the clamped edge the circumferential curvature b_{22} is zero, which implies that:

$$M^{22} = \nu\, M^{11}$$

The previous solution makes it possible to evaluate the validity of the hypotheses used in the establishment of linearised bending equations. In particular, it has been assumed that the influence of normal forces on flexion is negligible (which makes it possible to neglect the term $b \cdot N$ in the transverse equilibrium).

However, during bending, the plate passes from a state where the Gaussian curvature is zero to a state where it is equal to $\dfrac{w'w''}{r}$, therefore non-zero, except at the clamped edge and on the circle of the radius $\dfrac{R\sqrt{3}}{3}$ where $w'' = 0$. Such a transformation can not be done without extension (see §3.3.4.1). Therefore, membrane forces appear.

It is possible to evaluate the membrane forces by admitting that the solution w established above remains a sufficient approximation.

Given the particular conditions of loading, the tangential displacement $\underline{\xi} = u a_1 + v a_2$ is such that v is zero. Assuming further that, because of the lateral blocking, u is negligible vis-à-vis w, the strain tensor is reduced, by (7.27), to:

$$\varepsilon_{11} = \frac{w'^2}{2}; \quad \varepsilon_{12} = \varepsilon_{22} = 0$$

From which:

$$N^{11} = K\frac{w'^2}{2} = \frac{K}{2}\left\{\frac{pR^3}{16D}\left[\left(\frac{r}{R}\right)^3 - \left(\frac{r}{R}\right)\right]\right\}^2; \quad N^{22} = \nu\, N^{11}$$

Membrane energy can be compared to flexural energy. The displacement of the centre of the plate in linearised theory is:

$$q = \frac{pR^4}{64D}$$

So, with (7.69):

$$\mathcal{F}_f = \frac{1}{2} 2\pi D \int_0^R \left[w''^2 + \frac{w'^2}{r^2} + 2\nu \frac{w'w''}{r} \right] r\, dr = \frac{1}{2} 2\pi D \frac{32q^2}{3R^2}$$

$$\mathcal{F}_m = \frac{1}{2} 2\pi K \int_0^R \left(\frac{w'^2}{2} \right)^2 r\, dr = \frac{1}{2} 2\pi K \frac{32q^4}{105R^2}$$

from which:

$$\frac{\mathcal{F}_m}{\mathcal{F}_f} = \frac{K}{D} \frac{q^2}{35} = \frac{12}{35} \frac{q^2}{e^2} \qquad (7.70)$$

The ratio of energies is proportional to $\left(\frac{q}{e}\right)^2$, which is verified in other situations of loading or shape of the plate. In conclusion, the hypotheses used in classical theory are only realistic if the displacement remains small in front of the thickness. Indeed, if this is not the case, there is an "inverted vault effect" and the transverse forces p are balanced by both membrane forces and bending moments. This point is detailed in Section 7.3.2.4. It should be noted that:

- this effect is partly related to the fact that tangent movements are prevented. If they are free, the effect is less accentuated, the displacement u tends to decrease the deformation ε_{11};
- nevertheless, according to Gauss' theorem, the extension can not be zero, even if the horizontal displacements are free. This situation, related to the two-dimensional character of the plate, is not encountered in the case of a straight beam. The two-dimensional character of the plate is the main explanation for the existence of a vault effect when the plate bends.

7.2.1.5 Simply supported plate loaded in the centre

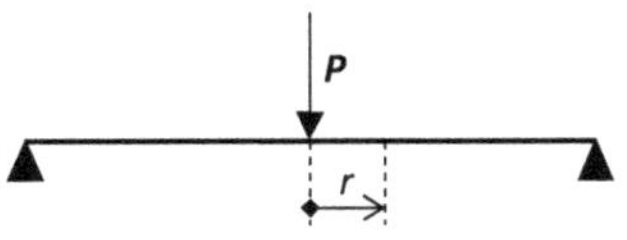

Figure 7.9

The concentrated charge brings a singularity (Figure 7.9). Indeed, the shear force at the distance r is written:

$$Q^1 = -\frac{P}{2\pi r}$$

In comparison with the expression (7.68), it becomes:

$$A = \frac{P}{2\pi D}$$

Moreover, C is zero because w is finite in $r = 0$.

It remains to express the conditions at the boundary $r = R$:

$$\begin{cases} w(R) = \dfrac{P}{2\pi D}\left(\dfrac{R^2}{4} \ln R - \dfrac{R^2}{4} \right) + BR^2 + F = 0 \\ M^{11}(R) = \dfrac{P}{2\pi}\left(\dfrac{1}{2} \ln R + \dfrac{1}{4} \right) + 2BD + \dfrac{P\nu}{2\pi}\left(\dfrac{1}{2} \ln R - \dfrac{1}{4} \right) + 2\nu\, BD = 0 \end{cases}$$

hence the displacement:

$$w(r)=\frac{PR^2}{8\pi D}\left[\frac{r^2}{R^2}\ln\frac{r}{R}+\frac{1}{2}\frac{3+\nu}{1+\nu}\left(1-\frac{r^2}{R^2}\right)\right]$$

and:

$$\begin{cases} M^{11}(r)=\dfrac{P}{4\pi}(1+\nu)\ln\dfrac{r}{R} \\ M^{22}(r)=\dfrac{P}{4\pi}\left[(1+\nu)\ln\dfrac{r}{R}-(1-\nu)\right] \end{cases}$$

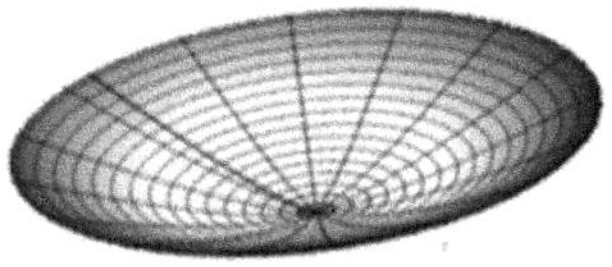

On the edge:

$$M^{22}(R)=-\frac{P(1-\nu)}{4\pi}$$

which is not zero because the circumferential curvature is not zero on the edge. Moments tend to be infinity under concentrated load.

7.2.1.6 *Plate simply supported, uniformly loaded*

The shear force is obtained simply by the equilibrium of the disk with radius r:

$$Q^1=-\frac{pr}{2}$$

By (7.68), the constant A is zero. The solution is then given by (7.64), with $A = C = 0$ (w is also finite in the centre). It becomes, expressing the conditions to the boundary $r = R$:

$$\begin{cases} w(R)=\dfrac{pR^4}{64D}+BR^2+F=0 \\ M^{11}(R)=0 \Rightarrow \dfrac{pR^2}{16}(3+\nu)+2BD(1+\nu)=0 \end{cases}$$

hence the solution:

$$w(r)=\frac{pR^4}{64D}\left[\left(\frac{r}{R}\right)^4-2\frac{3+\nu}{1+\nu}\left(\frac{r}{R}\right)^2+\frac{5+\nu}{1+\nu}\right]$$

At the centre:

$$\begin{cases} w(0)=\dfrac{pR^4}{64D}\dfrac{5+\nu}{1+\nu} \\ M^{11}(0)=M^{22}(0)=-\dfrac{pR^2}{16}(3+\nu) \end{cases}$$

which is approximately $-\frac{pR^2}{5}$. For this value, the maximum stress is equal, according to (7.45), to:

$$\sigma_{\max} = -\frac{6M}{e^2} = \frac{6}{5}p\frac{R^2}{e^2}$$

The ratio $\frac{\sigma_{\alpha\beta}}{\sigma_{33}}$ is therefore of the order of $\left(\frac{R}{e}\right)^2$, which justifies the hypothesis of plane stresses.

On the edge:

$$\begin{cases} M^{11}(R) = 0 \\ M^{22}(R) = -\frac{pR^2}{8}(1-\nu) \end{cases}$$

7.2.1.7 Bottom of a tank

The bottom of a tank with a cylindrical skirt is a circular plate resting on the ground. Under the effect of filling by the fluid, the bottom is subjected to the uniform weight p of fluid, the linear moment m transmitted by the skirt on the periphery and the linear weight q of the skirt.

If the soil is stiff enough so that its deformability can be neglected, the bottom rises from the ground in the vicinity of the periphery (Figure 7.10). The outer edge is supported on the ground by the weight of the skirt, but the corresponding reaction q does not intervene otherwise in the bending of the bottom. Let a be the radius of the disk in contact with the ground.

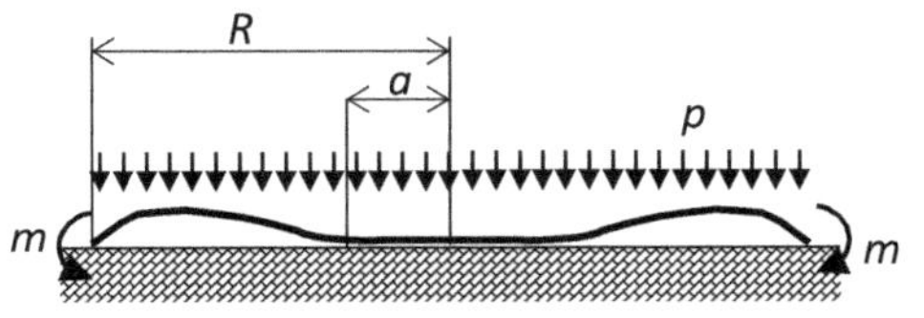

Figure 7.10

The general solution of the local equation is given by (7.64). In the unlifted central part, the load p (here opposite to the displacement) passes directly into the ground; displacement, bending moments and shear forces are therefore null in this region. The boundary conditions are:

$$w(a) = \frac{dw}{dr}(a) = 0;\ M^{11}(a) = 0$$

$$w(R) = 0;\ M^{11}(R) = -m$$

The shear force is not zero in $r = a$, because there is a linear density of reaction concentrated on the corresponding circle, as it appears hereinafter. Five boundary conditions are needed because a is unknown, in addition to the four integration constants A, B, C, and F. This leads to the system of five equations:

$$(i)\quad w(a) = -\frac{pa^4}{64D} + A\left(\frac{a^2}{4}\ln a - \frac{a^2}{4}\right) + Ba^2 + C\ln a + F = 0$$

$$(ii)\quad w'(a) = -\frac{pa^3}{16D} + A\left(\frac{a}{2}\ln a - \frac{a}{4}\right) + 2Ba + \frac{C}{a} = 0$$

$$(iii)\quad M^{11}(a) = D\left[-\frac{pa^2}{16D}(3+\nu) + \frac{A}{2}(1+\nu)\ln a + \frac{A}{4}(1-\nu) + 2B(1+\nu) - \frac{C}{a^2}(1-\nu)\right] = 0$$

$$(iv)\quad w(R) = -\frac{pR^4}{64D} + A\left(\frac{R^2}{4}\ln R - \frac{R^2}{4}\right) + BR^2 + C\ln R + F = 0$$

$$(v)\quad M^{11}(R) = D\left[-\frac{pR^2}{16D}(3+\nu) + \frac{A}{2}(1+\nu)\ln R + \frac{A}{4}(1-\nu) + 2B(1+\nu) - \frac{C}{R^2}(1-\nu)\right] = -m$$

B is calculated with *(ii)* as a function of A and C, which makes it possible to simplify $M^{11}(a)$ in *(iii)*, then express C and B as a function of A. The system can thus be reduced to two equations expressing $M^{11}(R)$ and $w(R)-w(a)$:

$$\frac{pR^2}{16D}\left(1-\frac{a^2}{R^2}\right)(3+\nu) + \frac{pR^2}{16D}\frac{a^2}{R^2}\left(1-\frac{a^2}{R^2}\right)(1-\nu) - \frac{A}{4}(1-\nu)\left(1-\frac{a^2}{R^2}\right) + \frac{A}{2}(1+\nu)\ln\frac{a}{R} = \frac{m}{D}$$

$$\frac{pR^4}{16D}\left[\frac{1}{4}\left(1-\frac{a^4}{R^4}\right) - \frac{a^2}{R^2}\left(1-\frac{a^2}{R^2}\right) - \frac{a^4}{R^4}\ln\frac{a}{R}\right] + \frac{AR^2}{4}\left[\left(1+\frac{a^2}{R^2}\right)\ln\frac{a}{R} + \left(1-\frac{a^2}{R^2}\right)\right] = 0$$

in which remain only A and a as unknown. Introducing:

- an adimensional parameter of loading $\mu = \dfrac{16m}{pR^2}$,
- the dimensionless constant $\alpha = \dfrac{16D}{pR^2}A$,
- the dimensionless length $\chi = \dfrac{a}{R}$,

the two equations are rewritten in adimensional form:

$$(1-\chi^2)(3+\nu) + \chi^2(1-\chi^2)(1-\nu) - \frac{\alpha}{4}(1-\nu)(1-\chi^2) + \frac{\alpha}{2}(1+\nu)\ln\chi = \mu$$

$$\left[\frac{1}{4}(1-\chi^4) - \chi^2(1-\chi^2) - \chi^4\ln\chi\right] + \frac{\alpha}{4}\left[(1+\chi^2)\ln\chi + (1-\chi^2)\right] = 0$$

By eliminating α:

$$\frac{\left[\frac{1}{4}\left(1-\chi^4\right)-\chi^2\left(1-\chi^2\right)-\chi^4\ln\chi\right]\left[\left(1-\nu\right)\left(1-\chi^2\right)-2\left(1+\nu\right)\ln\chi\right]}{\left(1+\chi^2\right)\ln\chi+\left(1-\chi^2\right)}$$

$$+\left(1-\chi^2\right)\left(3+\nu\right)+\chi^2\left(1-\chi^2\right)\left(1-\nu\right)=\mu$$

Figure 7.11 shows the relationship between χ and μ for $\nu = 0{,}2$.

The uprising takes place for $\chi \le 1$ and starts for the limit value $\mu = 0$, so as soon as a moment $m > 0$ is applied.

The support of the bottom on the ground is reduced to a point where $\chi = 0$, then $\alpha = 0$ and $\mu = \dfrac{5+\nu}{2}$. When μ is greater than this value, the bottom is fully raised.

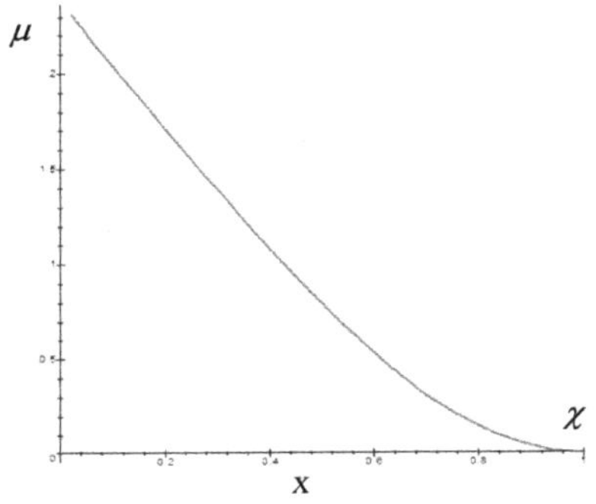

Figure 7.11

Shear force is given by (7.68):

$$Q^1(r)=\frac{pr}{2}-A\frac{D}{r} \quad\Rightarrow\quad \frac{16Q^1}{pR}=\frac{8r}{R}+\frac{\frac{1}{4}\left(1-\chi^4\right)-\chi^2\left(1-\chi^2\right)-\chi^4\ln\chi}{\left(1+\chi^2\right)\ln\chi+\left(1-\chi^2\right)}\frac{R}{r}$$

In the case of a partial uplift, there is a concentrated reaction density along the circle $r = a$, equal to the weight of fluid applied to the raised part of the bottom, minus the additional reaction in $r = R$. The latter is obtained from the expression of the shear force in R:

$$q=\frac{pR}{2}+\frac{pR}{16}\frac{\frac{1}{4}\left(1-\chi^4\right)-\chi^2\left(1-\chi^2\right)-\chi^4\ln\chi}{\left(1+\chi^2\right)\ln\chi+\left(1-\chi^2\right)}$$

whose result is worth $2\pi Rq$.

In $r = a$, the linear density of the reaction ρ is therefore:

$$\rho=\frac{1}{2a}\left[\left(R^2-a^2\right)p-2Rq\right]$$

If the central support is reduced to a point ($\chi = 0$), $q = pR/2$ and $\rho = 0$: the entire weight of fluid raised is balanced by the peripheral reaction q and there is no concentrated reaction in the centre.

If the tank rests on elastic soil, a common simplifying assumption is to assume that the soil response is distributed and proportional to the displacement, the proportionality factor k (also called ballast coefficient) being constant. In this case, Equation (7.63) is replaced by:

$$\frac{D}{r}\frac{d}{dr}\left\{r\frac{d}{dr}\left[\frac{1}{r}\frac{d}{dr}\left(r\frac{dw}{dr}\right)\right]\right\}+kw=p$$

Here, $w_0 = p/k$ is a particular solution of the equation and it is thus a question of finding a general solution of the equation without second member which makes it possible to respect the boundary conditions. (7.64) suggests looking for it as a combination of functions $A_n r^n$ and $B_n r^n \ln r$. Moreover, w, w', M^{11} must be finite in the centre, so the terms $n = 1$ and $n = 3$ are null. B_0 is also zero. The constant term A_0 can be taken as 0, as included in the particular solution. The solution of the equation is thus sought in the form:

$$w(r) = A_2 r^2 + \sum_{n=4}^{+\infty} A_n r^n + B_2 r^2 \ln r + \sum_{n=4}^{+\infty} B_n r^n \ln r$$

By transferring this form of solution to the equation without a second member, it becomes:

$$\sum_{n=4}^{+\infty} n(n-2)\left\{n(n-2)A_n + B_n\left[4n-4+n(n-2)\ln r\right]\right\} r^{n-4}$$

$$+\frac{k}{D}\left(A_2 r^2 + B_2 r^2 \ln r + \sum_{n=4}^{+\infty}\left[A_n + B_n \ln r\right] r^n\right) = 0$$

which leads to the following relations by matching the terms of the same power:

$$\begin{cases} n^2(n-2)^2 A_n + 4n(n-2)(n-1)B_n + \dfrac{k}{D}A_{n-4} = 0 \\ (n-2)^2 n^2 B_n + \dfrac{k}{D}B_{n-4} = 0 \end{cases}$$

$$\Rightarrow \begin{cases} A_n = \dfrac{k}{D}\dfrac{1}{n^2(n-2)^2}\left[\dfrac{4(n-1)B_{n-4}}{(n-2)n} - A_{n-4}\right] \\ B_n = -\dfrac{k}{D}\dfrac{B_{n-4}}{n^2(n-2)^2} \end{cases}$$

These recurrence relations are applicable from n = 4, except for 5 and 7, which leads to:

$$A_5 = B_5 = A_7 = B_7 = 0$$

added to:

$$A_0 = B_0 = A_1 = B_1 = A_3 = B_3 = 0$$

Finally, the coefficients A_{4n+2} and B_{4n+2} $(n \geq 1)$ are expressed as functions of A_2 and B_2. All other terms are null. The solution of the equation is therefore written:

$$w(r) = \frac{p}{k} + A_2 r^2 + B_2 r^2 \ln r + \sum_{n=1}^{+\infty} A_{4n+2} r^{4n+2} + \sum_{n=1}^{+\infty} B_{4n+2} r^{4n+2} \ln r$$

If the soil is flexible enough, the bottom does not uplift; this is the hypothesis that is taken here. Then, A_2 and B_2 are determined by the two boundary conditions in $r = R$.

The actions applied at the bottom of the tank are the weight p of fluid, applied over the entire surface and taken into account in the local equation, the weight of the cylindrical vertical skirt, which results in a vertical density of reaction q applied to the outer edge, and a bending moment m due to bending of the skirt. q and m are independent and the two loading cases can therefore be applied separately; the solution is then obtained by the superposition principle. The corresponding conditions are written:

$$m = -M_{11}(R) = -D\left(w''(R) + \nu\frac{w'(R)}{R}\right)$$

$$= -D\left[\begin{array}{l} 2A_2(1+\nu) + B_2\left(2+(1+\nu)(1+2\ln R)\right) + \sum_{n=1}^{+\infty}(4n+2)(4n+1+\nu)A_{4n+2}R^{4n} \\ \qquad + \sum_{n=1}^{+\infty}\left[8n+3+\nu+(4n+2)(4n+1+\nu)\ln R\right]B_{4n+2}R^{4n} \end{array}\right]$$

$$q = -Q^1 = D\frac{d}{dr}\left[\frac{1}{r}\frac{d}{dr}\left(r\frac{dw}{dr}\right)\right]_{r=R}$$

$$= D\left\{\begin{array}{l} \frac{4B_2}{R} + \sum_{n=1}^{+\infty}4n(4n+2)^2 A_{4n+2}R^{4n-1} + \\ \sum_{n=1}^{+\infty}\left[(12n+2)(4n+2)+4n(4n+2)^2\ln R\right]B_{4n+2}R^{4n-1} \end{array}\right\}$$

where all the coefficients are expressed as functions of A_2 and B_2 by the relations of recurrence. In practice, the coefficients are of the order of $\frac{1}{n^2}\left(\frac{kR^4}{D}\right)^n$ decreasing rapidly only if the soil is very flexible compared to the bottom.

Note: if the coefficients are increasing, it is easier to express the solution using Bessel functions, cf. §7.2.1.9.

Example: The skirt of the tank is supposed to be placed at its foot on a foundation that is sufficiently rigid so that its vertical displacement is negligible and only a moment m is transmitted to the bottom. By retaining only the first terms A_6 and B_6 of the series, the boundary conditions in R are written:

$$w(R) = \frac{p}{k} + A_2R^2 + B_2R^2\ln R + \frac{kR^6}{576D}\left(\frac{5}{6}B_2 - A_2\right) - B_2\frac{kR^6}{576D}\ln R = 0$$

$$\frac{m}{D} = -\left[\begin{array}{l} 2A_2(1+\nu) + B_2\left(2+(1+\nu)(1+2\ln R)\right) + 6(5+\nu)\frac{kR^4}{576D}\left(\frac{5}{6}B_2 - A_2\right) \\ \qquad - B_2\left[11+\nu+6(5+\nu)\ln R\right]\frac{kR^4}{576D} \end{array}\right]$$

The importance of the complementary terms is therefore dependent on $\frac{kR^4}{576D}$, compared to 1. If the soil is very stiff, the situation is similar to that previously examined in the case of infinitely stiff soil; this is the most common situation. In the opposite case, taking ν = 0 to simplify, it becomes:

$$w(r)=\frac{p}{k}\left\{1-\left[1+\frac{1}{2}\ln R-\frac{mkR^2}{4pD}\ln R\right]\frac{r^2}{R^2}+\left(\frac{1}{2}-\frac{mkR^2}{4pD}\right)\frac{r^2}{R^2}\ln r\right\}$$

m is of the order of pRe (see §8.2.3.1), which makes it possible to evaluate the relative order of magnitude of the different terms, in particular the relative influence of p and m in the displacement.

7.2.1.8 Any loading

Any load satisfies the conditions of periodicity in θ and can therefore be decomposed into Fourier series:

$$p(r,\theta)=p_0(r)+\sum_{j=1}^{\infty}\left[p_{sj}(r)\sin j\theta+p_{cj}(r)\cos j\theta\right] \tag{7.71}$$

The first term corresponds to an axisymmetric load which solutions have been treated in the preceding paragraphs. It is therefore sufficient to examine the solution of a mode of order j in cos jθ, obtained by transferring this load into Equation (7.55). The solution in sin jθ is analogous. It becomes:

$$\frac{1}{r}\frac{d}{dr}\left\{r\frac{d}{dr}\left[\Upsilon_j\right]\right\}-\frac{j^2}{r^2}\left[\Upsilon_j\right]=\frac{p_j}{D}$$

with: (7.72)

$$\Upsilon_j(r)=\frac{1}{r}\frac{d}{dr}\left(r\frac{d}{dr}w_j\right)-\frac{j^2}{r^2}w_j$$

The solutions of the equation without a second member are looked for in the form $w=r^\alpha$ which leads successively to $\Upsilon=\left(\alpha^2-j^2\right)r^{\alpha-2}$, then to the characteristic equation:

$$\left(\alpha^2-j^2\right)\left[\left(\alpha-2\right)^2-j^2\right]=0$$

which solutions are – j, + j, j + 2, – j + 2. The general solution of the equation without a second member is therefore:

$$\forall j>1,\qquad w_{1j}(r,\theta)=\left(A_{cj}r^{-j}+B_{cj}r^{j}+C_{cj}r^{-j+2}+D_{cj}r^{j+2}\right)\cos j\theta \tag{7.73}$$

In the case j = 1, two solutions of the characteristic equation are equal to 1 and the general solution is written:

$$j=1, \qquad w_{11}(r,\theta)=\left(A_{c1}r^{-1}+B_{c1}r+C_{c1}r\ \ln r+D_{c1}r^{3}\right)\cos\theta \tag{7.74}$$

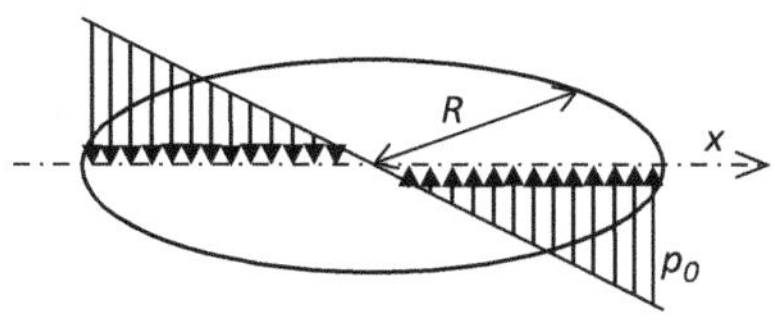

Figure 7.12

The solutions in sin jθ and sin θ are analogous to (7.73) and (7.74) respectively. It remains to determine a particular solution to Equation (7.72), which depends on the particular form of the functions $p_j(r)$.

Example: Statically variable antisymmetric loading (Figure 7.12).

Such loading corresponds for example to the variable part of pressure of fluid or soil.

Since the x-axis originates from the centre of the circle, the load is written as:

$$p(x)=p_0\frac{x}{R}$$

Either, in polar coordinates:

$$p(r,\theta)=p_0\frac{r}{R}\cos\theta$$

Looking for a particular solution in the form $w_2(r,\theta)=Kr^{\alpha}\cos\theta$, (7.72) is rewritten:

$$K\left(\alpha^2-1\right)\left[(\alpha-2)^2-1\right]r^{\alpha-4}=p_0\frac{r}{R}$$

which shows that the solution is completely determined with $\alpha=5$. The general solution to the problem is finally written:

$$\begin{aligned}w(r,\theta)=\frac{p_0}{192R}r^5\cos\theta&+\left(A_{c1}r^{-1}+B_{c1}r+C_{c1}r\ \ln r+D_{c1}r^3\right)\cos\theta\\&+\left(A_{s1}r^{-1}+B_{s1}r+C_{s1}r\ \ln r+D_{s1}r^3\right)\sin\theta\\&+\sum_{j=2}^{\infty}\left(A_{cj}r^{-j}+B_{cj}r^{j}+C_{cj}r^{-j+2}+D_{cj}r^{j+2}\right)\cos j\theta\\&+\sum_{j=2}^{\infty}\left(A_{sj}r^{-j}+B_{sj}r^{j}+C_{sj}r^{-j+2}+D_{sj}r^{j+2}\right)\sin j\theta\end{aligned}$$

The coefficients are determined by the boundary conditions. In the case of a disc without opening, the displacement and the moments are finite in the centre of the circle, which makes it possible to eliminate all the coefficients A_{cj}, A_{sj}, C_{cj}, and C_{sj}, with the exception of C_{c2} and C_{s2}. If the disc is supported on its edge, it should be written:

$$\begin{aligned}w(R,\theta)=\frac{p_0R^4}{192}\cos\theta+\left(B_{c1}R+D_{c1}R^3\right)\cos\theta+\left(B_{s1}R+D_{s1}R^3\right)\sin\theta+C_{c2}\cos 2\theta\\+C_{s2}\sin 2\theta+\sum_{j=2}^{\infty}\left(B_{cj}R^{j}+D_{cj}R^{j+2}\right)\cos j\theta+\sum_{j=2}^{\infty}\left(B_{sj}R^{j}+D_{sj}R^{j+2}\right)\sin j\theta\\=0\quad\forall\theta\end{aligned}$$

which implies that:

$$\frac{p_0R^4}{192}+\left(B_{c1}R+D_{c1}R^3\right)=0$$

$$C_{c2}=C_{s2}=0$$

$$B_{cj}R^{j}+D_{cj}R^{j+2}=0,\ \ j\geq 2$$

$$B_{sj}R^{j}+D_{sj}R^{j+2}=0,\ \ j\geq 1$$

If the edge is simply supported, the moment M^{rr} is zero on the edge, that is, according to (7.57):

$$M^{rr}=D\left[\partial^2_{r^2}w+\frac{\nu}{r^2}\left(\partial^2_{\theta^2}w+r\,\partial_r w\right)\right]_{r=R}=0\qquad\forall\theta$$

For example, the contribution of the term in cos jθ to M^{rr} is, in R:

$$j(j-1)B_{cj}R^{j-2}+(j+2)(j+1)D_{cj}R^{j}+\nu\left[-j^2B_{cj}R^{j-2}-j^2D_{cj}R^{j}+jB_{cj}R^{j-2}+(j+2)D_{cj}R^{j}\right]=0$$

either:

$$j(j-1)(1-\nu)B_{cj}+\left[(j+2)(j+1)+\left(-j^2+j+2\right)\nu\right]D_{cj}R^2=0$$

which shows that the two coefficients are zero and, by extension to the other terms, that all terms other than cos θ are zero. It remains:

$$M^{rr}=D\left[(20+4\nu)\frac{p_0R^2}{192}+(6+2\nu)D_{c1}R\right]\cos\theta=0\ \ \forall\theta$$

The two conditions in $r = R$ thus make it possible to obtain two equations to calculate the two remaining coefficients:

$$\frac{p_0R^4}{192}+B_{c1}R+D_{c1}R^3=0$$

$$(20+4\nu)\frac{p_0R^2}{192}+(6+2\nu)D_{c1}R=0$$

Finally, the solution is written:

$$w(r,\theta)=\frac{p_0R^4}{192D}\left(\frac{r^5}{R^5}+\frac{7+\nu}{3+\nu}\frac{r}{R}-\frac{10+2\nu}{3+\nu}\frac{r^3}{R^3}\right)\cos\theta$$

From where, according to (7.56), (7.57), and (7.58):

$$\begin{cases} M_{11} = \dfrac{p_0 R^2}{192}\left[\left((5+\nu)\dfrac{4r^3}{R^3} - (10+2\nu)\dfrac{2r}{R}\right)\right]\cos\theta \\ M_{22} = \dfrac{p_0 R^2}{192}\left[\dfrac{4r^3}{R^3}(1+5\nu) - \dfrac{2r}{R}\dfrac{(1+3\nu)(10+2\nu)}{3+\nu}\right]\cos\theta \\ M_{12} = (1-\nu)\dfrac{p_0 R^2}{192}\left[-\dfrac{4r^3}{R^3} + \dfrac{10+2\nu}{3+\nu}\dfrac{2r}{R}\right]\sin\theta \end{cases}$$

$$\begin{cases} Q_1 = -\dfrac{p_0 R}{24}\left(\dfrac{9r^2}{R^2} - \dfrac{10+2\nu}{3+\nu}\right)\cos\theta \\ Q_2 = \dfrac{p_0 R}{24}\left(\dfrac{3r^2}{R^2} - \dfrac{10+2\nu}{3+\nu}\right)\sin\theta \end{cases}$$

$$q^3 = -\frac{p_0 R}{4}\cos\theta$$

On the edge, the reaction q^3 is not equal to the shear force Q_1, since the torsion moment M_{12} varies along the edge. The resultant of the reaction along the edge is zero, as the resultant of the pressure applied to the plate. However, the integral of the difference $r(s) = q_3(s) - Q_1(s)$ is not zero along any partial segment of the edge (see §4.2.1.5).

7.2.1.9 *Vibrations of a circular plate*

In the case of free vibrations, Equation (7.55) is replaced by:

$$\frac{1}{r}\partial_r\left\{r\partial_r\left[\frac{1}{r}\partial_r(r\,\partial_r w) + \frac{1}{r^2}\partial^2_{\theta^2} w\right]\right\} + \frac{1}{r^2}\partial^2_{\theta^2}\left[\frac{1}{r}\partial_r(r\,\partial_r w) + \frac{1}{r^2}\partial^2_{\theta^2} w\right] + \frac{\rho}{D}\frac{\partial^2 w}{\partial t^2} = 0 \quad (7.75)$$

and it is natural to look for harmonic solutions in time and periodic in θ in the form:

$$w(r,\theta,t) = W(r)\cos n\theta\, e^{i\omega t} \quad (7.76)$$

which leads, by replacing in (7.75), to the equation of the eigenmodes:

$$\frac{1}{r}\frac{d}{dr}\left\{r\frac{d}{dr}\left[\frac{1}{r}\frac{d}{dr}\left(r\frac{dW}{dr}\right) - \frac{n^2}{r^2}W\right]\right\} - \frac{n^2}{r^2}\left(\frac{1}{r}\frac{d}{dr}\left(r\frac{dW}{dr}\right) - \frac{n^2}{r^2}W\right) - \frac{\rho\omega^2}{D}W = 0 \quad (7.77)$$

either, by developing:

$$\frac{d^4W}{dr^4} + \frac{2}{r}\frac{d^3W}{dr^3} - \frac{1+2n^2}{r^2}\frac{d^2W}{dr^2} + \frac{1+2n^2}{r^3}\frac{dW}{dr} + \left(\frac{n^4 - 4n^2}{r^4} - \frac{\rho\omega^2}{D}\right)W = 0 \quad (7.78)$$

which solutions are Bessel functions of order n.

Example: axisymmetric modes.
In this case, n = 0 and Equation (7.78) is simplified in:

$$\frac{d^4W}{dr^4}+\frac{2}{r}\frac{d^3W}{dr^3}-\frac{1}{r^2}\frac{d^2W}{dr^2}+\frac{1}{r^3}\frac{dW}{dr}-\frac{\rho\omega^2}{D}W=0 \tag{7.79}$$

which general solution is:

$$W(r)=A\,J_0(\lambda r)+B\,I_0(\lambda r) \tag{7.80}$$

with:

$$\lambda^4=\frac{\rho\omega^2}{D} \tag{7.81}$$

and where J_0 is the Bessel function and I_0 the modified Bessel function, of first kind and order 0. The two coefficients *A* and *B* are determined by the boundary conditions. If the plate is simply supported on its edge:

$$\begin{aligned} W(R)&=A\,J_0(\lambda R)+B\,I_0(\lambda R)=0\\ W''(R)&=A\,\lambda^2 J_0''(\lambda R)+B\lambda^2\,I_0''(\lambda R)=0 \end{aligned} \tag{7.82}$$

and to be a movement:

$$\begin{aligned} f(\lambda R)&=J_0(\lambda R)I_0''(\lambda R)-I_0(\lambda R)J_0''(\lambda R)\\ &=-J_0(\lambda R)J_0''(i\lambda R)-J_0(i\lambda R)J_0''(\lambda R)=0 \end{aligned} \tag{7.83}$$

This equation, which can be reevaluated using the functions J_1 and J_2 taking into account the properties of the Bessel functions, admits an infinite countable number of positive solutions λ_m, determining the pulsations ω_m by (7.81). The function $f(\lambda R)$ is plotted in Figure 7.13(a); its first three zeros are 2,1080, 5,4188, and 8,5920. Each value of λ_m is associated with an eigenmode:

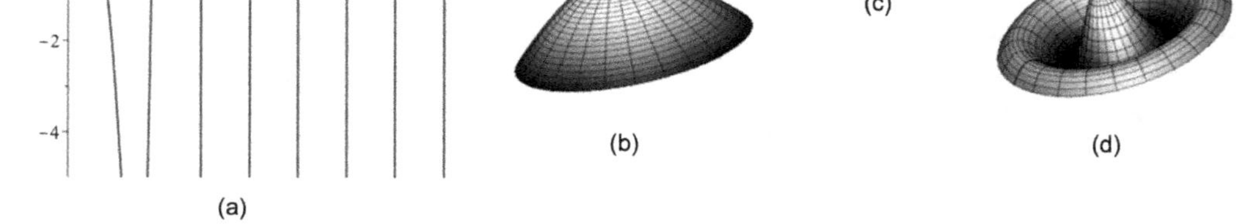

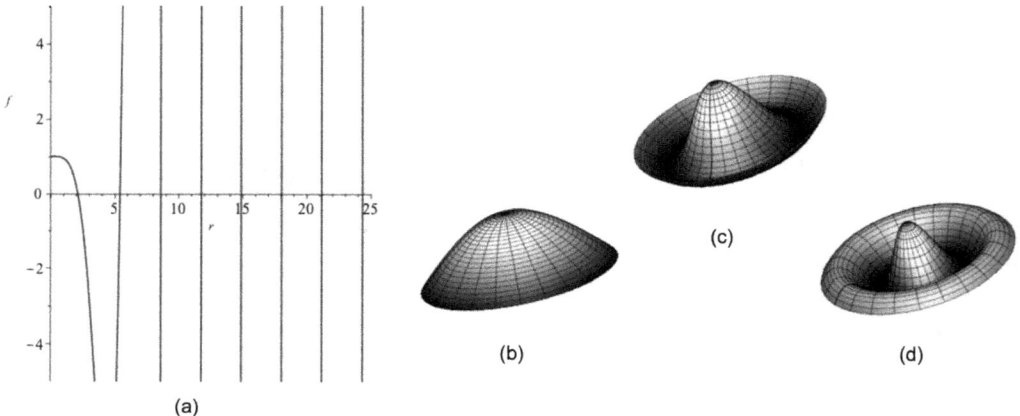

Figure 7.13

$$W_m(r) = J_0(\lambda_m r) - \frac{J_0(\lambda_m R)}{I_0(\lambda_m R)} I_0(\lambda_m r) \tag{7.84}$$

The first three axisymmetric eigenmodes are shown in Figure 7.13(b) to (d).

7.2.2 Rectangular plates

7.2.2.1 Isotropic plate supported on all four sides; Navier's solution

The problem is treated in orthonormal coordinates, in which the relations (7.33), (7.34), (7.37), and (7.44) are applied in particular. Boundary conditions are the conditions of simple supports along the entire length of the four sides (Figure 7.14):

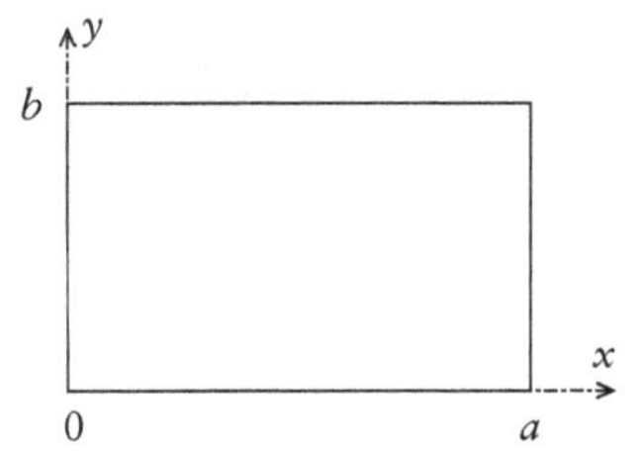

Figure 7.14

- $w = 0$ (zero displacement) (7.85)
- $\frac{\partial^2 w}{\partial x^2} = \frac{\partial^2 w}{\partial y^2} = 0$ (zero bending moment).

To solve the problem of the bending of such a plate subjected to transverse loads, it is natural to look for a solution developed in Fourier double series (Navier[5], 1820): any function f (x, y) intervening in the problem is expressed in the form:

$$f(x,y) = \sum_{m=1}^{\infty}\sum_{n=1}^{\infty} f_{mn} \sin\frac{m\pi x}{a}\sin\frac{n\pi y}{b} \tag{7.86}$$

Indeed, the functions:

$$\phi_{mn}(x,y) = \sin\frac{m\pi x}{a}\sin\frac{n\pi y}{b} \tag{7.87}$$

are orthogonal to the scalar product on the rectangle:

$$\int_0^a \int_0^b \phi_{mn}(x,y)\,\phi_{pq}(x,y)\,dxdy = \frac{ab}{4}\delta_m^p \delta_n^q \tag{7.88}$$

and form a basis of the space of summable square functions on the rectangle. The components of f are then expressed by:

$$f_{mn} = \frac{4}{ab}\int_0^a \int_0^b f(x,y)\,\phi_{mn}(x,y)\,dxdy \tag{7.89}$$

On the other hand, the functions ϕ_{mn} are eigenfunctions of the Laplacian and verify the relation:

$$\Delta\phi_{mn} = -\pi^2\left(\frac{m^2}{a^2} + \frac{n^2}{b^2}\right)\phi_{mn} \tag{7.90}$$

The decomposition (7.86) can be applied to the displacement $w(x, y)$ and to the load $p(x, y)$:

[5] Claude Louis Marie Henri Navier; French engineer and physicist (1785–1836).

$$w(x,y)=\sum_{m=1}^{\infty}\sum_{n=1}^{\infty} w_{mn}\sin\frac{m\pi x}{a}\sin\frac{n\pi y}{b}$$

$$p(x,y)=\sum_{m=1}^{\infty}\sum_{n=1}^{\infty} p_{mn}\sin\frac{m\pi x}{a}\sin\frac{n\pi y}{b}$$

This last double series is uniformly convergent if p vanishes on the edges. In this form, the displacement verifies the boundary conditions. The coefficients p_{mn} are determined by the relations (7.89). By (7.37) and considering the conditions of orthogonality of the functions ϕ_{mn}, w_{mn} verifies in the problems of statics the relation:

$$w_{mn}=\frac{1}{\pi^4\left(\frac{m^2}{a^2}+\frac{n^2}{b^2}\right)^2}\frac{p_{mn}}{D}$$

Moments are deduced by (7.33):

$$M_{xx}=-\frac{1}{\pi^2}\sum_{m=1}^{\infty}\sum_{n=1}^{\infty}\frac{\left(\frac{m^2}{a^2}+\nu\frac{n^2}{b^2}\right)p_{mn}}{\left(\frac{m^2}{a^2}+\frac{n^2}{b^2}\right)^2}\sin\frac{m\pi x}{a}\sin\frac{n\pi y}{b}$$

$$M_{yy}=-\frac{1}{\pi^2}\sum_{m=1}^{\infty}\sum_{n=1}^{\infty}\frac{\left(\nu\frac{m^2}{a^2}+\frac{n^2}{b^2}\right)p_{mn}}{\left(\frac{m^2}{a^2}+\frac{n^2}{b^2}\right)^2}\sin\frac{m\pi x}{a}\sin\frac{n\pi y}{b}$$

$$M_{xy}=\frac{1-\nu}{\pi^2 ab}\sum_{m=1}^{\infty}\sum_{n=1}^{\infty}\frac{mnp_{mn}}{\left(\frac{m^2}{a^2}+\frac{n^2}{b^2}\right)^2}\cos\frac{m\pi x}{a}\cos\frac{n\pi y}{b}$$

The shear forces by (7.34):

$$Q_x=\frac{1}{\pi a}\sum_{m=1}^{\infty}\sum_{n=1}^{\infty}\frac{mp_{mn}}{\left(\frac{m^2}{a^2}+\frac{n^2}{b^2}\right)}\cos\frac{m\pi x}{a}\sin\frac{n\pi y}{b}$$

$$Q_y=\frac{1}{\pi b}\sum_{m=1}^{\infty}\sum_{n=1}^{\infty}\frac{np_{mn}}{\left(\frac{m^2}{a^2}+\frac{n^2}{b^2}\right)}\sin\frac{m\pi x}{a}\cos\frac{n\pi y}{b}$$

and finally, by (7.40), the reaction on the edge $x = a$:

$$q=\frac{1}{\pi a}\sum_{m=1}^{\infty}\sum_{n=1}^{\infty}\frac{mp_{mn}}{\left(\frac{m^2}{a^2}+\frac{n^2}{b^2}\right)}\left[\frac{m^2}{a^2}+(2-\nu)\frac{n^2}{b^2}\right]\cos\frac{m\pi x}{a}\sin\frac{n\pi y}{b}$$

Example: Plate under uniform load.

If p is constant, it becomes:

$$p_{mn} = \begin{cases} \dfrac{16p}{\pi^2 \mathrm{mn}} & \text{if m and n are odd} \\ 0 & \text{if m or n is even} \end{cases}$$

In the case of a square plate, the deflection in the centre is:

$$w\left(\frac{a}{2},\frac{a}{2}\right) = \frac{16pa^4}{\pi^6 D} \underbrace{\sum\sum}_{\substack{m \text{ and } n \\ \text{odd}}} \frac{1}{\mathrm{mn}\left(\mathrm{m}^2+\mathrm{n}^2\right)^2}$$

and the bending moments in the centre:

$$M^{xx} = M^{yy} = -\frac{16pa^2}{\pi^4} \underbrace{\sum\sum}_{\substack{m \text{ and } n \\ \text{odd}}} \frac{\left(\mathrm{m}^2+\nu\mathrm{n}^2\right)}{\mathrm{mn}\left(\mathrm{m}^2+\mathrm{n}^2\right)^2}$$

An approximation of about 2,5% of the deflection in the centre is obtained by keeping only the first term in the series:

$$w\left(\frac{a}{2},\frac{a}{2}\right) = \frac{4pa^4}{\pi^6 D} \approx 0{,}00416\frac{pa^4}{D}$$

and for the moments in the centre:

$$M^{xx} = M^{yy} = -\frac{4pa^2(1+\nu)}{\pi^4} \approx -\frac{1+\nu}{24{,}35}pa^2$$

That is to say, for $\nu = 0{,}2$, approximately $\dfrac{pa^2}{20}$.

In a corner, only M^{xy} is non-zero:

$$M_{xy} = \frac{16pa^2(1-\nu)}{\pi^4} \underbrace{\sum\sum}_{\substack{m \text{ and } n \\ \text{odd}}} \frac{1}{\left(\mathrm{m}^2+\mathrm{n}^2\right)^2}$$

$$\approx \frac{4pa^2(1-\nu)}{\pi^4} \approx \frac{pa^2}{30}$$

It has been shown in §4.2.2 that there is a concentrated reaction at the four corners of the plate, normal to the plate and equal to $-2\,M_{xy}$. This result can be found in the present example.

Let q be the reaction density along the edges (Figure 7.15). Boundary conditions (7.7) indicate that:

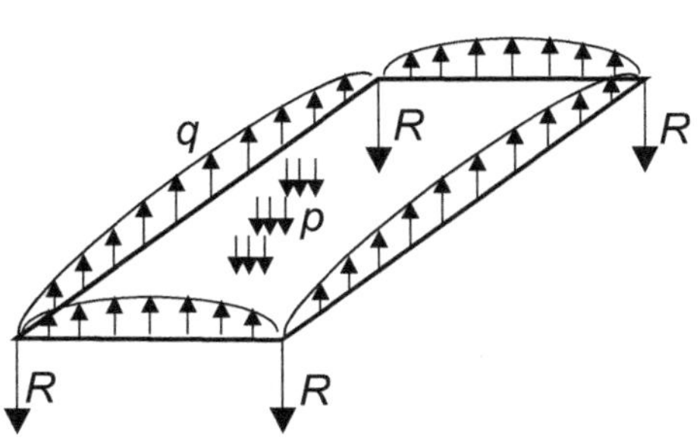

Figure 7.15

$$Q^{v}-\frac{\partial M^{vt}}{\partial s}=q$$

either, by integrating this relationship on the edge of the rectangle:

$$\oint Q\cdot v\,ds-2\left[M^{xy}(0,0)-M^{xy}(a,0)+M^{xy}(a,b)-M^{xy}(0,b)\right]=\oint q ds$$

where the discontinuities of the tangent to the edge have been taken into account in the integration of $\frac{\partial M^{vt}}{\partial s}$. On the other hand, according to equilibrium Equation (7.6):

$$\oint Q\cdot v\,ds=\int_0^a\int_0^b \operatorname{div}Q\,dx\,dy=-\int_0^a\int_0^b p\,dx\,dy=-pab$$

By considering the sign of w''_{xy} at the four corners, it appears that the signs of M^{xy} are the same on the same diagonal of the rectangle, opposite on the other diagonal. The absolute values are the same, which allows the conclusion of the announced result. This result is independent of the constitutive law of the plate, since obtained from equilibrium equations. It is to be compared with the interpretation of the boundary condition given in §7.1.1.3: in the corners, the two contiguous facets are orthogonal and the reactions created by M^{xy} can not be self-balancing.

In cases where the plate is simply laid unilaterally, the concentrated reactions can not develop and the corners rise: the solution indicated above can no longer be applied (the problem becomes non-linear) and, in particular, M^{xy} moments are null in the corners.

The Kirchhoff condition (7.7) adopted here along the edge has been verified experimentally for thin plates, notably with the existence of concentrated reactions in the corners. Nevertheless, in situations of intermediate thickness where it is not necessary to consider the distortion energy and where the Kirchhoff–Love hypotheses are used for the entire plate, the Kirchhoff condition may be less well respected if the rotation around v (thus equal to $\frac{\partial w}{\partial s}$) is zero along the edge. The reaction then directly balances the shear force (Poisson's condition) and there is no reaction in the corners (see §7.4.4 under Reissner–Mindlin's hypotheses).

The choice of one or the other condition mainly influences the calculation of the reactions and the internal forces in the vicinity of the support. The Kirchhoff condition must be used along the free edges or if the sections can turn on the support, for example in the case of elastic supports.

Example: Concentrated load P placed at the point of coordinates (ξ,η).

In this case:

$$P_{mn}=\frac{4P}{ab}\phi_{mn}(\xi,\eta)$$

The displacement at any point (x, y) under a unit concentrated load defines the influence function:

$$W(x,y,\xi,\eta)=\frac{4}{abD}\sum_{m=1}^{\infty}\sum_{n=1}^{\infty}\frac{1}{\pi^4\left(\frac{m^2}{a^2}+\frac{n^2}{b^2}\right)^2}\phi_{mn}(\xi,\eta)\,\phi_{mn}(x,y) \tag{7.91}$$

symmetrical with respect to couples *(x, y)* and *(ξ, η)* (Maxwell–Betti theorem).

In the case where a load *P* is applied to the centre and the deflection calculated in the centre, it becomes:

$$P\,W\left(\frac{a}{2},\frac{b}{2},\frac{a}{2},\frac{b}{2}\right)=\frac{4P}{abD}\sum_{m\ \text{odd}}\sum_{n\ \text{odd}}\frac{1}{\pi^4\left(\frac{m^2}{a^2}+\frac{n^2}{b^2}\right)^2}$$

In the case of a square plate, a good approximation (3,5%) of the deflection in the centre is obtained by retaining the first four terms:

$$W \approx 0{,}01121\frac{Pa^2}{D}$$

The influence function *W* makes it possible to obtain the displacement *w* by superposition, for any load *p(x, y)*.

$$w(x,y)=\int_0^a\int_0^b W(x,y,\xi,\eta)\,p(\xi,\eta)\,d\xi d\eta \tag{7.92}$$

Moments due to the influence function of displacements are obtained from the expression (7.91), for example for M^{xx}:

$$M^{xx}(x,y,\xi,\eta)=-\frac{4}{ab}\sum_{m=1}^{\infty}\sum_{n=1}^{\infty}\frac{\frac{m^2}{a^2}+\nu\frac{n^2}{b^2}}{\pi^2\left(\frac{m^2}{a^2}+\frac{n^2}{b^2}\right)^2}\phi_{mn}(\xi,\eta)\,\phi_{mn}(x,y) \tag{7.93}$$

An expression such as (7.93) defines an influence surface of the moment which can be used for any load; the moment at a point can be obtained by an integration such that (7.92). This method was commonly used before the finite element method was widely disseminated.

Example: Vibrations.

The equation of free movements is obtained from (7.38):

$$D\Delta\Delta w+\rho\,\ddot{w}=0 \tag{7.94}$$

where $\ddot{w}$ denotes the acceleration. The eigen movements are of the form:

$$w(x,y,t)=W(x,y)e^{i\omega t} \tag{7.95}$$

where the function *W* verifies:

$$D\Delta\Delta W-\rho\,\omega^2 W=0 \tag{7.96}$$

The solutions of this equation to the eigenfunctions of the operator ΔΔ are the functions ϕ_{mn}. The pulsations are deduced:

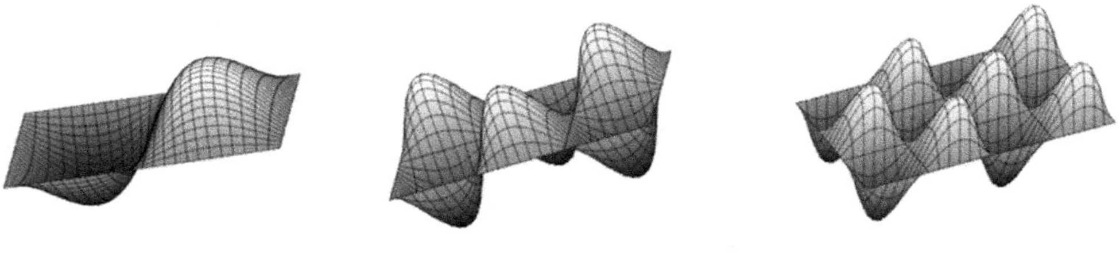

Figure 7.16

$$\omega_{mn}^{\ 2} = \frac{D}{\rho}\pi^2\left(\frac{m^2}{a^2}+\frac{n^2}{b^2}\right)^2$$

The first pulsation is obtained for (m, n) = (1, 1) and is:

$$\omega_{11} = \pi^2\left(\frac{1}{a^2}+\frac{1}{b^2}\right)\sqrt{\frac{D}{\rho}}$$

Any free movement is then written in the form:

$$w(x,y,t)=\sum_{m,n}\left(A_{mn}\cos\omega_{mn}t+B_{mn}\sin\omega_{mn}t\right)\phi_{mn}(x,y) \tag{7.97}$$

where coefficients A_{mn} and B_{mn} are determined by initial conditions.

Figure 7.16 illustrates the modes (2, 1), (3, 2), and (4, 3), in the case where the ratio of the lengths of the sides is equal to 2.

7.2.2.2 Isotropic plate supported on two opposite edges; Lévy solution

The plate is simply supported on two opposite edges $x = 0$ and $x = a$; the conditions on the other two edges depend on the nature of the support conditions: free, simply supported or clamped. In such a situation, it is natural to look for a solution developed in a series of terms respecting the simply supported edge conditions (Maurice Lévy[6] – 1899) :

$$w(x,y)=\sum_{m=1}^{\infty}Y_m(y)\sin\frac{m\pi x}{a} \tag{7.98}$$

The load density can be decomposed into a double Fourier series as before. The equation of equilibrium is then written:

$$D\sum_{m=1}^{\infty}\left(Y_m^{\ \mathrm{IV}}-2\frac{m^2\pi^2}{a^2}Y_m^{\ \prime\prime}+\frac{m^4\pi^4}{a^4}Y_m\right)\sin\frac{m\pi x}{a}=\sum_{m=1}^{\infty}\sum_{n=1}^{\infty}p_{mn}\sin\frac{n\pi y}{b}\sin\frac{m\pi x}{a}$$

The equations for the Y_m functions are obtained by noting that the functions $\sin\frac{m\pi x}{a}$ are orthogonal to each other on the domain $[0, a]$:

---

[6] Maurice Lévy, French engineer (1838–1910).

$$D\left(Y_m^{\mathrm{IV}}-2\frac{m^2\pi^2}{a^2}Y_m''+\frac{m^4\pi^4}{a^4}Y_m\right)=\sum_{n=1}^{\infty}p_{mn}\sin\frac{n\pi y}{b} \qquad (7.99)$$

This approach can be simplified if a particular solution w_1 of the equilibrium equation is known. It must verify:

$$D\Delta\Delta w_1 = p \qquad (7.100)$$

and the boundary conditions on the simply supported edges.

One way to obtain such a solution is to consider the plate cut in strips parallel to the x-axis, in flexion between the two opposed simply supported edges. This makes it possible to calculate the bending moment M^{xx} in each band as the bending moment $M(x)$ of a simply supported beam. The curvature b_{xx} is then deduced by $b_{xx}=\frac{M^{xx}}{D}$ (the cylindrical bending of a plate of infinite width, see §1.2.4.1), then w_1 by integration. The function w_1 thus obtained is acceptable if it satisfies the equation (7.100), which is the case in particular if p is independent of y.

Once determined, the function w_1 is also decomposed, for convenience, into a Lévy series:

$$w_1(x,y)=\sum_{m=1}^{\infty}W_m(y)\sin\frac{m\pi x}{a} \qquad (7.101)$$

with:

$$W_m(y)=\frac{2}{a}\int_0^a w_1(x,y)\sin\frac{m\pi x}{a}dx \qquad (7.102)$$

For example, for a load density $p(y) = \alpha y + \beta$:

$$M(y)=\frac{1}{2}p(y)x(x-a)$$

$$w_1(x)=\frac{p(y)}{24D}\left(x^4-2ax^3+a^3x\right) \qquad (7.103)$$

which verifies Equation (7.100).

If such a particular solution can be obtained, the general solution of the equation of equilibrium is in the form: $w = w_1 + w_2$, where w_2 satisfies the equation of equilibrium without a second member and the conditions on the simply supported edges. w must then respect the conditions on the other two edges.

More specifically, regarding w_2:

$$w_2(x,y)=\sum_{m=1}^{\infty}Y_m(y)\sin\frac{m\pi x}{a} \qquad (7.104)$$

with, for any value of m:

$$Y_m^{\mathrm{IV}}-2\frac{m^2\pi^2}{a^2}Y_m''+\frac{m^4\pi^4}{a^4}Y_m=0 \qquad (7.105)$$

The general solution of this last equation is written as:

$$Y_m(y) = a_m \mathrm{ch}\frac{m\pi y}{a} + b_m \frac{m\pi y}{a}\mathrm{ch}\frac{m\pi y}{a} + c_m \mathrm{sh}\frac{m\pi y}{a} + d_m \frac{m\pi y}{a}\mathrm{sh}\frac{m\pi y}{a} \tag{7.106}$$

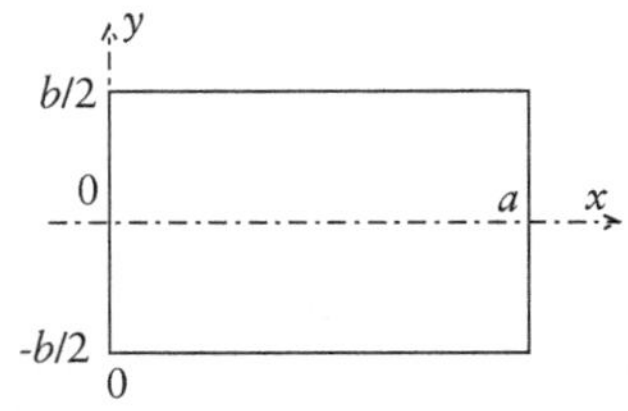

Figure 7.17

In cases where the load density is symmetrical about a median axis x, there is some interest in choosing this axis as the origin on the y-axis (Figure 7.17); then the boundary conditions not yet verified are to be expressed in $y = \pm b/2$.

If, furthermore, the boundary conditions are identical on these two edges, the deformation is symmetrical with respect to the x-axis and only the even terms are kept in the expression of Y_m:

$$Y_m(y) = a_m \mathrm{ch}\frac{m\pi y}{a} + d_m \frac{m\pi y}{a}\mathrm{sh}\frac{m\pi y}{a} \tag{7.107}$$

Example: Case of uniform load density in symmetric bending.

The function w_1 is written as:

$$w_1(x) = \frac{p}{24D}\left(x^4 - 2ax^3 + a^3 x\right)$$

hence the general solution of the equation in w:

$$w(x,y) = \frac{p}{24D}\left(x^4 - 2ax^3 + a^3 x\right) + \frac{pa^4}{D}\sum_{m=1}^{\infty}\left[A_m \mathrm{ch}\frac{m\pi y}{a} + D_m \frac{m\pi y}{a}\mathrm{sh}\frac{m\pi y}{a}\right]\sin\frac{m\pi x}{a} \tag{7.108}$$

where for convenience the term $\frac{pa^4}{D}$ has been particularised in the expression of the constants.

The relation (7.108) can be in a more useful form by developing w_1 in series. It becomes:

$$w(x,y) = \frac{pa^4}{D}\sum_{m=1}^{\infty}\left[\frac{2\left[1-(-1)^m\right]}{\pi^5 m^5} + A_m \mathrm{ch}\frac{m\pi y}{a} + D_m \frac{m\pi y}{a}\mathrm{sh}\frac{m\pi y}{a}\right]\sin\frac{m\pi x}{a} \tag{7.109}$$

The solution of the problem also satisfies the conditions on the edges $y = \pm b/2$, which makes it possible to determine the coefficients A_m and D_m.

The bending moments are obtained from the expression (7.108) of w:

$$M^{xx} = \frac{p}{2}x(x-a) - pa^2\pi^2(1-\nu)\sum_{m=1}^{\infty} m^2 \left[\begin{array}{l} A_m \mathrm{ch}\frac{m\pi y}{a} \\ +D_m\left(-\frac{2\nu}{1-\nu}\mathrm{ch}\frac{m\pi y}{a} + \frac{m\pi y}{a}\mathrm{sh}\frac{m\pi y}{a}\right)\end{array}\right]\sin\frac{m\pi x}{a}$$

$$M^{yy} = \nu \frac{p}{2} x(x-a) + pa^2\pi^2(1-\nu)\sum_{m=1}^{\infty} m^2 \begin{bmatrix} A_m \operatorname{ch}\frac{m\pi y}{a} \\ +D_m\left(\frac{2}{1-\nu}\operatorname{ch}\frac{m\pi y}{a} + \frac{m\pi y}{a}\operatorname{sh}\frac{m\pi y}{a}\right) \end{bmatrix} \sin\frac{m\pi x}{a}$$

$$M^{xy} = pa^2\pi^2(1-\nu)\sum_{m=1}^{\infty} m^2\left[(A_m + D_m)\operatorname{sh}\frac{m\pi y}{a} + D_m\frac{m\pi y}{a}\operatorname{ch}\frac{m\pi y}{a}\right]\cos\frac{m\pi x}{a}$$

(7.110)

and shear forces by (7.35):

$$Q^x = -p\left(x - \frac{a}{2}\right) - 2pa\pi^3\sum_{m=1}^{\infty} m^3 D_m \operatorname{ch}\frac{m\pi y}{a}\cos\frac{m\pi x}{a}$$

$$Q^y = -2pa\pi^3\sum_{m=1}^{\infty} m^3 D_m \operatorname{sh}\frac{m\pi y}{a}\sin\frac{m\pi x}{a}$$

(7.111)

then the normal reactions on the edges by (7.7), resulting in a relation such that (7.40):

- edge $x = 0$ (and in a similar way in $x = a$). The rectangle is travelled in the trigonometrical sense, so along this edge, the curvilinear abscissa is $-y$ and the normal vector is –i.

$$q = -Q^x + \frac{\partial M^{xy}}{\partial y}$$

$$= -\frac{pa}{2} + pa\pi^3\sum_{m=1}^{\infty} m^3 \begin{bmatrix} (1-\nu)A_m \operatorname{ch}\frac{m\pi y}{a} \\ +D_m\left(2(2-\nu)\operatorname{ch}\frac{m\pi y}{a} + (1-\nu)\frac{m\pi y}{a}\operatorname{sh}\frac{m\pi y}{a}\right) \end{bmatrix} \quad (7.112)$$

- edges $y = b/2$ (and in a similar way in $y = -b/2$). Along this edge, the curvilinear abscissa is $-x$ and the normal vector is +j.

$$q = Q^y - \frac{\partial M^{xy}}{\partial x}$$

$$= pa\pi^3\sum_{m=1}^{\infty} m^3 \begin{Bmatrix} (1-\nu)A_m \operatorname{sh}\alpha_m \\ +D_m\left[-(1+\nu)\operatorname{sh}\alpha_m + (1-\nu)\alpha_m \operatorname{ch}\alpha_m\right] \end{Bmatrix} \sin\frac{m\pi x}{a} \quad (7.113)$$

with:

$$\alpha_m = \frac{m\pi b}{2a} \quad (7.114)$$

The bending moments M^{xx} and M^{yy} reach their extreme values either in the middle of the clamped edges, or in the centre of the plate, where they are worth:

$$x=\frac{a}{2} \quad y=0 \quad \begin{cases} M^{xx}=-\dfrac{pa^2}{8}-pa^2\pi^2(1-\nu)\displaystyle\sum_{m\text{ odd}}\mathrm{m}^2(-1)^{(\mathrm{m}-1)/2}\left(A_m-D_m\frac{2\nu}{1-\nu}\right) \\ M^{xy}=0 \\ M^{yy}=-\nu\dfrac{pa^2}{8}+pa^2\pi^2(1-\nu)\displaystyle\sum_{m\text{ odd}}^{\infty}\mathrm{m}^2(-1)^{(\mathrm{m}-1)/2}\left(A_m+D_m\frac{2}{1-\nu}\right) \end{cases} \tag{7.115}$$

and M^{xy} is extremal in the corners, where it is worth:

$$M^{xy}\left(0,\pm\frac{b}{2}\right)=pa^2\pi^2(1-\nu)\sum_{m=1}^{\infty}\mathrm{m}^2\left[(A_m+D_m)\mathrm{sh}\alpha_m+D_m\alpha_m\mathrm{ch}\alpha_m\right] \tag{7.116}$$

Finally, the shear forces and the distributed reactions are extreme in the middle of the edges.

- In the case where the two edges $y = \pm b/2$ are simply supported, the conditions are written:

$$y=\pm\frac{b}{2};\quad \forall x:\begin{cases} w=0 \\ \dfrac{\partial^2 w}{\partial x^2}=0 \end{cases}$$

which leads, from the expression (7.109) of w, the functions $\sin\dfrac{\mathrm{m}\pi x}{a}$ being independent, to relations:

$$\begin{cases} A_m\mathrm{ch}\alpha_m+D_m\alpha_m\mathrm{sh}\alpha_m=-\dfrac{2\left[1-(-1)^{\mathrm{m}}\right]}{\pi^5\mathrm{m}^5} \\ A_m\mathrm{ch}\alpha_m+D_m(2\mathrm{ch}\alpha_m+\alpha_m\mathrm{sh}\alpha_m)=0 \end{cases}$$

The coefficients are zero if m is even and are worth, when m is odd:

$$\begin{cases} A_m=-\dfrac{2}{\pi^5\mathrm{m}^5}\dfrac{2+\alpha_m\mathrm{th}\alpha_m}{\mathrm{ch}\alpha_m} \\ D_m=\dfrac{2}{\pi^5\mathrm{m}^5}\dfrac{1}{\mathrm{ch}\alpha_m} \end{cases} \tag{7.117}$$

The ratio b/a, via α_m, determines the numerical values of the coefficients A_m and D_m, thus of the displacements. The Poisson's ratio is involved in the expression of moments. A numerical study of the above results shows that, for values of b/a greater than 5, the deflection and the moments in the centre are comparable, with an error of less than 1%, to those of an infinitely long plate in cylindrical bending. For a ratio of 3, the difference is of the order of 6%.

In the case of the square plate, retaining only the first term of the series, the deflection in the centre is:

$$w\left(\frac{a}{2},\frac{a}{2}\right)=\frac{5}{384}\frac{pa^4}{D}\approx 1{,}37123\frac{2pa^4}{\pi^5 D}\approx 0{,}00406\frac{pa^4}{D}$$

which is an approximation of less than 0,2%. This result, similar to that obtained in §7.2.2.1 by the Navier method, shows that the convergence of the Lévy series is very fast. The moments in the centre are obtained by (7.115):

$$\begin{matrix} x=\dfrac{a}{2} \\ y=0 \end{matrix}\quad \begin{cases} M^{xx}=-\dfrac{pa^2}{8}+\dfrac{2pa^2}{\pi^3}\displaystyle\sum_{m\text{ odd}}^{\infty}\dfrac{(-1)^{(m-1)/2}}{m^3\text{ch}\alpha_m}\left[2+(1-\nu)\alpha_m\text{th}\alpha_m\right] \\ M^{xy}=0 \\ M^{yy}=-\nu\dfrac{pa^2}{8}+\dfrac{2pa^2}{\pi^3}\displaystyle\sum_{m\text{ odd}}^{\infty}\dfrac{(-1)^{(m-1)/2}}{m^3\text{ch}\alpha_m}\left[2\nu-(1-\nu)\alpha_m\text{th}\alpha_m\right] \end{cases}$$

with $\alpha_m=\dfrac{m\pi}{2}$

or, keeping only the first term:

$$\begin{cases} M^{xx}\approx-\dfrac{pa^2}{8}+\dfrac{2pa^2}{\pi^3\text{ch}\frac{\pi}{2}}\left[2+(1-\nu)\dfrac{\pi}{2}\text{th}\dfrac{\pi}{2}\right] \\ M^{yy}\approx-\nu\dfrac{pa^2}{8}+\dfrac{2pa^2}{\pi^3\text{ch}\frac{\pi}{2}}\left[2\nu-(1-\nu)\dfrac{\pi}{2}\text{th}\dfrac{\pi}{2}\right] \end{cases}$$

which, for $\nu=0{,}2$, are worth $-\dfrac{pa^2}{22{,}75}$ and $-\dfrac{pa^2}{22{,}55}$ respectively. The equality of moments M^{xx} and M^{yy} in the centre of the plate implies that:

$$\sum_{m\text{ odd}}^{\infty}\frac{(-1)^{(m-1)/2}}{m^3\text{ch}\frac{m\pi}{2}}\left[1+\frac{m\pi}{2}\text{th}\frac{m\pi}{2}\right]=\frac{\pi^3}{32}$$

- In the case where the two edges $y = \pm b/2$ are clamped, the conditions are written:

$$y=\pm\frac{b}{2};\quad \forall x:\begin{cases} w=0 \\ \dfrac{\partial w}{\partial x}=0 \end{cases}$$

it comes, for odd m (the coefficients are zero if m is even):

$$\begin{cases} A_m \text{ch}\alpha_m + D_m \alpha_m \text{sh}\alpha_m = -\dfrac{4}{\pi^5 \text{m}^5} \\ A_m \text{sh}\alpha_m + D_m \left(\text{sh}\alpha_m + \alpha_m \text{ch}\alpha_m \right) = 0 \end{cases}$$

from which:

$$\begin{cases} A_m = -\dfrac{4}{\pi^5 \text{m}^5} \dfrac{\text{sh}\alpha_m + \alpha_m \text{ch}\alpha_m}{\alpha_m + \text{sh}\alpha_m \text{ch}\alpha_m} \\ D_m = \dfrac{4}{\pi^5 \text{m}^5} \dfrac{\text{sh}\alpha_m}{\alpha_m + \text{sh}\alpha_m \text{ch}\alpha_m} \end{cases} \tag{7.118}$$

The fixed-end moment is maximum in the middle of the clamped edges, where it is worth:

$$\begin{aligned} M^{yy}\left(\frac{a}{2}, \pm\frac{b}{2}\right) &= -\nu \frac{pa^2}{8} \\ &\quad + pa^2\pi^2(1-\nu) \sum_{m \text{ odd}} \text{m}^2 (-1)^{(\text{m}-1)/2} \begin{bmatrix} A_m \text{ch}\alpha_m \\ + D_m \left(\alpha_m \text{sh}\alpha_m + \dfrac{2}{1-\nu} \text{ch}\alpha_m \right) \end{bmatrix} \\ &= -\nu \frac{pa^2}{8} + \frac{4pa^2}{\pi^3} \sum_{m \text{ odd}} \frac{(-1)^{(\text{m}-1)/2}}{\text{m}^3} \frac{(1+\nu)\text{sh}\alpha_m \text{ch}\alpha_m - (1-\nu)\alpha_m}{\alpha_m + \text{sh}\alpha_m \text{ch}\alpha_m} \end{aligned} \tag{7.119}$$

In the case of the square plate, the deflection in the centre is worth, retaining only the first term in the series:

$$w\left(\frac{a}{2}, \frac{a}{2}\right) = \frac{5}{384} \frac{pa^4}{D} \approx 0{,}8499 \frac{4pa^4}{\pi^5 D} \approx 0{,}00191 \frac{pa^4}{D}$$

- In the case where the two edges $y = \pm b/2$ are free, the conditions are written:

$$y = \pm\frac{b}{2}; \quad \forall x: \begin{cases} M^{yy} = 0 \\ q = Q^y + \dfrac{\partial M^{xy}}{\partial x} = 0 \end{cases}$$

It becomes:

$$\begin{cases} (1-\nu) A_m \text{ch}\alpha_m + D_m \left[(1-\nu)\alpha_m \text{sh}\alpha_m + 2\text{ch}\alpha_m \right] = \dfrac{4\nu}{\pi^5 \text{m}^5} \\ (1-\nu) A_m \text{sh}\alpha_m + D_m \left[(1-\nu)\alpha_m \text{ch}\alpha_m - (1+\nu)\text{sh}\alpha_m \right] = 0 \end{cases}$$

from which:

$$\begin{cases} A_m = \dfrac{4\nu}{\pi^5 m^5} \dfrac{(1-\nu)\alpha_m \text{ch}\alpha_m - (1+\nu)\text{sh}\alpha_m}{(1-\nu)^2 \alpha_m - (3-2\nu-\nu^2)\text{sh}\alpha_m \text{ch}\alpha_m} \\ D_m = -\dfrac{4\nu}{\pi^5 m^5} \dfrac{(1-\nu)\text{sh}\alpha_m}{(1-\nu)^2 \alpha_m - (3-2\nu-\nu^2)\text{sh}\alpha_m \text{ch}\alpha_m} \end{cases} \tag{7.120}$$

Example: Hydrostatic pressure case.

The above results can easily be extended to other load cases symmetrical with respect to the x-axis. For example, for hydrostatic pressure $p = p_0 \dfrac{x}{a}$:

$$M(x) = \frac{p_0 x}{6a}\left(x^2 - a^2\right)$$

$$w_1(x) = \frac{p_0 x}{360aD}\left(3x^4 - 10a^2x^2 + 7a^4\right)$$

which breaks down into:

$$w_1(x) = \frac{2p_0 a^4}{D\pi^5} \sum_{m\ impair} \frac{1}{\text{m}^5} \sin\frac{\text{m}\pi x}{a}$$

which makes it possible to modify the second members of the equations (7.117), (7.118), or (7.120).

Before the generalisation of the finite element method, the Lévy solution was widely used for the calculation of slab bridges.

7.2.2.3 Plates of other shapes

For plates supported on their periphery with shapes other than the circular and rectangular shapes discussed in the paragraphs above, analytical solutions can be found by taking a transverse displacement w of the form:

$$w(x,y) = \vartheta(x,y) f(x,y) \tag{7.121}$$

where $\vartheta\ (x, y) = 0$ is the equation of the edge of the plate and f a function to be determined. In this case, the support conditions are automatically verified. To determine f, it is necessary to write the Germain equation and the boundary conditions other than the support condition.

Example: Elliptical plate clamped on its edge (Figure 7.18).

The edge is a standard ellipse, with orthogonal x- and y-axes. In that case:

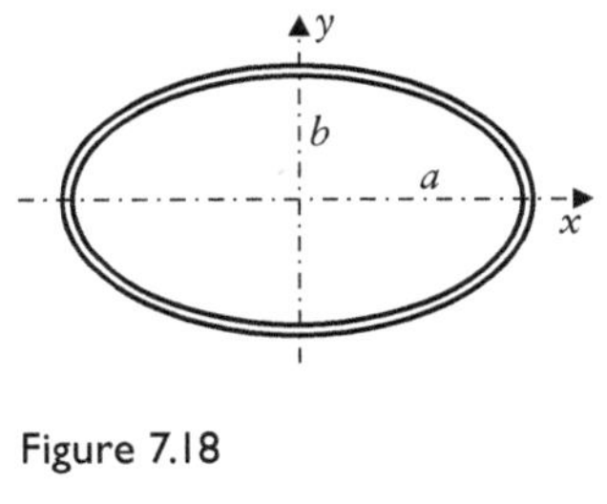

Figure 7.18

$$\vartheta(x,y) = 1 - \frac{x^2}{a^2} - \frac{y^2}{b^2}$$

By differentiation of (7.111), it becomes:

$$\Delta\Delta w = -4\left(\frac{1}{a^2} + \frac{1}{b^2}\right) \bullet \Delta f + \left(1 - \frac{x^2}{a^2} - \frac{y^2}{b^2}\right) \bullet \Delta\Delta f$$

$$-8\left(\frac{x}{a^2}\partial_x \Delta f + \frac{y}{b^2}\partial_y \Delta f + \frac{1}{a^2}\partial^2_{x^2} f + \frac{1}{b^2}\partial^2_{y^2} f\right)$$

In the case where the plate is loaded by a constant density q, only constant terms must remain in the previous expression and one way to respect this condition is to verify the (not necessary) relations:

$$\Delta\Delta w = -4\left(\frac{1}{a^2} + \frac{1}{b^2}\right) \bullet \Delta f - 8\left(\frac{1}{a^2}\partial^2_{x^2} f + \frac{1}{b^2}\partial^2_{y^2} f\right) = \frac{q}{D}$$

$$\Delta\Delta f = 0$$

$$\frac{x}{a^2}\partial_x \Delta f + \frac{y}{b^2}\partial_y \Delta f = 0$$

which can be respected when the two second derivatives are constant. Moreover, f is zero on the edge and can be taken proportional to ϑ, which second derivatives are constant. By referring to $\Delta\Delta w$ expression, it comes:

$$w(x,y) = \frac{q}{8D\left(\frac{3}{a^4} + \frac{3}{b^4} + \frac{2}{a^2 b^2}\right)}\left(1 - \frac{x^2}{a^2} - \frac{y^2}{b^2}\right)^2 \qquad (7.122)$$

The plate is clamped on the edge, and the other condition at the edge is written:

$$\text{grad } w \text{ grad } \vartheta = 0$$

with:

$$\text{grad } \vartheta = \begin{cases} -\dfrac{2x}{a^2} \\ -\dfrac{2y}{b^2} \end{cases} \qquad \text{grad } w = \frac{q\left(1 - \frac{x^2}{a^2} - \frac{y^2}{b^2}\right)}{8D\left(\frac{3}{a^4} + \frac{3}{b^4} + \frac{2}{a^2 b^2}\right)}\begin{cases} -\dfrac{2x}{a^2} \\ -\dfrac{2y}{b^2} \end{cases}$$

The clamping condition is therefore verified on the edge $\vartheta = 0$.

Example: Plate in the shape of an equilateral triangle resting on its edge (Figure 7.19). The centre of gravity is taken for origin. The equation of the edge is:

$$\vartheta(x,y) = \left(x + \frac{a}{3}\right)\left(y + \frac{x}{\sqrt{3}} - \frac{2a}{3\sqrt{3}}\right)\left(y - \frac{x}{\sqrt{3}} + \frac{2a}{3\sqrt{3}}\right)$$

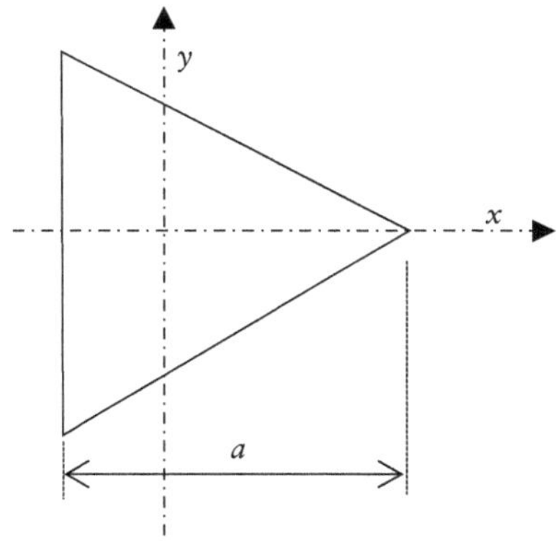

Figure 7.19

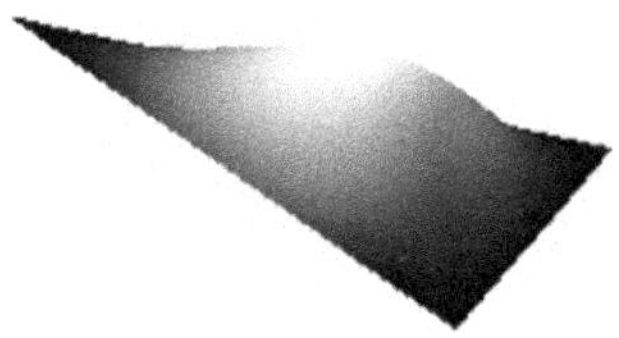

Following the same path as in the previous example and replacing the clamping condition at the edge with a zero moment condition, there comes the expression of the displacement for a uniform loading q:

$$w(x,y) = \frac{q}{64aD}\left(x^3 - 3xy^2 - ax^2 - ay^2 + \frac{4a^3}{27}\right)$$

$$\left(\frac{4a^2}{9} - x^2 - y^2\right)$$

7.2.3 Orthotropic plate

The equation of bending is Equation (7.53). The relations (7.51) and (7.52) define the constitutive law and make it possible to express the mechanical boundary conditions. Compared to the previous paragraph, only the conditions of the free edge are modified.

The resolution methods used are the same as for an isotropic plate and depend on the nature of the conditions at the edge.

Example: Plate supported on its four sides.
The Navier solution can be used. Let $\mathcal{L}$ be the operator:

$$\mathcal{L} = D_x \frac{\partial^4}{\partial x^4} + 2H \frac{\partial^4}{\partial x^2 \partial y^2} + D_y \frac{\partial^4}{\partial y^4} \tag{7.123}$$

H is defined in §7.1.3.4. The functions ϕ_{mn} defined by (7.68) are eigenfunctions of the operator $\mathcal{L}$ and verify the relation:

$$\mathcal{L}(\phi_{mn}) = \left[D_x \frac{m^4\pi^4}{a^4} + 2H \frac{m^2 n^2 \pi^4}{a^2 b^2} + D_y \frac{n^4 \pi^4}{b^4}\right]\phi_{mn} \tag{7.124}$$

which makes it possible to obtain the expression of the components w_{mn} of the displacement in the basis of the ϕ_{mn} according to the components of the loading, obtained by (7.89):

$$w_{mn} = \left[D_x \frac{m^4\pi^4}{a^4} + 2H \frac{m^2 n^2 \pi^4}{a^2 b^2} + D_y \frac{n^4 \pi^4}{b^4}\right]^{-1} p_{mn} \tag{7.125}$$

Example: Plate supported on two opposite sides, free on the other two sides.
This case is important in practice because it allows calculating the transverse bending in the isostatic bays of girder bridges (Guyon[7], 1946).

Lévy's method is used. The solution w_1 is obtained in the same way as in the isotropic case, the stiffness considered being D_x. The homogeneous solution w_2 is decomposed into a Lévy series (7.98) and the associated $Y_m(y)$ functions satisfy the differential equations:

$$Y_m^{IV} - 2\chi m^2 \vartheta^2 Y_m'' + m^4 \vartheta^4 Y_m = 0 \tag{7.126}$$

with:

$$\chi = H\sqrt{D_x D_y} \ ; \quad \vartheta = \frac{\pi}{a}\sqrt[4]{\frac{D_x}{D_y}} \tag{7.127}$$

By posing on the other hand:

$$\rho = \vartheta\sqrt{\frac{1+\chi}{2}} \ ; \quad \upsilon = \vartheta\sqrt{\frac{1-\chi}{2}} \tag{7.128}$$

the solution of the equation above can be written, in the usual cases where $\chi \leq 1$:

$$\begin{aligned} Y_m = e^{m\rho y}\left[A_m \cos(m\upsilon y) + B_m \sin(m\upsilon y)\right] \\ + e^{-m\rho y}\left[C_m \cos(m\upsilon y) + D_m \sin(m\upsilon y)\right] \end{aligned} \tag{7.129}$$

When the loading is symmetrical with respect to the x-axis, B_m and D_m are zero.

Solutions can develop with particular load cases as in §7.2.2.2, substituting (7.129) for (7.106).

7.3 PLATE BENDING TAKING INTO ACCOUNT MEMBRANE FORCES

7.3.1 Equilibrium equations

The equilibrium equations have been established in Euler variables: they apply in the present configuration, in which the load generates the internal actions. The hypothesis / H2 / (small displacements) allowed, in the previous paragraph, to express the equilibrium equations and the boundary conditions in the plane reference configuration. In particular, in this configuration, the curvature is zero and the term $b_{\alpha\beta}N^{\alpha\beta} + b_{\alpha\gamma}b_\gamma{}^\beta M^{\alpha\gamma}$ is neglected in the third equilibrium equation. But it happens that in certain situations (notably for the study of buckling), the term $b_{\alpha\beta}N^{\alpha\beta}$ cannot be neglected. The example of §7.2.1.4 has shown that this term is no longer negligible since the transverse displacement is of the order of the thickness of the plate. This term is generally preponderant when compared to $b_{\alpha\gamma}b_\gamma{}^\beta M^{\alpha\gamma}$, which means that:

$$R\,N^{\alpha\beta} >> M^{\alpha\beta} \tag{7.130}$$

[7] Yves Louis Marie Francisque Guyon, French engineer (1899–1975).

where R is the smallest radius of curvature of the deformed plate. This hypothesis is made later in this paragraph.

When the transverse displacement is no more smaller than the thickness, geometrical nonlinearities (membrane effect) must be taken into account

In these conditions, equilibrium equations are written:

$$\begin{aligned} &D_\alpha N^{\alpha\beta} + p^\beta - \rho\gamma^\beta = 0 \\ &D_\beta D_\alpha M^{\alpha\beta} - b_{\alpha\beta}N^{\alpha\beta} - p^3 + \rho\gamma^3 = 0 \end{aligned} \qquad (7.131)$$

The boundary conditions remain unchanged. (7.127) can be rewritten in intrinsic form:

$$\begin{aligned} &\mathrm{div}N + p - \rho\gamma = 0 \\ &\mathrm{div}\ \mathrm{div}M - b \bullet N - p^3 + \rho\gamma^3 = 0 \end{aligned}$$

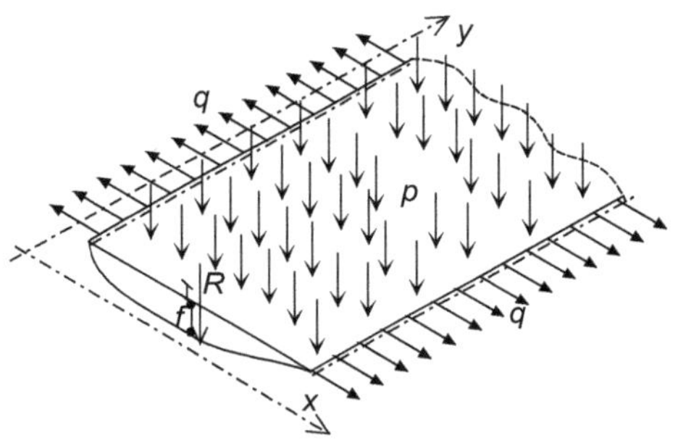

Figure 7.20

The orders of magnitude to justify the hypothesis (7.130) can be assessed using the example of §1.2.4.1. The infinite plate (Figure 7.20) is subjected to a lateral force density q along x and to a transverse loading p. f and R respectively denote the maximum deflection and the radius of the deformation in the centre. At first order:

$$N^{xx} = q\ ;\quad M^{xx} = \frac{p\ell^2}{8}$$

In the third equilibrium equation, cf. (4.97), div div M is of the order of p and the other terms are of the following orders of magnitude:

$$b \bullet N \approx \frac{q}{R};\ b \bullet b \bullet M \approx \frac{p\ell^2}{8R^2}$$

Assuming an order of magnitude of $b \bullet N \approx \frac{q}{R}$ comparable to p, this term must be taken into account in the equilibrium equation. This circumstance is due either to the existence of an external loading q or a reaction due to the blocking of the supports in the x direction, or more generally because of the variation of Gaussian curvature. By writing the moment in the centre in the deformed position, the influence of q appears naturally: $M^{xx} = \frac{p\ell^2}{8} - f\,q.$

In this hypothesis, by equating the deformed plate with a parabola, $\frac{\ell^2}{8R}$ is equal to f, the term $b \bullet b \bullet M \approx \frac{pf}{R}$ is negligible vis-à-vis div div M, but also vis-à-vis $b \bullet N$. So it can be neglected in the third equilibrium equation.

7.3.2 Isotropic elastic plates

7.3.2.1 Equations of flexion with membrane forces

The equations of flexion taking into account the membrane forces are obtained, for an isotropic linearly elastic plate, by reporting the constitutive laws (7.31) and (7.32) in the equilibrium equations (7.121), taking into account the expression of the curvature (7.29). In Cartesian orthonormal coordinates, it becomes:

$$\begin{aligned}
&\frac{\partial N^{xx}}{\partial x}+\frac{\partial N^{xy}}{\partial y}+p^{x}-\rho\frac{\partial^2 u}{\partial t^2}=0\\
&\frac{\partial N^{xy}}{\partial x}+\frac{\partial N^{yy}}{\partial y}+p^{y}-\rho\frac{\partial^2 v}{\partial t^2}=0\\
&\frac{\partial^2 M^{xx}}{\partial x^2}+2\frac{\partial^2 M^{xy}}{\partial x\partial y}+\frac{\partial^2 M^{yy}}{\partial y^2}-N^{xx}\frac{\partial^2 w}{\partial x^2}-2N^{xy}\frac{\partial^2 w}{\partial x\partial y}-N^{yy}\frac{\partial^2 w}{\partial y^2}-p^3+\rho\frac{\partial^2 w}{\partial t^2}=0
\end{aligned}\tag{7.132}$$

that it is necessary to complete by the constitutive laws, taking into account the expressions (7.27) of the in-plane deformations:

$$\begin{aligned}
&N^{xx}=K\left[\frac{\partial u}{\partial x}+\nu\frac{\partial v}{\partial y}+\frac{1}{2}\left(\frac{\partial w}{\partial x}\right)^2+\frac{\nu}{2}\left(\frac{\partial w}{\partial y}\right)^2\right]\\
&N^{yy}=K\left[\frac{\partial v}{\partial y}+\nu\frac{\partial u}{\partial x}+\frac{1}{2}\left(\frac{\partial w}{\partial y}\right)^2+\frac{\nu}{2}\left(\frac{\partial w}{\partial x}\right)^2\right]\\
&N^{xy}=K\left(\frac{1-\nu}{2}\right)\left(\frac{\partial u}{\partial y}+\frac{\partial v}{\partial x}+\frac{\partial w}{\partial x}\frac{\partial w}{\partial y}\right)
\end{aligned}\tag{7.133}$$

$$\begin{aligned}
&M^{xx}=D\left(\frac{\partial^2 w}{\partial x^2}+\nu\frac{\partial^2 w}{\partial y^2}\right)\\
&M^{yy}=D\left(\frac{\partial^2 w}{\partial y^2}+\nu\frac{\partial^2 w}{\partial x^2}\right)\\
&M^{xy}=D(1-\nu)\frac{\partial^2 w}{\partial x\partial y}
\end{aligned}\tag{7.134}$$

The system thus obtained is non-linear.

Two special cases play an important role in practice:

- A membrane prestress N_0 exists in the reference state and the variation of the membrane forces corresponds to negligible energy. Equation (7.132) applies with $N = N_0$.
- There is no membrane prestressing. Membrane forces come from the change of equilibrium and are due to the variation of the Gaussian curvature.

7.3.2.2 *Rectangular plate supported on four sides subject to an action in its plane*

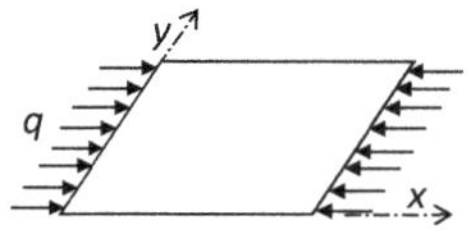

Figure 7.21

The plate is subjected to a lateral load q directed along the x-axis (q is negative in Figure 7.21). The membrane prestress tensor is:

$$N_0 = \begin{pmatrix} q & 0 \\ 0 & 0 \end{pmatrix}$$

According to the constitutive law, the variation of the tensor of the membrane forces is of the second order and is supposed negligible vis-à-vis q.

The equilibrium equation is then reduced to:

$$D\,\Delta\Delta w - q\frac{\partial^2 w}{\partial x^2} = p$$

Using the Navier method, the equation verified by the displacement components is written:

$$\forall(m,n): \quad \left[\pi^4\left(\frac{\mathrm{m}^2}{a^2}+\frac{\mathrm{n}^2}{b^2}\right)^2 D + q\frac{\mathrm{m}^2\pi^2}{a^2}\right] w_{mn} = p_{mn}$$

from where w_{mn} is obtained.

- If q is positive (traction), the stiffness of the plate is increased.
- If q is negative (compression), the stiffness decreases. It is clear that the displacement components can become infinite for some (discrete) values of q. Then there is **instability**. The most interesting value in practice is the smallest in absolute value, given by:

$$-q_{cr} = \min_{(m,n)} \frac{\pi^2\left(\frac{\mathrm{m}^2}{a^2}+\frac{\mathrm{n}^2}{b^2}\right)^2 Da^2}{\mathrm{m}^2}$$

which is obviously obtained for n = 1. There is therefore to study the minimum of the coefficient:

$$\lambda_m = \left(\mathrm{m} + \frac{1}{\mathrm{m}}\frac{a^2}{b^2}\right)^2$$

which provides, depending on the ratio $\frac{a}{b}$, the value of m for which λ_m is minimal and the value of the minimum. The associated function ϕ_{m1} is the buckling mode. This minimum value is enveloped inferiorly by the expression:

$$\lambda_{\min} = 4\frac{a^2}{b^2} \Rightarrow -q_{cr} = \pi^2\frac{D}{a^2}\min_m \lambda_m = 4\pi^2\frac{D}{b^2}$$

This approach to the study of instability is similar to Euler's for beams.

7.3.2.3 Föppl-Von Kármán equations

It is a question of rewriting the equations (7.132) to (7.134) for the problems of statics, in the case where there is no tangent load: $p^1 = p^2 = 0$. The equations of tangent equilibrium are reduced to:

$$\frac{\partial N^{xx}}{\partial x} + \frac{\partial N^{xy}}{\partial y} = 0$$
$$\frac{\partial N^{xy}}{\partial x} + \frac{\partial N^{yy}}{\partial y} = 0 \tag{7.135}$$

The elimination of the tangent displacement of components (u, v) in equations (7.27) leads to the condition of compatibility, obtained also by writing directly (3.113):

$$\frac{\partial^2 \varepsilon_{xx}}{\partial y^2} + \frac{\partial^2 \varepsilon_{yy}}{\partial x^2} - 2\frac{\partial^2 \varepsilon_{xy}}{\partial x \partial y} = \left(\frac{\partial^2 w}{\partial x \partial y}\right)^2 - \frac{\partial^2 w}{\partial x^2}\frac{\partial^2 w}{\partial y^2}$$

either, using the constitutive laws relating to the membrane forces:

$$\frac{\partial^2}{\partial y^2}\left(N^{xx} - \nu\, N^{yy}\right) + \frac{\partial^2}{\partial x^2}\left(N^{yy} - \nu\, N^{xx}\right) - 2(1-\nu)\frac{\partial^2 N^{xy}}{\partial x \partial y} = Ee\left[\left(\frac{\partial^2 w}{\partial x \partial y}\right)^2 - \frac{\partial^2 w}{\partial x^2}\frac{\partial^2 w}{\partial y^2}\right]$$

which is simplified considering (7.135):

$$\frac{\partial^2 N^{xx}}{\partial y^2} + \frac{\partial^2 N^{yy}}{\partial x^2} - 2\frac{\partial^2 N^{xy}}{\partial x \partial y} = Ee\left[\left(\frac{\partial^2 w}{\partial x \partial y}\right)^2 - \frac{\partial^2 w}{\partial x^2}\frac{\partial^2 w}{\partial y^2}\right] \tag{7.136}$$

Moreover, by noting $p^3 = p$ and introducing the bending constitutive law into the transverse equilibrium of (7.132), it comes:

$$D\,\Delta\Delta w - N^{xx}\frac{\partial^2 w}{\partial x^2} - N^{yy}\frac{\partial^2 w}{\partial y^2} - 2N^{xy}\frac{\partial^2 w}{\partial x \partial y} - p = 0 \tag{7.137}$$

Equations (7.135) to (7.137) give four relationships for four unknowns. These are the equations of Föppl[8] (1907) and Von Kármán[9] (1910).

They can be reduced by introducing the stress function F such that:

$$N^{xx} = \frac{\partial^2 F}{\partial y^2} \qquad N^{yy} = \frac{\partial^2 F}{\partial x^2} \qquad N^{xy} = -\frac{\partial^2 F}{\partial x\, \partial y} \tag{7.138}$$

Equation (7.135) is identically verified.

Equation (7.136) is then in the form:

$$\Delta\Delta F = Ee\left[\left(\frac{\partial^2 w}{\partial x \partial y}\right)^2 - \frac{\partial^2 w}{\partial x^2}\frac{\partial^2 w}{\partial y^2}\right] \tag{7.139}$$

and Equation (7.137):

[8] August Otto Föppl, German mechanician (1854–1924).

[9] Theodore von Kármán, Hungarian-American mathematician and physicist (1881–1963).

$$D\,\Delta\Delta w-\frac{\partial^2 F}{\partial y^2}\frac{\partial^2 w}{\partial x^2}-\frac{\partial^2 F}{\partial x^2}\frac{\partial^2 w}{\partial y^2}+2\frac{\partial^2 F}{\partial x\partial y}\frac{\partial^2 w}{\partial x\partial y}-p=0 \tag{7.140}$$

The two Equations (7.139) and (7.140) have two unknown scalar functions w and F. They are non-linear and must therefore be solved by numerical methods or approximate analytical methods.

These equations can be rewritten in intrinsic form, introducing the curvature tensor instead of the displacement w:

$$\begin{aligned}&\Delta\Delta F = Ee\,\hat{b}\\&D\,\Delta(\mathrm{TR}\,b)-b_{\alpha\beta}e^{\alpha\lambda}e^{\beta\mu}D_\lambda(\mathrm{grad}F)_\mu-p=0\end{aligned} \tag{7.141}$$

where the membrane forces are obtained as a function of F by a general expression analogous to (6.13).

7.3.2.4 Plate in axisymmetric bending

The objective is to highlight the influence of membrane forces on flexural stiffness.

In polar coordinates and assuming that the different quantities do not depend on θ :

$$N^{11}=\frac{F'}{r};\quad N^{22}=F''$$

and, with (7.60):

$$b_{11}=w'';\quad b_{22}=\frac{w'}{r}$$

Equation (7.141) becomes, using relations (7.60) and (7.63):

$$\begin{aligned}&\frac{1}{r}\frac{d}{dr}\left\{r\frac{d}{dr}\left[\frac{1}{r}\frac{d}{dr}\left(r\frac{dF}{dr}\right)\right]\right\}=-Eew''\frac{w'}{r}\\&D\frac{1}{r}\frac{d}{dr}\left\{r\frac{d}{dr}\left[\frac{1}{r}\frac{d}{dr}\left(r\frac{dw}{dr}\right)\right]\right\}=\frac{1}{r}\left(w''F'+F''w'\right)+p\end{aligned} \tag{7.142}$$

The first equation suggests that if w is an even (odd) function, then F is even (odd).

If p is constant, the second equation indicates that the solutions are even. By developing w and F:

$$w(r)=\sum_{k=0}^{\infty}A_{2k}r^{2k}\qquad F(r)=\sum_{k=1}^{\infty}B_{2k}r^{2k} \tag{7.143}$$

then:

$$\Delta\Delta w=\sum_{n=1}^{\infty}16(\mathrm{n}+1)^2\,\mathrm{n}^2A_{2n+2}r^{2\mathrm{n}-2}$$

And a similar formula for $\Delta\Delta F$. Equation (7.132) becomes:

$$\sum_{n=1}^{\infty} 16(\mathrm{n}+1)^2 \mathrm{n}^2 B_{2n+2}\, r^{2\mathrm{n}-2} = -Ee\left(\sum_{i=1}^{\infty} 2\mathrm{i}(2\mathrm{i}-1) A_{2i} r^{2\mathrm{i}-2}\right)\left(\sum_{j=1}^{\infty} 2\mathrm{j}\, A_{2j} r^{2\mathrm{j}-2}\right)$$

$$= -Ee\sum_{n=1}^{\infty}\left(\sum_{k=1}^{n} 4\mathrm{k}(2\mathrm{k}-1)(\mathrm{n}-\mathrm{k}+1)\, A_{2k}A_{2n-2k+2}\right) r^{2\mathrm{n}-2} \tag{7.144}$$

$$D\sum_{n=1}^{\infty} 16(\mathrm{n}+1)^2 \mathrm{n}^2 A_{2n+2}\, r^{2\mathrm{n}-2} = p + 8\sum_{n=1}^{\infty} \mathrm{n}\left(\sum_{k=1}^{n} \mathrm{k}(\mathrm{n}-\mathrm{k}+1)\, A_{2k}B_{2n-2k+2}\right) r^{2\mathrm{n}-2}$$

where the second member of the second equation is calculated from the expression:

$$F'w' = \left(\sum_{i=1}^{\infty} 2\mathrm{i}\, A_{2i} r^{2\mathrm{i}-1}\right)\left(\sum_{j=1}^{\infty} 2\mathrm{j}\, B_{2j} r^{2\mathrm{j}-1}\right)$$

$$= 4\sum_{n=1}^{\infty}\left(\sum_{k=1}^{n} \mathrm{k}(\mathrm{n}-\mathrm{k}+1)\, A_{2k}B_{2n-2k+2}\right) r^{2\mathrm{n}}$$

These expressions must be verified for all values of r, so there is equality of the monomials coefficients, either:

- for n = 1:

$$16B_4 = -Ee(A_2)^2$$
$$64DA_4 = p + 8A_2B_2 \tag{7.145}$$

- for n > 1:

$$4(\mathrm{n}+1)^2 \mathrm{n}^2 B_{2n+2} = -Ee\sum_{k=1}^{n} \mathrm{k}(2\mathrm{k}-1)(\mathrm{n}-\mathrm{k}+1)\, A_{2k}A_{2n-2k+2}$$
$$2D(\mathrm{n}+1)^2 \mathrm{n}^2 A_{2n+2} = \mathrm{n}\left(\sum_{k=1}^{n} \mathrm{k}(\mathrm{n}-\mathrm{k}+1)\, A_{2k}B_{2n-2k+2}\right) \tag{7.146}$$

These expressions make it possible to determine all the coefficients as functions of A_2 and B_2. A_0 is then determined to eliminate a general translation, for example:

$$w(R) = A_0 + \sum_{k=1}^{\infty} A_{2k} R^{2\mathrm{k}} = 0 \tag{7.147}$$

It is then a question of expressing the boundary conditions. If the plate is clamped on its edge:

$$\frac{dw}{dr}(R)=\sum_{k=1}^{\infty}2kA_{2k}R^{2k-1}=0 \tag{7.148}$$

$$u(R)=0 \tag{7.149}$$

To calculate the radial displacement $u(r)$, it is necessary to introduce the constitutive law again. By (7.26):

$$\frac{du}{dr}=\frac{1}{Ee}\left(N^{11}-\nu N^{22}\right)-\frac{1}{2}\left(\frac{dw}{dr}\right)^2$$

that is to say:

$$\frac{du}{dr}=\frac{1}{Ee}\left(\sum_{i=1}^{\infty}2i\,B_{2i}r^{2i-2}-\nu\sum_{i=1}^{\infty}2i(2i-1)B_{2i}r^{2i-2}\right)-\frac{1}{2}\left(\sum_{i=1}^{\infty}2i\,A_{2i}r^{2i-2}\right)\left(\sum_{j=1}^{\infty}2j\,A_{2j}r^{2i-2}\right)$$

or again:

$$\frac{du}{dr}=\frac{2}{Ee}\sum_{n=1}^{\infty}n\left[1-\nu(2n-1)\right]B_{2n}r^{2i-2}-2\sum_{n=1}^{\infty}\left(\sum_{k=1}^{n}k(n+1-k)A_{2k}A_{2n+2-2k}\right)r^{2n}$$

Taking into account the condition $u(0) = 0$, the integration of the previous relation leads to:

$$u(r)=\frac{2}{Ee}\sum_{n=1}^{\infty}\frac{n}{2n-1}\left[1-\nu(2n-1)\right]B_{2n}r^{2n-1}$$
$$-\sum_{n=1}^{\infty}\frac{2n}{2n+1}\left(\sum_{k=1}^{\infty}k(n+1-k)A_{2k}A_{2n+2-2k}\right)r^{2n+1} \tag{7.150}$$

It is interesting to compare this result with that obtained in §7.2.1.4 in the case of small displacements. To that aim, w is developed to the fourth order: N is of the order of $(w')^2$, so of degree 6 and F must be of degree 8 so that the developments are compatible:

$$w=A_0+A_2r^2+A_4r^4$$

$$F=B_2r^2+B_4r^4+B_6r^6+B_8r^8$$

Local equations (7.146) lead to:

$$B_4=-\frac{Ee}{16}\left(A_2\right)^2\;;\;B_6=-\frac{Ee}{18}A_2A_4$$

$$B_8 = -\frac{Ee}{48}(A_4)^2 \; ; \; A_4 = \frac{1}{64D}(p + 8A_2B_2)$$

and the boundary conditions to:

$$w(R) = A_0 + A_2R^2 + A_4R^4 = 0$$

$$w'(R) = 0 \qquad \Rightarrow \qquad A_2 + 2A_4R^2 = 0$$

$$u(R) = \frac{2}{Ee}\left[(1-\nu)B_2R + \frac{2}{3}(1-3\nu)B_4R^3 + \frac{3}{5}(1-5\nu)B_6R^5 + \frac{4}{7}(1-7\nu)B_8R^7\right]$$

$$-\left[\frac{2}{3}(A_2)^2R^3 + \frac{8}{5}A_2A_4R^5 + \frac{12}{7}(A_4)^2R^7\right] = 0$$

namely a total of seven equations for the seven unknown coefficients. A_0 intervenes only in the fifth relation, which makes it possible to calculate this constant. Within a constant, A_2 is the main term of displacement.

By posing:

$$x = A_2R \quad ; \quad \lambda = \frac{32D}{EeR^2} = \frac{8}{3(1-\nu^2)}\frac{e^2}{R^2} \quad ; \quad \Pi = \frac{pR}{Ee}$$

and by elimination between the fourth and the sixth relations:

$$A_4R^3 = -\frac{x}{2} = \frac{R^2}{64D}(pR + 8xB_2)$$

It becomes:

$$B_2 = -\frac{1}{8x}(pR + \lambda Eex)$$

hence the equation in x, replacing the different coefficients by their value as a function of x:

$$-\frac{1}{4x}(\Pi + \lambda x)(1-\nu) + x^2\left[-\frac{1}{12}(1-3\nu) + \frac{1}{30}(1-5\nu) - \frac{1}{168}(1-7\nu) - \frac{2}{3} + \frac{4}{5} - \frac{3}{7}\right] = 0$$

then:

$$\frac{59-21\nu}{42-42\nu}x^3 + \lambda x + \Pi = 0$$

x is the leading term of the deflection and Π is the load. x can be developed with respect to Π. It becomes:

$$x \approx -\frac{\Pi}{\lambda} + \frac{59-21\nu}{42-42\nu}\frac{\Pi^3}{\lambda^4} + O(\Pi^3)$$

Hence an approximation of the deflection and membrane forces:

$$f = w(0) - w(R) = -A_2R^2 - A_4R^4 = -x\frac{R}{2} \approx \frac{pR^2}{64D} - \frac{59-21\,\nu}{42-42\,\nu}\frac{p^3R^{10}}{2\times(32D)^4}$$

$$N^{rr}(0) = N^{\theta\theta}(0) = 2B_2 = -\frac{1}{4x}(pR + \lambda Eex)$$

$$\approx -\frac{pR}{4}\frac{59-21\nu}{42-42\nu}\frac{\Pi^3}{\lambda^4} = -\frac{59-21\nu}{42-42\nu}\frac{p^4R^{12}Ee}{4(32D)^4}$$

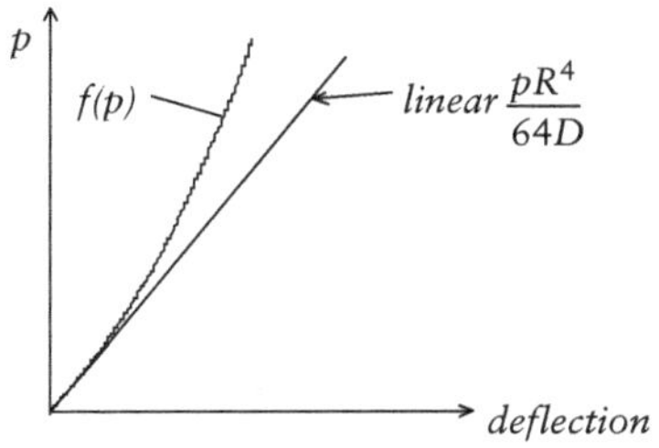

Figure 7.22

The deflection f is to be compared with that obtained in the hypothesis of small displacements, ie $\frac{pR^4}{64D}$ (Figure 7.22). The linearised theory gives a satisfactory result for small values of Π. Beyond, the stiffness of the plate increases due to the appearance of membrane forces not taken into account in the linearised theory.

7.3.2.5 Plate in cylindrical bending

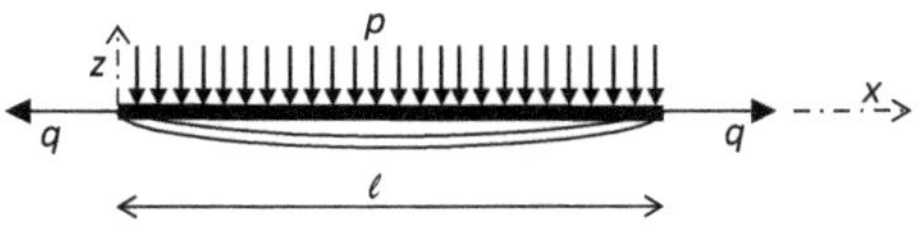

Figure 7.23

It has been previously demonstrated that the bending of plates generates membrane forces, as soon as the deflection cannot be considered much smaller than the thickness. This effect is due to the existence of a double curvature in the deformed state ($K \neq 0$) and/or the boundary conditions. To demonstrate this second cause, independently of the first, consider a cylindrical bending state producing a simple curvature, for example, the infinite plate of §1.2.4.1, uniformly loaded, the edges being supported but free in rotation (fixed articulations).

Given the infinite character of the plate, the horizontal reaction per unit length is a constant tension q. If $w(x)$ denotes the displacement in z, the bending moment M^{xx} is worth (Figure 7.23):

$$M^{xx} = D\frac{\partial^2 w}{\partial x^2} = -\frac{p}{2}(\ell - x)x + qw(x) \qquad (7.151)$$

where $w(x)$ and q are unknown.

If $u(x)$ denotes the displacement in the x direction, the axial deformation is:

$$\varepsilon_{xx} = u' + \frac{w'^2}{2}$$

This problem can be tackled by writing the Föppl–Von Kármán equations, which are reduced to:

$$\frac{\partial N^{xx}}{\partial x} = 0$$

$$D\frac{\partial^4 w}{\partial x^4} - N^{xx}\frac{\partial^2 w}{\partial x^2} = p \tag{7.152}$$

and necessarily, considering the middle section:

$$N^{xx} = q$$

To write the boundary conditions $u(0) = u(\ell) = 0$, the expression below must be integrated:

$$u' = \varepsilon_{xx} - \frac{w'^2}{2} = \frac{q}{K} - \frac{w'^2}{2}$$

which leads to the relationship:

$$\int_0^\ell \frac{w'^2}{2}\,dx = \frac{q\ell}{K} \tag{7.153}$$

Thus, the problem is entirely posed by the two equations (7.152) and (7.153), with unknowns q and $w(x)$. In this particular case, the expression (7.151) of M^{xx} results from the double integration of (7.152). By proceeding again to a double integration, it becomes:

$$w(x) = A\text{sh}\alpha x + B\text{ch}\alpha x + \frac{p}{2q}\left(x^2 + Cx + C'\right)$$

with $\alpha^2 = \dfrac{q}{D}$.

Expressing the conditions of support in $x = 0$

$$\left.\begin{aligned} w''(0) &= \alpha^2 B + \frac{p}{q} = 0 \\ w(0) &= B + \frac{pC'}{2q} = 0 \end{aligned}\right\} \Rightarrow B = -\frac{p}{q\alpha^2}\ ;\ C' = \frac{2}{\alpha^2}$$

It becomes:

$$w(x) = A\text{sh}\alpha x - \frac{p}{q\alpha^2}\text{ch}\alpha x + \frac{p}{2q}\left(x^2 + Cx + \frac{2}{\alpha^2}\right) \tag{7.154}$$

then, by expressing the boundary conditions in $x = \ell$:

$$w''(\ell) = A\alpha^2\text{sh}\alpha\ell - \frac{p}{q}\text{ch}\alpha\ell + \frac{p}{q} = 0$$

$$w(\ell) = A\text{sh}\alpha\ell - \frac{p}{q\alpha^2}\text{ch}\alpha\ell + \frac{p}{2q}\left(\ell^2 + C\ell + \frac{2}{\alpha^2}\right) = 0$$

finally:

$$w(x) = -\frac{p}{q\alpha^2}\left[\left(1-\text{ch}\alpha\ell\right)\frac{\text{sh}\alpha x}{\text{sh}\alpha\ell} + \text{ch}\alpha x - \frac{\alpha^2}{2}\left(x^2 - \ell x + \frac{2}{\alpha^2}\right)\right]$$

Then:

$$w'(x) = -\frac{p}{q\alpha}\left[\left(1-\text{ch}\alpha\ell\right)\frac{\text{ch}\alpha x}{\text{sh}\alpha\ell} + \text{sh}\alpha x - \alpha\left(x - \frac{\ell}{2}\right)\right]$$

Posing $\lambda = \alpha\,\ell$, the integral (7.153) is written:

$$\int_0^\ell \frac{w'^2}{2}dx = \frac{1}{2\alpha}\left(\frac{p}{q\alpha}\right)^2 \int_0^\lambda \left[\text{sh}\xi + \left(1-\text{ch}\lambda\right)\frac{\text{ch}\xi}{\text{sh}\lambda} - \xi + \frac{\lambda}{2}\right]^2 d\xi$$

that is to say:

$$\frac{q\ell}{K} = \frac{p^2}{2q^2\alpha^3} f(\lambda)$$

with:

$$f(\lambda) = \frac{2\lambda}{\left(e^\lambda + 1\right)^2} - \frac{2\lambda + 10}{e^\lambda + 1} + \frac{\lambda^3}{12} - 2\lambda + 5$$

In this equation, only q is unknown and can be calculated. It can be reduced to an equation in λ by expressing α and q according to λ:

$$\mu^2 = \frac{\lambda^9}{6f(\lambda)} \approx \frac{3360}{17}\lambda^2 + \frac{34720}{867}\lambda^4 + O(\lambda^5)$$

where:

$$\mu = \frac{p\ell^4}{De}$$

is a parameter that depends only on the data of the problem and is proportional to the load. The equation above makes it possible to calculate numerically λ for a given value of μ (Figure 7.24 (a)), then q using $q = \alpha^2 D = \frac{\lambda^2 D}{\ell^2}$. q vanishes when the load p vanishes.

The overall stiffness of the plate can be expressed by the deflection in the centre:

$$w\left(\frac{\ell}{2}\right) = -\frac{\mu e}{\lambda^4}\left[\frac{\left(1 - e^{\lambda/2}\right)^2}{1 + e^\lambda} - \frac{\lambda^2}{8}\right]$$

which can be compared to the deflection obtained when membrane forces are neglected:

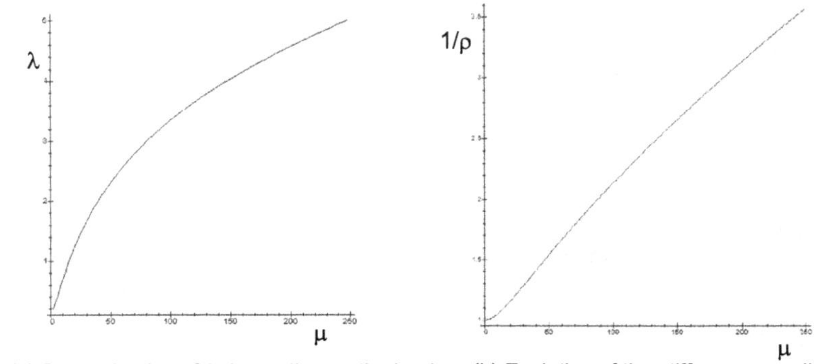

(a) Determination of λ depending on the load *(b) Evolution of the stiffness according to the loading*

Figure 7.24

$$w\left(\frac{\ell}{2}\right) = -\frac{5p\ell^4}{384D} = -\frac{5e\mu}{384}$$

The ratio ρ between the first value of the deflection and the second makes it possible to evaluate the evolution of the apparent stiffness of the plate, when the membrane force is taken into account, as the inverse of ρ (Figure 7.24 (b)):

$$\rho = \frac{384}{5\lambda^4}\left[\frac{\left(1-e^{\lambda/2}\right)^2}{\left(1+e^{\lambda}\right)} - \frac{\lambda^2}{8}\right]$$

This expression deserves to be developed in the neighbourhood of 0:

$$\rho \approx 1 + \frac{61}{600}\lambda^2 + O\left(\lambda^2\right)$$

which shows the stiffening of the plate due to the membrane force. The above results assume the small displacements, that is to say $\frac{5e\mu}{384\,\ell} = \frac{5p\ell^3}{384D} << 1$ or $\mu << \frac{384\ell}{5e}$. If the edges are clamped, it is necessary to start from the expression (7.154) and to express the conditions $w'(0) = w'(\ell) = 0$ or, alternatively, to add the fixed-end moment M_0 to (7.151). The calculations are not pursued here; they are detailed in (Timoshenko).

7.4 REISSNER–MINDLIN PLATES

7.4.1 Equilibrium equations

7.4.1.1 Scope

In cases where the thickness/span ratio is greater than 1/20 for an isotropic plate or for orthotropic plates sensitive to shear force distortion (see §5.6.1.2), consideration should be given to using the kinematics of Reissner-Mindlin rather than that of Kirchhoff. Nevertheless, in what follows, the hypothesis of thin plates is preserved, which assumes that the thickness/span ratio remains less than about ¼.

The hypothesis of small displacements is also preserved, so that, compared to the developments of §7.1, only the hypothesis /H4/ is replaced by the hypothesis /H5/ of Reissner–Mindlin, the other assumptions being the same.

7.4.1.2 Equilibrium equations and boundary conditions

The equilibrium equations given by relations (7.2) to (7.6) are applicable.

Boundary conditions must be changed from (7.7). Indeed, in Reissner–Mindlin's theory, reactions directly balance internal actions, following (4.70):

$$\begin{cases} N^{\alpha i}\nu_\alpha = q^i \\ M^{\alpha\beta}\nu_\alpha = m^\beta \end{cases} \tag{7.155}$$

7.4.1.3 Local stresses

Relations (7.8) to (7.23) remain true.

7.4.2 Deformations

7.4.2.1 Displacement

The displacement field is defined by relations (7.24) or (7.25). It is supplemented by the vector θ of the middle plane, characterising the rotation of the normal material segment (Figure 7.25). The displacement at any point in the thickness of the plate is given by the relations (5.61) and (5.62):

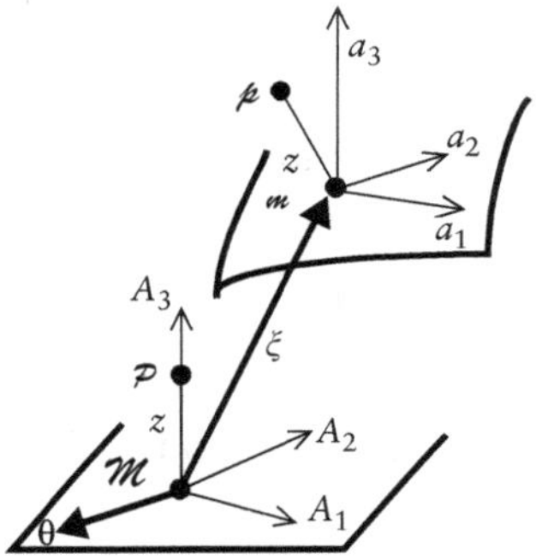

Figure 7.25

$$\mathcal{P}\mathscr{p} = \bar{\xi}(\mathcal{P}) = \xi(\mathcal{M}) + x^3(\mathbf{d}-\mathbf{k}) = \xi(\mathcal{M}) + \theta(\mathcal{M}) \wedge \mathcal{MP} \tag{7.156}$$

θ is expressed by its components:

$$\theta = \theta^\alpha\left(x^1, x^2\right) A_\alpha \tag{7.157}$$

in any coordinates of the plane. In an orthonormal Cartesian coordinate system:

$$\theta = \theta^x(x,y)\,\mathbf{i} + \theta^y(x,y)\,\mathbf{j} \tag{7.158}$$

Finally, the displacement is written according to the components:

$$\bar{\xi}(\mathcal{P}) = \left(\xi^\gamma - z\varepsilon_\delta{}^\gamma\theta^\delta\right)A_\gamma + w\,\mathbf{k} \tag{7.159}$$

either, in an orthonormal Cartesian coordinate system:

$$\bar{\xi}(\mathcal{P}) = \left(u + z\theta^y\right)\mathbf{i} + \left(v - z\theta^x\right)\mathbf{j} + w\,\mathbf{k} \tag{7.160}$$

It is convenient to introduce the vector ϑ such that $\vartheta = \mathbf{k} \times \theta$, as this allows rewriting the displacement in the simplest form:

$$\boxed{\bar{\xi}(\mathcal{P}) = \xi(\mathcal{M}) - z\vartheta(\mathcal{M}) = \left(\xi^{\alpha} - z\vartheta^{\alpha}\right)A_{\alpha} + w\mathbf{k}} \tag{7.161}$$

7.4.2.2 Deformations

The deformations can be obtained from the results of §5.3.2.4. Nevertheless, it is useful to establish them directly taking into account the assumptions /H1/, /H2/, /H3/ and /H5/.

The expressions (7.159) and (7.161) of the displacement in $\mathcal{P}$ make it possible to calculate the vectors of the local basis in p:

$$g_{\alpha} = A_{\alpha} + \left(D_{\alpha}\xi^{\gamma} - z\mathcal{E}_{\delta}{}^{\gamma} D_{\alpha}\theta^{\delta}\right)A_{\gamma} + \partial_{\alpha} w\, A_3$$

$$= A_{\alpha} + \left(D_{\alpha}\xi^{\gamma} - zD_{\alpha}\vartheta^{\gamma}\right)A_{\gamma} + \partial_{\alpha} w\, A_3$$

$$g_3 = A_3 - \mathcal{E}_{\delta}{}^{\gamma}\theta^{\delta} A_{\gamma} = A_3 - \vartheta^{\gamma} A_{\gamma}$$

Finally, following the same procedure as in §7.1.2.2, neglecting the quadratic terms of displacement and rotation with respect to the corresponding terms of the first degree:

$$\bar{\varepsilon}_{\alpha\beta} = \frac{1}{2}\left(g_{\alpha\beta} - G_{\alpha\beta}\right) = \varepsilon_{\alpha\beta} - z\kappa_{\alpha\beta}$$

where $\varepsilon_{\alpha\beta}$ is given by (7.26) and:

$$\boxed{\kappa_{\alpha\beta} = \frac{1}{2}\left(\mathcal{E}_{\delta\beta} D_{\alpha}\theta^{\delta} + \mathcal{E}_{\delta\alpha} D_{\beta}\theta^{\delta}\right) = \frac{1}{2}\left(D_{\alpha}\vartheta_{\beta} + D_{\beta}\vartheta_{\alpha}\right)} \tag{7.162}$$

in a more general form:

$$\boxed{\kappa = \frac{1}{2}\mathfrak{b} \otimes \mathbb{1}\left(\text{grad}\ \vartheta + \text{grad}^{T}\vartheta\right)} \tag{7.163}$$

then the distortion:

$$\bar{\varepsilon}_{\alpha 3} = \frac{1}{2}\left(g_{\alpha 3} - G_{\alpha 3}\right) = \frac{1}{2}\left(\partial_{\alpha} w - \theta^{\gamma}\mathcal{E}_{\gamma\alpha}\right) = \frac{1}{2}\left(\partial_{\alpha} w - \vartheta_{\alpha}\right) \tag{7.164}$$

In orthonormal Cartesian coordinates, $\varepsilon_{\alpha\beta}$ is given by (7.27) and:

$$\begin{aligned} \kappa_{xx} &= -\partial_x\theta^{y} = \partial_x\vartheta^{x} \\ \kappa_{yy} &= \partial_y\theta^{x} = \partial_y\vartheta^{y} \\ \kappa_{xy} &= \frac{1}{2}\left(\partial_x\theta^{x} - \partial_y\theta^{y}\right) = \frac{1}{2}\left(\partial_x\vartheta^{y} + \partial_y\vartheta^{x}\right) \end{aligned} \tag{7.165}$$

$$\begin{aligned} \gamma_x &= 2\bar{\varepsilon}_{x3} = \partial_x w + \theta^{y} = \partial_x w - \vartheta_x \\ \gamma_y &= 2\bar{\varepsilon}_{y3} = \partial_y w - \theta^{x} = \partial_y w - \vartheta_y \end{aligned} \tag{7.166}$$

The distortion is written in intrinsic form:

$$\boxed{\gamma = \text{grad } w + \theta \times \mathbf{k} = \text{grad } w - \vartheta} \tag{7.167}$$

7.4.3 Linear elastic plate

For a homogeneous and isotropic plate, the constitutive laws are given by (7.31) and (7.32), in which b is replaced by κ given by (7.162). In orthonormal Cartesian coordinates:

$$\begin{aligned}
M_{xx} &= D\left(\kappa_{xx} + \nu\,\kappa_{yy}\right) = D\left(-\partial_x\theta^y + \nu\,\partial_y\theta^x\right) = D\left(\partial_x\vartheta^x + \nu\,\partial_y\vartheta^y\right) \\
M_{yy} &= D\left(\kappa_{yy} + \nu\,\kappa_{xx}\right) = D\left(\partial_y\theta^x - \nu\,\partial_x\theta^y\right) = D\left(\partial_y\vartheta^y + \nu\,\partial_x\vartheta^x\right) \\
M_{xy} &= D\left(1-\nu\right)\kappa_{xy} = \frac{D}{2}\left(1-\nu\right)\left(\partial_x\theta^x - \partial_y\theta^y\right) = \frac{D}{2}\left(1-\nu\right)\left(\partial_x\vartheta^y + \partial_y\vartheta^x\right)
\end{aligned} \tag{7.168}$$

For shear force, the constitutive law is given by (5132), either:

$$\boxed{Q_\alpha = kGe\gamma_\alpha = kGe\left(\partial_\alpha w - \theta^\gamma \varepsilon_{\gamma\alpha}\right)} \tag{7.169}$$

or:

$$\boxed{Q = kGe\left(\text{grad } w - \vartheta\right)}$$

In orthonormal Cartesian coordinates:

$$\begin{aligned}
Q_x &= kGe\left(\partial_x w + \theta^y\right) = kGe\left(\partial_x w - \vartheta_x\right) \\
Q_y &= kGe\left(\partial_y w - \theta^x\right) = kGe\left(\partial_y w - \vartheta_y\right)
\end{aligned} \tag{7.170}$$

7.4.4 Equation of isotropic elastic plates

7.4.4.1 Equations of flexion

The equations of flexion of a Reissner–Mindlin isotropic elastic plate are obtained by reporting the constitutive laws (7.168) and (7.170) in the third equation of equilibrium (7.2), from where, by adding the first equation (7.6):

$$\begin{aligned}
&D\Delta\left(\text{div}\vartheta\right) - p^3 + \rho\frac{\partial^2 w}{\partial t^2} = 0 \\
&Q = kGe\left(\text{grad } w - \vartheta\right) \\
&Q + \text{div}M = 0
\end{aligned} \tag{7.171}$$

The first equation is easily demonstrated in orthonormal Cartesian coordinates. After expressing the third line using the second, there remain three equations for three unknowns. In orthonormal Cartesian coordinates, the three unknowns $w, \vartheta^x, \vartheta^y$ verify the equations:

$$
\begin{aligned}
&D\left(\frac{\partial^2}{\partial x^2}+\frac{\partial^2}{\partial y^2}\right)\left(\partial_x\vartheta^x+\partial_y\vartheta^y\right)-p^3+\rho\frac{\partial^2 w}{\partial t^2}=0\\
&D\left(\partial^2_{x^2}\vartheta^x+\frac{1}{2}(1-\nu)\partial^2_{y^2}\vartheta^x+\frac{1}{2}(1+\nu)\partial^2_{xy}\vartheta^y\right)+kGe\left(\frac{\partial w}{\partial x}-\vartheta^x\right)=0\\
&D\left(\frac{1}{2}(1+\nu)\partial^2_{xy}\vartheta^x+\frac{1}{2}(1-\nu)\partial^2_{x^2}\vartheta^y+\partial^2_{y^2}\vartheta^y\right)+kGe\left(\frac{\partial w}{\partial y}-\vartheta^y\right)=0
\end{aligned}
\tag{7.172}
$$

To these local equations, it is necessary to add boundary conditions on the edges, which are of two types:

- the kinematic conditions:
  - supported edge : $w = 0$,
  - clamped edge : $\vartheta^\nu = 0$ (ν being normal to the edge).

Indeed, the rotation operator $\mathcal{R}_{\mathcal{M}}$ such that $\mathcal{R}_{\mathcal{M}}(\mathbf{k})=\mathbf{d}$ is determined by (5.64). It is reduced to identity if $\theta\times\mathbf{k}=-\vartheta=0$. To express a clamping condition for the rotation in direction ν, it is necessary to write

$$(\theta\times\mathbf{k})\cdot\nu=-\vartheta^\nu=0.$$

- the mechanical conditions: these are the conditions (7.155) where the forces are expressed as functions of the displacement. For example, in flexion, for an edge perpendicular to i:

$$
\begin{aligned}
&\frac{D}{2}(1-\nu)\left(\partial_x\vartheta^y+\partial_y\vartheta^x\right)=m^x\\
&D\left(\partial_x\vartheta^x+\nu\,\partial_y\vartheta^y\right)=m^y
\end{aligned}
\tag{7.173}
$$

$$
-D\left(\partial^2_{x^2}\vartheta^x+\frac{1}{2}(1-\nu)\partial^2_{y^2}\vartheta^x+\frac{1}{2}(1+\nu)\partial^2_{xy}\vartheta^y\right)=q
\tag{7.174}
$$

In the case of an articulated rectilinear edge, these conditions are reduced to (the equation of the edge being x = cst) :

- $w = 0$ (zero displacement),
- $\partial_x\vartheta^x+\nu\,\partial_y\vartheta^y=0$ (zero moment) (7.175)

In addition, the support of the edge is continuous: $\dfrac{\partial w}{\partial y}=0$.

The resolution of equations (7.172), associated with the boundary conditions, makes it possible to obtain the displacements and the rotations, then the bending moments and the shear forces by (7.168) and (7.170). The reactions on the edge are then obtained by (7.155).

Example: Plate in cylindrical bending.
This is the example discussed in §1.2.4.1. Displacement and rotations do not depend on y. By the third Equation (7.172), ϑ^y is zero. The first equation then reduces to:

$$D\frac{\partial^3\vartheta^x}{\partial x^3} - p = 0$$

which general solution is:

$$\vartheta^x = \frac{p}{D}\frac{x^3}{6} + A\frac{x^2}{2} + Bx + C$$

By reporting to the second equation (7.172):

$$\frac{\partial w}{\partial x} = \vartheta^x - \frac{D}{kGe}\partial^2_{x^2}\vartheta^x = \frac{p}{D}\frac{x^3}{6} + A\frac{x^2}{2} + Bx + C - \frac{D}{kGe}\left(\frac{p}{D}x + A\right)$$

From where:

$$w = \frac{p}{D}\frac{x^4}{24} + A\frac{x^3}{6} + B\frac{x^2}{2} + Cx - \frac{D}{kGe}\left(\frac{p}{D}\frac{x^2}{2} + Ax\right) + C'$$

w and $\frac{\partial\vartheta^x}{\partial x}$ are zero in 0 and ℓ, which gives four relationships to determine the four integration constants. It finally comes, noting $\lambda = \frac{D}{kGe\ell^2}$:

$$w = \frac{p\ell^4}{24D}\left(\frac{x^4}{\ell^4} - 2\frac{x^3}{\ell^3} + \frac{x}{\ell}\right) + \frac{\lambda p\ell^4}{2D}\frac{x}{\ell}\left(1 - \frac{x}{\ell}\right)$$

$$\vartheta^x = \frac{p\ell^3}{24D}\left(4\frac{x^3}{\ell^3} - 6\frac{x^2}{\ell^2} + 1\right)$$

This solution is reduced to the KL solution when the plate is infinitely stiff in shear ($\lambda = 0$). Taking into account the shear deformation increases the displacement.

Bending moments identical to those obtained in §1.2.4.1 by equilibrium considerations are given by the constitutive laws (7.168). The shear force becomes:

$$Q_x = kGe(\partial_x w - \vartheta_x) = -\frac{p\ell}{2}\left(2\frac{x}{\ell} - 1\right)$$

$$Q_y = kGe(\partial_y w - \vartheta_y) = 0$$

in accordance with equilibrium conditions.

7.4.4.2 Looking for solutions

More generally, it is a question of finding the solutions to the system of equations (7.172) in the case of a change of equilibrium. The previous example suggests looking for a solution in the form:

$$w = W + \varpi$$
$$\vartheta = \operatorname{grad} W \tag{7.176}$$

where W is the solution to the Kirchhoff problem and ϖ a function to be determined. Then:

$$Q = kGe(\operatorname{grad} w - \vartheta) = kGe \operatorname{grad}\varpi \tag{7.177}$$

Transferring (7.177) in the equilibrium Equation (7.6), then eliminating p^3 using the first equation of (7.171), it comes, with $\upsilon = \dfrac{Gke}{D}$:

$$\operatorname{grad}(\Delta W + \upsilon\varpi) = 0 \tag{7.178}$$

which makes it possible to obtain ϖ by integration. Such a decomposition is the solution of the problem if it makes it possible to respect the boundary conditions. With $k = 5/6$, ϖ is worth $\dfrac{5(1-\nu)}{e^2}$.

Example: Rectangular plate supported on all four sides.
Navier's solution for a Kirchhoff–Love plate was determined in §7.2.2.1. (7.178) then makes it possible to determine ϖ:

$$\varpi(x,y) = \Upsilon_0 + \frac{\pi^2}{\upsilon}\sum_{m=1}^{\infty}\sum_{n=1}^{\infty} w_{mn}\left(\frac{m^2}{a^2} + \frac{n^2}{b^2}\right)\sin\frac{m\pi x}{a}\sin\frac{n\pi y}{b}$$

Since the displacement must be zero on the edge $x = 0$, the constant Υ_0 is zero. The bending moment must also be zero along the edges. For example, along an edge $x = \text{cst}$, the condition:

$$\partial_x\vartheta^x + \nu\,\partial_y\vartheta^y = \frac{\partial^2 W}{\partial x^2} + \nu\frac{\partial^2 W}{\partial y^2} = 0$$

is identically verified. Finally, the solution is written:

$$w(x,y) = \frac{1}{D\pi^2}\sum_{m=1}^{\infty}\sum_{n=1}^{\infty}\frac{p_{mn}}{\left(\frac{m^2}{a^2} + \frac{n^2}{b^2}\right)}\left(\frac{1}{\pi^2\left(\frac{m^2}{a^2} + \frac{n^2}{b^2}\right)} + \frac{1}{\upsilon}\right)\sin\frac{m\pi x}{a}\sin\frac{n\pi y}{b}$$

$$\vartheta^x(x,y) = \frac{1}{D\pi^3 a}\sum_{m=1}^{\infty}\sum_{n=1}^{\infty}\frac{m p_{mn}}{\left(\frac{m^2}{a^2} + \frac{n^2}{b^2}\right)^2}\cos\frac{m\pi x}{a}\sin\frac{n\pi y}{b} \tag{7.179}$$

$$\vartheta^y(x,y) = \frac{1}{D\pi^3 b}\sum_{m=1}^{\infty}\sum_{n=1}^{\infty}\frac{n p_{mn}}{\left(\frac{m^2}{a^2} + \frac{n^2}{b^2}\right)^2}\sin\frac{m\pi x}{a}\cos\frac{n\pi y}{b}$$

p_{mn} is determined by (7.89) depending on the load.

ϑ being equal to the gradient of the Kirchhoff displacement W, by (7.33) and (7.168) the moments (thus the shear forces) obtained in Reissner–Mindlin theory are identical to those obtained in Kirchhoff theory. Both theories give different results for displacements and support reactions.

Indeed, the reaction along the edges is given by (7.174). For example, along an edge x = cst:

$$q = \frac{1}{\pi a} \sum_{m=1}^{\infty} \sum_{n=1}^{\infty} \frac{m p_{mn}}{\left(\frac{m^2}{a^2} + \frac{n^2}{b^2} \right)} \sin \frac{n \pi y}{b}$$

The resultant of the reaction q along such an edge is worth:

$$\int_0^b q dy = \frac{1}{\pi a} \sum_{m=1}^{\infty} \sum_{n=1}^{\infty} \frac{m p_{mn}}{\left(\frac{m^2}{a^2} + \frac{n^2}{b^2} \right)} \int_0^b \sin \frac{n \pi y}{b} dy = \frac{b}{\pi^2 a} \sum_{m=1}^{\infty} \sum_{n=1}^{\infty} \frac{m p_{mn} \left[1 - (-1)^n \right]}{n \left(\frac{m^2}{a^2} + \frac{n^2}{b^2} \right)}$$

In the case of a uniform load (§7.2.2.1):

$$p_{mn} = \begin{cases} \dfrac{16p}{\pi^2 mn} & \text{if m and n are odd} \\ 0 & \text{if m or n are even} \end{cases}$$

Noting that the Riemann series $\sum_{n=1}^{\infty} \frac{1}{n^2}$ is worth $\frac{\pi^2}{6}$, it is easy to show that the series restricted to odd integers is worth $\frac{\pi^2}{8}$. The resultant of the reaction q on the edge is deduced:

$$\oint q ds = \frac{64pab}{\pi^4} \left(\sum_{m \text{ odd}} \frac{1}{m^2} \right) \left(\sum_{n \text{ odd}} \frac{1}{n^2} \right) = pab$$

and balances the load. There is therefore no concentrated reaction in the corners, unlike the result obtained in the Kirchhoff–Love model.

In the particular case of a square plate with a uniform load, the displacement in the centre is:

$$w\left(\frac{a}{2}, \frac{a}{2} \right) = W\left(\frac{a}{2}, \frac{a}{2} \right) + \frac{16pa^2}{D\upsilon\pi^4} \underbrace{\sum \sum}_{m \text{ et } n \text{ odd}} \frac{1}{mn(m^2 + n^2)}$$

to compare to the value obtained in §7.2.2.1. Keeping only the first term in the series, the displacement is increased by $\frac{8pa^2}{D\upsilon\pi^4}$, or by a proportion equal to $\frac{2}{\upsilon a^2}$ due to the contribution of the shear force strain proportional to $\frac{e^2}{a^2}$.

The moments and shear forces are the same as in Kirchhoff's theory whenever decomposition (7.176) is possible while respecting the boundary conditions. When a condition relates to the reactions (for example on a free edge), Kirchhoff's solution W does not respect the boundary condition.

Example: Rectangular plate on opposite simple supports and with two other free edges.

The general form of the solution in Kirchhoff's theory where only the simple support conditions are met is given by (7.108):

$$W(x,y)=\frac{p}{24D}\left(x^4-2ax^3+a^3x\right)+\frac{pa^4}{D}\sum_{m=1}^{\infty}\left[A_m\mathrm{ch}\frac{m\pi y}{a}+D_m\frac{m\pi y}{a}\mathrm{sh}\frac{m\pi y}{a}\right]\sin\frac{m\pi x}{a}$$

the coefficients of the Levy decomposition being given by (7.120). If the decomposition (7.176) is adopted, ϖ is determined by:

$$\varpi=\Upsilon_0-\frac{1}{\upsilon}\Delta W=\Upsilon_0-\frac{p}{\upsilon D}\left[\frac{1}{2}\left(x^2-ax\right)+2\pi^2a^2\sum_{m=1}^{\infty}D_m m^2\mathrm{ch}\frac{m\pi y}{a}\sin\frac{m\pi x}{a}\right]$$

To respect the conditions of simple support, Υ_0 is null. So:

$$w(x,y)=\frac{p}{24D}\left(x^4-2ax^3+a^3x\right)+\frac{pa^4}{D}\sum_{m=1}^{\infty}\left[A_m\mathrm{ch}\frac{m\pi y}{a}+D_m\frac{m\pi y}{a}\mathrm{sh}\frac{m\pi y}{a}\right]\sin\frac{m\pi x}{a}$$

$$-\frac{p}{\upsilon D}\left[\frac{1}{2}\left(x^2-ax\right)+2\pi^2a^2\sum_{m=1}^{\infty}D_m m^2\mathrm{ch}\frac{m\pi y}{a}\sin\frac{m\pi x}{a}\right]$$

The values of the coefficients of the decomposition can not be determined by (7.120) to respect the free edge conditions and the decomposition (7.176) therefore does not make it possible to determine the solution. Nevertheless, the general solution above allows obtaining the solution to the problem by writing the conditions of the free edge, which leads to different values of the coefficients of the decomposition. So rotations are determined by:

$$\vartheta^x=\frac{p}{2D}\left(x^2-ax\right)+\frac{\pi pa^3}{D}\sum_{m=1}^{\infty}m\left(A_m\mathrm{ch}\frac{m\pi y}{a}+D_m\frac{m\pi y}{a}\mathrm{sh}\frac{m\pi y}{a}\right)\cos\frac{m\pi x}{a}$$

$$\vartheta^y=\frac{\pi pa^3}{D}\sum_{m=1}^{\infty}m\left[A_m\mathrm{sh}\frac{m\pi y}{a}+D_m\left(\mathrm{sh}\frac{m\pi y}{a}+\frac{m\pi y}{a}\mathrm{ch}\frac{m\pi y}{a}\right)\right]\sin\frac{m\pi x}{a}$$

The bending moment M^{yy} is obtained by (7.168):

$$M^{yy}=-\nu\frac{p}{2}(2x-a)+\pi^2pa^2\sum_{m=1}^{\infty}m^2\left(\begin{array}{l}A_m(1+\nu)\mathrm{ch}\dfrac{m\pi y}{a}\\+D_m\left(2\mathrm{ch}\dfrac{m\pi y}{a}+(1+\nu)\dfrac{m\pi y}{a}\mathrm{sh}\dfrac{m\pi y}{a}\right)\end{array}\right)\sin\frac{m\pi x}{a}$$

and the shear force Q^y by (7.177):

$$Q^y = kGe\frac{\partial \varpi}{\partial y} = -2\pi^3 pa\sum_{m=1}^{\infty} D_m \mathrm{m}^3 \mathrm{sh}\frac{m\pi y}{a}\sin\frac{m\pi x}{a}$$

The free edge condition is written as:

$$y = \pm\frac{b}{2};\quad \forall x: \begin{cases} M^{yy} = 0 \\ q = Q^y = 0 \end{cases}$$

which implies that the coefficients D_m are zero. It stays:

$$-\nu\frac{p}{2}(2x-a) + \pi^2 pa^2(1+\nu)\sum_{m=1}^{\infty} A_m \mathrm{m}^2 \mathrm{ch}\alpha_m \sin\frac{m\pi x}{a} = 0\quad \forall x$$

which allows calculating coefficients A_m.

As a result, the moments and shear forces here are different from those obtained in Kirchhoff's solution.

MAIN RESULTS

EQUATIONS OF MOVEMENT

Local equations
$$\begin{cases} \operatorname{div} N + p - \rho\gamma = 0 \\ \operatorname{div}\operatorname{div} M - b \cdot N - p^3 + \rho\gamma^3 = 0 \end{cases}$$

Boundary conditions
$$\begin{cases} N^{\alpha\beta} v_\alpha = q^\beta \\ Q^\alpha v_\alpha - \dfrac{\partial}{\partial s}\left(m^v + M^{vt}\right) = q^3 \\ M^{vv} = m^t \end{cases}$$

DEFORMATIONS

$$\varepsilon_{\alpha\beta} = \frac{1}{2}\left(D_\alpha \xi_\beta + D_\beta \xi_\alpha + \partial_\alpha w\, \partial_\beta w\right)$$

$$\kappa_{\alpha\beta} = b_{\alpha\beta} \approx \partial^2_{\alpha\beta} w - \Gamma^\gamma_{\alpha\beta} \partial_\gamma w$$

LINEAR ELASTIC BEHAVIOUR

$$N = K\left[(1-\nu)\,\varepsilon + \nu\, \mathrm{TR}\varepsilon\, G\right]$$

$$M = D\left[(1-\nu)\, b + \nu\, \mathrm{TR}b\, G\right]$$

$$Q = -D \operatorname{grad} \Delta w$$

GERMAIN EQUATION

$$D\, \Delta\Delta w = p - \rho \frac{\partial^2 w}{\partial t^2}$$

ELASTIC THIN PLATE OF KIRCHHOFF

Elastic flexural potential

$$\mathcal{F}_f = \frac{1}{2} D \iint_{\mathcal{P}} \left[\left(\frac{\partial^2 w}{\partial x^2}\right)^2 + \left(\frac{\partial^2 w}{\partial y^2}\right)^2 + 2\nu \frac{\partial^2 w}{\partial x^2}\frac{\partial^2 w}{\partial y^2} + 2(1-\nu)\left(\frac{\partial^2 w}{\partial x \partial y}\right)^2 \right] dx\, dy$$

Local stresses

$$\sigma^{\alpha\beta}(\mathcal{P}) = -\frac{12}{e^3} z\, M^{\alpha\beta} \quad \sigma^{vz} = \frac{6Q^v}{e}\left[\frac{1}{4} - \left(\frac{z}{2}\right)^2\right]$$

EXERCISES

Exercise 7.1 Grid of beams

A bridge deck is made up of a grid of orthogonal girders in the x and y directions. In each direction, the set of beams consists of homogeneous elastic T-beams. The spacing of the x-direction beams is d_x, their flexural stiffness is EI_x and their torsional stiffness GK_x (similarly for the beams in the y-direction). The beams are sufficiently close and numerous for it to be possible to assume a regular behaviour. Establish the bending equation on transverse displacement w (similar to Huber's equation) under the effect of a transverse loading p.

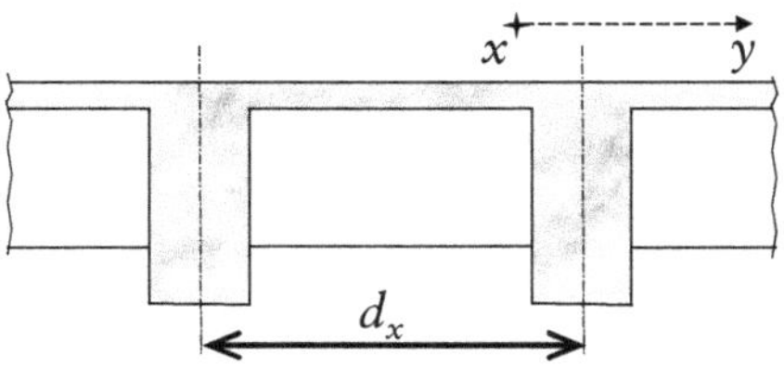

Figure 7.26

Answer: Since the beams are close together, the strain energy can be assumed to be continuous and regular by reducing it to the unit width of the beams. The rotation of a beam is equal to the derivative of the displacement w in the transverse direction. Finally, by adding the energy of the two orthogonal systems of beams:

$$\mathcal{T} = \frac{1}{2}\iint_{\mathcal{P}} \left[\frac{EI_x}{d_x}\left(\frac{d^2w}{dx^2}\right)^2 + \left(\frac{GK_x}{d_x} + \frac{GK_y}{d_y}\right)\left(\frac{d^2w}{dxdy}\right)^2 + \frac{EI_y}{d_y}\left(\frac{d^2w}{dy^2}\right)^2 \right] dxdy$$

By application of the PVP, there comes the equation relating to the displacement under the effect of p, analogous to the Huber Equation (7.53) therefore admitting the same methods of resolution:

$$\frac{EI_x}{d_x}\frac{d^4w}{dx^4} + \left(\frac{GK_x}{d_x} + \frac{GK_y}{d_y}\right)\frac{d^4w}{dx^2dy^2} + \frac{EI_y}{d_y}\frac{d^4w}{dy^4} = p$$

Exercise 7.2 Orthotropic plate

A rectangular plate simply supported on its edges (Figure 7.14) has an orthotropic constitutive law given by relation (7.51). Calculate:

1. its first pulsation;
2. the minimum critical load when the plate is subjected to a lateral load q directed along x (Figure 7.21).

Note: these results are particularly useful for a ribbed plate, cf. Exercise 7.1.

Answer:

1. The first mode is still $\phi_{1,1}$, with the associated pulsation:

$$\omega_{1,1} = \frac{\pi^4}{\rho e}\left[\frac{D_x}{a^4} + 2\frac{D' + D_{xy}}{a^2b^2} + \frac{D_y}{b^4}\right]$$

2. The minimum critical value also depends on the ratio of the different stiffnesses. n is always equal to 1, but the value of m which minimises the critical load is $\mathrm{m} = \frac{a}{b}\sqrt[4]{\frac{D_y}{D_x}}$, from which:

$$\left(-q_{cr}\right)_{\min} = \frac{2\pi^2}{b^2}\left(\sqrt{D_x D_y} + D' + D_{xy}\right)$$

which reduces to the value found in §7.3.2.2 when the plate is isotropic.

Exercise 7.3 Buckling of a plate subjected to shear in its plane

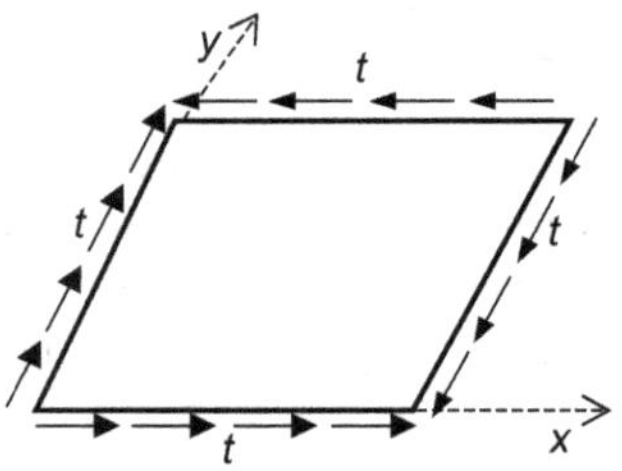

Figure 7.27

A plate panel is subjected to pure shear t. To assess the critical shear causing buckling, the panel is modelled as a square plate with a side of length a, simply supported on its edges (Figure 7.27). Establish a method to find the critical shear and the associated buckling mode in the linear approach.

Answer: The buckling equation is written as:

$$\mathrm{D}\Delta\Delta w - 2t\frac{\partial^2 w}{\partial x \partial y} = 0 \qquad \text{(i)}$$

Taking into account the boundary conditions, the space of kinematically admissible displacements is generated by the functions $\phi_{mn}(x,y)$ introduced in §7.2.2.1, then:

$$w(x,y) = \sum_{m=1}^{\infty}\sum_{n=1}^{\infty} w_{mn} \sin\frac{\mathrm{m}\pi x}{a}\sin\frac{\mathrm{n}\pi y}{a}$$

The functions $\psi_{mn}(x,y) = \cos\frac{\mathrm{m}\pi x}{a}\cos\frac{\mathrm{n}\pi y}{a}$ should also be introduced, they allow to rewrite the equation (i) of buckling:

$$\sum_{m,n}\left[\frac{D\pi^4}{a^4}\left(\mathrm{m}^2+\mathrm{n}^2\right)^2\phi_{m,n} - 2t\frac{\pi^2}{a^2}\mathrm{mn}\,\psi_{m,n}\right]w_{m,n} = 0 \qquad \text{(ii)}$$

The coefficients $w_{m,n}$ verifying (ii) also verify the relations obtained by carrying out dot products of (ii) with ϕ_{mn} and ψ_{mn}. The functions ϕ_{mn} and ψ_{mn} are orthogonal over the period, equal to $2a$, but on a the dot products are not all zero:

$$\left\langle \phi_{mn}, \psi_{pq} \right\rangle = \iint_{\mathcal{D}} \sin\frac{\mathrm{m}\pi x}{a}\sin\frac{\mathrm{n}\pi y}{a}\cos\frac{\mathrm{p}\pi x}{a}\cos\frac{\mathrm{q}\pi y}{a}\,dxdy$$

$$= \frac{4a^2}{\pi^2}\frac{\mathrm{mn}}{\left(\mathrm{m}^2-\mathrm{p}^2\right)\left(\mathrm{n}^2-\mathrm{q}^2\right)}\gamma_{mn}^{pq}$$

where γ_{mn}^{pq} is equal to 1 if m + p and n + q are odd, 0 otherwise. The other products are given by (88).

By making the dot product of equation (ii) by ψ_{pq}, it comes:

$$\sum_{m,n}\left[\left(m^2+n^2\right)^2\frac{mn}{\left(m^2-p^2\right)\left(n^2-q^2\right)}\gamma_{mn}^{pq}\right]w_{m,n}-\lambda pq\, w_{p,q}=0 \qquad \text{(iii)}$$

where $\lambda=\dfrac{a^2}{8D}t=\dfrac{3a^2\left(1-\nu^2\right)}{2e^3}t$ is the reduced shear.

An approximate solution of (iii) can be sought in a kinematically admissible subspace of finite dimension. The system (iii) obtained is then a problem with eigenvalues λ et eigenvectors X containing the components $w_{m,n}$ of displacement. The smallest eigenvalue gives the critical shear. (iii) can be solved numerically by limiting it to a subspace of finite dimension n_b and is then rewritten:

$$\left[C\right]X-\lambda X=0$$

where [C] is the matrix of dimensions $n_b \times n_b$ whose components are:

$$C_{mn}^{pq}=\frac{\left(m^2+n^2\right)^2}{\left(m^2-p^2\right)\left(n^2-q^2\right)}\frac{mn}{pq}\gamma_{mn}^{pq}$$

Taking $n_b = 25$ with the first values of m and n up to 5, the value of λ stabilises at around 35, therefore $t_{cr}\approx 35\dfrac{8D}{a^2}$. The value of λ increases with n_b, so the value obtained is on the safe side. The mode of instability obtained is plotted in Figure 7.28, where only the modes with the highest participation were retained: (m, n) = (1,1), (3,1), (1,3), (4,4), (5,1), (1,5), (5,3), and (3,5). The figure shows the very strong influence of the first three terms.

Exercise 7.4 Vibrations of a Reissner-Mindlin plate

Determine the modes and pulsations of a rectangular Reissner-Mindlin plate simply supported on its four sides (Figure 7.14). Study the evolution of the first pulsation as a function of the relative thickness of a square plate of side a.

Answer: Taking into account the results of §7.4.4.2, it is natural to seek eigenmodes in the form of displacement and components of rotation:

$$W_{mn}\left(x,y\right)=\sin\frac{m\pi x}{a}\sin\frac{n\pi y}{b}$$

$$\Theta_{mn,x}\left(x,y\right)=\alpha_{mn}\cos\frac{m\pi x}{a}\sin\frac{n\pi y}{b}$$

$$\Theta_{mn,y}\left(x,y\right)=\beta_{mn}\sin\frac{m\pi x}{a}\cos\frac{n\pi y}{b}$$

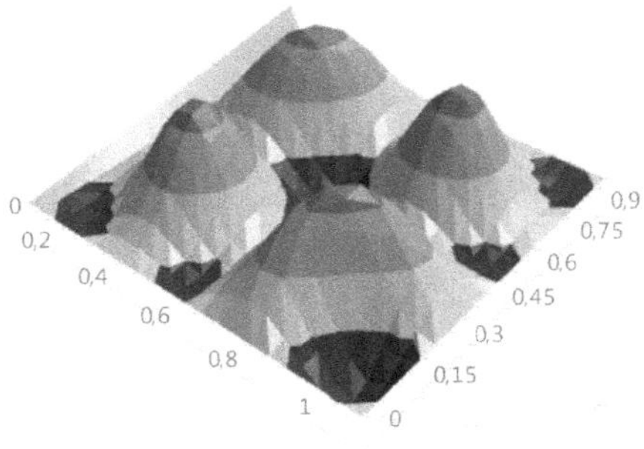

Figure 7.28

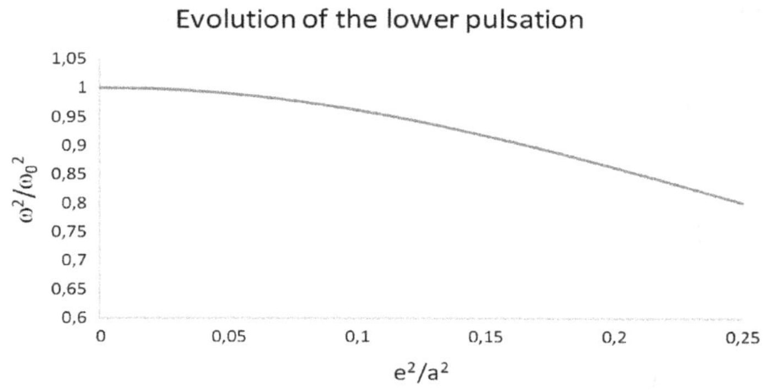

Figure 7.29

By reporting in the equilibrium Equation (7.172):

$$-D\left(\alpha_{mn}\frac{m\pi}{a}+\beta_{mn}\frac{n\pi}{b}\right)\left(\left(\frac{m\pi}{a}\right)^2+\left(\frac{n\pi}{b}\right)^2\right)-\rho\omega_{mn}{}^2=0$$

$$-D\left(\alpha_{mn}\left(\frac{m\pi}{a}\right)^2+\frac{1}{2}\alpha_{mn}(1-\nu)\left(\frac{n\pi}{b}\right)^2+\frac{1}{2}(1+\nu)\beta_{mn}\frac{m\pi}{a}\frac{n\pi}{b}\right)+kGe\left(\frac{m\pi}{a}-\alpha_{mn}\right)=0$$

$$-D\left(\frac{1}{2}(1+\nu)\alpha_{mn}\frac{m\pi}{a}\frac{n\pi}{b}+\frac{1}{2}(1-\nu)\left(\frac{m\pi}{a}\right)^2\beta_{mn}+\beta_{mn}\left(\frac{n\pi}{b}\right)^2\right)+kGe\left(\frac{n\pi}{b}-\beta_{mn}\right)=0$$

The last two relations make it possible to determine the relative amplitudes of the components of the rotation, then the pulsation of rank (m, n):

$$\alpha_{mn}=-\frac{\frac{m\pi}{a}\times\frac{kGe}{D}}{\left[\frac{kGe}{D}+\left(\frac{m\pi}{a}\right)^2+\left(\frac{n\pi}{b}\right)^2\right]}\;;\;\beta_{mn}=-\frac{\frac{n\pi}{b}\times\frac{kGe}{D}}{\left[\frac{kGe}{D}+\left(\frac{m\pi}{a}\right)^2+\left(\frac{n\pi}{b}\right)^2\right]}$$

$$\omega_{mn}{}^2=\frac{D}{\rho}\frac{\left(\left(\frac{m\pi}{a}\right)^2+\left(\frac{n\pi}{b}\right)^2\right)^2}{1+\frac{e^2}{5(1-\nu^2)}\left(\left(\frac{m\pi}{a}\right)^2+\left(\frac{n\pi}{b}\right)^2\right)}$$

For the first pulsation of a square plate, the evolution is given in Figure 7.27.

This result confirms that taking into account the shear strain only provides a weak correction for slenderness higher than five.

Exercise 7.5 Buckling of a Reissner–Mindlin plate

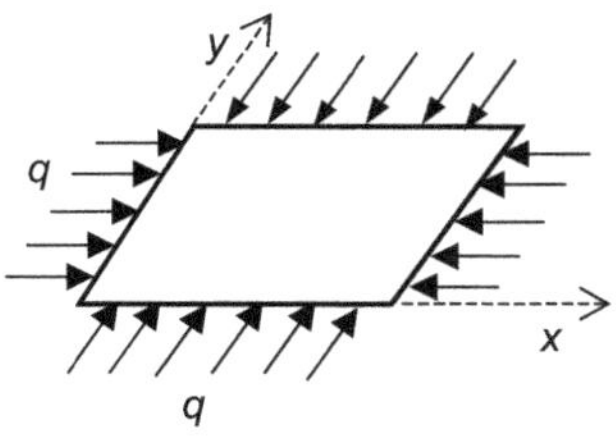

Figure 7.30

A rectangular Reissner-Mindlin plate is subjected to a density of force q on its four edges (Figure 7.28). Calculate the critical value of q causing the instability of the plate. Study the evolution of this critical value as a function of the relative thickness of the plate.

Answer:
The buckling modes are similar to the vibration modes in Exercise 7.4. The minimum critical value of q is obtained for m = n = 1:

$$q_{cr} = -D\pi^2\left(\frac{1}{a^2}+\frac{1}{b^2}\right)\frac{1}{1+\dfrac{\pi^2 e^2 / a^2}{5\left(1-\nu^2\right)}\left(1+\dfrac{a^2}{b^2}\right)}$$

For $a = b$, the variation is the same as for the pulsation in Exercise 7.4. Shear strain decreases the critical force.

Chapter 8

Shells in bending

The purpose of this chapter is the study of shells in bending.

The main theories of shells are based on the assumptions of Kirchhoff–Love or Reissner–Mindlin. A common assumption is that of thin shells, which has been presented in Chapter 5. The simplest approximation concerning the kinematics of motion is that of small displacements presented in the preceding chapters; it can no longer be adopted when the displacement is not negligible compared to the shell thickness. The approximations of Love or Donnell differ from that resulting from the complete linearisation of displacement-strain relations; they make it possible to obtain analytical solutions in certain situations and to address numerous stability problems. The kinematics corresponding to these various approximations and the associated equilibrium equations are presented in the first section.

The application to the cases of cylindrical thin shells and thin shells of revolution is presented in the next sections, where classic examples of resolution and some stability problems are presented. The analytical solutions developed make it possible to highlight the shells' behaviour, in particular the relative orders of magnitude of the different terms composing the solutions.

The chapter concludes with the study of the flexibility of the elbows of pipes, which shows how all the concepts introduced in this course are used in practice.

8.1 MAIN THEORIES AND APPROXIMATIONS

8.1.1 Kinematics of transformation

8.1.1.1 General

The equations for the movement of shells have been established in Chapters 4 and 5, with the help of the results of Chapter 3:

- in Chapter 4, for equilibrium equations, either with a "forces" approach, Equations (4.66) and (4.68), or with an energy approach, Equation (4.97), with the associated boundary conditions. Equation (4.97) takes into account the Kirchhoff–Love hypotheses, but they can be generalised by Equation (4.175) of §4.5.
- Chapter 5, regarding the relationship between three-dimensional stresses and generalised stresses, with relations (5.9) and (5.10), and constitutive laws.
- the equations (3.97) and (3.101) allow the expression of the relations between the variables of surface deformation and displacements.

DOI: 10.1201/9780429440403-8

The different "theories" of shells exposed in the literature are in fact approximations of the equations mentioned above. It seems fair to reserve the term "theory" for developments based on kinematic assumptions in thickness (Kirchhoff–Love or Reissner–Mindlin) and to keep the term "approximation" for developments based on considerations of the order of magnitude to neglect certain terms. This distinction, although arbitrary, makes it possible to classify the various methods better.

There may be a question of the practical interest of analysing these approximations in greater detail, since the finite element method discussed in the next chapter makes it possible to solve the shell problems without the need to simplify the equations. In fact, in addition to the historical interest of the various approaches, they allow the proper evaluation of the orders of magnitude of the different terms and the fields of application of the approximations granted, which is essential for a good understanding of the physical behaviour of shells. Second, they lead to a certain number of analytical solutions, which also contribute to a good understanding of the behaviour and make it possible to validate numerical solutions. Finally, they serve as a guide to an adequate formulation of the finite elements of shells.

A curved shell partially resisting by membrane effect, an important choice concerns the possibility of neglecting either the membrane energy or the flexional energy. There is no general answer to this question. Rayleigh (1881), to calculate the pulsations of a hemisphere, assumes that the deformations are essentially inextensional. Love (1888) suggests, on the contrary, that membrane effects are predominant. In reality, the situation is different depending on whether the shell is open or closed or whether kinematic conditions (support or embedding) are imposed on the shell. Lamb[1] and Basset[2] (1890) showed in the case of the cylinder the existence of a "bending limit layer" along the edges (see §8.2.3.1) where the solution evolves from preponderant bending to the membrane solution.

8.1.1.2 Indicators

The ratio e/R of the thickness to the smallest radius of curvature is a good indicator for the validity of various hypotheses:

- If $\frac{e}{R} < \frac{1}{1000}$, flexion can be neglected and membrane theory is applicable.
- If $\frac{e}{R} < \frac{1}{20}$, the shell is thin and the Kirchhoff–Love hypotheses are applicable.
- If $\frac{e}{R} > \frac{1}{20}$, the shell is thick and changes in thickness are no longer negligible.

The above limits should be considered as orders of magnitude.

A second indicator consists of the ratio of the transverse displacement to the thickness: if this ratio is significantly lower than 1, the displacements can be considered small and the deformation–displacement relationship can be linearised. In the opposite case, the geometric non-linearities must be taken into account. This is the subject of the below sections, where different approximations are adopted depending on the situation.

[1] Sir Horace Lamb, British mathematician and mechanician (1849–1934).

[2] Alfred Barnard Basset, British mathematician and physicist (1854–1930).

Note that if the displacement is small compared to the thickness, on the one hand, and that, on the other hand, the shell is thin, it is possible to consider that the displacements are of the first order, but that, divided by R, they are of the second order, which is the subject of the development in §8.1.1.6 and in §8.1.1.8.

8.1.1.3 Expression of deformations

The deformations of the middle surface are expressed by relations (3.95), (3.97), (3.98), (3.99) and (3.101) which are recalled here:

$$\varepsilon_{\alpha\beta} = \frac{1}{2}\left(\mathcal{A}_{\alpha\beta} + \mathcal{A}_{\beta\alpha} + \mathcal{A}_{\alpha}{}^{\gamma}\mathcal{A}_{\beta\gamma} + \mathcal{B}_{\alpha}\mathcal{B}_{\beta}\right) \tag{8.1}$$

$$\begin{aligned}\kappa_{\alpha\beta} &= \left(\partial_{\alpha}\mathcal{A}_{\beta}{}^{\gamma} + \Gamma^{\gamma}_{\alpha\mu}\mathcal{A}_{\beta}{}^{\mu} + \Gamma^{\gamma}_{\alpha\beta} - B_{\alpha}{}^{\gamma}\mathcal{B}_{\beta}\right)\mathcal{N}_{\gamma} \\ &\quad + \left(\partial_{\alpha}\mathcal{B}_{\beta} + B_{\alpha\gamma}\mathcal{A}_{\beta}{}^{\gamma}\right)\cos\omega - B_{\alpha\beta}\left(1-\cos\omega\right)\end{aligned} \tag{8.2}$$

with:

$$\mathcal{A}_{\alpha}{}^{\beta} = \nabla_{\alpha}\xi^{\beta} - \zeta B_{\alpha}{}^{\beta} \quad ; \quad \mathcal{B}_{\alpha} = B_{\alpha}{}^{\beta}\xi_{\beta} + \partial_{\alpha}\zeta$$

In the transformation due to the displacement field, $\mathcal{A}$ is representative of the elongation of the basis vectors tangent to the surface and $\mathcal{B}$ is representative of the rotation of the normal.

$$\begin{aligned}\mathcal{N}_{\alpha} &= \sqrt{\frac{\hat{A}}{\hat{a}}}\left(-\mathcal{B}_{\alpha} + \delta^{\lambda\mu}_{12}\mathcal{B}_{\lambda}\mathcal{A}_{\mu}{}^{\beta}\tilde{e}_{\gamma\alpha}\right) \\ \cos\omega &= \left(1+\Lambda\right)\sqrt{\frac{\hat{A}}{\hat{a}}} \\ \hat{a} &= \det\left(A_{\alpha\beta} + 2\varepsilon_{\alpha\beta}\right) \\ \Lambda &= \mathcal{A}_{\beta}{}^{\beta} + \det\left(\mathcal{A}_{\alpha}{}^{\beta}\right)\end{aligned} \tag{8.3}$$

$$\mathcal{N}_{\beta}\left(\delta^{\beta}_{\alpha} + \mathcal{A}_{\alpha}{}^{\beta}\right) + \mathcal{B}_{\alpha}\cos\omega = 0$$

and the rotation of the normal is given by (3.98):

$$a_3 = \mathcal{N}^{\alpha}A_{\alpha} + A_3\cos\omega \tag{8.4}$$

In the Kirchhoff–Love hypothesis, the local deformation is given by equation (5.37):

$$\bar{\varepsilon}\left(\mathcal{P}\right) = \varepsilon - x^3\kappa + \left(x^3\right)^2\tau \tag{8.5}$$

8.1.1.4 Usual theories

The two main theories have been discussed in Chapter 5: the Kirchhoff–Love and Reissner–Mindlin theories. In both cases, the segment consisting of the material points situated on the normal to the surface in the reference situation remains rectilinear and of constant length in the transformation.

- In Kirchhoff–Love's theory, this segment remains normal to the surface during .evolution. In this case, the distortion energy due to the shear force is zero.
- In Reissner–Mindlin's theory, this segment is free to rotate independently of the surface, which makes it possible to approximate the distortion energy due to the shear force.

It is possible to further improve the model by taking into account a variation in the length of the segment (variation of shell thickness), either by assuming null stress (Flügge, Naghdi), or by avoiding this last simplification (Reissner, Naghdi). Some indications are given on the "global" approach of these theories in §4.5. It is obviously possible to introduce other kinematic hypotheses, for example by assuming a variable distortion in the thickness. It should be borne in mind, however, that a complex shell model where the number of unknown fields would be too large would lose interest over a three-dimensional approach, which can be modelled using finite elements with a satisfactory approximation.

In the rest of this chapter, the Kirchhoff–Love hypotheses are the only ones considered.

8.1.1.5 Order of linearisation according to x^3

The approximations presented below relate mainly to the expression of the deformations. The linearisation of the lowest degree is that corresponding to the hypothesis of thin shells. It leads in particular to retaining only the linear term in x^3 in the expression of the three-dimensional deformation $\overline{\varepsilon}(\mathcal{P})$, which is the minimum to correctly represent bending:

$$\overline{\varepsilon}(\mathcal{P}) = \varepsilon - x^3\kappa \tag{8.6}$$

By retaining only the constant term, the equations would reduce to those of the membrane theory.

The expression of the generalised stresses as a function of the deformations by the usual elastic laws gives rise to a term of the type $\left(1-\dfrac{x^3}{R}\right)^{-1}$, cf. §5.6.2.2. The **first-order** approximations retain only the first term of development in x^3 : $1+\dfrac{x^3}{R}$: or, for the classical approximation of Love, only the value 1, which leads to the relations established in Chapter 5. On the other hand, some authors (Flügge, Lüre, Byrne) retain a complementary term (see 5.6.2.3):

$$\left(1-\frac{x^3}{R_1}\right)^{-1} \approx 1+\frac{x^3}{R_1}+\left(\frac{x^3}{R_1}\right)^2 \tag{8.7}$$

or proceed to complete development (Vlasov, Naghdi). Of course, in these cases, the equations are not linear and are therefore of a more delicate use.

8.1.1.6 Hypothesis of small deformations and small rotations

The idea developed here is to find out which kinematics hypotheses can be adopted, leading to neglecting certain deformation terms in a coherent way, to allow approximations, while remaining in conformity with the physics of evolutions.

The hypothesis of small displacements presented in Chapters 4 and 5 has the advantage of allowing a complete linearisation of the equations, subject to a linear constitutive law. But it turns out too strong for some applications, where the displacement can not be considered small compared to the thickness.

A weaker hypothesis actually consists of two approximations:

- The first approximation given in this hypothesis is that of **small deformations** of the middle surface: the variations of angles and lengths inscribed on the middle surface remain small. This assumption is reasonable in the case where bending is preponderant, which is for example the case of plates under transverse loading. In this approximation, all components of the middle surface strain tensor are assumed to be of the same order, at least of the first order.

A limit of the hypothesis of this approximation is the **inextensional transformation hypothesis** (Rayleigh), in which the in-plane strain tensor remains zero during evolution.

- The second approximation in this hypothesis is that of **small rotations**, which does not necessarily imply the hypothesis of small displacements. Here, the rotation of the normal vector is supposed to be infinitely small in the first order.

Before examining the consequences of this hypothesis on the kinematics of the shell, it is useful to come back to the one-dimensional case, that is to say, that of the beams or arches limited to plane arches. This allows specifying of the method followed, in a simpler mathematical framework.

The vector **T** (or **N**) is the tangent (or normal) vector to the arch in its initial position, and the vector **t** (or **n**) is the tangent vector to the arch (or normal) in its current position (Figure 8.1).

The displacement has components $u(s)$ (tangential) and $w(s)$ (normal), where s denotes the curvilinear abscissa. A simple calculation makes it possible to obtain the expression of the deformation:

$$\varepsilon = u' - \frac{v}{R} + \frac{1}{2}\left[\left(u' - \frac{v}{R}\right)^2 + \left(v' + \frac{u}{R}\right)^2\right]$$

where R is the radius of curvature of the arch in its initial position. If s' denotes the curvilinear abscissa along the deformed arch:

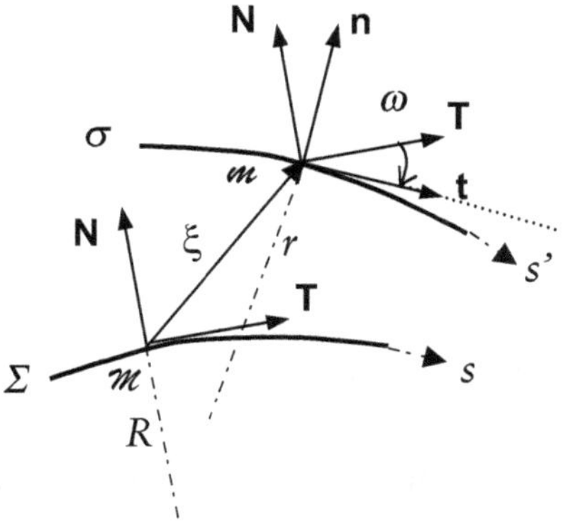

Figure 8.1

$$\left(\frac{ds'}{ds}\right)^2 = 1 + 2\varepsilon$$

The normal to the deformed arc is in the Frenet coordinate system associated with the initial arch is:

$$\mathbf{n} = \frac{ds}{ds'}\left[\left(1 + u' - \frac{v}{R}\right)\mathbf{N} - \left(v' + \frac{u}{R}\right)\mathbf{T}\right]$$

and the angle of rotation ω of the normal is characterised by:

$$\cos\omega = \mathbf{n}\cdot\mathbf{N} = \frac{ds}{ds'}\left(1 + u' - \frac{v}{R}\right) = \mathbf{t}\cdot\mathbf{T}$$

$$\sin\omega = -\mathbf{n}\cdot\mathbf{T} == \frac{ds}{ds'}\left(v' + \frac{u}{R}\right) = \mathbf{t}\cdot\mathbf{N}$$

With these formulas established, it is appropriate to express the two approximations:

- deformation ε is assumed to be small, at least of the first order. So:

$$\frac{ds}{ds'} \approx 1 - \varepsilon + \frac{3}{2}\varepsilon^2$$

- the rotation is supposed to be small, of the first order. So:

$$\sin\omega = \frac{ds}{ds'}\left(v' + \frac{u}{R}\right) \approx \omega \quad\Rightarrow\quad v' + \frac{u}{R} \approx \omega$$

is of the first order and:

$$\cos\omega = \frac{ds}{ds'}\left(1 + u' - \frac{v}{R}\right) \approx 1 - \frac{\omega^2}{2}$$

Consequently:

$$\varepsilon - \left(u' - \frac{v}{R}\right) = \frac{1}{2}\left[\left(u' - \frac{v}{R}\right)^2 + \left(v' + \frac{u}{R}\right)^2\right]$$

is of the second order.

At this stage, two coherent hypotheses can be taken:

- ε and $u' - \frac{v}{R}$ are of the first order:

$$\varepsilon \approx u' - \frac{v}{R}$$

to the second order. This hypothesis corresponds to complete linearisation: all the components of the displacement are of the same order.

- ε and $u' - \frac{v}{R}$ are of the second order:

$$\varepsilon \approx u' - \frac{v}{R} + \frac{1}{2}\left(v' + \frac{u}{R}\right)^2$$

In this hypothesis, $v' + \frac{u}{R}$ is of the first order, and $u' - \frac{v}{R}$ is of the second order.

In both cases, the curvature variation is:

$$\chi \approx \frac{d\omega}{ds} \approx v'' + \frac{d}{ds}\left(\frac{u}{R}\right)$$

and is of the first order.

These are the two types of assumptions usually taken for arches. For example, in the case of straight beams (R infinite):

- in the first hypothesis, ε, u', and v'' are of the first order.
- in the second hypothesis, v'' is of the first order, ε and u' are of the second order. It is this last hypothesis which gives the best results in the case of preponderant bending.

Returning to the case of shells, the reasoning is the same, it is to extend it to the two-dimensional situation. The approximation of small rotations is expressed by the relations:

$$\begin{aligned}
&(a_3, A_3) = \omega \\
&(a_3, A_\alpha) = \frac{\pi}{2} + \phi_\alpha \\
&(a_\alpha, A_3) = \frac{\pi}{2} + \phi'_\alpha \\
&(A_\alpha, A_\beta) = \gamma_{\alpha\beta} \\
&(a_\alpha, A_\beta) = \gamma_{\alpha\beta} + \phi_{\alpha\beta}
\end{aligned} \tag{8.8}$$

where angles ω, ϕ_α, ϕ'_α and $\phi_{\alpha\beta}$ are small vis-à-vis 1. ω is assumed to be of the **first-order** in limited developments. It should be noted that these different angles are interrelated, since three components are sufficient to define the rotation vector. $\gamma_{\alpha\beta}$ is a finite angle of the reference geometry, introduced here for convenience and therefore not small.

Using a limited development:

$$a_3 \cdot A_3 = \cos\omega \approx 1 - \frac{\omega^2}{2}$$

Noting, using (3.16) and (3.20), that:

$$\hat{a} = \det\left(A_{\alpha\beta} + 2\varepsilon_{\alpha\beta}\right) = \frac{1}{2}\tilde{e}^{\alpha\gamma}\tilde{e}^{\beta\delta}\left(A_{\alpha\beta} + 2\varepsilon_{\alpha\beta}\right)\left(A_{\gamma\delta} + 2\varepsilon_{\gamma\delta}\right)$$

$$\approx \frac{1}{2}\tilde{e}^{\alpha\gamma}\tilde{e}^{\beta\delta}\left(A_{\alpha\beta}A_{\gamma\delta} + 2A_{\alpha\beta}\varepsilon_{\gamma\delta} + 2A_{\gamma\delta}\varepsilon_{\alpha\beta}\right)$$

$$\approx \hat{A} + 2\tilde{e}^{\alpha\gamma}\tilde{e}^{\beta\delta}A_{\alpha\beta}\varepsilon_{\gamma\delta} = \hat{A}\left(1 + 2A^{\gamma\delta}\varepsilon_{\gamma\delta}\right)$$

$$\approx \hat{A}\left(1 + 2\,\mathrm{TR}(\varepsilon)\right)$$

relationship (8.3) makes it possible to write:

$$1 + \Lambda = \left[\frac{\hat{a}}{\hat{A}}\right]^{1/2}\cos\omega \approx 1 - \frac{\omega^2}{2} + \mathrm{TR}(\varepsilon) \tag{8.9}$$

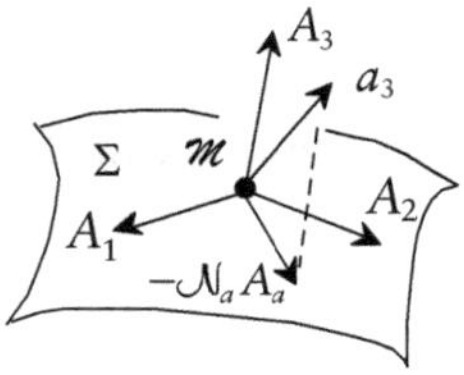

Figure 8.2

which shows that is of the order of ε (first or second order); consequently, $\mathcal{A}_{\alpha\beta}$ is of the same order by (8.1).

Relationships (8.3) and (8.8) can also be used to write (Figure 8.2):

$$a_3 \bullet A_\alpha = \mathcal{N}_\alpha = -\left\|A_\alpha\right\|\sin\phi_\alpha \approx -\left\|A_\alpha\right\|\varphi_\alpha$$

By (8.4), projecting a_3 on the initial tangent plane:

$$\sin\omega = \left(\mathcal{N}_\alpha\mathcal{N}^\alpha\right)^{1/2} \approx \omega$$

which is of the first order. So $\mathcal{N}$ is of the first order, which means that ϕ_α is also of the first order.

These results obtained, the last relation (8.3) shows that $\mathcal{N}_\alpha \approx -\mathcal{B}_\alpha$ is a first-order equality if ε is of the first order, to the second order, or a second-order equality if ε is of the second order, to the third order. This is also a relationship between ϕ_α and ϕ'_α. Indeed, relations (3,95) and (8.8) make it possible to obtain the approximation:

$$a_\alpha \bullet A_3 = \mathcal{B}_\alpha = -\left\|a_\alpha\right\|\sin\phi'_\alpha \approx -\left\|A_\alpha\right\|\left(1 + \varepsilon_{\alpha\alpha}\right)\phi'_\alpha \approx -\left\|A_\alpha\right\|\phi'_\alpha$$

It comes then $\phi_\alpha + \phi'_\alpha \approx 0$ and ϕ'_α is also of the first order. $\mathcal{N}$ and $\mathcal{B}$ being of the first order, (7.2) allows us to deduce that $\kappa_{\alpha\beta}$ is of the first order. Then:

$$a_\alpha \bullet A_\beta = A_{\alpha\beta} + \mathcal{A}_{\alpha\beta} = \left\|a_\alpha\right\|\left\|A_\beta\right\|\cos\left(\gamma_{\alpha\beta} + \phi_{\alpha\beta}\right)$$

$$\approx \left\|A_\alpha\right\|\left\|A_\beta\right\|\left(1 + \varepsilon_{\alpha\alpha}\right)\left[\left(1 - \frac{\left(\phi_{\alpha\beta}\right)^2}{2}\right)\cos\gamma_{\alpha\beta} - \phi_{\alpha\beta}\sin\gamma_{\alpha\beta}\right]$$

$$\approx A_{\alpha\beta} + A_{\alpha\beta}\left[\varepsilon_{\alpha\alpha} - \frac{\left(\phi_{\alpha\beta}\right)^2}{2} - \phi_{\alpha\beta}\left(1 + \varepsilon_{\alpha\alpha}\right)\tan\gamma_{\alpha\beta}\right]$$

$$\Rightarrow \quad a_\alpha \bullet A_\beta \approx A_{\alpha\beta} + A_{\alpha\beta}\left[\varepsilon_{\alpha\alpha} - \phi_{\alpha\beta}\tan\gamma_{\alpha\beta}\right]$$

ε and $\mathcal{A}$ are of the same order, as it has been seen previously, and $\phi_{\alpha\beta}$ is therefore of an order greater than or equal to that of ε.

At this point, it is useful to summarise the results:

ω, $\mathcal{N}_\alpha \approx -\mathcal{B}_\alpha$, $\phi_\alpha = -\phi'_\alpha$ and $\kappa_{\alpha\beta}$ are of first order; $\mathcal{A}_{\alpha\beta}$, Λ and $\phi_{\alpha\beta}$ are of the order of ε, which is not yet specified. So:

$$a_3 \approx A_3 - \mathcal{B}^\alpha A_\alpha$$

$1 + \Lambda$ is given by (8.9), it follows, by (8.1):

$$\mathrm{TR}(\varepsilon) = \mathrm{TR}(\mathcal{A}) + \frac{1}{2}\mathcal{A}_\alpha{}^\gamma \mathcal{A}^\alpha{}_\gamma + \frac{1}{2}\mathcal{B}_\alpha \mathcal{B}^\alpha$$

The above relationships are respected, to a higher order, when one of the following two consistent assumptions is made:

a) ε is of the first order and given by:

$$\varepsilon_{\alpha\beta} \approx \frac{1}{2}\left(\mathcal{A}_{\alpha\beta} + \mathcal{A}_{\beta\alpha}\right) \tag{8.10}$$

or by (3.102) depending on the displacement. $\mathcal{A}_{\alpha\beta} = \nabla_\alpha \xi_\beta - \zeta\, B_{\alpha\beta}$ and $\mathcal{B}_\alpha = B_\alpha{}^\beta \xi_\beta + \partial_\alpha \zeta$ are of the first order; then, using (3.95) and (3.107):

$$\begin{aligned} a_\alpha &= \left(\delta_\alpha{}^\beta + \mathcal{A}_\alpha{}^\beta\right) A_\beta + \mathcal{B}_\alpha A_3 \\ \kappa_{\alpha\beta} &\approx \nabla_\alpha \mathcal{B}_\beta + B_{\alpha\gamma} \mathcal{A}_\beta{}^\gamma \end{aligned} \tag{8.11}$$

In κ, the first term is the derivative of the rotation of the normal, the only one in the case of a shallow surface; the second term is due to the combined effect of deformation and curvature; for example, in the uniform increase w of the radius of a circle, the deformation is w/R and the variation of curvature - w/R^2. Also, for shallow shells, the second term may be neglected.

b) ε is of the second order, $\mathcal{A}_{\alpha\beta} = \nabla_\alpha \xi_\beta - \zeta\, B_{\alpha\beta}$ is of the second order and $\mathcal{B}_\alpha = B_\alpha{}^\beta \xi_\beta + \partial_\alpha \zeta$ is of the first order. Then:

$$a_\alpha = A_\alpha + \mathcal{B}_\alpha A_3$$

and:

$$\varepsilon_{\alpha\beta} \approx \frac{1}{2}\left(\mathcal{A}_{\alpha\beta} + \mathcal{A}_{\beta\alpha} + \mathcal{B}_\alpha \mathcal{B}_\beta\right) \tag{8.12}$$

either, depending on the displacement and to the fourth order:

$$\varepsilon_{\alpha\beta} \approx \frac{1}{2}\left(\nabla_\alpha \xi_\beta + \nabla_\beta \xi_\alpha - 2\zeta B_{\alpha\beta} + \left(B_\alpha{}^\lambda \xi_\lambda + \partial_\alpha \zeta\right)\left(B_\beta{}^\lambda \xi_\lambda + \partial_\beta \zeta\right)\right) \tag{8.13}$$

The expression of ε is therefore nonlinear. At the first order, neglecting the second order:

$$\kappa_{\alpha\beta} \approx \nabla_\alpha \mathcal{B}_\beta = \partial_\alpha \mathcal{B}_\beta - \Gamma^\gamma_{\alpha\beta}\mathcal{B}_\gamma$$
$$\approx \nabla_\alpha B_{\beta\gamma}.\xi^\gamma + B_{\beta\gamma}\nabla_\alpha \xi^\gamma + \nabla_\alpha \partial_\beta \zeta \tag{8.14}$$

This expression is linear. On the other hand, $\kappa_{\alpha\beta}$ obtained in this approximation is not symmetrical, which is a serious inconvenience, since κ and the associated tensor of moments would not be symmetrical, which would be contrary to the result of the general theory. However, in the first order, $\mathcal{A}$ is zero, so:

$$B_{\beta\gamma}\nabla_\alpha \xi^\gamma \approx B_{\beta\gamma}B_\alpha{}^\gamma \zeta$$

or, by the symmetry of the third fundamental form C:

$$B_{\beta\gamma}\nabla_\alpha \xi^\gamma \approx B_{\alpha\gamma}B_\beta{}^\gamma \zeta = B_{\alpha\gamma}\nabla_\beta \xi^\gamma$$

Considering also the Mainardi–Codazzi equality (3.80), the symmetry of $\kappa_{\alpha\beta}$ is thus ensured by (8.14) in the approximation, to the second order, and it is possible to retain a symmetrical expression:

$$\kappa_{\alpha\beta} \approx \frac{1}{2}\left(\nabla_\alpha \mathcal{B}_\beta + \nabla_\beta \mathcal{B}_\alpha\right) \tag{8.15}$$

An equivalent expression can be obtained from (8.14), replacing $B_{\beta\gamma}\nabla_\alpha \xi^\gamma$ with its equivalent above:

$$\kappa_{\alpha\beta} \approx \frac{1}{2}\left(\nabla_\alpha \partial_\beta \zeta + \nabla_\beta \partial_\alpha \zeta\right) + \nabla_\alpha B_{\beta\gamma}.\xi^\gamma + B_\alpha{}^\gamma B_{\beta\gamma}\zeta \tag{8.16}$$

In conclusion, in the approximation of small deformations and small rotations, the expression of $\varepsilon_{\alpha\beta}$ is of the first order or of the second order and is not linear in the latter case, whereas $\kappa_{\alpha\beta}$ is of the first order and linear, but with different expressions depending on the order of ε. From these orders of magnitude, two types of approximations are generally retained, neglecting the term $B_{\alpha\gamma}\mathcal{A}_\beta{}^\gamma$ in front of $\nabla_\alpha \mathcal{B}_\beta$. This last assumption is coherent in the case of the second order assumption, it is acceptable for shallow shells in the first order assumption; nevertheless, it leads to an expression of nonsymmetrical κ, it is preferable in this case to use the symmetrical expression (8.15). The validity of this assumption is discussed in §8.2.1.8 in the case of cylinders.

In addition, it appears in (8.16) that neglecting $B_{\alpha\gamma}\mathcal{A}_\beta{}^\gamma$ leads to not considering the term $B_{\alpha\gamma}\nabla_\beta \xi^\gamma$ while $B_{\beta\gamma}\nabla_\alpha \xi^\gamma$ is conserved, which may seem contradictory, but it is indeed the deformation which is of the second order, not this particular term.

8.1.1.7 Love approximation

It is a second-order approach where the expression of the deformation $\varepsilon_{\alpha\beta}$ is linearised, but where the expression of κ in the second-order approach is retained:

$$\varepsilon_{\alpha\beta} \approx \frac{1}{2}\left(\nabla_\alpha \xi_\beta + \nabla_\beta \xi_\alpha - 2\zeta \mathsf{B}_{\alpha\beta}\right) \tag{8.17}$$

$\kappa_{\alpha\beta}$ is given by (8.15), $B_{\alpha\gamma}, \mathcal{A}_{\beta}^{\gamma}$ being neglected compared to $\nabla_{\alpha}, \mathcal{B}_{\beta}$, which amounts to considering that the tangent components and ζ / R_{α} are of the same order of magnitude and small compared to ζ, which is physically acceptable in certain situations, for example when the transverse displacement is preponderant and the curvature sufficiently small (shallow shells). There are however many situations where the components of the displacement are of the same order of magnitude and, to the first order, it is then necessary to return to the linearised approximation. The advantage of the Love simplification is to lead to a fully linearised expression of the deformations.

8.1.1.8 Donnell approximation

This approximation is still based on the assumption of small deformations and small rotations. Donnell's approximation applies to shells with a low curvature (shallow shells), itself considered first-order; it is particularly useful for the study (in the second order) of the stability of this type of shell. In this case, $\mathcal{A}$ and the tangent components ξ^{α} of the displacement are of the second order. These components are small vis-a-vis the thickness:

$$\left\| B_{\alpha\beta}\ \xi^{\alpha} \right\| << \frac{e}{R_1} << 1$$

The deformation is then of the second order. $\mathcal{B}$ being of the first order, ζ is necessarily of the first order and may be of the order of the thickness. Thus, the participation of displacements ξ^{α} in the tangent plane is small vis-à-vis the normal displacement. So:

$$\mathcal{B}_{\alpha} = \partial_{\alpha}\zeta$$

In this case, deformations and variations in curvature are reduced to:

$$\varepsilon_{\alpha\beta} \approx \frac{1}{2}\left(\nabla_{\alpha}\xi_{\beta} + \nabla_{\beta}\xi_{\alpha} - 2\zeta \mathrm{B}_{\alpha\beta} + \partial_{\alpha}\zeta \cdot \partial_{\beta}\zeta\right) \tag{8.18}$$

$$\kappa_{\alpha\beta} \approx \nabla_{\alpha}\partial_{\beta}\zeta \tag{8.19}$$

This is the level of approximation used in §8.3 to study the case of plates subjected to a membrane force influencing flexion. It corresponds, in one dimension, to the classical hypotheses of the theory of straight or little curved beams in the second order.

8.1.1.9 Classical theory of shells in bending

In the case of a shell of moderate curvature, the various components of the displacement are assumed of the same order and small vis-à-vis the thickness; the deformation is then of the second order. The **classical linearised approximation of shells in bending** uses linearised expressions (8.17) and (8.19). It can be obtained also as a special case of Love and Donnell approximations. The classical approximation uses expressions that are not quite identical to the expressions obtained by completely linearising ε and κ from formulas (3.102) and (3.108): the terms related to the initial curvature are neglected in the expression of the variation of curvature, the approximation thus applies to shallow shells with small curvature.

8.1.2 Equilibrium equations in the main approximations

8.1.2.1 General

The local equations (4.97) and the boundary conditions (4.99) are expressed on the deformed configuration and therefore depend on the metric (via the differentiation) and the curvature of the deformed surface, which are unknown. These can nevertheless be expressed as a function of the metric and the curvature of the reference configuration and the displacement field.

The most common simplification is to express equilibrium equations in the reference configuration, which greatly simplifies the problem. This simplification can be adopted in the context of the small disturbances hypothesis and is therefore acceptable to solve most problems of change of equilibrium. Nevertheless, it can not be adopted whenever the change in geometry influences the behaviour of the shell, which is the case for example in the study of instability phenomena.

A simplification of the equilibrium equations can then be obtained by introducing into their expressions the different levels of approximation considered in the preceding sections. This can be done directly by replacing the characteristics of the deformed geometry with those of the reference geometry and displacement in the equilibrium equations. But the simplest method is to write the PVP by virtualising the expression of the deformation variables obtained in the different approximations: it is similar to that used in Chapter 4.

The deformation obtained depends on the assumptions of kinematics taken into account, the equilibrium equations obtained therefore differ from one approximation to another.

8.1.2.2 Small deformations and small rotations (first order)

It is a question of expressing the local equations and boundary conditions, in the first-order approximation, that is to say from the deformations given by (8.10) and (8.11). Indeed, the deformations were particularised and the equations are modified consequently, compared to the complete linearisation. To simplify the presentation, only the problem of change of equilibrium (static) is developed, the addition of the inertial actions does not show any difficulty, and the outer surface actions are only force densities, which is the most common case.

Given the symmetry of N and M, locally, the virtual power of deformation is written:

$$
\begin{aligned}
\mathsf{N}\bullet\overset{*}{\varepsilon}+\mathsf{M}\bullet\overset{*}{\kappa} &= \mathsf{N}^{\alpha\beta}\,\overset{*}{\mathcal{A}}_{\alpha\beta}+\mathsf{M}^{\alpha\beta}\left(\nabla_\alpha\,\overset{*}{\mathcal{B}}_\beta+B_{\alpha\gamma}\,\overset{*}{\mathcal{A}}_\beta{}^{\gamma}\right)=\left(\mathsf{N}^{\alpha\gamma}+\mathsf{M}^{\alpha\beta}B_\beta{}^{\gamma}\right)\overset{*}{\mathcal{A}}_{\alpha\gamma}+\mathsf{M}^{\alpha\beta}\nabla_\alpha\,\overset{*}{\mathcal{B}}_\beta \\
&=\left(\mathsf{N}^{\alpha\gamma}+B_\beta{}^{\gamma}\mathsf{M}^{\alpha\beta}\right)\left(\nabla_\alpha\,\overset{*}{\xi}_\gamma-B_{\alpha\gamma}\,\overset{*}{\zeta}\right)+\mathsf{M}^{\alpha\beta}\nabla_\alpha\left(B_\beta{}^{\gamma}\,\overset{*}{\xi}_\gamma+\partial_\beta\,\overset{*}{\zeta}\right) \\
&=-\nabla_\alpha\left(\mathsf{N}^{\alpha\gamma}+B_\beta{}^{\gamma}\mathsf{M}^{\alpha\beta}\right)\overset{*}{\xi}_\gamma-B_\beta{}^{\gamma}\nabla_\alpha\mathsf{M}^{\alpha\beta}\,\overset{*}{\xi}_\gamma+\nabla_\beta\nabla_\alpha\mathsf{M}^{\alpha\beta}\cdot\overset{*}{\zeta}-B_{\alpha\gamma}\left(\mathsf{N}^{\alpha\gamma}+B_\beta{}^{\gamma}\mathsf{M}^{\alpha\beta}\right)\overset{*}{\zeta} \\
&\quad+\nabla_\alpha\left[\left(\mathsf{N}^{\alpha\gamma}+B_\beta{}^{\gamma}\mathsf{M}^{\alpha\beta}\right)\overset{*}{\xi}_\gamma\right]+\nabla_\alpha\left[B_\beta{}^{\gamma}\mathsf{M}^{\alpha\beta}\,\overset{*}{\xi}_\gamma\right]+\nabla_\alpha\left[\mathsf{M}^{\alpha\beta}\partial_\beta\,\overset{*}{\zeta}\right]-\nabla_\beta\left[\nabla_\alpha\mathsf{M}^{\alpha\beta}\,\overset{*}{\zeta}\right]
\end{aligned}
\tag{8.20}
$$

After the application of Stokes' theorem, the virtual power of deformation is written, the integral being expressed on the reference surface:

$$\overset{*}{\mathcal{W}}_f = \int_{\Sigma} \left(\mathsf{N} \bullet \overset{*}{\varepsilon} + \mathsf{M} \bullet \overset{*}{\kappa} \right) d\Sigma$$

$$= -\int_{\Sigma} \left\{ \begin{aligned} &\left[\nabla_{\alpha} \left(\mathsf{N}^{\alpha\gamma} + B_{\beta}{}^{\gamma} \mathsf{M}^{\alpha\beta} \right) + B_{\beta}{}^{\gamma} \nabla_{\alpha} \mathsf{M}^{\alpha\beta} \right] \overset{*}{\xi}_{\gamma} \\ &+ \left[B_{\alpha\gamma} \left(\mathsf{N}^{\alpha\gamma} + B_{\beta}{}^{\gamma} \mathsf{M}^{\alpha\beta} \right) - \nabla_{\beta} \nabla_{\alpha} \mathsf{M}^{\alpha\beta} \right] \overset{*}{\zeta} \end{aligned} \right\} d\Sigma + \int_{\partial\Sigma} \left\{ \begin{aligned} &\left(\mathsf{N}^{\nu\gamma} + 2\mathsf{M}^{\nu\beta} B_{\beta}{}^{\gamma} \right) \overset{*}{\xi}_{\gamma} - \nabla_{\beta} \mathsf{M}^{\beta\nu} \bullet \overset{*}{\zeta} \\ &+ \mathsf{M}^{\nu\beta} \bullet \partial_{\beta} \overset{*}{\zeta} \end{aligned} \right\} ds_0 \tag{8.21}$$

The last term of the edge integral is rewritten, as it was seen in §4.2.2.1:

$$\int_{\partial\Sigma} \left[-\frac{\partial \mathsf{M}^{\nu t}}{\partial s} \overset{*}{\zeta} + \mathsf{M}^{\nu\nu} \partial_{\nu} \overset{*}{\zeta} \right] ds_0 \tag{8.22}$$

The virtual power of external actions is deduced from (4.87) and expressed on the reference configuration:

$$\overset{*}{\mathcal{W}}_e = \int_{\Sigma} \left(p^{\gamma} \overset{*}{\xi}_{\gamma} + p^{3} \overset{*}{\zeta} \right) d\Sigma + \int_{\partial\Sigma} \left\{ \begin{aligned} &\left(q^{\gamma} - m^{\nu} B_{t}{}^{\gamma} + m^{t} B_{\nu}{}^{\gamma} \right) \overset{*}{\xi}_{\gamma} \\ &+ \left(q^{3} + \frac{\partial m^{\nu}}{\partial s} \right) \overset{*}{\zeta} + m^{t} \frac{\partial \overset{*}{\zeta}}{\partial \nu} \end{aligned} \right\} ds_0 \tag{8.23}$$

where the actions are broken down in the reference configuration coordinate system.

Using relations (8.22) to (8.23), the PVP leads to:

- Local equations:

$$\boxed{\begin{aligned} &\nabla_{\alpha} \left(\mathsf{N}^{\alpha\gamma} + B_{\beta}{}^{\gamma} \mathsf{M}^{\alpha\beta} \right) + B_{\beta}{}^{\gamma} \nabla_{\alpha} \mathsf{M}^{\alpha\beta} + p^{\gamma} = 0 \\ &B_{\alpha\gamma} \left(\mathsf{N}^{\alpha\gamma} + B_{\beta}{}^{\gamma} \mathsf{M}^{\alpha\beta} \right) - \nabla_{\beta} \nabla_{\alpha} \mathsf{M}^{\alpha\beta} + p^{3} = 0 \end{aligned}} \tag{8.24}$$

either, noting:

$$\mathsf{Q}^{\beta} = -\nabla_{\alpha} \mathsf{M}^{\alpha\beta} \tag{8.25}$$

$$\boxed{\begin{aligned} &\nabla_{\alpha} \left(\mathsf{N}^{\alpha\gamma} + B_{\beta}{}^{\gamma} \mathsf{M}^{\alpha\beta} \right) - B_{\beta}{}^{\gamma} \mathsf{Q}^{\beta} + p^{\gamma} = 0 \\ &B_{\alpha\gamma} \left(\mathsf{N}^{\alpha\gamma} + B_{\beta}{}^{\gamma} \mathsf{M}^{\alpha\beta} \right) + \mathsf{Q}^{\beta} + p^{3} = 0 \end{aligned}} \tag{8.26}$$

- Boundary conditions:

$$
\boxed{\begin{array}{l}
\mathsf{N}^{\nu\gamma} + 2B_\beta{}^\gamma \mathsf{M}^{\nu\beta} - \left(q^\gamma - m^\nu B_t{}^\gamma + m^t B_\nu{}^\gamma\right) = 0 \\[2ex]
\dfrac{\partial \mathsf{M}^{\nu t}}{\partial s} + \nabla_\beta \mathsf{M}^{\beta\nu} + q^3 + \dfrac{\partial m^\nu}{\partial s} = 0 \\[2ex]
\mathsf{M}^{\nu\nu} - m^t = 0
\end{array}} \tag{8.27}
$$

B.M terms are often negligible, especially in the case of shallow shells.

The relations obtained are to be compared with (4.97) and (4.99) established in the general framework. They differ in that it is the curvature of the reference position that is involved.

8.1.2.3 Small deformations and small rotations (second order)

It is a matter of expressing the equilibrium equations and the boundary conditions in the context of the approximation in §8.1.1.6, while neglecting the term $B_{\alpha\gamma}\mathcal{A}_\beta{}^\gamma$ in the expression of the variation of curvature, as is done in particular in the Love, Donnell and the second-order approximations.

The generalised stresses N and M are given in Euler variables. The deformation variables are given in a current configuration according to the characteristics of the reference configuration and the displacement by the expressions (8.12) and (8.15) in the second-order approximation; their virtual velocities are written:

$$
\overset{*}{\varepsilon}_{\alpha\beta} \approx \frac{1}{2}\left(\begin{array}{l}
\nabla_\alpha \overset{*}{\xi}_\beta + \nabla_\beta \overset{*}{\xi}_\alpha - 2\overset{*}{\zeta} B_{\alpha\beta} \\
+\left(B_\alpha{}^\mu \overset{*}{\xi}_\mu + \partial_\alpha \overset{*}{\zeta}\right)\left(B_\beta{}^\lambda \xi_\lambda + \partial_\beta \zeta\right) + \left(B_\alpha{}^\lambda \xi_\lambda + \partial_\alpha \zeta\right)\left(B_\beta{}^\mu \overset{*}{\xi}_\mu + \partial_\beta \overset{*}{\zeta}\right)
\end{array}\right) \tag{8.28}
$$

$$
\overset{*}{\kappa}_{\alpha\beta} \approx \frac{1}{2}\left[\nabla_\alpha\left(\partial_\beta \overset{*}{\zeta} + B_{\beta\gamma}\overset{*}{\xi}{}^\gamma\right) + \nabla_\beta\left(\partial_\alpha \overset{*}{\zeta} + B_{\alpha\gamma}\overset{*}{\xi}{}^\gamma\right)\right] \tag{8.29}
$$

These are the components of $\overset{*}{\varepsilon}$ and $\overset{*}{\kappa}$ in the coordinate system of the reference surface.

The PVP is used in the same way as in §8.1.2.2, noting that the integrals are expressed on the reference surface and the variation of surface area, of the second order, is neglected. It finally becomes:

- Local equations:

$$
\boxed{\begin{array}{l}
\nabla_\alpha \mathsf{N}^{\alpha\gamma} + B_\beta{}^\gamma \nabla_\alpha \mathsf{M}^{\alpha\beta} - \mathsf{N}^{\alpha\beta} B_\alpha{}^\gamma\left(B_\beta{}^\mu \xi_\mu + \partial_\beta \zeta\right) + p^\gamma = 0 \\[2ex]
\mathsf{N}^{\alpha\beta} B_{\alpha\beta} + \nabla_\alpha\left[\mathsf{N}^{\alpha\beta}\left(B_\beta{}^\mu \xi_\mu + \partial_\beta \zeta\right)\right] - \nabla_\beta \nabla_\alpha \mathsf{M}^{\alpha\beta} + p^3 = 0
\end{array}} \tag{8.30}
$$

either, noting:

$$Q^{\beta} = -\nabla_{\alpha} M^{\alpha\beta} + N^{\alpha\beta}\left(B_{\alpha}{}^{\mu}\xi_{\mu} + \partial_{\alpha}\zeta\right) \tag{8.31}$$

$$\boxed{\begin{aligned} &\nabla_{\alpha} N^{\alpha\gamma} - B_{\beta}{}^{\gamma} Q^{\beta} + p^{\gamma} = 0 \\ &N^{\alpha\beta} B_{\alpha\beta} + \nabla_{\alpha} Q^{\alpha} + p^{3} = 0 \end{aligned}} \tag{8.32}$$

- Boundary conditions:

$$\boxed{\begin{aligned} &N^{\nu\gamma} + M^{\nu\beta} B_{\beta}{}^{\gamma} - \left(q^{\gamma} - m^{\nu} B_{t}{}^{\gamma} + m^{t} B_{\nu}{}^{\gamma}\right) = 0 \\ &N^{\nu\beta}\left(B_{\beta}{}^{\mu}\xi_{\mu} + \partial_{\beta}\zeta\right) - \nabla_{\beta} M^{\beta\nu} - \left(q^{3} + \frac{\partial m^{\nu}}{\partial s}\right) - \frac{\partial M^{\nu t}}{\partial s} = 0 \\ &M^{\nu\nu} - m^{t} = 0 \end{aligned}} \tag{8.33}$$

and the third condition (second line) is rewritten:

$$Q^{\nu} = q^{3} + \frac{\partial m^{\nu}}{\partial s} + \frac{\partial M^{\nu t}}{\partial s} \tag{8.34}$$

The equilibrium equations and the boundary conditions thus formulated are written in the deformed state, the latter being described in an approximate manner by linearisation of the deformation. Generalised stresses are described in Euler variables. These equations, where the generalised stresses and the displacements are coupled, can only be solved by introducing the constitutive laws, which must therefore be expressed also in Euler variables. The set of relations thus obtained are obviously non-linear and can be solved analytically only in a few simple cases. For this purpose, they can be a little simplified by introducing the Love or Donnell approximations.

The expression for $\kappa_{\alpha\beta}$ given by (8.16) can also be used to obtain the equations and boundary conditions, in which case the virtual variation of bending energy is rewritten:

$$M^{\alpha\beta}\overset{*}{\kappa}_{\alpha\beta} \approx M^{\alpha\beta}\left[\nabla_{\alpha}\partial_{\beta}\overset{*}{\zeta} + \nabla_{\alpha}B_{\beta\gamma}.\overset{*}{\xi}{}^{\gamma} + B_{\alpha}{}^{\gamma}B_{\beta\gamma}\overset{*}{\zeta}\right]$$

$$\approx M^{\alpha\beta}\nabla_{\alpha}B_{\beta\gamma}.\overset{*}{\xi}{}^{\gamma} + \left(\nabla_{\beta}\nabla_{\alpha}M^{\alpha\beta} + M^{\alpha\beta}B_{\alpha}{}^{\gamma}B_{\beta\gamma}\right)\cdot\overset{*}{\zeta} + \nabla_{\alpha}\left[M^{\alpha\beta}\partial_{\beta}\overset{*}{\zeta}\right] - \nabla_{\beta}\left[\nabla_{\alpha}M^{\alpha\beta}\cdot\overset{*}{\zeta}\right]$$

and following the same process, the local equations:

$$\begin{aligned} &\nabla_{\alpha}N^{\alpha\gamma} - M^{\alpha\beta}\nabla_{\alpha}B_{\beta}{}^{\gamma} - N^{\alpha\beta}B_{\alpha}{}^{\gamma}\left(B_{\beta}{}^{\mu}\xi_{\mu} + \partial_{\beta}\zeta\right) + p^{\gamma} = 0 \\ &B_{\alpha\beta}N^{\alpha\beta} + \nabla_{\alpha}\left[N^{\alpha\beta}\left(B_{\beta}{}^{\mu}\xi_{\mu} + \partial_{\beta}\zeta\right)\right] - \left(\nabla_{\beta}\nabla_{\alpha}M^{\alpha\beta} + M^{\alpha\beta}B_{\alpha}{}^{\gamma}B_{\beta\gamma}\right) + p^{3} = 0 \end{aligned} \tag{8.35}$$

and the boundary conditions:

$$\begin{aligned}
&\mathsf{N}^{\nu\gamma} - \left(q^\gamma - m^\nu B_t^{\ \gamma} + m^t B_\nu^{\ \gamma}\right) = 0 \\
&\mathsf{N}^{\nu\beta}\left(B_\beta^{\ \mu}\xi_\mu + \partial_\beta\zeta\right) - \nabla_\beta \mathsf{M}^{\beta\nu} - \left(q^3 + \frac{\partial m^\nu}{\partial s}\right) - \frac{\partial \mathsf{M}^{\nu t}}{\partial s} = 0 \\
&\mathsf{M}^{\nu\nu} - m^t = 0
\end{aligned} \tag{8.36}$$

Equations (8.35) and (8.36) are equivalent to (8.30) and (8.33), in second-order approximation; they can lead to slightly different results. The difference between the two systems of equations concerns the term $B \cdot B \cdot \mathsf{M}$, which occurs with different signs in the third equation, due to different deformation assumptions. This term, which does not intervene in linearised approximation (cf. §8.1.2.4), is thus related to the assumed kinematics; it is usually negligible. If this should not be the case, this would mean that the assumptions are not suitable and the problem should therefore be solved without simplifying assumptions.

The results obtained in the different approximations are given in the following paragraphs, without the demonstrations, these being similar to that established above.

8.1.2.4 Linearised approximation

In linearised approximation, displacements are assumed of the first order, without distinction between components. The equilibrium equations are then expressed on the reference position and are obtained by neglecting the displacement–dependent terms in (8.30) and (8.33):

- Local equations:

$$\boxed{\begin{aligned}
&\nabla_\alpha \mathsf{N}^{\alpha\gamma} + B_\beta^{\ \gamma}\nabla_\alpha \mathsf{M}^{\alpha\beta} + p^\gamma = 0 \\
&\mathsf{N}^{\alpha\beta}B_{\alpha\beta} - \nabla_\beta\nabla_\alpha \mathsf{M}^{\alpha\beta} + p^3 = 0
\end{aligned}} \tag{8.37}$$

- Boundary conditions:

$$\boxed{\begin{aligned}
&\mathsf{N}^{\nu\gamma} + \mathsf{M}^{\nu\beta}B_\beta^{\ \gamma} - \left(q^\gamma - m^\nu B_t^{\ \gamma} + m^t B_\nu^{\ \gamma}\right) = 0 \\
&\nabla_\beta \mathsf{M}^{\beta\nu} + \left(q^3 + \frac{\partial m^\nu}{\partial s}\right) + \frac{\partial \mathsf{M}^{\nu t}}{\partial s} = 0 \\
&\mathsf{M}^{\nu\nu} - m^t = 0
\end{aligned}} \tag{8.38}$$

8.1.2.5 Love approximation

In Love's kinematics, equilibrium equations reduce to:

- Local equations:

$$\boxed{\begin{aligned}
&\nabla_\alpha \mathsf{N}^{\alpha\gamma} + B_\beta^{\ \gamma}\nabla_\alpha \mathsf{M}^{\alpha\beta} + p^\gamma = 0 \\
&\mathsf{N}^{\alpha\beta}B_{\alpha\beta} - \nabla_\beta\nabla_\alpha \mathsf{M}^{\alpha\beta} + p^3 = 0
\end{aligned}} \tag{8.39}$$

- Boundary conditions:

$$\boxed{\begin{aligned}&\mathsf{N}^{\nu\gamma}+\mathsf{M}^{\nu\beta}B_\beta{}^\gamma-\left(q^\gamma-m^\nu B_t{}^\gamma+m^t B_\nu{}^\gamma\right)=0\\&\nabla_\beta\mathsf{M}^{\beta\nu}+\left(q^3+\frac{\partial m^\nu}{\partial s}\right)+\frac{\partial\mathsf{M}^{\nu t}}{\partial s}=0\\&\mathsf{M}^{\nu\nu}-m^t=0\end{aligned}} \tag{8.40}$$

The expression of the deformations being linear, the Love equations involve only the reference position: the displacement does not intervene and, in particular, there is no coupling due to ζ in the third equation. As a result, equilibrium equations are linear. They are identical to those obtained in linearised approximation.

8.1.2.6 Donnell approximation

The kinematics is defined by the relations (8.18) and (8.19). Equilibrium equations become:

- Local equations:

$$\boxed{\begin{aligned}&\nabla_\alpha\mathsf{N}^{\alpha\gamma}+p^\gamma=0\\&\mathsf{N}^{\alpha\beta}B_{\alpha\beta}+\nabla_\alpha\left(\mathsf{N}^{\alpha\beta}\partial_\beta\zeta\right)-\nabla_\beta\nabla_\alpha\mathsf{M}^{\alpha\beta}+p^3=0\end{aligned}} \tag{8.41}$$

- Boundary conditions:

$$\boxed{\begin{aligned}&\mathsf{N}^{\nu\gamma}-\left(q^\gamma-m^\nu B_t{}^\gamma+m^t B_\nu{}^\gamma\right)=0\\&\mathsf{N}^{\nu\beta}\partial_\beta\zeta-\nabla_\beta\mathsf{M}^{\beta\nu}-\left(q^3+\frac{\partial m^\nu}{\partial s}\right)-\frac{\partial\mathsf{M}^{\nu t}}{\partial s}=0\\&\mathsf{M}^{\nu\nu}-m^t=0\end{aligned}} \tag{8.42}$$

The Donnell approximation has the advantage of the first two very simple equations, which allows the resolution by a stress function in certain cases. The third equation introduces a non-linearity due to the presence of displacement. It is this approximation which has been used in §7.3 to study the influence of the membrane forces on the bending of plates. In particular, it allows an approach to the study of instabilities (see §8.2.2.2), but is mainly applicable to shallow shells.

8.1.2.7 Classical approximation

Since the deformation variables are completely linearised from their Donnell approximation expression, the equilibrium equations are written:

- Local equations:

$$\boxed{\begin{aligned}&\nabla_\alpha\mathsf{N}^{\alpha\gamma}+p^\gamma=0\\&\mathsf{N}^{\alpha\beta}B_{\alpha\beta}-\nabla_\beta\nabla_\alpha\mathsf{M}^{\alpha\beta}+p^3=0\end{aligned}} \tag{8.43}$$

- Boundary conditions:

$$\boxed{\begin{aligned}&\mathsf{N}^{\nu\gamma}-\left(q^{\gamma}-m^{\nu}B_t{}^{\gamma}+m^{t}B_{\nu}{}^{\gamma}\right)=0\\&\nabla_{\beta}\mathsf{M}^{\beta\nu}+\left(q^{3}+\frac{\partial m^{\nu}}{\partial s}\right)+\frac{\partial \mathsf{M}^{\nu t}}{\partial s}=0\\&\mathsf{M}^{\nu\nu}-m^{t}=0\end{aligned}} \tag{8.44}$$

The equations above have all the advantages of the Love and Donnell equations, of which they are an approximation. They are further simplified in membrane theory or in classical thin plate theory. Here, as in the case of small perturbations, which is a special case, the classical approximation makes it possible to express equilibrium on the reference position. But it applies to shallow shells; if this is not the case, the linearised approximation or the Love approximation should be used at first order; they differ only in the kinematics.

8.1.3 Classical resolution methods

8.1.3.1 Displacement method

As is usual in the mechanics of solids, the equations of motion can be solved in the different approximations by taking the displacement field as unknown.

In the Kirchhoff–Love hypotheses, there are 17 unknowns: displacement fields (3), membrane forces and bending moments (6), deformation and variation of curvature of the middle surface (6), and density and temperature. The equations are 17 in number: equations of motion (3), relations between displacement and deformation fields (6), constitutive laws (6), continuity equation (conservation of mass), and equation of heat. The unknowns can be reduced to 5: displacement, density and temperature; the equations are then reduced to 5: equations of motion, continuity equation, and the heat equation, replacing the generalised stresses by the deformations (via the constitutive laws), then by displacements, in the equations of motion.

In the usual cases of isothermal or adiabatic transformations, in small displacements, the unknowns and the equations are reduced to 3: the components of displacement. If the geometry of the shell and the kinematics of its movement allow simplifications, these equations can sometimes be solved by an analytical method. It is, however, much faster in general to solve the problem of displacements by minimising the total potential energy. This is also the method used most often to solve equations numerically. Examples are given in §8.2.

In the Reissner–Mindlin hypotheses, two unknowns are added to the previous ones: the two rotational components of the normal, but two equations are added by application of the PVP.

8.1.3.2 Stress method

This is the method of taking the generalised stresses as unknown (that is, 6 unknowns). This method is a priori less easy to handle, because there are more unknowns and it is necessary to express compatibility conditions via constitutive laws (the equivalent of Beltrami[3] equations for three-dimensional media).

[3] Eugenio Beltrami, Italian mathematician (1835–1900).

However, in the case of the Donnell approximation, it is possible to envisage the use of a stress function for the membrane forces. Indeed, the first equation in Equation (8.41):

$$\nabla_\alpha N^{\alpha\gamma} + p^\gamma = 0 \tag{8.45}$$

are similar to those obtained in plane elasticity, which suggests introducing an Airy function ψ such that:

$$N^{\alpha\beta} = a^{\alpha\beta}\left(\phi + \Delta\psi\right) - a^{\alpha\gamma} a^{\beta\delta} \nabla_\gamma \nabla_\delta \psi \tag{8.46}$$

where ϕ is the surfacic potential of external tangential actions:

$$p = -\text{grad}\ \phi \tag{8.47}$$

That is:

$$p^\alpha = -a^{\alpha\beta}\ \nabla_\beta \phi \tag{8.48}$$

Here, to express the differentiation on the surface and the actions p^α, the geometry of the surface can be assimilated to that of the reference surface, because the displacements in the tangent plane and the deformations are small.

In elasticity, the function ψ introduced in (8.46) makes it possible to respect the compatibility condition of the plane stresses in the tangent plane (the equations obtained in the tangent plane are similar to those obtained in plane elasticity). It is only suitable if the relation (8.45) is satisfied, that is if:

$$a^{\alpha\beta} \nabla_\alpha \left(\nabla^\gamma \nabla_\gamma \psi\right) = \nabla_\alpha \left(\nabla^\alpha \nabla^\beta \psi\right)$$

However, this equality is only possible if the order of the differentiations can be reversed, which is the case only if the surface is developable (see §3.2.2.2).

In conclusion, the Airy function method can only be applied if the tangential displacements are small and if the middle surface is developable. Nevertheless, it provides an approximate solution in the case of shallow shells.

8.2 CIRCULAR CYLINDRICAL THIN SHELLS

8.2.1 Deformations and equilibrium equations

8.2.1.1 Geometry and displacement

The geometry of the cylinder was introduced in 6.2.1. It is a developable surface.

In Figure 8.3, the vector a_3 is directed to the centre and the radius of curvature is positive.

The displacement is decomposed in the orthonormal local coordinate system in the form:

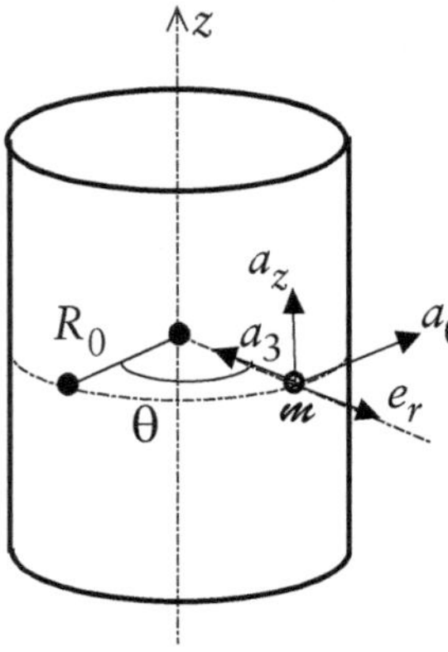

Figure 8.3

Photo 8.1: Grain silos consisting of vertical cylinders

$$\xi = u(z,s)A_z + v(z,s)A_s + w(z,s)A_3 \tag{8.49}$$

8.2.1.2 Calculation of deformations

The strain tensor and the curvature variation tensor are calculated from the expression of the vectors of the natural basis on the deformed surface (§3.3.3.1):

$$\begin{aligned} a_z &= \frac{\partial m}{\partial z} = \left(1+\frac{\partial u}{\partial z}\right)A_z + \frac{\partial v}{\partial z}A_s + \frac{\partial w}{\partial z}A_3 \\ a_s &= \frac{\partial m}{\partial s} = \frac{\partial u}{\partial s}A_z + \left(1+\frac{\partial v}{\partial s}-\frac{w}{R}\right)A_s + \left(\frac{\partial w}{\partial s}+\frac{v}{R}\right)A_3 \end{aligned} \tag{8.50}$$

hence the surface strain tensor:

$$\boxed{\begin{aligned} \varepsilon_{zz} &= \frac{\partial u}{\partial z} + \frac{1}{2}\left[\left(\frac{\partial u}{\partial z}\right)^2 + \left(\frac{\partial v}{\partial z}\right)^2 + \left(\frac{\partial w}{\partial z}\right)^2\right] \\ \varepsilon_{zs} &= \frac{1}{2}\left[\frac{\partial u}{\partial s} + \frac{\partial v}{\partial z} + \frac{\partial u}{\partial s}\frac{\partial u}{\partial z} + \frac{\partial v}{\partial z}\left(\frac{\partial v}{\partial s}-\frac{w}{R}\right) + \frac{\partial w}{\partial z}\left(\frac{\partial w}{\partial s}+\frac{v}{R}\right)\right] \\ \varepsilon_{ss} &= \frac{\partial v}{\partial s} - \frac{w}{R} + \frac{1}{2}\left[\left(\frac{\partial u}{\partial s}\right)^2 + \left(\frac{\partial v}{\partial s}-\frac{w}{R}\right)^2 + \left(\frac{\partial w}{\partial s}+\frac{v}{R}\right)^2\right] \end{aligned}} \tag{8.51}$$

The curvature tensor of the deformed surface is obtained by (3.25). successively:

$$a_z \times a_s = \left[1 + \frac{\partial u}{\partial z} + \frac{\partial v}{\partial s} - \frac{w}{R} - \frac{\partial v}{\partial z}\frac{\partial u}{\partial s} + \frac{\partial u}{\partial z}\left(\frac{\partial v}{\partial s} - \frac{w}{R}\right)\right] A_3$$
$$+\left[-\frac{\partial w}{\partial z} + \frac{\partial v}{\partial z}\left(\frac{\partial w}{\partial s} + \frac{v}{R}\right) - \frac{\partial w}{\partial z}\left(\frac{\partial v}{\partial s} - \frac{w}{R}\right)\right] A_z$$
$$+\left[-\frac{\partial w}{\partial s} - \frac{v}{R} + \frac{\partial w}{\partial z}\frac{\partial u}{\partial s} - \frac{\partial u}{\partial z}\left(\frac{\partial w}{\partial s} + \frac{v}{R}\right)\right] A_s$$

which allows us to calculate the normal vector in the deformed state:

$$a_3 = \frac{a_z \times a_s}{\|a_z \times a_s\|} \quad \text{with}$$

$$\|a_z \times a_s\| = \sqrt{\hat{a}} = \sqrt{\begin{aligned}&\left[1 + \frac{\partial u}{\partial z} + \frac{\partial v}{\partial s} - \frac{w}{R} - \frac{\partial v}{\partial z}\frac{\partial u}{\partial s} + \frac{\partial u}{\partial z}\left(\frac{\partial v}{\partial s} - \frac{w}{R}\right)\right]^2 \\ &+\left[-\frac{\partial w}{\partial z} + \frac{\partial v}{\partial z}\left(\frac{\partial w}{\partial s} + \frac{v}{R}\right) - \frac{\partial w}{\partial z}\left(\frac{\partial v}{\partial s} - \frac{w}{R}\right)\right]^2 \\ &+\left[-\frac{\partial w}{\partial s} - \frac{v}{R} + \frac{\partial w}{\partial z}\frac{\partial u}{\partial s} - \frac{\partial u}{\partial z}\left(\frac{\partial w}{\partial s} + \frac{v}{R}\right)\right]^2\end{aligned}} \tag{8.52}$$

Then:

$$\begin{aligned}\frac{\partial^2 \boldsymbol{m}}{\partial z^2} &= \frac{\partial^2 u}{\partial z^2} A_z + \frac{\partial^2 v}{\partial z^2} A_s + \frac{\partial^2 w}{\partial z^2} A_3 \\ \frac{\partial^2 \boldsymbol{m}}{\partial z \partial s} &= \frac{\partial^2 u}{\partial z \partial s} A_z + \left(\frac{\partial^2 v}{\partial z \partial s} - \frac{1}{R}\frac{\partial w}{\partial z}\right) A_s + \left(\frac{\partial^2 w}{\partial z \partial s} + \frac{1}{R}\frac{\partial v}{\partial z}\right) A_3 \\ \frac{\partial^2 \boldsymbol{m}}{\partial s^2} &= \frac{\partial^2 u}{\partial s^2} A_z + \left(\frac{\partial^2 v}{\partial s^2} - \frac{2}{R}\frac{\partial w}{\partial s} - \frac{v}{R^2}\right) A_s + \left(\frac{1}{R} + \frac{\partial^2 w}{\partial s^2} + \frac{2}{R}\frac{\partial v}{\partial s} - \frac{w}{R^2}\right) A_3\end{aligned} \tag{8.53}$$

From where finally:

$$\kappa_{zz} = \frac{1}{\sqrt{\hat{a}}}\left\{\begin{aligned}&\frac{\partial^2 w}{\partial z^2} + \left[\frac{\partial u}{\partial z} + \frac{\partial v}{\partial s} - \frac{w}{R} - \frac{\partial v}{\partial z}\frac{\partial u}{\partial s} + \frac{\partial u}{\partial z}\left(\frac{\partial v}{\partial s} - \frac{w}{R}\right)\right]\frac{\partial^2 w}{\partial z^2} \\ &+\left[-\frac{\partial w}{\partial z} + \frac{\partial v}{\partial z}\left(\frac{\partial w}{\partial s} + \frac{v}{R}\right) - \frac{\partial w}{\partial z}\left(\frac{\partial v}{\partial s} - \frac{w}{R}\right)\right]\frac{\partial^2 u}{\partial z^2} \\ &+\left[-\frac{\partial w}{\partial s} - \frac{v}{R} + \frac{\partial w}{\partial z}\frac{\partial u}{\partial s} - \frac{\partial u}{\partial z}\left(\frac{\partial w}{\partial s} + \frac{v}{R}\right)\right]\frac{\partial^2 v}{\partial z^2}\end{aligned}\right\}$$

$$\kappa_{zs}=\frac{1}{\sqrt{\hat{a}}}\left\{\begin{aligned}&\frac{\partial^2 w}{\partial z\partial s}+\frac{1}{R}\frac{\partial v}{\partial z}+\left[\frac{\partial u}{\partial z}+\frac{\partial v}{\partial s}-\frac{w}{R}-\frac{\partial v}{\partial z}\frac{\partial u}{\partial s}+\frac{\partial u}{\partial z}\left(\frac{\partial v}{\partial s}-\frac{w}{R}\right)\right]\left(\frac{\partial^2 w}{\partial z\partial s}+\frac{1}{R}\frac{\partial v}{\partial z}\right)\\&+\left[-\frac{\partial w}{\partial z}+\frac{\partial v}{\partial z}\left(\frac{\partial w}{\partial s}+\frac{v}{R}\right)-\frac{\partial w}{\partial z}\left(\frac{\partial v}{\partial s}-\frac{w}{R}\right)\right]\frac{\partial^2 u}{\partial z\partial s}\\&+\left[-\frac{\partial w}{\partial s}-\frac{v}{R}+\frac{\partial w}{\partial z}\frac{\partial u}{\partial s}-\frac{\partial u}{\partial z}\left(\frac{\partial w}{\partial s}+\frac{v}{R}\right)\right]\left(\frac{\partial^2 v}{\partial z\partial s}-\frac{1}{R}\frac{\partial w}{\partial z}\right)\end{aligned}\right\} \tag{8.54}$$

$$\kappa_{ss}=\frac{1}{\sqrt{\hat{a}}}\left\{\begin{aligned}&\frac{1}{R}+\frac{\partial^2 w}{\partial s^2}+\frac{3}{R}\frac{\partial v}{\partial s}-\frac{2w}{R^2}+\frac{1}{R}\frac{\partial u}{\partial z}+\left(\frac{\partial u}{\partial z}+\frac{\partial v}{\partial s}-\frac{w}{R}\right)\left(\frac{\partial^2 w}{\partial s^2}+\frac{2}{R}\frac{\partial v}{\partial s}-\frac{w}{R^2}\right)\\&+\left[-\frac{\partial v}{\partial z}\frac{\partial u}{\partial s}+\frac{\partial u}{\partial z}\left(\frac{\partial v}{\partial s}-\frac{w}{R}\right)\right]\left(\frac{1}{R}+\frac{\partial^2 w}{\partial s^2}+\frac{2}{R}\frac{\partial v}{\partial s}-\frac{w}{R^2}\right)\\&+\left[-\frac{\partial w}{\partial z}+\frac{\partial v}{\partial z}\left(\frac{\partial w}{\partial s}+\frac{v}{R}\right)-\frac{\partial w}{\partial z}\left(\frac{\partial v}{\partial s}-\frac{w}{R}\right)\right]\frac{\partial^2 u}{\partial s^2}\\&+\left[-\frac{\partial w}{\partial s}-\frac{v}{R}+\frac{\partial w}{\partial z}\frac{\partial u}{\partial s}-\frac{\partial u}{\partial z}\left(\frac{\partial w}{\partial s}+\frac{v}{R}\right)\right]\left(\frac{\partial^2 v}{\partial s^2}-\frac{2}{R}\frac{\partial w}{\partial s}-\frac{v}{R^2}\right)\end{aligned}\right\}-\frac{1}{R}$$

The above expressions, applicable to finite transformations, have limited practical interest in solving problems, given their high non-linearity. Nevertheless, they allow the evaluation of second-order approximations for the study of stability. Their development shows that the calculation effort required to obtain them on the basis of the methods given in Chapter 3 is reasonable. It should be noted however that here the geometry and associated differentiation are simple.

For axisymmetric loading such that v is zero (no general torsion around the axis of revolution), the displacement is independent of s and the deformation variables are reduced to:

$$\begin{aligned}&\varepsilon_{zz}=\frac{\partial u}{\partial z}+\frac{1}{2}\left[\left(\frac{\partial u}{\partial z}\right)^2+\left(\frac{\partial w}{\partial z}\right)^2\right]\\&\varepsilon_{zs}=0\\&\varepsilon_{ss}=-\frac{w}{R}+\frac{1}{2}\left(\frac{w}{R}\right)^2\end{aligned} \tag{8.55}$$

$$\begin{aligned}&\kappa_{zz}=\frac{1}{\sqrt{\hat{a}}}\left[\frac{\partial^2 w}{\partial z^2}+\left(\frac{\partial u}{\partial z}-\frac{w}{R}-\frac{w}{R}\frac{\partial u}{\partial z}\right)\frac{\partial^2 w}{\partial z^2}+\left(-\frac{\partial w}{\partial z}+\frac{w}{R}\frac{\partial w}{\partial z}\right)\frac{\partial^2 u}{\partial z^2}\right]\\&\kappa_{zs}=0\\&\kappa_{ss}=\frac{1}{\sqrt{\hat{a}}}\left[\frac{1}{R}-\frac{2w}{R^2}+\frac{1}{R}\frac{\partial u}{\partial z}-\frac{w}{R^2}\left(\frac{\partial u}{\partial z}-\frac{w}{R}\right)-\frac{w}{R}\frac{\partial u}{\partial z}\left(\frac{1}{R}-\frac{w}{R^2}\right)\right]-\frac{1}{R}\end{aligned} \tag{8.56}$$

with:

$$\sqrt{\tilde{a}} = \sqrt{\left(1+\frac{\partial u}{\partial z}-\frac{w}{R}-\frac{w}{R}\frac{\partial u}{\partial z}\right)^2+\left(-\frac{\partial w}{\partial z}+\frac{w}{R}\frac{\partial w}{\partial z}\right)^2} \tag{8.57}$$

8.2.1.3 Hypothesis of small deformations and small rotations

By applying the results of §8.1.1.6, the surface deformations are written:

- At the first order:

$$\boxed{\begin{aligned} \varepsilon_{zz} &= \frac{\partial u}{\partial z} \\ \varepsilon_{zs} &= \frac{1}{2}\left(\frac{\partial u}{\partial s}+\frac{\partial v}{\partial z}\right) \\ \varepsilon_{ss} &= \frac{\partial v}{\partial s}-\frac{w}{R} \end{aligned}} \tag{8.58}$$

$$\boxed{\begin{aligned} \kappa_{zz} &= \frac{\partial^2 w}{\partial z^2} \\ \kappa_{zs} &= \frac{\partial^2 w}{\partial z \partial s}+\frac{1}{R}\frac{\partial v}{\partial z} \\ \kappa_{ss} &= \frac{\partial^2 w}{\partial s^2}+\frac{2}{R}\frac{\partial v}{\partial s}-\frac{w}{R^2} \end{aligned}} \tag{8.59}$$

- At the second order:

$$\boxed{\begin{aligned} \varepsilon_{zz} &= \frac{\partial u}{\partial z}+\frac{1}{2}\left(\frac{\partial w}{\partial z}\right)^2 \\ \varepsilon_{zs} &= \frac{1}{2}\left[\frac{\partial u}{\partial s}+\frac{\partial v}{\partial z}+\frac{\partial w}{\partial z}\left(\frac{\partial w}{\partial s}+\frac{v}{R}\right)\right] \\ \varepsilon_{ss} &= \frac{\partial v}{\partial s}-\frac{w}{R}+\frac{1}{2}\left(\frac{\partial w}{\partial s}+\frac{v}{R}\right)^2 \end{aligned}} \tag{8.60}$$

The application of (8.13) leads to the same expressions (8.59) of the curvature variation. Nevertheless, the alternative expression (8.16) of the variation of curvature also makes it possible to write, in the second order:

$$\boxed{\kappa_{ss} = \frac{\partial^2 w}{\partial s^2}+\frac{w}{R^2}} \tag{8.61}$$

Note that in (8.59) κ_{ss} can be written:

$$\kappa_{ss} = \frac{\partial}{\partial s}\left(\frac{\partial w}{\partial s} + \frac{v}{R}\right) + \frac{1}{R}\left(\frac{\partial v}{\partial s} - \frac{w}{R}\right) \tag{8.62}$$

where the first term is the derivative along s of the rotation of the normal and the second depends directly on the elongation of the circle; the second can be neglected in second-order approximation. The torsion κ_{zs} is the z derivative of the rotation of the normal when following the circle. These are the expressions that should be retained in first-order approximation, but they can be simplified in the case of shallow shells.

The expression (8.50) of the basis vectors makes it possible to introduce the transformation (3.95):

$$\mathcal{A}_z^{\ z} = \frac{\partial u}{\partial z};\ \mathcal{A}_z^{\ s} = \frac{\partial v}{\partial z} \qquad \mathcal{B}_z = \frac{\partial w}{\partial z}$$

$$\mathcal{A}_s^{\ z} = \frac{\partial u}{\partial s};\ \mathcal{A}_s^{\ s} = \frac{\partial v}{\partial s} - \frac{w}{R} \qquad \mathcal{B}_s = \frac{\partial w}{\partial s} + \frac{v}{R}$$

from which:

$$\nabla_z \mathrm{B}_z = \frac{\partial^2 w}{\partial z^2};\ \nabla_s \mathrm{B}_z = \frac{\partial^2 w}{\partial z \partial s};\ \nabla_z \mathrm{B}_s = \frac{\partial^2 w}{\partial z \partial s} + \frac{1}{R}\frac{\partial v}{\partial z};\ \nabla_s \mathrm{B}_s = \frac{\partial^2 w}{\partial s^2} + \frac{1}{R}\frac{\partial v}{\partial s}$$

This allows us to find the linearised expressions (8.58) and (8.59) of the deformations by (3.102) and (3.107).

To compare the first- and second-order approaches, it is necessary to express the consistency of the orders of magnitude, with the following hypotheses, compatible with relations (8.59):

- v, w and their derivatives are of the first order;
- $\frac{\partial u}{\partial z}$, $\frac{\partial u}{\partial s} + \frac{\partial v}{\partial z}$ and $\frac{\partial v}{\partial s} - \frac{w}{R}$ are of the second order, since ε is of second order, $\frac{\partial v}{\partial z}$ is also of the second order, since $\mathcal{A}_z^{\ s}$ is also of second order and can therefore be neglected in κ_{zs}.

An additional simplification is carried out in the case of shallow shells (cf. §8.2.1.4 and 8.2.1.5).

When the displacement is axisymmetric, the generatrices remain in a radial plane and, in this direction, as in a straight beam, the longitudinal displacement u is of an order greater than the transverse displacement w, whereas, in the direction s, as in a ring, the two displacements v and w are of the same order of magnitude.

8.2.1.4 Love approximation

The surface deformations are of the second order, but their linearised expression (8.58) is used, with the symmetric expression (8.59) of the curvature variations.

8.2.1.5 Donnell approximation

In Donnell's approximation to shallow shells, tangential displacements are assumed of the second order. Deformations and variations of curvature obtained by (8.60) and (8.59) are reduced to:

$$\begin{aligned}
&\varepsilon_{zz} = \frac{\partial u}{\partial z} + \frac{1}{2}\left(\frac{\partial w}{\partial z}\right)^2 && \kappa_{zz} = \frac{\partial^2 w}{\partial z^2} \\
&\varepsilon_{zs} = \frac{1}{2}\left[\frac{\partial u}{\partial s} + \frac{\partial v}{\partial z} + \frac{\partial w}{\partial z}\frac{\partial w}{\partial s}\right] && \kappa_{zs} = \frac{\partial^2 w}{\partial z \partial s} \\
&\varepsilon_{ss} = \frac{\partial v}{\partial s} - \frac{w}{R} + \frac{1}{2}\left(\frac{\partial w}{\partial s}\right)^2 && \kappa_{ss} = \frac{\partial^2 w}{\partial s^2}
\end{aligned} \tag{8.63}$$

In practice, this approximation is questionable for a closed cylinder, when it can not be considered shallow, because it does not give a correct result in the direction s, for which it is better to use (8.59) and (8.60). On the other hand, it can be used for the stability analysis of a low curvature cylindrical roofing, cf. §8.2.4.3.

8.2.1.6 Classical approximation

This is a linearised approximation in which the linearised expression (8.58) of the surface deformation and the expression (8.63) of the curvature variation are taken into account. It assumes that the shell is shallow and therefore does not generally apply to a closed cylinder, except of a large radius.

8.2.1.7 Inextensional transformation

In the hypothesis of inextensional flexion introduced by Lord Rayleigh, the three components of the surface strain tensor are zero. It follows in all cases that the Gaussian curvature, zero before transformation, is preserved. This is written, in the deformed cylinder, by (8.63):

$$k = \frac{\partial^2 w}{\partial z^2}\left(\frac{1}{R} + \frac{\partial^2 w}{\partial s^2} + \frac{1}{R}\frac{\partial v}{\partial s}\right) - \left(\frac{\partial^2 w}{\partial s \partial z}\right)^2$$

where only the term of the first order is retained. Then $\frac{\partial^2 w}{\partial z^2} = 0$. By considering the relations (8.58), and expressing that the deformations are zero, it also becomes:

$$\frac{\partial u}{\partial z} = 0\,; \quad \frac{\partial u}{\partial s} + \frac{\partial v}{\partial z} = 0\,; \quad \frac{\partial v}{\partial s} - \frac{w}{R} = 0$$

By integrating these equations:

$$u = f(s)\,; \quad v = -f'(s)z + g(s)\,; \quad w = -R\,f''(s)z + R\,g'(s)$$

These solutions are periodic in s, with a period $2\pi R$.

It is clear that the form of the solution is rather restrictive and that it does not allow to respect any boundary conditions. Thus, it is not possible to fix the cylinder longitudinally ($u = 0$) in the context of this hypothesis. If the cylinder is of length h, it is not possible to fix it transversely at both ends ($w = 0$). Thus, a cylinder supported by stiffening rings spaced from h can not be in inextensional bending.

f and g are decomposed into Fourier series. Considering a term of decomposition, it is clear that u and w, on the one hand, and v, on the other hand, are in quadrature:

$$u = f(s) = A_n \cos\left(n\frac{s}{2\pi R}\right);\; g(s) = B_n \sin\left(n\frac{s}{2\pi R}\right)$$

$$v = \left(A_n \frac{n}{2\pi R} z + B_n\right) \sin\left(n\frac{s}{2\pi R}\right) \tag{8.64}$$

$$w = \left[A_n R\left(\frac{n}{2\pi R}\right)^2 z + B_n \frac{n}{2\pi}\right] \cos\left(n\frac{s}{2\pi R}\right)$$

If the transverse displacement w is fixed at $z = 0$, B_n is zero. Thus, any non-zero transverse displacement can be imposed on the free edge $z = h$, but then the displacement field is entirely determined. On the free edge, the ratio between the transverse displacement w and the other components of displacement is in h/R, which confirms its preponderance in slender cylinders. Figure 8.4 shows the pattern of displacement for $n = 2$ and $n = 4$. Inextensional transformations involve only bending energy, but the shell must have enough free edges for them to occur. This explains the much greater flexibility of open shells compared to closed shells. This greater flexibility gives them greater sensitivity to instabilities. To improve their behaviour from this point of view, the free edges must be stiffened.

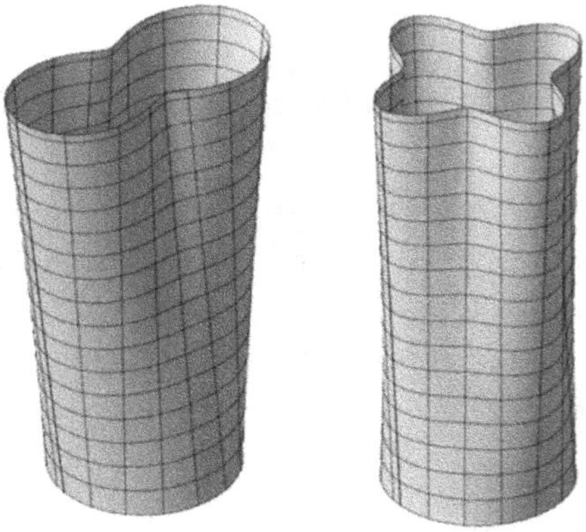

Figure 8.4

This also suggests, in the case of open shells, looking for modes of inextensional vibration, which was Lord Rayleigh's original intention (cf. §8.2.3.2).

8.2.1.8 Equilibrium equations in first-order approximation

It is a question of establishing the equilibrium equations and the boundary conditions within the framework of the hypothesis of small deformations and small rotations, at the first order. For the second order, cf. §8.2.4.1.

At the first order, the deformations are given by (8.58) and (8.59). Two approaches are then possible, depending on whether the terms due to the deformation in the expression (8.62) of the curvature variation are neglected or not.

When the complete expression of κ is considered, relation (8.59) applies. Then, the equilibrium equations and boundary conditions are obtained by particularising (8.24) and (8.27) to the geometry of the cylinder:

- Equilibrium equations:

$$\begin{aligned}
&\partial_z N^{zz} + \partial_s N^{zs} + p^z = 0 \\
&\frac{\partial}{\partial z}\left(N^{zs} + \frac{2M^{zs}}{R}\right) + \frac{\partial}{\partial s}\left(N^{ss} + \frac{2M^{ss}}{R}\right) + p^s = 0 \\
&\partial^2_{z^2} M^{zz} + 2\partial^2_{zs} M^{zs} + \partial^2_{s^2} M^{ss} - \frac{1}{R}\left(N^{ss} + \frac{M^{ss}}{R}\right) + p^3 = 0
\end{aligned} \tag{8.65}$$

- Boundary conditions along two edges in z for a cylinder closed in s:

$$\begin{aligned}
&N^{zz} - q^z = 0 \\
&N^{sz} + \frac{2M^{zs}}{R} - \left(q^s - \frac{m^z}{R}\right) = 0 \\
&\partial_z M^{zz} + 2\partial_s M^{zs} + q^3 + \partial_s m^z = 0 \\
&M^{zz} - m^s = 0
\end{aligned} \tag{8.66}$$

The terms due to deformation in the expression (8.62) can often be neglected, for example in the case of shallow shells where in addition v/R can also be neglected. The deformations are then given by (8.58) and (8.9), keeping the symmetry of κ. The equilibrium equations and the boundary conditions are obtained by eliminating the terms in M/R. It becomes:

- Equilibrium equations:

$$\begin{aligned}
&\partial_z N^{zz} + \partial_s N^{zs} + p^z = 0 \\
&\partial_z N^{sz} + \partial_s N^{ss} + p^s = 0 \\
&\partial^2_{z^2} M^{zz} + 2\partial^2_{zs} M^{zs} + \partial^2_{s^2} M^{ss} - \frac{N^{ss}}{R} + p^3 = 0
\end{aligned} \tag{8.67}$$

- Boundary conditions, only the second condition is modified by deleting 2 M^{zs}/R.

Between the two approaches, the equations and boundary conditions present two differences:

- the first is that $N^{ss} + M^{ss}/R$ (resp. $N^{zs} + M^{zs}/R$) is simplified to N^{ss} (resp. N^{zs}) in (8.65), neglecting the term $B_{\alpha\gamma}\mathcal{A}_\beta{}^\gamma$;
- the second is due to the fact that v/R is neglected, which leads to the deletion of the derivatives of M^{ss}/R and M^{zs}/R in (8.65) and the boundary conditions.

The two hypotheses, therefore, lead to the same consequences, which is consistent.

The Love and Donnell approximations neglect bending moment participation in first-order approximation, which assumes small curvatures, an assumption not necessarily acceptable for closed cylinders; it is therefore necessary to examine the influence of these terms.

The orders of magnitude can be specified by considering the case of a cylinder under internal pressure undergoing a constant displacement w (see also the example of §8.2.3.1). The application of equations (8.67) without taking into account the moment leads to the solution:

$$w = \frac{pR^2}{K} \ ; \ N^{ss} = pR \ ; \ M^{ss} = -D\frac{w}{R^2} = -\upsilon pR^2 \text{ with } \upsilon = \frac{D}{KR^2} = \frac{e^2}{12R^2}$$

The application of the complete equations (8.65) leads to:

$$w = \frac{pR^2}{K}(1-\upsilon)^{-1} \ ; \ N^{ss} = pR(1-\upsilon)^{-1} \ ; \ M^{ss} = -\upsilon(1-\upsilon)^{-1} pR^2$$

In the case of a thin shell, υ is negligible compared to 1 and the two solutions are equivalent. Also, the condition to neglect it does not depend only on the curvature, but also on thickness, and, in practice, taking into account the bending moments in the equations is to be considered for the thick shells of small radius.

The resolution of (8.67) by neglecting M^{ss}/R and M^{sz}/R compared to the corresponding membrane forces makes it possible to evaluate the value of these moments and to thus judge their importance.

8.2.1.9 Constitutive laws

The constitutive law of bending for a Hooke shell is written, with the assumptions taken:

$$\begin{aligned}
N^{zz} &= N_0^{zz} + K\left[\frac{\partial u}{\partial z} + \frac{1}{2}\left(\frac{\partial w}{\partial z}\right)^2 + \nu\left(\frac{\partial v}{\partial s} - \frac{w}{R} + \frac{1}{2}\left(\frac{\partial w}{\partial s} + \frac{v}{R}\right)^2\right)\right] \\
N^{zs} &= N_0^{zs} + \frac{K(1-\nu)}{2}\left[\frac{\partial u}{\partial s} + \frac{\partial v}{\partial z} + \frac{\partial w}{\partial z}\left(\frac{\partial w}{\partial s} + \frac{v}{R}\right)\right] \\
N^{ss} &= N_0^{ss} + K\left[\frac{\partial v}{\partial s} - \frac{w}{R} + \frac{1}{2}\left(\frac{\partial w}{\partial s} + \frac{v}{R}\right)^2 + \nu\left(\frac{\partial u}{\partial z} + \frac{1}{2}\left(\frac{\partial w}{\partial z}\right)^2\right)\right]
\end{aligned} \tag{8.68}$$

In first-order approximation, only the linear terms are to be retained in (8.68).

$$\begin{aligned}
M^{zz} &= M_0^{zz} + D\left[\frac{\partial^2 w}{\partial z^2} + \nu\,\kappa_{ss}\right] \\
M^{zs} &= M_0^{zs} + D(1-\nu)\,\kappa_{sz} \\
M^{ss} &= M_0^{ss} + D\left(\kappa_{ss} + \nu\frac{\partial^2 w}{\partial z^2}\right)
\end{aligned} \tag{8.69}$$

κ_{ss} is given by (8.59), but can also be expressed by (8.61) at the second order. κ_{sz} is given by (8.59) at the first order and holds $\frac{\partial^2 w}{\partial z \partial s}$ at the second order.

In a second-order approximation, to express the third equation of (8.65) or (8.67) in the function of the displacement, it is useful to note the following expression obtained by transferring the constitutive laws (8.69) (excluding prestresses):

$$\frac{\partial^2 M^{zz}}{\partial z^2} + 2\frac{\partial^2 M^{zs}}{\partial s \partial z} + \frac{\partial^2 M^{ss}}{\partial s^2} = D\left[\frac{\partial^4 w}{\partial z^4} + 2\frac{\partial^4 w}{\partial z^2 \partial s^2} + \frac{\partial^4 w}{\partial s^4} + \left(\nu\frac{\partial^2}{\partial z^2} + \frac{\partial^2}{\partial s^2}\right)\left(\frac{w}{R^2}\right)\right] \tag{8.70}$$

Note that the last term of (8.70) is equal to $\frac{M^{ss}}{R^2}$.

8.2.1.10 Form of solutions for closed cylinders

The equations expressed in displacement are obtained by replacing the generalised stresses in the equilibrium equations (§8.2.1.8) with the constitutive laws (§8.2.1.9). The form of the solutions has already been discussed in §8.2.1.7. The solutions are functions of the dimensionless variables $\chi = \frac{z}{h}$ and $\frac{s}{R}$ for a cylinder of radius R and of finite length h, or of z and $\frac{s}{R}$ for a cylinder without edge. For cylinders closed in the circumferential direction (tubes), the displacement and the stresses are periodic with period $2\pi R$, so can be decomposed into Fourier series in s, for example for the transverse displacement w:

$$w(z,s) = A(z) + \sum_{m=1}^{\infty} f_m(\chi)\sin\frac{ms}{R} + \sum_{m=1}^{\infty} g_m(\chi)\cos\frac{ms}{R} \tag{8.71}$$

The functions f_m and g_m have the dimensions of a length and are generally considered first order. They are determined on the one hand by deferring these expressions in the equilibrium equations, on the other hand by expressing the boundary conditions on the edges in z. The other displacement components can be decomposed in the same way.

Moreover, within the framework of the second-order approximation, the displacements are linked by the first-order equalities: $\frac{\partial v}{\partial s} - \frac{w}{R} = 0$ and $\frac{\partial u}{\partial s} + \frac{\partial v}{\partial z} = 0$, which lead to:

$$\begin{aligned} v(z,s) &= B(z) - \sum_{m=1}^{\infty} \frac{f_m(\chi)}{m}\cos\frac{ms}{R} + \sum_{m=1}^{\infty} \frac{g_m(\chi)}{m}\sin\frac{ms}{R} \\ u(z,s) &= C(z) + \frac{R}{h}\sum_{m=1}^{\infty} \frac{f'_m(\chi)}{m^2}\sin\frac{ms}{R} + \frac{R}{h}\sum_{m=1}^{\infty} \frac{g'_m(\chi)}{m^2}\cos\frac{ms}{R} \end{aligned} \tag{8.72}$$

It is necessary to block the cylinder so that it is not free in space, therefore to block six degrees of freedom, for example, the displacement of two distinct points, which leads to conditions on the functions A, B, and C which, here, can be taken as null without affecting the generality.

κ_{ss} is given by (8.53) in which, for a mode of order m, for example: $w_m = f_m(\chi)\sin\frac{ms}{R}$.

So $\frac{\partial^2 w_m}{\partial s^2} = -\frac{m^2}{R^2} w_m$. This term is predominant for high-order modes compared to $\frac{w_m}{R^2}$ and the latter can be neglected, but it is not the case for low-order modes, in particular the first one. So, in general, $\frac{w}{R^2}$ or $\frac{v}{R}$ should not be neglected with respect to $\frac{\partial^2 w}{\partial s^2}$ in (8.53), in second-order approximation for a closed cylinder.

In conclusion, the shape of the displacements given by (8.72) therefore shows that, for the modes with a high circumferential wave number m, the radial displacement w is preponderant and the variation of circumferential curvature can be taken equal to $\frac{\partial^2 w}{\partial s^2}$. This is not the case for a low wave number.

8.2.2 Resolution methods

The equations below are established within the framework of the Donnell approximation applicable in particular to shallow shells. The classical approximation is a simplification usable to the first order.

8.2.2.1 Equilibrium equations

In Donnell's approximation, the local equations (8.41) are reduced to:

$$\begin{aligned} &\partial_\alpha N^{\alpha\gamma} + p^\gamma = 0 \\ &\frac{N^{22}}{R} + \partial_\alpha\left(N^{\alpha\beta}\partial_\beta w\right) - \partial_\beta\partial_\alpha M^{\alpha\beta} + p^3 = 0 \end{aligned} \tag{8.73}$$

where the index *1* is relative to z and the index *2* to s. In classical approximation, the term $\partial_\alpha\left(N^{\alpha\beta}\partial_\beta w\right)$ is not taken into account.

8.2.2.2 Airy function method

Given Donnell's approximation, there is an Airy function defined by (8.46), which is reduced in the coordinate system (a_z, a_s), to:

$$\begin{aligned} &N^{11} = \phi + \partial^2_{22}\psi \\ &N^{22} = \phi + \partial^2_{11}\psi \\ &N^{12} = -\partial^2_{12}\psi \end{aligned} \tag{8.74}$$

The transverse equilibrium (8.73) is then written, the terms in ψ eliminating themselves:

$$\frac{N^{22}}{R} + a^{\alpha\beta}\ \partial_\alpha\phi\ \partial_\beta w + N^{\alpha\beta}\partial^2_{\alpha\beta} w - \partial^2_{\alpha\beta} M^{\alpha\beta} = -p^3 \tag{8.75}$$

Otherwise, after (8.19):

$$\kappa_{\alpha\beta} = \partial_\alpha\partial_\beta w \tag{8.76}$$

In the case where the shell is linearly elastic, homogeneous and isotropic:

$$M = D\left[(1-\nu)\,\kappa + \nu\,\mathrm{TR}(\kappa)\,a\right]$$

It becomes, as in the case of plates:

$$\partial_\alpha \partial_\beta M^{\alpha\beta} = D\Delta\Delta w \tag{8.77}$$

Given this relationship, by replacing the membrane forces by their expression (8.74) and by posing $p^3 = p$ (p is positive when directed towards the axis of the cylinder):

$$\boxed{\begin{aligned}&\frac{1}{R}\left(\partial^2_{11}\psi + \phi\right) + a^{\alpha\beta}\partial_\alpha\phi\,\partial_\beta w + \phi\,\Delta w - D\Delta\Delta w\\ &\qquad + \partial^2_{11}\psi\bullet\partial^2_{22}w + \partial^2_{22}\psi\bullet\partial^2_{11}w - 2\partial^2_{12}\psi\bullet\partial^2_{12}w + p = 0\end{aligned}} \tag{8.78}$$

The deformations are given by the relations (8.63). The elimination of u and v between these relations makes it possible to obtain the relation of compatibility (see also §3,3.4):

$$\begin{aligned}&2\left(\partial^2_{22}\varepsilon_{11} + \partial^2_{11}\varepsilon_{22} - 2\partial^2_{12}\varepsilon_{12}\right)\\ &\qquad = \partial^2_{11}\left[-2\frac{w}{R} + \left(\partial_2 w\right)^2\right] + \partial^2_{22}\left[\left(\partial_1 w\right)^2\right] - 2\partial^2_{12}\left(\partial_1 w\cdot\partial_2 w\right)\end{aligned} \tag{8.79}$$

and, by the constitutive law:

$$\varepsilon = \frac{1}{Ee}\left[(1+\nu)\,N - \nu\,\mathrm{TR}(N)\,a\right]$$

it becomes:

$$\begin{aligned}&\partial^2_{22}\varepsilon_{11} + \partial^2_{11}\varepsilon_{22} - 2\partial^2_{12}\varepsilon_{12}\\ &\qquad = \frac{1}{Ee}\left[\partial^2_{22}\left(N^{11} - \nu\,N^{22}\right) + \partial^2_{11}\left(N^{22} - \nu\,N^{11}\right) - 2(1+\nu)\,\partial^2_{12}N^{12}\right]\end{aligned}$$

either, by use of (8.74):

$$\partial^2_{22}\varepsilon_{11} + \partial^2_{11}\varepsilon_{22} - 2\partial^2_{12}\varepsilon_{12} = \frac{1}{Ee}\left[(1-\nu)\Delta\varphi + \Delta\Delta\psi\right] \tag{8.80}$$

Combining (8.72) and (8.73), it becomes:

$$\boxed{\frac{1}{Ee}\left[(1-\nu)\Delta\phi + \Delta\Delta\psi\right] = -\partial^2_{11}\left(\frac{w}{R}\right) - \partial^2_{11}w\bullet\partial^2_{22}w + \left(\partial^2_{12}w\right)^2} \tag{8.81}$$

which is the equivalent of the Beltrami equations obtained in elasticity. The problem is thus completely governed by Equations (8.78) and (8.81) in ψ and w. This system is not linear. It must be associated with the boundary conditions (8.42) obtained in the Donnell approximation.

When the membrane forces are due mainly to external loads (first-order) and their variations due to the change of Gaussian curvature (which is of the second order), they can be neglected vis-à-vis the principal terms and in the current case where ϕ is zero, Equations (8.78) and (8.81) are rewritten:

$$D\Delta\Delta w - \frac{1}{R}\partial^2_{11}\psi - N^{11}\,\partial^2_{11}w - N^{22}\,\partial^2_{22}w - 2N^{12}\,\partial^2_{12}w = p$$
$$\frac{1}{Ee}\Delta\Delta\psi = -\partial^2_{11}\left(\frac{w}{R}\right) - \partial^2_{11}w \bullet \partial^2_{22}w + \left(\partial^2_{12}w\right)^2 \tag{8.82}$$

By eliminating ψ between the two equations, it becomes:

$$\boxed{D\Delta^4 w - \Delta^2\left(N^{\alpha\beta}\,\partial^2_{\alpha\beta}w\right) + \frac{Ee}{R^2}\partial^4_{1111}w = \Delta^2 p} \tag{8.83}$$

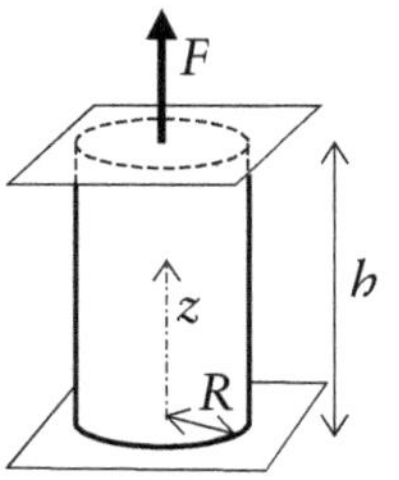

Figure 8.5

This is the Donnell equation (1933), established to study the stability of shallow cylindrical shells. This equation can not be used in cases where the membrane forces are due solely to bending, because then the change of Gaussian curvature can no longer be neglected in (8.81).

Example: Axisymmetric buckling of a circular cylinder.

A cylindrical shell undergoes an axial force F, which is distributed uniformly around its circumference by means of two rigid plates (Figure 8.5). The boundary conditions are such that the radial displacements are blocked at the bottoms. The potential ϕ of external surface actions is zero. At equilibrium, the membrane forces are:

$$N^{11} = \frac{F}{2\pi R} = q$$
$$N^{22} = N^{12} = 0$$

An axisymmetric buckling is characterised by a radial displacement $w(z)$. As this is a longitudinal bending, so, in the direction of zero curvature, the Donnell approximation can be used. Here, it is preferable to introduce the Airy function ψ associated with the variation of the membrane forces, such as:

$$N^{11} = q + \partial^2_{22}\psi \;;\; N^{22} = \partial^2_{11}\psi$$

The symmetry of revolution implies that:

$\partial^2_{12}\psi = \partial^3_{222}\psi = 0$ and $\partial_2 w = 0$

and $\Delta\Delta\psi$ reduces to $\partial^4_{1111}\psi$ which, according to (8.81), is equal to $-\frac{Ee}{R}\partial^2_{11}w$. By double integration, taking into account that N^{22} does not depend on θ:

$$\partial^2_{11}\psi = -\frac{Ee}{R}w$$

The first equation of (8.68) is then reduced, to the second order:

$$D\frac{d^4w}{dz^4}-q\frac{d^2w}{dz^2}+\frac{Ee}{R^2}w=0 \tag{8.84}$$

At both edges, the boundary conditions are reduced to $w = 0$ and $w" = 0$ (the moment being nul, given the first condition).

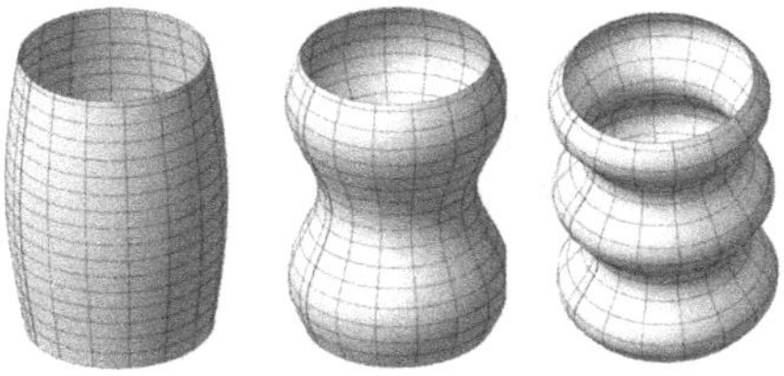

Figure 8.6

Functions such that sin αz are solutions of the equation, provided that sin $\alpha h = 0$, that is, $\alpha h = n\pi$. By reporting to equation (8.84), it becomes :

$$q\alpha^2=-\left(D\alpha^4+\frac{Ee}{R^2}\right)$$

so q takes the discrete values:

$$q_n=-\left(D\frac{n^2\pi^2}{h^2}+\frac{Ee}{R^2}\frac{h^2}{n^2\pi^2}\right) \tag{8.85}$$

for which there is a non-zero deformation, that is to say, a mode of buckling in $\sin\frac{n\pi}{h}z$ (Figure 8.6). Instability occurs for the smallest absolute value of q_n (which is compression in all cases). It is q_1 when:

$$\frac{h^2}{eR}\leq\frac{\pi}{\sqrt{12\left(1-\nu^2\right)}}\ (\approx 1)$$

For larger values of this ratio, a lower bound value of $|q_{cr}|$ is given by $\dfrac{Ee^2}{R\sqrt{3\left(1-\nu^2\right)}}$.

8.2.3 Some classical solutions

8.2.3.1 Cylinder in axisymmetric bending

Here, the membrane forces are due to bending and the effect of prestressing is negligible. In axisymmetric bending, the assumptions of the classical approximation can be retained, the bending occurring in the direction of the null curvature; Equations (8.78) and (8.81) are linearised and are reduced to:

$$-\frac{1}{R}\left(\partial_{11}^2\psi+\phi\right)+D\Delta\Delta w-p=0 \tag{8.86}$$

$$\frac{1}{Ee}\left[(1-\nu)\Delta\phi+\Delta\Delta\psi\right]=-\partial_{11}^2\left(\frac{w}{R}\right) \tag{8.87}$$

In the current case where ϕ is zero, the elimination of ψ leads to:

$$D\Delta^4 w+\frac{Ee}{R^2}\partial_{1111}^4 w=\Delta\Delta p$$

or:

$$\Delta^4 w+4\lambda^4\partial_{1111}^4 w=\frac{1}{D}\Delta\Delta p \tag{8.88}$$

with:

$$\boxed{\frac{1}{\lambda^4}=\frac{R^2e^2}{3\left(1-\nu^2\right)}} \tag{8.89}$$

The equation fits into:

$$w^{\mathrm{IV}}+4\lambda^4 w=\frac{p}{D} \tag{8.90}$$

which general solution is, in the case where p is constant:

$$w=\frac{pR^2}{Ee}+e^{\lambda z}\left(A_1\sin\lambda z+B_1\cos\lambda z\right)+e^{-\lambda z}\left(A_2\sin\lambda z+B_2\cos\lambda z\right) \tag{8.91}$$

If p is not constant, simply change the first term, which is a particular solution of (8.90).

Example: pipe under pressure.

This approach makes it possible to find solutions in the case of pipes containing a fluid under pressure and subjected to conditions respecting the axisymmetry. If, for example, such a pipe is clamped in a stiff ring (Figure 8.7):

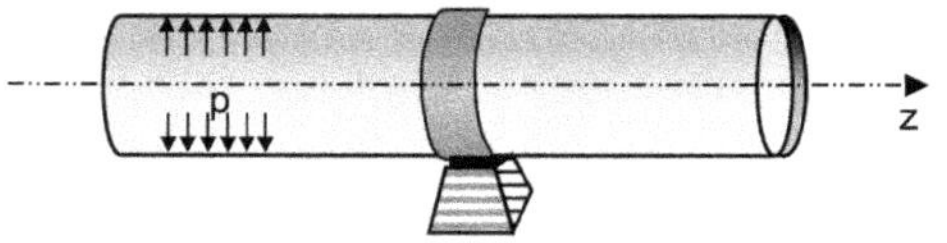

Figure 8.7

- at the embedment: $w(0) = w'(0) = 0$,
- far from embedding, the displacement must remain finite, so $A_1 = B_1 = 0$.

The solution is written in this case:

$$w=\frac{pR^2}{Ee}\left[1-e^{-\lambda z}\left(\sin\lambda z+\cos\lambda z\right)\right] \tag{8.92}$$

When z tends to infinity, w tends to $\frac{pR^2}{Ee}$, which is the solution obtained in membrane theory when $N^{zz} = 0$. The singularity introduced by the ring decreases rapidly while moving away, the wavelength of the oscillation and its decay being related to $1/\lambda$, that is to say to the characteristic length $\sqrt{\text{Re}}$.

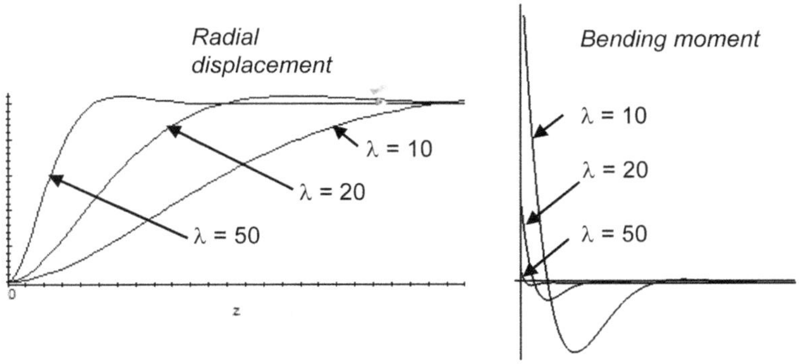

Figure 8.8

Bending therefore exists only in the vicinity of the singularity; the membrane theory gives a satisfactory solution at a distance significantly greater than the characteristic length.

Figure 8.8 shows the evolution of the radial displacement and the bending moment as a function of z, for different values of λ. For a pipe of a given radius, the bending wavelength and the amplitude of the moment decrease when λ increases, ie when the thickness decreases.

Note: The energy approach makes it possible to compare in a simple way the classical solution (of linearised Donnell's equations) with the solution obtained by directly linearising the deformations. In axisymmetric bending and in linearised approximation, the linearised deformations are given by:

$$\varepsilon_{22} = -\frac{w}{R}\,;\ \kappa_{11} = w''\,;\ \kappa_{22} = -\frac{w}{R^2}$$

If the cylinder is free in the longitudinal direction, N^{zz} is zero and $\varepsilon_{11} = -\nu\,\varepsilon_{22}$. So:

$$\mathcal{F} = \frac{1}{2}2\pi R\int_0^{\ell}\left\{Ee\left(\frac{w}{R}\right)^2 + D\left[\left(w''\right)^2 + \left(\frac{w}{R^2}\right)^2 - 2\nu\, w''\frac{w}{R^2}\right]\right\}dz$$

The virtual power of deformation is written:

$$\overset{*}{\mathcal{F}} = 2\pi R\int_0^{\ell}\left\{Ee\left(\frac{w}{R}\right)\left(\frac{\overset{*}{w}}{R}\right) + D\left[w''\,\overset{*}{w}'' + \left(\frac{w}{R^2}\right)\left(\frac{\overset{*}{w}}{R^2}\right) - \nu\left(\overset{*}{w}''\frac{w}{R^2} + w''\frac{\overset{*}{w}}{R^2}\right)\right]\right\}dz$$

and the virtual power of pressure:

$$\overset{*}{\mathcal{W}}_e = 2\pi R\int_0^\ell p\overset{*}{w}\,dz$$

The application of the PVP leads, after the necessary integrations by parts, to the local equation:

$$D\left(\frac{d^4w}{dz^4} - \frac{2\nu}{R^2}\frac{d^2w}{dz^2} + \frac{w}{R^4}\right) + \frac{Ee}{R^2}w = p \tag{8.93}$$

to be compared to (8.90).

$\frac{D}{R^4}$ is negligible vis-a-vis $\frac{Ee}{R^2}$ in the hypothesis of thin shells. Using the solution (8.92), $\frac{Ee}{R^2}w$ is of the order of p, whereas $D\frac{w''}{R^2}$ is of the order of $D\frac{p\lambda^2}{Ee}$, so of $p\frac{e}{R}$, which justifies to neglect it also. The equation obtained is thus reduced to Equation (8.88), which amounts to neglecting $\frac{w}{R^2}$ with respect to w'' in the bending energy (Donnell approximation, justified in the longitudinal direction, cf. §8.2.1.10).

It should also be noted that Equation (8.90) is analogous to the equation of a beam with continuous elastic supports. The elastic supports are here formed by "rings" working in extension and opposing the longitudinal flexion. In the approximation leading to (8.90), the cylinder behaves like an orthogonal network of longitudinal lamellae and rings.

If the pipe is blocked in the longitudinal direction, ε_{11} is zero and $N^{11} = \nu N^{22}$. Membrane energy is then reduced to:

$$\mathcal{F}_m = \frac{1}{2}2\pi R\int_\sigma\left[K\left(\frac{w}{R}\right)^2\right]dz$$

and Equation (8.83) remains applicable, but with λ defined by:

$$\frac{1}{\lambda^4} = \frac{R^2e^2}{3}$$

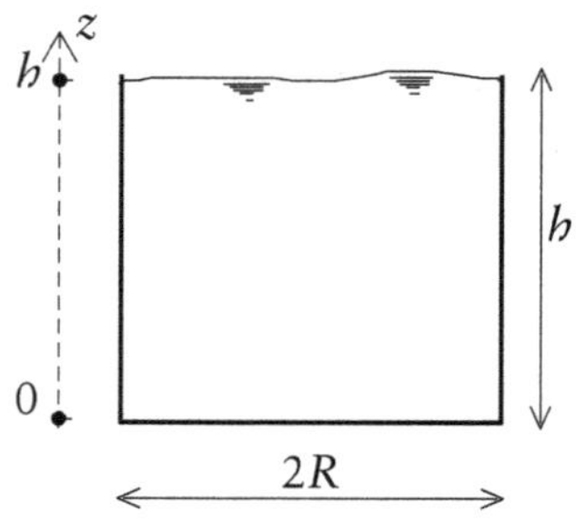

Figure 8.9

Example: Cylindrical wall of a tank with a rigid bottom.

The tank is cylindrical with radius R and height h (Figure 8.9). It is filled with fluid to the brim. The problem has been addressed in membrane theory in 6.2.1.5. The solution obtained then constitutes a particular solution of the bending Equation (8.90), therefore:

$$w = -\frac{\rho g R^2}{Ee}(h-z) + e^{\lambda z}(A_1\sin\lambda z + B_1\cos\lambda z) + e^{-\lambda z}(A_2\sin\lambda z + B_2\cos\lambda z)$$

The top edge $z = h$ is a free edge where (in accordance with Donnell's approximation): $M^{zz} = D\frac{\partial^2 w}{\partial z^2} = 0$; $q^3 = D\frac{\partial^3 w}{\partial z^3} = 0$, resulting in:

$$e^{\lambda h}\left(-B_1\sin\lambda h + A_1\cos\lambda h\right) + e^{-\lambda h}\left(B_2\sin\lambda h - A_2\cos\lambda h\right) = 0$$

$$e^{\lambda h}\left(-\left(A_1 + B_1\right)\sin\lambda h + \left(A_1 - B_1\right)\cos\lambda h\right)$$
$$+ e^{-\lambda h}\left(\left(A_2 - B_2\right)\sin\lambda h + \left(A_2 + B_2\right)\cos\lambda h\right) = 0$$

The bottom being sufficiently stiff in its plane, the radial displacement is assumed to be zero in $z = 0$, either:

$$-\frac{\rho g R^2 h}{Ee} + B_1 + B_2 = 0$$

The last condition depends on how the wall is tied to the bottom. In the case where the bottom is very stiff, the wall can be assumed clamped in it and the rotation is zero:

$$\frac{dw}{dz}(0) = \frac{\rho g R^2}{Ee} + \lambda\left(A_1 + B_1 + A_2 - B_2\right) = 0$$

The four conditions above make it possible to determine the four coefficients and the solution is entirely determined. This solution can usefully be put in adimensional form w/h according to the adimensional parameter:

$$\Lambda = \lambda h = \frac{h}{\sqrt{Re}}\sqrt[4]{3\left(1-\nu^2\right)} \approx 1{,}3\frac{h}{\sqrt{Re}}$$

which depends essentially on the geometry, and the dimensionless parameter of loading $P = \frac{\rho g R^2}{Ee}$. It can therefore be expressed as the ratio ω of displacement to loading $\omega = \frac{w}{Ph} = \frac{Eew}{\rho g R^2 h}$ as a function of $\zeta = \frac{z}{h}$ and of the parameter Λ:

$$\omega = -\left(1-\zeta\right) + e^{\Lambda\zeta}\left(\alpha_1\sin\Lambda\zeta + \beta_1\cos\Lambda\zeta\right) + e^{-\Lambda\zeta}\left(\alpha_2\sin\Lambda\zeta + \beta_2\cos\Lambda\zeta\right) \qquad \text{(i)}$$

The bending moment is also put in an adimensional form $\mu = \frac{M^{zz}h}{DP} = \frac{d^2\omega}{d\zeta^2}$, with:

$$\frac{d^2\omega}{d\zeta^2} = 2\Lambda^2 e^{\Lambda\zeta}\left(-\beta_1\sin\Lambda\zeta + \alpha_1\cos\Lambda\zeta\right) + 2\Lambda^2 e^{-\Lambda\zeta}\left(\beta_2\sin\Lambda\zeta - \alpha_2\cos\Lambda\zeta\right) \qquad \text{(ii)}$$

Three adimensional constants are expressed as functions of the fourth β_1 by the first three boundary conditions (unless $\cos\Lambda = 0$, in which case the constants are expressed differently):

$$\beta_2 = -\beta_1 + 1$$

$$\alpha_1 = \frac{1}{2e^{2\Lambda}\cos^2\Lambda}\left\{\beta_1\left[1 + e^{2\Lambda}\left(1 + 2\sin\Lambda\cos\Lambda\right)\right] - 1\right\} \qquad \text{(iii)}$$

$$\alpha_2 = \frac{1}{2\cos^2\Lambda}\left\{\beta_1\left(1 + e^{2\Lambda} - 2\cos\Lambda\sin\Lambda\right) - 1 + 2\sin\Lambda\cos\Lambda\right\}$$

The adimensional rotation in $\zeta = 0$ is given by:

$$\frac{d\omega}{d\zeta}(\zeta = 0) = 1 + \Lambda(\alpha_1 + \beta_1 + \alpha_2 - \beta_2) \qquad \text{(iv)}$$

The embedding condition of the wall in the bottom is then written:

$$1 + \Lambda(\alpha_1 + \beta_1 + \alpha_2 - \beta_2) = 0$$

and allows to calculate β_1:

$$\beta_1 = \frac{1 + e^{-2\Lambda} - \dfrac{2}{\Lambda}\cos^2\Lambda + 2(\cos\Lambda - \sin\Lambda)\cos\Lambda}{2 + 4\cos^2\Lambda + e^{-2\Lambda} + e^{2\Lambda}}$$

then the other coefficients by (iii).

Photo 8.2 : Cylindrical tanks in a farm.

When Λ goes to infinity, $\beta_1 e^{\Lambda}$ and $\alpha_1 e^{\Lambda}$ tend to 0, β_2 and α_2 to 1. In practice, the limit values can be considered as reached for values of Λ greater than 25, which is generally the case for metal tanks. Then the solution is similar to (8.92), that is to say, it is not

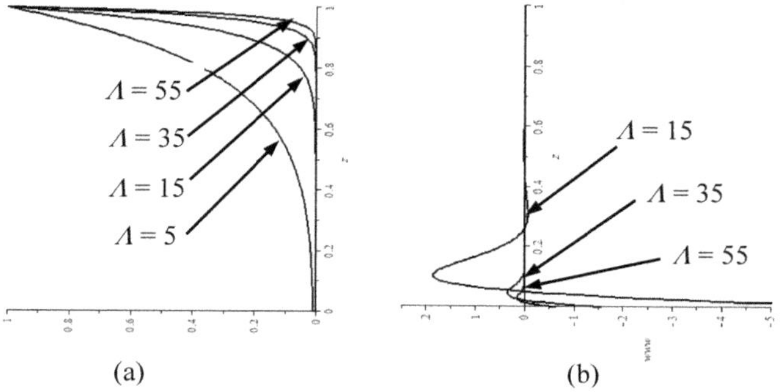

Figure 8.10

influenced by the upper edge: **bending remains localised at the bottom of the tank.** In the case of thicker concrete tanks, such values of Λ may not be reached and the exact calculation must be made.

Displacement and dimensionless moment are given in Figure 8.10 as functions of ζ and for four (or three) values of Λ. On the curves (a), the displacements are normalised with respect to their maximum value at the free end; it increases rapidly with Λ, so with the flexibility of the wall. The moments are given by the curves (b). The moments at embedment and the length of the flexion wave decrease when Λ increases. For a given reservoir volume, Λ increases when the thickness decreases.

The moment at wall embedment is proportional to $\frac{D\Lambda^2}{Ee}$, so to e. In elasticity, the stresses are proportional to $\frac{M}{e^2}$, therefore inversely proportional to e. Thus, in this situation, an increase in thickness is much less effective in reducing stress than in the case of a beam, for example. However, this conclusion is only partial, since it is also necessary to consider the vertical force due to the weight of the higher structures.

Example: Cylindrical tank with a deformable bottom.

In this example, the wall and the bottom are made of the same material, but may have different thicknesses, respectively e and e'. The bottom is supposed to rest on a rigid floor or raft. It is a question of connecting the solution established above for the wall with that obtained in §7.2.1.7 for the circular bottom, in which m denotes the moment per unit length applied to the periphery of the bottom.

For the wall, the assumption of rigidity of the bottom is no longer retained, it is appropriate to express continuity with the bottom. This transmits a linear moment m to the cylinder, which provides the condition:

$$M^{zz}(0) = -m = D\frac{d^2w}{dz^2}(z=0) = 2\lambda^2(A_1 - A_2)$$

which replaces the embedding condition and makes it possible to determine the coefficients as functions of m. In adimensional form, with the notation of the previous example:

$$\mu_c = \frac{mh}{DP} = -\frac{d^2\omega}{d\zeta^2}(\zeta=0) = -2\Lambda^2(\alpha_1 - \alpha_2) \tag{v}$$

from which:

$$\beta_1 = \frac{1}{1+4e^{2\Lambda}\sin\Lambda\cos\Lambda - e^{4\Lambda}}\left[-\mu_c\frac{e^{2\Lambda}\cos^2\Lambda}{\Lambda^2} + 1 - e^{2\Lambda} + 2e^{2\Lambda}\sin\Lambda\cos\Lambda\right] \tag{vi}$$

The solution is entirely determined in the wall by (iii) and (iv) as a function of m. Displacements and dimensionless moments are deduced by (i) and (ii).

For the bottom, according to §7.2.1.7, by introducing the shape parameter $\Upsilon = h/R$, the dimensionless coordinate $\tau = r/R$ and the thickness ratio $\varepsilon = e/e'$, the vertical displacement is written in the adimensional form:

$$\Omega(\tau) = \frac{w(r)}{PR} = -\frac{\Lambda^4\varepsilon^3}{16\Upsilon^3}\tau^4 + \alpha\left(\frac{\tau^2}{4}\ln\tau - \frac{\tau^2}{4}\right) + \beta\tau^2 + \gamma\ln\tau + \varphi \tag{a}$$

with:

$$\alpha = \frac{AR}{P};\ \beta = \frac{BR}{P};\ \gamma = \frac{C}{PR};\ \varphi = \left(\frac{\alpha}{4} + \gamma\right)\ln R + \frac{F}{PR} \tag{b}$$

hence the slope:

$$\frac{w'(r)}{P} = \Omega'(\tau) = -\frac{\Lambda^4\varepsilon^3}{4\Upsilon^3}\tau^3 + \alpha\frac{\tau}{2}\left(\ln\tau - \frac{1}{2}\right) + 2\beta\tau + \frac{\gamma}{\tau}$$

which value in $r = R$ is:

$$\frac{w'(R)}{\mathrm{P}} = \Omega'(\tau = 1) = -\frac{\Lambda^4\varepsilon^3}{4\Upsilon^3} - \frac{\alpha}{4} + 2\beta + \gamma \tag{c}$$

and the radial moment M^{rr} becomes, in the adimensional form:

$$\frac{M^{rr}(r)R}{D'P} = \left[\begin{array}{c} -\frac{\Lambda^4\varepsilon^3}{4\Upsilon^3}(3+\nu)\tau^2 + \frac{\alpha}{2}\left[(1+\nu)\ln\tau + (1+\nu)\ln R + \frac{1}{2}(1-\nu)\right] \\ +2\beta(1+\nu) - \frac{\gamma}{\tau^2}(1-\nu) \end{array}\right]$$

At the point $r = a$, either $\tau = \chi$, the moment is zero and can therefore be subtracted from the expression above:

$$\frac{M^{rr}(r)R}{D'P} = \left[\begin{array}{c} -\frac{\Lambda^4\varepsilon^3}{4\Upsilon^3}(3+\nu)(\tau^2 - \chi^2) + \frac{\alpha}{2}\left[(1+\nu)\ln\tau - (1+\nu)\ln\chi\right] \\ -\gamma(1-\nu)\left(\frac{1}{\tau^2} - \frac{1}{\chi^2}\right) \end{array}\right] \tag{d}$$

either at the junction with the cylinder:

$$\mu_f = \frac{mR}{D'P} = -\frac{M^{rr}(R)R}{D'P} = \frac{\Lambda^4\varepsilon^3}{4\Upsilon^3}(3+\nu)(1-\chi^2) + \frac{\alpha}{2}(1+\nu)\ln\chi + \gamma(1-\nu)\left(1 - \frac{1}{\chi^2}\right) \tag{e}$$

The raft is in contact with the central part of radius a, the different conditions, explained in §7.2.1.7, are written:

$$\Omega(\chi) = -\frac{\Lambda^4\varepsilon^3}{16\Upsilon^3}\chi^4 + \alpha\left(\frac{\chi^2}{4}\ln\chi - \frac{\chi^2}{4}\right) + \beta\chi^2 + \gamma\ln\chi + \varphi = 0 \tag{f}$$

$$\Omega'(\chi) = -\frac{\Lambda^4\varepsilon^3}{4\Upsilon^3}\chi^3 + \alpha\frac{\chi}{2}\left(\ln\chi - \frac{1}{2}\right) + 2\beta\chi + \frac{\gamma}{\chi} = 0 \tag{g}$$

$$\Omega(1) = -\frac{\Lambda^4\varepsilon^3}{16\Upsilon^3} - \frac{\alpha}{4} + \beta + \varphi = 0 \tag{h}$$

The relations (e), (f), (g), and (h) make it possible to determine the four constants as functions of m and χ. The displacement and the bending moment are then given by (a) and (b).

To determine m, it is necessary to express the continuity of the rotation at the junction between the bottom and the wall, using the expressions (iv) and (c), which is written $\Omega'(\tau)=\omega'(\xi)$. The problem is thus completely solved according to the adimensional parameters χ, ν, ε, Λ, and Υ.

Proceeding as in §7.2.1, m can be expressed in terms of χ:

$$\frac{16m}{pR^2}=\frac{4\Upsilon^3}{\Lambda^4\varepsilon^3}\mu_f=\frac{4\Upsilon^2}{\Lambda^4}\mu_c$$

$$=\frac{\left[\frac{1}{4}\left(1-\chi^4\right)-\chi^2\left(1-\chi^2\right)-\chi^4\ln\chi\right]\left[\left(1-\nu\right)\left(1-\chi^2\right)-2\left(1+\nu\right)\ln\chi\right]}{\left(1+\chi^2\right)\ln\chi+\left(1-\chi^2\right)} \tag{i}$$

$$+\left(1-\chi^2\right)\left(3+\nu\right)+\chi^2\left(1-\chi^2\right)\left(1-\nu\right)$$

which simplifies the resolution[4]. This highlights the fact that the solution does not depend linearly on the loading, because of the bottom uplift: it is physically obvious that the uplift parameter depends on the loading P, which is implicit in solving the equations; indeed, the relation (i) makes it possible to calculate the reduced moment μ_f, which leads to the dimensionless loading P by (v).

In the dimensionless solution, the material intervenes a little with ν and weakly in Λ and thus has only little influence. On the other hand, the loading parameter P is directly related to the specific weight of the fluid and the membranar stiffness.

The geometry, by the adimensional parameters Υ and Λ, intervenes in an essential way in the solution, as it was seen previously. The ratio ε of thicknesses (which determines the ratio ε^3 of stiffness to flexion) also plays an important role in the solution. More precisely, it appears by (i) that, in the bottom, the solution depends mainly on the composite non-dimensional parameter $\frac{\Lambda^4\varepsilon^3}{\Upsilon^3}pe$, whereas Λ and the composite non-dimensional parameter $\frac{\Upsilon^2}{\Lambda^4}$ are important in the cylinder.

8.2.3.2 Cylindrical vibrations

In this type of axisymetrical movement, the displacement of a circular section perpendicular to z is independent of z which assumes that there is no boundary condition restricting the displacement (v, w):

$$\xi=u(z,s)A_z+v(s)A_s+w(s)A_3 \tag{8.94}$$

Since the bending occurs in the direction s tangent to the circular section of radius R, the hypotheses of low curvature can not be taken and it is, therefore, necessary to use the linearised deformations. These are written:

[4] Indeed, being taken as parameter of the calculation, (i) determines μ_c, which leads to β_1 then α_1, α_2 and β_2 and finally to the rotation by (iv). α, β, γ and φ are then determined by (c), (f), (g), and (h).

$$\varepsilon_{zz} = \frac{\partial u}{\partial z} \qquad \kappa_{zz} = \kappa_{zs} = 0$$

$$\varepsilon_{zs} = \frac{1}{2}\left(\frac{\partial u}{\partial s}\right) \qquad \kappa_{ss} = \frac{\partial^2 w}{\partial s^2} + \frac{2}{R}\frac{\partial v}{\partial s} - \frac{w}{R^2}$$

$$\varepsilon_{ss} = \frac{\partial v}{\partial s} - \frac{w}{R}$$

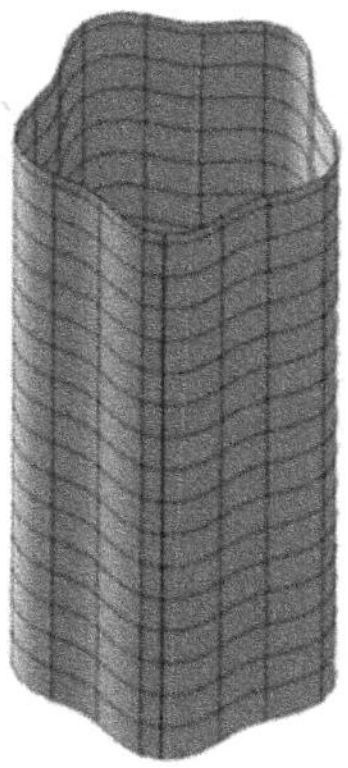

The displacement u depends on the boundary conditions at the ends of the cylinder. If the cylinder is blocked in z, u is zero and the deformation is purely plane (infinite cylinder). If on the contrary, the edges are free: $\varepsilon_{zz} = \nu\, \varepsilon_{ss}$ and $\varepsilon_{zs} = 0$, u is then an affine function of z. In both cases $\varepsilon_{zs} = 0$.

v and w are now expressed as functions of θ. If the shell is thin, the deformation is essentially due to bending and, following Lord Rayleigh, it is useful to analyse the pure bending modes. The hypothesis of inextensional transformation $\varepsilon_{ss} = \varepsilon_{zz} = 0$ makes it possible to write as:

$$w = \frac{\partial v}{\partial \theta} \quad \Rightarrow \quad \kappa_{ss} = \frac{1}{R^2}\left(\frac{\partial^2 w}{\partial \theta^2} + w\right)$$

In this hypothesis, the two components of the displacement v and w are of the same order of magnitude and in quadrature.

The elastic potential for a unit length of a cylinder becomes:

$$\mathcal{F} = \frac{1}{2}\int_0^{2\pi} D\left[\frac{1}{R^2}\left(\frac{\partial^2 w}{\partial \theta^2} + w\right)\right]^2 R d\theta$$

For the application of PVP:

$$\overset{*}{\mathcal{F}} = \frac{D}{R^3}\int_0^{2\pi}\left(\frac{\partial^2 w}{\partial \theta^2} + w\right)\left(\frac{\partial^2 \overset{*}{w}}{\partial \theta^2} + \overset{*}{w}\right)d\theta = \frac{D}{R^3}\int_0^{2\pi}\left(\frac{\partial^4 w}{\partial \theta^4} + 2\frac{\partial^2 w}{\partial \theta^2} + w\right)\overset{*}{w}\, d\theta$$

$$\overset{*}{\mathcal{W}}_j = -\int_0^{2\pi}\left(\ddot{v}\,\overset{*}{v} + \ddot{w}\,\overset{*}{w}\right)\rho R d\theta$$

where ρ is the surfacic mass. Since the two components of displacement are linked in the hypothesis of isometry, the last expression suggests expressing everything in terms of v, then:

$$\overset{*}{\mathcal{F}} = \frac{D}{R^3}\int_0^{2\pi}\left(\frac{\partial^5 v}{\partial \theta^5} + 2\frac{\partial^3 v}{\partial \theta^3} + \frac{\partial v}{\partial \theta}\right)\frac{\partial \overset{*}{v}}{\partial \theta} d\theta$$

$$\overset{*}{\mathcal{W}}_j = -\int_0^{2\pi}\left(\ddot{v}\,\overset{*}{v} + \frac{\partial \ddot{v}}{\partial \theta}\frac{\partial \overset{*}{v}}{\partial \theta}\right)\rho R d\theta$$

from where, by proceeding to a final integration by parts, the equation of the free movements:

$$\frac{D}{R^4}\left(\frac{\partial^6 v}{\partial\theta^6}+2\frac{\partial^4 v}{\partial\theta^4}+\frac{\partial^2 v}{\partial\theta^2}\right)-\rho\left(\ddot{v}-\ddot{v}''\right)=0$$

The periodic solution given by (8.64) and (8.65) can be sought in the complex form $v(\theta,t)=C_n e^{i(\omega_n t+n\theta)}$. This makes it possible to obtain the pulsations:

$$-\frac{D}{R^4}\left(n^2-1\right)^2 n^2+\rho\omega_n^{\,2}\left(1+n^2\right)=0$$

$$\boxed{\omega_n^{\,2}=\frac{D}{\rho R^4}\frac{\left(n^2-1\right)^2 n^2}{n^2+1}} \tag{8.95}$$

Note that n = 1 corresponds to a rigid translation, so to a zero pulsation. The n = 2 mode is an ovalisation deformation of the cylinder (Figure 8.11).

If the hypothesis of inextensional motion is not taken, the boundary conditions have a small influence on the membrane energy. If the edges are free, the elastic potential is written as:

$$\mathcal{F}=\frac{1}{2}\int_0^{2\pi}\left\{\frac{Ee}{R^2}\left[\frac{\partial v}{\partial\theta}-w\right]^2+\frac{D}{R^4}\left[\left(\frac{\partial^2 w}{\partial\theta^2}+2\frac{\partial v}{\partial\theta}-w\right)\right]^2\right\}Rd\theta$$

hence the virtual power:

$$\overset{*}{\mathcal{F}}=\int_0^{2\pi}\left\{\frac{Ee}{R^2}\left(\frac{\partial v}{\partial\theta}-w\right)\left(\frac{\partial \overset{*}{v}}{\partial\theta}-\overset{*}{w}\right)+\frac{D}{R^4}\left(\frac{\partial^2 w}{\partial\theta^2}+2\frac{\partial v}{\partial\theta}-w\right)\left(\frac{\partial^2 \overset{*}{w}}{\partial\theta^2}+2\frac{\partial \overset{*}{v}}{\partial\theta}-\overset{*}{w}\right)\right\}Rd\theta$$

$$=\int_0^{2\pi}\left\{\begin{array}{l}\left[-\frac{Ee}{R^2}\left(\frac{\partial^2 v}{\partial\theta^2}-\frac{\partial w}{\partial\theta}\right)-\frac{2D}{R^4}\left(\frac{\partial^3 w}{\partial\theta^3}+2\frac{\partial^2 v}{\partial\theta^2}-\frac{\partial w}{\partial\theta}\right)\right]\overset{*}{v}\\ +\left[-\frac{Ee}{R^2}\left(\frac{\partial v}{\partial\theta}-w\right)+\frac{D}{R^4}\left(\frac{\partial^4 w}{\partial\theta^4}+2\frac{\partial^3 v}{\partial\theta^3}-2\frac{\partial v}{\partial\theta}-2\frac{\partial^2 w}{\partial\theta^2}+w\right)\right]\overset{*}{w}\end{array}\right\}Rd\theta$$

If the cylinder were blocked in the longitudinal direction, the membrane energy term should be divided by $1-\nu^2$.

In the case of free edges, the equations of the free movements are written:

$$\left[\frac{Ee}{R^2}\left(\frac{\partial^2 v}{\partial\theta^2}-\frac{\partial w}{\partial\theta}\right)+\frac{2D}{R^4}\left(\frac{\partial^3 w}{\partial\theta^3}+2\frac{\partial^2 v}{\partial\theta^2}-\frac{\partial w}{\partial\theta}\right)\right]-\rho\ddot{v}=0$$

$$\left[-\frac{Ee}{R^2}\left(\frac{\partial v}{\partial\theta}-w\right)+\frac{D}{R^4}\left(\frac{\partial^4 w}{\partial\theta^4}+2\frac{\partial^3 v}{\partial\theta^3}-2\frac{\partial v}{\partial\theta}-2\frac{\partial^2 w}{\partial\theta^2}+w\right)\right]+\rho\ddot{w}=0$$

Given the shape of the equations and the periodicity conditions, it makes sense to look for solutions of v and w in quadrature, for example:

$$v_n(\theta,t) = A_n e^{i\omega_n t}\cos(n\theta)$$

$$w_n(\theta,t) = B_n e^{i\omega_n t}\sin(n\theta)$$

Hence, replacing in the equations of motion:

$$\left[\frac{Ee}{R^2}\left(-n^2A_n - nB_n\right) + \frac{2D}{R^4}\left(-n^3B_n - 2n^2A_n - nB_n\right)\right] + \rho\omega_n{}^2A_n = 0$$

$$\left[-\frac{Ee}{R^2}\left(-nA_n - B_n\right) + \frac{D}{R^4}\left(n^4B_n + 2n^3A_n + 2nA_n + 2n^2B_n + B_n\right)\right] - \rho\omega_n{}^2B_n = 0$$

or, reordering:

$$\left(-n^2 - 4kn^2 + x\right)A_n - \left(n + 2n^3k + 2kn\right)B_n = 0$$

$$\left(n + 2n^3k + 2nk\right)A_n + \left(1 + n^4k + 2n^2k + k - x\right)B_n = 0$$

with $k = \dfrac{D}{EeR^2} = \dfrac{1}{12\left(1-\nu^2\right)}\dfrac{e^2}{R^2}$ and $x = \dfrac{\rho\omega_n{}^2R^2}{Ee} = \dfrac{\bar{\rho}\omega_n{}^2R^2}{E}$.

where $\bar{\rho}$ is the volumic mass. The pulsations are determined by writing that the movement exists, which implies that the system of equations is singular. The determinant of the system of equations in A_n, B_n is zero:

$$\left(-n^2 - 4kn^2 + x\right)\left(1 + n^4k + 2n^2k + k - x\right) + \left(n + 2n^3k + 2kn\right)^2 = 0$$

x is the solution of the equation of the second degree:

$$x^2 - \left(1 + n^2 + 4kn^2 + k\left(n^2+1\right)^2\right)x + n^2k\left(n^2-1\right)^2 = 0 \tag{8.96}$$

for which two positive solutions $x_{n,1}$ and $x_{n,2}$ exist whatever the data. Two pulsations are thus obtained for each value of n.

For the modes where the bending energy can be neglected, especially when the bending inertia is low, e/R very small, the equations of motion are reduced to:

$$\frac{\partial^2 v}{\partial\theta^2} - \frac{\partial w}{\partial\theta} - \frac{\bar{\rho}R^2}{E}\ddot{v} = 0$$

$$-\frac{\partial v}{\partial\theta} + w + \frac{\bar{\rho}R^2}{E}\ddot{w} = 0$$

and are independent of the thickness e of the shell. Indeed, membrane stiffness and surfacic mass depend linearly on the thickness; the corresponding pulsations do not depend on it either. The modes are of the same shape as above and the pulsations are then given by:

$$\omega_n{}^2 = \left(n^2+1\right)\frac{E}{\rho R^2} \tag{8.97}$$

associated with the modes defined by $\left(A_n = n\,;\, B_n = 1\right)$. For each value of n, a null solution exists, associated with the mode $\left(A_n = 1\,;\, B_n = -n\right)$.

These two extreme hypotheses (negligible extension or bending) highlight the role of the thickness in relation to the dimensions of the cylinder. The lowest pulsation between those given by (8.95) and (8.97), for n = 2, is extension when $\frac{D}{\rho R^4}\frac{36}{5} > 5\frac{E}{\rho R^2}$, that is $\frac{e^2}{R^2} > \frac{25}{3}\left(1-\nu^2\right)$, flexion otherwise. So, if the shell is not really thin, the extension plays a significant role. This also highlights the need to couple the two energy sources in the equations.

The relative importance of these two components is also highlighted in the case of thin cylinders. by taking k = 0 in (8.96). The constant term being then zero in the characteristic Equation (8.96), it is clear that one of the solutions is zero: when the thickness of the shell tends towards 0, the lowest pulsation tends towards 0 and the highest towards (8.97).

This is also highlighted by developing the solutions of (8.96) with respect to k. By sticking to the first order in k, the root of Equation (8.96) is:

$$\sqrt{\Delta} \approx 1+n^2+\frac{-n^6+11n^4+5n^2+1}{\left(1+n^2\right)}k$$

and the main term of the smallest solution is worth:

$$x_{n,1} \simeq \frac{n^2\left(n^2-1\right)^2}{\left(1+n^2\right)}k$$

and gives the inextensional pulsation (8.95), for which it is easy to verify that the mode is $\left(A_n = 1\,;\, B_n = -n\right)$.

The main term of the largest solution is worth:

$$x_{n,2} \simeq 1+n^2$$

and corresponds to the purely extensional pulsation (8.97). Thus, in the case of thin shells, when the thickness is small relative to the radius R, for the mode of order n, the lowest pulsation is close to the purely flexional pulsation (8.95) and tends to 0 with the thickness; the highest pulsation is close to the purely extensional pulsation (8.97), which is independent of the thickness. This justifies Rayleigh's hypothesis for the search for pulsations due to bending and gives a suitable approximation for all modes. Figure 8.11 shows the two series of modes for the values of n from 2 to 5. The shapes of the two modes corresponding to the same value of n are quite different, in particular, because the extension modes do not exhibit flexural stiffness. The greater flexural flexibility thus demonstrated also confirms the results obtained for the study of the axial stability of the cylinder (see §8.2.4.2).

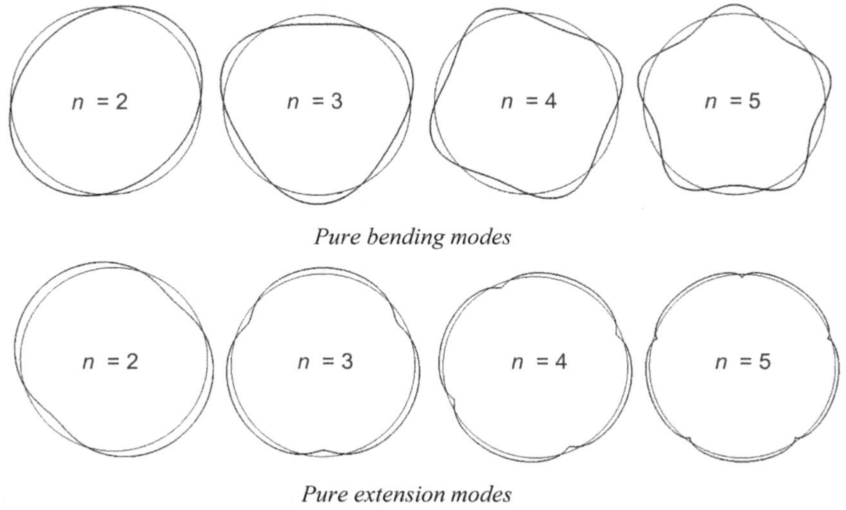

Figure 8.11

8.2.3.3 Cylindrical vault

A roof element consists of a cylinder supported on a flexible tympanum at each end. This problem has been addressed in §6.2.2.1, where a membrane solution has been established. It has been shown that it is difficult for the membrane solution to meet all the boundary conditions. Here, there is no longitudinal edge beam. In addition, the opening angle of the cylinder section is $\theta_0 << \frac{\pi}{2}$ and the curvature of the cylinder is small. This geometry corresponds to a usual situation and makes it possible to consider the shell as shallow, so to apply the Donnell assumptions.

Boundary conditions are:

- in $\theta = \pm\theta_0$: $N^{12} = N^{22} = M^{22} = q^2 = 0$;
- in $z = \pm\ell$: $N^{11} = 0$; $w = v = 0$.

The equations relating to the displacement field can be obtained either by reporting the constitutive laws in equilibrium equations, this is the approach used in §8.2.2.1, or from the strain energy; this is the second way that is followed here. It is a question of finding the Equation (8.76), in classical approximation, by the energy method. Starting from relations (8.49) for ε and (8.56) for κ, the strain energy is written as:

$$\mathcal{F} = \frac{K}{2}\left\{\begin{aligned} &\int_\sigma \left[\left(\frac{\partial u}{\partial z}\right)^2 + \frac{1-\nu}{2}\left(\frac{\partial u}{R\partial\theta} + \frac{\partial v}{\partial z}\right)^2 + \frac{1}{R^2}\left(\frac{\partial v}{\partial\theta} - w\right)^2 + \frac{2\nu}{R}\frac{\partial u}{\partial z}\left(\frac{\partial v}{\partial\theta} - w\right)\right] dS \\ &+ \frac{e^2}{12}\int_\sigma \left[\left(\frac{\partial^2 w}{\partial z^2}\right)^2 + \left(\frac{\partial^2 w}{R^2\partial\theta^2}\right)^2 + \frac{2(1-\nu)}{R^2}\left(\frac{\partial^2 w}{\partial z\partial\theta}\right)^2 + \frac{2\nu}{R^2}\frac{\partial^2 w}{\partial z^2}\frac{\partial^2 w}{\partial\theta^2}\right] dS \end{aligned}\right\}$$

By virtualisation of the strain energy:

$$\overset{*}{\mathcal{F}} = K\left\{\begin{aligned}&\int_\sigma\left[\begin{aligned}&\frac{\partial u}{\partial z}\frac{\partial \overset{*}{u}}{\partial z}+\frac{1-\nu}{2}\left(\frac{\partial u}{R\partial\theta}+\frac{\partial v}{\partial z}\right)\left(\frac{\partial \overset{*}{u}}{R\partial\theta}+\frac{\partial \overset{*}{v}}{\partial z}\right)+\frac{1}{R^2}\left(\frac{\partial v}{\partial\theta}-w\right)\left(\frac{\partial \overset{*}{v}}{\partial\theta}-\overset{*}{w}\right)\\&+\frac{\nu}{R}\frac{\partial \overset{*}{u}}{\partial z}\left(\frac{\partial v}{\partial\theta}-w\right)+\frac{\nu}{R}\frac{\partial u}{\partial z}\left(\frac{\partial \overset{*}{v}}{\partial\theta}-\overset{*}{w}\right)\end{aligned}\right]dS\\&+\frac{e^2}{12}\int_\sigma\left[\begin{aligned}&\frac{\partial^2 w}{\partial z^2}\frac{\partial^2 \overset{*}{w}}{\partial z^2}+\frac{\partial^2 w}{R^2\partial\theta^2}\frac{\partial^2 \overset{*}{w}}{R^2\partial\theta^2}+\frac{2(1-\nu)}{R^2}\frac{\partial^2 w}{\partial z\partial\theta}\frac{\partial^2 \overset{*}{w}}{\partial z\partial\theta}\\&+\frac{\nu}{R^2}\frac{\partial^2 \overset{*}{w}}{\partial z^2}\frac{\partial^2 w}{\partial\theta^2}+\frac{\nu}{R^2}\frac{\partial^2 w}{\partial z^2}\frac{\partial^2 \overset{*}{w}}{\partial\theta^2}\end{aligned}\right]dS\end{aligned}\right\}$$

then integrations by parts and application of the PVP, it becomes the coupled equations allowing to determine the three components of the displacement field:

$$\begin{aligned}&\frac{\partial^2 u}{\partial z^2}+\frac{1-\nu}{2R^2}\frac{\partial^2 u}{\partial\theta^2}+\frac{1+\nu}{2R}\frac{\partial^2 v}{\partial z\partial\theta}-\frac{\nu}{R}\frac{\partial w}{\partial z}=-\frac{p^1}{K} && (a)\\&\frac{1-\nu}{2}\frac{\partial^2 v}{\partial z^2}+\frac{1}{R^2}\frac{\partial^2 v}{\partial\theta^2}+\frac{1+\nu}{2R}\frac{\partial^2 u}{\partial\theta\,\partial z}-\frac{1}{R^2}\frac{\partial w}{\partial\theta}=-\frac{p^2}{K} && (b)\\&-\frac{1}{R^2}\left(\nu R\frac{\partial u}{\partial z}+\frac{\partial v}{\partial\theta}-w\right)+\frac{e^2}{12}\left[\frac{\partial^4 w}{\partial z^4}+\frac{\partial^4 w}{R^4\partial\theta^4}+\frac{2}{R^2}\frac{\partial^4 w}{\partial z^2\partial\theta^2}\right]=\frac{p^3}{K} && (c)\end{aligned}\qquad(8.98)$$

The membrane solution given in §6.2.2.1 is a particular solution of this system of equations.

Differentiating (8.98-a) on the one hand twice with respect to z, on the other hand twice with respect to $R\theta$, and subtracting the two expressions obtained, multiplied by the appropriate coefficients to eliminate v from the relation (8.98-b), itself previously differentiated with respect to z and $R\theta$, it comes:

$$\begin{aligned}&\frac{\partial^4 u}{\partial z^4}+2\frac{\partial^4 u}{R^2\partial\theta^2\partial z^2}+\frac{\partial^4 u}{R^4\partial\theta^4}=\Delta\Delta u\\&\qquad=-\frac{\partial^3 w}{R^3\partial\theta^2\partial z}+\frac{\nu}{R}\frac{\partial^3 w}{\partial z^3}+\frac{1+\nu}{1-\nu}\frac{1}{K}\frac{\partial^2 p^2}{R\partial\theta\,\partial z}-\frac{1}{K}\frac{\partial^2 p^1}{\partial z^2}-\frac{2}{1-\nu}\frac{1}{K}\frac{\partial^2 p^1}{R^2\partial\theta^2}\end{aligned}\qquad(8.99)$$

Similarly, by differentiating twice (8.98-b) on the one hand with respect to z, on the other hand with respect to $R\theta$, and subtracting the two expressions obtained from the relation (8.98-a) derived with respect to z and $R\theta$, eliminating u, it becomes:

$$\begin{aligned}&\frac{\partial^4 v}{\partial z^4}+2\frac{\partial^4 v}{R^2\partial\theta^2\partial z^2}+\frac{\partial^4 v}{R^4\partial\theta^4}=\Delta\Delta v\\&=\frac{2+\nu}{R}\frac{\partial^3 w}{R\partial\theta\,\partial z^2}+\frac{1}{R}\frac{\partial^3 w}{R^3\partial\theta^3}+\frac{1+\nu}{1-\nu}\frac{\partial^2}{R\partial\theta\,\partial z}\frac{p^1}{K}-\frac{2}{1-\nu}\frac{\partial^2}{\partial z^2}\frac{p^2}{K}-\frac{\partial^2}{R^2\partial\theta^2}\frac{p^2}{K}\end{aligned}\qquad(8.100)$$

Taking the double Laplacian of (8.98-c):

$$-\frac{1}{R^2}\left(\nu R\Delta\Delta\frac{\partial u}{\partial z}+\Delta\Delta\frac{\partial v}{\partial \theta}-\Delta\Delta w\right)+\frac{e^2}{12}\Delta^4 w=\Delta\Delta\frac{p^3}{K} \tag{101}$$

Finally, by deriving (8.99) with respect to z and (8.93) with respect to θ and by reporting the expressions obtained in (8.101), it comes the equation (of Donnell or Jenkins or Vlassov) bearing only on w, to compare with (8.76):

$$\boxed{\frac{e^2}{12}\Delta^4 w+\frac{1-\nu^2}{R^2}\frac{\partial^4 w}{\partial z^4}=\frac{1}{K}\left[\Delta\Delta p^3+\frac{1}{R}\left(\begin{array}{l}\dfrac{\partial^3 p^1}{R^2\partial\theta^2\partial z}-\nu\dfrac{\partial^3 p^1}{\partial z^3}\\ -(2+\nu)\dfrac{\partial^3 p^2}{R\partial\theta\,\partial z^2}-\dfrac{\partial^3 p^2}{R^3\partial\theta^3}\end{array}\right)\right]} \tag{8.102}$$

This equation of the eight order makes it possible to determine w. u and v are then given by equations (8.99) and (8.100). With respect to (8.76), the membrane forces do not influence the displacement since here the theory is linearised. The second member has been completed by external actions tangent to the surface.

Example: Vault under self-weight.
The vault being subjected to its self-weight, the components of the external action density are worth (see 6.2.2.1):

$$p^2=\varpi R\sin\theta\,;\;\; p^3=\varpi R\cos\theta$$

The second member of (8.102) is then worth $\frac{2\varpi}{KR^3}\cos\theta$. A particular solution of the equation is given by:

$$w_0=\frac{12R^8}{e^2}\frac{2\varpi}{KR^3}\cos\theta=\frac{2\varpi R^5}{D\left(1-\nu^2\right)}\cos\theta$$

A complete solution in $\cos\theta$ does not allow to respect the boundary conditions in θ_0. It is, therefore, necessary to look for the solutions of the equation without a second member by decomposing them into Fourier series following θ, or, taking symmetry into account, as a combination of elementary functions:

$$w_n(z,\theta)=W_n(z)\cos n\theta$$

including the term corresponding to $n=0$. By referring to (8.102), there comes the equation verified by the function W_n for $n>0$:

$$\frac{\partial^8 W_n}{\partial z^8}-4\frac{n^2}{R^2}\frac{\partial^6 W_n}{\partial z^6}+6\frac{n^4}{R^4}\frac{\partial^4 W_n}{\partial z^4}-4\frac{n^6}{R^6}\frac{\partial^2 W_n}{\partial z^2}+\frac{n^8}{R^8}W_n+4\lambda^4\frac{\partial^4 W_n}{\partial z^4}=0 \tag{8.103}$$

with λ given by (8.89). The general solution of this equation is obtained in the form e^{rz}, where r is a complex number, which satisfies the characteristic equation:

$$\left(\frac{R^2r^2}{n^2}-1\right)^4+4\left(\frac{\lambda R}{n}\right)^4\left(\frac{Rr}{n}\right)^4=0$$

which can be reduced to the four complex equations of the second degree:

$$\left(\frac{Rr}{n}\right)^2-1+(\varepsilon+i\varepsilon')\frac{\lambda R}{n}\left(\frac{Rr}{n}\right)=0$$

- ε and ε' taking the values +1 and –1. The solutions of the four characteristic equations are constructed by noting:

$$\mu_n=\frac{\lambda R}{2n};\quad \chi_n=\sqrt{\sqrt{1+4\mu_n^{\ 4}}+2\mu_n^{\ 2}}\ ;\quad \tau_n=\sqrt{\sqrt{1+4\mu_n^{\ 4}}-2\mu_n^{\ 2}}\ ;$$

$$\alpha_n=\frac{n}{R}\left[\mu_n-\frac{1}{2}(\chi_n+\tau_n)\right];\quad \beta_n=\frac{n}{R}\left[\mu_n-\frac{1}{2}(\chi_n-\tau_n)\right];$$

$$\gamma_n=\frac{n}{R}\left[\mu_n+\frac{1}{2}(\chi_n+\tau_n)\right];\quad \delta_n=\frac{n}{R}\left[\mu_n+\frac{1}{2}(\chi_n-\tau_n)\right]$$

and finally, the eight solutions in r are written:

$$-\alpha_n-i\beta_n\ ;\ \alpha_n-i\beta_n\ ;\ -\alpha_n+i\beta_n\ ;\ \alpha_n+i\beta_n\ ;\ -\gamma_n-i\delta_n\ ;\ \gamma_n-i\delta_n\ ;\ -\gamma_n+i\delta_n\ ;\ \gamma_n+i\delta_n$$

In the case of a thin arch, μ_1 is of the order of 3 or 5, for example. μ_n decreases rapidly with n and tends towards 0 when n increases.

When n tends to infinity, χ_n is equivalent to $1+\mu_n^{\ 2}$ and τ_n to $1-\mu_n^{\ 2}$, which therefore both tend towards 1. So the second term of α_n and γ_n tends to infinity with n, whereas β_n and δ_n tend towards $\lambda/2$. Conversely, when μ_n is large enough, χ_n tends toward $2\mu_n$ and τ_n towards 0, but this asymptotic situation is not achieved in practice.

We must then look at the case $n = 0$. The differential equation verified by W_0 is:

$$\frac{\partial^8 W_0}{\partial z^8}+4\lambda^4\frac{\partial^4 W_0}{\partial z^4}=0$$

which solution is easy considering the results obtained in §8.2.3.1. Finally comes the general solution of (8.103):

$$\begin{aligned}\mathrm{w}(\theta,z)=&\frac{2\varpi R^5}{D(1-\nu^2)}\cos\theta+\mathrm{A}z^3+\mathrm{B}z^2+\Gamma z+\mathrm{K}\\&+e^{\lambda z}(A_0\sin\lambda z+B_0\cos\lambda z)+e^{-\lambda z}(C_0\sin\lambda z+D_0\cos\lambda z)\\&+\sum_{n=1}^{\infty}\left[\begin{array}{l}e^{\alpha_n z}(A_n\sin\beta_n z+B_n\cos\beta_n z)+e^{-\alpha_n z}(C_n\sin\beta_n z+D_n\cos\beta_n z)\\+e^{\gamma_n z}(E_n\sin\delta_n z+F_n\cos\delta_n z)+e^{-\gamma_n z}(G_n\sin\delta_n z+H_n\cos\delta_n z)\end{array}\right]\cos n\theta\end{aligned}\tag{8.104}$$

(8.104) being the general solution of (8.102) for the loading considered on $\left]-\pi,+\pi\right]$ is also the solution in the subspace $\left[-\theta_0,\theta_0\right]$.

The solution (8.104) is simplified by noting that only the even terms should be retained to respect the symmetry with respect to $x = 0$ and that w is zero in $x = \pm\ell$ for any value of θ:

$$\mathrm{w}(\theta,z) = b\left(z^2-\ell^2\right) + a_0\left(\frac{\mathrm{sh}\lambda z\,\sin\lambda z}{\mathrm{sh}\lambda\ell\,\sin\lambda\ell}-1\right) + b_0\left(\frac{\mathrm{ch}\lambda z\,\cos\lambda z}{\mathrm{ch}\lambda\ell\,\cos\lambda\ell}-1\right)$$
$$+\sum_{n=1}^{\infty}\left[\begin{array}{l} a_n\left(\dfrac{\mathrm{sh}\alpha_n z\sin\beta_n z}{\mathrm{sh}\alpha_n\ell\sin\beta_n\ell}-1\right)+b_n\left(\dfrac{\mathrm{ch}\alpha_n z\cos\beta_n z}{\mathrm{ch}\alpha_n\ell\cos\beta_n\ell}-1\right) \\ +e_n\left(\dfrac{sh\gamma_n z\sin\delta_n z}{\mathrm{sh}\gamma_n\ell\sin\delta_n\ell}-1\right)+f_n\left(\dfrac{ch\gamma_n z\cos\delta_n z}{\mathrm{ch}\gamma_n\ell\cos\delta_n\ell}-1\right)\end{array}\right]\cos n\theta \qquad (8.105)$$

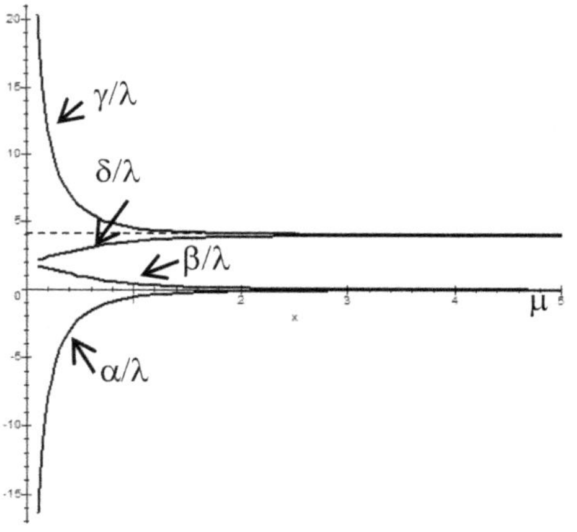

Figure 8.12

According to (8.100), v is odd in θ and even in z. According to (8.99), u is even in θ and odd in z. The complete analytical solution is not very affordable and non-competitive with respect to a finite element solution (see Chapter 9), but the following interpretation clarifies the flexional behaviour of the vault.

To deepen the physical interpretation of (8.105), it is interesting to express α_n,β_n, γ_n, and δ_n according to μ_n, assuming that this last parameter is continuous, that is μ in Figure 8.12

When μ_n tends to 0, α_n tends towards $-\infty$, γ_n towards $+\infty$, β_n, and δ_n towards 2λ. This corresponds to high values of n or, alternatively, to low values of λR, then of $\sqrt{\frac{R}{e}}$. This last situation is to be removed for a thin shell, it corresponds to shortwaves in θ. In direction z, the undulations then have a half-wavelength tending towards:

$$l = \frac{\pi}{2\lambda} = \frac{\pi\sqrt{\mathrm{Re}}}{2\sqrt[4]{3\left(1-\nu^2\right)}} \approx 1{,}21\sqrt{\mathrm{Re}}$$

independent of n. The amplitudes $e^{\pm\alpha_n z}$ and $e^{\pm\gamma_n z}$ of these waves tend to 0 or infinity with n and z. They therefore correspond to solutions that are either incompatible with the boundary conditions in $z = \pm\ell$, or have a small contribution to the result.

When μ_n tends towards infinity (in practice for $\mu_n = 3$ limits are almost reached), α_n and β_n tend towards 0, γ_n and δ_n towards 4λ. This corresponds to low values of n for a thin shell. In this case, the solution is the superposition of weakly damped long waves providing a contribution comparable to a constant and to waves of half-length:

$$L = \frac{\pi}{4\lambda} \approx 0{,}60\sqrt{\text{Re}}$$

independent of n and with amplitude $e^{\pm 4\lambda z}$. These observations make it possible to obtain an approximation of the solution in the case of thin shells. In the "intermediate" situations of μ_n, all the terms must be taken into account in the solution.

For example, for $R/e = 25$, $\mu_n = 3$ is obtained for $n = 1$. μ_n takes values lower than 0,1 for n greater than 3. We must therefore keep the terms of orders 2 and 3 corresponding to "intermediate" values of μ_n. This is the common situation with thin shells.

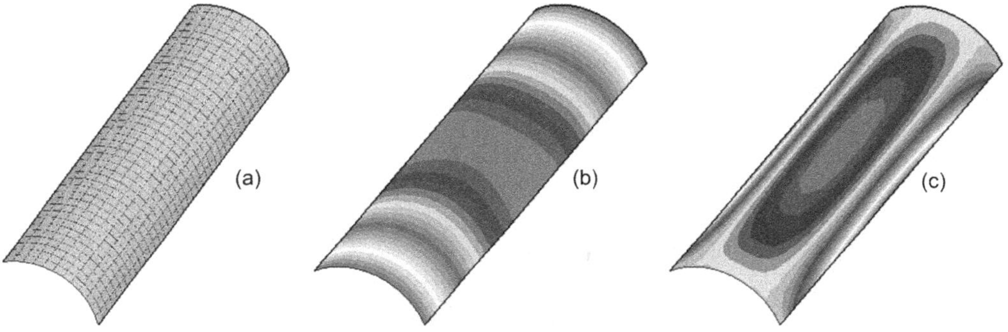

Figure 8.13

The finite element method discussed in the next chapter provides a numerical solution directly by minimising the potential energy which terms have been established above. The vault of which the model is shown in Figure 8.13(a) is such that $R/e = 25$, $\lambda R = 6{,}5$, and $\lambda \ell = 20{,}8$. The vault is closed by diaphragms at its ends (displacements in the plane of the diaphragm blocked) and free to move in the longitudinal direction of the vault. In a one-dimensional model, the vault would be likened to a beam on two simple supports at its ends. It is subject to its self-weight.

Figure 8.13(b) shows the contour lines of displacements, which are essentially vertical. Near the supports, Navier–Bernoulli–Euler's hypothesis is approximately verified, notably because of the presence of the tympanum. On the other hand, in the central part, displacements depend significantly on θ: the section z = cst is deformed. Figure 8.14(a) shows the deformation of the central section (deformed in bold lines, the average displacement of the section was eliminated).

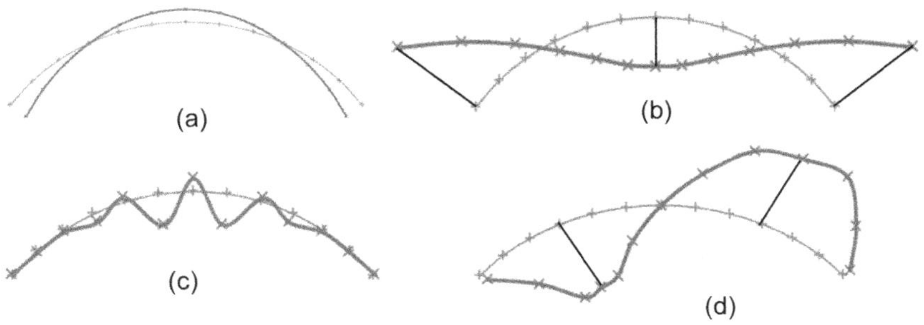

Figure 8.14

Figure 8.13(c) shows the amplitude level curves of longitudinal stresses N^{zz}. It highlights the dependence in θ and z established using the analytical solution. In the central part, the longitudinal stresses are tensions at the bottom and compressions at the top (see Figure 8.14(b) the distribution of the longitudinal stresses in the central section), which is also given by the beam model.

The transverse stresses $N^{\theta\theta}$ (see Figure 8.14(c)) are eminently variable in θ; they show the influence of the term $n = 3$. Figure 8.14(d) shows the distribution of $N^{\theta z}$ shear stresses on the bearing section. The beam model does not give sufficiently precise results, the section being open and therefore deformable, and it is preferable here to use the shell model.

Note: in Figure 8.14(c) and (d), the stresses are magnified by a factor of 10 with respect to Figure 8.14(b).*

Photo 8.3: Sheds made up of pieces of cylinders arranged to facilitate the passage of light

8.2.4 Instability of cylindrical shells

8.2.4.1 Equilibrium in second-order approximation

In second-order approximation, the deformations are given by (8.54) and (8.55). The equilibrium equations are obtained by particularising (8.30) to the geometry of the cylinder:

$$\begin{aligned}
&\partial_z N^{zz} + \partial_s N^{zs} + p^z = 0\\
&\partial_z\left(N^{zs} + \frac{M^{zs}}{R}\right) + \partial_s\left(N^{ss} + \frac{M^{ss}}{R}\right) - \frac{N^{sz}}{R}\partial_z w - \frac{N^{ss}}{R}\left(\frac{v}{R} + \partial_s w\right) + p^s = 0\\
&\partial^2_{z^2} M^{zz} + 2\partial^2_{zs} M^{zs} + \partial^2_{s^2} M^{ss} - \frac{N^{ss}}{R} - \partial_z\left[N^{zz}\partial_z w + N^{zs}\left(\frac{v}{R} + \partial_s w\right)\right]\\
&\qquad - \partial_s\left[N^{sz}\partial_z w + N^{ss}\left(\frac{v}{R} + \partial_s w\right)\right] - p^3 = 0
\end{aligned} \tag{8.106}$$

Moreover, in the case of a sufficiently thin shell, the terms M^{sz}/R and M^{ss}/R are negligible in the second equation.

For a closed cylinder limited by the edges $z = 0$ and $z = h$, they are supplemented by the boundary conditions on the edges, from (8.27):

$$N^{zz} - q^z = 0$$

$$N^{sz} + \frac{2M^{zs}}{R} - \left(q^s - \frac{m^z}{R} \right) = 0$$

$$\partial_z M^{zz} + 2\partial_s M^{zs} + \partial_z M^{zs} - N^{zz}\partial_z \mathrm{w} - N^{zs}\left(\frac{v}{R} + \partial_s w \right) + q^3 + \partial_s m^z = 0 \tag{8.107}$$

$$M^{zz} - m^s = 0$$

These equations are sufficient to deal with most problems, with the order of magnitude assumptions set out in §8.2.1.3.

But the alternative form (8.35) can also be used, which leads to equivalent local equations:

$$\partial_z N^{zz} + \partial_s N^{zs} + p^z = 0$$

$$\partial_z N^{zs} + \partial_s N^{ss} - \frac{N^{sz}}{R}\partial_z w - \frac{N^{ss}}{R}\left(\frac{v}{R} + \partial_s w \right) + p^s = 0$$

$$\partial^2_{z^2} M^{zz} + 2\partial^2_{zs} M^{zs} + \partial^2_{s^2} M^{ss} + \frac{M^{ss}}{R^2} - \frac{N^{ss}}{R} - \partial_z \left[N^{zz}\partial_z w + N^{zs}\left(\frac{v}{R} + \partial_s w \right) \right] \tag{8.108}$$

$$- \partial_s \left[N^{sz}\partial_z w + N^{ss}\left(\frac{v}{R} + \partial_s w \right) \right] - p^3 = 0$$

The constitutive law for a Hooke shell is given by (8.57) and (8.58).

Note that, considering the general equations (4.97), in the presence of prestresses N_0 and M_0, if they are not themselves small, $b_{\alpha\beta} N_0^{\alpha\beta}$ and $b_{\alpha\beta} b_\gamma{}^\beta M_0{}^{\alpha\gamma}$ can be comparable (cf. example of §8.2.1.8 where they are of the order of p), except if the shell is thin and shallow. The influence of the moment might not be negligible if the prestressing moment is not.

8.2.4.2 Study of buckling of closed cylinders

8.2.4.2.1 Use of the second-order approximation

In linear buckling theory, the variations of membrane forces due to the buckling deformation are neglected compared to the prestressing terms. The prestresses balance the external actions; it is a question of seeking if there is an equilibrium different from the equilibrium obtained at the first order, this one is taken as a reference state. The equations, therefore, relate to the possible equilibrium change, the external actions being unchanged. A first realistic approach consists in assuming the second-order approximation on the deformations, therefore a quasi-inextensional bending deformation. The first two equations are of the second order (they make it possible to evaluate the variations of normal force according to the displacement); only the last equilibrium equation, therefore, remains at the first order; it is obtained from (8.106) or (8.108).

As noted in §8.1.2.3, these equations differ in the linear part from those obtained in the first-order approximation by the role played by the term M^{ss}/R^2, which intervenes differently. Indeed, different kinematic hypotheses lead to different "current" configurations, and therefore to different equilibriums. It is therefore legitimate to neglect this term in the equations for the search for instability, except if there is a moment of prestress: it is not the final equilibrium which is sought, it cannot be obtained with linear equations, but the initiation of the instability movement.

8.2.4.2.2 Non-axisymmetric buckling of the cylinder under axial loading

It is a question of taking again the example of §8.2.2.2 for non-axisymmetric modes, by using the results from Figure 8.5, in second-order approximation first.

The cylinder being simply supported laterally on its two edges, starting from the general form of the solution given in §8.2.1.10 by (8.64) and (8.65), w can be sought in the form of a sinusoid in z. Thus, the following movements respect these conditions:

$$w(z,s)=\sum_{m,n}\sin\frac{n\pi z}{h}\left(A_{n,m}\sin\frac{ms}{R}+B_{n,m}\cos\frac{ms}{R}\right)$$

$$v(z,s)=F(z)+\sum_{m,n}\frac{1}{m}\sin\frac{n\pi z}{h}\left(-A_{n,m}\cos\frac{ms}{R}+B_{n,m}\sin\frac{ms}{R}\right) \tag{8.109}$$

$$u(z,s)=-F'(z)s+G(z)+\sum_{m,n}\frac{n\pi R}{m^2h}\cos\frac{n\pi z}{h}\left(A_{n,m}\sin\frac{ms}{R}+B_{n,m}\cos\frac{ms}{R}\right)$$

F' must be zero to respect the periodicity in s and, by fixing v at a point, $F = 0$. By hypothesis $\frac{\partial u}{\partial z}=0$, up to the second order, whatever s, G' is, therefore, zero and by fixing G at a point, $G = 0$.

It is already noted that, in a mode (m, n), $u_{m,n}$ is of the form: $u_{m,n}=\frac{n}{m^2\kappa}\cos\frac{n\pi z}{h}\sin\frac{ms}{R}$ and is therefore small compared to the displacements in the transverse plane, on the one hand for slender cylinders, on the other hand when the number of circumferential waves increases. Conversely, it increases when the number of longitudinal waves increases: more folds in this direction imply more displacement.

(8.106) reduces to:

$$\frac{\partial^2 M^{zz}}{\partial z^2}+2\frac{\partial^2 M^{zs}}{\partial z\partial s}+\frac{\partial^2 M^{ss}}{\partial s^2}-q\frac{\partial^2 w}{\partial z^2}=0 \tag{8.110}$$

It becomes, by expressing the flexural behaviour by (8.63):

$$\frac{\partial^4 w}{\partial z^4}+2\frac{\partial^4 w}{\partial z^2\partial s^2}+\frac{\partial^4 w}{\partial s^4}-\frac{q}{D}\frac{\partial^2 w}{\partial z^2}+\left(\nu\frac{\partial^2}{\partial z^2}+\frac{\partial^2}{\partial s^2}\right)\frac{w}{R^2}=0 \tag{8.111}$$

The equation sought can also be obtained by applying the PVP, by expressing the second-order deformations as functions of w, which leads to the total potential:

$$\mathcal{T}+\mathcal{V}=\int_0^h\int_0^{2\pi R}\frac{1}{2}q\left[\left(\frac{\partial w}{\partial z}\right)^2\right]+\frac{1}{2}D\left[\begin{array}{l}(1-\nu)\left[\left(\frac{\partial^2 w}{\partial z^2}\right)^2+\left(\frac{\partial^2 w}{\partial s^2}+\frac{w}{R^2}\right)^2+2\left(\frac{\partial^2 w}{\partial z\partial s}\right)^2\right]\\ +\nu\left(\frac{\partial^2 w}{\partial z^2}+\frac{\partial^2 w}{\partial s^2}+\frac{w}{R^2}\right)^2\end{array}\right]ds\,dz$$

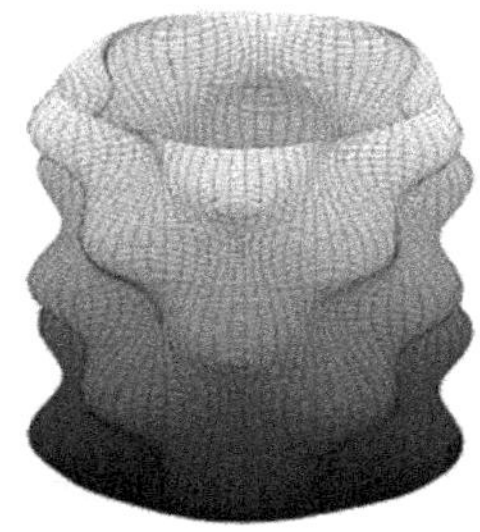

The equation obtained in this way expresses the equilibrium Equation (8.108) and therefore differs from (8.111) due to the term M^{ss}/R^2.

Equation (8.111) differs from (8.73) in particular by the absence of the term in w linked to the circumferential extension; here, the second-order deformation hypotheses mean that the cylinder does not undergo any extension in the s direction, up to the second order, so this term disappears as soon as the cylinder "creases" in the circumferential direction: in this hypothesis, circumferential extension is negligible compared to bending. The second-order approach considered here is quasi-inextensional. Thus, the approach leading to (8.73) is appropriate only for axisymmetric deformations where such bending does not exist.

The mode $w_{m,n}=\sin\frac{n\pi z}{h}\sin\frac{ms}{R}$ is a solution of the displacement equation. It verifies $w=w''=0$ on both edges. The condition of null moment: $\frac{\partial^2 w}{\partial z^2}+\nu\left(\frac{\partial^2 w}{\partial s^2}+\frac{w}{R^2}\right)=0$ is also verified.

Figure 8.15 shows the possible instability modes for (m = 1, n = 1), (m = 2, n = 1), and (m = 8, n = 1).

The case m = 1 is a special case, since the circle does not deform, and therefore its curvature variation $\frac{\partial^2 w}{\partial s^2}+\frac{w}{R^2}$ is zero. This corresponds well to the hypotheses of the beam

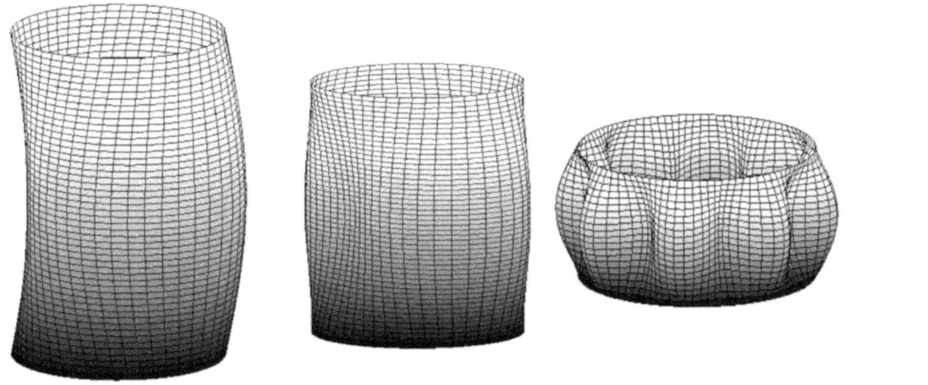

Figure 8.15

theory, with an undeformable section. The only difference here is that the energy corresponding to the bending of the walls can be evaluated (longitudinal bending and torsion). Also, the total potential with bending of the walls alone is reduced to:

$$\mathcal{T}+\mathcal{V}=\int_0^h\int_0^{2\pi R}\frac{1}{2}q\left[\left(\frac{\partial w}{\partial z}\right)^2\right]+\frac{1}{2}D\left[\left(\frac{\partial^2 w}{\partial z^2}\right)^2+2(1-\nu)\left(\frac{\partial^2 w}{\partial z\partial s}\right)^2\right]ds\,dz$$

The corresponding critical loading would then be obtained from the Rayleigh ratio:

$$-q_{crit}=D\frac{\int_0^h\int_0^{2\pi R}\left[\left(\frac{\partial^2 w}{\partial z^2}\right)^2+2(1-\nu)\left(\frac{\partial^2 w}{\partial z\partial s}\right)^2\right]ds\,dz}{\int_0^h\int_0^{2\pi R}\left[\left(\frac{\partial w}{\partial z}\right)^2\right]ds\,dz}=\frac{\pi^2 D}{h^2}\left[1+2(1-\nu)\kappa^2\right] \tag{8.112}$$

with κ the slenderness of the cylinder walls equal to $\kappa = h/\pi R$. To cause instability, q must be a compression: the second member is positive. For a given length h, the critical loading increases when R decreases, the influence of the torsional moment decreasing with the radius.

The critical Euler loading for a tubular beam simply supported at both ends is $-q_{crit}=\frac{1}{2\pi R}\frac{\pi^2 E\pi R^3 e}{h^2}=\frac{\pi^2 EeR^2}{2h^2}=\frac{Ee}{2\kappa^2}$ and completely differs from the value found above. The tube undergoes membrane stresses due to beam bending, whereas (8.112) is obtained by considering only the bending of the walls and is in a ratio e^2/R^2 with the critical Euler loading. The wall shell bending approach therefore only provides corrective action in this case, which is generally negligible, except for thick cylinders that are not very slender. The second-order approach, therefore, does not apply here, the membrane terms being preponderant. This highlights the fact that the second-order approach has limitations, which will be evaluated in what follows.

For the modes involving the deformation of the circle (for example ovalisation, for $m = 2$), the displacements $v_{n,m}$ and $u_{n,m}$ associated with $w_{n,m}$ in the circumferential bending are therefore given by:

$$v_{m,n}=-\frac{1}{m}\sin\frac{n\pi z}{h}\cos\frac{ms}{R}\;;\;u_{m,n}=\frac{n}{m^2\kappa}\cos\frac{n\pi z}{h}\sin\frac{ms}{R}$$

By deferring in the equation the expression of w for a mode $w_{m,n}$, it becomes:

$$\left[\left(\frac{n\pi}{h}\right)^2+\left(\frac{m}{R}\right)^2\right]^2-\frac{1}{R^2}\left(\nu\left(\frac{n\pi}{h}\right)^2+\left(\frac{m}{R}\right)^2\right)+\frac{q}{D}\left(\frac{n\pi}{h}\right)^2=0$$

then the critical force associated with the $w_{m,n}$ mode:

$$-q_{m,n}=\frac{D\pi^2}{h^2}\left[n^2+\kappa^2\left(2m^2-\nu\right)+\kappa^4\frac{m^2}{n^2}\left(m^2-1\right)\right] \tag{8.113}$$

The Poisson's ratio does not play a favourable role, so it should not be neglected.

The minimum value is obtained for $m = 2$ and $n = 1$; the corresponding critical value is given by:

$$-q_{2,1} = \frac{D\pi^2}{h^2}\left[1+\kappa^2(8-\nu)+12\kappa^4\right] \tag{8.114}$$

The corresponding loading increases with κ, for a given length h of the cylinder; indeed, decreasing the diameter also increases the stiffness with respect to the ovalisation.

Conversely, for a given diameter, the increase in length causes a reduction in the critical loading. For small values of κ corresponding to rather compact cylinders, there is very little difference between the critical values associated with the first modes in m and the deformation which is established can thus be very influenced by the geometrical imperfections (see below).

Photo 8.4: Non-axisymmetric buckling of a cylindrical shaft

Equation (8.114) can be compared to the lower bound value given by (8.74), equal to $\frac{D}{R^2}4\sqrt{3(1-\nu^2)}\frac{R}{e}$. For thin shells ($e << R$), the value given by (8.114) is much smaller, which highlights the importance of the kinematic assumptions made: the existence of waves in the circumferential direction introduces flexibility by bending, which takes **precedence in deformation.** This is the underlying assumption in the second-order approach, but it may be faulty.

Indeed, for compact (not slender) cylinders, the result is reversed, for example, when $\frac{D\pi^2}{h^2}\left[1+8\kappa^2\right] > \frac{D}{R^2}4\sqrt{3(1-\nu^2)}\frac{R}{e}$, i.e. $1+8\left(\frac{h}{\pi R}\right)^2 > \pi^2\frac{h^2}{Re}4\sqrt{3(1-\nu^2)}$, a condition which depends on the "thinness" e/R, but also on the slenderness κ. The term on the left in the inequality is related to the pure bending of the cylinder walls, while the one on the right is related to the bending wavelength. Also, when the cylinder is not thin and slender (short wavelength), it is no longer possible to neglect the membrane energy (see below).

The presence of an imperfection changes the way the cylinder deforms. By keeping the preponderant bending energy hypothesis, consider for example an imperfection proportional to the third mode in z: $w_0 = e\sin\frac{3\pi z}{h}\sin\frac{s}{R}$. In this case, the displacement with respect to the perfect geometry can be put in the form $w = w_0 + \bar{w}$. Whichever way the shape w_0 was obtained, it constitutes the new reference position.

This initial defect induces bending in the cylinder when the axial loading is applied, such that $M_0^{zz} = -qw_0$. q is applied to the cylinder with the defect, therefore, neglecting the variation in membrane force induced by this bending, Equation (8.110) becomes:

$$\frac{\partial^4\bar{w}}{\partial z^4}+2\frac{\partial^4\bar{w}}{\partial z^2\partial s^2}+\frac{\partial^4\bar{w}}{\partial s^4}+\frac{1}{R^2}\frac{\partial^2\bar{w}}{\partial s^2}+\left(\frac{\nu}{R^2}-\frac{q}{D}\right)\frac{\partial^2\bar{w}}{\partial z^2}-\frac{q}{D}\frac{\partial^2 w_0}{\partial z^2}=0$$

$\bar{w}$ is decomposed as in (8.109). The total displacement defining the deformed geometry (compared to the perfect cylinder) is written as:

$$w = \sum_{m,n} W_{m,n} \sin\frac{n\pi z}{h} \sin\frac{ms}{R} + w_0 \sin\frac{3\pi z}{h} \sin\frac{s}{R}$$

and the components $W_{m,n}$ must make it possible to respect the equation. In this specific case, the defect only has a component in mode (*1*, *3*), and the component $W_{1,3}$ according to this mode verifies:

$$W_{1,3}\left\{\left[\left(\frac{3\pi}{h}\right)^2+\left(\frac{1}{R}\right)^2\right]^2-\frac{1}{R^2}\left(\frac{1}{R}\right)^2-\left(\frac{\nu}{R^2}-\frac{q}{D}\right)\left(\frac{3\pi}{h}\right)^2\right\}+\frac{qe}{D}\left(\frac{3\pi}{h}\right)^2=0$$

that is $W_{1,3} = e\dfrac{q}{q_{1,3}-q}$. The total displacement according to this mode is therefore equal to

$e\left(1-\dfrac{q}{q_{1,3}}\right)^{-1}$ and increases progressively from e when q increases and tends towards infinity for the critical value $q_{1,3}$ corresponding to this mode, determined previously.

The other components have the same expression as determined above, and are thus null before the critical value associated with each mode, infinite for these values.

For the defect considered, the instability still appears for the critical value associated with the mode representing the defect. The first buckling mode is then superimposed on the displacement due to the imperfection determined above. In reality, the additional displacement due to the buckling mode is not suddenly infinite when the critical value is reached, but the linear method developed here does not make it possible to calculate its evolution. Nevertheless, this approach makes it possible to highlight the fact that a defect generates a bending which can lead to yielding or to a rupture even before reaching the first critical value.

Similar reasoning can be held for a circumferential defect, which explains why the number of visible waves in this direction is random for a compact cylinder, the critical values being close.

When the defect has any shape, it can be broken down into a double series, which makes it possible to generalise the previous result. Then each mode undergoes a progressive deformation depending on the ratio between q and $q_{m,n}$. The total displacement results from the superposition of all the modal components, the one causing the instability becoming preponderant as q increases.

The above approach applies well to thin and slender shells where the effects of extension can be neglected. The study of cylindrical vibrations carried out in §8.2.3.2 has shown that, if these slenderness assumptions are no longer verified, there is a coupling between flexion and extension. The usual approach to solve this problem is that of Donnell, from his Equation (8.83).

If the cylinder is free to deform longitudinally, the variation of total normal force applied in this direction to the cylinder is zero, but N^{zz} may not be local, because the buckling causes an overall bending of the cylinder considered like a beam with axis z, therefore a local deformation ε_{zz}, which results in a longitudinal displacement u varying according to s. It was noted earlier that this displacement can be considered small compared to the others, especially when the number of transverse waves increases.

Also, the membrane deformation may not be negligible, both longitudinally and transversely. The deformations are then given by (8.58), except for ε_{zz}, given by (8.60) to keep the second-order terms, and the curvature variations by (8.59).

It is advisable to first respect the first two equilibrium equations which are taken in the simplified form (8.67), neglecting the terms in M/R. Inspired by the previous solution, N^{ss} is taken proportional to $\sin\frac{n\pi z}{h}\sin\frac{ms}{R}$ under the same boundary conditions. The integration of these two equations makes it possible to obtain the other two membrane forces, and then the membrane deformations by the inverse constitutive law. The three deformation components are expressed as functions of the three displacement components by (8.58), which makes it possible to calculate these three components. Without detailing this calculation here, a solution, up to a coefficient, is given by:

$$w = R\left(1+\frac{m^2}{n^2}\kappa^2\right)^2 \sin\frac{n\pi z}{h}\sin\frac{ms}{R} \qquad \varepsilon_{ss} = -\left(1-\nu\frac{m^2}{n^2}\kappa^2\right)\sin\frac{n\pi z}{h}\sin\frac{ms}{R}$$

$$v = -\frac{m}{n^2}\kappa^2 R\left(2+\nu+\frac{m^2}{n^2}\kappa^2\right)\sin\frac{n\pi z}{h}\cos\frac{ms}{R} \qquad \varepsilon_{zz} = -\left(\frac{m^2}{n^2}\kappa^2-\nu\right)\sin\frac{n\pi z}{h}\sin\frac{ms}{R} \qquad (8.115)$$

$$u = \frac{\kappa R}{n}\left(\frac{m^2}{n^2}\kappa^2-\nu\right)\cos\frac{n\pi z}{h}\sin\frac{ms}{R} \qquad \varepsilon_{sz} = -\frac{m}{n}\kappa(1+\nu)\cos\frac{n\pi z}{h}\cos\frac{ms}{R}$$

It is noted that, for this solution to be valid, the longitudinal displacement u must be left free, which is zero only on average in $z = 0$. If this is not the case, it is the general form of the solution which is to be called into question; it must then take the form of a double series.

It should also be noted that this solution is not compatible with the non-deformability of circles for $m = 1$. The instability of the Euler tube necessarily implies a deformation of the section.

The curvature variations are then obtained by (8.59). Note that in Donnell's approach (8.63), the v/R terms are neglected, but it was seen in §8.2.1.10 that this is not justified if the curvature is not small and for a small number of transverse waves. Then the complete calculation can be made from the solution above, which leads to the following expressions of the variations of curvature:

$$\kappa_{zz} = -\frac{1}{R}n^2\frac{1}{\kappa^2}\left(1+\frac{m^2}{n^2}\kappa^2\right)^2\sin\frac{n\pi z}{h}\sin\frac{ms}{R}$$

$$\kappa_{zs} = \frac{m}{R}\left(1+\frac{m^2}{n^2}\kappa^2\right)^2\left[\frac{n}{\kappa}-\frac{\kappa}{n}\left(2+\nu+\frac{m^2}{n^2}\kappa^2\right)\left(1+\frac{m^2}{n^2}\kappa^2\right)^{-2}\right]\cos\frac{n\pi z}{h}\cos\frac{ms}{R}$$

$$\kappa_{ss} = \frac{1}{R}\left(1+\frac{m^2}{n^2}\kappa^2\right)^2\left[-\left(1+m^2\right)+2\frac{m^2}{n^2}\kappa^2\left(2+\nu+\frac{m^2}{n^2}\kappa^2\right)\left(1+\frac{m^2}{n^2}\kappa^2\right)^{-2}\right]\sin\frac{n\pi z}{h}\sin\frac{ms}{R}$$

(8.116)

Another aspect to examine is the work of q in the second order terms of ε_{zz}, which are equal to:

$$\frac{\partial w}{\partial z}=\frac{n}{\kappa}\left(1+\frac{m^2}{n^2}\kappa^2\right)^2\cos\frac{n\pi z}{h}\sin\frac{ms}{R}$$

$$\frac{\partial u}{\partial z}=\left(\frac{m^2}{n^2}\kappa^2-\nu\right)\cos\frac{n\pi z}{h}\sin\frac{ms}{R}$$

The ratio between the first and the second terms is of the order of $\frac{m^2}{n}\kappa$ and is not necessarily negligible for very compact cylinders. For the latter, the hypothesis of small rotations is no longer justified and it is then necessary to take the complete expression of ε_{zz} given by (8.51).

These results make it possible to give a more precise expression of the critical loading corresponding to this mode of displacement obtained from the Rayleigh quotient:

$$-q_{crit}=\frac{\left\{\begin{array}{l}\frac{1}{2}K\int\limits_{C}\left[\varepsilon_{ss}{}^2+\varepsilon_{zz}{}^2+2\nu\varepsilon_{ss}\varepsilon_{zz}+2(1-\nu)\varepsilon_{sz}{}^2\right]dS\\ \qquad+\frac{1}{2}D\int\limits_{C}\left[\kappa_{ss}{}^2+\kappa_{zz}{}^2+2\nu\kappa_{ss}\kappa_{zz}+2(1-\nu)\kappa_{sz}{}^2\right]dS\end{array}\right\}}{\frac{1}{2}\int\limits_{C}\left[\left(\frac{\partial w}{\partial z}\right)^2+\left(\frac{\partial u}{\partial z}\right)^2\right]dS} \tag{8.117}$$

By replacing and rearranging the membrane term, it becomes:

$$\begin{aligned}-q_{crit}=&\frac{\frac{1}{2}Ee\left[1+\frac{m^2}{n^2}\kappa^2\right]^2}{\frac{n^2}{\kappa^2}\left(1+\frac{m^2}{n^2}\kappa^2\right)^4+\left(\frac{m^2}{n^2}\kappa^2-\nu\right)^2}\\&+\frac{\frac{1}{2}\frac{D}{R^2}\left(1+\frac{m^2}{n^2}\kappa^2\right)^4\left[\begin{array}{l}\left[-(1+m^2)+2\frac{m^2}{n^2}\kappa^2\left(2+\nu+\frac{m^2}{n^2}\kappa^2\right)\left(1+\frac{m^2}{n^2}\kappa^2\right)^{-2}\right]^2+\left[n^2\frac{1}{\kappa^2}\right]^2\\ +2\nu\left[-(1+m^2)+2\frac{m^2}{n^2}\kappa^2\left(2+\nu+\frac{m^2}{n^2}\kappa^2\right)\left(1+\frac{m^2}{n^2}\kappa^2\right)^{-2}\right]n^2\frac{1}{\kappa^2}\\ +2(1-\nu)m^2\left[\frac{n}{\kappa}-\frac{\kappa}{n}\left(2+\nu+\frac{m^2}{n^2}\kappa^2\right)\left(1+\frac{m^2}{n^2}\kappa^2\right)^{-2}\right]^2\end{array}\right]}{\frac{n^2}{\kappa^2}\left(1+\frac{m^2}{n^2}\kappa^2\right)^4+\left(\frac{m^2}{n^2}\kappa^2-\nu\right)^2}\end{aligned} \tag{8.118}$$

The critical mode (m, n) causing the instability depends on the geometric characteristics κ and e/R of the cylinder.

This result is compared to that obtained with the simplifying Donnell assumptions:

$$-q_{crit} = Een^2\kappa^2\left(n^2 + m^2\kappa\right)^{-2} + \frac{\pi^4 D}{n^2 h^2}\left(n^2 + m^2\kappa\right)^2 \tag{8.119}$$

For the membrane term, the difference essentially comes from $\left(\frac{\partial u}{\partial z}\right)^2$ which may not be negligible for very compact cylinders.

For the bending term, the difference comes from the simplified expression (8.63) of the curvatures.

In conclusion, Donnell's result is a useful simplification for not-too-compact cylinders.

The second-order approach has the advantage of being more simple, but can only be used for slender cylinders. For example, (8.119) becomes for $n = 1$ and $m = 2$:

$$-q_{crit} = Ee\frac{\kappa^2}{\left(1+4\kappa\right)^2} + \frac{\pi^4 D}{h^2}\left(1+4\kappa\right)^2$$

which can be compared to (8.114). There is a small difference in the bending term, which comes from Donnell's simplifying assumption. The first term can be neglected if:

$$\frac{1}{\left(1+4\kappa\right)^2}\frac{h^2}{eR}\sqrt{1-\nu^2} << \frac{\pi^3}{\sqrt{12}} \approx 9 \tag{8.120}$$

For large values of κ, this condition is written approximately $\frac{R}{e} << 15$. For the second-order approach to be valid, it is not enough for the cylinder to be slender, it must also be very thin.

8.2.4.2.3 Torsional buckling of the cylinder

The same cylinder is subjected to a global torsion C, applied on the lower and upper edges by a uniform shear N_0^{sz} such that $N_0^{sz} = T = \frac{C}{2\pi R^2}$. The state of membrane prestress is given by:

$$\overline{\overline{N_0}} = \begin{pmatrix} 0 & T \\ T & 0 \end{pmatrix}$$

respects the first-order equilibrium equations and boundary conditions on stresses. In this problem, the first-order tangential displacement v can be fixed on one of the edges, but not on the other, to allow the application of the torsion torque. Once the torsion is applied, subsequent buckling conditions can be chosen based on the behaviour of the edges.

In the second-order approximation, the equilibrium Equation (8.106) is written:

$$\frac{\partial^2 M^{zz}}{\partial z^2} + 2\frac{\partial^2 M^{zs}}{\partial z \partial s} + \frac{\partial^2 M^{ss}}{\partial s^2} - \frac{T}{R}\frac{\partial v}{\partial z} - 2T\frac{\partial^2 w}{\partial s \partial z} = 0$$

By replacing in the bending equation:

$$\frac{\partial^4 w}{\partial z^4}+2\frac{\partial^4 w}{\partial z^2 \partial s^2}+\frac{\partial^4 w}{\partial s^4}+\left(\nu\frac{\partial^2}{\partial z^2}+\frac{\partial^2}{\partial s^2}\right)\left(\frac{w}{R^2}\right)-\frac{T}{D}\left(\frac{1}{R}\frac{\partial v}{\partial z}+2\frac{\partial^2 w}{\partial s \partial z}\right)=0$$

In z, the solution of this equation can no longer consist of simple sines, it is necessary to use the more general form given by (8.71) and (8.72), hence replacing in the equation:

$$\sum_{p=1}^{\infty}\left[\begin{array}{l}\left(\frac{1}{h^4}f_p^{\text{IV}}-\frac{1}{h^2}\frac{2p^2-\nu}{R^2}f_p''+\frac{p^2\left(p^2-1\right)}{R^4}f_p+\frac{T}{DRh}\frac{2p^2-1}{p}g_p'\right)\sin\frac{ps}{R}\\ +\left(\frac{1}{h^4}g_p^{\text{IV}}-\frac{1}{h^2}\frac{2p^2-\nu}{R^2}g_p''+\frac{p^2\left(p^2-1\right)}{R^4}g_p-\frac{T}{DRh}\frac{2p^2-1}{p}f_m'\right)\cos\frac{ps}{R}\end{array}\right]=0$$

Performing the dot product of this expression by sin ms and cos ms, for a fixed value of m, then making the results adimensional by setting:

$$\rho=\frac{R}{h}\;;\;\tau=\frac{TR^3}{Dh\rho}\;;\;\chi=\frac{z}{h}$$

it becomes:

$$\rho^4 f_m^{\text{IV}}-\left(2m^2-\nu\right)\rho^2 f_m''+m^2\left(m^2-1\right)f_m+\tau\rho\frac{2m^2-1}{m}g_m'=0$$

$$\rho^4 g_m^{\text{IV}}-\left(2m^2-\nu\right)\rho^2 g_m''+m^2\left(m^2-1\right)g_m-\tau\rho\frac{2m^2-1}{m}f_m'=0$$

Thus, the problem is solved independently for each value of m. Fixing m, a solution of the system obtained is sought in the form:

$$\begin{pmatrix}f_m\\ g_m\end{pmatrix}=\begin{pmatrix}F_m\\ G_m\end{pmatrix}e^{r\chi}$$

From where, by deferring in the system of equations, and by posing $x=\rho r$:

$$\begin{pmatrix}x^4-\left(2m^2-\nu\right)x^2+m^2\left(m^2-1\right) & \frac{\tau}{\rho}\frac{2m^2-1}{m}x\\ -\frac{\tau}{\rho}\frac{2m^2-1}{m}x & x^4-\left(2m^2-\nu\right)x^2+m^2\left(m^2-1\right)\end{pmatrix}\begin{pmatrix}F_m\\ G_m\end{pmatrix}=0$$

For this system to have a non-zero solution, its determinant must be zero:

$$\det=\left[x^4-\left(2m^2-\nu\right)x^2+m^2\left(m^2-1\right)\right]^2+\left[\left(\frac{\tau}{\rho}\right)\frac{2m^2-1}{m}x\right]^2=0$$

Below a value of τ which depends on m, that is, τ_{0m}, there are eight complex solutions formed from two real parts α_1 and α_2 and an imaginary part β, all taken positive, of such so that the solution can be written:

$$\begin{pmatrix} f_m \\ g_m \end{pmatrix} = \begin{pmatrix} F_m \\ G_m \end{pmatrix} \begin{Bmatrix} \left[A_{m1}\text{ch}(\alpha_1\chi) + A_{m2}\text{sh}(\alpha_1\chi) + A_{m3}\text{ch}(\alpha_2\chi) + A_{m4}\text{sh}(\alpha_2\chi)\right]\cos\beta\chi \\ +\left[B_{m1}\text{ch}(\alpha_1\chi) + B_{m2}\text{sh}(\alpha_1\chi) + B_{m3}\text{ch}(\alpha_2\chi) + B_{m4}\text{sh}(\alpha_2\chi)\right]\sin\beta\chi \end{Bmatrix}$$

The two components are the same functions of χ, so we can, for each value of m, choose the orientation of the radial axis perpendicular to z to keep only one of the components; this orientation varies from one value of m to another. Then the solution is sought for f_m alone.

With respect to the median plane of the cylinder, the solutions are symmetric in w and v, so it is simpler to take this median plane as the origin of z and to write the boundary conditions in $z = d$, i.e. $\chi = \frac{z}{d} = \pm 1$. Then the solution is of the form:

$$f_m(\chi) = \left[A_{m1}\text{ch}(\alpha_1\chi) + A_{m3}\text{ch}(\alpha_2\chi)\right]\cos\beta\chi + \left[B_{m2}\text{sh}(\alpha_1\chi) + B_{m4}\text{sh}(\alpha_2\chi)\right]\sin\beta\chi$$

u is antisymmetric and vanishes in the middle of the cylinder. Normally, the longitudinal displacement u is zero on one edge and free on the other; here, it is equivalent up to a general translation. Ultimately:

$$\text{u}_m(z,s) = \frac{R}{m^2 d} f_m'(\chi)\sin\frac{ms}{R}$$

It is then necessary to express the boundary conditions. The kinematic conditions relate to the displacements fixed on the edges. If w is taken as zero, so is v, given the expression of the form of the solution, these are two coherent conditions.

On the edges, as the displacement w is zero, the reaction q^3 is not. Similarly, since v is zero, q^s is not. From (8.107), it remains:

$$M^{zz} = D\left[\frac{\partial^2 w}{\partial z^2} + \nu\left(\frac{\partial^2 w}{\partial s^2} + \frac{w}{R^2}\right)\right]_{z=\pm d} = 0 \quad \Rightarrow \quad \left[f_m'' + \nu\frac{m^2+1}{R^2} f_m\right]_{\chi=\pm 1} = 0$$

Then:

$$\begin{aligned} f_m''(1) = {} & A_m\left[(\alpha_1^2 - \beta^2)\text{ch}(\alpha_1)\cos\beta - 2\alpha_1\beta\text{sh}(\alpha_1)\sin\beta\right] \\ & + B_m\left[(\alpha_2^2 - \beta^2)\text{ch}(\alpha_2)\cos\beta - 2\alpha_2\beta\text{sh}(\alpha_2)\sin\beta\right] \\ & + C_m\left[2\alpha_1\beta\text{ch}(\alpha_1)\cos\beta + (\alpha_1^2 - \beta^2)\text{sh}(\alpha_1)\sin\beta\right] \\ & + D_m\left[2\alpha_2\beta\text{ch}(\alpha_2)\cos\beta + (\alpha_2^2 - \beta^2)\text{sh}(\alpha_2)\sin\beta\right] = 0 \end{aligned}$$

There are therefore two equations relating to the four coefficients A_m, B_m, C_m, and D_m, with a zero second member, so non-zero solutions are possible. They depend on the roots

α_1, α_2, and β, which themselves depend on τ, which leads to values of τ allowing instability. The lowest gives the critical value of torsion, for a given mode m. The value sought corresponds to the critical values obtained, considering all the modes m.

The combination of two of the functions above makes it possible to find a buckling mode, obtained for the value of τ for which the corresponding 2×2 determinant is zero. For $m = 2$, the one that results in the lowest value of τ corresponds to the deformation:

$$\operatorname{ch}(\alpha_1\chi)\cos(\beta\chi) - 13{,}89 \times \operatorname{sh}(\alpha_2\chi)\sin(\beta\chi)$$

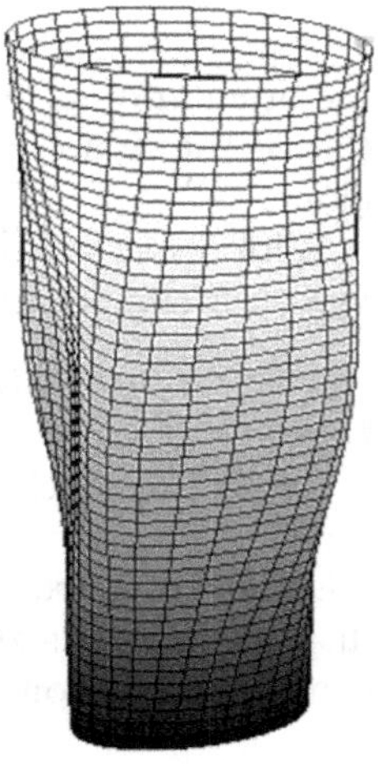

Figure 8.16

with $\alpha_1 = 3{,}069$, $\alpha_2 = 0{,}552$, and $\beta = 0{,}928$; see Figure 8.16 including the initial torsion. τ is then 4,836. To this critical value of τ corresponds the critical value of shear:

$$T_{cr} = \frac{D}{R^2}\tau_{cr} = 4{,}836\frac{D}{R^2} \tag{8.121}$$

It should be noted that this displacement is associated with strong boundary conditions, namely that a displacement in the plane of the edges is zero, which corresponds approximately to the situation of a tin can with edges closed by a rigid membrane. For this mode, the critical value does not depend on the height. A similar search can be performed for other values of m.

As in the case of vertical pressure, for thick and not slender cylinders, the membrane term brings a non-negligible additional rigidity. The same approach as in the case of the axial compression can be followed, but here, it is necessary to consider a complete decomposition in the Fourrier series to allow the respect of the boundary conditions on the two edges. Nevertheless, each circumferential mode m can be treated separately and, for each value of m, an approximate solution can be found by taking a finite number of modes in n.

An approximation can also be obtained by the Rayleigh quotient by admitting the same deformation as that obtained in the second-order approximation. These calculations are not continued here.

8.2.4.3 Introduction to the stability of a cylindrical vault

The main difference is that these vaults, often made up of portions of juxtaposed cylinders, are open shells. Subject to the shallow shells assumption, Donnell's assumptions may be used. To study vault stability, the effect of membrane forces must be introduced in (8.101). The displacement w_0 at first-order equilibrium is due to the loading appearing on the right-hand side of (8.102). The membrane prestress balances the external loading and, neglecting the variations of membrane forces due to the displacement w with respect to the first order equilibrium position (as to establish (8.81)), it becomes:

$$\boxed{D\Delta^4 w - \Delta^2\left(N^{\alpha\beta}\,\partial^2_{\alpha\beta} w\right) - \frac{Ee}{R^2}\partial^4_{z^4} w = 0} \tag{8.122}$$

This is Equation (8.90) already established by a slightly different method, here without a second member for the search for instability. It can also be obtained by the energy approach by considering the displacements with respect to the position of equilibrium obtained above.

The self-weight induces a prestress $N^{\alpha\beta}$ which varies greatly as a function of θ and z; when the boundary conditions lend themselves to it, this distribution can be given by the membrane solution of §6.2.2.4, otherwise, it is necessary to proceed as in §8.2.3.3. A numerical solution of (8.122) can then be obtained. Nevertheless, the search for the buckling modes numerically from (8.122) is not concurrent compared to a resolution by the finite element method (cf. Chapter 9), subject to an adequate treatment of the terms of the second order. It should also be noted that bending is not negligible in the shell, under self-weight, and that there is therefore a moment of prestress for the search for instability, which does not appear in (8.122).

8.3 THIN SHELLS OF REVOLUTION

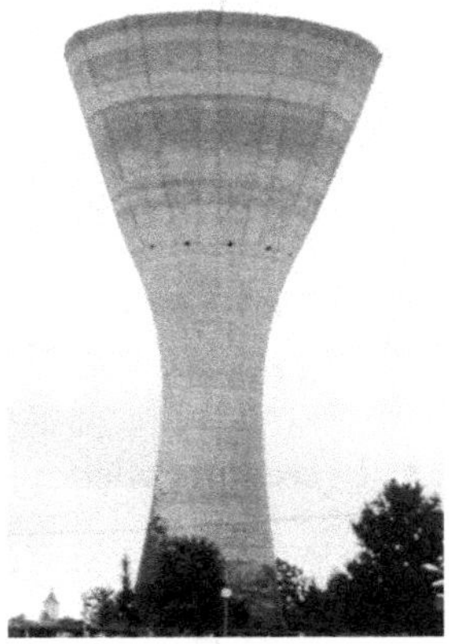

Photo 8.5: Drinking water tank

8.3.1 Deformations

8.3.1.1 Geometry

The geometry of the shells of revolution has been specified in §6.3.1.1 (Figure 8.17). The basis (A_1, A_2) is orthonormal and principal of curvature, the index *1* being relative to the curvilinear abscissa along the generatrix, also parameterised by the angle ϕ, the index *2* to the curvilinear abscissa along the circle of radius r, also parameterised by the angle θ.

8.3.1.2 Deformations in small axisymmetric transformation

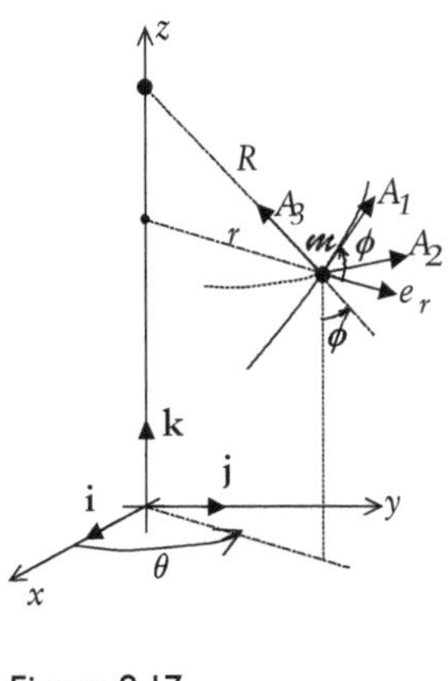

Figure 8.17

The following developments are limited to the case of axisymmetric loadings for which:

$$\xi = u(s)A_1 + w(s)A_3 \tag{8.123}$$

At the current point m of the deformed surface, the local basis is:

$$a_1 = \frac{dm}{ds} = \left(1 + u' - \frac{w}{R_s}\right)A_1 + \left(w' + \frac{u}{R_s}\right)A_3$$

$$a_2 = \frac{dm}{rd\theta} = \left(1 + \frac{u\cos\phi - w\sin\phi}{r}\right)A_2$$

which allows expressing of the transformation (3.95):

$$\mathcal{A}_1^{\ 1} = u' - \frac{w}{R_s}\ ;\ \mathcal{A}_1^{\ 2} = 0 \qquad \mathcal{B}_1 = w' + \frac{u}{R_s}$$

$$\mathcal{A}_2^{\ 1} = 0\ ;\ \mathcal{A}_2^{\ 2} = \frac{u\cos\phi - w\sin\phi}{r} \qquad \mathcal{B}_2 = 0$$

from which:

$$\nabla_1\mathcal{B}_1 = \frac{d}{ds}\left(w' + \frac{u}{R_s}\right);\ \nabla_2\mathcal{B}_1 = \nabla_1\mathcal{B}_2 = 0;\ \nabla_2\mathcal{B}_2 = \left(w' + \frac{u}{R_s}\right)\frac{\cos\phi}{r}$$

This leads to the expression of the deformation variables under the assumption of small deformations and small rotations:

- In first-order approximation:

$$\begin{aligned} \varepsilon_{11} &= u' - \frac{w}{R_s} \\ \varepsilon_{22} &= \frac{u\cos\phi - w\sin\phi}{r} \end{aligned} \tag{8.124}$$

$$\begin{aligned} \kappa_{11} &= \frac{1}{R_s}\left(u' - \frac{w}{R_s}\right) + w'' + \frac{d}{ds}\left(\frac{u}{R_s}\right) \\ \kappa_{22} &= \frac{1}{r}\left(\frac{u\cos\phi - w\sin\phi}{r}\right)\sin\phi + \frac{1}{r}\left(w' + \frac{u}{R_s}\right)\cos\phi \end{aligned} \tag{8.125}$$

- In second-order approximation:

$$\begin{aligned} \varepsilon_{11} &= u' - \frac{w}{R_s} + \frac{1}{2}\left(w' + \frac{u}{R_s}\right)^2 \\ \varepsilon_{22} &= \frac{u\cos\phi - w\sin\phi}{r} \end{aligned} \tag{8.126}$$

with the symmetric expression of κ:

$$\begin{aligned} \kappa_{11} &= \frac{d}{ds}\left(w' + \frac{u}{R_s}\right) \\ \kappa_{22} &= \left(w' + \frac{u}{R_s}\right)\frac{\cos\phi}{r} = \left(w' + \frac{u}{R_s}\right)\frac{\cot\phi}{R_\theta} \end{aligned} \tag{8.127}$$

The normal vector is given by:

$$a_3 \approx A_3 - \left(w' + \frac{u}{R_s}\right)A_1$$

The rotation $\boldsymbol{\omega}$ of the normal is obtained by (3.104):

$$\boldsymbol{\omega}\times A_3 = -\left(w' + \frac{u}{R_s}\right)A_1 \quad \Rightarrow \quad \boldsymbol{\omega} = -\left(w' + \frac{u}{R_s}\right)A_2 = -\gamma\, A_2$$

In Love approximation, the deformations are linearised, given by (8.124), and the curvature variations by (8.127).

In Donnell's approximation, the deformations are of second order and reduced to:

$$\varepsilon_{11} = u' - \frac{w}{R_s} + \frac{1}{2}(w')^2$$
$$\varepsilon_{22} = \frac{u\cos\phi - w\sin\phi}{r} \tag{8.128}$$

and the variations of curvature are simplified in:

$$\kappa_{11} = w''$$
$$\kappa_{22} = \frac{w'}{r}\cos\phi \tag{8.129}$$

Finally, in classical approximation, the deformations are given by (8.124) and the curvature variations by (8.129).

8.3.2 Resolution methods

The transformations considered in this section are axisymmetric.

8.3.2.1 Equilibrium equations

The following developments are made as part of Love's approximation. After the expression of the connection coefficients, the equilibrium equations are reduced to:

$$\frac{dN^{11}}{ds} + \frac{\cos\phi}{r}\left(N^{11} - N^{22}\right) + \frac{1}{R_s}Q^1 + p^1 = 0$$
$$-\frac{N^{11}}{R_s} - \frac{N^{22}\sin\phi}{r} + \frac{dQ^1}{ds} + \frac{Q^1\cos\phi}{r} + p^3 = 0$$

with:

$$Q^1 = -\left[\frac{dM^{11}}{ds} + \frac{\cos\phi}{r}\left(M^{11} - M^{22}\right)\right]$$
$$Q^2 = 0$$

They can also be written:

$$\frac{1}{R_s}\frac{dN^{11}}{d\phi} + \frac{\cot\phi}{R_\theta}\left(N^{11} - N^{22}\right) + \frac{1}{R_s}Q^1 + p^1 = 0$$
$$-\frac{N^{11}}{R_s} - \frac{N^{22}}{R_\theta} + \frac{1}{R_s}\frac{dQ^1}{d\phi} + \frac{Q^1\cot\phi}{R_\theta} + p^3 = 0 \tag{8.130}$$

The edge is defined by the two circles of abscissa s_1 and s_2, where:

$$N^{11} + \frac{M^{11} + m^{v}}{R_s} - q^{1} = 0$$

$$Q^{1} = q^{3} \tag{8.131}$$

$$M^{11} = m^{t}$$

8.3.2.2 Homogeneous and isotropic elastic shell

Given the expression of the deformations (8.124) and (8.127), in small transformations, the constitutive laws are written:

$$N^{11} = K\left(u' - \frac{w}{R_s} + \nu\frac{u\cos\phi - w\sin\phi}{r}\right) = K\left(\frac{1}{R_s}\frac{du}{d\phi} - \frac{w}{R_s} + \nu\frac{u\cot\phi - w}{R_\theta}\right)$$

$$N^{22} = K\left(\frac{u\cos\phi - w\sin\phi}{r} + \nu\left(u' - \frac{w}{R_s}\right)\right) = K\left(\frac{u\cot\phi - w}{R_\theta} + \nu\left(\frac{1}{R_s}\frac{du}{d\phi} - \frac{w}{R_s}\right)\right)$$

$$M^{11} = D\left[\frac{d}{ds}\left(w' + \frac{u}{R_s}\right) + \nu\left(w' + \frac{u}{R_s}\right)\frac{\cos\phi}{r}\right] = D\left[\frac{1}{R_s}\frac{d\gamma}{d\phi} + \nu\,\gamma\frac{\cot\phi}{R_\theta}\right]$$

$$M^{22} = D\left[\left(w' + \frac{u}{R_s}\right)\frac{\cos\phi}{r} + \nu\frac{d}{ds}\left(w' + \frac{u}{R_s}\right)\right] = D\left[\gamma\frac{\cot\phi}{R_\theta} + \nu\frac{1}{R_s}\frac{d\gamma}{d\phi}\right] \tag{8.132}$$

$$Q^{1} = -D\left\{\begin{aligned}&\frac{d^2}{ds^2}\left(w' + \frac{u}{R_s}\right) + \nu\frac{d}{ds}\left[\left(w' + \frac{u}{R_s}\right)\frac{\cos\phi}{r}\right]\\ &+(1-\nu)\frac{\cos\phi}{r}\left[\frac{d}{ds}\left(w' + \frac{u}{R_s}\right) - \left(w' + \frac{u}{R_s}\right)\frac{\cos\phi}{r}\right]\end{aligned}\right\}$$

$$= -D\left\{\begin{aligned}&\frac{1}{R_s}\frac{d}{d\phi}\left(\frac{1}{R_s}\frac{d\gamma}{d\phi}\right) + \nu\frac{1}{R_s}\frac{d}{d\phi}\left(\gamma\frac{\cot\phi}{R_\theta}\right)\\ &+(1-\nu)\left(\frac{1}{R_s}\frac{d\gamma}{d\phi} - \gamma\frac{\cot\phi}{R_\theta}\right)\frac{\cot\phi}{R_\theta}\end{aligned}\right\}$$

where is introduced γ related to the rotation of the normal vector:

$$\gamma = \frac{1}{R_s}\left(\frac{dw}{d\phi} + u\right)$$

8.3.2.3 Resolution from equilibrium equations

The problem can be solved by introducing the constitutive laws (8.132) into equilibrium Equation (8.130). Equation (8.130) admits a particular solution, that is obtained in membrane theory, cf. 6.3. To deal with bending, it is therefore sufficient to look for the solutions of the equation without a second member ($p^1 = p^3 = 0$) and to adapt the boundary conditions accordingly. In the case where the thickness of the shell is constant:

$$
\begin{aligned}
&\frac{1}{R_s}\frac{d}{d\phi}\left(\frac{1}{R_s}\frac{du}{d\phi}-\frac{w}{R_s}+\nu\frac{u\cot\phi-w}{R_\theta}\right) \\
&\qquad +\frac{\cot\phi}{R_\theta}(1-\nu)\left(\frac{1}{R_s}\frac{du}{d\phi}-\frac{w}{R_s}-\frac{u\cot\phi-w}{R_\theta}\right)+\frac{1}{KR_s}Q^1=0 \\
&-\frac{K}{R_s}\left(\frac{1}{R_s}\frac{du}{d\phi}-\frac{w}{R_s}+\nu\frac{u\cot\phi-w}{R_\theta}\right)-\frac{K}{R_\theta}\left(\frac{u\cot\phi-w}{R_\theta}+\nu\left(\frac{1}{R_s}\frac{du}{d\phi}-\frac{w}{R_s}\right)\right) \\
&\qquad +\frac{1}{R_s}\frac{d}{d\phi}\left(Q^1\right)+\frac{Q^1\cot\phi}{R_\theta}=0
\end{aligned}
\tag{8.133}
$$

where Q_1 is given by (8.132). The solutions of these equations are discussed in the next paragraph, in the particular case of the sphere.

8.3.2.4 Resolution by the energy approach

The strain energy is evaluated in Love approximation from the expressions of the deformations (8.117) and (8.120):

$$
\begin{aligned}
\mathcal{F} &= \frac{1}{2}2\pi\int_{s1}^{s2} K\left\{
\begin{aligned}
&(1-\nu)\left[\left(u'-\frac{w}{R_s}\right)^2+\left(\frac{u\cos\phi-w\sin\phi}{r}\right)^2\right] \\
&+\nu\left(u'-\frac{w}{R_s}+\frac{u\cos\phi-w\sin\phi}{r}\right)^2
\end{aligned}
\right\}ds \\
&+\frac{1}{2}2\pi\int_{s1}^{s2} D\left\{
\begin{aligned}
&(1-\nu)\left[\left(w''+\frac{d}{ds}\left(\frac{u}{R_s}\right)\right)^2+\left(\frac{1}{r}\left(w'+\frac{u}{R_s}\right)\cos\phi\right)^2\right] \\
&+\nu\left(w''+\frac{d}{ds}\left(\frac{u}{R_s}\right)+\frac{1}{r}\left(w'+\frac{u}{R_s}\right)\cos\phi\right)^2
\end{aligned}
\right\}ds
\end{aligned}
\tag{8.134}
$$

which makes it possible to obtain two equations in u and w by minimizing the total potential. Although the establishment of the two equilibrium equations presents no particular difficulty, it is a little tedious to develop them if the shape of the particular geometry of the shell is not specified (which is done in the example below).

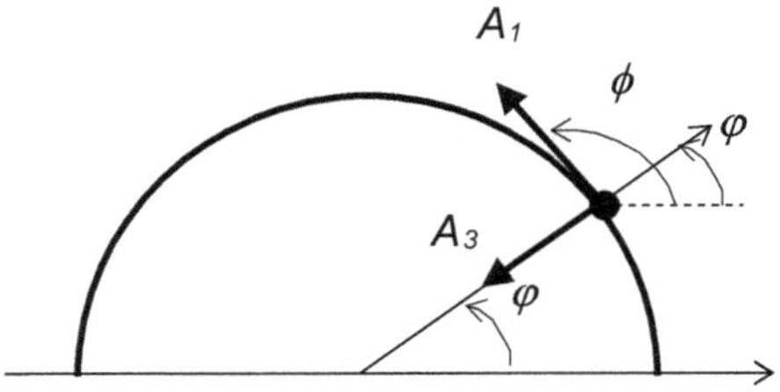

Figure 8.18

Example: Spherical dome under internal pressure.

In this case, spherical coordinates are used. The latitude φ is related to the angle ϕ previously used by $\phi = \varphi + \frac{\pi}{2}$. The vector A_1 is tangent to the meridian, the vector A_2 to the parallel and in this case the normal vector A_3 is directed towards the centre of the sphere (Figure 8.18). The displacement is written according to the latitude:

$$\xi(\varphi) = u(\varphi)A_1 + w(\varphi)A_3$$

The vectors of the natural basis are, in the deformed position:

$$a_\varphi = \left(1 + \frac{u' - w}{R}\right)A_\varphi + (w' + u)A_3$$

$$a_\theta = \left(1 - \frac{w + u\,tg\varphi}{R}\right)A_\theta$$

and in small displacements:

$$a_3 = A_3 - \frac{w' + u}{R^2}A_\varphi$$

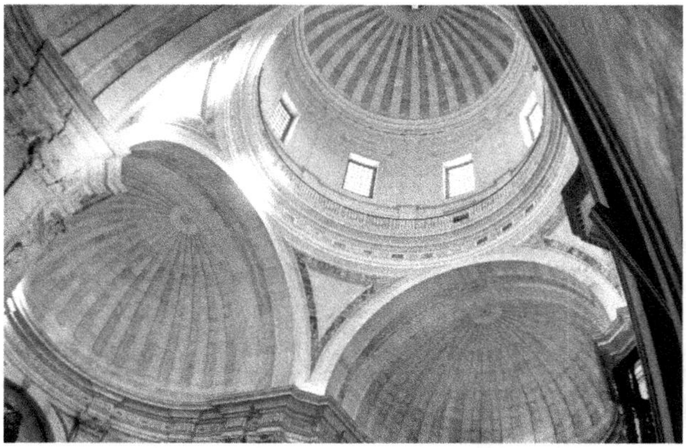

Photo 8.6: Spherical cupolas (Lisbon, Portugal)

The equilibrium equation (8.123) is rewritten:

$$\frac{dN^{11}}{d\varphi} - \left(N^{11} - N^{22}\right)\tan\varphi + Q^1 = 0$$
$$-\left(N^{11} + N^{22}\right) + \frac{dQ^1}{d\varphi} - Q^1\tan\varphi + pR = 0 \qquad (8.135)$$

Linearised deformation variables become, in the associated orthonormal basis:

$$\tilde{\varepsilon}_{\varphi\varphi} = \frac{u' - w}{R} \qquad \tilde{\kappa}_{\varphi\varphi} = \frac{w'' - w + 2u'}{R^2} = \frac{w'' + u'}{R^2} + \frac{\tilde{\varepsilon}_{\varphi\varphi}}{R}$$
$$\tilde{\varepsilon}_{\theta\theta} = -\frac{w + u\tan\varphi}{R} \qquad \tilde{\kappa}_{\theta\theta} = \frac{-(2u + w')\tan\varphi - w}{R^2} = -\frac{u + w'}{R^2}\tan\varphi + \frac{\tilde{\varepsilon}_{\theta\theta}}{R} \qquad (8.136)$$

In the expression of the bending energy, terms in $\frac{D}{R^2}\tilde{\varepsilon}^2$ which are small (in a ratio of the order of $\frac{e^2}{R^2}$), with respect to the terms $K\tilde{\varepsilon}^2$ coming from the membrane energy, appear.

This justifies neglecting the terms $\tilde{\varepsilon}$ in the expression of the variations of curvature. This is the Love approximation. Under these conditions, the strain energy is reduced to:

$$\mathcal{F} = \frac{1}{2} 2\pi \int_{s_1}^{s_2} K \left\{ \begin{aligned} &(1-\nu)\left[\left(\frac{u'-w}{R}\right)^2 + \left(\frac{w+u\tan\varphi}{R}\right)^2\right] \\ &+\nu\left(\frac{u'-w}{R} - \frac{w+u\tan\varphi}{R}\right)^2 \end{aligned} \right\} R^2 \cos\varphi \, d\varphi$$

$$+\frac{1}{2} 2\pi \int_{s_1}^{s_2} D \left\{ \begin{aligned} &(1-\nu)\left[\left(\frac{w''+u'}{R^2}\right)^2 + \left(\frac{u+w'}{R^2}\tan\varphi\right)^2\right] \\ &+\nu\left(\frac{w''+u'}{R^2} - \frac{u+w'}{R^2}\tan\varphi\right)^2 \end{aligned} \right\} R^2 \cos\varphi \, d\varphi$$

This expression can be simplified by posing:

$$u + w' = \gamma \tag{8.137}$$

where γ is related to the rotation $\boldsymbol{\omega}$ of the normal and intervenes as the only variable in flexion. Indeed, by (3.105)[5] :

$$\boldsymbol{\omega} = \frac{w'+u}{R} A_1 \times A_3 = -\frac{\gamma}{R} A_2$$

This allows to rewrite (8.136) in the form:

$$\begin{aligned} \tilde{\varepsilon}_{\varphi\varphi} &= \frac{\gamma' - w'' - w}{R} & \tilde{\kappa}_{\varphi\varphi} &= \frac{\gamma'}{R^2} \\ \tilde{\varepsilon}_{\theta\theta} &= -\frac{w + (\gamma - w')\tan\varphi}{R} & \tilde{\kappa}_{\theta\theta} &= -\frac{\gamma}{R^2}\tan\varphi \end{aligned} \tag{8.138}$$

It becomes:

$$\mathcal{F} = \frac{1}{2} 2\pi \int_{\varphi_1}^{\varphi_2} K \left\{ \begin{aligned} &(1-\nu)\left[(\gamma' - w'' - w)^2 + (w + (\gamma - w')\tan\varphi)^2\right] \\ &+\nu\left[\gamma' - w'' - 2w - (\gamma - w')\tan\varphi\right]^2 \end{aligned} \right\} \cos\varphi \, d\varphi$$

$$+\frac{1}{2} 2\pi \int_{\varphi_1}^{\varphi_2} \frac{D}{R^2} \left\{ (1-\nu)\left[\gamma'^2 + (\gamma\tan\varphi)^2\right] + \nu(\gamma' - \gamma\tan\varphi)^2 \right\} \cos\varphi \, d\varphi$$

[5] For the sake of simplification, the definition of γ used here differs slightly from that used in 8.3.2.2.

The potential of (internal) pressure is written:

$$\mathcal{V}_e = -2\pi\int_{\varphi_1}^{\varphi_2} p\, w\, R^2 \cos\varphi\, d\varphi$$

By virtualisation of the elastic potential:

$$\overset{*}{\mathcal{F}} = 2\pi\int_{\varphi_1}^{\varphi_2} K\left\{\begin{array}{l}(1-\nu)\left[\begin{array}{l}(\gamma'-w''-w)\left(\overset{*}{\gamma}{}'-\overset{*}{w}{}''-\overset{*}{w}\right)\\[2ex] +\left(w+(\gamma-w')\tan\varphi\right)\left(\overset{*}{w}+\left(\overset{*}{\gamma}-\overset{*}{w}{}'\right)\tan\varphi\right)\end{array}\right]\\[4ex] +\nu\left[\gamma'-w''-2w-(\gamma-w')\tan\varphi\right]\left[\overset{*}{\gamma}{}'-\overset{*}{w}{}''-2\overset{*}{w}-\left(\overset{*}{\gamma}-\overset{*}{w}{}'\right)\tan\varphi\right]\end{array}\right\}\cos\varphi\, d\varphi$$

$$+2\pi\int_{\varphi_1}^{\varphi_2}\frac{D}{R^2}\left\{(1-\nu)\left[\gamma'\overset{*}{\gamma}{}'+\gamma\overset{*}{\gamma}\,(\tan\varphi)^2\right]+\nu\,(\gamma'-\gamma\tan\varphi)\left(\overset{*}{\gamma}{}'-\overset{*}{\gamma}\tan\varphi\right)\right\}\cos\varphi\, d\varphi$$

After integrations by parts and application of the PVP, it comes the two local equations:

$$(1-\nu)\left(\begin{array}{l}(-\gamma'+w''+w)\cos\varphi+\dfrac{d^2}{d\varphi^2}\left[(-\gamma'+w''+w)\cos\varphi\right]\\[2ex] +\left[w+(\gamma-w')\tan\varphi\right]\cos\varphi+\dfrac{d}{d\varphi}\left\{\left[w+(\gamma-w')\tan\varphi\right]\sin\varphi\right\}\end{array}\right)$$

$$+\nu\left(\begin{array}{l}-2\left[\gamma'-w''-2w-(\gamma-w')\tan\varphi\right]\cos\varphi\\[2ex] -\dfrac{d^2}{d\varphi^2}\left\{\left[\gamma'-w''-2w-(\gamma-w')\tan\varphi\right]\cos\varphi\right\}\\[2ex] -\dfrac{d}{d\varphi}\left\{\left[\gamma'-w''-2w-(\gamma-w')\tan\varphi\right]\sin\varphi\right\}\end{array}\right)=\frac{pR^2}{K}\cos\varphi$$

$$K\left\{\begin{array}{l}(1-\nu)\left(\dfrac{d}{d\varphi}\left[(-\gamma'+w''+w)\cos\varphi\right]+\left[w+(\gamma-w')\tan\varphi\right]\sin\varphi\right)\\[3ex] +\nu\left(\begin{array}{l}-\dfrac{d}{d\varphi}\left\{\left[\gamma'-w''-2w-(\gamma-w')\tan\varphi\right]\cos\varphi\right\}\\[2ex] -\left[\gamma'-w''-2w-(\gamma-w')\tan\varphi\right]\sin\varphi\end{array}\right)\end{array}\right\} \tag{8.139}$$

$$+\frac{D}{R^2}\left\{\begin{array}{l}(1-\nu)\left[-\dfrac{d}{d\varphi}(\gamma'\cos\varphi)+\gamma(\tan\varphi)^2\cos\varphi\right]\\[3ex] -\nu\left(\dfrac{d}{d\varphi}\left[(\gamma'-\gamma\tan\varphi)\cos\varphi\right]+(\gamma'-\gamma\tan\varphi)\sin\varphi\right)\end{array}\right\}=0$$

To obtain the boundary conditions, it is necessary to express the virtual power of the reactions on the edges, namely a density of force q carried by A_3 and a moment density m carried by A_2:

$$\overset{*}{\mathcal{W}}_r = 2\pi R\left(q\overset{*}{w} - m\frac{\overset{*}{\gamma}}{R}\right)$$

hence the boundary conditions, using the edge terms resulting from the integration by parts of $\overset{*}{\mathcal{T}}$:

$$(1-\nu)(\gamma' - w'' - w) + \nu\left[\gamma' - w'' - 2w - (\gamma - w')\tan\varphi\right] = N^{11} = 0$$

$$K\left[\begin{matrix}(1-\nu)\left[\dfrac{d}{d\varphi}\left[(\gamma' - w'' - w)\cos\varphi\right] - \left(w + (\gamma - w')\tan\varphi\right)\sin\varphi\right] \\ +\nu\left[\begin{matrix}\dfrac{d}{d\varphi}\left[\left(\gamma' - w'' - 2w - (\gamma - w')\tan\varphi\right)\cos\varphi\right] \\ -\left(\gamma' - w'' - 2w - (\gamma - w')\tan\varphi\right)\sin\varphi\end{matrix}\right]\end{matrix}\right] = qR$$

$$\begin{aligned} &K\left\{(1-\nu)(\gamma' - w'' - w) + \nu\left[\gamma' - w'' - 2w - (\gamma - w')\tan\varphi\right]\right\}\cos\varphi \\ &\qquad + \frac{D}{R^2}\left\{(1-\nu)\gamma' + \nu(\gamma' - \gamma\tan\varphi)\right\}\cos\varphi = -m \end{aligned} \tag{8.140}$$

The generalised stresses are then obtained from the constitutive laws:

$$N^{11} = K\left[\frac{u' - w}{R} - \nu\frac{w + u\tan\varphi}{R}\right]$$

$$N^{22} = K\left[-\frac{w + u\tan\varphi}{R} + \nu\frac{u' - w}{R}\right]$$

$$M^{11} = \frac{D}{R^2}\left[w'' + u' - \nu(u + w')\tan\varphi\right] = \frac{D}{R^2}\left[\gamma' - \nu\gamma\tan\varphi\right]$$

$$M^{22} = \frac{D}{R^2}\left[-(u + w')\tan\varphi + \nu(w'' + u')\right] = \frac{D}{R^2}\left[-\gamma\tan\varphi + \nu\gamma'\right]$$

$$\begin{aligned} Q^1 &= -\frac{D}{R^3}\left[w''' + u'' + \nu\frac{d}{d\varphi}\left((u + w')\cot\varphi\right) - (1-\nu)\left(w'' + u' + (u + w')\tan\varphi\right)\tan\varphi\right] \\ &= -\frac{D}{R^3}\left[\gamma'' + \nu\frac{d}{d\varphi}(\gamma\cot\varphi) - (1-\nu)(\gamma' + \gamma\tan\varphi)\tan\varphi\right] \end{aligned} \tag{8.141}$$

A particular solution is obtained by taking $u = 0$ and w constant. In this case, γ is zero and the bending energy is zero: it is the membrane solution $w = \frac{pR^2(1-\nu)}{2Ee}$. The solution of the problem can thus be obtained by adding to the membrane solution a self-balanced flexional solution allowing to respect the boundary conditions.

Equilibrium equation (8.139) with boundary conditions (8.140) does not admit a known analytical solution and must therefore be solved numerically. However, transformations lead to a simpler form.

Equation (8.139) is rewritten, by (8.136):

$$
\begin{aligned}
&-(1-\nu)\left(\tilde{\varepsilon}_{\varphi\varphi}\cos\varphi+\frac{d^2}{d\varphi^2}\left[\tilde{\varepsilon}_{\varphi\varphi}\cos\varphi\right]+\tilde{\varepsilon}_{\theta\theta}\cos\varphi+\frac{d}{d\varphi}\left\{\tilde{\varepsilon}_{\theta\theta}\sin\varphi\right\}\right)\\
&-\nu\left(2\left[\tilde{\varepsilon}_{\varphi\varphi}+\tilde{\varepsilon}_{\theta\theta}\right]\cos\varphi+\frac{d^2}{d\varphi^2}\left\{\left[\tilde{\varepsilon}_{\varphi\varphi}+\tilde{\varepsilon}_{\theta\theta}\right]\cos\varphi\right\}+\frac{d}{d\varphi}\left\{\left[\tilde{\varepsilon}_{\varphi\varphi}+\tilde{\varepsilon}_{\theta\theta}\right]\sin\varphi\right\}\right)=\frac{pR}{K}\cos\varphi\\
&-KR\left\{\begin{array}{l}(1-\nu)\left(\dfrac{d}{d\varphi}\left[\tilde{\varepsilon}_{\varphi\varphi}\cos\varphi\right]+\tilde{\varepsilon}_{\theta\theta}\sin\varphi\right)\\[2mm]+\nu\left(\dfrac{d}{d\varphi}\left\{\left[\tilde{\varepsilon}_{\varphi\varphi}+\tilde{\varepsilon}_{\theta\theta}\right]\cos\varphi\right\}+\left[\tilde{\varepsilon}_{\varphi\varphi}+\tilde{\varepsilon}_{\theta\theta}\right]\sin\varphi\right)\end{array}\right\}\\
&\qquad+\frac{D}{R^2}\left\{\begin{array}{l}(1-\nu)\left[-\dfrac{d}{d\varphi}\left(\gamma'\cos\varphi\right)+\gamma\left(\tan\varphi\right)^2\cos\varphi\right]\\[2mm]-\nu\left(\dfrac{d}{d\varphi}\left[\left(\gamma'-\gamma\tan\varphi\right)\cos\varphi\right]+\left(\gamma'-\gamma\tan\varphi\right)\sin\varphi\right)\end{array}\right\}=0
\end{aligned}
\tag{8.142}
$$

This suggests using the secondary unknown:

$$
\begin{aligned}
V&=\frac{R}{\cos\varphi}\left\{\begin{array}{l}(1-\nu)\left(\dfrac{d}{d\varphi}\left[\tilde{\varepsilon}_{\varphi\varphi}\cos\varphi\right]+\tilde{\varepsilon}_{\theta\theta}\sin\varphi\right)\\[2mm]+\nu\left(\dfrac{d}{d\varphi}\left\{\left[\tilde{\varepsilon}_{\varphi\varphi}+\tilde{\varepsilon}_{\theta\theta}\right]\cos\varphi\right\}+\left[\tilde{\varepsilon}_{\varphi\varphi}+\tilde{\varepsilon}_{\theta\theta}\right]\sin\varphi\right)\end{array}\right\}\\
&=R\left\{(1-\nu)\left(\frac{d\tilde{\varepsilon}_{\varphi\varphi}}{d\varphi}+\left(\tilde{\varepsilon}_{\theta\theta}-\tilde{\varepsilon}_{\varphi\varphi}\right)\tan\varphi\right)+\nu\frac{d\left(\tilde{\varepsilon}_{\varphi\varphi}+\tilde{\varepsilon}_{\theta\theta}\right)}{d\varphi}\right\}\\
&=\frac{R}{K}\left[\frac{d\tilde{N}^{11}}{d\varphi}-\left(\tilde{N}^{11}-\tilde{N}^{22}\right)\tan\varphi\right]\\
&=\frac{1}{\cos\varphi}\frac{d}{d\varphi}\left[\left(\gamma'-w''-w\right)\cos\varphi\right]+\nu\left[\gamma'-w''-w\right]\sin\varphi\\
&\qquad-\frac{R}{\cos\varphi}\left[w+\left(\gamma-w'\right)\tan\varphi+\nu\frac{d}{d\varphi}\left\{\left[w+\left(\gamma-w'\right)\tan\varphi\right]\cos\varphi\right\}\right]\sin\varphi
\end{aligned}
\tag{8.143}
$$

which allows us to rewrite the second equation of equilibrium (8.142):

$$\boxed{\mathcal{D}(\gamma)+\nu\gamma+\frac{KR^2}{D}V=0} \tag{8.144}$$

where $\mathcal{D}$ refers to the differential operator (from Meissner[6], 1913):

$$\mathcal{D}(\gamma)=\gamma''-\gamma'\tan\varphi-\gamma\tan^2\varphi \tag{8.145}$$

According to the first equilibrium equation (8.135), V is connected to Q^1 by:

$$V=-\frac{Q^1R}{K} \tag{8.146}$$

With the definitions (8.137) and (8.143), V and γ are homogeneous to a length.

The first equilibrium equation (8.142) becomes:

$$-(1+\nu)\left(\tilde{\varepsilon}_{\varphi\varphi}+\tilde{\varepsilon}_{\theta\theta}\right)-\frac{1}{R\cos\varphi}\frac{d(V\cos\varphi)}{d\varphi}=\frac{pR}{K} \tag{8.147}$$

and is the second equilibrium equation (8.135). By eliminating w from (8.138), it becomes:

$$\frac{d\tilde{\varepsilon}_{\theta\theta}}{d\varphi}-\left(\tilde{\varepsilon}_{\theta\theta}-\tilde{\varepsilon}_{\varphi\varphi}\right)\tan\varphi=-\frac{\gamma}{R}$$

and then reporting this relation to the second expression (8.143) of V:

$$V=R\frac{d\left(\tilde{\varepsilon}_{\varphi\varphi}+\tilde{\varepsilon}_{\theta\theta}\right)}{d\varphi}+(1-\nu)\gamma \tag{8.148}$$

This makes it possible to express the second equilibrium equation (8.142) as a function of V and γ only:

$$(1+\nu)\left[\frac{V}{R}-(1-\nu)\frac{\gamma}{R}\right]+\frac{d}{d\varphi}\left[\frac{1}{R\cos\varphi}\frac{d(V\cos\varphi)}{d\varphi}\right]=0$$

which is rewritten, by introducing the differential operator $\mathcal{D}$ (8.145):

$$\boxed{\mathcal{D}(V)+\nu V-\left(1-\nu^2\right)\gamma=0} \tag{8.149}$$

Equations (8.144) and (8.149) constitute the two local equations of the problem (H. Reissner[7], 1912), from which V can be eliminated:

$$\boxed{\mathcal{D}\mathcal{D}(\gamma)+2\nu\mathcal{D}(\gamma)+\left(\left(1-\nu^2\right)\frac{KR^2}{D}+\nu^2\right)\gamma=0} \tag{8.150}$$

The last term ν^2 can be neglected in (8.150).

[6] Ernst Meissner, Swiss mathematician (1883–1939).

[7] Hans Jacob Reissner, German aeronautical engineer (1874–1967).

The generalisation of this approach to a shell of revolution of any form can be found in {Calladyne}.

Thus, the general form of γ and V is determined by (8.150), then (8.144). Then, w can be calculated by the last relation (8.143) and finally u by (8.137).

Boundary conditions are now in a simpler form:

$$M^{11} = \frac{D}{R^2}\left[\gamma' - \nu\,\gamma\,\tan\varphi\right] = m$$
$$Q^1 = -\frac{KV}{R} = \frac{D}{R^3}\left(\mathcal{D}(\gamma) + \nu\gamma\right) = q \tag{8.151}$$

The first condition (8.140) contributes to calculating the displacements once determined V and γ. Most often, the kinematic links relate to u and w and it is, therefore, necessary to solve all the variables to apply all the boundary conditions.

This system of equations, accompanied by all the boundary conditions, can be solved numerically. The shear force is then determined by (8.146), and the bending moments by the constitutive laws (8.141).

Operationally, this approach has made great progress in solving shell problems in the twentieth century. It is no longer competitive since the development of the finite element method (Chapter 9). It nevertheless retains a great interest in understanding the orders of magnitude of the different terms.

8.3.2.5 Cylindrical tank with a hemispherical bottom

This section provides a complete solution to the problem of the pressure vessel already discussed in §6.1.2.3. It has been demonstrated that the membrane solution is unsatisfactory in the vicinity of the connection of the bottom on the cylinder where bending due to discontinuity of curvature occurs. To simplify the presentation, the thicknesses are taken identically in both parts of the tank, although this is not the most optimal design. The generalisation to different thicknesses is done without difficulty.

Approximations make it possible to simplify the equations established in §8.2.3.1 and §8.3.2.4.

The length of the cylindrical portion is assumed to be sufficiently large relative to the bending waves ($\ell >> \sqrt{\text{Re}}$) for the cylinder to be considered infinite. The membrane solution given in §6.1.2.3 is supplemented by bending, according to formula (8.84), which allows the bottom effect to be taken into account:

$$w = \frac{(2-\nu)pR^2}{2Ee} + e^{-\lambda z}\left(A_2 \sin\lambda z + B_2 \cos\lambda z\right)$$

where λ is given by (8.89) and z is null at the connection and positive when moving away from it. The conditions at $z = 0$ are written considering Donnell's assumptions in the cylinder, resulting in:

Photo 8.7 : Cylindrical tank with hemispherical bottom

$$M^{zz} = D\frac{\partial^2 w}{\partial z^2} = -2\lambda^2 D A_2 = m$$

$$Q^z = -D\frac{\partial^3 w}{\partial z^3} = -2\lambda^3 D(A_2 + B_2) = q$$

The bending moment m and the shear force q are identical on both sides of the connection and are determined by writing the continuity of displacement and rotation.

Finally, the transverse displacement is worth in the cylinder:

$$w_c = \frac{(2-\nu)pR^2}{2Ee} + \frac{1}{2\lambda^2 D} e^{-\lambda z}\left[-m\sin\lambda z + \left(m - \frac{q}{\lambda}\right)\cos\lambda z\right]$$

On the hemisphere side, the equations in §8.3.2.4 can be simplified by considering the Geckeler[8] approximation. It consists in neglecting the last two terms in the expression (8.145) of the differential operator $\mathcal{D}$. This is justified if the geometry, represented here by $\frac{\tan\varphi}{R}$ characterising the rotation of the basis vectors, varies little over a flexural wavelength. Taking inspiration from the case of the cylinder and anticipating the result, it is estimated by the quantity $\sqrt{\text{Re}}$, which leads to the condition:

$$\tan\varphi << \sqrt{\frac{R}{e}}$$

which is respected for moderate values of φ. For example, for a slenderness of 100, the condition can be considered respected for $\varphi < 60°$. The orders of magnitude can also be verified on the result.

Note 1: In the example presented here, this approximation is particularly justified since it involves writing connection conditions in the neighbourhood of $\varphi = 0$.
Note 2: Near the top of the hemisphere, where the above approximation cannot be adopted, the shallow shell conditions are fulfilled and another solution may therefore be sought as part of the Donnell approximation.

On the physical point of view, referring to (8.145), the Geckeler approximation consists in neglecting the variation of circumferential curvature $\tilde{\kappa}_{\theta\theta}$ in front of $\tilde{\kappa}_{\varphi\varphi}$ and in considering that the circular deformation $\tilde{\varepsilon}_{\theta\theta}$ is reduced to $-\frac{w}{R}$, both hypotheses being necessarily true in the vicinity of the connection.

Note: The hypothesis is not relevant for studying the unstable scalloping in θ of pressure vessels, which involves non-axisymmetric displacements.

[8] Josef Geckeler, German physicist (1897–1952)

If, as in the case of the cylinder, bending waves appear (which is verified *in fine* on the shape of the solution), noting s the curvilinear abscissa in the longitudinal direction ($s = R\varphi$), the oscillatory part of the solution is of the form $\sin\lambda s = \sin\lambda R\varphi$, for which the second derivative is in the form $\lambda^2 R^2 \sim \frac{R}{e}$. γ'' (resp. V'') is therefore predominant with respect to $\nu\,\gamma$ (resp. $\nu\, V$) and Equations (8.144) and (8.149) are simplified in:

$$\gamma'' + \frac{KR^2}{D} V = 0 \qquad (8.152)$$

$$V'' - \left(1 - \nu^2\right)\gamma = 0$$

Under these conditions, Equation (8.150) is reduced to:

$$\gamma^{IV} + 4\lambda^4 R^4 \gamma = 0 \qquad (8.153)$$

with λ given by (8.87) and the simplified boundary conditions in $\varphi = 0$:

$$\gamma' = \frac{mR^2}{D} \;;\; \gamma'' = \frac{qR^3}{D} \qquad (8.154)$$

The general solution of (8.153) is:

$$\gamma = e^{\lambda R\varphi}\left(C_1 \sin\lambda R\varphi + D_1 \cos\lambda R\varphi\right) + e^{-\lambda R\varphi}\left(C_2 \sin\lambda R\varphi + D_2 \cos\lambda R\varphi\right)$$

This justifies the hypothesis on flexural wavelength. The boundary conditions (8.154) are not sufficient to completely determine the solution. To do this, it should be considered that the bending waves must dampen away from the connection, eliminating the first terms. The conditions (8.154) are then written:

$$\lambda R\left(C_2 - D_2\right) = \frac{mR^2}{D} \;;\; -2\lambda^2 R^2 C_2 = \frac{qR^3}{D}$$

and finally:

$$\gamma = -\frac{Re^{-\lambda R\varphi}}{D\lambda}\left(\frac{q}{2\lambda}\sin\lambda R\varphi + \left(\frac{q}{2\lambda} + m\right)\cos\lambda R\varphi\right) \qquad (8.155)$$

$$V = -\frac{D}{KR^2}\gamma'' = -\frac{\lambda Re^{-\lambda R\varphi}}{K}\left(\frac{q}{\lambda}\cos\lambda R\varphi - \left(\frac{q}{\lambda} + 2m\right)\sin\lambda R\varphi\right)$$

The fourth relationship (8.143) is reduced to:

$$V = \gamma'' - w''' - w'$$

This shows with (8.152) that γ'' is preponderant with respect to V and, considering that w' is small in front of w''' for the same reason as established above, it becomes:

$$w'''' + \frac{KR^2}{D} V \approx 0 \Rightarrow w''' = \gamma'' \Rightarrow w' = \gamma \tag{8.156}$$

This means that u is negligible in front of w', which is coherent with the other hypotheses, provided that, defined to within a constant, u was taken zero at the connection (in §6.1.2.3, u is called v). This latter convention also makes it possible to eliminate in w the solution in $\sin\varphi$ appearing in the membrane solution. w is then obtained by integration of (8.156), that is, considering the particular (membrane) solution:

$$w_s = \frac{pR^2(1-\nu)}{2Ee} - \frac{1}{2D\lambda^2} e^{-\lambda R\varphi}\left(m \sin\lambda R\varphi - \left(\frac{q}{\lambda} + m\right)\cos\lambda R\varphi\right)$$

In the cylinder and the sphere, the normal vectors a_3 are directed inwards, so the displacements w have been obtained with the same sign. On the other hand, a_z in the cylinder and a_1 in the sphere are opposite, so the rotations are of opposite signs. At the connection, the continuity of displacement and rotation is thus expressed by the two relations:

$$w_c(z=0) = w_s(\varphi=0) \Rightarrow \frac{(2-\nu)pR^2}{2Ee} + \frac{\left(m - \frac{q}{\lambda}\right)}{2\lambda^2 D} = \frac{pR^2(1-\nu)}{2Ee} + \frac{\left(\frac{q}{\lambda} + m\right)}{2D\lambda^2}$$

$$w_c'(0) = -\frac{\gamma(0)}{R} \Rightarrow -\frac{2m - \frac{q}{\lambda}}{2\lambda D} = \frac{\frac{q}{2\lambda} + m}{D\lambda}$$

from which:

$$q = \frac{pR^2}{2Ee} D\lambda^3 = \frac{p}{8\lambda} \; ; \; m = 0$$

Thus, the bending moment is zero at the connection. This result makes it possible to express the transverse displacement according to the parameter $\varpi = \frac{pR^2}{Ee}$.

- Displacement in the cylinder: $\frac{w_c}{\varpi} = 1 - \frac{\nu}{2} - \frac{1}{4} e^{-\lambda z} \cos\lambda z$
- Displacement in the sphere: $\frac{w_s}{\varpi} = \frac{1}{2}\left[(1-\nu) + \frac{1}{2} e^{-\lambda R\varphi} \cos\lambda R\varphi\right]$

At the connection, the displacement is worth $\frac{3-2\nu}{4}\varpi$ and is intermediate between the displacements of the cylinder and the sphere given by the membrane theory.

Regarding the bending moment:

- Moment in the cylinder: $M_c = Dw_c'' = -\frac{p}{8\lambda^2} e^{-\lambda z} \sin\lambda z$
- Moment in the sphere: $M_s = \frac{D}{R^2} w_s'' = \frac{p}{8\lambda^2} e^{-\lambda R\varphi} \sin\lambda R\varphi$

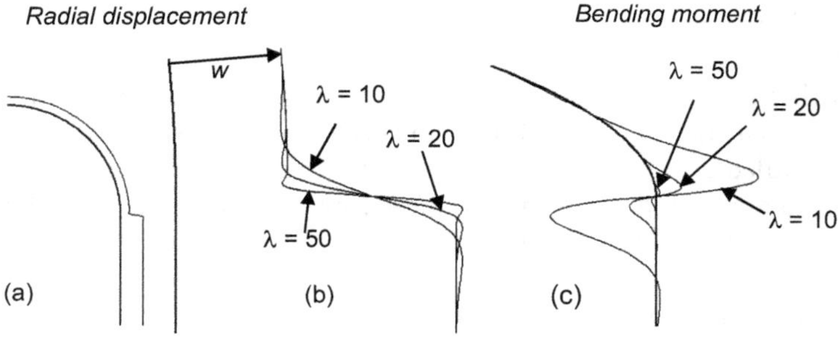

Figure 8.19

It should be noted that the Geckeler simplification transforms the sphere into a quasi-cylinder from a geometric point of view, for bending, while retaining its membrane stiffness.

Figure 8.19(a) shows the normal displacement, enlarged in Figure 8.19(b) in the vicinity of the connection. Figure 8.19(c) shows the moment in the vicinity of the connection, where the strong influence of the parameter λ appears. The bending wavelength and the moment amplitude decrease as λ increases, that is, when the thickness decreases, for a given radius.

8.3.3 Introduction to any loading

When the loading and, consequently, the displacement can not be considered axisymmetric, the displacement is written in the most general form:

$$\xi = u(s,\sigma)A_1 + v(s,\sigma)A_2 + w(s,\sigma)A_3 \tag{8.157}$$

where σ is the curvilinear abscissa along A_2. The vectors of the natural basis in the deformed position are deduced:

$$a_1 = A_1 + \frac{\partial u}{\partial s}(s,\sigma)A_1 + u(s,\sigma)\frac{\partial A_1}{\partial s} + \frac{\partial v}{\partial s}(s,\sigma)A_2$$

$$+ v(s,\sigma)\frac{\partial A_2}{\partial s} + \frac{\partial w}{\partial s}(s,\sigma)A_3 + w(s,\sigma)\frac{\partial A_3}{\partial s}$$

$$= \left(1 + \frac{\partial u}{\partial s} - \frac{w}{R}\right)A_1 + \frac{\partial v}{\partial s}A_2 + \left(\frac{\partial w}{\partial s} + \frac{u}{R}\right)A_3$$

$$a_2 = A_2 + \frac{\partial u}{\partial \sigma}(s,\sigma)A_1 + u(s,\sigma)\frac{\partial A_1}{\partial \sigma} + \frac{\partial v}{\partial \sigma}(s,\sigma)A_2$$

$$+ v(s,\sigma)\frac{\partial A_2}{\partial \sigma} + \frac{\partial w}{\partial \sigma}(s,\sigma)A_3 + w(s,\sigma)\frac{\partial A_3}{\partial \sigma}$$

$$= \frac{\partial u}{\partial \sigma}A_1 + \left(1 - \frac{u}{R}\tan\varphi + \frac{\partial v}{\partial \sigma} + \frac{v}{R}\tan\varphi - \frac{w}{R}\right)A_2 + \left(\frac{v}{R} + \frac{\partial w}{\partial \sigma}\right)A_3$$

from which:

$$\mathcal{A}_s^{\,s} = \frac{\partial u}{\partial s} - \frac{w}{R}; \quad \mathcal{A}_\sigma^{\,s} = \frac{\partial u}{\partial \sigma} \qquad \mathcal{B}_s = \frac{\partial w}{\partial s} + \frac{u}{R}$$

$$\mathcal{A}_s^{\,\sigma} = \frac{\partial v}{\partial s}; \qquad \mathcal{A}_\sigma^{\,\sigma} = -\frac{u}{R}\tan\varphi + \frac{\partial v}{\partial \sigma} + \frac{v}{R}\tan\varphi - \frac{w}{R} \qquad \mathcal{B}_\sigma = \frac{v}{R} + \frac{\partial w}{\partial \sigma}$$

then:

$$\nabla_s \mathcal{B}_s = \frac{\partial^2 w}{\partial s^2} + \frac{1}{R}\frac{\partial u}{\partial s}; \quad \nabla_s \mathcal{B}_\sigma = \frac{1}{R}\left(\frac{\partial u}{\partial s} + \frac{\partial w}{\partial \sigma}\tan\varphi\right);$$

$$\nabla_\sigma \mathcal{B}_s = \frac{\partial^2 w}{\partial s \partial \sigma} + \frac{1}{R}\frac{\partial u}{\partial \sigma}; \quad \nabla_\sigma \mathcal{B}_\sigma = \frac{\partial^2 w}{\partial \sigma^2} + \frac{1}{R}\frac{\partial u}{\partial \sigma}$$

hence the linearised deformation variables:

$$\varepsilon_{ss} = \frac{\partial u}{\partial s} - \frac{w}{R} \qquad \kappa_{ss} = \frac{\partial^2 w}{\partial s^2} + \frac{1}{R}\frac{\partial u}{\partial s} + \frac{1}{R}\left(\frac{\partial u}{\partial s} - \frac{w}{R}\right)$$

$$\varepsilon_{\sigma\sigma} = -\frac{u}{R}\tan\varphi + \frac{\partial v}{\partial \sigma} + \frac{v}{R}\tan\varphi - \frac{w}{R} \qquad \kappa_{\sigma\sigma} = \frac{\partial^2 w}{\partial \sigma^2} + \frac{1}{R}\frac{\partial u}{\partial \sigma} + \frac{1}{R}\left(-\frac{u}{R}\tan\varphi + \frac{\partial v}{\partial \sigma} + \frac{v}{R}\tan\varphi - \frac{w}{R}\right)$$

$$\varepsilon_{s\sigma} = \frac{1}{2}\left(\frac{\partial u}{\partial \sigma} + \frac{\partial v}{\partial s}\right) \qquad \kappa_{s\sigma} = \frac{1}{2}\left[\frac{1}{R}\left(\frac{\partial u}{\partial \sigma} + \frac{\partial u}{\partial s} + \frac{\partial w}{\partial \sigma}\tan\varphi\right) + \frac{\partial^2 w}{\partial s \partial \sigma}\right] \tag{8.158}$$

The deformation energy is deduced from this, which makes it possible to establish the three equations of motion relating to the three displacement components by application of the PVP. The search for analytical solutions is not competitive with the search for the displacement field minimising the total potential by numerical methods, including the application of the finite element method which is the subject of the next chapter, and also the Ritz and Galerkin methods.

8.4 EXAMPLE OF APPLICATION: FLEXION OF A PIPE ELBOW

The theory of beams is usually applied to the calculation of pipelines. In the straight parts, the stiffnesses at the extension ES and at the flexion EI of the annular section give correct results. In the elbows, on the other hand, experience shows that the bending stiffness of the annular section is lower than predicted by the beam theory, due to the ovalisation of the section.

A practical method to avoid using the shell model in the curved part is to correct the bending beam constitutive law, that is, in linear elasticity:

$$M = \frac{EI}{k}\chi$$

where:

- χ is the variation of curvature of the centreline of the beam;
- k is a coefficient of flexibility greater than 1.

It is therefore necessary to evaluate the coefficient k by a shell model. The method used consists of comparing the strain energy in the beam model to that obtained in the shell model, with simplifying assumptions. Given the difference in "fineness" between the two models, the equivalence of the energies can not lead to the same value of k for all the load cases considered. In the Von Kármán[9] solution developed below, the case used to express the equivalence is the case of uniform bending (constant bending moment, of axis normal to the plane of the elbow, supposed horizontal for convenience of presentation).

Photo 8.8 : Pipes in an industrial plant

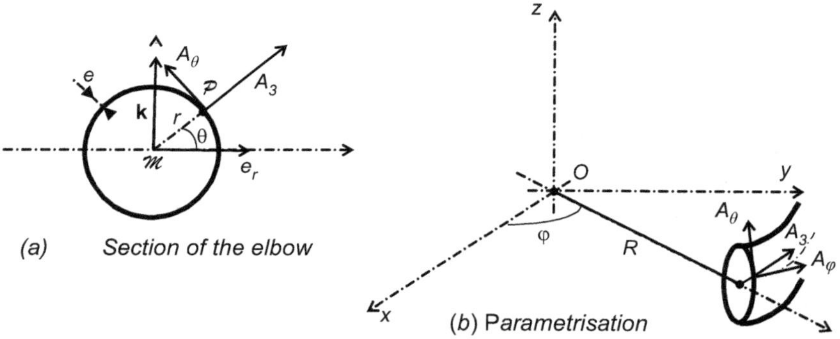

Figure 8.20

a. The geometry of the elbow (Figure 8.20).

The elbow is a torus of revolution, in which a small radius r is much smaller than the large radius R ($r << R$). The position of the circular section is marked by the angle φ in the horizontal plane and the current point $\mathcal{P}$ of the section by the angle θ of the vector radius with the horizontal.

[9] Théodore Von Kármán, Hungarian-American mathematician, aerospace engineer and physicist (1881–1963); a pioneer in aerodynamics, has also contributed to various advances in the field of elastic plates and shells.

In the parameterisation (φ, θ), the coordinates of the current point $\mathcal{P}$ are:

$$\mathcal{P}\begin{cases}(R+r\cos\theta)\cos\varphi\\(R+r\cos\theta)\sin\varphi\\r\sin\theta\end{cases}$$

from where the natural coordinate system:

$$A_\varphi=\begin{cases}-(R+r\cos\theta)\sin\varphi\\(R+r\cos\theta)\cos\varphi\\0\end{cases}\qquad A_\theta=\begin{cases}-r\sin\theta\cos\varphi\\-r\sin\theta\sin\varphi\\r\cos\theta\end{cases}\qquad A_3=\begin{cases}\cos\theta\cos\varphi\\\cos\theta\sin\varphi\\\sin\theta\end{cases}$$

in which the curvature tensor is written:

$$B_{\bullet\bullet}=\begin{pmatrix}-(R+r\cos\theta)\cos\theta & 0\\0 & -r\end{pmatrix}$$

and the Riemannian connection coefficients:

$$\Gamma^{\theta}_{\varphi\varphi}=\frac{R+r\cos\theta}{r}\sin\theta\qquad\Gamma^{\varphi}_{\varphi\theta}=-\frac{r\sin\theta}{R+r\cos\theta}$$

A_1 and A_2 designate the unit vectors associated respectively with A_φ and A_θ. e_r denotes the unit vector associated with $O\mathcal{M}$, radius of the centreline.

b. The transformation.

The following assumptions are made for a simplified approach to the problem:

- Under the effect of a bending moment of a vertical axis, the elbow bends in its plane.
- Sections φ = cst (sections of the equivalent beam) remain normal to the centreline during the transformation (Navier–Bernoulli–Euler hypothesis).
- The displacement of a point $\mathcal{P}$ of the elbow is composed of (Figure 8.19):
  - o the displacement of the centreline of the equivalent beam. This displacement $\xi_1(\mathcal{P})$ is itself composed of:
  - o the displacement $\xi(\mathcal{M})=u(\varphi)A_1+\delta(\varphi)e_r$ of the centre $\mathcal{M}$ of the section in the horizontal plane of the centreline of the beam, where e_r designates the vector normal to the circle forming the centreline,
  - o the z-axis rotation ψ(φ) of the beam.

In this fictional beam transform, p' is the transformed point of $\mathcal{P}$.

- The ovalisation of the section, represented by a displacement:

$$\xi_2=vA_2+wA_3$$

expressed in the reference system (small movements) (Figure 8.21).

- The hypothesis of small displacements: the displacement ξ_2 due to the ovalisation remains small compared to the thickness. Deformations are linearised in the first order.

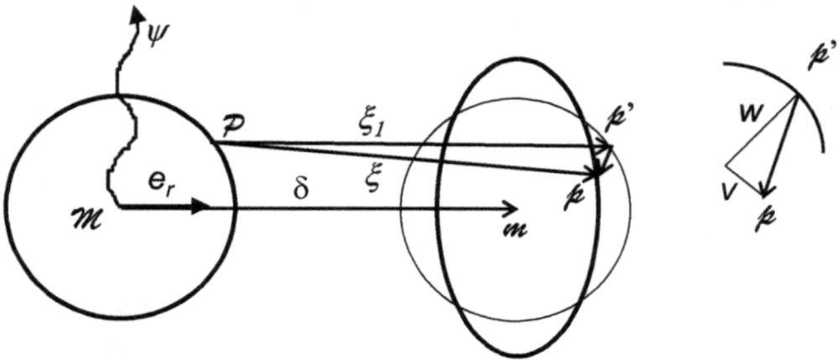

Figure 8.21

- The radial displacement δ (of the centreline of the beam) is constant: from the point of view of the beam, the moment (thus the variation of curvature) is constant.
- The effects of the connection between the straight parts and the elbow are not considered: the part of the elbow considered in the calculation is located at the middle of the elbow where the flexion can be considered as axisymmetric about the axis z: v and w only depend on θ.
- The axial deformation of the beam centreline is negligible (pure bending).

c. The deformation fields of the shell.

The above hypotheses make it possible to express the displacement $\xi(\mathcal{P}) = \overrightarrow{\mathcal{P}p}$ of any point of the shell in the reference coordinate system(A_1, A_2, A_3):

$$\xi(\mathcal{P}) = \underbrace{uA_1 + \delta(A_3\cos\theta - A_2\sin\theta)}_{\xi(\mathcal{M})} + \underbrace{\psi r\ A_1\cos\theta}_{\psi\,\mathbf{k}\wedge\overrightarrow{\mathcal{MP}}} + \underbrace{v\ A_2 + w\ A_3}_{\xi_2}$$

$$= (u + \psi r\cos\theta)A_1 + (v - \delta\sin\theta)A_2 + (w + \delta\cos\theta)A_3$$

Then in the coordinate system $(A_\varphi, A_\theta, A_3)$:

$$\xi(\mathcal{P}) = \underbrace{\frac{u + \psi r\cos\theta}{R + r\cos\theta}}_{\xi^\varphi} A_\varphi + \underbrace{\frac{v - \delta\sin\theta}{r}}_{\xi^\theta} A_\theta + \underbrace{(w + \delta\cos\theta)}_{\zeta} A_3$$

The deformations of the shell are then calculated in the natural coordinate system:

$$\varepsilon_{\varphi\varphi} = \nabla_\varphi \xi_\varphi - \zeta\ B_{\varphi\varphi}$$

$$= \partial_\varphi\left[(R + r\cos\theta)^2 \times \frac{u + \psi r\cos\theta}{R + r\cos\theta}\right] + \frac{R + r\cos\theta}{r}\sin\theta\ r(v - \delta\sin\theta)$$

$$+ (w + \delta\cos\theta)(R + r\cos\theta)\cos\theta$$

$$\varepsilon_{\theta\theta} = \nabla_\theta \xi_\theta - \zeta\ B_{\theta\theta} = \partial_\theta\left[r^2 \frac{v - \delta\sin\theta}{r}\right] + r(w + \delta\cos\theta)$$

$$\kappa_{\theta\theta} = \nabla_\theta\left(\partial_\theta\zeta + B_{\theta\theta}\xi^\theta\right) + B_{\theta\theta}\left(\partial_\theta\xi^\theta - \zeta B_\theta{}^\theta\right)$$

$$= \partial_\theta\left[\partial_\theta\left(w + \delta\cos\theta\right) - r\frac{v - \delta\sin\theta}{r}\right] - r\left[\partial_\theta\left(\frac{v - \delta\sin\theta}{r}\right) + \frac{1}{r}\left(w + \delta\cos\theta\right)\right]$$

then in the orthonormal coordinate system:

$$\varepsilon_{11} = \frac{\varepsilon_{\varphi\varphi}}{\left(R + r\cos\theta\right)^2} = \frac{u' + \psi' r\cos\theta - v\sin\theta + w\cos\theta + \delta}{R + r\cos\theta}$$

$$\varepsilon_{22} = \frac{\varepsilon_{\theta\theta}}{r^2} = \frac{v' + w}{r}$$

$$\kappa_{22} = \frac{\kappa_{\theta\theta}}{r^2} = \frac{w'' - 2v' - w}{r^2}$$

d. The deformation fields of the beam.

The displacement of the centreline of the beam is written as:

$$\xi(\mathcal{M}) = \frac{u}{R}A_\varphi + \delta e_r$$

hence its axial deformation:

$$\varepsilon_{\varphi\varphi} = \frac{de_r}{d\varphi}\cdot\frac{d}{d\varphi}\left(\frac{u}{R}A_\varphi + \delta e_r\right) = R\left(u' + \delta\right)$$

and its variation of curvature:

$$\chi = \frac{1}{R}\frac{d\psi}{d\varphi}$$

The latter is constant by hypothesis, so $\psi' = \chi R = \text{cst}$.

By hypothesis again, the deformation of the centreline is zero, so $u' + \delta = 0$.

Moreover, given the assumption on the orders of magnitude of the radii:

$$R + r\cos\theta \approx R$$

These results make it possible to simplify the expression of the deformation ε_{11} of the shell:

$$\varepsilon_{11} = \frac{\chi Rr\cos\theta - v\sin\theta + w\cos\theta}{R}$$

e. The strain energy of the elbow.

The strain energy of the elbow is obtained from its expression in the shell model, making the following complementary hypotheses:

- The shell is thin.
- Sections φ = cst are inextensible ($\varepsilon_{22}=0$): ovalisation bending is preponderant.
- Longitudinal folds are neglected ($\kappa_{11}=0$).
- w is a periodic function of θ, only the first term of its Fourier series decomposition is retained: $w = a\cos 2\theta$, which corresponds well to the studied ovalisation.

These hypotheses lead to the following expressions of the deformation variables:

$$\varepsilon_{22}=0 \;\Rightarrow\; v' = -w = -a\cos 2\theta \;\Rightarrow\; v = -\frac{a}{2}\sin 2\theta$$

$$\varepsilon_{11} = \chi r\cos\theta + \frac{a}{R}\left(\frac{1}{2}\sin\theta\sin 2\theta + \cos\theta\cos 2\theta\right) = \chi r\cos\theta + \frac{a}{R}\cos^3\theta$$

$$\kappa_{22} = -\frac{3a}{r^2}\cos 2\theta$$

Hence the strain energy per unit length of elbow:

$$\mathcal{F} = \frac{1}{2}\frac{Ee}{1-\nu^2}\int_0^{2\pi}\left(\varepsilon_{11}{}^2 + \frac{e^2}{12}\kappa_{22}{}^2\right) r\,d\theta$$

$$= \frac{1}{2}\frac{Eer}{1-\nu^2}\int_0^{2\pi}\left(\left(\chi r\cos\theta + \frac{a}{R}\cos^3\theta\right)^2 + \frac{e^2}{12}\left(\frac{3a}{r^2}\cos 2\theta\right)^2\right)d\theta$$

$$= \frac{1}{2}\frac{\pi Eer}{1-\nu^2}\left[\left(\chi r + \frac{3a}{4R}\right)^2 + a^2\left(\frac{1}{16R^2} + \frac{3e^2}{4r^4}\right)\right]$$

f. The solution.

The equations governing the behaviour of the elbow are obtained by application of the PVP:

$$\overset{*}{\mathcal{F}} = \frac{\pi Eer}{1-\nu^2}\left[\left(\chi r + \frac{3a}{4R}\right) r\overset{*}{\chi} + \left(\chi r + \frac{3a}{4R}\right)\frac{3}{4R}\overset{*}{a} + a\overset{*}{a}\left(\frac{1}{16R^2} + \frac{3e^2}{4r^4}\right)\right]$$

$$\overset{*}{\mathcal{W}}_e = M\overset{*}{\chi}$$

$$\overset{*}{\mathcal{F}} = \overset{*}{\mathcal{W}}_e \quad \forall\left(\overset{*}{a},\overset{*}{\chi}\right) \Rightarrow \begin{cases}\left(\chi r + \dfrac{3a}{4R}\right)\dfrac{3}{4R} + a\left(\dfrac{1}{16R^2} + \dfrac{3e^2}{4r^4}\right) = 0 \\[2ex] \dfrac{\pi Eer^2}{1-\nu^2}\left(\chi r + \dfrac{3a}{4R}\right) = M\end{cases}$$

and finally:

$$M = \frac{\pi E e r^3}{1-\nu^2}\frac{1+12\lambda^2}{10+12\lambda^2}\chi = \frac{EI}{k}\chi$$

where $\lambda = \frac{\mathrm{Re}}{r^2}$ and $\mathrm{I} = \pi e r^3$. The value of k is thus determined:

$$k = \left(1-\nu^2\right)\frac{10+12\lambda^2}{1+12\lambda^2}$$

Stresses are also determined from the shell model:

$$\sigma^{11} = E\varepsilon_{11} = k\frac{Mr}{\mathrm{I}}\left[\cos\theta - \frac{3}{10+12\lambda^2}\left(\cos 3\theta + 3\cos\theta\right)\right]$$

while the uncorrected beam model would have given $\sigma^{11} = \frac{Mr}{\mathrm{I}}\cos\theta$. σ^{11} is zero in $\theta = \pm\frac{\pi}{2}$, but in $\theta = 0$ or π, ie along the most stressed lines, its value is amplified by the coefficient:

$$\tau_1 = \left(1-\nu^2\right)\frac{12\lambda^2 - 2}{1+12\lambda^2}$$

The stress σ^{22} does not appear in the beam model. In the shell model, it is worth:

$$\sigma^{22} = -E\kappa_{22}\, z = E\frac{3a}{r^2} z\cos 2\theta = -\frac{1}{r}\frac{18R}{5+6\lambda^2}\frac{kM}{\mathrm{I}} z\cos 2\theta$$

which maximum value:

$$\max\left(\left|\sigma^{22}\right|\right) = \lambda\frac{9}{5+6\lambda^2}\frac{kMr}{\mathrm{I}}$$

can be compared to the maximum stress σ^{11} obtained in the beam model and is therefore deduced by the application of the coefficient:

$$\tau_2 = \left(1-\nu^2\right)\frac{18\lambda}{1+12\lambda^2}$$

k, τ_1, and τ_2 depend a little on the Poisson's ratio, but especially on λ which depends only on the geometry of the elbow. Their variations are given in Figure 8.22 for $\nu = 0$.

When λ tends towards 0, k tends towards $10\left(1-\nu^2\right)$: the elbow is very flexible vis-à-vis the ovality and the longitudinal stress is multiplied by 2 while changing of sign. σ^{22} is equal

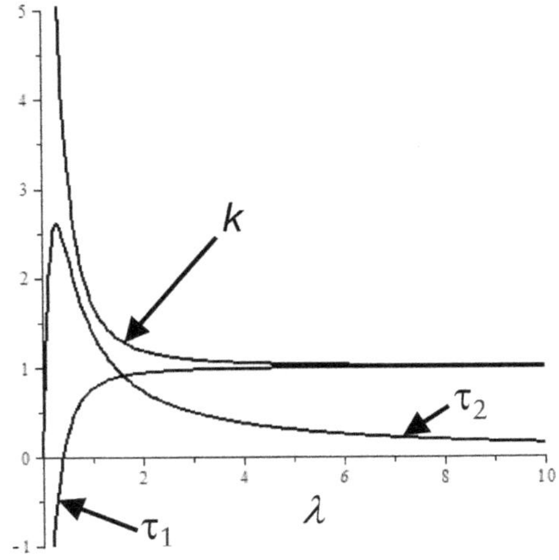

Figure 8.22

Photo 8.9: Water towers (Villejuif, France)

to 0 for $\lambda = 0$, but increases very fast with λ.

Note that σ^{11} and σ^{22} have comparable values for $\lambda = 1$.

When λ tends to infinity, k and σ^{11} tend towards $1-\nu^2$, σ^{22} towards 0: the section of the elbow becomes rigid, the only difference with the beam model being the intervention of the Poisson's ratio.

In conclusion, the uncorrected beam model is only valid for large values of λ. Below this, longitudinal stress is significantly affected by ovalisation. The circumferential stress is not negligible for intermediate values of λ.

MAIN RESULTS

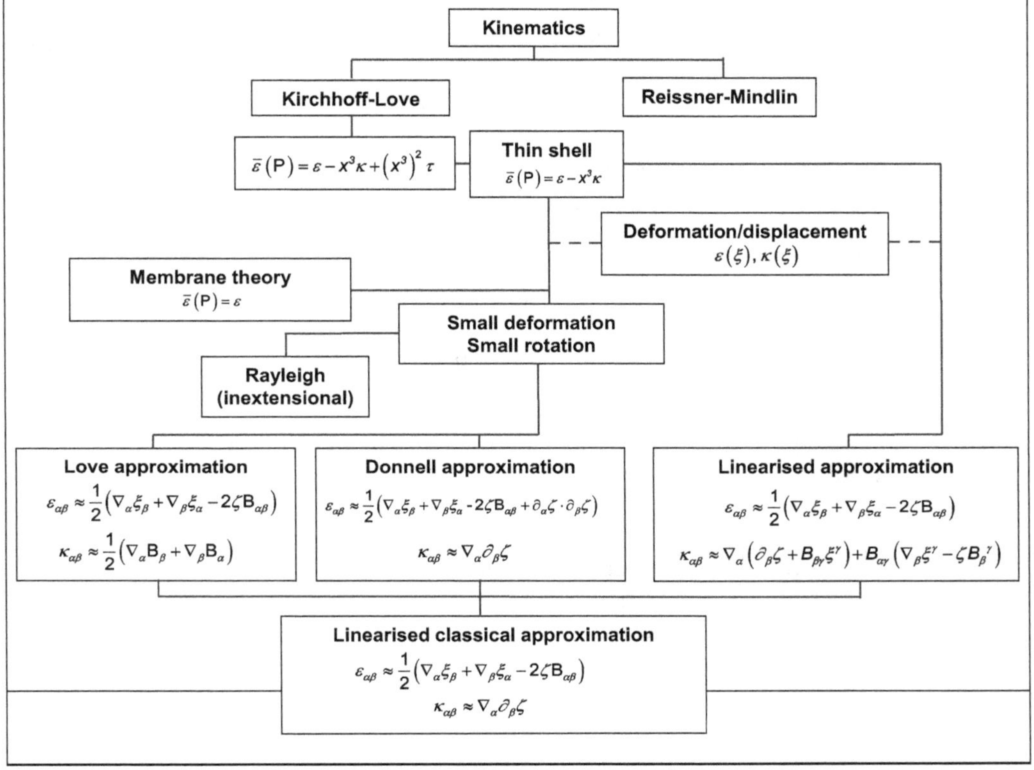

EXERCISES

Exercise 8.1 Cone deformation

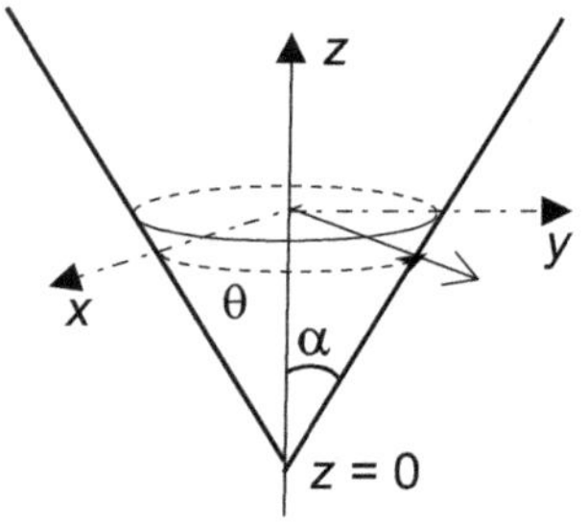

Figure 8.23

A cone (Figure 8.23) is parametrised by the angle θ along each horizontal circle of radius r and z, the vertical coordinate.

The natural basis and the corresponding orthonormal basis are associated with this parameterisation. By noting respectively s and σ the curvilinear abscissas along the horizontal circle at height z and along a straight generatrix, the displacement is written:

$$\xi = \zeta^{\theta}(\theta, z) A_{\theta} + \zeta^{z}(\theta, z) A_{z} + w(\theta, z) A_{3}$$

$$= v(\theta, z) A_{s} + u(\theta, z) A_{\sigma} + w(\theta, z) A_{3}$$

Calculate without approximation the strain tensor in the natural base.

Express the strain tensor and the curvature variation tensor in the orthonormal basis, assuming small transformations:

- in first-order linearised approximation;
- in Love's approximation;
- in Donnell approximation;
- in linearised classical approximation.

Answer:

The geometry of the cone was the subject of Exercise 3.1.

$$\mathcal{A}_{\theta}{}^{\theta} = \partial_{\theta}\xi^{\theta} + \frac{1}{z}\xi^{z} + \frac{\cos^2\alpha}{z\sin\alpha} w \; ; \; \mathcal{A}_{z}{}^{z} = \partial_{z}\xi^{z}$$

$$(3.95) \quad \mathcal{A}_{\theta}{}^{z} = \partial_{\theta}\xi^{z} - z\sin^2\alpha\, \xi^{\theta} \; ; \; \mathcal{A}_{z}{}^{\theta} = \partial_{z}\xi^{\theta} + \frac{1}{z}\xi^{\theta}$$

$$\mathcal{B}_{\theta} = -z\sin\alpha\, \xi^{\theta} + \partial_{\theta} w \; ; \; \mathcal{B}_{z} = \partial_{z} w$$

$$\varepsilon_{\theta\theta} = z^2\tan^2\alpha\left[\mathcal{A}_{\theta}{}^{\theta} + \frac{1}{2}\left(\mathcal{A}_{\theta}{}^{\theta}\right)^2\right] + \frac{1}{2}\frac{1}{\cos^2\alpha}\left(\mathcal{A}_{\theta}{}^{z}\right)^2 + \frac{1}{2}\left(\mathcal{B}_{\theta}\right)^2$$

$$(3.97) \quad \varepsilon_{\theta z} = \frac{1}{2}\left[\frac{1}{\cos^2\alpha}\left(\mathcal{A}_{\theta}{}^{z} + \mathcal{A}_{\theta}{}^{z}\mathcal{A}_{z}{}^{z}\right) + z^2\tan^2\alpha\left(\mathcal{A}_{z}{}^{\theta} + \mathcal{A}_{\theta}{}^{\theta}\mathcal{A}_{z}{}^{\theta}\right) + \mathcal{B}_{\theta}\mathcal{B}_{z}\right]$$

$$\varepsilon_{zz} = \frac{1}{\cos^2\alpha}\left[\mathcal{A}_{z}{}^{z} + \frac{1}{2}\left(\mathcal{A}_{z}{}^{z}\right)^2\right] + \frac{1}{2}z^2\tan^2\alpha\left(\mathcal{A}_{z}{}^{\theta}\right)^2 + \frac{1}{2}\left(\mathcal{B}_{z}\right)^2$$

First-order linearised tensors:

$$\varepsilon_{ss} = \frac{1}{\tan\alpha}\partial_\theta\left(\frac{v}{z}\right) + \cos\alpha\,\frac{u}{z} + \frac{\cos^2\alpha}{\sin\alpha}\frac{w}{z}$$

$$(3.102)\quad \varepsilon_{s\sigma} = \frac{1}{2}\left[\frac{1}{\tan\alpha}\partial_\theta\left(\frac{u}{z}\right) - \cos\alpha\frac{v}{z} + \cos\alpha \bullet \partial_z v\right]$$

$$\varepsilon_{\sigma\sigma} = \cos\alpha \bullet \partial_z u$$

$$\kappa_{ss} = \frac{1}{z^2\tan^2\alpha}\begin{bmatrix}\partial^2_{\theta^2}w + z\sin^2\alpha\,\partial_z w - w\cos^2\alpha \\ -2\cos\alpha\,\partial_\theta v - u\sin\alpha\cos\alpha\end{bmatrix}$$

$$(3.107)\quad \kappa_{s\sigma} = \frac{\cos^2\alpha}{z\sin\alpha}\left[\partial^2_{z\theta}w - \partial_\theta\left(\frac{w}{z}\right) - \partial_z v \bullet \cos\alpha + \frac{v}{z}\cos\alpha\right] = \kappa_{\sigma s}$$

$$\kappa_{\sigma\sigma} = \cos^2\alpha\,\partial^2_{z^2}w$$

Love's approximation = ① + ③

$$\kappa_{ss} = \frac{1}{z^2\tan^2\alpha}\left[\partial^2_{\theta^2}w + z\sin^2\alpha\,\partial_z w - \partial_\theta v \bullet \cos\alpha\right]$$

$$(8.13)\quad \kappa_{s\sigma} = \frac{\cos^2\alpha}{z\sin\alpha}\left[\partial^2_{z\theta}w - \partial_\theta\left(\frac{w}{z}\right) + \frac{v}{z}\cos\alpha\right]$$

$$\kappa_{\sigma\sigma} = \cos^2\alpha\,\partial^2_{z^2}w$$

Donnell approximation = ④ + ⑤

$$\varepsilon_{ss} = \frac{1}{z\tan\alpha}\left(\partial_\theta v + u\sin\alpha + w\cos\alpha\right) + \frac{1}{2z^2\tan^2\alpha}\left(\partial_\theta w\right)^2$$

$$(8.15)\quad \varepsilon_{s\sigma} = \frac{1}{2}\left[\frac{\partial_\theta u}{z\tan\alpha} - \frac{v\cos\alpha}{z} + \partial_z v\cos\alpha\right] + \frac{1}{2}\frac{\cos^2\alpha}{z\sin\alpha}\partial_\theta w\,\partial_z w$$

$$\varepsilon_{\sigma\sigma} = \partial_z u\cos\alpha + \frac{1}{2}\left(\partial_z w\cos\alpha\right)^2$$

$$\kappa_{ss} = \frac{1}{z^2\tan^2\alpha}\left[\partial^2_{\theta^2}w + z\sin^2\alpha\,\partial_z w\right]$$

$$(8.16)\quad \kappa_{s\sigma} = \frac{\cos^2\alpha}{z\sin\alpha}\left(\partial^2_{\theta z}w - \frac{1}{z}\partial_\theta w\right)$$

$$\kappa_{\sigma\sigma} = \cos^2\alpha\,\partial^2_{z^2}w$$

Classical linear approximation = ① + ⑤

Exercise 8.2 Inextensional eigenmodes of a cone

A cone is fixed at its vertex located at $z = 0$ (Figure 8.23). Characterise the displacement field associated with a small inextensional motion and comment on the validity of this hypothesis.

Calculate the natural modes and pulsations of a truncated cone bounded by elevations $z = h_1$ and $z = h_2$ in the same inextensional movement with bending only (therefore verifying the associated conditions).

Answer:
The strain and curvature variation tensors were established in Exercise 8.1. To calculate the modes of the cone, the first-order linearisation ① + ② is used.

The three deformations of the system ① are null, which leads to the general form of the displacement (taking into account the null displacement in $z = 0$):

$$u(\theta,z)=0 \ ; \ v(\theta,z)=B(\theta)z \ ; \ w(\theta,z)=-\frac{1}{\cos\alpha}B'(\theta)z$$

The deformation remains a cone and is characterised by the deformation of the circle at each elevation z. This deformation depends on the single function $B(\theta)$, to be determined.

For this deformation, the tensor of variation of curvature ② is rewritten:

$$\kappa_{ss}=-\frac{\cos\alpha}{z\sin^2\alpha}\left[B'''(\theta)+B'(\theta)\right] \ ; \ \kappa_{s\sigma}=0 \ ; \ \kappa_{\sigma\sigma}=0$$

The variation in curvature of the horizontal circle tends towards infinity when z tends towards 0. Indeed, the radial displacement w being zero at the top, the radius of the circle remains equal to zero as it is initially. But it is a discontinuity of the tangent to the surface which requires special treatment.

The upper boundary conditions at $z = h$ cannot be arbitrary to satisfy the inextension condition. The displacement u must be zero and the displacements v and w free. If the Poisson's ratio is zero, the moment $M_{\sigma\sigma}$ is zero and the edge is articulated, free to move in the plane (a_s, a_3).

The application of the PPV leads to the equation in B:

$$-\kappa\left(B^{\mathrm{VI}}+2B^{\mathrm{IV}}+B''\right)+\mu\left(-\frac{\ddot{B}''}{\cos^2\alpha}+\ddot{B}\right)=0$$

For a truncated cone limited by elevations h_1 and h_2, the modes and associated pulsations respecting the conditions of inextension are:

$$X_n\begin{cases}u_n=0\\ v_n=z\sin(n\theta)\\ w_n=-\dfrac{nz}{\cos\alpha}\cos(n\theta)\end{cases} \qquad \omega_n^2=\frac{\kappa}{\mu}\frac{n^2\left(n^2-1\right)^2}{\dfrac{n^2}{\cos^2\alpha}+1}, \text{ with : } \begin{aligned} &p_n=n\pi\\ &\kappa=\frac{D}{\sin^3\alpha}\ln\left(\frac{h_2}{h_1}\right)\\ &\mu=\rho\frac{\sin\alpha}{4\cos^2\alpha}\left(h_2^{\,4}-h_1^{\,4}\right)\end{aligned}$$

Exercise 8.3 Buckling of a vacuum box.

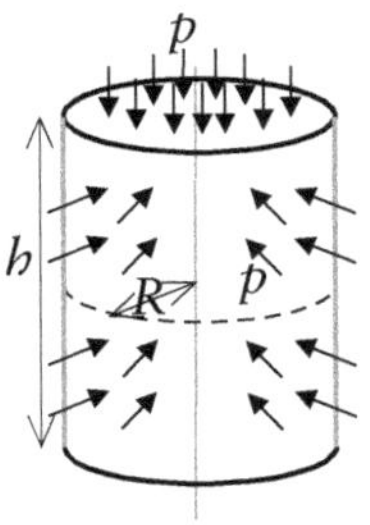

Figure 8.24

A cylindrical vacuum box of radius R and height h is subjected to a constant pressure p all around (Figure 8.24), acting on the cylindrical wall and on the bottoms. Assuming that the cylindrical shell constituting the box is thin (υ small) and slender, establish the membrane solution taken for the prestressed state, then look for the critical buckling value and the associated mode.

For the search for buckling, to simplify the calculations, the cylindrical part is supposed to be simply supported on its bottoms and second-order approximation is applied.

Answer:
The equilibrium equations admit the membrane solution:

$$\mathrm{N}_0^{zz} = -\frac{pR}{2} \;;\; \mathrm{N}_0^{ss} = -pR$$

This state can only be established if w is free at the bottom, so it is only an approximation, there is actually a first-order bending in the vicinity of the edges, but the effect of the flexural prestress is neglected.

To determine the buckling mode, Equation (8.106) reduces to:

$$\frac{\partial^2 \mathrm{M}^{zz}}{\partial z^2} + 2\frac{\partial^2 \mathrm{M}^{zs}}{\partial z \partial s} + \frac{\partial^2 \mathrm{M}^{ss}}{\partial s^2} + \frac{pR}{2}\frac{\partial^2 w}{\partial z^2} + pR\frac{\partial}{\partial s}\left(\frac{v}{R} + \partial_s w\right) = 0$$

Expressing the bending behaviour, it becomes the equation in w and v:

$$\frac{\partial^4 w}{\partial z^4} + 2\frac{\partial^4 w}{\partial z^2 \partial s^2} + \frac{\partial^4 w}{\partial s^4} + \left(\frac{1}{2}\frac{pR}{D} + \frac{\nu}{R^2}\right)\frac{\partial^2 w}{\partial z^2} + \left(\frac{pR}{D} + \frac{1}{R^2}\right)\frac{\partial^2 w}{\partial s^2} + \frac{p}{D}\frac{\partial v}{\partial s} = 0$$

The modes are the same as for the cylinder considered in §8.2.4.2.2, it becomes:

$$(n\mu)^4 + 2(n\mu)^2 m^2 + m^4 - \left(\frac{1}{2}\lambda + \nu\right)(n\mu)^2 - (\lambda + 1)m^2 + \lambda = 0$$

with $\lambda = \dfrac{pR^3}{D}$ and $\mu = \dfrac{\pi R}{h} = \dfrac{1}{\kappa}$.

hence the value of λ allowing a deformation according to this mode:

$$\lambda_{m,n} = \frac{\left((n\mu)^2 + m^2\right)^2 - \nu(n\mu)^2 - m^2}{\frac{1}{2}(n\mu)^2 + m^2 - 1}$$

The critical value is obtained for the couple (m, n) which minimises $\lambda_{m,n}$, that is (1, 1). Ultimately :

$$p_{cr} = \frac{2D}{R^3} \frac{\left(\mu^2+1\right)^2 - \nu\mu^2 - 1}{\mu^2}$$

The shape of the mode was given in §8.2.4.2.2.

For very slender cylinders ($\mu \to 0$), a lower value is given by $p_{cr} \approx \frac{2D}{R^3}(2-\nu)$

For a very short cylinder ($\mu = 1$) : $p_{cr} \approx \frac{2D}{R^3}(3-\nu)$

Exercise 8.4 Axisymmetric buckling of a cone

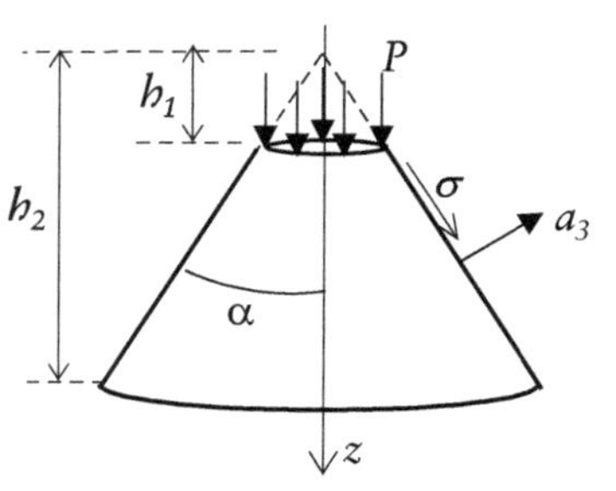

Figure 8.25

The objective is to determine the critical buckling force of a portion of the cone subjected to the state of membrane equilibrium determined in exercise 6.1, according to an axisymmetric mode, under the sole action of the force density P distributed over the edge of the upper opening (Figure 8.25). The cone is included between the elevations h_1 and h_2 where the boundary conditions allow the establishment of the membrane equilibrium; at the top and at the base, a rigid ring makes it possible to balance the horizontal reaction so that the total reaction is tangent to the cone; the cone is supported at the base and free to deform in the direction of the generatrices.

1. The displacement field is defined as in Exercise 8.1, give the tensors of deformation and variation of curvature in second-order approximation, for any displacement.
2. Apply the results of 1. if the displacement is axisymmetric. Write the axisymmetric buckling equation using the energy method.

Answer :

ε and κ were determined in Exercise 8.1. $\mathcal{A}$, therefore $\partial_z u$, $\partial_z v$, $\partial_\theta v + u \sin\alpha + w \cos\alpha$ and $\partial_\theta u - v \tan^2\alpha$ are of the second order. $\mathcal{B}$, therefore $\partial_z w$ and $-v \cos\alpha + \partial_\theta w$, are of the first order.

1. In second-order approximation, they reduce to:

$$\varepsilon_{ss} = \frac{1}{z \tan\alpha}\left(\partial_\theta v + u \sin\alpha + w \cos\alpha\right) + \frac{1}{2}\left(\frac{-v \cos\alpha + \partial_\theta w}{z \tan\alpha}\right)^2$$

$$\varepsilon_{s\sigma} = \frac{1}{2}\left[\frac{\partial_\theta u}{z \tan\alpha} - \frac{v \cos\alpha}{z} + \partial_z v \cos\alpha + \frac{\cos^2\alpha}{z \sin\alpha}\partial_z w\left(-v \cos\alpha + \partial_\theta w\right)\right]$$

$$\varepsilon_{\sigma\sigma} = \partial_z u \cos\alpha + \frac{1}{2}\left(\partial_z w \cos\alpha\right)^2$$

$$\kappa_{ss} = \frac{1}{z^2 \tan^2\alpha}\left[\partial^2_{\theta^2} w + z\sin^2\alpha\, \partial_z w - \partial_\theta v \bullet \cos\alpha\right]$$

$$\kappa_{s\sigma} = \frac{\cos^2\alpha}{z\sin\alpha}\left[\partial^2_{z\theta} w - \partial_\theta\left(\frac{w}{z}\right) + \frac{v}{z}\cos\alpha\right]$$

$$\kappa_{\sigma\sigma} = \cos^2\alpha\, \partial^2_{z^2} w$$

2. In an axisymmetric displacement, the displacement v is zero, as well as the partial derivatives with respect to θ. The cone is free to deform along the generatrices in the buckling transformation from the membrane state, therefore $\varepsilon_{\sigma\sigma} = -\nu\varepsilon_{ss}$. In the orthonormal physical coordinate system:

$$\varepsilon_{ss} = \frac{u\sin\alpha + w\cos\alpha}{z\tan\alpha} \;;\; \varepsilon_{\sigma\sigma} = u'\cos\alpha + \frac{1}{2}\left(w'\cos\alpha\right)^2$$

$$\kappa_{ss} = \frac{\cos^2\alpha}{z} w' \;;\; \kappa_{\sigma\sigma} = w''\cos^2\alpha$$

The membrane state under the effect of the loading P was determined in Exercise 6.1: $N = N^{\sigma\sigma} = -\dfrac{h_1 P}{z\cos\alpha}$

This supposes that a rigid ring beam blocks the horizontal displacement on the two edges and resumes a horizontal reaction making it possible to balance N. This compression constitutes the prestress which works in the deformation $\varepsilon_{\sigma\sigma}$ of the generatrix.

The cone is free in the vertical direction: $\varepsilon_{\sigma\sigma} = -\nu\varepsilon_{ss}$, which makes it possible to simplify the expression of the elastic potential.

The elastic potential and the potential of P are written:

$$\mathcal{F} = -2\pi h_1 P\tan\alpha \int_{h_1}^{h_2}\left(\frac{u'}{\cos\alpha} + \frac{1}{2}w'^2\right)dz + \frac{1}{2}\frac{2\pi Ee}{\sin\alpha}\int_{h_1}^{h_2}\frac{\left(u\sin\alpha + w\cos\alpha\right)^2}{z}dz$$

$$+\frac{1}{2}2\pi D\cos^2\alpha\sin\alpha\int_{h_1}^{h_2}\left(\frac{w'^2}{z} + z\,w''^2 + 2\nu\, w'\, w''\right)dz$$

$$\mathcal{V} = 2\pi h_1\tan\alpha\, P\left[u(h_1)\cos\alpha - w(h_1)\sin\alpha\right]$$

The rings of the free edges prevent horizontal displacements on the edges, therefore, for example in h_1:

$$u(h_1)\sin\alpha + w(h_1)\cos\alpha = 0 \;\Rightarrow\; u(h_1)\cos\alpha - w(h_1)\sin\alpha = \frac{u(h_1)}{\cos\alpha}$$

The PVP is applied to a field of kinematically admissible virtual velocities. It comes :

The conditions on the edges:

$$D\cos^2\alpha\left[\left(z\,w''\right)'-\frac{w'}{z}\right]_{h_1}^{h_2}+\frac{h_1P}{\cos\alpha}\left[w'\right]_{h_1}^{h_2}=0$$

$$\left[z\,w''+\nu\,w'\right]_{h_1}^{h_2}=0$$

The first condition connects the shear force to the projection of the membrane force in the cone on the perpendicular to the deformed middle surface. The second expresses that the bending moment in the meridian direction is zero.

The local equation:

$$\left(z\,w''\right)''-\left(\frac{w'}{z}\right)'+\frac{h_1P}{D\cos^3\alpha}w''=0$$

The local equation makes it possible to seek the eigenvalues, making it possible to determine the critical forces P. But here the operator is nonlinear, so the solutions can be searched numerically.

Chapter 9

Shell finite elements

The purpose of this chapter is the application of the finite element method to the resolution of plate and shell problems. The approach used is the displacement method; mixed formulations are not discussed here.

The principles of the finite element method are recalled in the first section, where the notations used later are also introduced.

The second section is devoted to the plate elements and the third to the shell elements. The principles for making finite elements of plates and shells and the main difficulties encountered in their development or use are discussed

The formulation of some common finite elements is established as a demonstrative application.

9.1 THE FINITE ELEMENTS METHOD

9.1.1 Purpose of the chapter

Finite elements were originally developed to solve shell problems in the aircraft industry. For more than 50 years, many efforts have been made to formulate reliable and efficient general shell elements and new elements continue to be developed.

The object here is not to go into the details of the method, nor to describe all the elements available in the literature, which are in great numbers. It is rather to identify the main properties of the large families of finite elements, highlighting the possible difficulties related to each family. Some elements of the different families are shown, allowing appreciation of the differences existing between them; nevertheless, no detail of their construction is given, this being in the abundant specialised literature. In particular, only the finite elements used in the displacement method are discussed; elements that can be used in a mixed formulation are detailed in (Zienkiewicz and Taylor) or (Batoz and Dhatt), for instance.

9.1.2 The principles of the finite element method

The main steps of solving a problem of solid mechanics by the finite element method are described in this paragraph, where, to simplify the presentation, only the case of conservative and linearly elastic systems in small displacements (first order) is considered.

The Rayleigh–Ritz method can serve as a general framework for the finite element method; so it is approached through this.

Let $u(\mathcal{M},t)$ the field of displacement of the structure under the action of external forces; u belongs to the functional space $\mathcal{U}$ of the kinematically admissible displacement fields. To

DOI: 10.1201/9780429440403-9

the search of the exact solution u in $\mathcal{U}$ is substituted the search for an approximation in a subspace $\mathcal{U}_n$ of $\mathcal{U}$, $\mathcal{U}_n$ being of finite dimension n.

The space $\mathcal{U}_n$ is generated by n independent functions ϕ_r of $\mathcal{U}_n$. A field of displacement in $\mathcal{U}_n$ is thus written:

$$v \in \mathcal{U}_n: \quad v(\mathcal{M},t) = q^r(t)\,\phi_r(\mathcal{M}) \tag{9.1}$$

The functions ϕ_r being chosen *a priori*, it remains to determine the n functions $q^r(t)$.

In a static problem of change of equilibrium under a distribution of forces $f(\mathcal{M})$, q^r are scalars independent of time. In this case, the potential of external actions $\mathcal{V} = -\int_\Sigma f(\mathcal{M}) \cdot u(\mathcal{M})\, dV$ admits in $\mathcal{U}_n$ an approximation:

$$\mathcal{V} = -\int_\Sigma f(\mathcal{M}) \cdot \left[q^r\,\phi_r(\mathcal{M})\right] dV = -q^r \int_\Sigma \mathrm{f}(\mathcal{M}) \cdot \phi_r(\mathcal{M})\, dV = -q^r f_r \tag{9.2}$$

with:

$$f_r = \int_\Sigma f(\mathcal{M}) \cdot \phi_r(\mathcal{M})\, dV \tag{9.3}$$

$\mathcal{V}$ can also be written in matrix form:

$$\mathcal{V} = -\,Q^T F \tag{9.4}$$

where:

Q is the **displacement vector** containing the unknown "parameters" q^r;
F is the **force vector** containing the generalised actions f_r.
Q and F are of dimension n.

In small displacements (in first-order theory), in the case of a three-dimensional solid, the three-dimensional strain tensor $\bar{\varepsilon}(\mathcal{M})$ is expressed as a function of the displacement $u(\mathcal{M})$ by a linear differential operator $\mathcal{D}$:

$$\bar{\varepsilon}(\mathcal{M}) = \mathcal{D}\left[u(\mathcal{M})\right]$$

More precisely:

$$\mathcal{D}[u] = \frac{1}{2}\left(\mathrm{grad}\, u + \mathrm{grad}^T u\right) \tag{9.5}$$

If the solid is linearly elastic, Λ denoting the tensor of elasticity, the elastic potential of the system, linearised to the second order with respect to an unloaded natural state, is written:

$$\mathcal{F} = \frac{1}{2}\int_\Sigma \bar{\varepsilon}(\mathcal{M}) \cdot \Lambda \cdot \bar{\varepsilon}(\mathcal{M})\, dV = \frac{1}{2}\int_\Sigma \mathcal{D}(u) \cdot \Lambda \cdot \mathcal{D}(u)\, dV \tag{9.6}$$

and admits in $\mathcal{U}_n$ the approximation:

$$\mathcal{F} = \frac{1}{2} q^r q^s \int_{\Sigma} \mathcal{D}(\phi_r) \bullet \Lambda \bullet \mathcal{D}(\phi_s)\, dV = \frac{1}{2} k_{rs} q^r q^s \qquad (r,s) \in [1,n] \times [1,n] \tag{9.7}$$

either, in matrix notation:

$$\mathcal{F} = \frac{1}{2} Q^T K Q \tag{9.8}$$

where the square matrix K is the elastic stiffness matrix of the structure, of components:

$$k_{rs} = \int_{\Sigma} \mathcal{D}(\phi_r) \bullet \Lambda \bullet \mathcal{D}(\phi_s)\, dV \tag{9.9}$$

Since $\mathcal{F}$ is a strain energy, the matrix K is a definite positive if the system is stable.

Note: the system is suitably fixed, since the functions ϕ_r are kinematically admissible.

The solution u minimises the total potential $\mathcal{F} + \mathcal{V}$ in $\mathcal{U}$. An approximate solution u_n of u is obtained by minimizing $\mathcal{F} + \mathcal{V}$, in $\mathcal{U}_n$, that is to say by seeking u such that:

$$[\mathcal{F} + \mathcal{V}](u_n) = \underbrace{\min}_{v \in \mathcal{U}_n} \{[\mathcal{F} + \mathcal{V}](v)\} \tag{9.10}$$

which is equivalent to looking for Q such that:

$$\frac{1}{2} Q^T K Q - Q^T F$$

is minimum, hence the equation:

$$\boxed{KQ = F} \tag{9.11}$$

which is usually invertible given the assumptions[1].

To solve a problem of free or forced vibrations by the same method, it is necessary to discretise the kinetic energy:

$$\mathcal{C} = \frac{1}{2} \int_{\Sigma} \dot{u}(\mathcal{M},t)^2 \rho\, dV \approx \frac{1}{2} \int_{\Sigma} \left[\dot{q}^r(t)\, \phi_r(\mathcal{M})\right]^2 \rho\, dV = \frac{1}{2} \dot{q}^r \dot{q}^s \int_{\Sigma} \phi_r(\mathcal{M}) \phi_s(\mathcal{M}) \rho\, dV \tag{9.12}$$

either, in matrix form:

$$\mathcal{C} = \frac{1}{2} \dot{Q}^T M \dot{Q} \tag{9.13}$$

where the square matrix M is the mass matrix of the system, of components:

$$m_{rs} = \int_{\Sigma} \phi_r(\mathcal{M}) \phi_s(\mathcal{M}) \rho\, dV \tag{9.14}$$

[1] The stiffness matrix may not be invertible in (rare) cases of models with few under-integrated elements.

By applying the Lagrange equations, there comes the matrix equation of motion, where F can depend on time:

$$\boxed{M\ddot{Q}+KQ=F} \tag{9.15}$$

The same procedure is applied for all elastic structures, including plates and shells, and formally leads to the same formulation.

The degree of approximation in the discretisation depends essentially on the choice of the functions $\phi_r(\mathcal{M})$of the basis.

In the finite element method, a large number of functions $\phi_r(\mathcal{M})$ with a "small" support are used. Indeed, for two functions $\phi_r(\mathcal{M})$ and $\phi_s(\mathcal{M})$ with disjoint supports, the component k_{sr} of the stiffness matrix and m_{sr} of the mass matrix are zero. With an adequate numbering of the functions ϕ_r, the matrix K can be in the form of a band matrix, with a band as narrow as possible, which favours the numerical resolution of the equation to the displacement Q.

To achieve this objective, the structure Σ is divided into subdomains of given geometry. In each subdomain, the displacement of any point is expressed as a function of a finite number of parameters; in practice, these parameters are displacements, or spatial derivatives of displacements (rotations, curvatures...), in points of the subdomain. The support of the function ϕ consists of elements connected to a node.

9.1.3 Formulation of a finite element

A finite element is constituted (Figure 9.1):

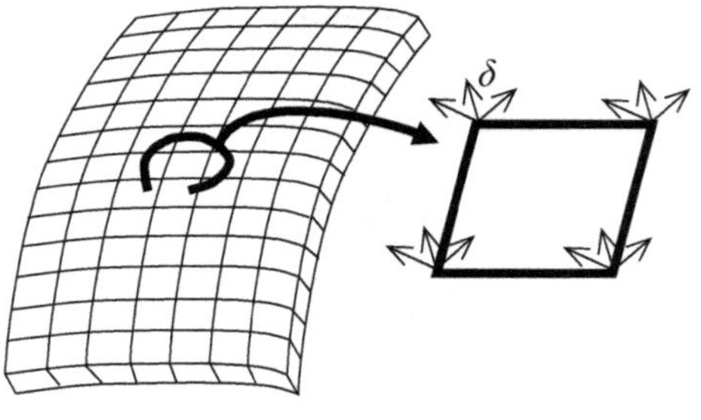

Figure 9.1

- a closed **geometric domain**, of generally simple (but not necessarily plane) shape: triangular or quadrangular for a two-dimensional structure;
- particular points in finite number, called **nodes**; they are most often the apexes of the domain, the midpoints of the sides, and have the characteristic of interior points;
- **parameters** consisting of displacements, rotations and their derivatives at the nodes of the element; these are the unknowns of the problem, in finite numbers;
- an **interpolation function** making it possible to express *a priori* the displacements and rotations at any point of the element as functions of the parameters.

The set of geometric domains associated with the finite elements covers the entire modelled structure without holes or overlaps (unless the elements can be superimposed). On the other hand, in the case of shells (curved), when the domains can not exactly "stick" to the middle surface, an approximation on the geometry can be used (see §9.3.2).

Let $[\delta]$ be the vector containing the parameters of an element (displacements, rotations, curvatures at the nodes of the element), generally expressed in a local coordinate system associated with the element.

$\left[w(\mathcal{M})\right]$ is the approximation of the displacement field within an element, and $\left[N(\mathcal{M})\right]$ the matrix expressing the interpolation function. In a coordinate system (x, y, z) of $\mathbb{R}^3$:

$$\left[w(x,y,z)\right]=\left[N(x,y,z)\right]\left[\delta\right] \tag{9.16}$$

Let [ε] be the vector containing the characteristic deformations of the model: in three-dimensional theory, it contains the components of the strain tensor; in the theory of plane beams, it contains the deformation of the axis and its variation of curvature; in shell theory, it contains the components of the tensors of deformation and of variation of curvature of the middle surface and possibly the distortions.

In small displacements, these deformations are deduced from displacements by a linear differential operator and thus parameters $\left[\delta\right]$ by a matrix $[B]$ deduced from $[N]$ by differentiation operations:

$$\left[\varepsilon(x,y,z)\right]=\left[B(x,y,z)\right]\left[\delta\right] \tag{9.17}$$

Example: Beam in extension.

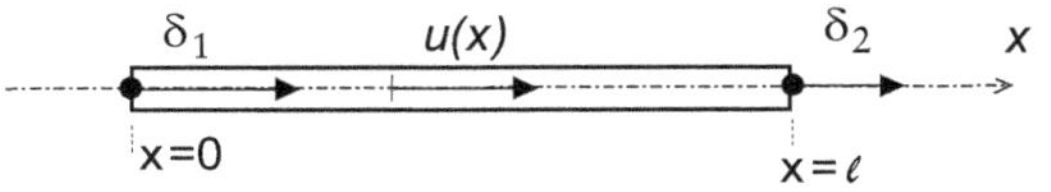

Figure 9.2

Let be a straight beam of axis x, in pure extension; $u(x)$ is the axial displacement at abscissa x. The simplest finite element comprises only two nodes at both ends; δ_1 and δ_2 are the displacements of the two nodes of the element (Figure 9.2).

A good approximation of u(x) is given by an affine function of x, because it is the exact solution in static when the element undergoes forces only at its ends:

$$u(x)=\delta_1+(\delta_2-\delta_1)\frac{x}{\ell}=\begin{bmatrix}1-\frac{x}{\ell} & \frac{x}{\ell}\end{bmatrix}\begin{bmatrix}\delta_1\\ \delta_2\end{bmatrix}$$

Here:

$$\left[N\right]=\begin{bmatrix}1-\frac{x}{\ell} & \frac{x}{\ell}\end{bmatrix}$$

The beam element of length ℓ, its end nodes, the parameters δ_1 and δ_2 and the matrix $[N]$ representative of the interpolation function constitute a finite element of a beam in pure extension.

In small displacements:

$$\varepsilon(x)=\frac{\partial u}{\partial x}=\left[\frac{\partial N}{\partial x}\right]\left[\delta\right]=\begin{bmatrix}-\frac{1}{\ell} & \frac{1}{\ell}\end{bmatrix}\begin{bmatrix}\delta_1\\ \delta_2\end{bmatrix}$$

then:

$$\left[B\right]=\begin{bmatrix}-\frac{1}{\ell} & \frac{1}{\ell}\end{bmatrix}$$

9.1.4 Expression of elementary energies

In the hypothesis of a linearly elastic behaviour, the constitutive law is written in the form:

$$\left[\sigma\right]=\left[\sigma_0\right]+\left[D\right]\left[\varepsilon\right] \tag{9.18}$$

in which case the quadratic part of the elastic potential of the element is written:

$$\mathcal{F}_e = \frac{1}{2}\int_{\mathcal{E}} [\varepsilon]^T [\sigma]\, dV = \frac{1}{2}[\delta]^T \left(\int_{\mathcal{E}} [B]^T [D][B]\, dV\right)[\delta] = \frac{1}{2}[\delta]^T [k_e][\delta] \tag{9.19}$$

where:

$$[k_e] = \int_{\mathcal{E}} [B]^T [D][B]\, dV \tag{9.20}$$

is called an **elementary stiffness matrix.**

w denoting all the useful components of the displacement, the potential of the external forces applied to the element is written (*p* can be a distribution):

$$\mathcal{V}_e = -\int_{\mathcal{E}} p \cdot \mathrm{w}\, \mathrm{dV} = -[\delta]^T \left(\int_{\mathcal{E}} \mathrm{p} \cdot [\mathrm{N}]^T\, \mathrm{dV}\right)[\delta] = -[\delta]^T [f_e] \tag{9.21}$$

where $[f_e]$ is the **elementary force vector.**

In dynamics problems, the parameters depend on time and the kinetic energy is expressed from an adequate formulation of the masses and the displacements concerned. In simplified form:

$$\mathcal{C}_e = \frac{1}{2}\int_{\mathcal{E}} \left(\frac{\partial w}{\partial t}\right)^2 \rho\, dV = \frac{1}{2}[\dot{\delta}]^T \left(\int_{\mathcal{E}} [N]^T [N] \rho\, dV\right)[\dot{\delta}] = \frac{1}{2}[\dot{\delta}]^T [m_e][\dot{\delta}] \tag{9.22}$$

$[m_e] = \int_{\mathcal{E}} [N]^T [N] \rho\, dV$ is the **elementary mass matrix.**

This mass matrix is said to be **coherent** here because it is expressed with the same interpolation function as the stiffness matrix. The force vector and the mass matrix are not always coherent in the current finite elements, because a good discretisation of $\mathcal{V}$ and $\mathcal{C}$ does not necessarily require the same level of development of the interpolation functions as for $\mathcal{F}$, as it appears in the examples discussed later in this chapter.

Example: A beam in extension.

The extensional energy of the beam in linear elasticity is written:

$$\mathcal{F}_e = \frac{1}{2}\int_0^{\ell} \mathrm{ES}\varepsilon^2 dx = \frac{1}{2}\begin{bmatrix}\delta_1 & \delta_2\end{bmatrix} \left(\int_0^{\ell} \mathrm{ES}\begin{bmatrix} -\frac{1}{\ell} & \frac{1}{\ell}\end{bmatrix} \begin{bmatrix} -\frac{1}{\ell} \\ \frac{1}{\ell} \end{bmatrix} dx\right) \begin{bmatrix}\delta_1 \\ \delta_2\end{bmatrix}$$

Hence, in the case where *ES* is constant:

$$[k_e] = \begin{bmatrix} \frac{ES}{\ell} & -\frac{ES}{\ell} \\ -\frac{ES}{\ell} & \frac{ES}{\ell} \end{bmatrix}$$

For constant axial load density q:

$$\mathcal{V}_e = -\int_0^\ell q \cdot u(x)\,dx = -\begin{bmatrix}\delta_1 & \delta_2\end{bmatrix}\left(\int_0^\ell q\begin{bmatrix}1-\dfrac{x}{\ell}\\ \dfrac{x}{\ell}\end{bmatrix}dx\right)$$

from which:

$$[f_e] = \begin{bmatrix}\dfrac{q\ell}{2}\\ \dfrac{q\ell}{2}\end{bmatrix}$$

When the load distribution is not constant, the integral is simply recalculated.

For dynamics problems, kinetic energy is approximated by:

$$\mathcal{C}_e = \frac{1}{2}\int_0^\ell \left(\frac{\partial u}{\partial t}\right)^2 \rho S\,dx = \frac{1}{2}\begin{bmatrix}\dot{\delta}_1 & \dot{\delta}_2\end{bmatrix}\left(\int_0^\ell \rho S\begin{bmatrix}1-\dfrac{x}{\ell} & \dfrac{x}{\ell}\end{bmatrix}\begin{bmatrix}1-\dfrac{x}{\ell}\\ \dfrac{x}{\ell}\end{bmatrix}dx\right)\begin{bmatrix}\dot{\delta}_1\\ \dot{\delta}_2\end{bmatrix}$$

hence, if ρS is constant, and noting $m = \rho S\ell$ the mass of the bar:

$$[m_e] = \begin{bmatrix}\dfrac{m}{3} & \dfrac{m}{6}\\ \dfrac{m}{6} & \dfrac{m}{3}\end{bmatrix}$$

9.1.5 Expression of the energies of the complete system

The parameterisation $[\delta]$ of each element is usually expressed in local axes related to the element. On the other hand, the whole structure is described in global axes, in which are expressed the parameters q^i of the structure, corresponding to displacements (translations and rotations) of the nodes (or their derivatives) (Figure 9.3).

The parameters q^i must obviously be chosen so that any parameter δ can be expressed as a function of q^i. For a given element:

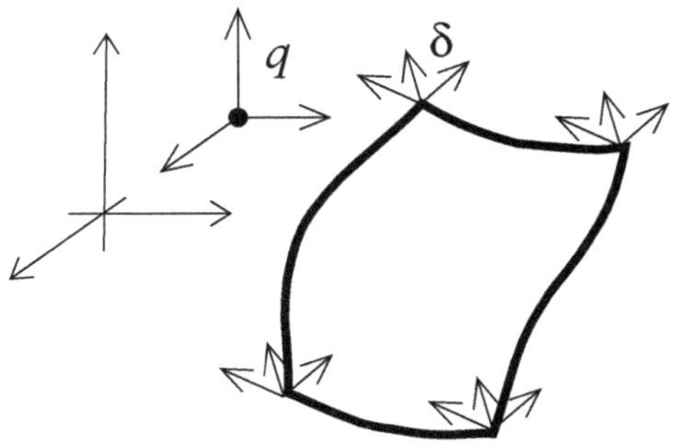

Figure 9.3

$$[\delta] = \begin{bmatrix}\delta^1\\ \delta^2\\ \bullet\\ \delta^p\end{bmatrix} = [R_e]\begin{bmatrix}q^1\\ q^2\\ \bullet\\ \bullet\\ \bullet\\ q^n\end{bmatrix} = [R_e][q] \qquad (9.23)$$

The matrix $[R_e]$ is the representative matrix of the application obtained by composing a projection of $\mathbb{R}^n$ in $\mathbb{R}^p$ and a rotation in $\mathbb{R}^p$ (p denoting the number of parameters of the element, n the number of parameters of the structure).

From then on, the elementary energies can be expressed according to the global parameterisation q:

$$\begin{aligned} \mathcal{F}_e &= \frac{1}{2}[\delta]^T[k_e][\delta] = \frac{1}{2}[q]^T[R_e]^T[k_e][R_e][q] \\ \mathcal{V}_e &= \frac{1}{2}[\delta]^T[f_e] = \frac{1}{2}[q]^T[R_e]^T[f_e] \\ \mathcal{C}_e &= \frac{1}{2}[\dot{\delta}]^T[m_e][\dot{\delta}] = \frac{1}{2}[\dot{q}]^T[R_e]^T[m_e][R_e][\dot{q}] \end{aligned} \tag{9.24}$$

The energies being additive, the energies of the complete structure are obtained by summing the elementary energies obtained above, now all expressed in the same parameterisation:

- Elastic potential:

$$\mathcal{F} = \sum_{\mathcal{E}} \mathcal{F}_e = \frac{1}{2}[q]^T[K][q] \tag{9.25}$$

where:

$$[K] = \sum_{\mathcal{E}} [R_e]^T[k_e][R_e] \tag{9.26}$$

is the **elastic stiffness matrix** of the structure.

- Potential of external actions:

$$\mathcal{V} = \sum_{\mathcal{E}} \mathcal{V}_e = -[q]^T[F] \tag{9.27}$$

where:

$$[F] = \sum_{\mathcal{E}} [R_e]^T[f_e] \tag{9.28}$$

is the **vector of external forces** applied to the structure.

- Kinetic energy:

$$\mathcal{C} = \sum_{\mathcal{E}} \mathcal{C}_e = \frac{1}{2}[\dot{q}]^T[M][\dot{q}] \tag{9.29}$$

where:

$$[M] = \sum_{\varepsilon} [R_e]^T [m_e][R_e] \tag{9.30}$$

is the **mass matrix** of the structure.

9.1.6 Equations of motion (displacement-based method)

The equations of motion are those obtained by the Principle of Virtual Powers from the expressions of the energies obtained above, as recalled in §9.1.1.

In order to obtain an invertible K matrix, it is necessary to eliminate the possible rigid movements by expressing the support conditions. Similarly, to obtain a regular mass matrix, the degrees of freedom without mass should be reduced if necessary. These reduction of parameters operations are not detailed here.

Finally, the matrix equation is written as (9.11) for change of equilibrium problems and (9.15) for vibration problems.

9.1.7 Finite element quality tests

The finite elements must respect a certain number of conditions of stability and/or of constitution making it possible to ensure the convergence of the method. The first condition concerns the stability of the finite element: the kernell of the elementary stiffness matrix must contain only rigid displacement modes (three modes for a plate in flexion, five modes for a shell) and no parasitic mode (called "internal"), which would not be associated with any strain energy. Rigid modes are eliminated by taking into account appropriate boundary conditions.

The finite element approximation is acceptable if it converges, ie if the approximated solution u_h tends to the exact solution u when the "size" h of the elements tends regularly to 0. This implies the convergence of the discretised form $\mathcal{W}_h$ of the total energy to its exact value $\mathcal{W}$. However $\mathcal{W}$ is a function of the displacement u and its derivatives up to the order m. The approximations of u in each finite element must therefore be able to correctly represent all its derivatives up to the order m. The interpolation functions adopted must in particular be able to represent a zero (rigid motion) or constant deformation state. More generally, to ensure good accuracy of the element, each of the terms involved in the expression of the total energy must be able to take a given constant value. Tests to verify the good convergence of an element are called **individual tests**.

The elements used are said **conforming** elements if the solution u obtained by discretisation is continuous, as well as its derivatives up to the order $m - 1$. This implies that the interpolation functions are continuous up to the order $m - 1$ within each element, but also that there is no discontinuity of u and its derivatives at the crossing of the boundary between two elements. The approximation is permissible if it uses conforming elements satisfying the individual convergence criteria.

When the approximations are not in conformity, it should be ascertained (analytically or numerically) that the influence of the discontinuities is null or negligible in the expression of the discretised energy $\mathcal{W}_h$. For this, it is necessary to make sure of the good representation of the constant terms of $\mathcal{W}$ on a set of elements (a mode of constant deformation must produce correct stresses). These tests on assemblies of elements are called **patch-tests**.

A correct solution of the dynamics problems rests notably on a good constitution of the elementary mass matrix. The **weighing test** makes it possible to verify that the mass matrix correctly restores the kinetic energy during a rigid movement. During such a movement (translation and/or overall rotation), the kinetic energy can be calculated from Koenig's second theorem. The value thus obtained is compared with that obtained by (9.22).

Several generic elements are presented in the next paragraphs. They may include variants intended to improve such or such a point, for example concerning the locking in shear or in membrane (cf §9.2.2.3 and §9.3.1).

9.2 PLATE ELEMENTS IN BENDING

9.2.1 The constitution of the plate elements

9.2.1.1 Determination of interpolation functions

The following paragraphs deal with plates in bending, the membrane effects being treated by elements of plane elasticity.

The approach described below relates to flexural plates in classical theory (see §7.1); in this case, the unknown displacement field is the transverse displacement w. In Reissner-Mindlin theory (cf. §7.4), it is necessary to add two components of rotation of the normal vector. The interpolation functions used are generally of a polynomial type:

$$[w]=[N][\delta]=[P][C]^{-1}[\delta] \tag{9.31}$$

where:

[P] is a vector containing monomials.

$$[P]^T=\begin{bmatrix}1 & x & y & x^2 & xy & y^2 & x^3 & y^3 & etc...\end{bmatrix}$$

[C] is a matrix of coefficients et [δ] the vector of the parameters containing the values of displacements at the nodes (translation and rotations).

The coefficients in the matrix [C] are determined by expressing the displacements (or their derivatives) at the nodes, either by calling [δ_i] the subvector of the parameters associated with the node i:

$$[P(x=x_i)][C]^{-1}[\delta]=[\delta_i] \tag{9.32}$$

There must be as many coefficients as there are parameters in the element. This often leads to having to make choices among the monomials of higher degree, essential choices for the good quality of the element.

The following paragraphs deal with flexural plates, and the membrane effects in plates being treated by two-dimensional elements.

9.2.1.2 Strain energy of a plate in flexion

In Reissner's theory, in bending without membrane forces and in small displacements, the displacement of the middle surface is essentially normal to the plate: (Figure 9.4)

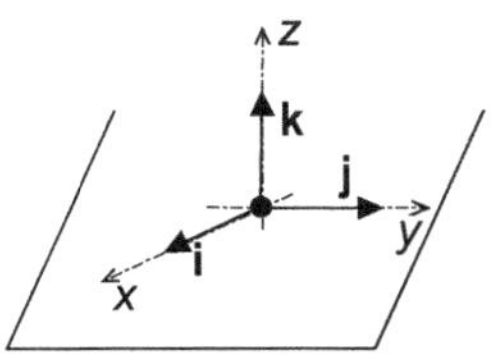

Figure 9.4

$$\xi = w\,\mathbf{k}$$

The normal vector is rotated during the transformation:

$$\theta_x\,\mathbf{i} + \theta_y\,\mathbf{j}$$

so that a point at distance z of the middle point undergoes a displacement:

$$\xi = \begin{pmatrix} u \\ v \\ w \end{pmatrix} = w\,\mathbf{k} + \left(\theta_x\,\mathbf{i} + \theta_y\,\mathbf{j}\right) \wedge z\,\mathbf{k} = \begin{pmatrix} z\theta_y \\ -z\theta_x \\ w \end{pmatrix} \tag{9.33}$$

The strain energy of a plate made of a homogeneous and isotropic elastic linear material is written:

$$\mathcal{F} = \frac{1}{2} D \left\{ \begin{aligned} &\int_e \left[\left(\frac{\partial \theta_x}{\partial x} \right)^2 - 2\nu \frac{\partial \theta_x}{\partial x} \frac{\partial \theta_y}{\partial y} + \left(\frac{\partial \theta_y}{\partial y} \right)^2 + \frac{1-\nu}{2} \left(\frac{\partial \theta_x}{\partial y} - \frac{\partial \theta_y}{\partial x} \right)^2 \right] dxdy \\ &+ \frac{6k(1-\nu)}{e^2} \int_e \left[\left(\frac{\partial w}{\partial x} + \theta_y \right)^2 + \left(\frac{\partial w}{\partial y} - \theta_x \right)^2 \right] dxdy \end{aligned} \right\} \tag{9.34}$$

where k is the shear correction factor.

For very thin isotropic plates, it is generally preferable to retain Kirchhoff's theory, in which case:

$$\begin{cases} \theta_x = \dfrac{\partial w}{\partial y} \\ \theta_y = -\dfrac{\partial w}{\partial x} \end{cases} \tag{9.35}$$

The strain energy is then reduced to:

$$\mathcal{F} = \frac{1}{2} D \left\{ \int_e \left[\left(\frac{\partial^2 w}{\partial x^2} \right)^2 + 2\nu \frac{\partial^2 w}{\partial x^2} \frac{\partial^2 w}{\partial y^2} + \left(\frac{\partial^2 w}{\partial y^2} \right)^2 + 2(1-\nu) \left(\frac{\partial^2 w}{\partial x \partial y} \right)^2 \right] dxdy \right\} \tag{9.36}$$

For a study where the displacements can not be considered as "small", it is necessary to superimpose on the plate bending element a plane element (plane stresses) allowing to take into account the membrane forces and their coupling with flexion.

9.2.1.3 Kirchhoff plate elements

In Kirchhoff's theory, energy reveals second derivatives of the transverse displacement w, which necessitates a continuity C^1 at the boundary of two elements, ie the continuity of

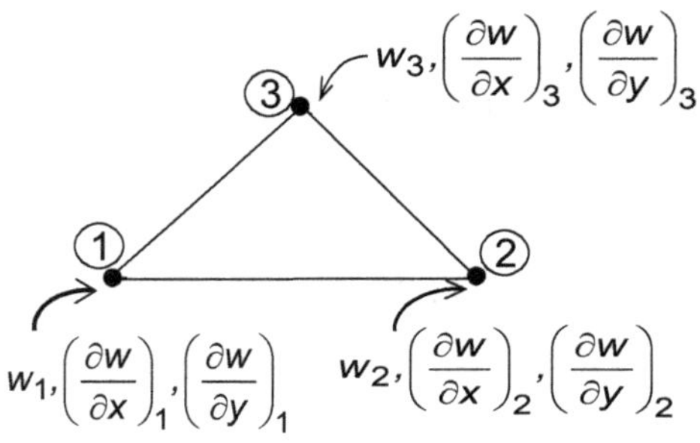

Figure 9.5

the displacement w and the rotation $\frac{\partial w}{\partial n}$, to obtain conforming elements, but this continuity is not easy to obtain.

In the first developed elements (triangles or quadrangles), the values of displacement and derivatives at nodes (apexes) were generally chosen as parameters (Figure 9.5).

Thus, the triangular element T3 has nine degrees of freedom and $w(x, y)$ can be discretised by means of an incomplete cubic polynomial, comprising 9 coefficients C_0 to C_8:

$$w(x,y) = C_0 + C_1x + C_2y + C_3x^2 + C_4xy + C_5y^2 + C_6x^3 + C_7\left(x^2y + xy^2\right) + C_8y^3$$

Note: A complete cubic polynomial has 10 coefficients. The choice made here maintains a symmetry in the monomials between x and y. Nevertheless, this formulation leads to difficulties for certain geometries. On the other hand, the incompleteness of the cubic polynomial loses interesting properties of convergence when the size of the elements tends towards 0. Other choices can be found in the literature to eliminate the "coefficient in excess", see (Zienkiewicz and Taylor).

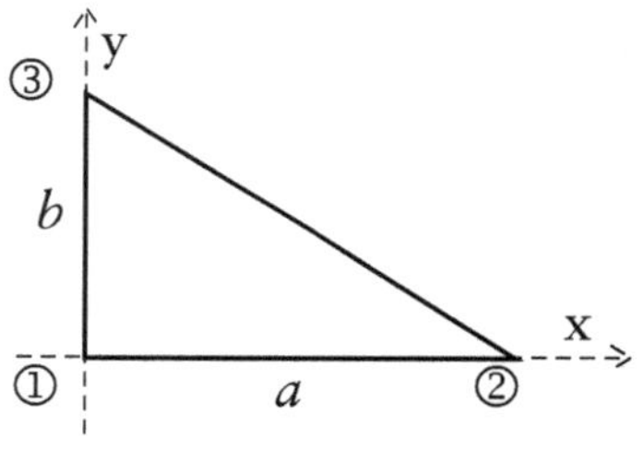

Figure 9.6

C^1 continuity is not verified with such a choice, as shown in the example below (Figure 9.6):

The continuity of $\frac{\partial w}{\partial y}$ alongside 1-2 imposes:

$$\frac{\partial w}{\partial y} = \left(1 - \frac{x}{a}\right)\left(\frac{\partial w}{\partial y}\right)_1 + \frac{x}{a}\left(\frac{\partial w}{\partial y}\right)_2$$

Similarly, the continuity of $\frac{\partial w}{\partial x}$ along 1-3 requires:

$$\frac{\partial w}{\partial x} = \left(1 - \frac{x}{b}\right)\left(\frac{\partial w}{\partial x}\right)_1 + \frac{x}{b}\left(\frac{\partial w}{\partial x}\right)_3$$

At node (①), it comes then:

$$\frac{\partial^2 w}{\partial x \partial y} = \begin{cases} \frac{1}{a}\left[\left(\frac{\partial w}{\partial y}\right)_2 - \left(\frac{\partial w}{\partial y}\right)_1\right] \\ \text{or} \\ \frac{1}{b}\left[\left(\frac{\partial w}{\partial x}\right)_3 - \left(\frac{\partial w}{\partial x}\right)_1\right] \end{cases}$$

expressions *a priori* different. There is no continuity of the torsional curvature at the node.

A first proposed remedy for ensuring the continuity C^1 between two elements consists in increasing the degree of the polynomial interpolation and consequently the number of degrees of freedom at the nodes or the number of nodes.

The simplest element thus constituted uses a polynomial of degree five and uses in addition the second derivatives $\frac{\partial^2 w}{\partial x^2}, \frac{\partial^2 w}{\partial y^2}, \frac{\partial^2 w}{\partial x \partial y}$ as degrees of freedom (Figure 9.7a). Nevertheless, this type of element can not give good results if the rigidity of the plate varies from one element to another, because there is not then continuity of the moments.

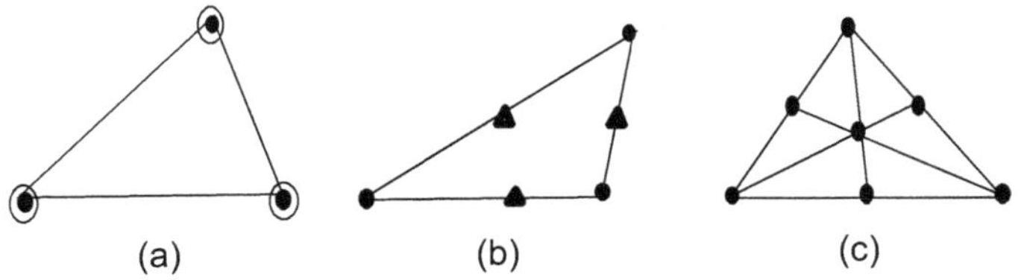

Degrees of freedom :

● $w, \frac{\partial w}{\partial x}, \frac{\partial w}{\partial y}$

⦿ $w, \frac{\partial w}{\partial x}, \frac{\partial w}{\partial y}, \frac{\partial^2 w}{\partial x^2}, \frac{\partial^2 w}{\partial y^2}, \frac{\partial^2 w}{\partial x \partial y}$

▲ $\frac{\partial w}{\partial \mathbf{n}}$

Figure 9.7

It is possible to introduce intermediate nodes on the sides, with, as nodal parameter, the normal rotation $\frac{\partial w}{\partial n}$ (Figure 9.7b), but in the case of quadrangles, this is only feasible for simple geometries, for example rectangular elements. In the triangular case, such an element can be formulated in the following way: to the interpolation function of a simple triangular element such as that of Figure 9.6 are added three interpolation functions each associated with one of the three additional parameters. Each of these additional functions corresponds to a zero displacement w on the three edges, to zero slopes $\frac{\partial w}{\partial n}$ on two edges and quadratic on the third. The setting value of $\frac{\partial w}{\partial n}$ in the middle of this third side makes it possible to calibrate the amplitude of the function.

It is also possible to increase the maximum degree of the polynomial by increasing the number of degrees of freedom (Figure 9.7c).

Another way to develop thin plate elements is to respect Kirchhoff's relations not anywhere in the element, but only at a small number of points (DK elements: "Discrete Kirchhoff"). Under these conditions, the strain energy is expressed as a function of w, θ_x and θ_y only a continuity C° is ensured between two elements, but the shear energy is neglected.

For example, a possible choice for θ_x and θ_y consists of fourth-degree polynomials and for w cubic interpolations on each side. Kirchhoff's relations are then imposed at certain points, for example, the apexes and the midpoints of the sides, so that the rotations θ_x and θ_y can be expressed as functions of the degrees of freedom at the nodes: w, $\dfrac{\partial w}{\partial x}$ and $\dfrac{\partial w}{\partial y}$.

In practice, this is only feasible on a reference geometry and an isoparametric transformation (see §9.3.3.1) is then used to move to the actual geometry of the element.

9.2.1.4 Reissner–Mindlin plate elements

It is necessary here to use the complete expression of the strain energy, since the term due to shear is not negligible compared to the term due to bending. Only a continuity C^0 is required between two elements, when the transverse displacement w and the two rotations θ_x and θ_y are adopted as parameters. The choice of interpolation functions depends on the number of nodes of the element.

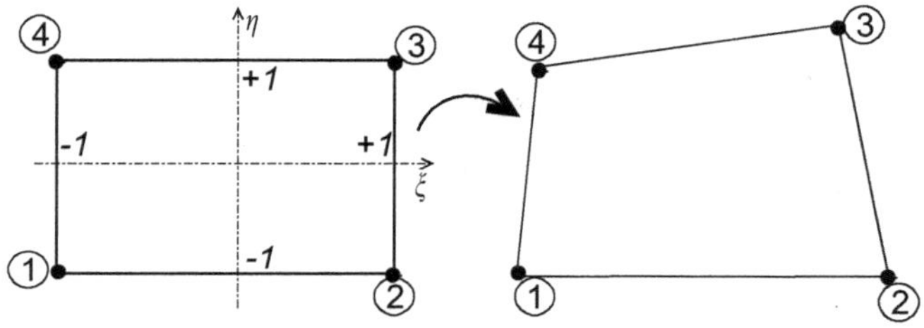

Figure 9.8

For example, a quadrangular element can be defined from an isoparametric transformation (see §9.3.3.1), where w, θ_x and θ_y are bilinear polynomials as a function of the reference coordinates (ξ,η), Figure 9.8.

An 8-node element can also be developed (Figure 9.9), using polynomials of the second degree in (ξ,η).

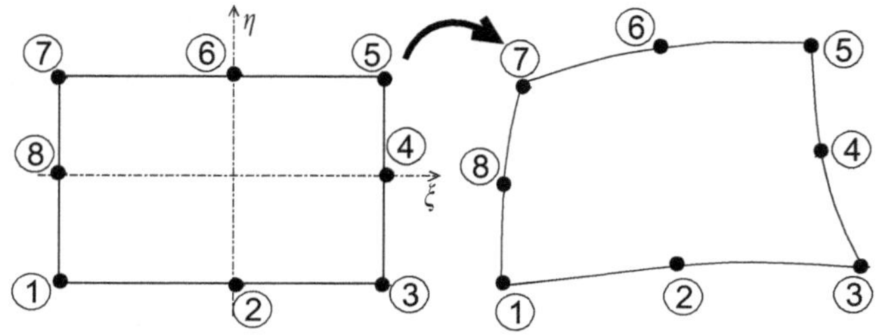

Figure 9.9

The elements presented above have, unfortunately, parasitic phenomena when they are used to model thin plates; indeed, when the thickness tends towards 0, these elements become abnormally stiff, because of a locking in transverse shear.

This phenomenon can be illustrated on a straight beam element taking into account the deformation due to transverse shear (Figure 9.10):

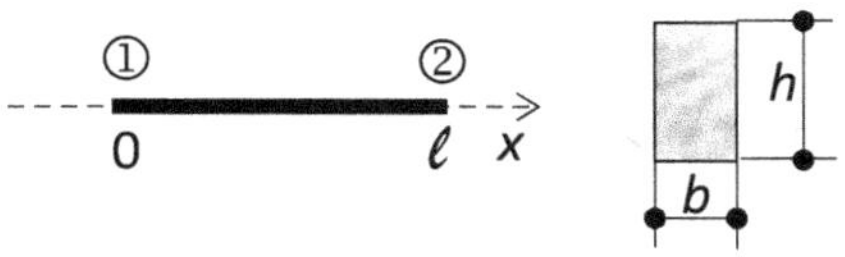

Figure 9.10

The deformation energy of the bending element is written as:

$$\mathcal{F} = \frac{1}{2}\frac{Ee^3b}{12}\int_0^\ell \left(\frac{d\theta}{dx}\right)^2 dx + \frac{1}{2}Gbek\int_0^\ell \left(\theta + \frac{dw}{dx}\right)^2 dx$$

where:

- G is the shear modulus,
- k is the shear correction factor,
- w is the deflection of the beam,
- θ is the rotation of the current section.

By choosing a linear interpolation for w and θ:

$$w(x) = w_1\left(1 - \frac{x}{\ell}\right) + w_2\frac{x}{\ell}$$

$$\theta(x) = \theta_1\left(1 - \frac{x}{\ell}\right) + \theta_2\frac{x}{\ell}$$

the deformations are written:

- Variation of curvature: $\chi = \dfrac{d\theta}{dx} = \dfrac{\theta_2 - \theta_1}{\ell}$
- Distorsion: $\gamma = \dfrac{dw}{dx} + \theta = \dfrac{w_2 - w_1}{\ell} + \theta_1\left(1 - \dfrac{x}{\ell}\right) + \theta_2\dfrac{x}{\ell}$

The distortion γ varies linearly along the beam. By integrating the previous expressions, the strain energy is written:

$$\mathcal{F} = \frac{1}{2}\frac{Ee^3b}{12}\frac{(\theta_2 - \theta_1)^2}{\ell}$$
$$+ \frac{1}{2}Gbek\ell\left[\left(\frac{w_2 - w_1}{\ell} + \theta_1\right)^2 + \left(\frac{w_2 - w_1}{\ell} + \theta_1\right)\left(\frac{\theta_2 - \theta_1}{\ell}\right)\ell + \left(\frac{\theta_2 - \theta_1}{\ell}\right)^2\frac{\ell^2}{3}\right]$$

In the case of thin plates ($e \to 0$), the strain energy must tend towards the bending energy (first term), which imposes that:

$$\frac{w_2 - w_1}{\ell} + \theta_1 = 0 \text{ and } \theta_2 - \theta_1 = 0$$

The first condition expresses that the Kirchhoff equation (Euler–Navier for the beams) is satisfied on the element on average; so it has a clear physical meaning. On the other hand, the second condition would entail the nullity of the variation of curvature and therefore of the energy of flexural deformation; it can not be verified.

As a result, the energy due to shear can not cancel out, which causes the beam to behave abnormally stiffly when its thickness decreases. The term due to shear energy appears as a penalty term, all the more so as the ratio $\frac{e}{\ell}$ is small. Several remedies have been proposed to remove transverse shear locking, for example:

- use of a selective reduced integration: the various contributions to the strain energy are numerically integrated, realising an integration as precise as possible of the bending terms and a reduced integration of the shear terms having the effect of "filtering" the monomials of high degree in the expression of distortions; relationships are then verified only on average.
- use of modified representations of distortions, eliminating terms that give rise to non-physical stresses.

9.2.2 An example of a Kirchhoff plate element construction: the rectangular four-node plate element

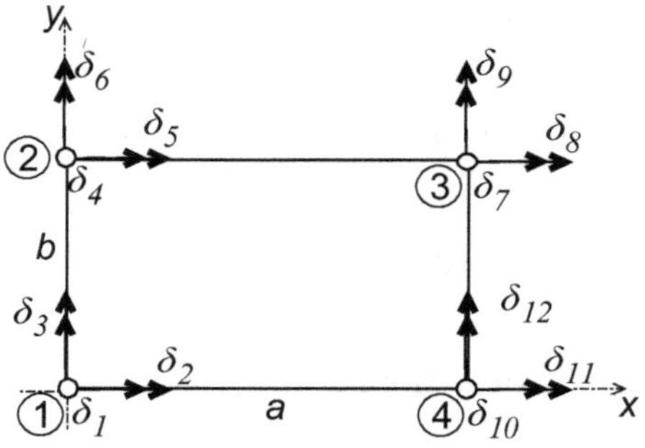

Figure 9.11

This is a rectangular element with twelve parameters (one translation and two rotations per node), see Figure 9.11. At each node, the vector [δ] contains the translation and the two rotations, for example at node 1:

$$\delta_1 = w(x=0;y=0) \; ; \; \delta_2 = -\frac{\partial w}{\partial y}(x=0;y=0) \; ; \; \delta_3 = \frac{\partial w}{\partial x}(x=0;y=0)$$

An incomplete Fourth Degree Polynomial with twelve coefficients is used:

$$w(x,y) = \alpha_1 + \alpha_2 x + \alpha_3 y + \alpha_4 x^2 + \alpha_5 xy + \alpha_6 y^2$$
$$+ \alpha_7 x^3 + \alpha_8 x^2 y + \alpha_9 xy^2 + \alpha_{10} y^3 + \alpha_{11} x^3 y + \alpha_{12} xy^3$$

which is written $w(x,y) = [P][\alpha]$ with:

$$[P] = \begin{bmatrix} 1 & x & y & x^2 & xy & y^2 & x^3 & x^2y & xy^2 & y^3 & x^3y & xy^3 \end{bmatrix}$$

$[\alpha\,]$ is the column vector of unknown coefficients α_i.

The expression of w, $\frac{\partial w}{\partial x}$ and $\frac{\partial w}{\partial y}$ at the nodes leads to:

$$[\delta] = [C]\cdot[\alpha]$$

where [C] is the matrix 12 × 12 of coefficients depending on nodal coordinates. In the present case:

$$\begin{bmatrix}
1 & 0 & 0 & 0 & 0 & 0 & 0 & 0 & 0 & 0 & 0 & 0 \\
0 & 0 & 1 & 0 & 0 & 0 & 0 & 0 & 0 & 0 & 0 & 0 \\
0 & -1 & 0 & 0 & 0 & 0 & 0 & 0 & 0 & 0 & 0 & 0 \\
1 & -b & 0 & 0 & -b^2 & 0 & 0 & 0 & 0 & b^3 & 0 & 0 \\
0 & 0 & 1 & 0 & 0 & b & 0 & -b^2 & 0 & 0 & -b^3 & 0 \\
0 & -1 & 0 & 0 & -2b & 0 & 0 & 0 & 0 & 3b^2 & 0 & 0 \\
1 & -b & a & a^2 & -b^2 & ab & a^3 & -ab^2 & a^2b & b^3 & -ab^3 & a^3b \\
0 & 0 & 1 & 2a & 0 & b & 3a^2 & -b^2 & 2ab & 0 & -b^3 & 3a^2b \\
0 & -1 & 0 & 0 & -2b & a & 0 & -2ab & a^2 & 3b^2 & -3ab^2 & a^3 \\
1 & 0 & a & a^2 & 0 & 0 & a^3 & 0 & 0 & 0 & 0 & 0 \\
0 & 0 & 1 & 2a & 0 & 0 & 3a^2 & 0 & 0 & 0 & 0 & 0 \\
0 & -1 & 0 & 0 & 0 & a & 0 & 0 & a^2 & 0 & 0 & a^3
\end{bmatrix}$$

It becomes:

$$[\alpha] = [C]^{-1} \cdot [\delta]$$

The matrix $[N]$ is then obtained by (9.31):

$$[N] = [P] \cdot [C]^{-1}$$

Either, by posing $\begin{cases} x = \xi a \\ y = \eta b \end{cases}$:

$$[N]^{\mathrm{T}} = \begin{bmatrix}
1 - \xi\eta - (3-2\xi)\xi^2(1-\eta) - (1-\xi)(3-2\eta)\eta^2 \\
(1-\xi)\eta(1-\eta)^2 b \\
-(1-\xi)^2 \xi(1-\eta) a \\
(1-\xi)(3-2\eta)\eta^2 + \xi(1-\xi)(1-2\xi)\eta \\
-(1-\xi)(1-\eta)\eta^2 b \\
-(1-\xi)^2 \xi\eta a \\
(3-2\xi)\xi^2\eta - \xi(1-\eta)(1-2\eta)\eta \\
-\xi(1-\eta)\eta^2 b \\
(1-\xi)\xi^2\eta a \\
(3-2\xi)\xi^2(1-\eta) + \xi(1-\eta)(1-2\eta)\eta \\
\xi\eta(1-\eta)^2 b \\
(1-\xi)\xi^2(1-\eta) a
\end{bmatrix} \qquad (9.37)$$

It should be noted that the element thus constituted is not conforming: $\partial_n w$ is not continuous on the edges, but it passes the patch-tests in Kirchhoff kinematics.

The mass matrix is obtained by applying (9.22). Two of the terms of the mass matrix are given by (9.38), as an example, with $m = \rho eab$ the mass of the element:

$$M_{ij} = \rho eab\int_0^1 d\xi\int_0^1 N_i N_j d\eta \Rightarrow \begin{cases} M_{8,8} = \dfrac{1}{315}mb^2 \\ M_{12,5} = -\dfrac{1}{900}mab \end{cases} \tag{9.38}$$

This coherent matrix withstands the weighing test.

The strain vector contains the curvatures and the stress vector the bending moments:

$$[\varepsilon] = \begin{bmatrix} \dfrac{\partial^2 w}{\partial x^2} \\ \dfrac{\partial^2 w}{\partial y^2} \\ 2\dfrac{\partial^2 w}{\partial x \partial y} \end{bmatrix} \quad [\sigma] = \begin{bmatrix} M_{xx} \\ M_{yy} \\ M_{xy} \end{bmatrix}$$

Factor two applied to the distortion term is introduced so that the dot product of the two vectors gives the correct expression for the energy.

[B] is therefore obtained by double differentiation of [N]:

$$[B]^{\mathrm{T}} = \begin{bmatrix}
-(1-2\xi)(1-\eta)\dfrac{6}{a^2} & -(1-2\eta)(1-\xi)\dfrac{6}{b^2} & -\left[1-6\xi(1-\xi)-6\eta(1-\eta)\right]\dfrac{2}{ab} \\
0 & -(2-3\eta)(1-\xi)\dfrac{2}{b} & -(1-4\eta+3\eta^2)\dfrac{2}{a} \\
(2-3\xi)(1-\eta)\dfrac{2}{a} & 0 & (1-4\xi+3\xi^2)\dfrac{2}{b} \\
-(1-2\xi)\eta\dfrac{6}{a^2} & (1-2\eta)(1-\xi)\dfrac{6}{b^2} & \left[1-6\xi(1-\xi)-6\eta(1-\eta)\right]\dfrac{2}{ab} \\
0 & -(1-\xi)(1-3\eta)\dfrac{2}{b} & \eta(2-3\eta)\dfrac{2}{a} \\
(2-3\xi)\eta\dfrac{2}{a} & 0 & -(1-4\xi+3\xi^2)\dfrac{2}{b} \\
(1-2\xi)\eta\dfrac{6}{a^2} & (1-2\eta)\xi\dfrac{6}{b^2} & \left[-1+6\xi(1-\xi)+6\eta(1-\eta)\right]\dfrac{2}{ab} \\
0 & -(1-3\eta)\xi\dfrac{2}{b} & -\eta(2-3\eta)\dfrac{2}{a} \\
(1-3\xi)\eta\dfrac{2}{a} & 0 & \xi(2-3\xi)\dfrac{2}{b} \\
(1-2\xi)(1-\eta)\dfrac{6}{a^2} & -\xi(1-2\eta)\dfrac{6}{b^2} & \left[-1+6\xi(1-\xi)+6\eta(1-\eta)\right]\dfrac{2}{ab} \\
0 & -\xi(2-3\eta)\dfrac{2}{b} & (1-4\eta+3\eta^2)\dfrac{2}{a} \\
(1-3\xi)(1-\eta)\dfrac{2}{a} & 0 & -\xi(2-3\xi)\dfrac{2}{b}
\end{bmatrix} \tag{9.38}$$

It remains to write the constitutive law, that is to say, the relation between [σ] and [ε], which reveals the elasticity matrix [D] calculated from (9.36):

$$\begin{bmatrix} M_{xx} \\ M_{yy} \\ M_{xy} \end{bmatrix} = \underbrace{\begin{bmatrix} D & D\nu & 0 \\ D\nu & D & 0 \\ 0 & 0 & D(1-\nu) \end{bmatrix}}_{[D]} \begin{bmatrix} \dfrac{\partial^2 w}{\partial x^2} \\ \dfrac{\partial^2 w}{\partial y^2} \\ 2\dfrac{\partial^2 w}{\partial x \partial y} \end{bmatrix}$$

which then makes it possible to calculate the stiffness matrix by application of (9.20). Posing $\beta = a/b$, two terms are given by (9.40), as an example:

$$K_{ij} = ab\int_0^1 d\xi \int_0^1 \sum_{k=1}^{3} B_{ik} \sum_{l=1}^{3} D_{kl} B_{lj} d\eta \Rightarrow \begin{cases} K_{2,2} = \dfrac{4}{3}\left[\dfrac{1}{\beta^2} + \dfrac{1-\nu}{5}\right] b^2 \\ K_{7,5} = \left[2\beta^2 + \dfrac{1-\nu}{5}\right] a \end{cases} \tag{9.40}$$

The complete matrices calculated as indicated above are given in (Przemieniecki).

9.2.3 Some examples of plate elements

9.2.3.1 Kirchhoff plate elements

Table 9.1 shows some examples of Kirchhoff plate elements with some of their properties. All elements of the table are conforming. The 12 ddl element developed in §9.2.3 and the 9 ddl triangular element discussed in §9.2.2.2 are commonly used, but are not conforming. They nevertheless have the correct behaviour with respect to the patch-test, subject to low distortion of the geometry.

The first elements of the table are expressed directly in Kirchhoff's formulation, the next ones are degenerate Reissner–Mindlin elements, for which the Kirchhoff relations are expressed in a finite number of points (discrete elements DK). The DK elements (DKT, DKQ, DKTP, and DKQP) have an additional parameter in the middle of each side, defining the amplitude of the variation of θ_s (rotation in the direction of the normal) along the side.

9.2.3.2 Reissner–Mindlin plate elements

Table 9.2 shows some examples of Reissner–Mindlin plate elements with some of their properties. The DST and DSQ elements have an additional parameter in the middle of each side, defining the amplitude of the variation of θ_s along the side.

These elements make it possible to take into account the energy of distortion and do not present more difficulty of mesh than the Kirchhoff elements, nor do they require the taking into account of more degrees of freedom. It is therefore tempting to use them for any type of plate. Nevertheless, it is necessary to properly choose the type of element if the plate is slender, to avoid the risk of locking in shear.

Table 9.1 Kirchhoff plate elements

Name	*Outline*	*Parameters (number)*	*Comments*
R16	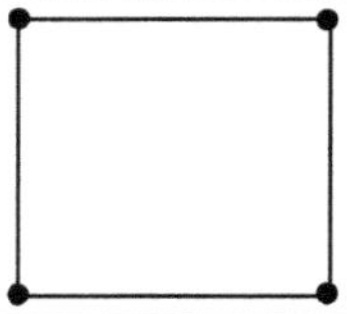	• $\left\{ w, \frac{\partial w}{\partial x} \right.$ $\left\{ \frac{\partial w}{\partial y}, \frac{\partial^2 w}{\partial x \partial y} \right.$ (16)	Incomplete polynomial of the sixth degree. The terms of [*N*] are products of two cubic Hermite functions. C^1 continuity.
T18	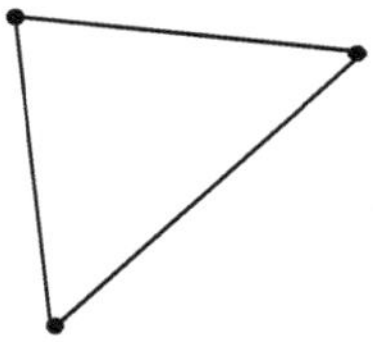	• $\left\{ w, \frac{\partial w}{\partial x}, \frac{\partial w}{\partial y} \right.$ $\left\{ \frac{\partial^2 w}{\partial x^2}, \frac{\partial^2 w}{\partial y^2}, \frac{\partial^2 w}{\partial x \partial y} \right.$ (18)	Incomplete polynomial of the fifth degree. $\partial_n w$ cubic on the sides. C^1 continuity.
DKT	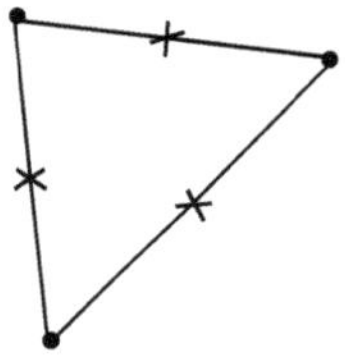	• $\left\{ w, \; \theta_x = \frac{\partial w}{\partial y}, \; \theta_y = -\frac{\partial w}{\partial x} \right.$ × θ_s (12)	Reissner elements with neglected distortion energy. The distortion is zero on the edges. Interpolations on θ are quadratic for DKT and cubic for DKQ. Along one side: • θ_s quadratic in s, • θ_n linear in s. C^0 continuity of θ_x and θ_y. Interpolations on *w* can be linear, quadratic or cubic.
DKQ	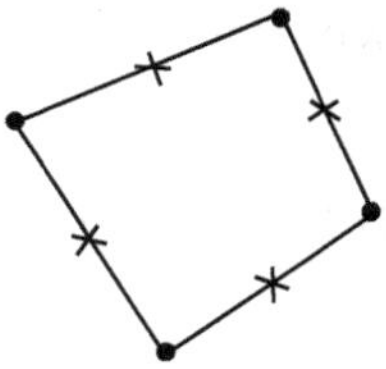	• $\left\{ w, \; \theta_x = \frac{\partial w}{\partial y}, \; \theta_y = -\frac{\partial w}{\partial x} \right.$ × θ_s (16)	
DKTP	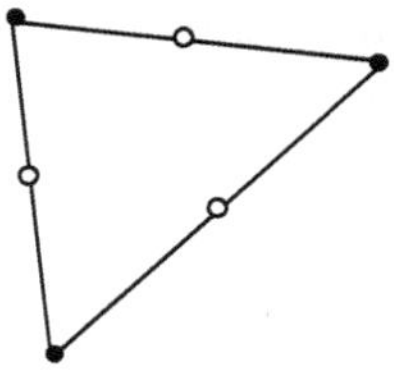	• $\left\{ w, \; \theta_x = \frac{\partial w}{\partial y}, \; \theta_y = -\frac{\partial w}{\partial x} \right.$ ○ w, θ_s (15)	Elements developed to be superimposed with a quadratic membrane element. They look like the previous ones, but the interpolations on θ are quadratic for DKTP and cubic for DKQP. Along one side: • θ_s cubic in s, • θ_n linear in s. The same elements exist with curvilinear edges.
DKQP	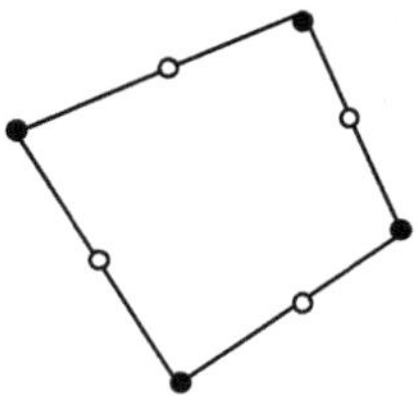	• $\left\{ w, \; \theta_x = \frac{\partial w}{\partial y}, \; \theta_y = -\frac{\partial w}{\partial x} \right.$ ○ w, θ_s (20)	

Table 9.2 Reissner–Mindlin plate elements

Name	*Outline*	*Parameters (number)*	*Comments*
T3		• $\begin{cases} w \\ \theta_x \\ \theta_y \end{cases}$	All of these elements have three parameters per node. w and θ being independent, lower degree interpolation functions can be taken independently for the three variables, depending on the number of nodes. The degree of interpolation therefore increases as a function of the number of nodes. T and Q elements are more numerous than those shown here. The quality of these elements is very variable, in particular because some of them exhibit a shear lock, such as Q4, Q8 and Q12. Q9 and Q16 converge properly, provided their geometry is not over-distorted.
T6	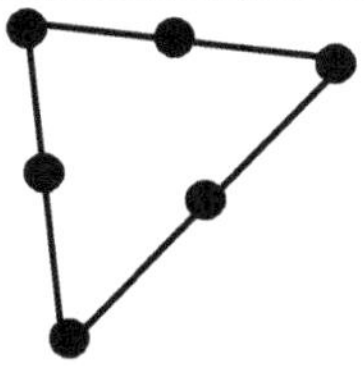		
Q4	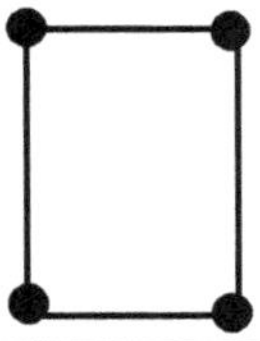		
Q9	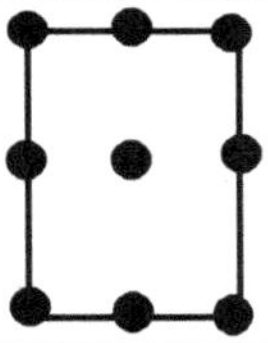		
Q16	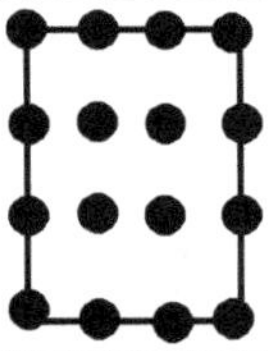		
Q4γ	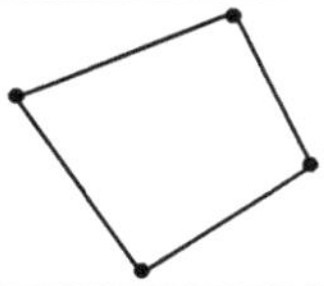	• $\begin{cases} w \\ \theta_x \\ \theta_y \end{cases}$ (12)	Bilinear approximation of the parameters. Along one side: constant CT No locking in CT.
DST	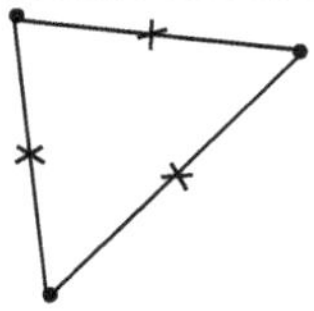	• $\begin{cases} w \\ \theta_x \\ \theta_y \end{cases}$ $\times\ \theta_s$ (12)	Element analogous to DKT. No locking in CT.
DSQ	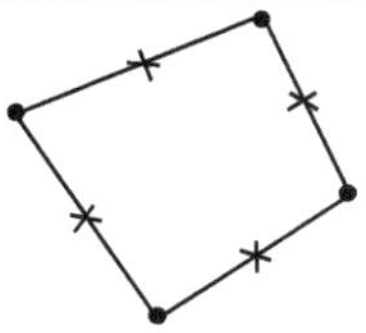	• $\begin{cases} w \\ \theta_x \\ \theta_y \end{cases}$ $\times\ \theta_s$ (16)	Element analogous to DKQ. No locking in CT.

9.3 SHELL ELEMENTS

9.3.1 General

Four major families of shell elements are discussed below:

- flat elements;
- intrinsic curved elements;
- parametric curved elements;
- degenerate massive isoparametric elements.

The elastic potential depends on the thickness e for the membrane and shear terms and e^3 for the bending terms. When e tends to 0 (thin shell), bending should not be underestimated compared to other terms. Shear locking has already been addressed for plates. A similar locking can occur in membrane but is generally less critical for a good accuracy of the results. To avoid this, it is necessary that the interpolations make it possible to obtain null membrane deformations without affecting bending. The techniques used are similar to those used for shear locking.

9.3.2 Flat shell elements

9.3.2.1 Properties of flat elements

The basic idea is to represent the shell by an assembly of plane facets. Each flat facet is modelled by a flat finite element. The big advantage of this modelling is to use simple geometry to model any shape; but it has certain disadvantages which are discussed below. A faceted mesh has an additional advantage if the shell has ribs or edge stiffeners, the connection of flat elements with beam elements then being easier.

For any geometry with double curvature, only the triangular elements allow a mesh sufficiently close to the geometry; the quadrangular elements are more suitable for single curvature geometries (cylinders, cones, ...) or shells consisting of a set of planes (buildings consisting of walls and floors, roofs in the assembly of plane faces, for example).

The first problem is that of the precision of the approximation of the geometry of the shell by an assembly of plane facets. If the dimensions of the facets are small enough vis-à-vis the dimensions of the shell (span and radius of curvature), the error is of the same order of magnitude as that produced by the finite element discretisation. It is thus quite clear that the good precision of the result depends directly on the mesh refinement.

A flat shell element is obtained by the superposition of a membrane element and a plate element (Figure 9.12).

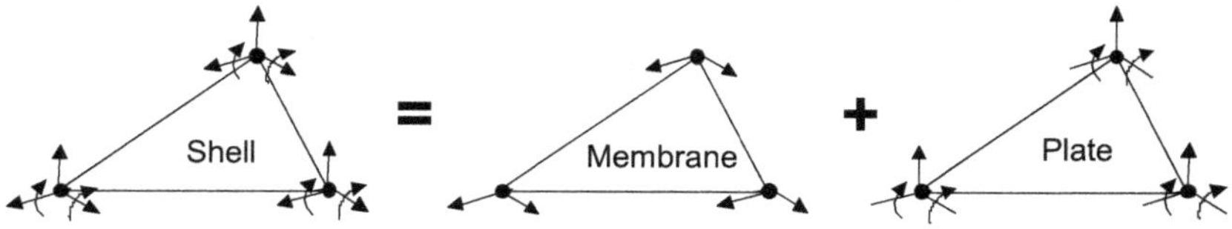

Figure 9.12

The parameters at each node are therefore at least the three components of the displacement and the two rotations which axes are in the plane of the element.

This superposition is not always possible because it supposes a decoupling between the membrane and flexural behaviours, which implies first-order kinematics and appropriate material behaviour, for example, isotropic and homogeneous properties in the thickness of the shell or orthotropic properties allowing decoupling.

In other cases, couplings appear, either by the kinematics (large displacements), or by the constitutive law. The flat element must then be formulated directly, without the superposition of two elements.

During the formulation of the flat elements, the difficulties inherent in the formulation of the plate elements remain. Added to this are the difficulties of connecting solutions to the boundary between two non-coplanar elements; indeed:

- when elements are coplanar, they have zero stiffness vis-à-vis the degree of freedom of rotation around the normal to their plane. This creates a parasitic mechanism, from which it is possible to get rid either by eliminating the unnecessary degree of freedom, or by introducing into the elementary stiffness matrices a very small diagonal term, associated with the corresponding degree of freedom. On the other hand, when the elements are not coplanar, it is necessary to keep the rotation perpendicular to each of the elements, with a risk of instability when the angles between elements are very small.
- when elements are not coplanar, the continuity of the displacements can be relatively poor, because of the differences of the degree of interpolation between the membrane displacements and the flexural displacements. This is illustrated by the following example (Figure 9.13): A right-angle connection between two elements can give rise to a gap.

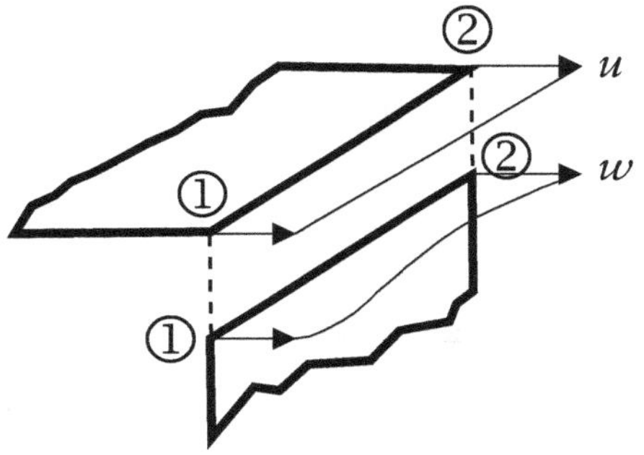

Figure 9.13

For the elements of lower degree, the membrane displacement is linear, while the flexional displacement is cubic along the common 1-2 side. Only the values at nodes 1 and 2 are identical:

$$\begin{cases} u_1 = w_1 \\ u_2 = w_2 \end{cases}$$

The solution here is to use elements with comparable membrane and bending interpolations along the edges.

- The approximation of a curved surface by an assembly of flat elements can give rise to parasitic bending moments in the elements, when the solution is mainly a membrane one. In this case, the remedy is to use curved elements.

9.3.2.2 Examples of flat elements

Flat elements are often constructed from 2D elements and flexural plate elements discussed in 9.2.4. Some examples are given in Table 9.3.

Table 9.3 Flat shell elements

Name	*T/Q*	*Membrane*	*Plate*	*Comments*
DKT18	T	CST	DKT	The CST element is the simplest triangular membrane element, with three nodes at the vertices and two displacements *u* and *v* per node. The "fictitious" parameter θ_3 is added to each vertex.
DST18	T	CST	DST	
DKT12	T	CST	DKT6	The same as above, but the elements DKT and DST are simplified by not considering θ_x and θ_y at the vertices.
DST12	T	CST	DST6	
DQK20	Q	Q4	DKQ	The element Q4 is the quadrangular membrane element with four nodes at the vertices and two displacements *u* and *v* per node, with quadratic interpolation.
DKT27	T	QST	DKT	The QST element is the triangular six-node membrane element at the vertices and at the midpoints of the sides and two displacements *u* and *v* per node, with quadratic interpolation.
DST27	T	QST	DST	
DKTP27	T	QST	DKTP	
DKQP32	Q	Q8	DKQP	The element Q8 is the isoparametric quadrangular membrane element with eight nodes at the vertices and the middle of the sides, with two displacements *u* and *v* per node and quadratic or cubic interpolation.
$Q4_\gamma 20$	Q	Q4	$Q4_\gamma$	

9.3.3 Parametric curved elements

9.3.3.1 Curved geometry of the elements

To generate complex-shaped elements, a geometric transformation defined in parametric form can be used: any point (x, y, z) of the element is then defined as a function of parameters (ξ, η) defined on a reference plane element, cf. Figure 9.14 in the case of $\mathbb{R}^2$. This is the usual representation of a surface by mapping $\mathbb{R}^2$ into $\mathbb{R}^3$.

In this case, the matrices $[N]$ and $[B]$ which intervene in the calculation of the elementary energies are expressed according to the parameters (ξ, η). The integrals are calculated over the parameter space and the volume element dV involves the determinant $\mathcal{J}$ of the Jacobian of the transformation x (ξ).

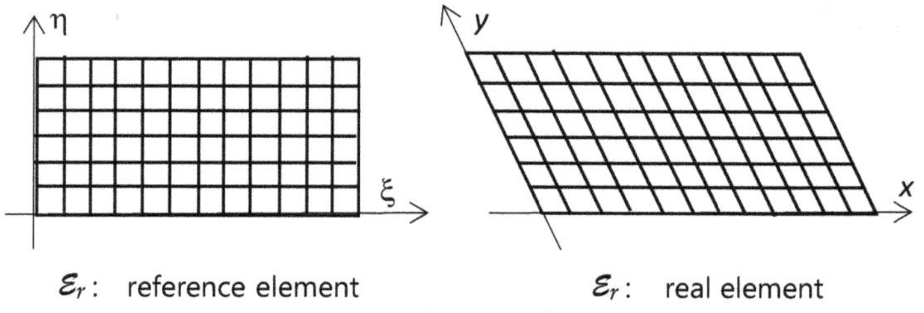

Figure 9.14

Hence for example:

$$[k_e] = \int_{\mathcal{E}_r} [B(\xi,\eta)]^T [D(\xi,\eta)][B(\xi,\eta)]\mathcal{J}(\xi,\eta)\, d\xi d\eta$$

$$[m_e] = \int_{\mathcal{E}_r} [N(\xi,\eta)]^T [N(\xi,\eta)]\mathcal{J}(\xi,\eta)\rho(\xi,\eta)\, d\xi d\eta \tag{9.41}$$

where the integrals are expressed on the reference elements.

The integrations are most often carried out numerically in the form (for example, for the stiffness matrix):

$$\left[k_e\right]=\sum_{1}^{n_I}\pi_i\left[k_\xi\left(\xi_i,\eta_i\right)\right] \tag{9.42}$$

where:

- n_I is the number of points (ξ_i, η_i) known as "integration points" (or Gauss points) ;
- $\left[k_\xi\left(\xi_i,\eta_i\right)\right]$ are the numerical values of $[k_\xi]$ at integration points;
- $\left[k_\xi(\xi,\eta)\right]=\left[B(\xi,\eta)\right]^T\left[D(\xi,\eta)\right]\left[B(\xi,\eta)\right]\ell(\xi,\eta)$
- π_i is the "weight" of point (ξ_i, η_i) in the integration scheme. The accuracy of the scheme depends on the number of Gauss points and the weights adopted.

The representation is said to be **isoparametric** if the geometric transformation $x(\xi_i)$ is of the same form as the displacement interpolation function:

$$\left[x\right]=\left[N(\xi,\eta)\right]\left[\xi\right] \tag{9.43}$$

$[N]$ is the matrix defined by (9.16).

This method has the advantage of giving the same orders of continuity and derivability for the geometry and the transformation of the medium (in particular at the border between two elements). When the element used is conforming, the continuity of the geometry is respected, without creating angular edges (Figure 9.15).

Example: A bar in extension.

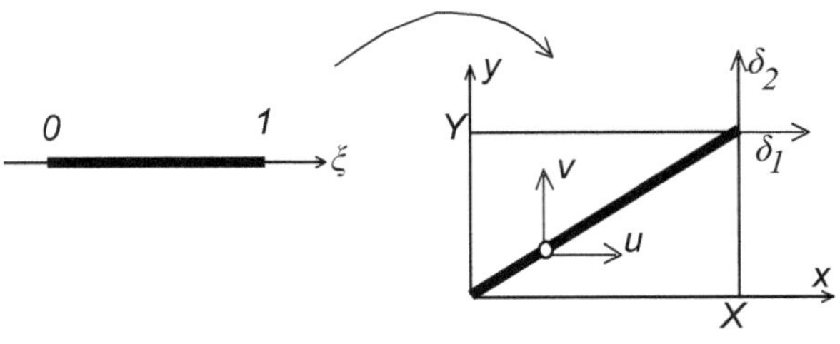

Figure 9.15

The isoparametric element is defined by:

$$\left[N\right]=\begin{bmatrix}\xi & 0\\ 0 & \xi\end{bmatrix}$$

It becomes:

$$\begin{bmatrix}x\\ y\end{bmatrix}=\left[N\right]\begin{bmatrix}X\\ Y\end{bmatrix} \text{ and } \begin{bmatrix}u\\ v\end{bmatrix}=\left[N\right]\begin{bmatrix}\delta_1\\ \delta_2\end{bmatrix}$$

Furthermore:

$$\ell=\left(X^2+Y^2\right)^{1/2}$$

9.3.3.2 Examples of parametric curved elements

Two examples of parametric elements (iso or not) are given in Table 9.4.

Table 9.4 Parametric curved elements

Name	*Outline*	*Parameters (number)*	*Comments*
SEMI-LOOF32		$\bullet \begin{cases} u \\ v \\ w \end{cases}$ $\times\ \theta_s$ (32)	Cartesian isoparametric description. Based on the discrete Kirchhoff semi-loof plate element with 16 dof. Quadratic interpolation of w and rotations.
curved DKT27	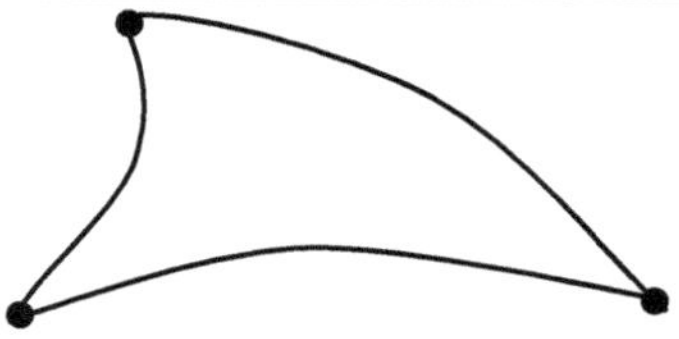	$\bullet \begin{cases} u \quad v \quad w \\ \frac{\partial u}{\partial x}\ \frac{\partial v}{\partial x}\ \frac{\partial w}{\partial x} \\ \frac{\partial u}{\partial y}\ \frac{\partial v}{\partial y}\ \frac{\partial w}{\partial y} \end{cases}$ (9)	Shallow mean surface. Constant curvatures. Cubic displacements, quadratic rotations. Discrete Kirchhoff. See (Batoz and Dhatt) for the detail of the element's constitution.

9.3.4 Shell elements of revolution

The intrinsic curved elements have a defined geometry, of truncated cone, cylindrical or spherical for example. They can follow a KL or RM kinematics. In practical terms, these are mainly shells of revolution elements.

9.3.4.1 Element of revolution in axisymmetric bending

Among the elements of revolution, it is particularly appropriate to distinguish those which are subjected to an axisymmetric loading which does not induce general torsion (Figure 9.16).

In this case, the deformations given by (8.124) and (8.127) in Love approximation, the membrane and bending tensors are diagonal in the main axes of curvature and can be represented by the vectors:

$$[\varepsilon] = \begin{pmatrix} \varepsilon_s \\ \varepsilon_\theta \\ \kappa_s \\ \kappa_\theta \end{pmatrix} = \begin{pmatrix} \dfrac{du}{ds} + \dfrac{w}{R_s} \\ \dfrac{u\cos\phi + w\sin\phi}{r} \\ \dfrac{d^2w}{ds^2} - \dfrac{d}{ds}\left(\dfrac{u}{R_s}\right) \\ \dfrac{\cos\phi}{r}\left(\dfrac{dw}{ds} - \dfrac{u}{R_s}\right) \end{pmatrix}$$

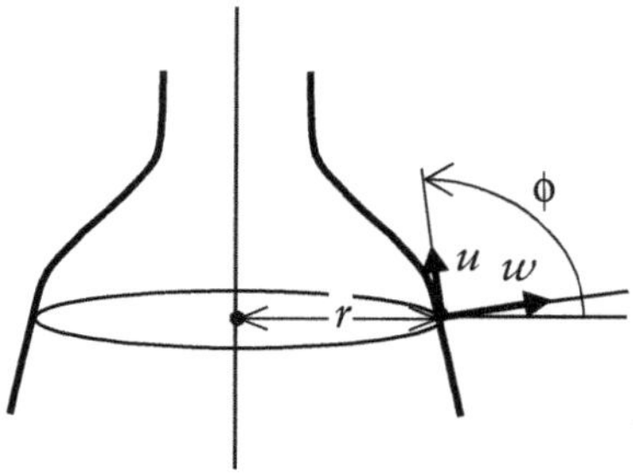

Figure 9.16

$$[\sigma] = \begin{pmatrix} N_{ss} \\ N_{\theta\theta} \\ M_{ss} \\ M_{\theta\theta} \end{pmatrix} \tag{9.44}$$

9.3.4.2 *An example of the construction of an axisymmetric membrane element: the isoparametric element with three nodes*

The membrane displacements (U,W) are expressed directly in the global system (r,z).

The geometry of the element is given by the parametric transformation (Figure 9.17):

$$\begin{pmatrix} r \\ z \end{pmatrix} = \underbrace{\begin{pmatrix} N_1 & N_2 & N_3 & 0 & 0 & 0 \\ 0 & 0 & 0 & N_1 & N_2 & N_3 \end{pmatrix}}_{[N(\xi)]} \begin{pmatrix} r_1 \\ r_2 \\ r_3 \\ z_1 \\ z_2 \\ z_3 \end{pmatrix}$$

with:

$$N_1 = -\frac{1}{2}\xi(1-\xi);\ N_2 = 1-\xi^2;\ N_3 = \frac{1}{2}\xi(1+\xi)$$

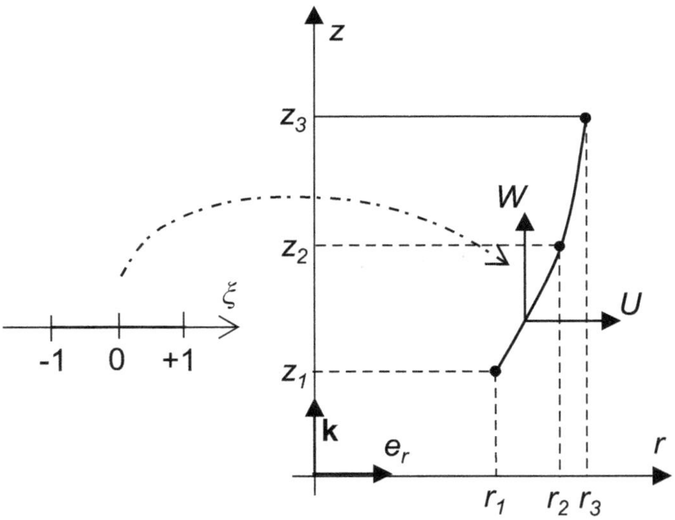

Figure 9.17

The vector a_ξ is worth:

$$a_\xi = \begin{pmatrix} \sum_{i=1}^{3} N_i' r_i \\ \sum_{i=1}^{3} N_i' z_i \end{pmatrix} = A\begin{pmatrix} \cos\phi \\ \sin\phi \end{pmatrix}$$

with:

$$N_1' = -\frac{1}{2}(1-2\xi);\ N_2' = -2\xi;\ N_3' = \frac{1}{2}(1+2\xi)$$

$$A = \left[\left(\sum_{i=1}^{3} N_i' r_i\right)^2 + \left(\sum_{i=1}^{3} N_i' z_i\right)^2\right]^{1/2}$$

The Jacobian is worth:

$$\mathcal{J}(\xi) = \sqrt{a_{\xi\xi} a_{\theta\theta}} = \left(\sum_{i=1}^{3} N_i r_i\right) A$$

Since the element is isoparametric, the displacements are obtained by the same interpolation function from the displacements at the three nodes: (Figure 9.17)

$$\begin{pmatrix} U \\ W \end{pmatrix} = \underbrace{\begin{pmatrix} N_1 & N_2 & N_3 & 0 & 0 & 0 \\ 0 & 0 & 0 & N_1 & N_2 & N_3 \end{pmatrix}}_{[N(\xi)]} \begin{pmatrix} U_1 \\ U_2 \\ U_3 \\ W_1 \\ W_2 \\ W_3 \end{pmatrix}$$

The strain in the direction of the generator is evaluated in the basis vector a_1, then normalised:

$$\varepsilon_{11} = \frac{1}{A^2}\varepsilon_{\xi\xi} = \frac{1}{2A^2}\left[\left(\frac{dm}{d\xi}\right)^2 - \left(\frac{d\mathcal{M}}{d\xi}\right)^2\right] \approx \frac{1}{A^2}\left(\frac{dr}{d\xi}e_r + \frac{dz}{d\xi}\mathbf{k}\right)\left(\frac{dU}{d\xi}e_r + \frac{dW}{d\xi}\mathbf{k}\right)$$

$$\varepsilon_{22} = \frac{U}{r}$$

hence, depending on the interpolation:

$$\varepsilon_{11} = \frac{1}{A^2}\left[\left(\sum_{i=1}^{3} N_i' r_i\right)\left(\sum_{j=1}^{3} N_j' U_j\right) + \left(\sum_{i=1}^{3} N_i' z_i\right)\left(\sum_{j=1}^{3} N_j' W_j\right)\right]$$

$$\varepsilon_{22} = \left(\sum_{j=1}^{3} N_j' U_j\right) \Big/ \left(\sum_{i=1}^{3} N_i' r_i\right)$$

then the matrix $\left[B(\xi)\right]$:

$$\left[B(\xi)\right]=\begin{pmatrix} \frac{1}{A^2}\left(\sum_{i=1}^{3}N_i'r_i\right)N_1' & \frac{1}{A^2}\left(\sum_{i=1}^{3}N_i'r_i\right)N_2' & \frac{1}{A^2}\left(\sum_{i=1}^{3}N_i'r_i\right)N_3' \\ N_1'\Big/\left(\sum_{i=1}^{3}N_i'r_i\right) & N_2'\Big/\left(\sum_{i=1}^{3}N_i'r_i\right) & N_3'\Big/\left(\sum_{i=1}^{3}N_i'r_i\right) & \cdots \end{pmatrix}$$

$$\cdots \begin{matrix} \frac{1}{A^2}\left(\sum_{i=1}^{3}N_i'z_i\right)N_1' & \frac{1}{A^2}\left(\sum_{i=1}^{3}N_i'z_i\right)N_2' & \frac{1}{A^2}\left(\sum_{i=1}^{3}N_i'z_i\right)N_3' \\ 0 & 0 & 0 \end{matrix}\Bigg)$$

In the case of a homogeneous and isotropic linearly elastic membrane, the matrix of elasticity coefficients is written as:

$$\left[D\right]=\begin{pmatrix} K & K\nu \\ K\nu & K \end{pmatrix}$$

But there is nothing against taking a variable thickness in the expression of $[D]$. At this stage, all the matrices expressed as a function of ξ occurring in the integrals (9.41) are determined. The integrations are carried out numerically according to a Gauss scheme.

9.3.4.3 Element of revolution in non-axisymmetric bending

Any loading p depends on both s and longitude θ. It can be decomposed into Fourier series in the form:

$$p(s,\theta)=\sum_{i=1}^{3}\left[p_i^0(s)+\sum_j p_i^{cj}(s)\cos j\theta+\sum_j p_i^{sj}(s)\sin j\theta\right]a_i \tag{9.45}$$

The zero-order term corresponds to axisymmetric bending and / or torsion (if the term p_2^0 exists). Each of the components of the decomposition is studied separately, considering the symmetry or antisymmetry in θ. The same type of elements as the axisymmetric elements can be used to study each j-order mode, however adding the distortion ε_{12} and torsion κ_{12} deformations.

9.3.4.4 Examples of elements of revolution

Table 9.5 shows examples of elements of revolution in axisymmetric bending with some of their properties.

Table 9.5 Elements of revolution in axisymmetric bending

Name	*Outline*	*Parameters (number)*	*Comments*
CAXI_K	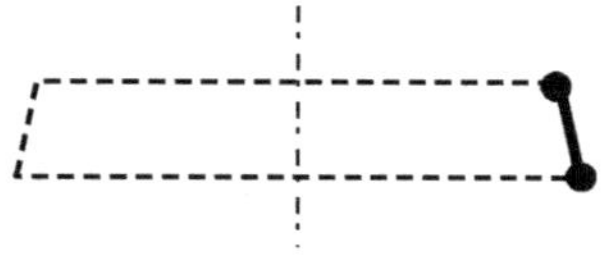	• $\begin{cases} u \\ w \end{cases}$ (4)	Axisymmetric truncated cone element (KL). $u = \alpha_0 + \alpha_1 s$ $w = \beta_0 + \beta_1 s + \beta_2 s^2 + \beta_3 s^3$ $\theta = \dfrac{dw}{ds}$
CAXI_L		• $\begin{cases} u \\ w \\ \theta \end{cases}$ or $\begin{cases} U \\ W \\ \theta \end{cases}$ (6)	Axisymmetric truncated cone element (RM). $\begin{cases} u = \alpha_0 + \alpha_1 s \\ w = \beta_0 + \beta_1 s \\ \theta = \gamma_0 + \gamma_1 s \end{cases}$ No risk of blockage in CT.
CAXI_Q	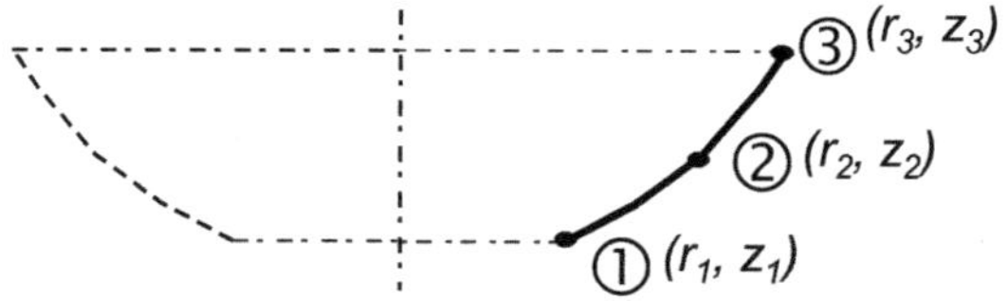	• $\begin{cases} u \\ w \\ \theta \end{cases}$ (9)	Axisymmetric isoparametric element (RM). Generalisation of the membrane element of §9.3.2.4. $\begin{cases} u = \sum_i N_i(\xi) u_i \\ w = \sum_i N_i(\xi) w_i \\ \theta = \sum_i N_i(\xi) \theta_i \end{cases}$ Possibility of variable thickness. No risk of blockage in CT.

9.3.5 Curved isoparametric elements (degenerated 3D-element)

Most isoparametric shell elements are degenerated 3D elements. These elements are particularly suited to the Reissner–Mindlin kinematics.

9.3.5.1 *General principle of the formulation of these elements*

The current point in the thickness of the shell is given by the parametric representation:

$$\bar{z}_j = \sum_{i=1}^{nd} N_i(\xi,\eta)\left[z_{ji} + \frac{1}{2}\zeta e_i n_{ji}\right] \tag{9.46}$$

where nd is the number of node, z_{ji} the j-th coordinate of node i, n_{ji} the j-th component of the normal vector a_3 at node i, e_i the thickness at node i and ζ the transverse coordinate varies from –1 to +1.

In the same way, the displacement is given at any point by the same interpolation:

$$[w] = [N(\xi,\eta,\zeta)][\delta] \tag{9.47}$$

$[w]$ is the vector displacement at any point of the shell, $[\delta]$ is the vector of the element's parameters, defined below for five parameters per node and $[N]$ the interpolation function defined here according to the parameters ξ, η and ζ, where appear the components of the two vectors of the local natural basis a_1 and a_2 associated with the parameters ξ and η:

$$[\delta]^{\mathrm{T}} = [\cdots \;\vdots u_i \quad v_i \quad w_i \quad \theta_{1i} \quad \theta_{2i} \vdots \;\cdots \; i = 1, nd] \tag{9.48}$$

$$[N] = \left[\cdots \left|\begin{matrix} N_i & 0 & 0 & -\frac{1}{2}\zeta N_i e_i a_{2xi} & \frac{1}{2}\zeta N_i e_i a_{1xi} \\ 0 & N_i & 0 & -\frac{1}{2}\zeta N_i e_i a_{2yi} & \frac{1}{2}\zeta N_i e_i a_{1yi} \\ 0 & 0 & N_i & -\frac{1}{2}\zeta N_i e_i a_{2zi} & \frac{1}{2}\zeta N_i e_i a_{1zi} \end{matrix}\right| \cdots \; i = 1, nd\right] \tag{9.49}$$

The matrix $[N]$ is of dimensions (3, $5 \times nd$).

The local strain tensor is calculated from this expression by (9.5). The integrals (9.41) are performed on the volume determined by the element. As the variation of the strains as functions of the thickness is linear, the integration in z can be reduced (two Gauss points) or even carried out analytically.

9.3.5.2 *Examples of isoparametric elements*

Table 9.6 shows examples of degenerated 3D isoparametric elements with some of their properties.

Table 9.6 Degenerated 3D isoparametric elements

Name	*Outline*	*Parameters (number)*	*Comments*
COQ6	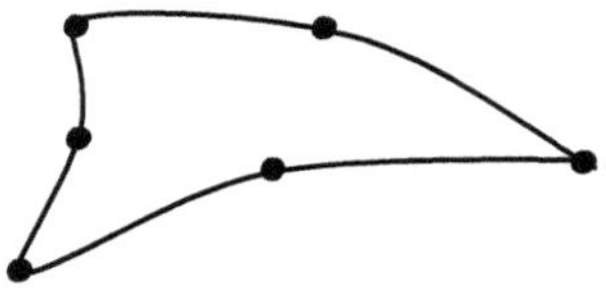	• $\begin{cases} u \\ v \\ w \\ \theta_1 \\ \theta_2 \end{cases}$ (30)	Element according to the method of §9.3.5.1, developed from the plane triangular element T6.
Q8H (hétérosis)	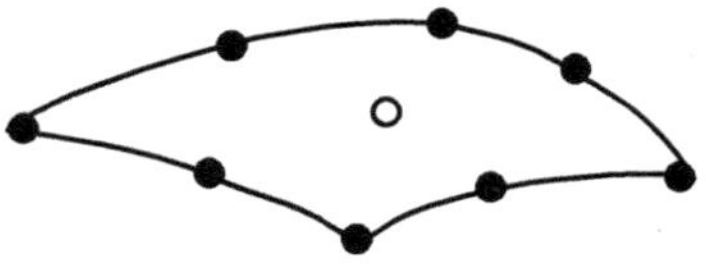	• $\begin{cases} u \\ v \\ w \\ \theta_1 \\ \theta_2 \end{cases}$ ○ $\begin{cases} \theta_1 \\ \theta_2 \end{cases}$ (42)	A sixth variable is added to the nodes if the normal vector rotates from one element to another. Different interpolation in displacement and rotation.
Q4γ24	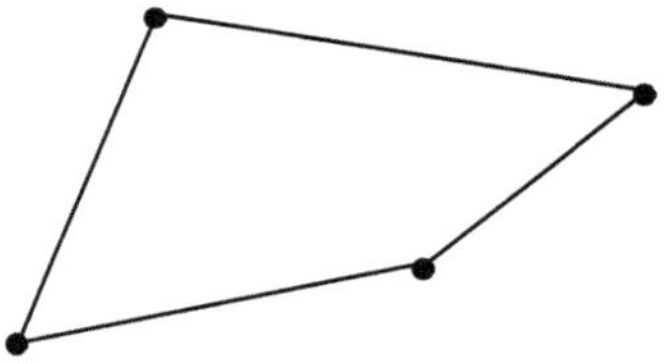	• $\begin{cases} u \\ v \\ w \\ \theta_1 \\ \theta_2 \\ \theta_3 \end{cases}$ (24)	Geometry: bilinear isoparametric (four non-coplanar nodes). Normal vectors are not continuous from element to element. Obtained by superimposing the membrane element Q4 and the bending element Q4γ. Quadratic deformation in z and numerical integration in the thickness. Special treatment of θ_3 to avoid singularity. The CT stiffness contains terms in e and e^3. No membrane or shear blockage.
SHB8	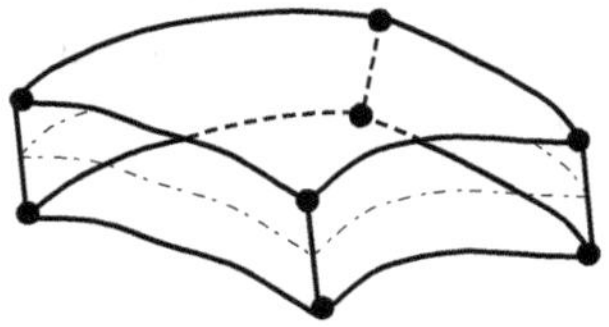	• $\begin{cases} u \\ v \\ w \end{cases}$ (24)	Element developed from the eight-node hexahedron. Numerical (five Gauss points) or analytical integration of strain energy in the thickness. Possibility of variable thickness. Possibility of connection with 3D elements. Element developed for nonlinear behaviours.

EXERCISE

Exercise 9.1

It is a question of constructing a finite element of rectangular plate, in Reissner-Mindlin theory (Figure 9.18).

1) The distribution of transverse shear stresses is given in the *a priori* form:

$$\sigma^{iz} = f(z)Q^i$$

By writing the displacement and strain fields in the thickness of the plate, on the Reissner-Mindlin assumption, show that the elastic strain energy can be in the form:

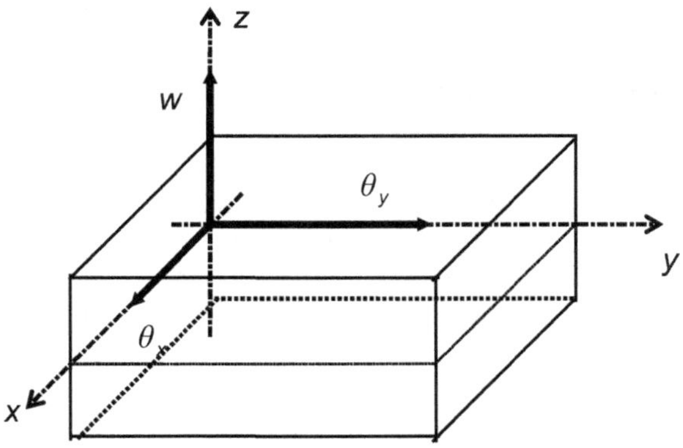

Figure 9.18

$$\mathcal{F} = \mathcal{F}_f + \mathcal{F}_c = \frac{1}{2}\int_\sigma [\varepsilon]^T [C_\varepsilon][\varepsilon] d\sigma + \frac{1}{2}\int_\sigma [\gamma]^T [C_\gamma][\gamma] d\sigma$$

where:

- $[\varepsilon]$ and $[\gamma]$ the mean surface strain vectors associated with bending and shear respectively;
- $\mathcal{F}_f$ is the bending energy associated with the assumption of plane stresses, expressed using the strain vector $[\varepsilon]$;
- $\mathcal{F}_c$ is the distortion energy, expressed using the distortion vector $[\gamma]$;
- $[C_\varepsilon]$ et $[C_\gamma]$ are matrices explaining the constitutive laws. Calculate these two matrices in the following cases:
  - shell made of a homogeneous and isotropic material, with parabolic shear distribution;
  - shell made of two steel sheets separated by a honeycomb, with constant shear in pieces.

2) The finite plate element to be constructed is rectangular, with sides *a* and *b*, parameterised by the local variables x and y.

The interpolation functions are the same for the three functions w, θ_x and θ_y, and they are assumed to be affine with respect to each variable. These functions are represented by a single matrix $[N]$, for example for displacement: (Figure 9.19)

$$w = [N(x,y)][W]$$

$$\text{with: } [W] = \begin{pmatrix} w_1 \\ w_2 \\ w_3 \\ w_4 \end{pmatrix}$$

and identical interpolations for the rotations.

The vector of the parameters [δ] of the element is composed successively for each node of the translation w and of the two rotations θ_x and θ_y, in the order of numbering of the nodes, ie 12 parameters.

Express the matrices [N] and [B] of the element in Reissner-Mindlin theory, then the term $K_{2,7}$ of the stiffness matrix, in the case of the homogeneous and isotropic material.

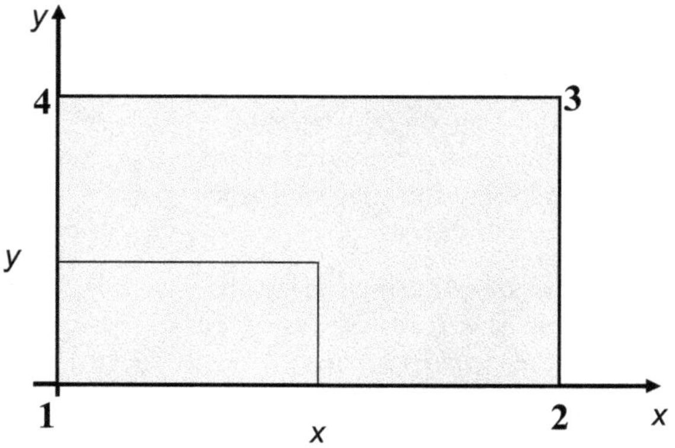

Figure 9.19

Answer:

1. See §7.4, for the Reissner–Mindlin kinematics. In pure bending, the displacement of any point in the thickness of the plate is written: $\xi = \left(u = z\theta_y; v = -z\theta_x; w\right)$, hence the local strain tensor:

$$\begin{aligned} \bar{\varepsilon}_{xx} &= z\frac{\partial\theta_y}{\partial x} \\ \bar{\varepsilon}_{xx} &= -z\frac{\partial\theta_x}{\partial y} \\ \bar{\varepsilon}_{xy} &= \frac{z}{2}\left(\frac{\partial\theta_y}{\partial y}+\frac{\partial\theta_x}{\partial x}\right) \end{aligned} \Rightarrow \left[\bar{\varepsilon}\right] = z\left[\varepsilon\right] \qquad \begin{aligned} \varepsilon_{xz} &= \frac{1}{2}\left(\theta_y+\frac{\partial w}{\partial x}\right) \\ \varepsilon_{yz} &= \frac{1}{2}\left(-\theta_x+\frac{\partial w}{\partial y}\right) \end{aligned} = \frac{1}{2}\left[\gamma\right]$$

In the cases where the constitutive law does not couple the membrane strains and the distortions, the corresponding strain energies can be calculated separately. For the pure bending energy, by introducing the local plane elasticity constitutive law:

$$\begin{bmatrix} \sigma_{xx} \\ \sigma_{yy} \\ \sigma_{xy} \end{bmatrix} = \left[\bar{C}_\varepsilon\right] \begin{bmatrix} \varepsilon_{xx} \\ \varepsilon_{yy} \\ 2\varepsilon_{xy} \end{bmatrix}$$

energy is written as:

$$\mathcal{F}_f = \frac{1}{2}\int_{\mathcal{V}}\left[\bar{\varepsilon}\right]^{\mathrm{T}}\left[\bar{C}_\varepsilon\right]\left[\bar{\varepsilon}\right]dV = \frac{1}{2}\int_{\Sigma}\left[\varepsilon\right]^{\mathrm{T}}\left[C_\varepsilon\right]\left[\varepsilon\right]dV \text{ with } \left[C_\varepsilon\right] = \int_{-e/2}^{e/2}\left[\bar{C}_\varepsilon\right]z^2dz$$

For the distortion energy, it should be noted beforehand that the shear stresses are zero on both sides of the shell, so $f\left(\pm\frac{e}{2}\right) = 0$, and that the shear stresses balance the shear force, therefore:

$$Q^\alpha = \int_{-e/2}^{e/2}\sigma^{\alpha z}dz = \int_{-e/2}^{e/2}f(z)Q^\alpha dz \Rightarrow \int_{-e/2}^{e/2}f(z)dz = 1$$

The local shear constitutive law is of the form:

$$[\tau]=\begin{bmatrix}\sigma_{xz}\\ \sigma_{yz}\end{bmatrix}=\left[\bar{C}_\gamma\right][\gamma]$$

To express the global law $[Q]=[C_\gamma][\gamma]$, the distribution of shear stresses in the thickness should be taken into account, the distortions being constant in the model, and therefore not locally representative (see §5.4.4.2). The energy is therefore expressed as a function of the stresses:

$$\mathcal{F}_c=\frac{1}{2}\int_{\mathcal{V}}[\gamma]^{\mathrm{T}}[\tau]dV=\frac{1}{2}\int_{\mathcal{V}}[\tau]^{\mathrm{T}}\left[\bar{C}_\gamma\right]^{-1}[\tau]dV$$

$$=[Q]^{\mathrm{T}}\underbrace{\left[\int_{-e/2}^{e/2}\left[\bar{C}_\gamma\right]^{-1}(f(z))^2\,dz\right]}_{[C_\gamma]^{-1}}[Q]=\frac{1}{2}\int_{\Sigma}[\gamma]^{\mathrm{T}}[C_\gamma][\gamma]dV$$

If the material is homogeneous and isotropic (see §9.2.2), the distribution of shear stresses is parabolic in the thickness (see §5.4.3.4) and:

$$[C_\varepsilon]=D\begin{bmatrix}1&\nu&0\\ \nu&1&0\\ 0&0&1-\nu\end{bmatrix}$$

$$[C_\gamma]^{-1}=\int_{-e/2}^{e/2}\frac{1}{G}\begin{bmatrix}1&0\\0&1\end{bmatrix}\left[\frac{3}{2e}\left(1-\frac{4z^2}{e^2}\right)\right]^2 dz\Rightarrow[C_\gamma]=\frac{5}{6}Ge\begin{bmatrix}1&0\\0&1\end{bmatrix}$$

In the case of the sandwich material (see §5.6.3.2), the thickness a of the sheets is small compared to the total thickness e. The sheets are homogeneous and isotropic and only balance the bending moments, the stiffness of the honeycomb being low.

The honeycomb alone is deformable with respect to shear and the shear stresses can be considered there to be little variable and therefore taken constant; so $f(z)=1/e$. The shear moduli can be different in both directions. Finally:

$$[C_\varepsilon]=\frac{E_s e^2 a}{2\left(1-\nu_s^2\right)}\begin{bmatrix}1&\nu_s&0\\ \nu_s&1&0\\ 0&0&1-\nu_s\end{bmatrix}\quad[C_\gamma]=\begin{bmatrix}eG_x&0\\0&eG_y\end{bmatrix}$$

2. In the case of an affine interpolation in a rectangular finite element, the displacement is written (Figure 9.19):

$$w=\alpha_1+\alpha_2 x+\alpha_3 y+\alpha_4 xy$$

By writing the equality of displacements with the parameters $[W]$ at the four vertices of the rectangle, it comes:

$$[N]=\begin{bmatrix}1-\xi-\eta+\xi\eta & \xi-\xi\eta & \xi\eta & \eta-\xi\eta\end{bmatrix} \text{ with } x=\xi a\,;\, y=\eta b$$

Since the interpolation is the same for the rotations, the matrix $[N]$ is the same for these parameters.

The matrix $[B]$ is the sum of two matrices which can be calculated separately for the bending and the distortions:

$$[B_\varepsilon]^T=\begin{bmatrix} 0 & 0 & 0 \\ 0 & \frac{1-\xi}{b} & \frac{1-\eta}{a} \\ -\frac{1-\eta}{a} & 0 & -\frac{1-\xi}{b} \\ 0 & 0 & 0 \\ 0 & \frac{\xi}{b} & -\frac{1-\eta}{a} \\ \frac{1-\eta}{a} & 0 & -\frac{\xi}{b} \\ 0 & 0 & 0 \\ 0 & -\frac{\xi}{b} & -\frac{\eta}{a} \\ \frac{\eta}{a} & 0 & \frac{\xi}{b} \\ 0 & 0 & 0 \\ 0 & -\frac{1-\xi}{b} & \frac{\eta}{a} \\ -\frac{\eta}{a} & 0 & \frac{1-\xi}{b} \end{bmatrix} \quad [B_\gamma]=\begin{bmatrix} -\frac{1-\eta}{a} & -\frac{1-\xi}{b} \\ 0 & -(1-\xi-\eta+\xi\eta) \\ 1-\xi-\eta+\xi\eta & 0 \\ \frac{1-\eta}{a} & -\frac{\xi}{b} \\ 0 & -(\xi-\xi\eta) \\ \xi-\xi\eta & 0 \\ \frac{\eta}{a} & \frac{\xi}{b} \\ 0 & -\xi\eta \\ \xi\eta & 0 \\ -\frac{\eta}{a} & \frac{1-\xi}{b} \\ 0 & -(\eta-\xi\eta) \\ \eta-\xi\eta & 0 \end{bmatrix}$$

The stiffness matrix is then obtained by:

$$K=\int_\Sigma\left([B_\varepsilon]^T[C_\varepsilon][B_\varepsilon]+[B_\gamma]^T[C_\gamma][B_\gamma]\right)dS$$

or, for the term (2,7):

$$\left([B_\varepsilon]^T[C_\varepsilon][B_\varepsilon]\right)_{2,7}=\sum_{j=1}\left([B_\varepsilon]^T\right)_{j,7}\sum_{i=1}^{3}\left([C_\varepsilon]\right)_{j,i}\left([B_\varepsilon]\right)_{i,2}$$

$$=D\begin{bmatrix}0\\0\\0\end{bmatrix}^T\begin{bmatrix}1 & \nu & 0\\ \nu & 1 & 0\\ 0 & 0 & 1-\nu\end{bmatrix}\left[\begin{bmatrix}0\\ \frac{1-\xi}{b}\\ \frac{1-\eta}{a}\end{bmatrix}\right]=0$$

$$\left([B_\gamma]^T[C_\gamma][B_\gamma]\right)_{2,7} = \sum_{j=1}^{2}\left([B_\gamma]^T\right)_{j,7}\sum_{i=1}^{2}\left([C_\gamma]\right)_{j,i}\left([B_\gamma]\right)_{i,2}$$

$$= \frac{5}{6}Ge\begin{bmatrix}\frac{\eta}{a}\\ \frac{\xi}{b}\end{bmatrix}^T\begin{bmatrix}1 & 0\\ 0 & 1\end{bmatrix}\begin{bmatrix}0\\ -(1-\xi-\eta+\xi\eta)\end{bmatrix}$$

$$= -\frac{5}{6}Ge\frac{\xi}{b}(1-\xi-\eta+\xi\eta)$$

Ultimately: $K_{2,7} = 0$.

The courageous reader will be able to calculate all the terms of the stiffness matrix by following the same method.

Reading tips

The reading advice below is voluntarily limited to documents that can constitute the first extension of this course, useful for the reader. Most of these works themselves contain a more detailed bibliography which refers to the fundamental texts referenced in the various chapters or to the many articles dealing with the subjects covered here.

Understanding the material covered in this course requires a solid knowledge of the mechanics of solids, of the theory of structures in particular. The energy method used is particularly developed in:

- Y. BAMBERGER "Mécanique de l'Ingénieur" Tomes 1 à 3. HERMANN. 1981. (in French)
- F. VOLDOIRE et Y. BAMBERGER « Mécanique des structures ». Presses de l'École nationale des ponts et chaussées. 2008. (in French)

The foundations of the theory of shells are particularly well exposed in:

- P.M. NÀGHDI "Foundations of elastic shell theory". in "Progress in Solid Mechanics" Vol.4. NORTH-HOLLAND PUBLISHING COMPANY. 1963. (in English)
- P.M. NÀGHDI "The Theory of Shells and Plates". Vol. VIa/2. in "Encyclopaedia of Physics". SPRINGER-VERLAG. 1972. (in English)
- W. FLUGGE "Theory of shells". in "Tensor analysis and Continuum Mechanics"; Chapter 9. SPRINGER-VERLAG. 1972. (in English)
- J. BARBE "Théorie des Coques". Cours de l'ENSTA. 1977. (in French)

A fundamental reference work that covers many of the topics discussed here:

- C. R. CALLADINE "Theory of shell structures". CAMBRIDGE UNIVERSITY PRESS. 1983. (in English)

Books that are easier to read on a mathematical level, although sometimes less rigorous, but where many examples can be found:

- P. L. GOULD « Analysis of plates and shells ». PRENTICE HALL. 1999. (in English)
- R. LHERMITE "Résistance des Matériaux". DUNOD. 1958. (in French)
- RAAMACHADRAN "Thin shells - Theory and problems". SANGAM BOOKS. 1993. (in English)

- S.TIMOSHENKO "Théorie des Plaques et Coques". LIBRAIRIE POLYTECHNIQUE BERANGER. 1951. (this edition is in French, but the original work is in English)

Books that present recent developments in the application of the finite element method, sometimes with practical examples. They are very useful for making the link between theory and practical applications, in particular for the proper use of finite elements.

- J. S. PRZEMIENIECKI. « Theory of matrix structural analysis ». DOVER PUBLICATIONS INC. 1985. (in English)
- O.C. ZIENKIEWICZ et R.L. TAYLOR « The finite element method – volume 2: Solid mechanics ». 5th edition. Butterworth Heinemann. 2000. (in English)
- J. L. BATOZ & G. DHATT "Modélisation des structures par éléments finis". HERMES. 1992. (in French)
- M. PRAT & all. "La modélisation des ouvrages". AFPC – Emploi des éléments finis en génie civil. HERMES. 1995. (in French)

Index

Q

R

S

T

V

W

For Product Safety Concerns and Information please contact our EU
representative GPSR@taylorandfrancis.com
Taylor & Francis Verlag GmbH, Kaufingerstraße 24, 80331 München, Germany

www.ingramcontent.com/pod-product-compliance
Lightning Source LLC
Chambersburg PA
CBHW060948210326
41598CB00031B/4755
* 9 7 8 1 0 3 2 4 5 1 7 1 8 *